Recruitment, Colonization and Physical–Chemical Forcing in Marine Biological Systems

Developments in Hydrobiology 132

Series editor
H. J. Dumont

Recruitment, Colonization and Physical–Chemical Forcing in Marine Biological Systems

Proceedings of the 32nd European Marine Biology Symposium, held in Lysekil, Sweden, 16–22 August 1997

Edited by

S. Baden, L.Pihl, R. Rosenberg, J.-O. Strömberg, I. Svane and P. Tiselius

Reprinted from Hydrobiologia, volumes 375/376 (1998)

Springer-Science+Business Media, B.V.

Library of Congress Cataloging-in-Publication Data

A C.I.P. Catalogue record for this book is available from the Library of Congress.

ISBN 978-90-481-5109-7 ISBN 978-94-017-2864-5 (eBook)
DOI 10.1007/978-94-017-2864-5

Printed on acid-free paper

Cover illustration: JTC design and Maria Granberg

Hydrobiologia **375/376**, v–vii, 1998.
S. Baden, L. Pihl, R. Rosenberg, J.-O. Strömberg, I. Svane & P. Tiselius (eds),
Recruitment, Colonization and Physical-Chemical Forcing in Marine Biological Systems.

Contents

Preface

The 32nd European Marine Biology Symposium was held in Lysekil, Sweden, between August 16th and 22nd 1997 and organized by the Kristineberg Marine Research Station. The symposium was attended by 280 participants from 29 countries contributing 65 oral presentations and 120 posters. The topics were, 'Recruitment and Colonization', and 'Physical and Chemical Forcing on Marine Biological Systems', partly reflecting the present research interests at Kristineberg.

The 32nd EMBS is the first in Sweden and the event marked at the same time the beginning of a new era in the history of Kristineberg Marine Research Station. Since the station was founded in 1877 by The Royal Swedish Academy of Sciences, a large number of scientists of many nationalities have visited Kristineberg and contributed significantly to the development of the marine sciences. In 1994, the station was thoroughly modernized and a new laboratory building completed. At the same time a significant investment in new laboratory equipment was made. The management of these new resources is jointly shared by the Royal Swedish Academy of Sciences and the Göteborg University. This development has transformed Kristineberg into one of Europe's largest and most modern marine laboratories now appointed to be a 'large scale facility' within EU – a support given to promote extended marine collaboration within Europe.

We are grateful to all participants, our sponsors (see below), staff and students for making the symposium a great success. The high quality presentations and posters, with contributions from many generations of marine biologists, made the discussions reach melting points of enthusiasm throughout the meeting. We were privileged by warm and sunny summer weather adding mellowness to the beautiful scenery of the Lysekil archipelago – a throughly unforgettable event in the spirit of the EMBS initiators.

We are convinced that this volume of *Hydrobiologia*, containing 33 major scientific contributions presented at the meeting, reflects the high standard of European marine science and will contribute to further advancement. We would like to thank the authors, the referees, the editor-in chief, and the publisher for excellent cooperation throughout the numerous working hours necessary to complete publication of these proceedings.

Thanks to you all from the editors:

SUSANNE BADEN
LEIF PHIL
RUTGER ROSENBERG
JARL-OVE STRÖMBERG
IB SVANE
PETER TISELIUS

The symposium was kindly sponsored by

Göteborg University Marine Research Center (GMF), The Royal Swedish Academy of Sciences, Swedish Environmental Protection Agency (NV), Swedish Natural Science Research Council (NFR), Swedish Council for Forestry and Agricultural Research (SJFR), Community of Lysekil, Bohus County Council, Provincial Government of Göteborg and Bohus län, Göteborg and Bohus län Vattenvårdsförbund, EU-Objective 2 (Region Fyrstad), Munkedals AB, Preem Environmental Foundation, AB Volvo Penta, Akzo Nobel Surface Chemistry AB, Neste Oxo AB, Borealis AB, Hydro Plast AB, Vattenfall Generation Services AB.

Preface

Hydrobiologia **375/376:** 1–21, 1998.
S. Baden, L. Pihl, R. Rosenberg, J.-O. Strömberg, I. Svane & P. Tiselius (eds),
Recruitment, Colonization and Physical–Chemical Forcing in Marine Biological Systems.

Larval supply and recruitment of benthic invertebrates: do larvae always disperse as much as we believe?

Christopher D. Todd
Gatty Marine Laboratory, School of Environmental & Evolutionary Biology, University of St Andrews, St Andrews, Fife, Scotland, KY16 8LB. E-mail: cdt@st-and.ac.uk; fax: 01334 463443

Abstract

The concept of larval supply, and the relative importance of pre- and post-settlement processes as structuring agents of marine benthic invertebrate assemblages and communities are discussed. Because pelagic larvae result in an essential decoupling of local reproduction from local larval recruitment, many marine populations should be considered demographically 'open'. Accordingly, factors affecting larval input to the benthos might be especially important in structuring both populations and communities, and even supercede the importance of post-settlement processes such as competition, predation and physical disturbance. Comprehensive studies on intertidal barnacles have provided an initial focus for field studies, but perhaps the greatest challenge facing marine ecologists lies in extrapolating beyond single-species populations to assess the relative importance of pre- and post-settlement processes for whole assemblages or communities.

Amongst rocky intertidal assemblages, larval supply processes do appear especially important in circumstances where recruitment levels are low: at higher recruitment levels, post-settlement processes are prominent. No comprehensive evaluation of the relative importance of pre- and post-settlement processes has been undertaken for soft sediment assemblages, but the prevailing consensus is that hydrodynamic factors, density dependent effects of food availability and post-settlement processes are of prime importance. For epifaunal 'fouling' assemblages, probably both pre- and post-settlement factors are very important, and illustrative data are presented to show that post-recruitment processes may mask even strong larval signals in later assemblage development. Short-term panel data showed that hard substratum epifaunal recruitment levels do not necessarily conform to simple measures of current velocity at a given site: application of a tidal flushing model did, however, assist in explaining the variation in larval recruitment rates in relation to water flux below an arbitrary critical velocity for larval settlement.

An indirect means of assessing realized larval dispersal – and hence geographic scale of population 'openness' – is available from analyses of population genetic differentiation. Data are presented for two species of epifaunal molluscs with contrasting larval types. The planktotrophic species conformed to expectations, in showing genetic homogeneity of populations over a 1600 km range, and therefore in displaying large-scale larval dispersal. The pelagic lecithotrophic species, however, revealed population differentiation on very small scales (< 10 km), even in highly dispersive environments with strong tidal currents (up to 3.6 m s^{-1}). The deduction is that contrary to actually facilitating dispersal, those pelagic lecithotrophic larvae are behaviourally constrained to minimize larval transport. The consequences of this to general deductions about 'openness' of epifaunal assemblages are discussed, in the knowledge that spatial domination there is attributable largely to three phyla which characteristically reproduce by means of short-term pelagic lecithotrophic larval stages.

Introduction

The formal recognition of soft sediment benthic communities, as more or less predictable species abundance entities, stems from the pioneering work of Petersen in the Baltic and near Atlantic. His initial objective was to survey and quantify the amount of food available to commercially important fish, but this progressively developed as much toward analyses of benthic assemblage structure, culminating in what were later termed the 'Petersen communities'. In his final review report of the 25 year programme (Petersen, 1918), under the subheading 'How the animal communities are maintained', he focused on the *Macoma*-community in illustrating how settlement and post-settlement processes may determine species presence/absence and abundance. That community group above all others seemed widespread and independent of sediment composition. He noted a strong inverse relationship between (predatory) *Amphiura* and bivalve mollusc densities, including high abundances of juvenile *Mactra* and *Tellina* in shallow and deeper sediments lacking *Amphiura*. He presumed that these juveniles would never reach adulthood because of anomalously low tides and occasionally severe wave action or low winter temperatures; the same factors were deduced to explain the absence of echinoderms. The *Macoma*-community species appeared well-adapted to these stressful environments, hence their numerical success there, but the question Petersen raised – and could not explain to his own satisfaction – was why were species of the *Macoma*-community absent from seemingly suitable deeper sediments which also lacked echinoderms?

Over the intervening 80 years, of course, a whole discipline in larval ecology has evolved and immense strides have been made in our understanding of larval developmental patterns and the subtleties and complexities of larval habitat selection and metamorphic behaviour (Young, 1990). Moreover, community ecology has undergone numerous paradigm shifts, from the perception of the importance of competition within the context of equilibrial models, to currently and widely-held views of the importance of stochastic processes and disturbance (of both physical and biological origins) and non-equilibrial models (Connell, 1978). But in particular reference to coral reef assemblage structure, it is of some interest to note that Knowlton & Jackson (1994) have pointed to modern advances in species taxonomy which might predicate a resurgence in interest in the importance of the equilibrial niche-diversification hypothesis.

Whilst acknowledging the advances and developments in these distinct fields of marine research, only in recent years have larval ecology and community ecology been successfully melded to provide a clearer understanding of the dynamics of whole communities. Although marine biologists have long believed that larval input has important consequences for species population dynamics, and community structure (Young, 1990), there is no doubt that since Petersen's landmark studies much emphasis has been placed on post-settlement processes – competition, predation, physical disturbance and their interaction (Table 1). Roughgarden et al. (1985) and Gaines & Roughgarden (1985) therefore achieved much in refocusing the interest of benthic community ecologists to explicitly addressing questions about the relative importance of larval settlement/recruitment (so-called 'supply-side') processes.

But of necessity, Roughgarden et al. (1985) and Gaines & Roughgarden (1985) modelled single-species populations in order to achieve the necessary level of quantitative detail and rigour in their analyses. One of the goals of ecology is an understanding of the processes that underlie the patterns recognised in nature and one of the major challenges facing benthic ecologists today is to extend our appreciation of supply-side processes beyond single-species populations to the level of whole assemblages and communities (e.g. Menge & Sutherland, 1987) across the full range of marine biotopes.

Marine communities and systems are unusually amenable to observation and experimental manipulation because, at least in intertidal and shallow water systems, the ecologically important gradients in physical factors are steep and the sizes/abundances and taxonomic diversity of organisms are tractable. The level of phyletic and divisional complexity in marine systems can be extremely high in comparison to terrestrial and freshwater systems, but just as this may confer an additional layer of ecological diversity or complexity on the system under scrutiny, there commonly is much convergence of disparate organisms resolving the same environmental challenge. For example, sublittoral epifaunal assemblages world-wide tend to be characterized by space-occupying species of the phyla Porifera, Cnidaria, Bryozoa and Urochordata; these predominantly are modular, rather than unitary, organisms with indeterminate growth form.

There are considerable logistic and practical difficulties in considering all macroscopic taxa, including meiofauna, in any benthic marine study. None the less, in order to achieve a full understanding of the relative importance of pre-settlement and post-settlement processes at the community level, an inclusive approach between algologists and zoologists and between vertebrate and invertebrate ecologists would seem expedient. Whereas much progress on pre-settlement processes has been made by invertebrate biologists, and especially those with interests in hard substratum systems, developments to a simi-

Table 1. Summary of pre-settlement, settlement and post-settlement processes and factors likely to be implicated in determining benthic community structure. The duration ranges are given in parentheses

PRE-SETTLEMENT PROCESSES and FACTORS [Seconds–Months]

1. Predation on larvae (in the water column)
2. Oceanographic influences (e.g. directional currents, upwelling, Ekman transport)
3. Local hydrographic effects (e.g. tidal current variation, residual drift, hyposaline runoff)
4. Larval 'quality' (are all larvae equal?)
5. Larval behaviour (e.g. vertical migration, responses to salinity)
6. Substratum electivity (making the 'correct' choice)
7. Benthic predation (e.g. inhibitory effects of adult suspension feeders)

SETTLEMENT and RECRUITMENT PROCESSES and FACTORS [Minutes–Hours]

1. Successful encounter of, or attachment to, the substratum
2. Responses to substratum cues (e.g. flow regime, particle size, bacteria, surface biofilms, chemical cues, substratum texture and orientation)
3. Detection of conspecifics or founders (gregarious species)
4. Detection/avoidance of superior competitors and predators
5. Responses to allelochemical agents
6. Successful completion of metamorphosis

POST-SETTLEMENT PROCESSES [Seconds–Decades]

1. Intra-specific competition
2. Inter-specific competition
3. Amensalism in soft sediments?
4. Biotic interactions (predation [including numerical responses], grazing, inadvertant killing by sedentary consumers, compensatory mortality)
5. Parasitism and disease
6. Physical disturbance (including displacement due to hydrodynamic factors)
7. Reproductive performance and success
8. Sublethal and lethal abiotic stresses
9. Density-related immigration/emigration of juveniles and adults

lar complexity in algology are not apparent (see, for example, Booth & Brosnan, 1995). Zoologists and algologists face strikingly similar demands in this regard, as illustrated by the range of invertebrate larval strategies and types of algal spores, both groups of which vary in their longevity, developmental status and dispersal potential. Similarly, theoretical and experimental advances attributable to studies on pelagic and demersal fish assemblages (e.g. lottery hypothesis, Sale, 1977; recruitment limitation hypothesis, Doherty, 1981) have encouraged and facilitated the productive transfer of ideas between vertebrate and invertebrate ecologists (see Caley et al., 1996).

This is a selective review of a range of specific issues, general features and current problems pertaining to the general question of the relative importance of pre-settlement (larval supply) and post-settlement processes to the dynamics of marine benthic invertebrate assemblages. The range of studies discussed is not intended to be comprehensive and there are numerous excellent recent reviews, including those by Ólafsson et al. (1994), Booth & Brosnan (1995) and Caley et al. (1996). Population studies on barnacles are a focal point of larval supply studies and warrant some detailed consideration in their own right. The focus then is extended to address intertidal rocky shore communities as whole systems; whilst acknowledging the obvious importance of macroalgae (see, for example, Menge et al., 1993), a detailed appraisal of these is beyond the present scope.

Because soft sediment benthic systems predominate throughout the World Ocean they provide probably the greatest difficulties and potential rewards to ecologists in the present context. The logistic and practical complexities in sampling at depth and of quantifying settlement are well known; furthermore, these assemblages embrace considerable ecological heterogeneity, especially in terms of energetic flux, productivity and life history variation among deep sea, neritic, estuarine and intertidal species' populations. But of all assemblage types, possibly the most tractable systems for addressing larval supply issues at the whole-assemblage level are hard substratum sublittoral or cryptic epifauna. Their advantages include the following: (1) the biotope is essentially two dimensional and can be readily photographed or censused, (2) the organisms will readily settle on artificial substrata, (3) in marked contrast to soft sediment biota, larval settlement and recruitment are relatively easy to quantify, (4) space is a common, potentially limiting, resource to most species, and (5) by definition, macroalgae are absent and this simplifies the ecological patterns and processes. As a means of illustrating two particular issues, data are included here from our own experimental work on epifaunal assemblages. First, selected data are presented from a panel study of sublittoral epifaunal succession which showed strong recruitment signals with post-recruitment effects also. Second, as part of a continuing panel study of epifaunal larval settlement and recruitment at the same location, the effects that variations in local flow regime can exert on realized larval supply *versus* potential larval supply are shown. The feature of especial interest there is whether or not larvae in the water column are capable of settling throughout a tidal cycle and therefore are 'available' to the benthos.

The final topic addressed is that of 'openness' or 'closedness' of benthic populations and assemblages, and the implications thereof to the relative importance of larval supply as a structuring influence. Analyses of the degree and scale of genetic differentiation of species' populations are a powerful indirect method of addressing this particular question (e.g. Grosberg & Levitan, 1992; Booth & Brosnan, 1995). The data included here were sought specifically in questioning whether or not the primary role of a pelagic larva to an otherwise benthic adult is dispersal. We quantified adult population differentiation patterns for two mollusc species with pelagic larvae, of contrasting developmental mode and planktonic longevity. Our expectations of larval transport potential for the pelagic lecithotrophic species were not upheld; realized larval dispersal distances revealed its populations to be demographically closed on surprisingly small spatial scales.

Larval types and dispersal of benthic invertebrates

Most benthic marine invertebrates reproduce by means of a planktonic larval stage which may be dispersed extensively in the water column before settling to the substratum and undergoing metamorphosis. The duration of the pelagic phase and the spatial scale of dispersal potential commonly are inferred from knowledge of the developmental strategy of the larva. Long-term planktotrophs – which require an obligatory and often extended (weeks to months) period of feeding, growth and morphological development towards the competent state – are presumed to be dispersed considerable distances. It is commonly stated that pelagic larvae are 'for' dispersal and that their ability to enable horizontal transport is the primary selective reason for their inclusion in the life cycle (reviewed by Strathmann, 1980, 1985; Jablonski & Lutz, 1983; Day & McEdward, 1984; Grahame & Branch, 1985; Knowlton & Keller, 1986; Pechenik, 1990; Willis & Oliver, 1990). Pelagic lecithotrophic larvae, by contrast, may be planktonic for only minutes to days before attaining metamorphic competence and are presumed to be dispersed over shorter, but none the less potentially significant, distances (Todd et al., 1998). All planktotrophic species and most lecithotrophic species undergo an obligatory post-hatching/post-release pelagic phase during which settlement and/or metamorphosis to the benthic state is not possible even if the larva were to encounter a suitable substratum. Pelagic dispersal is thereby enforced as larvae are uncontrollably exported, and the ability of at least some larvae to delay metamorphosis (Pechenik, 1990) in the absence of the appropriate benthic cue(s) may result in yet further dispersal (but see Jackson & Strathmann, 1981).

In the demographic analysis of single-species populations, the classic Lotka-Volterra approach is to quantify dynamics with births and immigrants as inputs and deaths and emigrants as outputs (see Roughgarden et al., 1985). Such an approach, replacing births by recruits, may not, however, be appropriate for the study of many marine benthic populations or assemblages because they do not conform to the requirement that the population be demographically closed. The inclusion of a long-term planktonic larval phase essentially decouples local adult reproductive output from local larval settlement (Bertness et al., 1992). Given the prevalence of pelagic larval strategies amongst shallow water marine benthic invertebrates, these systems may be argued to be characterised by widespread larval dispersal and conceptually considered to be demographically open (Underwood & Fairweather, 1989).

Equilibrial and non-equilibrial models of assemblage structure and the concept of larval supply

Over the past 25 years, marine benthic ecologists have emphasized non-equilibrial models of community organization and post-settlement processes, since the shift from the primacy of equilibrial models in the 1950s and 1960s (e.g. Connell, 1978). But the perceived importance of pre-settlement, or larval supply, processes in structuring benthic assemblages is not novel (Grosberg, 1982; Connell, 1985; Young, 1990), as has been extensively discussed in recent reviews (Underwood & Fairweather, 1989; Grosberg & Levitan, 1992; Caley et al., 1996). Following Underwood & Fairweather (1989), the term 'supply' is applied here as the rate and timing of larval settlement, or input to the benthos, because larvae of soft-sediment infaunal species do not necessarily attach, settle and rapidly metamorphose in a manner directly comparable to sessile hard substratum species.

Debates about larval supply centre upon the relative importance of pre-settlement and post-settlement processes in structuring benthic adult populations and assemblages. In evaluating the importance of larval supply for single-species populations, clear distinctions have to be made between (1) population limitation and regulation (Caley et al., 1996), (2) the importance of variation in settlement being associated with adult population fluctuations, and (3) whether or not density dependent or density independent post-settlement mortality and growth are implicated in determining adult abundances. A process is limiting if it adds individuals to, or subtracts them from, a local population. A recruitment-limited population is one which may increase only if there is an increase in larval input (Hughes, 1990) or for which recruitment rates are insufficient to fill all the available space (Sutherland, 1987). Conversely, regulation implies density-related (i.e. density dependent or inversely density dependent) increase or decrease in a rate process (such as *per capita* mortality or growth rates) with increasing or decreasing population density.

The importance of recruitment phenomena has long been implicated in tropical reef fish assemblage dynamics (e.g. Sale, 1977, 1990, 1991) and in fisheries management (Frank & Leggett, 1994). In marine benthos, most experimental studies on larval supply are for intertidal barnacles. The focus of interest has been on the relationships between variations in larval flux to the substratum, the density of settlement, recruitment processes (with particular reference to the effects of resident adult densities) and whether or not adult population size is regulated by larval supply (Caley et al., 1996). Ólafsson et al. (1994) have addressed the relevance of recruitment limitation to soft-sediment assemblages and particular aspects are discussed further below.

Settlement and recruitment

The term recruitment is variously applied by marine ecologists according to the field of interest (Frank & Leggett, 1994). In fisheries biology, for example, recruitment can refer both to (1) the number of individuals in a specific cohort that survive to harvestable size, and (2) the numbers of pelagic larvae that survive to the juvenile stage. Other definitions in the non-fisheries literature include the number of individuals of a specific cohort that survive to enter the reproductive population. Larval settlement of a sessile benthic invertebrate occurs when the larva has irreversibly attached to the substratum (and metamorphosed): recruitment generally refers to those settled individuals that have survived an arbitrary or subjective time period before being censused (Keough & Downes, 1982; Connell, 1985; Minchinton & Scheibling, 1993a; Booth & Brosnan, 1995; Caley et al., 1996). Whilst this is a convenient definition of settlement for sessile hard substratum species, it can be extended to many soft sediment infaunal species only with caution because these do not attach permanently and may not undergo cataclysmic metamorphosis. Despite being comparatively easy to define operationally (Sutherland, 1987), larval settlement generally is extremely difficult to quantify in the field (but see Raimondi, 1990), especially in soft sediments (Ólafsson et al., 1994); none the less, confusion between settlement and recruitment still persists in the benthic literature. As a problem, this may be all the more acute if measures of the rates and density of settlement are inferred from censuses of recruits at varying time intervals from settlement (Caffey, 1985; Gaines & Roughgarden, 1985; Gaines et al., 1985; Hughes, 1990; Farrell et al., 1991; Ólafsson et al., 1994), although this lessens if post-settlement mortality is density independent (Connell, 1985; cf Holm, 1990).

Studies of larval supply and barnacle populations

Recent studies of the importance of larval supply/recruitment to adult population limitation and regulation have focused on intertidal barnacles (e.g. Caffey, 1985; Connell, 1985; Gaines & Roughgarden, 1985; Gaines et al., 1985; Raimondi, 1990; Sutherland, 1990; Minchinton & Scheibling, 1991, 1993a,b; Bertness et al., 1992; Gaines & Bertness, 1992; Gaines & Bertness, 1993; Miron et al., 1995), especially on the Pacific and Atlantic coasts of North America, Central America, Australia and NW Europe. Barnacles are ideal in that (1) the benthic stage is sessile (there is no immigration/emigration of adults), (2) the larvae are readily caught and identifiable in plankton samples, (3) settlement of cyprids can be enumerated *in situ* on natural substrata, or on settlement panels, (4) adult populations commonly are space-dominating and ecologically important components of mid-low shore rocky intertidal assemblages, and (5) mid shore populations generally are accessible after each tidal settlement period.

Roughgarden et al. (1985) and Gaines & Roughgarden (1985) modelled settlement, growth after settlement and death in *Balanus glandula* in varying areas (0.01 m^2–100 m^2). Their major findings were that (1) growth (space occupation) was destabilizing because increased growth (especially in relation to the mortality rate) led to reduced space for settlement and (2) when settlement was low, the mortality rate of recruits was independent of free space. An important assumption here is that the number of larvae landing or settling per unit time is a direct function of the available free space (cf. Raimondi, 1990 on *Chthamalus anisopoma* in Baja California). Although settlement of cyprids on adult *B. glandula* shell plates generally occurs only when free space is almost filled, that assumption does probably also conceal other important sources of variation; there may be density-related facilitatory and inhibitory effects of previously-settled incumbents on subsequent settlement (e.g. Hills & Thomason, 1996) and no allowance was made for possible variations in substratum biofilm cues to larval settlement. Such cues might be especially influenced by resident adult density (Wieczorek & Todd, 1997), with contrasting biofilms associated with the primary rock substratum and secondary space (adult shell plates).

Minchinton & Scheibling (1993a) explicitly addressed related issues in considering how recruitment can be operationally over-estimated or underestimated from repeatedly cleared quadrats according to the duration of observations. If there is density-related facilitation of settlement (gregarious behaviour), the summed data from a series of contiguous short-term observations of settlement will underestimate settlement and recruitment, compared to season-long observations for other quadrats or panels; conversely, if there is density-related inhibition (spacing behaviour), settlement and recruitment will be over-estimated. These errors are compounded further if there are density-related variations in settlement behaviour (gregariousness at low densities and spacing behaviour at higher densities) and, especially so, given that settling cyprids may arrive in unpredictable massive pulses over a settlement season (e.g. Farrell et al., 1991). Minchinton & Scheibling (1993a) compared *Semibalanus balanoides* settlement on quadrats which were cleared regularly, with quadrats that were initially cleared at the start of the settlement season and left throughout. They found that the estimate of recruitment and post-settlement mortality rose when the sampling interval was reduced from 2 d to 1.3 d. Presumably reducing the sampling interval to a twice-daily (e.g. tidal) basis would elevate these measures yet further.

Notwithstanding possible density-related variation, allowance should also be made for the facilitatory effects of biofilms on daily settlement rate. Laboratory observations for *Elminius modestus* (Neal et al., 1996), and field data (Wieczorek & Todd, unpubl.) for *Semibalanus balanoides* in Scotland, show facilitation of settlement on substrata preconditioned with biofilms. Manual removal of cyprids and metamorphosed spat from quadrats censused daily (e.g. Minchinton & Scheibling, 1991) would allow the assessment of settlement in the presence of natural biofilms, but errors are possible because of the confounding (facilitatory) effects of any remaining attached skeletal fragments or traces on subsequent larval settlement. Bertness et al. (1992), for example, cleared settlement quadrats of cyprids daily by means of a toothbrush and seawater. Whilst effective in removing cyprid traces, presumably such repeated treatment will result in substratum biofilms distinct from those on uncleared quadrats. These are not easy experimental problems to resolve (Minchinton & Scheibling, 1993b), and biofilm effects on larval settlement cannot be ignored if an appropriate measure of field settlement rates is the objective: these might partly explain Minchinton & Scheibling's (1991) observation that the availability of cyprids and the rates of settlement on cleared quadrats were more highly correlated later than earlier in the settlement season (see also Hills & Thomason, 1996).

Connell (1985) concluded, from analyses of data for various geographic localities and several barnacle species, that larval supply (settlement) did affect surviving adult density, but only when settlement was low (see also Minchinton & Scheibling, 1991). When barnacles settle at very high densities, primary space can become occupied quickly and further shell growth of individuals is constrained upward. Such crowding forces groups to form distinct hummocks, with a columnar growth form and minimal attachment of individuals to the primary substratum. These hummocks are inherently unstable and especially vulnerable to both physical (wave crash) and biotic (predation) disturbance agents: elimination of hummock patches may result in density dependent mortality and low survivorship of recruits to adulthood. At extremely low settlement densities, however, Connell (1985) found mortality of the adult barnacles to be independent of density and possibly total for recruit cohorts. Numerous studies have shown that barnacle populations may be not only limited, but regulated, by recruitment (Caley et al., 1996) because, in relation to adult density, the *per capita* recruitment rate in small patches is likely to be lower for high density adult populations. Recruitment rates generally are inversely density dependent; that is, they decrease with increase in adult density, and the concomitant decreases in space availability.

Farrell et al. (1991) observed four 'pulses' of *Balanus glandula* and *Chthamalus* spp. cyprid recruitment over 6.5 months, separated by periods of very low or zero recruitment. A connection between strong upwelling and (offshore) Ekman transport and zero-low recruitment was apparent; recruitment pulses ended when strong equatorward (northerly) winds and upwelling resumed. Data such as these, and those of Wing et al. (1995) for urchin and crab settlement in relation to upwelling relaxation in California, provide compelling and powerful illustrations of the effects of large-scale oceanographic variations in retaining or returning competent larvae to inshore waters and their possible contact with appropriate settlement substrata. On a smaller scale, Gaines & Bertness (1992) showed the importance of local hydrographic processes in determining local variations on *Semibalanus balanoides* settlement within bays. For a nine year record of recruitment variation in Narragansett Bay, Rhode Island, most (77%) of the variation in recruitment was explained by simple measures (average flushing times) of the likelihood of larval export from the bay. Minchinton & Scheibling (1991) also found that on average 79% of the variance in *S. balanoides* recruit density at the end of the main settlement period in Nova Scotia was explained by larval abundances. It is clear, therefore, that oceanographic and hydrographic factors may – in certain localities and at certain times – be responsible for recruitment failure of whole year classes over wide geographic ranges. The converse pattern of strong year-class dominance is not unusual amongst soft sediment infaunal species (Ólafsson et al., 1994); but is this unequivocal evidence of the importance of larval supply to infaunal assemblages? Not necessarily, because it could be a reflection of the local absence of density dependent post-settlement predation and/or competition.

Despite the advances made in the numerous excellent studies on intertidal barnacles, including strong connections between larval density and settler density (e.g. Minchinton & Scheibling, 1991), numerous challenges remain for benthic ecologists. Of especial importance are (1) the realistic measurement of settlement for a wider range of species and localities, (2) the wider assessment of the importance of variation in larval supply to actual recruitment and (3) the assessment of recruitment and post-recruitment processes or events in driving overall assemblage/community structure. Pump and net samples of pelagic larvae can provide measures of larval concentration (e.g. Minchinton & Scheibling, 1991; Gaines & Bertness, 1993), but in the absence of flow data over the substratum these are of limited value in deriving reliable estimates of larval supply to given sites. Using a passive larval trap design, previously utilised by Yund et al. (1991) and Bertness et al. (1992) for estimating barnacle larval supply, Gaines & Bertness (1993) were successful in obtaining a time-integrated larval flux estimate for *Semibalanus balanoides* which resulted in a very high correlation ($r = 0.98$) between larval abundance and settlement: colonisation of a particular site clearly was a function of larval density in the water column and flow. Such approaches will be of obvious value to quantitative studies of a wider variety of ecologically important species, localities and habitats – both in hard substratum and soft sediment assemblages – but as the authors acknowledge, unfortunately these trap designs may not be suitable for all species and are functionally restricted to deployment in locations with minimal vertical component to flow; they will not operate in environments with breaking waves (but see also Setran, 1992).

Even seemingly strongly seasonal species of barnacles tend to reproduce and settle over more or less extended periods (e.g. Connell, 1985; Gaines et al., 1985; Wethey, 1985; Farrell et al., 1991; Bertness et al., 1992), and some species can become reproductive within 6–10 weeks of settlement (e.g. Raimondi, 1990; Sutherland, 1990). Thus, some individual recruits can become reproductive adults whilst others in that year class are still settling as larvae. If photographs or maps of individuals are not available, and if growth of recruits is sufficiently rapid that they become indistinguishable from old adults at an early age, it will remain extremely difficult to quantify recruitment of a particular cohort and to assess density dependence or independence of recruit mortality with respect to previously-settled adult density. The difficulty here is in distinguishing density-related processes which are attributable to the prior adult densities (or free space), and the density of the recruits themselves, given that adult densities will probably decline over the period and space may be progressively occupied by growing recruits.

Community-level analyses of the relative importance of larval supply as a determinant of rocky intertidal community structure

In their early studies of rocky intertidal communities in New England and Panama, Menge and co-workers indicated competition, predation and physical disturbance to be largely responsible for post-recruitment variation in abundance of ecological dominants. In subsequent reanalyses of their data, Menge (1991) undertook an explicit comparison of the relative importance of estimates of recruitment and the above post-recruitment processes, using multiple regression with the major space-occupiers (community dominants) as the dependent variables. The taxa included several species of barnacles, two species of bivalves, and foliose algae as a composite functional group. For the New England shores investigated, recruitment explained 11% of the variance (with predation and competition together explaining 50–78%); in Panama recruitment explained 39–87%, and competition plus predation accounted for only 8–10% of the variance. Hence, in concurrence with Connell's (1985) analysis, when recruitment density is low, it does seem to explain a disproportionate amount of the variance in adult abundance. At sites, or in seasons, when recruitment is high, competition and predation then are the more potent influences. Menge's (1991) retrospective analyses were necessarily confined to estimates of recruitment only. Future variance partitioning analyses of the interplay of larval supply in the strict sense and post-settlement processes at the community level will undoubtedly be even more informative.

Much of the discussion of larval supply effects has focused on sessile space-dominant species, but it is important to consider the special problems in extending such analyses to sedentary or mobile species which may predominate at least in terms of species richness. On many shores, mobile grazers and predators are both abundant and ecologically important as consumers. Many have planktotrophic larval stages and presumably their populations are demographically open on large scales. Mussels share the pre-eminence of barnacles in being major space dominants of the mid-low shore zones on many wave exposed temperate rocky shores in both hemispheres. Their ecological importance extends beyond their occupation of primary space and competitive ability, because they may comprise an important resource for various predators (especially gastropods, echinoderms and birds). Furthermore, they can be an important source of bound sediment for infauna beneath their clumps and also provide extensive secondary space and refuge for other sedentary and sessile invertebrate species (e.g. Lohse, 1993).

All intertidal mussels have planktotrophic larvae but, although sedentary to the extent of being almost functionally sessile, it is a characteristic feature that settled individuals also can readily relocate and migrate within shores (especially as juveniles) and pre-emptively close small disturbance-generated gaps by lateral movement (Seed & Suchanek, 1992). For *Mytilus edulis*, the primary settlement of larvae (particularly on filamentous algal substrata) and secondary upshore migration of plantigrades (to join juvenile/adult clumps) emphasizes the extent to which populations are highly dynamic, especially on NW European shores (see Suchanek, 1985). The initial larval supply to shoreline zones might be readily measurable using traps or settlement bags (e.g. Hunt & Scheibling, 1996), but the possibilities for assessing settlement and recruitment (in the strict sense), and the importance of mussel larval supply in determining variations in adult abundances, on given small census patches or panels seem problematic because of this migratory potential. Never the less, as described above, Menge's variance partitioning approach should be informative so long as abundance data are available from suitably large-scale sample areas within shores to allow for migration effects.

Studies on soft sediment assemblages

Soft sediment substrata spatially predominate at all latitudes and depths in the benthic marine environment. Several recent reviews concerning soft sediment infaunal assemblages have reiterated our perception of the prevailing importance of hydrodynamic factors (Butman, 1987; Snelgrove & Butman, 1994) and post-settlement processes (Wilson, 1991; Woodin, 1991; Ólafsson et al., 1994) as structuring agents. Whilst acknowledging that Rhoads & Young's (1970) 'trophic group amensalism' hypothesis has been subject to criticism and recent reappraisal (Snelgrove & Butman, 1994), Woodin's (1976) adjunct of the hypothesis to infaunal 'mobility' groupings remains illustrative in focusing on negative, or inhibitory, recruitment-related adult-larval interactions: she predicted bioturbators would smother settling larvae at the sediment surface and suspension feeders would filter them from the water column. Tube-builders at certain densities may stabilize the sediment and alter the local hydrodynamic regime near the sediment surface – perhaps to the extent of facilitating larval contact with the sediment surface (enhancing larval supply) – but in Woodin's view their major inhibitory potential is in the pre-emption of substratum space and obscuring of the sediment surface by defaecation. The latter can be especially complex in masking both positive (adult- and sediment-associated) settlement cues and negative cues (such as ammonium concentrations in the sediment; Woodin, 1991). None the less, the consensus is recognized of the likely primacy of local hydrodynamic regime in determining and modifying sediment structure, modulating animal-sediment relationships, and in determining larval supply to the sediment surface.

Ólafsson et al. (1994) acknowledged that recruitment can, and sometimes does, determine patterns of infaunal adult abundance. They reviewed the importance of larval supply (recruitment limitation) in soft sediments on the basis of Thorson's paradigm – that infaunal species with long-term planktotrophic development will show the greatest spatial and temporal variations in larval production, planktonic survivorship and settlement input – but found no compelling support for this. They concluded larval supply to be not generally limiting. Rather, the bulk of the literature suggests the importance of fluxes in food supply (intimately connected to hydrodynamic regime), post-settlement regulatory processes (e.g. density dependent post-settlement adult inhibition of recruit establishment), predation and disturbance. Ólafsson et al. (1994) generally favoured the incidence of strong density dependent effects – especially for infaunal juveniles – as evidence of food limitation perhaps overriding even variations in larval input as a primary structuring influence. Quantifying the settlement and recruitment of soft sediment infaunal invertebrates is a demanding task, but given such data, variance-partitioning analyses (Menge, 1991) might be a much more powerful means of assessing the relative importance of larval supply at the assemblage level.

Woodin (1991) emphasized a difference between hard and soft substrata in that settling larvae, post-settlement juveniles and adults in soft sediments all can reject substrata and relocate, albeit with varying costs and benefits to fitness. For sessile species on hard substrata, only the pre-settlement larva has this 'option' before irreversibly attaching, although sedentary mussels are a notable exception and mobile consumers are vagile over the substratum. Whilst much of the soft sediment literature focuses on positive (facilitation) cues and choices for the settling larva and post-settlement stages, Woodin (1991) contended that negative responses probably are prevalent amongst soft sediment assemblages. Similarly, in a review of the effects of microbial biofilms on epifaunal larval settlement, Wieczorek & Todd (1997) noted that published reports of facilitation far outweigh those of inhibition. This probably reflects a bias in the investigations undertaken, because just as it is important for settling larvae to respond positively to favourable microhabitats, so too should they avoid microhabitats of low or zero survivorship probability.

Hard substratum epifaunal 'fouling' assemblages

Cryptic infralittoral and sublittoral epifaunal 'fouling' assemblages differ from open rock intertidal communities in two fundamental ways: hard substratum epifaunal assemblages are characterized by (1) an essential absence of macroalgae and (2) space-dominant invertebrates typically are colonial and indeterminate in growth form, rather than solitary and determinate in growth form (Woodin & Jackson, 1979). Despite the numerous and extensive studies on epifaunal assemblage succession, I am aware of none that have explicitly assessed the relative importance of pre-settlement and post-settlement processes at the whole-assemblage level. Hughes (1990) did, however, undertake a detailed study of a sublittoral population of an encrusting bryozoan, *Celleporella pumicosa*, which included analyses of recruitment, colony growth and mortality from *in situ* photographs of natural substrata. Svane (1984) similarly quantified recruitment and mortality from *in situ* photographs for six sublittoral populations of the long-lived solitary ascidian *Ascidia mentula* over a 12 year period; overall he found recruitment to be dependent on adult density but that mortality was density independent.

Although reports on many epifaunal studies have included speculation on variability in larval supply, and recruitment success, as structuring processes, Turner & Todd (1993) found no comprehensive evidence from the literature as to the relative importance of pre-and post-settlement processes. Competition for primary space can be visibly obvious within these assemblages (e.g. Turner & Todd, 1994 and refs. therein) but recruitment of larvae on live colonies (secondary space) – even of species, such as cheilostomatid bryozoans, with hard skeletons – seems comparatively infrequent. None the less, the leathery tunic of some solitary ascidians, such as *Styela plicata* (e.g. Walters & Wethey, 1996) may be readily colonized by other epifaunal species. Predation can be partial or total, at the level of the colony, and may be a potent source of assemblage disturbance and release of both primary and secondary space. Suctorial predators may leave intact cheilostomatid bryozoan prey skeletons which then may be effectively oversettled by other taxa. Similarly, well-established massive corals such as *Porites* spp. may be apically damaged or abraded to the extent that a sufficiently large area of calcareous substratum is exposed to permit the recruitment of other branching coral species or tridacnid bivalves.

Successional processes must to an extent be driven by larval supply and recruitment processes, but available evidence for epifaunal assemblages suggests that these do not conform to the classical facilitation model (Connell & Slatyer, 1977; Greene & Schoener, 1982; Turner & Todd, 1993). Sutherland & Karlson (1977) observed assemblage change to be unpredictable in North Carolina, with the rate and direction of change being determined by variable larval recruitment, the longevity of the space-occupying adults and the capacity of the assemblage as a whole to 'resist' further larval recruitment (inhibition model; Connell & Slatyer, 1977). Most species in Sutherland & Karlson's study were short-lived; there was much turnover of primary space and most recruitment was strongly seasonal. Accordingly, the coincidence of seasonally varying available space and competent larvae determined new recruitment, but the intensity of post-recruitment processes may mask even strong larval signals in the later stages of assemblage development.

Our observations of the solitary ascidian, *Ascidiella scabra*, a prominent component of sublittoral cryptic epifaunal assemblages in western Scotland, are perhaps illustrative of such a pattern. At Clachan Seil (Argyll, west Scotland) *A. scabra* can attain 3–4 cm in length and live for up to 12–18 months, depending upon growth rate. This species shows strongly seasonal settlement (Todd & Turner, 1986), at consistently high densities (unpubl. data, 1981–1986), but apparently is dominant only in the earlier stages of assemblage development.

Clachan Seil is a narrow, sheltered tidal rapid of approximately 1 km length which is delimited by tidal sills at its northern and southern extremities (Figure 1). Spring tides ebb to the same level between the sills, irrespective of open coast fluctuations, and the substratum at the study site (60 m to the south of 'HighFRS'; Figure 1) remains sublittoral. Experimental panels of natural slate, of dimensions 15 × 15 × 1 cm, were immersed for continuous periods of up to 35 months. Each experimental panel was 'sandwiched' by bolting a lower panel and separating the two by spacer rods of either 25, 10 or 5 mm height in an attempt to mimic the effects of variation in vertical relief between the undersides of boulder surfaces and the bedrock substratum; the present illustrative data pertain to the undersurface of the upper panel in each pair and for the 25 mm treatment only. The prime feature of interest in these experiments was the effect of seasonal variation in the timing of initial immersion (= space availability) on subsequent epifaunal assemblage development.

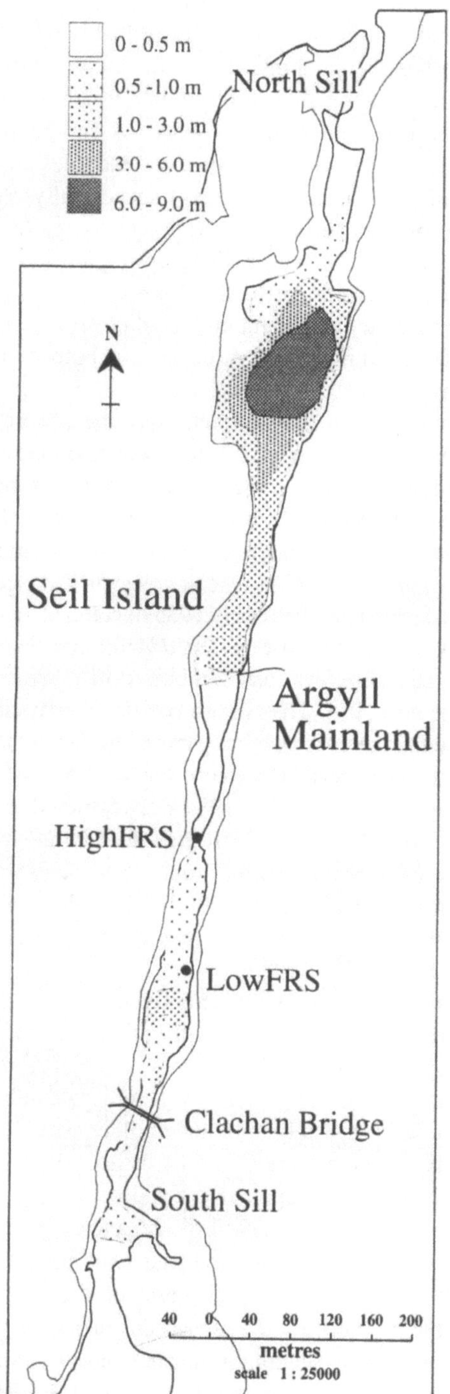

Figure 1. Sketch map of the tidal rapid at Clachan Seil, Argyll, W Scotland. The positions of the High Flow Rate Site (HighFRS) and Low Flow Rate Site (LowFRS) utilized in the study of larval recruitment to panel substrata are shown. The long-term (1983–86) epifaunal panel study (Figures 2 & 3) was undertaken at a site intermediate between the HighFRS and LowFRS.

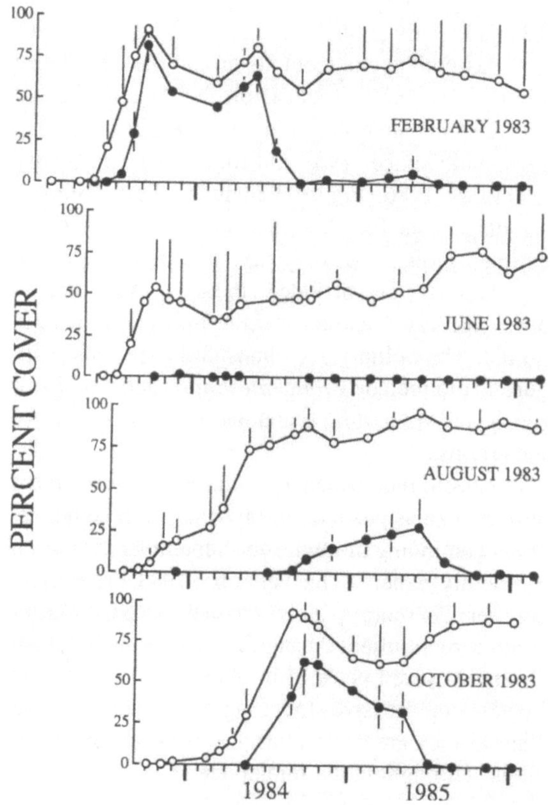

Figure 2. Total assemblage cover (back-transformed mean arcsine percentages) (open circles) and cover of *Ascidiella scabra* (closed circles) for the four separate experiments. Error terms are SE. Panel sets were first immersed at Clachan Seil (Figure 1) in February, June, August and October 1983.

Because the first immersion dates of the four panel sets were pre-determined and spanned eight months, these were treated as four separate experiments and the triplicated panels of the three gap sizes were accommodated in each of four frames placed within a 2 m radius.

Epifaunal abundances were recorded from identical repeat photographs of a permanently marked central 10 × 10 cm square on each panel, using 100 randomly-generated dots on an acetate overlay for each bimonthly census. Where possible all live sessile species beneath each sample dot were scored, resulting in some summed abundances exceeding 100%. Total percent cover (Figure 2) was derived from the hits for unoccupied bare slate. For the long-term experiments, panels were scanned with a binocular microscope at least every four months and all epifauna identified to species. The seasonality of recruitment was followed using new panels immersed and retrieved at each (approximately monthly) census date between February 1983 and January 1986. Figure 2 shows the trajectories of total assemblage percent cover and the abundance pattern of *Ascidiella scabra*, on the four triplicated sets of 25 mm panels first immersed in February, June, August and October 1983 and followed until January or February 1986. Figure 3 includes the partitioned assemblage data for all sessile taxa, summarized as functional groups, but only taxa which were present at $\geq$ 5% cover are included; there is, therefore, no plot for Solitary Ascidians for the June experiment in Figure 3. The Solitary Ascidians grouping (black fill, Figure 3) did include *Ciona intestinalis* and *Ascidiella aspersa*, but *A. scabra* accounted for > 98% of the observations.

Successful recruitment of *Ascidiella scabra* on the long-term panels peaked in May/June, with some settlement continuing through July-September at least on the monthly panels. The February 1983 experiment (hereafter 'February' panels) showed a strong pulse of *A. scabra* recruitment during the first spring/summer, with area covered peaking in September 1983; there followed a progressive sloughing of the individual ascidians as they grew to maturity over the ensuing nine months. Throughout the remainder of 1984 and 1985 the development of the epifauna stabilized, both in terms of area covered and functional group composition, but there was apparent inhibition of *A. scabra* recruitment to the February panels in both the 1984 and 1985 seasons (cf. August and October experiments). The lack of *A. scabra* recruitment to the February panels in spring/early summer of 1984 could have been due to either (i) the self-inhibitory effects of the established adult *A. scabra*, (ii) inhibition by other epifaunal species on those panels, or (iii) their combined effect. The possibility of an exclusive self-inhibitory effect of established *A. scabra* on settling conspecifics is not apparent, given the failure of *A. scabra* also to recruit to the February (and June) panels in spring 1985, a period in which adult *A. scabra* were virtually absent.

For the remaining three experiments there was essentially no recruitment of *Ascidiella scabra* during 1983. This was expected for both the August and October experiments, which were first immersed well after the peak *A. scabra* recruitment season. However, given that *A. scabra* recruitment was recorded for the June monthly panels in 1983, it was expected that the long-term June experiment would show at least modest *A. scabra* recruitment in that first season. Two possible explanations for the lack of *A. scabra* in 1983 on the June experiment panels (Figure 2) might be spatial heterogeneity of recruitment, or frame effects, but if so these would most probably have been manifest as a slight numerical reduction or increase in recruitment, compared to the monthly panels, rather than the observed virtual absence. A further possibility is that essentially only those tadpoles settling early in their season (i.e. May) successfully recruit and become established in the longer term. In May 1984 and 1985, however, there was again failure of *A. scabra* recruitment to the June experiment, implying that the incumbent epifaunal cover effectively inhibited recruitment of *A. scabra* in both of those seasons.

The panels in the October experiment were first immersed in early winter, after most taxa had completed recruitment in 1983. Space occupation by the following May (1984) was low but the incumbents at that time clearly failed there to inhibit recruitment of *Ascidiella scabra*. *A. scabra* cover on the October panels during late 1984/early 1985 developed to levels closely similar to those in 1983/84 for the February panels but there was total inhibition of *A. scabra* recruitment in 1985. The August panels latterly showed an assemblage development pattern similar to that for the October experiment. Recruitment of *A. scabra* to the August experiment in 1984 was restricted to relatively few individuals which grew to a large size. *A. scabra* recruitment across all four experiments was almost totally inhibited in 1985, despite the typically high numbers recorded on the monthly settlement panels in May/June 1985.

Notwithstanding the apparently anomalous failure of recruitment of *Ascidiella scabra* to the June experiment in 1983, the overall pattern was one of *A. scabra* recruitment being high on newly-immersed substrata and partially or totally inhibited in older assemblages: *A. scabra* larvae recruit essentially to primary substratum only and apparently avoid secondary space (unpubl. obs.). The marked discrepancies in recruitment of *A. scabra* across the four experiments were not, however, simply a reflection of differences in available free space in the censused area of each panel. The pattern was seemingly attributable to strong inhibitory influences on *A. scabra* larval settlement, possibly in some instances (e.g. 1984, February experiment; 1985, August and October experiments) by adult *A. scabra*, and/or by other established epifaunal taxa (e.g. 1985, February and June experiments). Inhibitory influences may include (i) epifauna-induced compositional changes in the biofilm of free primary

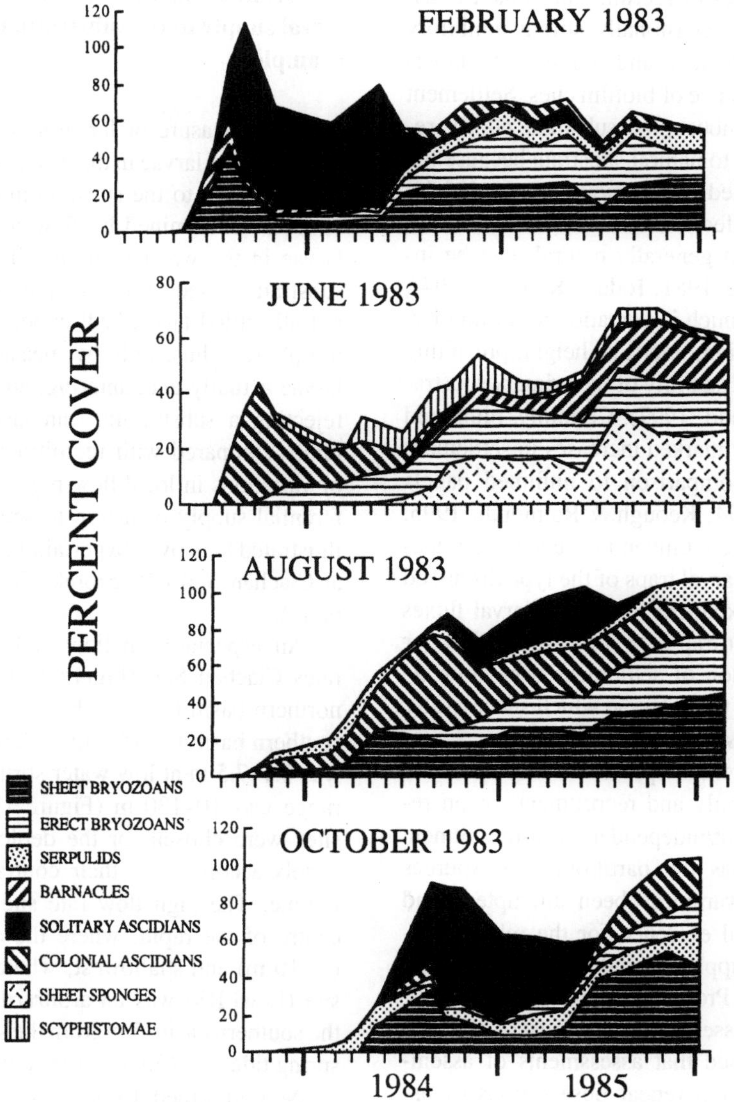

Figure 3. Development of epifaunal assemblages on the triplicated sets of natural slate panels in the four epifaunal experiments. Data are back-transformed mean arcsine percentages and are presented for taxa partitioned into eight functional groups. The black fill is for Solitary Ascidians, of which > 98% of the records were for *Ascidiella scabra*. Note ordinate scale changes for the June experiment.

space, (ii) age-related changes in biofilms, (iii) water column-mediated allelochemical agents emanating from incumbent taxa, or (iv) self-inhibition by adults arising from their space pre-emption and/or filtration of incoming tadpoles.

The early development of this Scottish epifaunal assemblage is therefore closely congruent with the mechanisms reported by Sutherland & Karlson (1977) for North Carolina. Larval supply can be very important in determining especially the early developmental stages of the assemblage, in relation to the seasonal timing of space availability. But general inhibitory and/or specific self-inhibitory post-recruitment events or processes may be potent modifiers of even very strong settlement/recruitment signals, as shown by the general failure of *Ascidiella scabra* to recruit to any of the experiments in the 1985 season.

A cautionary note in the same context as the previous remarks about recruitment to daily-cleared intertidal quadrats is perhaps warranted here. Short-term panel data of epifaunal settlement/recruitment in many studies (including our own) probably are especially unreliable, as indicators of recruitment to natural assemblages, given the occasionally strong inhibitory effects of other sessile taxa and our present understanding of the importance of biofilm cues. Settlement to new (unfilmed), unoccupied substrata poorly reflects larval responses to natural substrata because the latter will be biofilmed and variously occupied by resident taxa. The effects even of other incumbent new recruits, although generally neutral, can be inhibitory also (Grosberg, 1981; Todd & Keough, 1994). Biofilms can convey much information to settling larvae about the orientation, intertidal height, prevailing flow regime, 'age' and stability of the primary substratum (Wieczorek & Todd, 1997). Certainly, biofilmed panels generally show marked increases in settlement and recruitment in comparison to unfilmed panels (Todd & Keough, 1994; Keough & Raimondi, 1995, 1996). As a means of circumventing such experimental obstacles, passive larval traps of the type discussed above may provide excellent records of larval fluxes for particular taxa, but trap data still cannot allow for inhibition or facilitation of settlement behaviour affecting recruitment of those larvae which successfully encounter a given substratum.

I am not aware of multi-species data on epifaunal assemblage larval supply and recruitment, or on related density dependent/independent post-recruitment processes. Neither has the partitioning of sources of adult abundance variances been attempted, and therefore experimental evidence for the relative importance of larval supply to epifaunal assemblages remains unavailable. Probably both are very important in all epifaunal assemblages, but just as Connell & Sousa (1983) argued that assessments of assemblage 'stability' demand repeat observations of assemblages throughout at least one complete turnover of all individuals, so too can this criterion be applied to pre-settlement *versus* post-settlement comparisons. Realistically, however, such data will be practically extremely difficult to obtain, given the extended longevity of some epifaunal species. In Sweden, for example, the large solitary epifaunal ascidian, *Ascidia mentula*, survives for at least 7–8 years (Svane & Lundälv, 1981). There is, however, an additional consideration in such assessments, discussed further below, that epifaunal assemblages might not be demographically open in a manner comparable to some rocky intertidal populations and that this may temper the relative balance.

The relationship between larval concentration and larval supply to the substratum: an epifaunal example

The true measure of larval supply is not simply the abundance of larvae in the water column, but the actual flux of larvae to the substratum (Gaines & Bertness, 1993) as determined by flow velocities. But not all larvae in the water column will contact a particular substratum (see, for example, Butman, 1987) and not all settled larvae will choose to attach and metamorphose. Thus, only by measuring the numbers of larvae actually encountering, and then accepting and rejecting a substratum, can larval supply be accurately compared with recruitment input. The effects of variations in local flow regime in determining differential supply of larvae to settlement substrata are illustrated here by data pertaining to our ongoing study at Clachan Seil (Wieczorek, Todd & Cromey, ms in prep.).

An especially shallow and narrow region separates Clachan Seil (Figure 1) into a wide and deep northern basin ($\sim$ 85% by volume) and a shallower southern basin ($\sim$ 15% by volume). The depth range was 0.1–7.5 m at low water spring tide and the width range was 10–130 m (Figure 1). Two experimental sites were chosen for the deployment of settlement panels according to their contrasts in ambient flow regime. The high flow rate site (HighFRS) was the centre of the rapid, where the channel is narrowest ($\sim$ 10 m) and shallowest, whereas the low flow rate site (LowFRS) was an arbitrarily chosen location in the southern basin, 5 m from the shoreline at low water spring tide and 130 m to the south of the HighFRS.

New unfilmed 15 $\times$ 15 $\times$ 0.5 cm black perspex 'sandwiched' panels were exposed to larval settlement in three seasonal repeat experiments over a spring-neap tide cycle (28 d, August/September 1993; 31 d, October/November 1993; 32 d, May/June 1994) run contemporaneously at the two sites. Each repeat comprised 36 panels allocated to four treatments, but for convenience here consideration is given only to the summed abundances for all species, across all treatments, enumerated at the end of each experiment. The epifaunal data were the sums both of live recruits and all traces of dead settled individuals/colonies on each panel. Flow velocity data were obtained for both sites using an Interocean S4 electromagnetic current meter; flow recordings were obtained over two complete tidal cycles ($\sim$ 24 h) for both spring and neap tides at each site. From these flow data, and a low water spring tide bathymetric survey for the entire rapid, we have developed a simple flushing model using the mid-section method (Shaw, 1988) and which makes allowance for depth changes over the tidal cycle. The model predicted the average water volume and flushing time for the entire rapid (northern and southern basins) on a spring tide to be respectively 133 500 m^3 and 6.7 h; on a neap tide these parameters were predicted at 124 750 m^3 and 12.4 h. For comparative purposes, and for estimating larval fluxes to the experimental panels, we also modelled flow through a 5 cm thick 'slab' of water covering a 1 m^2 area of substratum at each of the two sites.

The complexities and asymmetries of the tidal patterns in this channel are clear from Figure 4; these arise from the partial retentive effects of the two tidal sills, which are emersed only at low water spring tides, and the occurrence of an amphidromic point 60 km to the southwest. The flow velocity data are plotted for the tidal cycle commencing with the northerly (flood) tide. This is of only a brief duration before the current equilibrates and reverses, with the tide protractedly ebbing to the south. The net flux of water through the entire rapid is from north to south.

Before undertaking the flow analysis, it was expected that settlement and recruitment would be greater at the HighFRS and that this would comprise the more favourable site for suspension-feeding sessile species: a greater flow was predicted to result in a greater flux of settling larvae to those panels than to the LowFRS. Contrary to expectations, however, recruitment to the LowFRS was consistently greater than to the HighFRS across all three repeat experiments, with multiples of $\times 1.43$, $\times 2.26$ and $\times 1.85$ more recruits. The overall multiple for all three repeats pooled was of 1.73 times as many larvae recruiting to the panels at the LowFRS than to the HighFRS. It is plausible that this differential response, which was not attributable to any particular species, reflects subtle larval responses to the development of the markedly different biofilms (reflecting the contrasting flow regimes) that we have observed at the two sites. In assessing that possibility, we presently are conducting reciprocal transfer experiments between the sites which permit analysis of the effects of the source of the initial biofilm. Preliminary results show no such effect: larvae simply respond to whether or not a substratum is biofilmed and recruit at rates characteristic for the two sites.

Although the S4 meter cannot provide a precise measure of flow across the actual substratum, it is likely that these data reflect real differences in larval supply to panel substrata at the two sites, despite these being exposed to similar concentrations of larvae. For the modelled 5 cm thick slab of water on a spring tide, the model predicted 1.5 times more flow per unit height of water column per unit width of the substratum at the HighFRS than at the LowFRS and 3.5 times more flow at the HighFRS than at the LowFRS on a neap tide. Because the available current meter data are only for extreme spring and neap tides, and the recruitment data are integrated over the entire spring-neap cycle, the assumption is made that the averages of the flushing model output (giving 2.0 times as much flow at the HighFRS than the LowFRS) will provide the best representation for the whole period.

A flow velocity of approximately ≥ 0.15 m s^{-1} probably results in general failure of many species' larvae to be able to settle and metamorphose, or for them to be resuspended and perhaps subject to bedload transport (e.g. Pawlik et al., 1991; Jonsson et al., 1991). Figure 4 shows that at the HighFRS, the current exceeds that critical velocity almost throughout the tidal cycle on both spring and neap tides. For the LowFRS, by contrast, only on neap tides is the current below that critical velocity for extended periods. By truncating the modelled data at that critical velocity, and assuming that flow rates < 0.15 m s^{-1} are suitable for settlement, a real estimate (in m^3 s^{-1} unit width^{-1} of the rapid) is obtainable from the revised flux model. On applying this condition, the model predicted 1.8 times more flow (= larval supply) over the LowFRS than the HighFRS; this was a reversal of our initial expectation for the sites, but a multiple which conforms closely to the 1.7 multiple observed for recruitment.

It is certain that other factors will contribute to the differential recruitment noted for these two sites. Although larval supply inevitably is a function of both larval concentration and water flux at a given site, larval settlement may in some localities (e.g. our HighFRS) be possible only over a very narrow 'window' in the tidal cycle. Evidently, contemporaneous larval trap data similar to those obtained by Yund et al. (1991) and Gaines & Bertness (1993) would be especially informative in this context, but higher flow rates do not necessarily confer higher larval supply or recruitment rates.

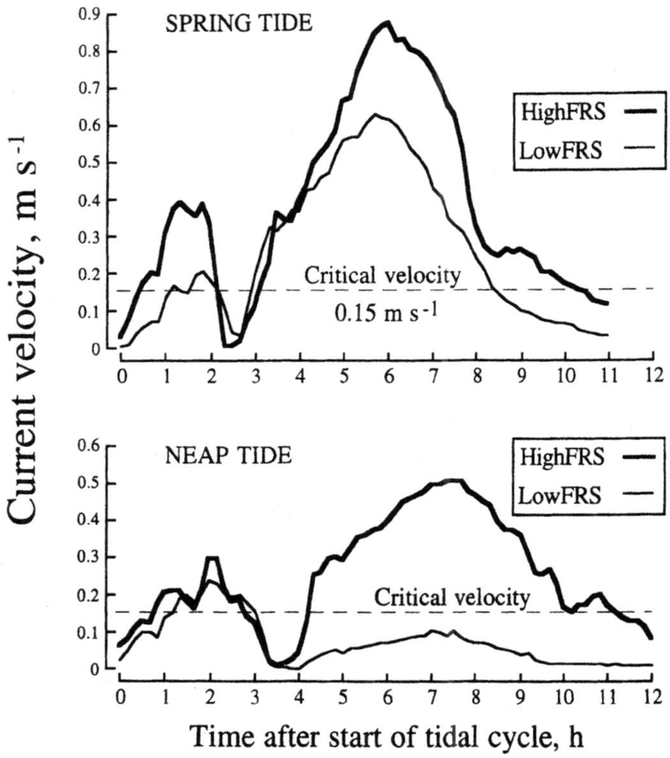

Figure 4. Current velocity data over complete tidal cycles on both spring and neap tides at the High Flow Rate Site (HighFRS) and Low Flow Rate Site (LowFRS), Clachan Seil (see Figure 1). These data were used in generating the simple flushing model for the whole rapid and the flux (= larval supply) predictions for the two epifaunal experimental sites. The arbitrarily applied critical flow velocity of 0.15 m s^{-1}, used in interpreting differential patterns of larval supply and recruitment at the two sites, is marked for each tidal curve (see text for further details).

Genetic evidence of realized dispersal by pelagic larvae and openness of benthic populations

Analyses of genetic heterogeneity of benthic populations provide an indirect means of assessing the scale of realized larval dispersal and whether or not populations are demographically open or closed over a given range. Most genetic studies for benthic marine invertebrates conform to the expectations that (1) planktonic larval transport confers migration and genetic cohesion between populations and (2) that species with non-pelagic development show marked interpopulation heterogeneity (Todd et al., 1998). In a comparative geographic study of genetic differentiation of intertidal populations of two species of epifaunal nudibranch molluscs (Todd et al., 1998), we investigated the population genetic consequences of the contrasting larval strategies of long-term planktotrophy (*Goniodoris nodosa*) and shorter term pelagic lecithotrophy (*Adalaria proxima*). Both species are strictly annual in life cycle and larvae hatch from benthic spawn masses as free-swimming veligers. Laboratory culture data indicated *G. nodosa* to be planktotrophic for up to ~ 3 months, whereas *A. proxima* larvae can metamorphose within 1–2 d of release but may delay metamorphosis for several weeks. On the basis of coastal currents we estimated larval transport potential at 10^2–10^3 km for *G. nodosa* and 10^1–10^2 km for *A. proxima*, and sampled respectively 11 and 19 populations for a single generation of both species around a 1600 km range of the northern British Isles (N Wales-Northern Ireland-Scotland-NE England). *G. nodosa* populations were predicted to show significant genetic differentiation on a scale of 10^3–10^4 km, in contrast to perhaps only 10^2–10^3 km for *A. proxima*. Thus, over the range sampled, we expected genetic homogeneity of *G. nodosa* populations, but perhaps an east-west coast differentiation for *A. proxima*.

F-statistic analyses (Weir, 1990) of the allelic (allozyme) data showed no evidence of differentiation for *Goniodoris nodosa* ($\theta = -0.003$) but *Adalaria proxima* over that same 1600 km range showed extremely high levels of differentiation ($\theta = 0.306$; 95% c.i. 0.179, 0.327). Supplementary analyses of nine additional populations of *A. proxima* over a 26 km range in western Scotland – around two islands characterised by highly dispersive surface tidal currents attaining 3.6 m s^{-1} – revealed similar levels of differentiation (θ = 0.221; 95% c.i. 0.132, 0.257) to the 1600 km macro-scale data: populations of *A. proxima* were differentiated on scales of < 10 km, with some indications of differentiation on scales as small as 10^2 m. Moreover, these latter data also revealed a significant inverse relationship between genetic differentiation and geographic distance. We have genetic data for up to 10 consecutive generations for certain populations of *A. proxima*. From those data, and field transplant experiments tracking introduced genetic markers between generations, we concluded that *A. proxima* populations are characterized by localized recruitment, founder effects, random genetic drift and occasional local extinction.

Despite reproduction by means of pelagic larvae, it is evident that typically there are minimal levels of larval migration in *Adalaria proxima* and local populations can be demographically closed on very small scales (10^2–10^3 m). The clear implication of these data is that possession of a pelagic lecithotrophic larval stage does not necessarily result in dispersal and colonisation potential: in the field, *A. proxima* larvae appear behaviourally adapted to effectively avoid dispersal in the water column and thereby generally recruit locally. These results highlight the pitfalls of our assuming the dispersal potential, extent of gene flow, scales of population genetic differentiation and openness of populations from knowledge only of the larval type or even laboratory observations of the duration of the pelagic larval stage.

Openness of epifaunal assemblages

Despite a wide array of investigations of epifaunal assemblage development and succession, detailed demographic analyses – which might allow assessments of the interplay between larval settlement, recruitment, post-settlement growth and mortality – are conspicuously lacking. In Hughes' (1990) photographic study of recruitment, growth and mortality of a sublittoral Californian population of the encrusting bryozoan, *Celleporella pumicosa*, recruitment of 120 colonies was noted over the 18 months of study. New recruits were visible when about two weeks old. At the start of the study, *C. pumicosa* averaged 206 colonies m^{-2} (6.1% cover) but the population steadily declined throughout, whereas the assemblage as a whole fluctuated seasonally. Brooded embryos of *C. pumicosa* were found only between October and May and yet recruitment was recorded year-round. There was no correlation between mortality rate of recruits and density of established *C. pumicosa* colonies (cf. Svane, 1984 for *Ascidia mentula*) or total benthic cover, but the possibility that post-settlement mortality over the first two weeks was density dependent cannot be discounted. Longer term data would be especially valuable, but the evidence indicated that this population, occupying a 250 m-long rockface, was demographically open although it may have been a 'marginal' declining component of a larger closed population. By contrast, empirical observational studies of the solitary ascidian *Ciona intestinalis* (Petersen & Svane, 1995) in Denmark strongly indicate that some epifaunal sublittoral populations on hard substrata may be demographically closed, as a result of very restricted larval dispersal. As for all ascidians, *C. intestinalis* reproduces by means of pelagic lecithotrophic tadpole larvae, but dispersal there appears especially constrained by at least some developing embryos failing to escape from sticky mucous egg strings spawned by the adult.

Despite the fuzziness of the ecosystem concept, even when applied to seemingly clearly-bounded systems such as coral reefs (Hatcher, 1997), of all sessile epifaunal benthic systems coral reefs might provide the ideal opportunity for assessing whole-assemblage openness: with the exception perhaps of extensive fringing reefs, coral reef systems commonly are comprised of multiple patches (discrete 'islands' on a wide range of spatial scales) and this can include even barrier and ribbon reefs. Coral larval dispersal and recruitment to the isolated Helix reef (Great Barrier Reef) were dependent both on water residence (flushing) time and local variations in circulation (Sammarco & Andrews, 1988, 1989; Sammarco, 1994), to the extent that reefs could 'self-seed'. The Helix reef sites with the lowest flushing rates had the highest recruitment. Although superficially similar in outcome to the differential levels of epifaunal recruitment at the Low *versus* High flow rate sites discussed above, it is important to emphasize that flushing times on those coral reefs were determined primarily by reef size (see also Black, 1993), not by flow velocities.

High levels of self-seeding might imply that coral assemblages are demographically relatively closed, but Sammarco & Andrews (1988, 1989) also found that a substantial proportion of planula larvae did settle at a distance (cf. *Adalaria proxima*, above) resulting in colonization and gene flow potential between neighbouring reefs (see also Fisk & Harriott, 1990). Although larval exchange between distant reefs remains uncertain, in a model simulation, Black (1994) found that patch reefs within a group could show surprisingly high levels of connectedness through larval interchange. Planulating corals (especially Pocilloporidae) release well-developed lecithotrophic larvae, whereas free-spawning species (e.g. Acroporidae, Poritidae, Faviidae) release unfertilized gametes. This alone confers a range of potential dispersal strategies on corals (Fisk & Harriott, 1990; Ayre & Dufty, 1994) although these patterns are far from simple even within families. Large species of Poritidae, for example, typically are broadcast spawners of gametes whereas smaller species tend to brood embryos and release planulae (Glynn, et al. 1994).

The few available genetic data indicate that some coral species' populations are largely demographically closed but the complexity of some coral species' reproductive strategies, and the pervasive ecological problem of scale, might render such assessments difficult. *Pocillopora damicornis*, for example, has been extensively studied and generally releases well-developed planula larvae (but see Glynn et al., 1991), which may be produced asexually or sexually (Benzie et al., 1995). This species can, however, also reproduce by fragmentation and has geographically varying and complex patterns of seasonal and lunar periodicities of reproduction (Tanner, 1996). Benzie et al. (1995) recorded significant differentiation of lagoonal and reef crest populations of *P. damicornis* at One Tree Island (Great Barrier Reef). Ayre & Dufty (1994) recorded extremely high levels of genetic variation and population subdivision for the pocilloporid *Seriatopora hystrix* at 12 reefs in the central Great Barrier Reef, indicating very restricted gene flow in this planulator. It may well be that pocilloporids typically display relatively closed populations whereas, on a similar spatial scale, populations of other families of corals will be relatively open, with the result that the assemblage as a whole lies intermediate to the notional extremes. Generalizations about coral reef assemblages are none the less problematic, not only because of the wide range of reef types, but also because of the geographic and ecological contrasts between Atlantic and Indo-Pacific reefs and the uneven distribution of studied locations (Connell, 1997).

With the obvious exception of cnidarians on coral reefs, temperate and tropical hard substratum epifaunal assemblages are characterized by the phyla Porifera, Bryozoa and Urochordata (Sutherland & Karlson, 1977) and this generalization appears to extend also to the Antarctic (Arntz et al., 1994). None of these phyla are especially speciose, with the Porifera (Nielsen, 1995) and Bryozoa (Boardman & Cheetham, 1987) comprising approximately 5000 and 4000 species respectively worldwide, and the benthic colonial and solitary ascidians the majority of the approximately 1250 species in the Urochordata (Nielsen, 1995). With the exception of five genera (30 species) of bryozoans – for which reproduction by means of longer-lived cyphonautes larvae has been recorded (Jackson & Coates, 1986) – all of the species in these epifaunal phyla reproduce by means of pelagic lecithotrophic larvae, the large majority of which are short-lived in the plankton. Short-distance dispersal is therefore almost certainly obligatory and given the above data for *Adalaria proxima* might in fact be minimal for many or most species. Thus, with specific reference to hard substratum epifaunal assemblages, widespread dispersal of juveniles may not be as common in shallow-water benthic habitats as suspected (cf. Underwood & Fairweather, 1989).

Every assemblage or community will lie at an intermediate point on the continuum of totally open to totally closed demography. In this regard epifaunal assemblages (which are characterized by species of clonal organisation and pelagic lecithotrophic larvae) are far separated from typical rocky shore invertebrate assemblages (commonly with planktotrophic larvae and characterized by species of unitary organisation). The latter assemblages, in some instances, may be especially responsive to variations in larval supply, whereas post-settlement processes might be prevalent in hard substratum epifaunal fouling assemblages. Whether this contrast arises simply from the differences in larval strategy or growth form remains open to future exploration. For both hard substratum epifaunal assemblages and coral reefs, Connell & Keough (1985) predicted that the main determinant of structure would be the effectiveness of surviving residents – adjacent to disturbance-generated patches – in recolonizing space. Surviving residents can influence the subsequent fate of patches by several means, including (i) vegetative growth/extension onto the free space, (ii) by affecting the local physical environment, (iii) by capturing and eating incoming larvae, and (iv) by the inhibitory influence of allelochemical agents on larval settlement. If vegetative extension and space preemption are especially prevalent, this will exaggerate further any tendency towards demographic closedness of epifaunal assemblages by mediating against the successful recruitment of pelagic larvae which may have immigrated from a distance.

Whilst long-term planktotrophic larvae clearly are subject to enforced dispersal during the obligatory (pre-competent) pelagic phase, our genetic data show that simple extrapolations of larval dispersal potential from laboratory observations of lecithotrophic larval longevity may be markedly erroneous. Our data argue against dispersal being an exclusive selective advantage in that category of larval type and certain marine benthic populations and assemblages might be demographically much more closed than suspected from knowledge of the larval developmental strategy alone.

Acknowledgements

My heartfelt thanks are due to my various colleagues and collaborators who have helped throughout. I thank especially Chris Cromey, Mick Keough, Walt Lambert, Neil MacDougall, Tom Pearson, Dave Raffaelli, John Thorpe, Cynthia Trowbridge, Stephanie Turner and Sabine Wieczorek. Financial support was provided by NERC in grant numbers GR3/1083, 8167 and 9951. To all I am grateful.

References

Arntz, W. E., T. Brey & V. A. Gallardo, 1994. Antarctic zoobenthos. Oceanogr. Mar. Biol. Ann. Rev. 32: 241–304.

Ayre, D. J. & S. Dufty, 1994. Evidence for restricted gene flow in the viviparous coral *Seriatopora hystrix* on Australia's Great Barrier Reef. Evolution 48: 1183–1201.

Benzie, J. A. H., A. Haskell & H. Lehmann, 1995. Variation in the genetic composition of coral (*Pocillopora damicornis* and *Acropora palifera*) populations from different reef habitats. Mar. Biol. 121: 731–739.

Bertness, M. D., S. D. Gaines, E. G. Stephens & P. O. Yund, 1992. Components of recruitment in populations of the acorn barnacle *Semibalanus balanoides* (Linnaeus). J. exp. mar. Biol. Ecol. 156: 199–215.

Black, K. P., 1994. Developments in our knowledge of dispersal on the Great Barrier Reef. In P. W. Sammarco & M. L. Heron (eds), The Bio-physics of Marine Larval Dispersal. American Geophysical Union, Washington DC: 159–192.

Black, K. P., 1993. The relative importance of local retention and inter-reef dispersal of neutrally buoyant material on coral reefs. Coral Reefs 12: 43–53.

Boardman, R. S. & A. H. Cheetham, 1987. Phylum Bryozoa. In R. S. Boardman, A. H. Cheetham & A. J. Rowell (eds), Fossil invertebrates. Blackwell Scientific Publications, Oxford: 497–549.

Booth, D. J. & D. M. Brosnan, 1995. The role of recruitment dynamics in rocky shore and coral reef fish communities. Adv. Ecol. Res. 26: 309–385.

Butman, C. A., 1987. Larval settlement of soft-sediment invertebrates: the spatial scales of pattern explained by active habitat selection and the emerging rôle of hydrodynamical processes. Oceanogr. Mar. Biol. Ann. Rev. 25: 113–165.

Caffey, H. M., 1985. Spatial and temporal variation in settlement and recruitment of intertidal barnacles. Ecol. Monogr. 55: 313–332.

Caley, M. J., M. H. Carr, M. A. Hixon, T. P. Hughes, G. P. Jones & B. A. Menge, 1996. Recruitment and the local dynamics of open marine populations. Ann. Rev. Ecol. Syst. 27: 477–500.

Connell, J. H., 1978. Diversity in tropical rain forests and coral reefs. Science 199: 1302–1310.

Connell, J. H., 1985. The consequences of variation in initial settlement vs. post-settlement mortality in rocky intertidal communities. J. exp. mar. Biol. Ecol. 93: 11–45.

Connell, J. H., 1997. Disturbance and recovery of coral assemblages. Coral Reefs 16, Suppl.: S101–S113.

Connell, J. H. & M. J. Keough, 1985. Disturbance and patch dynamics of subtidal marine animals on hard substrata. In S. T. A. Pickett & P. S. White (eds), The Ecology of Natural Disturbance and Patch Dynamics. Academic Press, London: 125–151.

Connell, J. H. & R. O. Slatyer, 1977. Mechanisms of succession in natural communities and their role in community stability and organization. Am. Nat. 111: 1119–1144.

Connell, J. H. & W. P. Sousa, 1983. On the evidence needed to judge ecological stability or persistence. Am. Nat. 121: 789–824.

Day, R. & L. McEdward, 1984. Aspects of the physiology and ecology of pelagic larvae of marine benthic invertebrates. In K. A. Steidinger & L. M. Walker (eds), Marine plankton life cycle strategies. CRC Press, Inc., Boca Raton, FL: 93–120.

Doherty, P. J., 1981. Coral reef fishes: recruitment-limited assemblages? Proc. Int. Coral Reef Symp., 4th. 2: 465–470.

Farrell, T. M., D. Bracher & J. Roughgarden, 1991. Cross-shelf transport causes recruitment to intertidal populations in central California. Limnol. Oceanogr. 36: 279–288.

Fisk, D. A. & V. J. Harriott, 1990. Spatial and temporal variation in coral recruitment on the Great Barrier Reef: implications for dispersal hypotheses. Mar. Biol. 107: 485–490.

Frank, K. T. & W. C. Leggett, 1994. Fisheries ecology in the context of ecological and evolutionary theory. Ann. Rev. Ecol. Syst. 25: 401–422.

Gaines, S. & J. Roughgarden, 1985. Larval settlement rate: a leading determinant of structure in an ecological community of the marine intertidal zone. Proc. Acad. nat. Sci. Philad. 82: 3707–3711.

Gaines, S., S. Brown & J. Roughgarden, 1985. Spatial variation in larval concentrations as a cause of spatial variation in settlement for the barnacle, *Balanus glandula*. Oecologia 67: 267–272.

Gaines, S. D. & M. D. Bertness, 1992. Dispersal of juveniles and variable recruitment in sessile marine species. Nature 360: 579–580.

Gaines, S. D. & M. Bertness, 1993. The dynamics of juvenile dispersal: why field ecologists must integrate. Ecology 74: 2430–2435.

Glynn, P. W., N. J. Gassman, C. M. Eakin, J. Cortes, D. B. Smith & H. M. Guzman, 1991. Reef coral reproduction in the eastern Pacific: Costa Rica, Panama, and Galapagos Islands (Ecuador): I. Pocilloporidae. Mar. Biol. 109: 355–368.

Glynn, P. W., S. B. Colley, C. M. Eakin, D. B. Smith, J. Cortes, N. J. Gassman, H. M. Guzman, J. B. Del Rosario & J. S. Feingold, 1994. Reef coral reproduction in the eastern Pacific: Costa

Rica, Panamá, and Galápagos Islands (Ecuador): I. Poritidae. Mar. Biol. 118: 191–208.

Grahame, J. & G. M. Branch, 1985. Reproductive patterns of marine invertebrates. Oceanogr. Mar. Biol. Ann. Rev. 23: 373–398.

Greene, C. H. & A. Schoener, 1982. Succession on marine hard substrata: a fixed lottery. Oecologia 55: 289–297.

Grosberg, R. K. & D. R. Levitan, 1992. For adults only? Supply-side ecology and the history of larval biology. Trends Ecol. Evol. 7: 130–133.

Grosberg, R. K., 1981. Competitive ability influences habitat choice in marine invertebrates. Nature 290: 700–702.

Grosberg, R. K., 1982. Intertidal zonation of barnacles: the influence of planktonic zonation of larvae on vertical distribution of adults. Ecology 63: 894–899.

Hatcher, B. G., 1997. Coral reef ecosystems: how much greater is the whole than the sum of the parts? Coral Reefs 16, Suppl.: S77–S91.

Hills, J.M. & J.C. Thomason, 1996. A multi-scale analysis of settlement density and pattern dynamics of the barnacle *Semibalanus balanoides*. Mar. Ecol. Prog. Ser. 138: 103–115.

Holm, E. R., 1990. Effects of density-dependent mortality on the relationship between recruitment and larval settlement. Mar. Ecol. Prog. Ser. 60: 141–146.

Hughes, T. P., 1990. Recruitment limitation, mortality, and population regulation in open systems: a case study. Ecology 71: 12–20.

Hunt, H. L. & R. E. Scheibling, 1996. Physical and biological factors influencing mussel (*Mytilus trossulus, M. edulis*) settlement on a wave-exposed rocky shore. Mar. Ecol. Prog. Ser. 142: 135–145.

Jablonski, D. & R. A. Lutz, 1983. Larval ecology of marine invertebrates: paleobiological implications. Biol. Rev. 58: 21–89.

Jackson, J. B. C. & A. G. Coates, 1986. Life cycles and evolution of clonal (modular) animals. Phil. Trans. r. Soc., Lond. 313: 7–22.

Jackson, G. A. & R. R. Strathmann, 1981. Larval mortality from offshore mixing as a link between precompetent and competent periods of development. Am. Nat. 118: 16–26.

Jonsson, P. R., C. André & M. Lindegarth, 1991. Swimming behaviour of marine bivalve larvae in a flume boundary-layer flow: evidence for near-bottom confinement. Mar. Ecol. Prog. Ser. 79: 67–76.

Keough, M. J. & B. J. Downes, 1982. Recruitment of marine invertebrates: the role of active larval choices and early mortality. Oecologia 54: 348–352.

Keough, M. J. & P. T. Raimondi, 1995. Responses of settling invertebrate larvae to bioorganic films: effects of different types of film. J. exp. mar. Biol. Ecol. 185: 235–253.

Keough, M. J. & P. T. Raimondi, 1996. Responses of settling marine invertebrate larvae to microbial films: effects of large-scale variation in films. J. exp. mar. Biol. Ecol. 207: 59–78.

Knowlton, N. & J. B. C. Jackson, 1994. New taxonomy and niche partitioning on coral reefs: jack of all trades or master of some? Trends Ecol. Evol. 9: 7–9.

Knowlton, N. & B. D. Keller, 1986. Larvae which fall far short of their potential: highly localized recruitment in an alpheid shrimp with extended larval development. Bull. mar. Sci. 39: 213–223.

Lohse, D. P., 1993. The importance of secondary substratum in a rocky intertidal community. J. exp. mar. Biol. Ecol. 166: 1–17.

Menge, B. A. & J. P. Sutherland, 1987. Community regulation: variation in disturbance, competition, and predation in relation to environmental stress and recruitment. Am. Nat. 130: 730–757.

Menge, B. A., 1991. Relative importance of recruitment and other causes of variation in rocky intertidal community structure. J. exp. mar. Biol. Ecol. 146: 69–100.

Menge, B. A., T. M. Farrell, A. M. Olson, P. van Tamelen & T. Turner, 1993. Algal recruitment and the maintenance of a plant mosaic in the low intertidal region of the Oregon coast. J. exp. mar. Biol. Ecol. 170: 91–116.

Minchinton, T. E. & R. E. Scheibling, 1991. The influence of larval supply and settlement on the population structure of barnacles. Ecology 72: 1867–1879.

Minchinton, T. E. & R. E. Scheibling, 1993a. Variations in sampling procedure and frequency affect estimates of recruitment of barnacles. Mar. Ecol. Prog. Ser. 99: 83–88.

Minchinton, T. E. & R. E. Scheibling, 1993b. Free space availability and larval substratum selection as determinants of barnacle population structure in a developing rocky intertidal community. Mar. Ecol. Prog. Ser. 95: 233–244.

Miron, G., B. Boudreau & E. Bourget, 1995. Use of larval supply in benthic ecology: testing correlations between larval supply and larval settlement. Mar. Ecol. Prog. Ser. 124: 301–305.

Neal, A. L., F. N. Simões & A. B. Yule, 1996. Interactions between shear rates and biofilms affecting exploratory behaviour by cyprids of *Elminius modestus* (Cirripedia). Mar. Biol. 127: 241–246.

Nielsen, C., 1995. Animal Evolution. Oxford University Press, Oxford. 467 pp.

Ólafsson, E. B., C. H. Peterson & W. G. Ambrose, 1994. Does recruitment limitation structure populations and communities of macro-invertebrates in marine soft sediments: the relative significance of pre- and post-settlement processes. Oceanogr. Mar. Biol. Ann. Rev. 32: 65–109.

Pawlik, J. R., C. A. Butman & V. R. Starczak, 1991. Hydrodynamic facilitation of gregarious settlement of a reef-building tube worm. Science, 251: 421–424.

Pechenik, J. A., 1990. Delayed metamorphosis by larvae of benthic marine invertebrates: does it occur? Is there a price to pay? Ophelia 32: 63–94.

Petersen, C. G. J., 1918. The sea bottom and its production of fish food. A survey of the work done in connection with valuation of the Danish waters from 1883–1917. Rep. Danish Biol. Stat. 25: 1–61.

Petersen, J. K. & I. Svane, 1995. Larval dispersal in the ascidian *Ciona intestinalis* (L.). Evidence for a closed population. J. exp. mar. Biol. Ecol. 186: 89–102.

Raimondi, P. T., 1990. Patterns, mechanisms, consequences of variability in settlement and recruitment of an intertidal barnacle. Ecol. Monogr. 60: 283–309.

Rhoads, D. C. & D. K. Young, 1970. The influences of deposit-feeding organisms on sediment stability and community trophic structure. J. mar. Res. 28: 150–178.

Roughgarden, J., Y. Isawa & C. Baxter, 1985. Demographic theory for an open marine population with space-limited recruitment. Ecology 66: 54–67.

Sale, P. F., 1977. Maintenance of high diversity in coral reef fish communities. Am. Nat. 111: 337–359.

Sale, P. F., 1990. Recruitment of marine species: is the bandwagon rolling in the right direction? Trends Ecol. Evol. 5: 25–27.

Sale, P. F., 1991. The ecology of fishes on coral reefs. Academic Press, London. 754 pp.

Sammarco, P. W., 1994. Larval dispersal and recruitment processes in Great Barrier Reef corals: analysis and synthesis. In P. W. Sammarco & M. L. Heron (eds), The Bio-physics of Marine Larval Dispersal. American Geophysical Union, Washington DC: 35–72.

Sammarco, P. W. & J. C. Andrews, 1988. Localized dispersal and recruitment in Great Barrier Reef corals: the Helix experiment. Science 239: 1422–1424.

Sammarco, P. W. & J. C. Andrews, 1989. The Helix experiment: differential localized dispersal and recruitment patterns in Great Barrier Reef corals. Limnol. Oceanogr. 34: 896–912.

Seed, R. & T. H. Suchanek, 1992. Population and community ecology of *Mytilus*. In E. Gosling (ed.), The Mussel *Mytilus*: Ecology, Physiology, Genetics, and Culture. Elsevier: 87–169.

Setran, A. C., 1992. A new plankton trap for use in the collection of rocky intertidal zooplankton. Limnol. Oceanogr. 37: 669–674.

Shaw, E. M., 1988. Hydrology in Practice, 2nd edition. Chapman & Hall, London: 97–103.

Snelgrove, P. V. R. & C. A. Butman, 1994. Animal-sediment relationships revisited: cause versus effect. Oceanogr. Mar. Biol. Ann. Rev. 32: 111–177.

Strathmann, R. R., 1980. Why does a larva live so long? Paleobiology 6: 373–376.

Strathmann, R. R., 1985. Feeding and non-feeding larval development and life history evolution in marine invertebrates. Ann. Rev. Ecol. Syst. 16: 339–361.

Suchanek, T. H., 1985. Mussels and their rôle in structuring rocky shore communities. In P. G. Moore & R. Seed (eds), The Ecology of Rocky Coasts. Hodder & Stoughton, London: 70–96.

Sutherland, J. P., 1987. Recruitment limitation in a tropical intertidal barnacle: *Tetraclita panamensis* (Pilsbry) on the Pacific coast of Costa Rica. J. exp. mar. Biol. Ecol. 113: 267–282.

Sutherland, J. P., 1990. Recruitment regulates demographic variation in a tropical intertidal barnacle. Ecology 71: 955–972.

Sutherland, J. P. & R. H. Karlson, 1977. Development and stability of the fouling community at Beaufort, North Carolina. Ecol. Monogr. 47: 425–446.

Svane, I., 1984. Observations on the long-term population dynamics of the perennial ascidian, *Ascidia mentula* O.F. Müller, on the Swedish west coast. Biol. Bull. 167: 630–646.

Svane, I. & T. Lundälv, 1981. Reproductive pattern and population dynamics of *Ascidia mentula* O.F. Müller on the Swedish west coast. J. exp. mar. Biol. Ecol. 50: 163–182.

Tanner, J. E., 1996. Seasonality and lunar periodicity in the reproduction of Pocilloporid corals. Coral Reefs 15: 59–66.

Todd, C. D. & M. J. Keough, 1994. Larval settlement in hard substratum epifaunal assemblages: a manipulative field study of the effects of substratum filming and the presence of incumbents. J. exp. mar. Biol. Ecol. 181: 159–187.

Todd, C. D. & S. J. Turner, 1986. Ecology of intertidal and sublittoral cryptic epifaunal assemblages: I. Experimental rationale and the analysis of larval settlement. J. exp. mar. Biol. Ecol. 99: 199–231.

Todd, C. D., W. J. Lambert & J. P. Thorpe, 1998. The genetic structure of intertidal populations of two species of nudibranch molluscs with planktotrophic and pelagic lecithotrophic larval stages: are pelagic larvae 'for' dispersal? J. exp. mar. Biol. Ecol. In Press.

Turner, S. J. & C. D. Todd, 1993. The early development of epifaunal assemblages on artificial substrata at two intertidal sites on an exposed rocky shore in St. Andrews Bay, N.E. Scotland. J. exp. mar. Biol. Ecol. 166: 251–272.

Turner, S. J. & C. D. Todd, 1994. Competition for space in encrusting bryozoan assemblages: the influence of encounter angle, site and year. J. mar. biol. Ass. U.K. 74: 603–622.

Underwood, A. J. & P. G. Fairweather, 1989. Supply-side ecology and benthic marine assemblages. Trends Ecol. Evol. 4: 16–20.

Walters, L. J. & D. S. Wethey, 1996. Settlement and early post-settlement survival of sessile marine invertebrates on topographically complex surfaces – the importance of refuge dimensions and adult morphology. Mar. Ecol. Prog. Ser. 137: 161–171.

Weir, B. S., 1990. Genetic Data Analysis. Methods for Discrete Population Data. Sinauer Associates, Sunderland, MA. 377 pp.

Wethey, D. S., 1985. Local and regional variation in settlement and survival in the littoral barnacle *Semibalanus balanoides* (L.): patterns and consequences. In P. G. Moore & R. Seed (eds), The Ecology of Rocky Coasts. Hodder & Stoughton, London: 194–202.

Wieczorek, S. K. & C. D. Todd, 1997. Inhibition and facilitation of settlement of epifaunal marine invertebrate larvae by microbial biofilm cues. Biofouling 12: 81–118.

Willis, B. L. & J. K. Oliver, 1990. Direct tracking of coral larvae: implications for dispersal studies of planktonic larvae in topographically complex environments. Ophelia 32: 145–162.

Wilson, W. H., 1991. Competition and predation in marine soft-sediment communities. Ann. Rev. Ecol. Syst. 21: 221–241.

Wing, S. R., J. L. Largier, L. W. Botsford & J. F. Quinn, 1995. Settlement and transport of benthic invertebrates in an intermittent upwelling region. Limnol. Oceanogr. 40: 316–329.

Woodin, S. A., 1976. Adult-larval interactions in dense infaunal assemblages: patterns of abundance. J. mar. Res. 34: 25–41.

Woodin, S. A., 1991. Recruitment of infauna: positive or negative cues? Am. Zool. 31: 797–807.

Woodin, S. A. & J. B. C. Jackson, 1979. Interphyletic competition between marine benthos. Am. Zool. 19: 1029–1043.

Young, C. M., 1990. Larval ecology of marine invertebrates: a sesquicentennial history. Ophelia 32: 1–48.

Yund, P. O., S. D. Gaines & M. D. Bertness, 1991. Cylindrical tube traps for larval sampling. Limnol. Oceanogr. 36: 1167–1177.

Hydrobiologia **375/376:** 23–34, 1998.
S. Baden, L. Pihl, R. Rosenberg, J.-O. Strömberg, I. Svane & P. Tiselius (eds),
Recruitment, Colonization and Physical–Chemical Forcing in Marine Biological Systems.

Recruitment in *Macoma balthica* after mild and cold winters and its possible control by egg production and shrimp predation

J. J. Beukema, P. J. C. Honkoop & R. Dekker
Netherlands Institute for Sea Research, P.O. Box 59, 1790 AB Den Burg, Texel, The Netherlands

Key words: recruitment, winter temperature, predators, tidal flats, Wadden Sea

Abstract

In several species of bivalve molluscs recruitment is more successful in a summer after a cold than after a mild winter. Results of a long-term monitoring program on Balgzand, a 50-km^2 tidal-flat area in the westernmost part of the Wadden Sea, included a significantly negative relationship between winter temperatures and subsequent densities of *Macoma balthica* recruits in August. We show that this correlation may have arisen from two winter-temperature related processes: 1. Numbers of eggs spawned in April (both per female and per m^2) were higher after cold than after mild winters. Results of experiments showed that this is due to higher body weights in late winter/early spring after cold than after mild winters, resulting from lower rates of weight loss at low than at high winter temperatures. 2. Densities of juvenile shrimps *Crangon crangon* on the tidal flats were significantly lower in springs after cold than after mild winters. Shrimps are important predators of post-larval just-settled *Macoma*. Numbers of *Macoma* recruits per spawned egg were generally between 0.001 and 0.0001 and were significantly higher at low than at high shrimp densities in late-spring. We conclude that the latter process was the decisive one.

Introduction

There is generally substantial year-to-year variation in reproductive success of marine benthic species. In several species of bivalve molluscs annual recruit densities have been assessed during long periods. Such data series indicate differences in spat densities amounting to two or three orders of magnitude: from (almost) complete recruitment failure in certain years to high densities of hundreds or thousands per m^2 in other years.

Such variability originates both (e.g. Möller, 1986) from abiotic environmental conditions and from biotic interrelations (such as predator-prey relationships). Among the abiotic factors, in particular the character of the preceding winter appears to play a role. Strong bivalve year classes originating after an unusually cold winter have been reported in several species, including cockles *Cerastoderma edule* (L.) by Thamdrup (1935), Kreger (1940), Kristensen (1957), Hancock (1973), Möller & Rosenberg (1983), Jensen & Jensen (1985), Yankson (1986), and Beukema (1982, 1992b); in mussels *Mytilus edulis* L. by Bahr (1950), Savage (1956), Jensen & Jensen (1985), Dijkema (1992), Beukema (1982, 1992b) and Young et al. (1996); in gaper clams *Mya arenaria* L. by Smidt (1951), Kühl (1955), Jensen & Jensen (1985), Möller (1986) and Beukema (1982, 1992b); and in the baltic tellinid *Macoma balthica* (L.) by Jensen & Jensen (1985), Desprez et al. (1991) and Beukema (1982, 1992b). As severe winters tend to occur simultaneously over vast areas, such winter-related variation in abundance is synchronized over large areas, including in some species e.g. the entire Wadden Sea (Beukema & Essink, 1986; Beukema et al., 1996). If recruitment failure occurs during the same prolonged period in more than one important species (as was the case in three successive years in both cockles and mussels in the Wadden Sea), the effects can be serious (Beukema, 1993a; Beukema & Cadée, 1996).

Among the biotic factors, densities of adults and abundance of predators are known to play important roles in determining recruitment success. The pres-

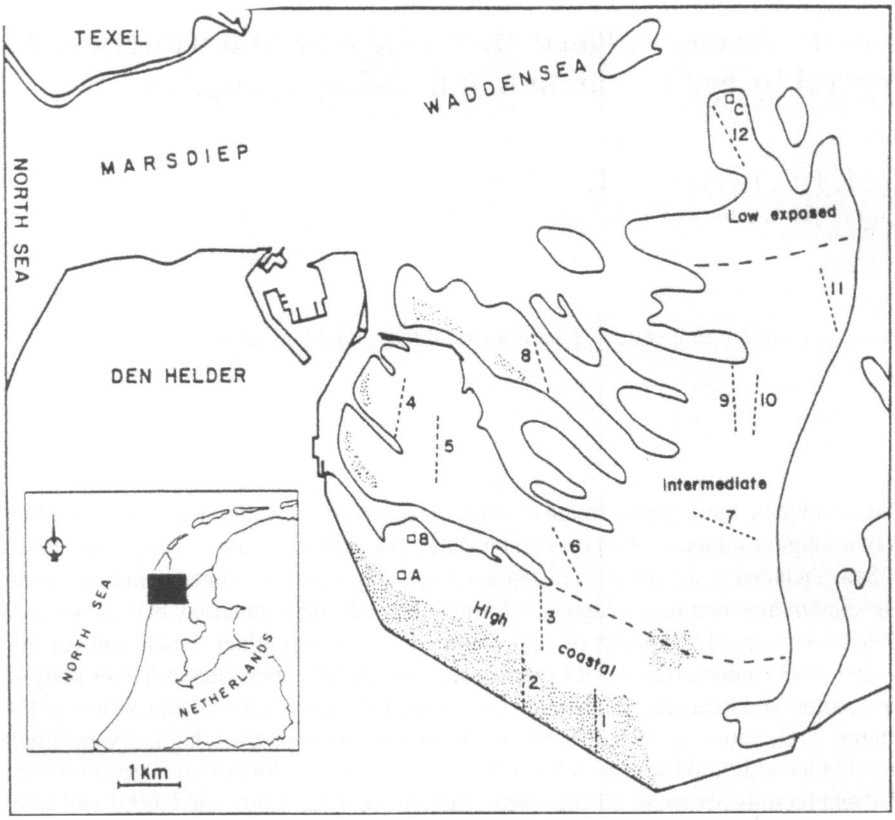

Figure 1. The Balgzand tidal-flat area, i.e. the area above low-water level, roughly indicated by the thin line. The positions of the 15 sampling stations (3 squares: A, B, and C, and the 12 transects: numbered 1 to 12) are indicated.

ence of high numbers of predators (such as shrimps and crabs) can severely reduce recruitment in several species of bivalves (examples can be found in Reise, 1985). Particularly shrimps *Crangon crangon* (L.) can act as important predators on *M. balthica* spat (Keus, 1986; Mattila et al., 1990). The role of adult abundance can vary from positive effects by higher numbers of spawned eggs to negative effects by cannibalism and competition for space (e.g. Bonsdorff et al., 1986, 1995).

The tellinid bivalve *Macoma balthica* (L.) is a suitable species to study in detail processes leading to success or failure of recruitment. Long-term data on abundance in the field are relatively reliable and precise due to its relatively even (Beukema, 1983) and widespread (Dankers & Beukema, 1982) distribution. It was found in high numbers in all years and at all sampling stations of a Wadden Sea tidal-flat monitoring program (Beukema, 1993b). It is also a suitable species for experimental manipulation.

In the present study, we combine long-term field data in the form of correlations (with e.g. winter temperature) with results of experiments to show which processes were involved in these correlative relationships and to prove that these relationships represented causative mechanisms.

Materials and methods

Field observations

All reported field observations were made at Balgzand, a 50-km^2 tidal-flat area in the westernmost part of the Wadden Sea, located at about 53° N and 5° E. A long-term monitoring program of the benthic macrofauna started in the early 1970s (since 1970 annual data are available on adult *M. balthica*, since 1973 also on recruits). The program includes twice-annual quantitative sampling (mostly in March and in August) at 15 permanent stations (see Figure 1): 12 transects of 1 km each (with 50 equally-spaced core samples) and 3 squares of 900 m^2 each (with at least 18 randomly taken core samples). Minimal sample size per station was 0.45 m^2 in summer and 0.90 m^2 in winter. The 15 stations are scattered over Balgzand and cover all types of sediment and the entire intertidal depth range, which runs from about +60 to about −80 cm compared with mean-tide level (MTL). See Beukema (1988) for details on the Balgzand fauna and Beukema (1993b) for the distribution of *M. balthica* over Balgzand. Numbers of adults are expressed as n per m^2. Shell lengths were measured in all individuals. Weights of the soft parts (body mass) were routinely determined in mm groups of equally-sized individuals as ash-free dry weight (AFDW). These weights are expressed as a mean 'condition factor' or body-mass index (BMI), defined as AFDW (in mg) divided by cubic shell length (in cm), thus $BMI = W/L^3$. The cores were sieved in the field on 1-mm mesh screens. Thus only those *Macoma* with a shell length of over 1.5 mm were retained. In the following, the 15-station average (expressed in n per m^2) of young-of-the-year of > 1.5 mm observed in August is denoted as recruitment. Because such data are available only from 1973 onwards, we use only data from the long-term program that were collected in the years 1973 up to and including 1996 in the present paper.

To study also the dynamics of the smaller post-larvae, weekly to monthly samples were taken in the May-September periods of five years (1987, 1988, 1994, 1995, and 1996) at a few Balgzand stations (near square B in Figure 1). These samples were sieved over 0.3-mm meshes, thus retaining all spat of over 0.5 mm length.

Numbers of eggs (and BMI) were determined in 1995 and 1996 by collecting high numbers of adults at the end of March at three Balgzand stations (near square B and transects 2 and 8 in Figure 1), differing strongly in intertidal height (they are described in Honkoop & Beukema, 1997). By a temperature shock, the animals were forced to release their eggs. Numbers and sizes of eggs spawned were estimated separately for each female. The spawned females were checked for (nearly) complete emptiness (for more details, see Honkoop & Van der Meer, 1997a).

Densities of shrimps *C. crangon* on the Balgzand tidal flats were estimated during the April-August periods of the years 1983 and following. Samples were taken mostly at about 10-d intervals at three neighbouring permanent stations (near squares A and B in Figure 1), located between 0.5 and 1.0 km from the coast at levels between 10 cm above and 40 cm below MTL. The samples were taken at low tide. The total of 120 cores per sampling date approximates 1 m^2. Shrimp abundance is expressed in mg AFDW per m^2. For further details on shrimp sampling and population dynamics see Beukema (1992a).

The air temperature data used were obtained from monthly weather reports (issued by KNMI, De Bilt, the Netherlands) which include data from a weather stations called De Kooy, located < 1 km from Balgzand.

The observed correlations were evaluated by Spearman rank tests.

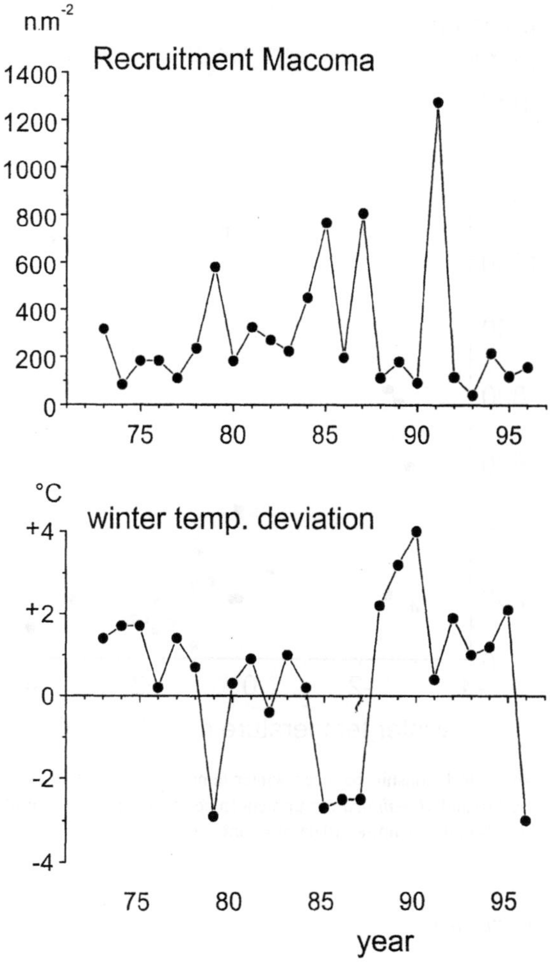

Figure 2. (a) Mean numbers per m^2 of *M. balthica* recruits (young-of-the-year, larger than 1.5 mm) as observed on Balgzand (means of 15 sampling stations) in August during 24 successive years (1973–1996, inclusive). (b) Mean winter (Jan-Febr-Mar) air temperatures, plotted in °C as deviations from the long-term average at a weather stations close to Balgzand.

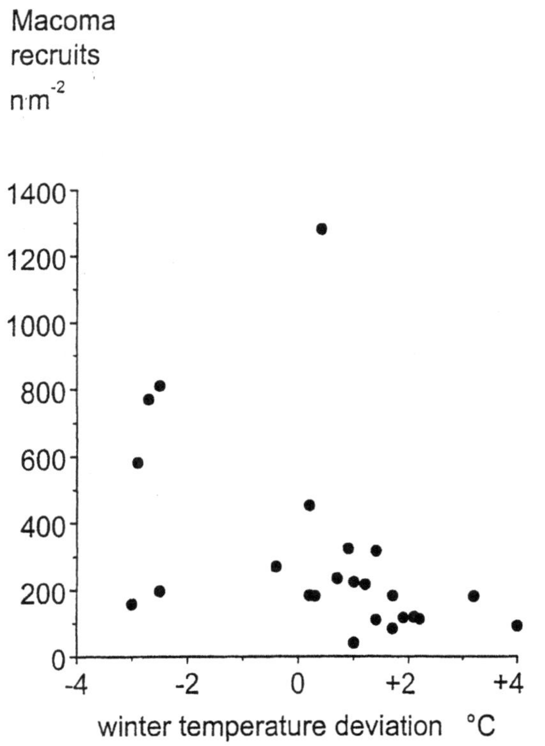

Figure 3. Relationship between winter temperatures (plotted as in Figure 2b) and *M. balthica* recruitment (n per m^2 of spat > 1.5 mm) in the subsequent summer (data of Figure 2a).

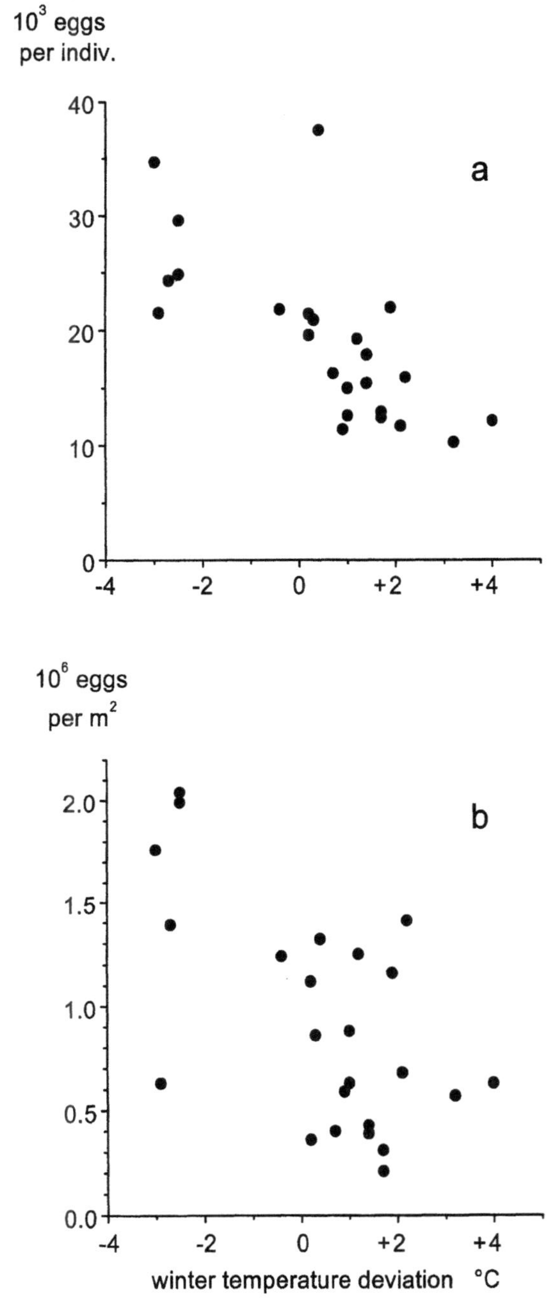

Figure 4. Dependence on winter temperature (plotted in °C as deviation from the long-term average) of (a) calculated numbers of eggs per *M. balthica* female of 15 mm length (in thousands) and (b) calculated numbers of *M. balthica* eggs per m^2 (in millions). Each point represents one of the 24 y 1973–1996, inclusive. See text for ways of calculation.

Experiments

To study in more detail relationships between winter temperature and BMI (and egg numbers and sizes), some thousands of Balgzand-collected adult (14–17 mm) *M. balthica* were kept for several months in large temperature-controlled basins (described in Honkoop & Beukema, 1997) which were continually fed with Wadden Sea water. In half of the basins, the water was cooled by 2.5 °C, to simulate a cold winter. Within the various basins, the animals were kept at two levels, representing intertidal (43% of time drained) and subtidal conditions and allowing the animals to feed about half or all of the time, respectively. The experiments were run in early 1994 and early 1995. During the experiments, BMI and state of the gonads were monitored. In April, groups of animals were forced to spawn and numbers and sizes of released eggs were determined in individual females (checked for complete spawning). For further details see Honkoop & van der Meer (1997a, 1997b).

Consumption of *M. balthica* spat by shrimps was studied both in the laboratory and in the field. In the laboratory, known numbers (mostly groups of 10 to 20) of well burrowed spat of known size were exposed to individual shrimps of known size. Numbers and sizes of disappeared spat were assessed and related to shrimp size. In the field, a few (2 to 4) shrimps of known size were introduced to 1-mm meshed cages that contained (in addition to the already present fauna) a known number (20 to 50) of colour-marked and measured spat (of up to 2 mm length). After two days, the numbers and sizes of the remaining marked and unmarked spat were determined. The total numbers of consumed spat of each size class were calculated under the assumption that marked and unmarked spat were equally vulnerable. These experimental results were compared with stomach contents (sizes and numbers of spat) of shrimps collected at Balgzand. For further details see Keus (1986).

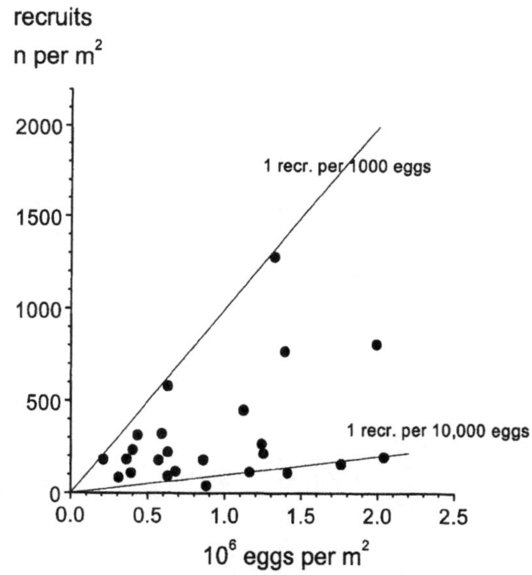

Figure 5. Relationship between the calculated numbers of eggs per m^2 (in millions) spawned by *M. balthica* females on Balgzand in April of each of the 24 y 1973–1996 and the resulting numbers per m^2 of recruits in August. The two lines indicate egg-to-recruit survival levels of 0.001 and 0.0001, respectively.

Results

Recruitment and winter temperatures

Recruitment in *M. balthica* varied strongly from year to year on Balgzand during the 1973–1996 period (Figure 2a). Annual mean numbers of spat ranged from 41 to 1280 per m^2 during the 24 y of observation. The average number per m^2 was 301 with a standard deviation of 291.

Strong year classes arose after three of the five cold winters (viz. those of 1979, 1985, and 1987), but also after the close-to- average winter of 1991 (compare Figures 2a, b). Recruitment was low after the three mildest winters (1988, 1989, and 1990). The relationship between winter temperature and subsequent recruitment was negative (Figure 3: $r = -0.62$, $n = 24$) and statistically significant ($p < 0.01$, Spearman rank test). In a simple linear regression, winter temperatures explained 36% of the variance in annual recruitment. Out of the 24 data points, 21 appear to fit well a linear relationship. The three odd points in Figure 3 represent the years 1986 and 1996 with cold winters and around-average recruitment and the year 1991 with high recruitment after a close-to-average winter.

Egg production and winter temperatures

As in other bivalves in the Wadden Sea (Zwarts, 1991), *M. balthica* generally loose less weight in cold than in mild winters (Honkoop & Beukema, 1997), resulting in higher BMI values just before spawning after a cold than after a mild winter (see figure 5A of Honkoop & Beukema, 1997). BMI values were found to be positively and linearly related to egg production: the number of eggs released per standard female (of 15 mm shell length) equals 7739 * BMI – 43314 (Honkoop & van der Meer, 1997a). Using this relationship, Honkoop et al. (1998) calculated the fecundity of standard females in each of the years 1970–1996 on Balgzand. In a linear regression, variation in winter temperature explained 68% of the variation in individual egg numbers. For the years of the 1973–1996 period, the relationship with winter temperatures was significantly negative (Figure 4a: Spearman $r = -0.74$, $n = 24$, $p < 0.001$).

As a next step, Honkoop et al. (1998) also calculated for each year the number of *M. balthica* eggs spawned per m^2, by multiplying the number of females per m^2 by the individual fecundity, taking into account the size distribution of the females in each year. They found a significant relationship with winter temperature, explaining 24% of the variance in egg numbers spawned per m^2. For the 1973–1996 period this relationship is shown in Figure 4b (Spearman $r = -0.44$, $n = 24$, $p < 0.05$).

As expected from the way of calculation, both the more than five-fold varying adult densities and the about four-fold varying numbers of eggs per fe-

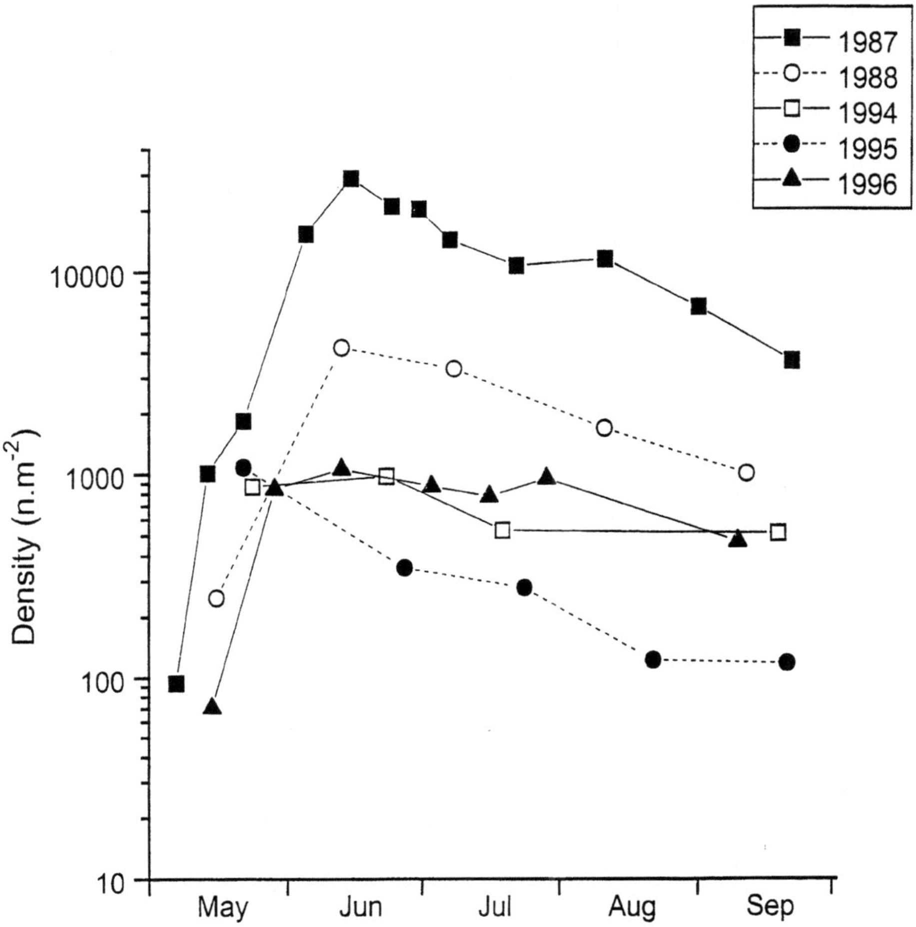

Figure 6. Logarithmic plots of density (n per m^2) changes of *M. balthica* postlarvae(of > 0.5 mm shell length) in May-September periods of five years (■ 1987, ○ 1988, □ 1994, ● 1995, and ▲ 1996) at station B on Balgzand (see Figure 1).

male strongly affected the estimated annual numbers of eggs per m^2 (Spearman $r = +0.66$ and $+0.65$, respectively, $n = 24$, $p < 0.01$). Together these factors explained 92% of the variance in egg numbers per m^2 (Honkoop et al., 1998), leaving only a low percentage for the year-to-year variation in size distribution of the females (which was indeed low).

Numbers of eggs and recruits

The high numbers (hundreds of thousands) of eggs spawned per m^2 in April were compared with the relatively low numbers (hundreds) of recruits per m^2 in August. On average, higher recruit numbers were found in years with higher egg numbers in the 1973–1996 period (Figure 5: Spearman $r = +0.29$, $n = 24$). However, this relationship was barely significant ($0.05 < p < 0.1$ in a one-tailed Spearman rank test) and could explain only a minor part (7%) of the year-to-year variation in recruitment. The higher numbers of recruits in years with high egg production were unrelated to the size of the adult stock ($r = -0.12$, $n = 24$), but significantly positively related to the numbers of eggs per female in the same year: $r = +0.55$ ($n = 24$, $p < 0.01$).

Mortality between the egg and recruit stages must be very high, as generally only one or a few (maximally about ten) per 10 000 eggs give rise to an August recruit (Figure 5). As at the start of benthic life, in May/June, maximal densities of postlarvae (of > 0.5 mm) were already reduced to thousands or tens of thousands per m^2 (Figure 6), most of this mortality will have occurred early in life, during the pelagic stages. Numbers of settled postlarvae declined further to hundreds or thousand in the course of summer (Figure 6). Note that the rate of these summer declines

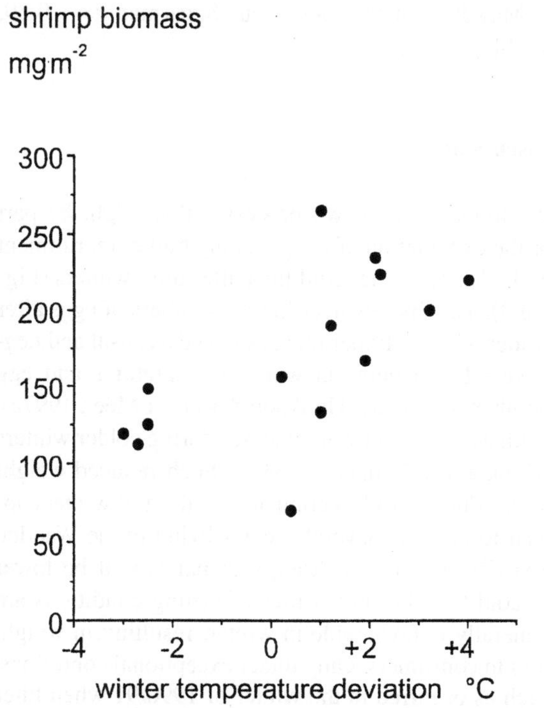

Figure 7. Relationship between winter temperatures (plotted as in Figure 2b) and mean shrimp biomass in the subsequent May/June periods of 14 y (1983–1996, inclusive). Biomass data in mg AFDW per m^2, means of 6 sampling dates and 3 stations.

were roughly equal (the lines of Figure 6 running roughly parallel from June to August/September). Thus, the numbers of postlarvae present in the sediment in June largely determined their abundance in August. This means that the year-to-year variation in year class strength was determined already shortly after settlement in May/June or even earlier. After June, the order of the strength of the different year classes (1987 > 1988 > 1996 > 1994 > 1995) did not change at this station. Though August recruit densities (as presented in Figure 2a) refer to spat of > 1.5 mm (by different sieve mesh sizes) and are therefore lower that the numbers shown in Figure 6 (including also smaller spat of > 0.5 mm), the order of the abundance in different years at this station was again identical (and similar to that of the 15-stations means presented in Figure 2a).

Abundance of shrimps and winter temperatures

At the time of settlement in May or early June, the only highly abundant predators on such small animals as postlarval bivalves appear to be juvenile shrimps. The other important predators on tidal-flat inhabiting

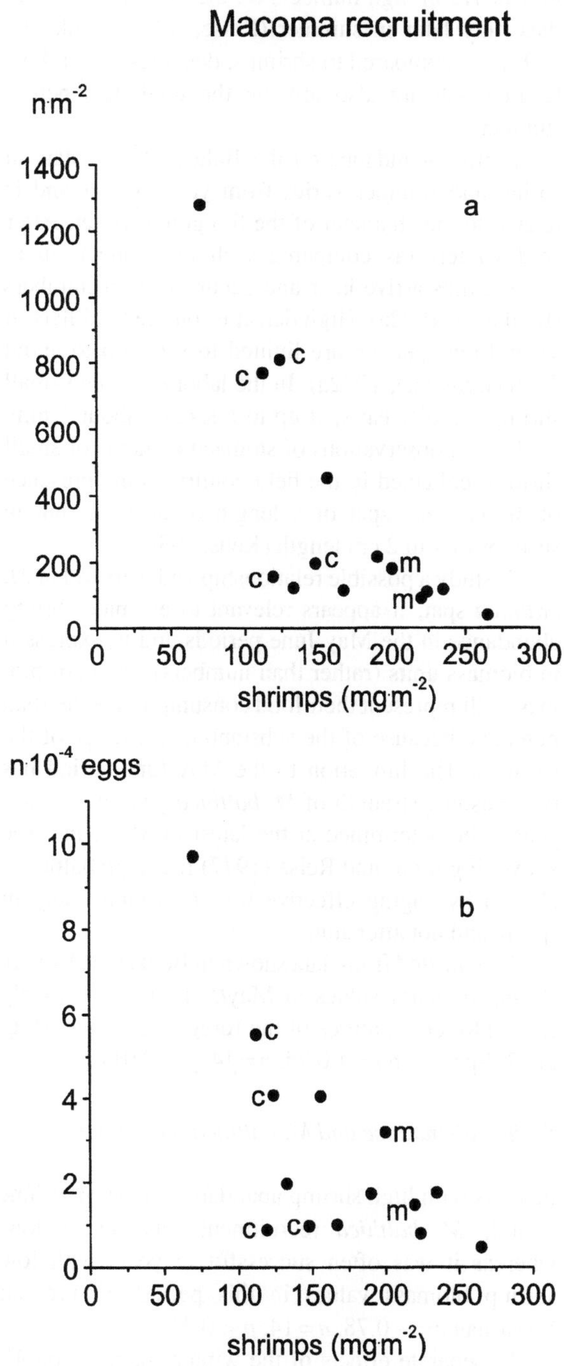

Figure 8. Relationship between shrimp biomass (mg AFDW per m^2) in May/June (data of Figure 7) with (a) density (n per m^2) of *M. balthica* recruits in August of the same year and (b) number of *M. balthica* recruits per 10 000 spawned eggs. One point for each of the 14 y 1983–1996, inclusive. Points representing recruitment after extreme winters (compare Figure 2b) indicated by either c (four extremely cold winters) or m (two extremely mild winters).

bivalve spat are shore crabs *Carcinus maenas* (L.), but they arrive in high numbers on the Wadden Sea tidal flats only later in summer (Reise, 1977; Beukema, 1991). As compared to shrimps, densities of benthos-feeding fish are also low on the tidal flats before summer.

Shrimp abundance on the Balgzand tidal flats in spring and summer varies from year to year and is related to the character of the foregoing winter. After cold winters (as compared with mild ones), juvenile shrimps arrive later and occur in lower numbers (Beukema, 1992a). High densities on the tidal flats of several tens per m^2 are limited to sizes up to about 2 cm (Beukema, 1992a). In the laboratory such small shrimps readily eat spat up to a size of about 1 mm. Additional observations of stomach contents of small shrimps collected in the field confirmed the presence of *M. balthica* spat of a length of up to 1 mm in shrimps of 1 to 2 cm length (Keus, 1986).

To study a possible relationship with survival of *M. balthica* spat, it appears relevant to estimate shrimp abundance in the May/June periods and to express it in biomass units (rather than numbers). Predator biomass will represent their food consumption better than numbers, because of the substantial size range of the shrimps. The limitation to the May/June period has two reasons: strength of *M. balthica* year classes appear to be determined at the latest in May/June (see above; Figure 6) and Reise (1977) found predator exclusion by caging effective for *M. balthica* only in spring and not after June.

As expected from data shown in Beukema (1992a), shrimp biomass values in May/June were positively related to temperatures of the foregoing winters (Figure 7: Spearman $r = +0.75$, $n = 14$, $p < 0.01$).

Shrimp abundance and M. balthica recruitment

In years with high shrimp abundance in the May/June period, *M. balthica* recruitment was always low, whereas it was often successful in years with low shrimp biomass values in this period (Figure 8a: Spearman $r = -0.78$, $n = 14$, $p < 0.01$).

To separate effects of the winter character on *M. balthica* recruitment via egg production (Figure 4b) from effects via shrimp abundance (Figure 7), the relationship between shrimp abundance and recruitment success has been plotted also for the numbers of recruits per 10 000 eggs (Figure 8b). The annual survival from eggs in April to recruits in August was significantly negatively correlated with shrimp biomass in May/June of the same year: Spearman $r = -0.59$, $n = 14$, $p < 0.05$.

Discussion

We found at least two processes that might be part of the explanation of the generally higher recruitment of *M. balthica* after cold than after mild winters (Figure 3). The first involves higher numbers of eggs after colder winters. Experiments showed a causal and negative relation between winter temperatures and egg number per female (Honkoop & van der Meer, 1997a), mediated by lower weight losses during colder winters (Honkoop & Beukema, 1997). Such reduced weight losses during cold as compared with mild winters appear to occur in several species living in the Wadden Sea (Zwarts, 1991). Energy demands will be lower in cold than in mild winters. Feeding conditions are generally unfavourable in winter, resulting in weight loss in consumers. Only under exceptional conditions, such as occurred in the winter of 1990/91 when filter feeders were extremely scarce, phytoplankton concentrations were high enough to keep bivalve body weights high and allow growth to start already in March (Beukema & Cadée, 1996). As a result, *M. balthica* fecundity was in spring 1991 higher than in any other year (see the odd high point in Figure 4a). In experiments (Honkoop & Beukema, 1997), longer feeding times also resulted in lower weight losses in several species of bivalves. Thus, more favourable energetic conditions in winter result in lower weight losses, higher body weights in spring, and higher reproductive outputs. In *M. balthica*, these higher outputs were expressed primarily in higher egg numbers per female (Figure 4a). Similar observations have been made in other bivalve species. Starved females of *Mytilus edulis* produced fewer and smaller eggs than fed ones (Bayne et al., 1975, 1978). In *Cerastoderma edule*, Honkoop & van der Meer (1997b) found more and larger eggs after winter periods with favourable than after such periods with unfavourable energetic conditions.

Higher numbers of eggs per female resulted in higher total numbers of eggs in the Balgzand population (Figure 4b). These higher numbers of eggs after cold as compared with mild winters, however, did not always generate higher numbers of recruits (Figure 5). Though recruitment was always low in years with low egg numbers, recruitment was highly variable in years with more than about 1 million eggs per m^2. Thus, egg numbers had little predictive value for recruitment success. Apparently, other processes had a more decisive influence on final year class strength.

Egg numbers per female showed a stronger positive correlation (a significant +0.55 versus a barely significant +0.29) with recruit densities than total egg numbers (resulting from number of females * number of eggs per female). Adult densities were found to be unrelated to recruit densities (if anything, the correlation was negative). Survival from egg to recruit stage was negatively related to adult densities. Maybe, high adult densities negatively affected survival between the egg and recruit stages, as observed in aquarium experiments with early bottom stages of *M. balthica* by Bonsdorff et al. (1986). Such negative relationships between adults and their progeny have been reported also in other bivalves, e.g. in the cockle *Cerastoderma edule* by André et al. (1993). An alternative explanation of low egg-to-recruit survival at high adult densities might be that negative interactions (such as competition for limiting resources) between larvae result in higher mortality at high densities of eggs and larvae.

Year class strength in *M. balthica* was determined before the end of June (Figure 6). During the period between spawning (in April) and late June the animals pass through three stages and experience widely different environmental conditions: the eggs and larvae are pelagic, the larvae settle on the bottom in mostly relatively deep water (low tidal flats and subtidal areas), whereas the postlarvae are mostly found on relatively high the tidal flats (Armonies & Hellwig-Armonies, 1992; Beukema, 1993b). During the pelagic stages, eggs and larvae can be transported by tidal currents over long distances. Even the postlarvae change their distribution pattern is this way (Beukema & De Vlas, 1989; Armonies & Hellwig-Armonies, 1992). In fact, the Balgzand population is an open one. Both washing out from the parental area and immigration can take place during all stages of the juvenile life. As most of the areas surrounding Balgzand are less densely populated by adults, washing out may prevail. Such transport processes might preclude the assessment of a stock-recruitment relationship as depicted in Figure 5. However, we think that the possibly distorting effects of migration of pelagic stages are limited, because the densities of *M. balthica* populations tend to fluctuate in a synchronised way over vast areas (Beukema & Essink, 1986; Desprez et al., 1991; Beukema et al., 1996). Moreover, the populations in the surrounding

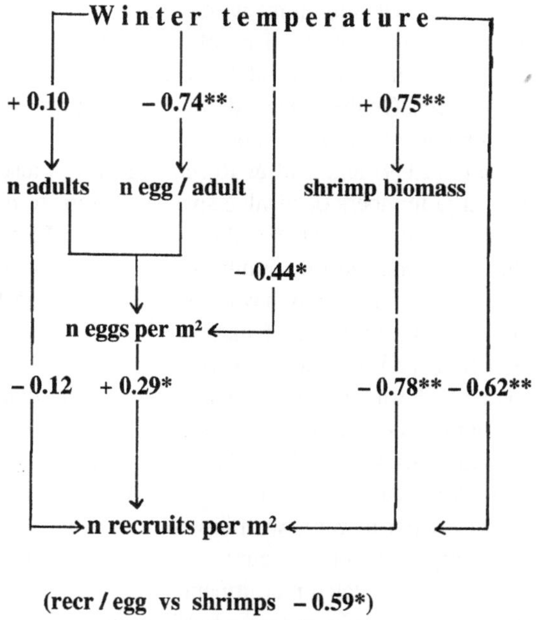

Figure 9. Summary of single rank correlations: Spearman-r values with * for $p < 0.05$ and ** for $p < 0.01$. Numbers of observation are either 14 (1983–1996, incl.) for correlations involving shrimp data or 24 for other correlations (1973–1996, incl.).

areas experienced similar winter temperatures as those on Balgzand.

We had no opportunity to study survival in larvae. Larval life will generally be more successful if a larva originates from a big than from a small egg. In oysters *Ostrea edulis*, Helm et al. (1973) observed low survival of larvae originating from low-quality eggs spawned by poorly fed adults. In *Mercenaria mercenaria* (L.) and *Astropecten irradians* (Lamarck), Kraeuter et al. (1982) observed survival to be reduced in larvae resulting from small eggs as compared with those from large eggs. In *Mytilus edulis*, Bayne et al. (1975) found lower growth rates in larvae from relatively small eggs. Moreover, such larvae showed relatively high proportions of deformations (Bayne, 1972). In *M. balthica* gametogenesis takes place already before the inset of winter and accordingly winter character does not affect egg sizes in this species (Honkoop & van der Meer, 1997a). Thus winter temperatures cannot affect larval survival in *M. balthica* via egg size variation.

After the larval stage, postlarvae of *M. balthica* settle in a hostile environment. Their main predators, juvenile shrimps, live in high numbers of several tens per m^2 on the tidal flats in spring and summer. After cold winters such high numbers are reached only in late May, but after mild winters already by late April (Beukema, 1992a). This offers another mechanism by which winter can affect recruitment success in *M. balthica.* On the Balgzand tidal flats, shrimps eat large numbers of *M. balthica* postlarvae (Keus, 1986). Plagmann (1939) and Pihl & Rosenberg (1984) also found large numbers of bivalve spat in shrimp stomachs. Raffaelli et al. (1989) did not, but according to these authors, the low numbers of bivalve spat in the shrimp stomachs may have reflected the low spat densities on the Ythan mudflats.

On Balgzand, biomass of shrimps, and thus their food demand, was in May/June higher after mild than after cold winters (Figure 7) and was negatively correlated with densities of *M. balthica* recruits in August (Figure 8a). Also survival from the egg to the recruit stage was better in years with low than in years with high shrimp abundance (Figure 8b). The large differences in shrimp-related recruitment success shown in Figure 8 point to an important role of shrimp predation in the determination of the year-to-year variation in year class strength of *M. balthica*. Such significant role of shrimp predation in bivalve recruitment has been shown also in Sweden by Möller & Rosenberg (1983), Pihl (1985) and Mattila et al. (1990). An important role of predators in the survival of *M. balthica* postlarvae and the determination of year-class strength has been suggested also by Bachelet (1986) and Bonsdorff et al. (1995).

To summarize (Figure 9): at least two processes might thus contribute to the observed better *M. balthica* recruitment in the Wadden Sea after cold than after mild winters: (1) body weight changes in adults determining egg numbers and (2) predation by shrimps on postlarvae. The path of winter temperatures to recruitment success via the egg numbers appears to be the least important. Particularly the relationship between egg numbers per m^2 and recruit densities was weak (Figure 5). Connections between winter temperatures and both fecundity (Figure 4a) and total egg production of the population (Figure 4b) were statistically significant, it's true, but a high number of eggs did certainly not guarantee a high number of recruits (Figure 5). On the other hand, the path via shrimp abundance appears to be more straightforward: shrimp biomass at the time of the critical early postlarval stage of *M. balthica* was significantly correlated with winter temperatures (Figure 7) and also with both subsequent recruit densities (Figure 8a) and survival of eggs to recruits (Figure 8b).

An obvious objection to the above reasoning is that, in a statistical sense, the two path ways were not completely independent. Both egg numbers in *M. balthica* and abundance of shrimps were significantly related to water temperatures in winter. High *M. balthica* egg numbers per m^2 (in early spring) occurred especially in years with low shrimp abundance (in late spring) (Spearman $r = -0.63$, $n = 14$, $p < 0.05$). Therefore, conventional analysis of variance should not be applied.

Nevertheless, there is some evidence that shrimp abundance affected recruitment in *M. balthica* in a direct way rather than acting via some other winter-related factor. The highest and lowest values of both recruitment and egg-to-recruit survival ratio were observed exactly in the years with the lowest and highest shrimp abundance, respectively, whereas neither of these two years was characterised by an extremely cold or mild winter (Figure 8). This suggests that shrimp abundance in May/June was a direct and important causative factor in the determination of recruitment success in *M. balthica* (and that it was not acting indirectly via some other winter-related factor). The relationship between *M. balthica* recruitment success and character of the foregoing winter may then be effected via the relationship between winter temperatures and shrimp abundance in spring.

References

André, C., P. R. Jonsson & M. Lindegarth, 1993. Predation on settling bivalve larvae by benthic suspension feeders: the role of hydrodynamics and larval behaviour. Mar. Ecol. Prog. Ser. 97: 183–192.

Armonies, W. & M. Hellwig-Armonies, 1992. Passive settlement of *Macoma balthica* spat on tidal flats of the Wadden Sea and subsequent migration of juveniles. Neth. J. Sea Res. 29: 371–378.

Bachelet, G., 1986. Recruitment and year-to-year variability in a population of *Macoma balthica* (L.). Hydrobiologia 142: 233–248.

Bahr, K., 1950. Die Ergebnisse der Miesmuscheluntersuchungen an der Ostfriesischen Küste in den Jahren 1947–1950. Veröff. Inst. Küst.-u. Binnenfisch. 68: 1–29.

Bayne, B. L., 1972. Some effects of stress in the adult on larval development of *Mytilus edulis*. Nature 237: 459.

Bayne, B. L., P. A. Gabbot & J. Widdows, 1975. Some effects of stress in the adults on the eggs and larvae of *Mytilus edulis*. J. mar. biol. Ass. UK 55: 675–689.

Bayne, B. L., D. L. Holland, M. N. Moore, D. M. Lowe & J. Widdows, 1978. Further studies on the effects of stress in the adult on the eggs of *Mytilus edulis*. J. mar. biol. Ass. U.K. 58: 825–841.

Beukema, J. J., 1982. Annual variation in reproductive success and biomass of the major macrozoobenthic species living in a tidal flat area of the Wadden Sea. Neth. J. Sea Res. 16: 37–45.

Beukema, J. J., 1988. An evaluation of the ABC-method (abundance/biomass comparison) as applied to macrozoobenthic communities living on tidal flats in the Dutch Wadden Sea. Mar. Biol. 99: 425–433.

Beukema, J. J., 1991. The abundance of shore crabs *Carcinus maenas* (L.) on a tidal flat in the Wadden Sea after cold and mild winters. J. exp. mar. Biol. Ecol. 153: 97–113.

Beukema, J. J., 1992a. Dynamics of juvenile shrimp *Crangon crangon* in a tidal-flat nursery of the Wadden Sea after mild and cold winters. Mar. Ecol. Prog. Ser. 83: 157–165.

Beukema, J. J., 1992b. Expected changes in the Wadden Sea benthos in a warmer world: lessons from periods with mild winters. Neth. J. Sea Res. 30: 73–79.

Beukema, J. J., 1993a. Increased mortality in alternative bivalve prey during a period when the tidal flats of the Dutch Wadden Sea were devoid of mussels. Neth. J. Sea Res. 31: 395–406.

Beukema, J. J., 1993b. Successive changes in distribution patterns as an adaptive strategy in the bivalve *Macoma balthica* (L.) in the Wadden Sea. Helgoländer Meeresunters. 47: 287–304.

Beukema, J. J. & G. C. Cadée, 1996. Consequences of the sudden removal of nearly all mussels and cockles from the Dutch Wadden Sea. P.S.Z.N. I: Mar. Ecol. 17: 279–289.

Beukema, J. J. & J. De Vlas, 1989. Tidal-current transport of thread-drifting postlarval juveniles of the bivalve *Macoma balthica* from the Wadden Sea to the North Sea. Mar. Ecol. Prog. Ser. 52: 193–200.

Beukema, J. J. & K. Essink, 1986. Common patterns in the fluctuations of macrozoobenthic species living at different places on tidal flats in the Wadden Sea. Hydrobiologia 142: 199–207.

Beukema, J. J., G. C. Cadée & H. Hummel, 1983. Differential variability in time and space of numbers in suspension and deposit feeding benthic species in a tidal flat area. Oceanol. Acta No Sp.: 21–26.

Beukema, J. J., K. Essink & H. Michaelis, 1996. The geographic scale of synchronized fluctuation patterns in zoobenthos populations as a key to underlying factors: climatic or man-induced. ICES J. Mar. Sci. 53: 964–971.

Bonsdorff, E., J. Mattila, C. Rönn & C.-S. Österman, 1986. Multidimensional interactions in shallow soft-bottom ecosystems; testing the competitive exclusion principle. Ophelia Suppl. 4: 37–44.

Bonsdorff, E., A. Norkko & C. Boström, 1995. Recruitment and population maintenance of the bivalve *Macoma balthica* (L.) – factors affecting settling success and early survival on shallow sandy bottoms. In: A. Eleftheriou et al. (eds), Biology and Ecology of Shallow Coastal Waters (Proc. 28th EMBS), Olsen & Olsen, Fredensborg: 253–260.

Dankers, N. & J. J. Beukema, 1982. Distributional patterns of macrozoobenthic species in relation to some environmental factors. In W. J. Wolff (ed.), Ecology of the Wadden Sea, Balkema, Rotterdam, 1: 69–103.

Desprez, M., G. Bachelet, J. J. Beukema, J.-P. Ducrotoy, K. Essink, J. Marchand, H. Michaelis, B. Robineau & J. G. Wilson, 1991. Dynamique des populations de *Macoma balthica* (L.) dans les estuaires du Nord-Ouest de l'Europe: première synthèse. In M. Elliott & J.-P. Ducrotoy (eds), Estuaries and Coasts: Spatial and Temporal Intercomparisons. Olsen & Olsen, Fredensborg: 159–166.

Dijkema, R., 1992. Spatfall and recruitment of mussels (*Mytilus edulis*) and cockles (*Cerastoderma edule*) on different locations along the European coast. ICES C.M. 1992/K:45 (Shellfish Committee).

Hancock, D. A., 1973. The relationship between stock and recruitment in exploited invertebrates. Rapp. P.-v. Réun. Cons. perm. Int. Explor. Mer 164: 113–131.

Helm, M. M., D. L. Holland & R. R. Stephenson, 1973. The effect of supplementary algal feeding of a hatchery breeding stock of *Ostrea edulis* L. on larval vigour. J. mar. biol. Ass. U.K. 53: 673–684.

Honkoop, P. J. C. & J. J. Beukema, 1997. Loss of body mass in winter in three intertidal bivalve species: an experimental and observational study of the interacting effects between water temperature, feeding time and feeding behaviour. J. exp. mar. Biol. Ecol. 212: 277–297.

Honkoop, P. J. C. & J. van der Meer, 1997a. Reproductive output of *Macoma balthica* populations in relation to winter-temperature and intertidal-height mediated changes of body mass. Mar. Ecol. Prog. Ser. 149: 155–162.

Honkoop, P. J. C. & J. van der Meer, 1997b. Experimentally induced effects of water temperature and immersion time on reproductive output of bivalves in the Wadden Sea. J. exp. mar. Biol. Ecol. 220: 227–246.

Honkoop, P. J. C., J. van der Meer, J. J. Beukema & D. Kwast, 1998. Does temperature-influenced egg production predict the recruitment in the bivalve *Macoma balthica*? Mar. Ecol. Prog. Ser. 164: 229–235.

Jensen, K. T. & J. N. Jensen, 1985. The importance of some epibenthic predators on the density of juvenile benthic macrofauna in the Danish Wadden Sea. J. exp. mar. Biol. Ecol. 89: 157–174.

Keus, B., 1986. De predatie van de garnaal (*Crangon crangon*) op het broed van het nonnetje (*Macoma balthica*). Interne Verslagen NIOZ, Texel 1986-5: 1–43.

Kraeuter, J. N., M. Castagna & R. van Dessel, 1982. Egg size and larval survival of *Mercenaria mercenaria* (L.) and *Argopecten irradians* (Lamarck). J. exp. mar. Biol. Ecol. 56: 3–8.

Kreger, D., 1940. On the ecology of *Cardium edule*. Archs néerl. Zool. 4: 157–200.

Kristensen, I., 1957. Differences in density and growth in a cockle population in the Dutch Wadden Sea. Archs néerl. Zool. 12: 351–453.

Kühl, H., 1955. Studien über die Klaffmuschel *Mya arenaria* 3. Das Junggut. Arch. Fisch. Wiss. 6: 33–44.

Mattila, J., E. B. 'Olafsson & A. Johansson, 1990. Predation effects of *Crangon crangon* on benthic infauna on shallow sandy bottoms – an experimental study from southern Sweden. In M. Barnes & R. N. Gibson (eds), Proc. 24th Eur. Mar. Biol. Symp., Aberdeen Univ. Press, Aberdeen: 503–516.

Möller, P., 1986. Physical factors and biological interactions regulating infauna in shallow boreal areas. Mar. Ecol. Prog. Ser. 30: 33–47.

Möller, P. & R. Rosenberg, 1983. Recruitment, abundance and production of *Mya arenaria* and *Cardium edule* in marine shallow waters, western Sweden. Ophelia 22: 33–55.

Pihl, L., 1985. Food selection and consumption of mobile epibenthic fauna in shallow marine areas. Mar. Ecol. Prog. Ser. 22: 169–179.

Pihl, L. & R. Rosenberg, 1984. Food selection and consumption of the shrimp *Crangon crangon* in some shallow marine areas in western Sweden. Mar. Ecol. Prog. Ser. 15: 159–168.

Plagmann, J., 1939. Ernährungsbiologie der Garnele (*Crangon vulgare* Fabr.). Helgoländer wiss. Meeresunters. 2: 113–162.

Raffaelli, D., A. Conacher, H. McLachlan & C. Emes, 1989. The role of epibenthic crustacean predators in an estuarine food web. Est. coast. Shelf Sci. 28: 149–160.

Reise, K., 1977. Predator exclusion experiments in an intertidal mud flat. Helgoländer wiss. Meeresunters. 30: 263–271.

Reise, K., 1985. Tidal flat ecology. An experimental approach to species interactions. Springer Verlag, Berlin, 191 pp.

Savage, R. E., 1956. The great spatfall of mussels (*Mytilus edulis* L.) in the river Conway estuary in spring 1940. Fish. Invest., London (2) 20: 1–22.

Smidt, E. L. B., 1951. Animal production in the Danish Wadden Sea. Meddr Kommn danm. Fisk.- og Havunders. (Serie Fiskeri) 11: 1–151.

Thamdrup, H. M., 1935. Beiträge zur Ökologie der Wattenfauna auf experimenteller Grundlage. Meddr Kommn danm. Fisk- og Havunders. (Serie Fiskeri) 10: 1–125.

Yankson, K., 1986. Reproductive cycles of *Cerastoderma glaucum* (Bruguière) and *C. edule* (L.) with special reference to the effects of the 1981–82 severe winter. J. moll. Stud. 52: 6–14.

Young, E. F., G. R. Bigg & A. Grant, 1996. A statistical study of environmental influences on bivalve recruitment in the Wash, England. Mar. Ecol. Prog. Ser. 143: 121–129.

Zwarts, L., 1991. Seasonal variation in body weight of the bivalves *Macoma balthica, Scrobicularia plana, Mya arenaria* and *Cerastoderma edule* in the Dutch Wadden Sea. Neth. J. Sea Res. 28: 231–245.

Hydrobiologia **375/376:** 35–47, 1998.
S. Baden, L. Pihl, R. Rosenberg, J.-O. Strömberg, I. Svane & P. Tiselius (eds),
Recruitment, Colonization and Physical–Chemical Forcing in Marine Biological Systems.

The spreading potential of polychaete larvae does not predict adult distributions; consequences for conditions of recruitment

Michel Bhaud
Université P. & M. Curie, Laboratoire Arago B.P. 44, 66651 Banyuls-sur-Mer, France

Key words: larval dispersion, success of settlement, ecological constraints, control of adult area, Polychaeta, Chaetopteridae, *Spiochaetopterus, Mesochaetopterus*

Abstract

We analysed the relationship between the area covered by each generation of larvae and the area of specific distribution containing adult forms. The specific area was defined as the area in which larvae could settle, juveniles could grow, and adults could reach maturity and produce a new generation of propagules. From observations of larval and adult distributions in precise situations, it appears that larvae of numerous species cover a much wider variety of habitats than those in which their adult forms can survive. The discrepancies between the initial area where larvae were found, or the area occupied by first benthic recruits, and the adult habitat, converge to show that ecological constraints take on more and more importance as the animal develops from the planktonic larval stage through the juvenile stage to adult maturity. In the same way, sterile populations express a shift in constraints intensity between the benthic stage and the reproductive phase. Based on Atlantic Ocean results, proof is provided here for the first time that the ability of larvae to cross an ocean does not bring about the panoceanic presence of adults. Finally, suitable conditions of recruitment do not necessarily imply the presence of adults, and conditions of colonisation are related to ecological constraints on adults. Consequently, the suggestion that species with a long planktonic larval development occupy a larger geographic area than species that spend only a short part of their life in the planktonic larval form does not seem to be justified.

Introduction

Numerous authors have suggested that the geographic distribution of benthic invertebrates is related to the type of larva present in the life cycle. Those species with a long planktonic larval development would intuitively be expected to occupy a larger geographic area than species that spend only a short part of their life as planktonic larvae. In this regard, an examination of the literature reveals divided opinions over the confinement of larvae to specific areas.

Gardiner (1904) discussed the duration of pelagic larval life within different organisms, and then considered the effect of the velocity of surface currents on their distribution. However, at that time, the duration of pelagic larval life was not well understood, in contrast to fifty-seven years later when the duration of this stage was known for 195 species. Thorson (1961) found that eighty per cent of all bottom-dwelling invertebrates with a pelagic larval stage have a planktonic life of less than 6 weeks. Thus, even if the length of this stage were doubled or if the larvae were transported by especially rapid currents, they would not have the slightest chance of crossing the larger ocean basins.

In several papers, Scheltema (1971, 1974) compared the same data, with the addition of a large number of low-latitude species, and concluded that the length of pelagic larval development and the velocities of ocean currents left no doubt that there was a reciprocal exchange of larvae between the eastern and western Atlantic. More recently, a difference in the distribution (i.e. absence of overlap of the respective areas) of larvae and adults of the same species was highlighted (Mileikovsky, 1961; Bhaud et al., 1978; Bhaud, 1993).

Hadfield & Strathmann (1996) noticed an absence of support for two assumptions: 1) that lecithotrophic larvae do not remain in the plankton long enough to effect genetic dispersal between populations over significant geological time and 2) that species with lecithotrophic development have narrow geographic ranges and are more vulnerable to extinction caused by localised environmental disasters. In addition, they suggested that if the correlation between lecithotrophic development and shorter geological survival was not due to limited larval dispersal, then other causal hypotheses should be examined.

From this short review it can be seen that there is a diversity of opinions as to the role of the dispersive period. The fundamental assumption, that the duration of a species' larval development is directly related to its geographic range, will be re-examined here in the light of the following questions. Does successful planktonic life and recruitment necessarily lead to the continuation of the life cycle? Even if larvae are disseminated over a large area, is their success secured in the areas that receive them?

Materials and methods

This work uses Polychaetes as its subject matter. This class of invertebrate is useful for comparisons in consideration of the fact that, among the main groups of marine invertebrates, Polychaetes exhibit the widest variety of reproductive strategies (Giangrande et al., 1994). The aim of the study is to identify relationships between physical structures in the environment and specific stages of the life cycle, taking advantage of differences in spatial occupation by successive developmental stages. To do this, results from personal work carried out in subantarctic, tropical and temperate areas are used (Bhaud; Bhaud & Duchêne, 1989, 1996). First of all, the result of larval dissemination on an oceanic scale (Atlantic and Pacific Oceans) is tested. There are two methods to test whether larvae really control the adult area by their ability to disseminate. 1) based on larvae when rigorous identification and known geographic distribution are available; 2) based on the adult form collected at two extremities of an ocean. In spite of several studies based on breeding living specimens belonging to the main genera *Chaetopterus*, *Spiochaetopterus*, *Phyllochaetopterus* and *Mesochaetopterus* (Bhaud, 1966), erroneous identification is always probable. Consequently, the second solution was chosen, including the description of adult morphology and the introduction of new criteria for differentiation. This process has been substantiated through detailed morphological analyses of the specialised seta borne by the A4 segment (fourth segment in the anterior region) of species belonging to the genus *Spiochaetopterus*.

Several specific cases based on a small spatial scale will now be described. In order to understand these specific cases, the mechanism of a boundary will be used, as proposed several years ago (Bhaud, 1983, 1989). Factors of specific distribution boundaries may be classified into two groups. The first set is linked to edaphic features (e.g. sediment for settlement), competitive features (when one species prevents the settlement of another), physical features (in reference to conditions of light) or chemical features (when, for instance, salinity is a limiting factor). All of these factors induce local boundaries and define the habitat. The second set of factors generates boundaries on a larger geographical scale. These factors belong to the climatic class and they help to define the specific area. Edaphic conditions may be suitable outside the climatic boundaries, which is why settlement (mainly influenced by the substratum type) may be successful but the following stage, for instance maturation to adulthood (primarily influenced by the seasonal variations in temperature), fails to develop.

Results

Polychaeta in the Atlantic Ocean with the Chaetopteridae family

This family is of interest through the presence of long-lived larvae in the life cycle. These larvae are pelagic with a very long planktonic life extending over several months. Nectochaete larvae are the most unusual larvae of the Polychaeta, with a large spherical body clearly divided into three regions: the anterior region, with a broad prostomium and peristomium, terminal mouth, paired eyes (1 to 3 pairs), short tentacular palps (on old larvae) and 7 to 9 setigerous segments with notosetae similar to those of adults; the middle region, with one or two prominent ciliary bands (mesotrochs) and indistinct segments; and the posterior region, which is tapered, with indistinct segments (but marked on old larvae by oncinal plates) and a terminal cylindrical appendage (Pettibone, 1982; p. 25).

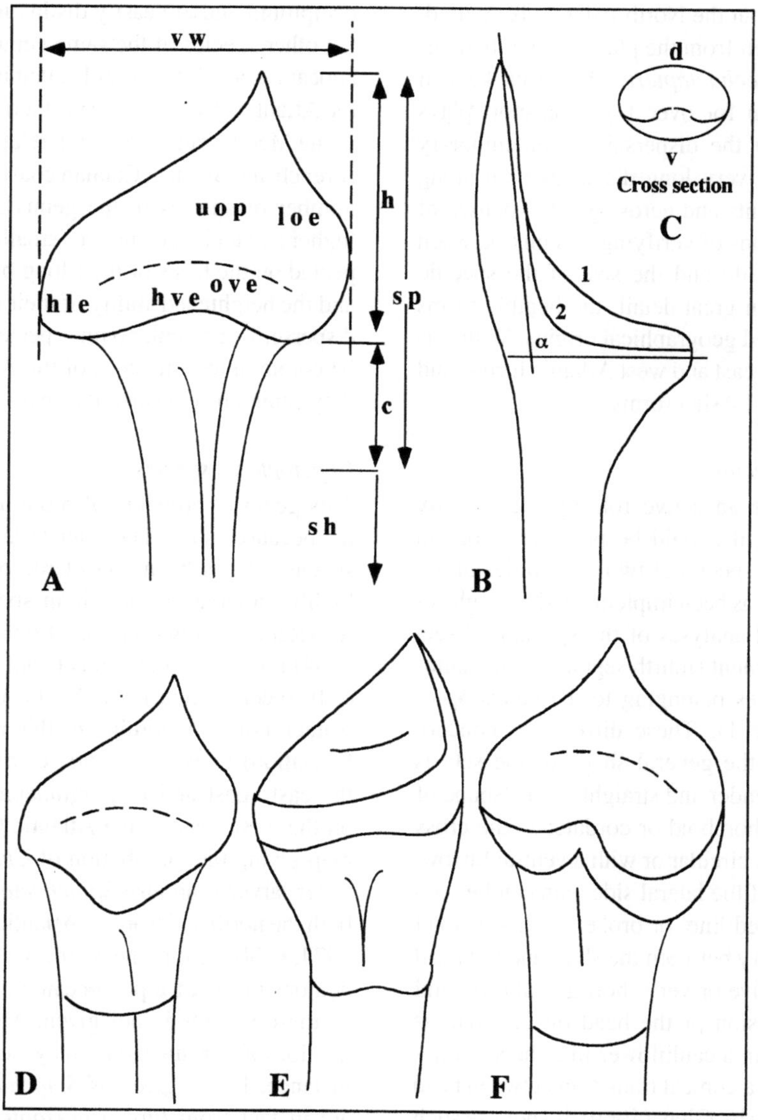

Figure 1. Illustration of new morphological characteristics on which is based the erection of a new species of the *Spiochaetopterus* genus. Illustrated features are based on the specialised seta borne by the fourth segment in the anterior region of the specimen. In A (ventral view), several parameters have been indicated: c: collar; d: dorsal side; dt: dorsal tip; h: head; hle: horizontal lateral edge; hve: horizontal ventral edge; loe: lateral oblique edge of the upper plan; sp: swollen part; sh: shaft; uop: upper oblique plane; v: ventral side; vw: ventral width. B is a lateral view showing the lateral oblique edge of the upper plan; α: angle between upper and lower oblique plans. Several setal features are illustrated: shape of upper oblique plane: rhomboid (D), subcircular (E), or cordate and bilobate and in this last case with equality (A) or inequality (F) of the two lobes; cross section of the shaft (C): circular (F) or with a ventral furrow (E); ventral edge of the lateral side plan: as an unbroken curved line (B1) or broken, with a distinct angle (B2); boundary between the shaft and the head: long and progressive (E) or very short (A); the lateral overflowing of the head on the shaft: present and marked as cauliflower-shaped (A) or absent but with a conical transition between head and shaft (D and E).

The capacity for dispersal of these larvae is significant. Scheltema (1971; p. 24 and 1974; p. 307) indicated that *Chaetopterus variopedatus* had a larval development of over 3 months and probably much longer; thus its dispersal throughout the North Atlantic is assured. Another species, taken from the plankton and belonging to the genus *Spiochaetopterus*, has been kept in a laboratory situation for over 1 y. Transport plays an important role in the dispersal of contemporary benthic species over very long distances both along coastlines of continents and across great expanses of open ocean. One means of verifying this link between the length of larval life and the size of the specific area is to examine, in great detail, the benthic forms from widely separated geographical origins. With this in mind, we compare east and west Atlantic forms, and also the European and Asian forms.

Spiochaetopterus species

If larvae really have an active role by their ability to disseminate, then it should be possible to obtain samples of the same species at two extremities of the ocean. This process has been implemented through detailed morphological analyses of the specialised seta borne by the A4 segment (fourth segment in the anterior region) of species belonging to the genus *Spiochaetopterus* (Figure 1). These differences concern several features: 1) the general shape of the seta is slightly curved or slender and straight; 2) the shape of the ventral edge is rhomboid or cordate; 3) the cross section of the shaft is circular or with a ventral furrow; 4) the ventral edge of the lateral side plan can be seen as an unbroken curved line, or broken with a distinct angle; 5) the boundary between the shaft and the head is long and progressive or very short and abrupt; and 6) the lateral protrusion of the head on the shaft is present and marked in a cauliflower-like shape, or absent and replaced by a conical transition between head and shaft. The review by Gitay (1969) did not use such a detailed description of hard structures.

In the Atlantic-Mediterranean biogeographic area, a comparison of 5 species of the genus *Spiochaetopterus* was undertaken (Bhaud, 1998) in which the species were divided into two groups. Group A, with boreal biogeographic affinity, consisted of *S. typicus* and *S. bergensis*. Group B, with temperate biogeographic affinity, was made up of *S. costarum*, *S. oculatus* and *S. solitarius*. *S. oculatus* is a distinct species located in the western part of the Atlantic Ocean, from Massachusetts to the Gulf of Mexico, and is clearly separate from the two eastern temperate species *S. costarum* and *S. solitarius*. In addition to differences observed in the morphology of specialised setae, the oncinal plates display another difference: these plates, on *S. oculatus*, are characterised by a straight posterior occipitium and are easily distinguished from those of ten other species of the same genus. *S. solitarius* also appears as a valid species located in the eastern part of the Atlantic Ocean (on the northern coast of Spain) and in the Mediterranean Sea in a relatively restricted area (French and Spanish Catalan coast). It appears that the number of species in the genus *Spiochaetopterus* is higher and their specific area smaller than might be expected on the basis of their long planktonic larval life and the heightened ability of their planktonic larvae to disperse. At present, no one species has been found to be common to both sides of the Atlantic, and as such they cannot be considered as pan-Atlantic species.

Chaetopterus species

This genus is probably of more interest than the latter because, since Hartman (1959) synonymised 25 species, it has been considered monotypic with one highly variable cosmopolitan species *C. variopedatus* (Renier). However recent observations (Petersen, 1984) reveal that the genus is not monotypic and that c. 10 species occur in the North Atlantic and adjacent waters alone. According to this author (personal information) *C. variopedatus*, *C. brevis* are located on the east coast and *C. pergamentaceus* and *C. sarsii* on the west coast of the Atlantic Ocean. In spite of a map giving the distribution of *Chaetopterus variopedatus* larvae (and also *Spiochaetopterus costarum*) in both the north and tropical Atlantic Ocean (Scheltema, 1974; p. 306) along the North Atlantic gyre, the direct demonstration of a pan-oceanic distribution of adults of these species is not given. Moreover, the identification of certain larvae may be contested and, for instance, larvae given as *Spiochaetopterus costarum* are in fact from *Phyllochaetopterus* sp. (Scheltema, 1974, Figure 7 and 8).

Polychaete larvae in the Pacific Ocean

This section concerns the comparison of larvae between distant areas in the Pacific Ocean. Differences in the frequencies at which Polychaeta larvae were found in regions of equal area led Scheltema (1988) to conclude that the eastern Pacific may act as a substantial barrier to the dispersion of many forms of larvae. For *Chaetosphaera* (Spionidae), *Mesotrocha* (Chaetopteridae) and *Mitraria* (Owenidae) taken to-

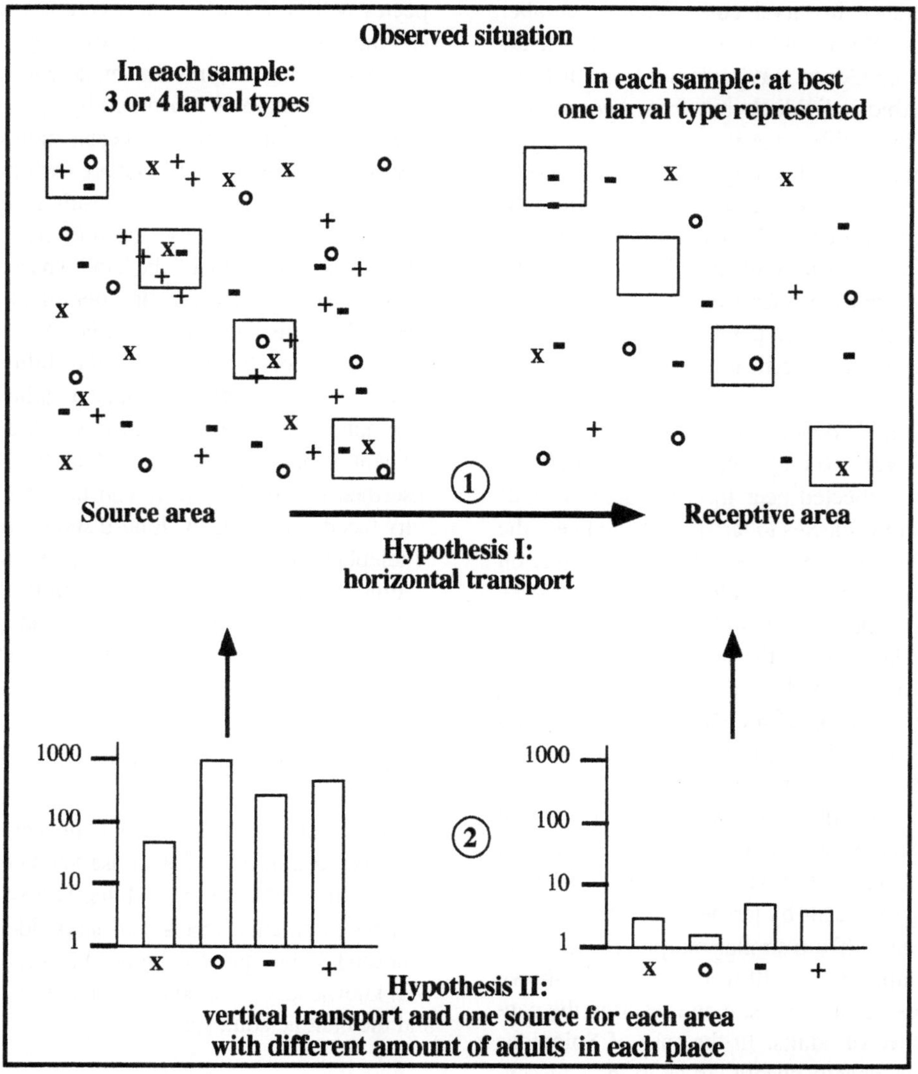

Figure 2. Two interpretations of differences in the density of larvae in two locations. Hypothesis 1 is based on one source and one reception area, and adopted when an age difference expressed by larval morphology is seen between the two locations. Hypothesis 2 assumes no link between the two locations; each one being both a source and a reception area. When no morphological difference, linked with development through a life cycle, can be demonstrated, larval spreading is probably more limited than previously thought.

gether, the percentage of stations where larvae are collected is 60 in the Central Pacific and only 28 in the Eastern Pacific. However, given that the quantitative argument is not sufficient to establish the success or failure of these larvae in crossing the Pacific, morphological and biological considerations must be integrated into an analysis of the situation.

In the case of *Mitraria* larvae (Oweniidae), several arguments can be advanced that could provide an explanation of their observed distribution without the help of transoceanic transport. (1) Larvae are observed at the same morphological stage; if there is really transportation from one area to another, then a morphological variation should be observed, which has not been the case. From recent observations in the Mediterranean Sea (Rey, 1995), *Mitraria* larvae begin metamorphosis in open water with the appearance of the juvenile body; from this stage, their positive flotation decreases and larvae leave the upper layers of the water. In neither cases was the initial stage of metamorphosis noted along with morphological differences that are indicative of differences in age. This result suggests that the larval source was not so far removed as might have been suggested by initial work.

(2) The difference in larval concentration (number per unit area) between the western and eastern areas of the Pacific Ocean may arise not from just one source of benthic individuals but from two sources characterised by a difference in the concentration of benthic organisms. In this way, instead of a model with one neritic source (localised in the western part of the Pacific Ocean) with horizontal transportation, a model with two partially bathyal sources could be substituted, where each source feeds the upper water layers of western or eastern parts of the Pacific Ocean with a predominance of vertical transport in each case (Figure 2).

This model, probably useful in the case of Owenidae, is even more applicable to chaetosphaera larvae. Larvae collected near the Marquesas Islands were of a different species to larvae collected near the Coco Islands, with at least one difference, visible on very fine original pictures (Scheltema 1988; Figure 3), in that the tip of the setae are either sharp or rounded. Consequently, larvae from the two sides of the Pacific Ocean differ in the morphology of their setae, which means that the crossing of the Pacific by these larval stages has yet to be demonstrated. As a general rule, chaetosphaera larvae do not have a large spatial extension: for 6 collection sites, it was not possible to identify a larval type common to any two sites (Bhaud et al., 1988; Bhaud & Lefèvre, 1986; Tachon, 1997). If the crossing of oceans by larvae is directly substantiated in the future (from similar morphology or from biochemical support), it will also need to be shown that these crossings are efficient and result in the pan-oceanic presence of adults. In the case of Spionidae adults that have chaetosphaera larvae in their life cycle, it is too early to tell, from benthic forms, whether these larvae had traversed the ocean successfully, as the link between larval and adult forms has yet to be established.

For *Mesotrocha* of Chaetopteridae, and in particular for the genus *Spiochaetopterus*, no detailed morphological descriptions of larvae are available. However, in the same way as for the Atlantic Ocean, the situation is more propitious when comparisons are based on observations of adult forms. At least two species have been identified on opposite coasts of the Pacific Ocean: *Spiochaetopterus miyukiae* (Nishi & Bhaud, in progress) from the Japanese coast, and *Spiochaetopterus pottsi*, from the eastern coast of the Pacific Ocean and previously considered to be a subspecies (Gitay, 1969). It is probable that several other species will be identified on each coast, as these are only the first results of a review of the genus.

Summarising the results on a wide geographic scale, the situation as illustrated by *Spiochaetopterus* in the Atlantic and Pacific Oceans shows that larvae may spread throughout a large area without implying that a larval presence should result in the presence of adults. More precisely, the presence of larvae in areas devoid of adults of the same species has yet to be demonstrated, because the identification of larval species by their morphology is not easy. It has only been established that, in spite of the ability of larvae to disseminate, the adults of a species inhabit a limited geographical area. Teleplanic larvae are thus not as efficient as has generally been thought. The situation described for the Pacific Ocean underlines the difficulty faced in proving the long-distance transportation of teleplanic larvae. A detailed morphological examination, perhaps based on the identification of new features, is required, both for larvae and adults before assuming that the same species is present in several distant areas.

The different kinds of boundary

It is only in the case of *Spiochaetopterus* species that a pronounced shift has been observed between the respective areas of larvae and adults. This shift is, in fact, only based on adult forms, and larval identification is contested. Consequently the need emerges to observe well known species in both the planktonic larval stage and the benthic adult stage.

Boundary of the specific area of Mesochaetopterus minutus

The species *Mesochaetopterus minutus* (Chaetopteridae) is as well known in its adult stage as in its larval stages. Planktonic larvae are big enough to attract attention when a live sample is sorted. Young larvae with one ciliated ring reach 600 μm long, while older larvae with two ciliated rings, approaching metamorphosis, may be 2500 μm long. Metamorphosis has been observed in larvae from the Mediterranean Sea (Bhaud, 1966) and the Indian Ocean (Bhaud unpublished observations). Adults of *Mesochaetopterus* sp. are the only chaetopterids that live in a sandy tube with a small diameter, of the order of 1 mm. Adult distribution of *M. minutus* is limited to upper levels (Bhaud, 1969, 1975, 1986; Bhaud et al., 1990; Gibbs, 1978; Day, 1961).

Table 1. Mesochaetopterus minutus. Details of collections of larval (L) and adult (A) specimens on a latitudinal section including the northern biogeographical boundary. The northernmost report of larvae was off Arcachon. The northernmost report of adults was Aveiro (Portugal).

Location	Observation of larvae (L) and adults (A)		References
	+: positive	–: negative	
Plymouth	L	–	Lebour, 1917
50°25′	A	–	Nelson-Smith et al., 1990
Roscoff	L	–	Bodo, 1965
48°45′	A	–	Cabioch et al., 1968
Southern Brittany	L	?	
47°40′	A	–	Glémarec, 1973
Arcachon	L	+	Bhaud et al., 1990
44°40′	A	–	Bachelet, 1987
Capbreton	L	+	Sorbe, 1990
43°30′	A	–	Amoureux, 1987
Aveiro (Portugal)	L	+	(personal observations)[1]
40°40′	A	+	Ravara, 1997
Atlantic Ocean	L	?	
25° N	A	+	Hartmann-Schröder, 1982
Atlantic Ocean	L	?	
20° N	A	+	Rosenfeldt, 1982

[1] From collection of plankton sampling made available by G. Marcano.

The length of larval life is estimated by recording change with time in the number and mean size of larval stocks collected successively (Bhaud, 1969). In the size range of larvae collected by the mesh of the plankton net (200 μm) the larval stage exceeds 3 weeks. However the actual time in the neritic province is certainly longer, because smaller stages must escape the net. Furthermore, in the oceanic province the time spent in the plankton is undoubtedly even longer (Bhaud et al., 1990).

On the Atlantic coast of France, adult specimens of this species have never been collected from the benthic communities of the continental shelf off northern Gascony (Glémarec, 1969), nor during numerous expeditions above the continental slope (Amoureux, 1987). In the Arcachon area, Cazaux (1970) and Bachelet (1987), who have collected extensively in the area, have never found adults, whereas planktonic samples from outside the Arcachon Basin contained many larvae (Sorbe, 1990; Bhaud et al., 1990). The northernmost station to have collected both larvae and adults is located close to Aveiro, Portugal (Ravara, 1997 and personal observation from a plankton collection made available by G. Marcano).

The case of *Mesochaetopterus minutus* demonstrates two elements useful in the discussion of specific range. 1) In the Atlantic Ocean there is a clear gap between the boundaries of adults and larvae (Table 1), probably due to the north-flowing currents which increase the area containing larvae. 2) Taking into account the general distribution of specimens in the inter-tropical zone and the reproductive period during the summer at the northern boundary, the optimum temperature appears relatively high, with a latitudinal range delimited to the north by waters which do not reach the lower boundary temperature in winter.

The transport of larvae by oceanic drift, particularly in a northward or southward direction, would account for the shift between larvae and adults. In this way, the Gulf Stream and its extension, the North Atlantic drift, could provide a means of larval transport from 30° N (Northern Florida) to 60° N (Shetland Islands). The same is true of the Peru and Benguela currents. Larvae are undoubtedly transported along these routes but the progression of the species involving a successful life cycle is questionable. At best, the larvae settle but the adults are unable to withstand the extremes of summer or winter temperatures for successful maturation, resulting in the formation of pseudo-populations (cf. illustration of specific boundaries on the Chilean Coasts: Brattström & Johanssen, 1983).

Thus, the spatial continuity of larval presence in the water mass does not necessarily lead to the continuity of the species with the adult benthic part of the life cycle. The occupation of a very large zone by sterile benthic populations regularly replenished by larvae confirms that environmental constraints are less severe on larvae than on adults. Although rare, examples of pseudo-populations or sterile populations have been identified in annelids (Mileikovsky, 1961; Bhaud et al., 1978), echinoderms (Gage & Tyler, 1981), and molluscs; in the latter case for both extant (Wells & Gray, 1960; Yamada, 1977) and fossil forms (Zinsmeister, 1974). The area of distribution of a given species, defined as the sum total of places where the complete life cycle is achieved, is therefore not necessarily identical to the area populated by any given dispersal stage.

Local boundary and the need of a spreading control
Observations of limited development, even after successful benthic recruitment, have been constructed, in the above sections, from examples integrating long planktonic larval life with long-distance transportation. The schema of an area compatible with continued development and more restricted than the total area held by larvae, is not only dependent on wide dispersion distance; it is also observed on a short spatial scale. Climatic constraints have an effect on short-distance transportation when, for instance, bathymetric conditions are considered (involving a thermocline). Edaphic constraints (nature of the bottom) are also important in a local situation. Finally, the hydrodynamic conditions controlling larval recruitment, and not adult reproduction, may also define the extent of the species. These situations are illustrated by 4 examples.

(1) A study of *Terebellides stroemi* (Bhaud et al., 1978) enabled factors connected with the nature of the substrate and hydrology to be distinguished. Variations in the sediment's properties did not account for the observed discontinuity in the quantitative distribution of the species. However, the bathymetric depth where a change in abundance was observed was related to the penetration of the thermocline in summer. During the spring, it was found that abrupt and unforeseeable rising temperatures were quite striking. In summer, the thermocline did not extend below a depth of 30 m, but the temperature at this depth temporarily rose to 16–19 °C. This summer warming of the superficial waters prevented the species from being present above a depth of 30–35 m. Thus, a clear faunal break in species distribution could be observed along a transect showing substrate continuity. In such situations larvae were required to 'choose' the appropriate substrate location in order to ensure their survival.

(2) A comparison of the vertical distribution of biogeographic units between Isefjord (Denmark) and Gullmar Fjord (Sweden), which differ in their topography, pointed to a control of species distribution by both temperature and bathymetry (Hannerz, 1956; Rasmussen, 1973; Bhaud et al., 1995). Isefjord, with its shallow bottom, was inhabited solely by species that tolerated relatively high summer temperatures as most of the benthic species reproduce during the summer. In contrast, Gullmar Fjord, much deeper, contained species of a more diversified biogeographical origin. As a result, the polychaete larvae in plankton collections displayed a bimodal seasonal pattern, with each mode corresponding to the reproduction of one of the two biogeographic stocks: winter breeding or boreal species, and summer breeding or temperate species. The dispersion of larvae belonging to one or the other of the two stocks of species, in a transverse direction relative to the coast, must be reduced if the life cycle is to be successfully completed. That is, larvae present during summer and originating from upper levels would not have survived until the adult stage if they had settled in deeper waters.

(3) Differences between adult and first benthic recruitment areas have been described in a recent survey of *Owenia fusiformis* undertaken in the Bay of Blanes, Spanish Catalan coast (Arnoux et al., 1995). Seasonal patterns relating to the abundance of specimens were characterised by a peak during the spring (recruitment period), a sharp decrease through summer, and lower densities during autumn and winter. A high density of small-sized individuals was observed at three stations during the recruitment period, which was followed by a period of high mortality. Adults were clearly detected at St 3, to a lesser extent at St 1 and had completely disappeared at St 2. The area covered by young benthic recruits was larger than that occupied by adults, and the settlement of larvae seemed to be associated with a wider range of sediments than was associated with adults. For the Bay of Banyuls (Verdier-Bonnet et al., 1997) the model shows that the fate of the larvae tracked by currents, appears to be determined early in the pelagic phase by being trapped in low current zones. This situation should be compared with that described for the Kerguelen archipelago (Bhaud & DuchÍne, 1989). Exceedingly violent hydrodynamic conditions prevent the maintenance, at least on the outer coastline of the archipelago, of species with planktonic larvae. Only

those species that have developed retentive processes can survive.

(4) *Eupolymnia nebulosa* (Terebellidae) is widely distributed in the Bay of Banyuls (Northwestern Mediterranean Sea) but is also present throughout the world. An English Channel population has also been studied, allowing different reproductive strategies to be compared: larvae developing in mucous egg-masses in the Mediterranean versus direct spawning in the English Channel. This difference appears to be linked to the extent of dissemination. In the Mediterranean, where the adults are distributed in a very limited narrow belt parallel to the coast, low larval dissemination appears to be relevant. On the other hand, in the English Channel, with maximal dissemination, larvae are still found in the area where adults are collected and restricted dissemination seems to be less importance (Arnoux et al., 1995).

These localised observations show an adult habitat of limited area in all the cases examined. The area occupied by larvae appears significantly larger than that inhabited by adults. For those species restricted to a narrow bathymetric belt, larvae may move in only a limited number of directions if their development is to be assured. Other directions, for instance towards the open sea, result in the failure of settlement or of a subsequent developmental stage. The consequence of reduced habitat area is the development of control processes for larval dissemination. This conclusion is only valid for zones characterised by a marked climatic gradient (mean and high latitudes). Under low latitudes, boundaries are far apart and a larva with a long planktonic life has the possibility to move off from the source without crossing geographic boundaries of reproduction and vegetative life. These remarks probably allow us to reconcile the opposing opinions of Thorson (1961) and Scheltema (1971, 1974) given in the introductory section. At low latitudes, planktonic larvae seem to have a real ability to disperse over long distances and to be successful at the end of their voyage, if only because the habitat boundaries are widely separated.

Discussion and conclusions

In the preceding sections dealing with spread on a large scale, it was shown that cases of supposed larval spread beyond identified boundaries (*Spiochaetopterus* sp.) were due to the inaccurate identification of the larvae. In turn, on a local scale, the path followed by larvae is known, which allows dispersion to be verified, and it is often found to be spatially restricted. In the life cycle of *Eupolymnia nebulosa* a biological structure, gelatinous egg masses, is interpreted as the main factor in the control of spreading. In addition, dissemination may be limited by the water-mass structure and specific behaviour of larvae, as with *Owenia fusiformis* (Verdier-Bonnet et al., 1997). Physical processes create structures within which biological mechanisms may proceed; they also influence the development of these biological mechanisms. In this context, the displacement of larvae is not only the result of the combined effects of advection of water masses and the larval lifespan. Advection does not always take place in a straight line, and hydrodynamic processes may form retentive structures leading to the development of specific larval behavioural patterns in connection with physical oceanographic processes. The result of this is the retention of pelagic phases within an appropriate geographic area, ensuring the continuation of the population (Boicourt, 1982; Christy & Stancyk, 1982; Rothschild, 1988; Sinclair, 1988; Lefèvre, 1990; Young & Chia, 1987 for review). Larvae do not disperse as much as is believed. Consequently, attempting to link the length of planktonic larval life with ocean current flow is meaningless. The traditional viewpoint (Thorson, 1961; Figure 3) connecting on the one hand, the number of a species spending a given number of weeks at the larval stage in the plankton, and on the other hand, the number of weeks taken by the strongest ocean-crossing surface currents to go from coast to coast, is given here as a symbol of an obsolete thought process. This pattern of calculation was used to estimate the length of the runs. That the situation might have been more complex due to factors such as the variability of oceanic currents, larval behaviour or variations in planktonic larval life was not considered. However, the combination of larval life-span and oceanic currents, and more subtly developmental pattern and specific area, is still very common in the literature (Scheltema, 1989; Jablonski, 1986; Hansen, 1980).

Two examples may be briefly discussed. Johannesson (1988) put forward 'The paradox of Rockall', asking: why is a brooding gastropod (*Littorina saxatilis*) more widespread than one having a planktonic larval dispersal stage (*L. littorea*)? The paradox was clear; the absence of a species with planktotrophic larvae on the small and isolated rocky cliff was not due to the larvae, which had a long planktonic life and were probably present close to the island in addition to being unsuccessful in the settling process. The explanation of negative collects of adults lies probably not in the reduced inflow of larvae but in the difficulty of maintaining larvae born there. This failure to reproduce was probably due to an absence of local retentive structures. An analogous situation has been described from the Kerguelen Archipelago (Bhaud & Duchêne, 1989), also due to frequent stormy conditions.

For their part, Wray & Raff (1991) found that a species with a long planktonic larval development period, *Heliocidaris tuberculata*, had spread on ocean currents from the coast of the Australian mainland (eastern coast from 29 to 38° S) to the North Island of New Zealand and several isolated islands including Norfolk, Kermadec and Lord Howe. Another species, with lecithotrophic larvae, *Heliocidaris erythrogramma*, had a range limited to the vicinity of the Australian continent (all the southern part from 25° S on the west coast to 27° S on the east coast). Clearly, the species with planktonic larvae has covered a much smaller distance than the species with direct development; the link between specific area and the developmental pattern was not particularly strong. Specific area is not necessarily a consequence of the ability of a species to spread. In the same way, the presence of a given species in several distant islands is not necessarily linked to the dispersal capacity of its larval stage. A large range of species with direct development is present in Pacific islands, spread over a large area (Bailey-Brock, 1985; Gibbs, 1971, 1972; Reish, 1968). Ecological constraints must be integrated alongside the dispersion process. The reduced dispersal area of the species with planktotrophic larvae following a N–S pattern on the eastern coast, with the large dispersal area of the species with lecithotrophic larvae following a E–W pattern, form a good illustration of ecological boundaries (Nilsson et al., 1977; Rochford, 1977; Reynolds, 1982). In the same way, the boundary, on the western part of the Australian Coast, of species with direct development is the consequence of a local ecological feature, probably related to the Leeuwin Current (Cresswell, 1991).

In order to end, temporarily, the question of the relationship – or not – between specific area and developmental patterns, it is useful to compare species with significantly different life cycles, for example holoplanktonic species (Furnestin, 1983; Wiebe & Boyd, 1978), benthic species (Bhaud, 1982; Mileikovsky, 1974), interstitial species (Westheide, 1977) and highly mobile species (Sund et al., 1981), together with species of different zoological levels: protozoa (Cifelli, 1976), invertebrates (Mileikovsky, 1974; Wiebe & Boyd, 1978; Bhaud, 1982; Furnestin, 1983), and vertebrates (Sund et al., 1981). Until now, the correlation between the potential for dissemination, from free species of holoplankton to epibenthic or interstitial species without planktonic larvae, and the size of specific area, has not been substantiated.

In conclusion, several results that have emerged from the present review are summarised in the following three points.

The situation illustrated with *Spiochaetopterus* in the Atlantic and Pacific Oceans shows that larvae may spread throughout a large area without implying that larval presence should result in the presence of adults. More precisely, the presence of larvae in areas devoid of adults of the same species has not yet been demonstrated because the identification of larval stages from their morphology is not sufficiently advanced. It has only been established that, in spite of the probable ability of larvae to disseminate, the adults of a species inhabit a limited geographical area. Teleplanic larvae are thus not as efficient (in controlling the adult area) as has generally been thought.

Observations on a smaller spatial scale show that successful planktonic life and recruitment does not necessarily lead to the continuation of the life cycle. In other words the adult area is not as deductible from the planktonic larval area as it is believed to be. Successful larval transport by the currents, or successful larval settlement on the substratum, is not necessarily followed by development to adulthood and the continuation of the life cycle. In other words, at the end of a long, or even a short, period of travel larvae may successfully settle, while subsequent stages fail to develop. Ecological constraints take on more and more importance as the animal develops from the planktonic larval stage through the juvenile stage to adult maturity. In the same way, sterile populations express a shift in constraint intensity between the benthic stage and the reproductive phase.

It seems more profitable to examine the significance of the lifespan of planktonic larvae alongside other traits, rather than its absolute value. The role of dissemination in the loss or the maintenance of individuals is probably connected with the size of the area suitable for larval reception (that is to say the area suitable for adults). In such a situation, a species with a reduced distribution area and a low capacity to disseminate, for instance *Eupolymnia nebulosa* in the Mediterranean Sea, would fare neither better nor worse than another species which leads a planktonic larval life for several months without going outside its adult distribution area. This similarity is possible either because the adult area is very large (as for *Mesochaetopterus minutus* under low latitudes) or because the larvae are subjected to anti-spreading structures. These structures may be of a biological nature as with *Eupolymnia nebulosa,* or connected to the structure of water masses as for *Owenia fusiformis*. Finally, larvae cover an area larger than that of adults, but always spread less than might be expected through water mass advection.

Acknowledgements

This work is a contribution to the French National Program on Determinants of Recruitment (PNDR France-GLOBEC) and was financed by grants from IFREMER and the CNRS to the author.

References

Amoureux, L., 1987. AnnÈlides polychËtes du talus continental europÈen. DonnÈes complÈmentaires concernant les campagnes 1970–1971–1972 de la 'Thalassa'. Données globales sur l'ensemble des six campagnes (1967 à 1973). Cah. Biol. mar. 28: 569–577.

Arnoux, S., M. Bhaud, C. Cazaux, J. H. Cha, J. C. Duchêne, P. Fraunié, G. Marcano, D. Martin, M. J. Nattero, C. Nozais, C. Rey, S. Pinedo & R. Sarda, 1995. Recruitment in the light of biological-physical interactions in coastal waters: results of PNDR action at the Arago Laboratory in 1994–1995. Vie Milieu 45: 85–105.

Bachelet, G., 1987. Processus de recrutement et role des stades juvèniles d'Invertébrés dans le fonctionnement des systèmes benthiques de substrat meuble en milieu intertidal estuarien. Thèse de doctorat Université de Bordeaux I: 478 pp.

Bailey-Brock, J. H., 1985. Polychaetes from Fijian coral reefs. Pac. Sci. 39: 195–220.

Bhaud, M., 1966. Etude du développement et de l'écologie de quelques larves de *Chaetopteridae*. Vie Milieu 17: 531–571.

Bhaud, M., 1969. Remarques systématiques et biogéographiques sur le genre *Mesochaetopterus* Potts 1914. Vie Milieu 20: 325–332.

Bhaud, M., 1975. Présence de *Mesochaetopterus minutus* (Annélide Polychéte, *Chaetopteridae*) dans la région de Banyuls-sur-Mer. Vie Milieu 25: 341–344.

Bhaud, M., 1982. Relations entre stratégies de reproduction et aire de répartition chez les Annélides Polychètes. Oceanol. Acta 5: 465–475.

Bhaud, M., 1983. Les limites des distributions géographiques chez quelques Invertébrés marins. Océanis 9: 317–336.

Bhaud, M., 1989. Rôle de la dissémination larvaire en paléobiogéographie réévalué à la lumière des données concernant l'époque actuelle. Bull. Soc. géol. France, V: 551–559.

Bhaud, M., 1993. Relationship between the type of larval development and the geographic range of marine species: complementary observations on gastropods. Oceanol. Acta 16: 191–198.

Bhaud, M., 1998. The species of the genus *Spiochaetopterus* (Polychaeta: Chaetopteridae) in the Atlantic-Mediterranean Biogeographic Area. Sarsia 83.

Bhaud, M., C. Bougnol & J. C. Duchêne, 1978. Observations sur la limite de répartition bathymétrique de la Polychète sédentaire *Terebellides stroemi* (Terebellidae) dans le Golfe du Lion. C.r. hebd. Séanc. Acad. Sci., Paris, Sér. D, 287: 947–950.

Bhaud, M. & M. Lefèvre, 1986. Les larves chétosphères des Polychètes Spionidae dans le Pacifique. Remarques sur l'identification des stades larvaires. Bull. Mus. natn. Hist. nat., Paris. 8 section A no. 3: 573–589.

Bhaud, M., C. Cazaux, C. Watson-Russell & M. Lefèvre, 1988. Description and identification of Polychaete larvae; their implications in current biological problems. Oceanis 13: 596–753.

Bhaud, M. & J. C. Duchêne, 1989. Biologie larvaire et stratégie de reproduction des Annélides Polychètes en province subantarctique. Colloque de Strasbourg, 1987. C.N.F.R.A.: 145–152.

Bhaud, M., C. Cazaux & M. H. Mathivat-Lallier, 1990. Métamorphose retardée chez les larves de Polychètes et modèle d'acquisition de la vie benthique. Oceanis 16: 207–223.

Bhaud, M., J. C. Duchêne, J. H. Cha & C. Nozais, 1995. Influence of temperature on the marine fauna: what can be expected from a climatic change? J. thermal Biol. 20: 91–104.

Bhaud, M. & J. C. Duchêne, 1996. Change from planktonic to benthic development: is life cycle evolution an adaptive answer to the constraints of dispersal? Oceanol. Acta 19: 335–346.

Bodo, F., 1965. Etude dynamique et variations saisonnières du plancton de la région de Roscoff. Cah. Biol. Mar. 6: 219–254.

Boicourt, W. C., 1982. Estuarine larval retention mechanisms on two scales. In V. S. Kennedy (eds), Estuarine Comparisons. Academic Press, New York: 445–456.

Brattström, H. & A. Johanssen, 1983. Ecological and regional zoogeography of the marine benthic fauna of Chile. Sarsia 68: 289–339.

Cabioch, L., J. P. L'Hardy & F. Rullier, 1968. Inventaire de la faune benthique de Roscoff. Annélides. Ed. Station Marine de Roscoff, 94 pp.

Cazaux, C., 1970. Recherche sur l'écologie et le développement larvaires des Polychètes de la région d'Arcachon. Thèse de Doctorat ès-Sciences Naturelles, Bordeaux, 2 tomes, 455 et 180 pp.

Christy, J. H. & S. E. Stancyk, 1982. Timing of larval production and flux of invertebrate larvae in a well-mixed estuary. In V. S. Kennedy (eds), Estuarine Comparisons. Academic Press, New York: 489–503.

Cifelli, R., 1976. The role of circulation in the parcelling and dispersal of north atlantic planktonic Foraminifera. In J. Gray & A. Boucot (ed.), 'Historical biogeography, Plate tectonics and the changing environment.' Oregon State Univ. Press: 417–426.

Cresswell, G. R., 1991. The Leeuwin Current-Observations and recent models. J. r. Soc. Western Australia 74: 1–14.

Day, J. H., 1961. The polychaete fauna of South Africa. part 6. Sedentary species dredged off Cape coasts with a few new records from the shore. J. Linn. Soc. zool. 44: 463–560.

Furnestin, M. L., 1983. Plancton et biogéographie: quelques exemples. Oceanis 9: 289–316.

Gage, J. D. & P. A. Tyler, 1981. Non-viable seasonal settlement of larvae of the upper bathyal brittle star *Ophiocten gracilis* in the Rockall trough abyssal. Mar. Biol. 64: 153–161.

Gardiner, J. S., 1904. Notes and observations on the distribution of the larvae of marine animals. Ann. Mag. Nat. Hist., ser. 7. 14: 403–410.

Giangrande, A., S. Geraci & G. Belmonte, 1994. Life-cycle and life-history diversity in marine invertebrates and implications in community dynamics. Oceanogr. Mar. Biol. annu. Rev. 32: 305–333.

Gibbs, P. E., 1971. The Polychaete fauna of the Solomon Islands. Bull. Brit. Mus. (Nat. History). Zoology. 21: 101–211.

Gibbs, P. E., 1972. Polychaete annelids from the Cook Islands. J. Zool. 168: 199–223.

Gibbs, P. E., 1978. Macrofauna of the intertidal sand flats on low wooded islands, northern Great Barrier Reef. Phil. Trans. r. Soc. Lond. B. 284: 81–97.

Gitay, A., 1969. A contribution to the revision of *Spiochaetopterus* (Chaetopteridae: Polychaeta). Sarsia 37: 9–20.

Glémarec, M., 1969. Les peuplements benthiques du plateau continental Nord-Gascogne. Thèse Doct. Etat, Univ. de Paris, 167 pp.

Glémarec, M., 1973. The benthic communities of the European North Atlantic continental shelf. Oceanogr. Mar. Biol. annu. Rev. 11: 263–289.

Hadfield, M. & M. Strathmann, 1996. Variability, flexibility and plasticityin life histories of marine invertebrates. Oceanol. Acta 19: 323–334.

Hannerz, L., 1956. Larval development of polychaete families *Spionidae Sars*, *Disomidae Mesnil* and *Poecilochaetidae* n. fam. in the Gullmar Fjord, Sweden. Zool. Bidr. 31: 1–204.

Hansen, A. T., 1980. Influence of larval dispersal and geographic distribution on species longevity in neogastropods. Paleobiology 6: 193–207.

Hartman, O., 1959. Catalogue of the Polychaetous Annelids of the World. Allan Hancock Foundation Publication, Occasional Papers, 23: 628 pp.

Hartmann Schröder, G., 1982. Die Polychaeten der Fahrt 36 (1975) von F. S. 'Meteor' in das Auftriebsgebiet vor Nordwest-Afrika. Meteor Forsch.-Ergebnisse, R D 35: 1–20.

Jablonski, D., 1986. Larval ecology and macroevolutionb in marine invertebrates. Bull. mar. Sci. 39: 565–587.

Johannesson, K., 1988. The paradox of Rockall: why is a brooding gastropod (*Littorina saxatilis*) more widespread than one having a planktonic larval dispersal stage (*L. littorea*)? Mar. Biol. 99: 507–513.

Lebour, P., 1917. Note on the inshore plankton of Plymouth. J. mar. biol. Ass. U.K. 26: 527–547.

Lefèvre, M., 1990. Interférence sur le recrutement benthique entre hydrodynamisme des masses d'eau et comportement larvaire. Océanis 16: 135–148.

Mileikovsky, S. A., 1961. Character and nature of deep-water populations of eurybathic benthic forms of invertebrates with pelagic larvae taking as an example the Polychaeta *Euphrosine borealis* Oersted 1843 from the North Atlantic. Okeanologiya 1: 679–687.

Mileikovsky, S. A., 1974. Types of larval development in marine bottom invertebrates: an integrated ecological scheme. Thalassia Jugosl. 10: 171–179.

Nelson-Smith, A., P. Knight-Jones & E. W. Knight-Jones, 1990. Annelida 201–306. In P. J. Hayward & J. S. Ryland (eds), The Marine Fauna of the British Isles and North-West Europe. Vol. 1: Introduction and Protozoans to Arthropods, Clarendon Press, Oxford.

Nilsson, C. S., J. C. Andrews & P. Scully-Power, 1977. Observations of eddy formation off East Australia. J. Phys. Oceanogr. 7: 659–669.

Petersen, M. E., 1984. *Chaetopterus variopedatus* (Renier) (Annelida: Polychaeta: Chaetopteridae): a species complex. What species are being used at MBL? Biol. Bull. 167: 513.

Pettibone, M. H., 1982. Annelida: Polychaeta. P. 3–43. In S. P. Parker (ed.), Synopsis and Classification of living organisms vol. 2. New York, McGraw Hill.

Rasmussen, E., 1973. Systematics and ecology of the Isefjord marine fauna (Denmark). Ophelia 11: 1–495.

Ravara, A. A., 1997. Estudo da Comunidade de Poliquetas da plataforma continental ao largo de Aveiro. Dissertaçao de Maestrado. Universidade de Aveiro, 142 pp.

Reish, D. J., 1968. *Polychaetous Annelids* of the Marshall Islands. Pacific Science 22: 208–231.

Rey, C., 1995. Originalité du recrutement benthique chez *Owenia fusiformis*, Polychaeata: dissémination larvaire et rapport avec le substrat. Rapport de Maitrise, Observatoire Océanologique de Banyuls, Laboratoire Arago, 43 pp.

Reynolds, R. W., 1982. A monthly averaged climatology of sea surface temperature. NOAA Tech. Rep. NWS 31, 35 pp

Rochford, D. J., 1977. Monitoring of Coastal Sea Temperatures around Australia. Search 8: 167–168.

Rosenfeldt, P., 1982. Die Polychaetender fahrt 44 (1977) von F. S. 'Meteor' in das Auftriebsgebiet vor Nordwest-Africa. Meteor Forsch.-Ergebnisse, R D 35: 43–53.

Rothschild, B. J., 1988. Dynamics of the sea: the Ecology of High Dimensionality Systems. In B. J. Rothschild (ed.), Toward a Theory on Biological-Physical Interactions in the World Ocean. Kluwer Academic Publishers, Dordrecht, Boston: 527–548.

Scheltema, R. S., 1971. The dispersal of the larvae of shoal-water benthic invertebrate species over long distances by ocean currents. In D. J. Crisps (ed.), Fourth European Marine Biology Symposium. London: Cambridge University Press: 7–28.

Scheltema, R. S., 1974. Relationship of dispersal to geographical distribution and morphological variation in the polychaete family *Chaetopteridae*. Thal. Jugosl. 10: 297–312.

Scheltema, R. S., 1988. Initial evidence for the transport of teleplanic larvae of benthic invertebrates across the east Pacific barrier. Biol. Bull. 174: 145–152.

Scheltema, R. S., 1989. Planktonic and non-planktonic development among prosobranch gastropods and its relationship to the geographic range of species. In J. S. Ryland & P. A. Tyler (eds), Reproduction, Genetics and Distribution of Marine Organisms, Olsen and Olsen, Fredensborg: 183–188.

Sinclair, M., 1988. Marine populations: an essay on population regulation and speciation. Washington Sea Grant Program, University of Washington Press, Seattle and London: 1–252.

Sorbe, J. C., 1990. Le canyon de Cap Breton: état des connaissances sur ce type exceptionnel de la marge atlantique européenne. VIII8 Curros de Verano en San Sebastian: Universidad del Pais Vasco 19–22 juillet 1989.

Sund, P. N., M. Blackburn & F. Williams, 1981. Tunas and their environment in the Pacific ocean: a review. Oceanogr. Mar. Biol. Annu. Rev., Londres 19: 443–512.

Tachon, O., 1997. Reproduction des Invertébrés benthiques sous hautes latitudes: étude d'une campagne de prélèvements planctoniques en Terre Adélie. Rapport de Maitrise, Observatoire Océanologique de Banyuls, Laboratoire Arago, 38 pp.

Thorson, G., 1961. Length of pelagic larval life in Marine bottom invertebrates as related to larval transport by ocean currents. In M. Sears (ed.), Oceanography. AAAS. no. 67: 455–474.

Verdier-Bonnet, C. F. Carlotti, C. Rey & M. Bhaud, 1997. A model of larval dispersion coupling wind-driven currents and vertical larval behaviour: application to the recruitment of the Annelid *Owenia fusiformis* in Banyuls Bay, France. Mar. Ecol. Prog. Ser. 160: 217–231.

Wells, H. W. & I. E. Gray, 1960. The seasonal occurence of *Mytilus edulis* on the Carolina coast as a result of transport around Cape Hatteras. Biol. Bull. 119: 550–559.

Westheide, W., 1977. The geographical distribution of interstitial polychaetes. Mikrofauna Meeresboden 61: 287–302.

Wiebe, P. H. & S. H. Boyd, 1978. Limits of Nematoscelis megalops in north-western Atlantic in relation to Gulf Stream cold ring. I. Horizontal and vertical distribution. J. mar. Res. 36: 119–141.

Wray, G. A. & R. A. Raff, 1991. The evolution of developmental strategy in marine Invertebrates. TREE 6: 45–50.

Yamada, S. B., 1977. Geographic range limitation of the intertidal gastropods *Littorina sithana* and *L. planaxii*. Mar. Biol. 39: 61–65.

Young, C. M. & F. S Chia, 1987. Abundance and distribution of pelagic larvae as influenced by predation, behavior, and hydrographic factors. In A. C. Giese, J. S. Pearse & V. B. Pearse (eds), Reproduction of Marine Invertebrates, Vol. 9. Pacific Grove: Boxwood Press: 385–463.

Zinsmeister, W. J., 1974. A new interpretation of thermally anomalous molluscan assemblages of the California Pleistocene. J. Paleont. 48: 84–94.

Hydrobiologia **375/376:** 49–58, 1998.
S. Baden, L. Pihl, R. Rosenberg, J.-O. Strömberg, I. Svane & P. Tiselius (eds),
Recruitment, Colonization and Physical–Chemical Forcing in Marine Biological Systems.

Regrowth of kelp and colonization of epiphyte and fauna community after kelp trawling at the coast of Norway

Hartvig Christie[1,*], Stein Fredriksen[2] & Eli Rinde[3]
[1] *Norwegian institute for nature research, P.O. Box 736, N-0105 Oslo, Norway*
[2] *Department of Biology, Section for Marine Botany, University of Oslo, P.O. Box 1069 Blindern, N-0316 Oslo, Norway*
[3] *Norwegian institute for nature research, P.O. Box 736, N-0105 Oslo, Norway*
* *E-mail: hartvig.christie@ninaosl.ninaniku.no*

Key words: kelp, kelp trawling, recovery, epiphytes, holdfast fauna

Abstract

The kelp *Laminaria hyperborea* is regularly harvested along the Norwegian coast. Kelp trawling is regulated by restricting this to every 5th year in specified areas. The kelp plants form dense forests, 1–2 m high, and house a large number of epiphytes and associated invertebrates. Kelp, epiphytes, and holdfast (hapteron) fauna were sampled at two different regions in untrawled kelp forest and at sites trawled different number of years ago. We have examined the rate of kelp regrowth after trawling, and in what time scale the associated flora and fauna colonize the trawled areas. The trawl removed all adult kelp plants (the canopy plants), while small understorey kelp plants were left undisturbed. These recruits, given improved light conditions, made the new generation of canopy-forming kelp plants, exceeding a height of 1 m within 2–3 y. The recruitment pattern of the kelp ensures maintenance of kelp forest dominance despite repeated trawling. Both percent cover, abundance and number of epiphytic species increased with time post trawling, but epiphytic communities were not totally re-established before the next trawling episode. Colonization of most species of fauna inhabiting the kelp holdfast were found as early as one year after trawling, but increasing size of the habitat by age of kelp gave room for increasing numbers of both individuals and species. Slow colonization rate by some species might be due to low dispersal potential. Due to a higher maximum age and size of kelp plants in the northernmost region studied, restoration of both kelp and kelp forest community was slower there.

Introduction

Kelp dominate on rocky subtidal coasts of temperate waters (Mann, 1982), making diverse kelp forest communities. The persistence of this community will depend on the stability properties of the kelp. Along European coasts, kelp forests are dominated by *Laminaria hyperborea* (Gunn.) Foslie, and are known as habitats for a diverse flora and fauna. Its stiff and upright stipe provide substrate for a number of macroalgae and sessile animals. More than 40 species of epiphytic algae can be found, of which the rhodophytes dominate (Sørlie, 1994). A rich fauna of mobile invertebrates has also been found associated with this kelp, and specially the fauna inhabiting the holdfasts (hapterons) have been studied (Jones, 1971; Moore, 1973, 1986; Schultze et al., 1990). The kelp is a resource exploited by alginate industries, and ca. 170 000 tonnes are harvested (trawled) annually in Norway. The regrowth of kelp plants in trawled areas is obviously of interest to the alginate industry, and also in an ecological context because the dynamics of recovery of the kelp forest ecosystem depend on the recovery of kelp as a keystone species. However, recolonization of such disturbed systems will not only depend on recovery time of the habitat (kelp) itself,

Table 1. Different sites sampled in the two areas Rogaland and Smøla. –: not found.

	Rogaland	Smøla
Untrawled	x	x
Newly trawled	x	x
Trawled 1 year ago	x	x
Trawled 2 years ago	x	–
Trawled 3 years ago	x	x
Trawled 4 years ago	x	x
Trawled 6 years ago	–	x

but also depend on how the associated flora and fauna are able to colonize and re-establish the system.

The adult *L. hyperborea* plants form 1–2 m high kelp forests at the wave-exposed coasts of Norway (Kain, 1971a; Sjøtun et al., 1993). Kelp trawling is carried out between 58° and 64° N where shallow rocky bottoms extend several km far out off the main coastline. The trawling is regulated and decisions regarding time interval of trawling is generally based on unpublished data of kelp regrowth provided by the kelp industry itself. However, newer independent reports have contributed to prolonging the trawling intervals from 4 to 5 y in 1992 (Sivertsen, 1991; Rinde et al., 1992). The potential for dispersal and recruitment of *L. hyperborea* is high (Fredriksen et al., 1995) and a mature kelp forest normally contains an understorey vegetation of recruits ready to take over when adults are removed (Svendsen, 1972; Røv et al., 1990). According to industry capture records, the kelp forest recover in biomass by subsequent trawling. However, the restored kelp forest may differ in size and demography compared to untrawled kelp forests. Moreover, other macroalgae have played important parts in the succession in kelp forest restoration after disturbances (e.g. Dayton et al., 1984, 1992; Oshurkov & Ivanjushina, 1993). In Norwegian waters, Leinaas & Christie (1996) found *L. hyperborea* to be abundant first after 3–4 y in the succession of re-establishing kelp forest after removal of sea urchins in an overgrazed area.

Kelp trawling may affect the occurrence of the associated flora and fauna in several ways. If the composition of associated species depend on the size, age and density of the host plants, kelp trawling may, by changing the demography and size distribution of the kelp forest, indirectly affect the kelp forest community. The epiphytic algal composition have been found to be affected by age of plants (Whittick, 1983) and by shading (Harkin, 1981). Associated flora and fauna composition may change with kelp species (Schultze et al., 1990), and the proportion of other kelp species in the recovery process may thus affect the composition of associated flora and fauna. Colonization of the invertebrate species to the trawled plots will also depend on the dispersal ability of the different species. Species with pelagic larval dispersal (supply side ecology, e.g. Roughgarden et al., 1987) may have advantages in colonizing the trawled areas compared to species with brooding or other forms of direct development of juveniles.

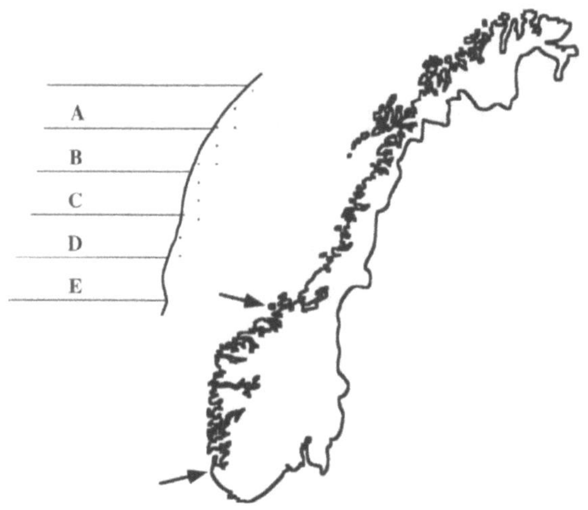

Figure 1. Map of Norway showing the two sampling sites. The figure also includes an illustration of how the coast line is divided into sectors (one nautical mile each), each permitted to trawling every 5th year. Trawling is performed one year in sector A, next year in sector B etc.

Sjøtun et al. (1993) found that the kelp forest structure (average age and size of population) vary with latitude along the coastline. Thus, recovery patterns found in one area may not be valid for the recovery at other parts of the coastal region.

In this study we focus on regrowth of the kelp forest and the return of epiphytes and invertebrate fauna in holdfasts in areas trawled a different number of years ago. For a study of possible latitudinal differences, similar studies have been done in a southern and in a northern region where kelp trawling is permitted in Norway.

Materials and methods

The sampling took part at a northern region; west of Smøla island just north of 63° N, in August 1991, and at a southern region; in Rogaland just south of 59° N, in August 1993 (Figure 1). Both areas have been important for kelp trawling for many years, and provide neighbouring latitudinal sectors of 1 nautical mile trawled at different years before any sampling (Figure 1). The regulation of trawling frequency of every 5th year give the opportunity to find sampling sites of newly trawled, tracks trawled different number of years ago, as well as untrawled kelp forest, all within short distance.

All sampling were obtained by SCUBA diving. For a comparison between the sampling sites, same wave exposure (strongly exposed, but some sheltered behind the outermost skerries) and depth (5 m below low water level) was chosen. Due to difficulties by working in the exposed waters off the Norwegian west coast, not all sampling was possible to fulfill according to the planned design. Table 1 shows the kelp forest sites sampled in the two areas.

For a description of kelp vegetation structure, a 1 × 1 m frame was dropped randomly above the kelp canopy, and the number of plants were counted in 6 replicate frames. The plants were grouped in canopy forming and understorey plants. The canopy forming plants in three of the six quadrates were harvested for individual measurements of stipes length and age determination (method from Kain, 1963).

Of the harvested plants from each site, 5 (Smøla) or 15 (Rogaland) were randomly picked out for analysis of epiphytes. The harvested plants were analysed on land soon after sampling (for visually occurring epiphytes). The epiphytes generally start to colonize the basal part of the stipes (the oldest part) and with time develop an increasing coherent cover upwards. The proportion of epiphytic cover of the stipes was measured. The stipes were then divided into three equal parts, the upper, middle, and lower stipe. Epiphytic macroalgae and macroinvertebrates were identified and semiquantitatively recorded for abundance (1 – present, 2 – common, 3 – dominating) on each part of the stipe and on the holdfast. The abundance records were added up for each plant, giving a possibility for comparison of relative abundance of epiphytes, and also illustrating the amount (volume) of the epiphytes.

Samples of 20 (Smøla) and 15 (Rogaland) randomly chosen kelp holdfasts were collected by first cutting off the stipe, then carefully loosening the holdfast using a knife and sealing each holdfast in a plastic bag. The holdfast volumes were measured according to Jones (1971), cut to pieces, washed and sieved (250 μm) for retaining associated fauna. A total quantitative and qualitative analysis of macrofauna is very time consuming, and only three replicate holdfasts from selected sites were analyzed. Holdfasts from untrawled kelp forest and from one year after trawling were chosen from each area, and holdfasts from 4 and 6 y after trawling from Smøla and 3 y after trawling from Rogaland were also analyzed. The invertebrates were individually counted and identified to species or higher taxonomic groups.

This study included two independent samplings, in 1991 from Smøla and in 1993 from Rogaland. The

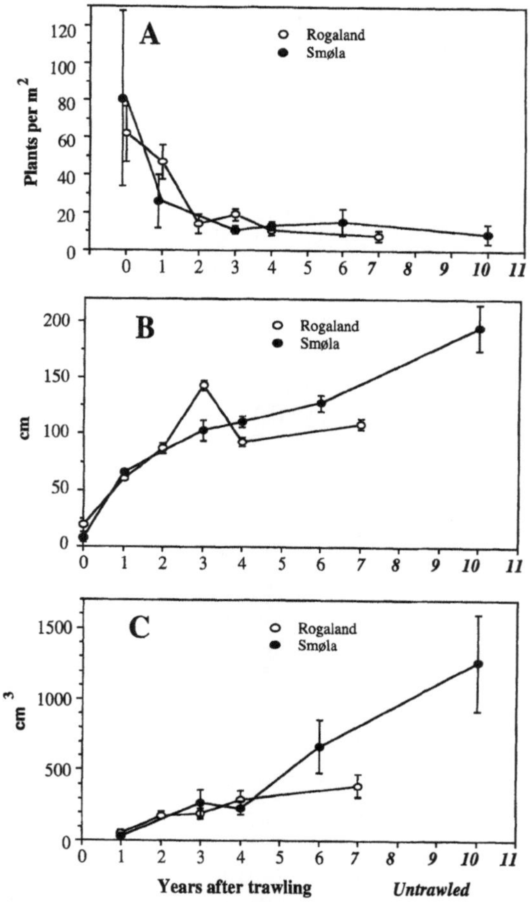

Figure 2. Kelp forest structure different number of years post trawling compared to untrawled forest at the two investigated sites Rogaland and Smøla (where untrawled kelp forest had an average age of 7 and 10 y respectively). Mean values are given with 95% confidence limits. A. Number of canopy plants per m^2, B. Canopy height (stipes length), C. Holdfast volume of canopy plants.

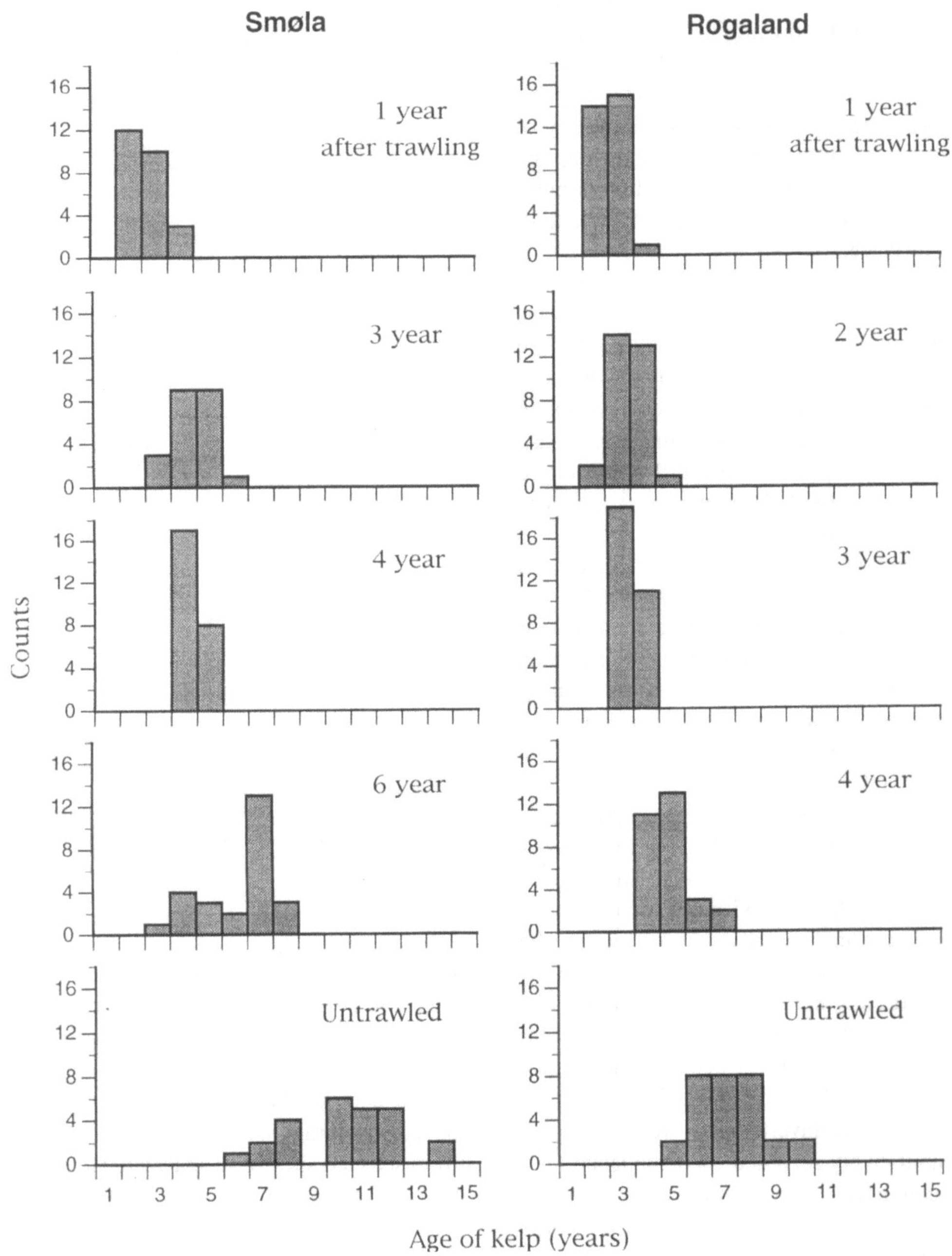

Figure 3. Age frequency distribution of canopy forming kelp plants in trawled and untrawled areas from the Rogaland and Smøla sites.

Table 2. Mean density of understorey plants (no/m^2 ± one standard error) in the trawled sites at Smøla and Rogaland. – no data from this site.

Years after trawling	1	2	3	4	Untrawled
Smøla	46(38)	–	71(25)	52(29)	83 (11)
Rogaland	23(7)	47(25)	9(7)	25(9)	34 (4)

aim was to draw a comparison between the two areas, and thus the second study was designed similar to the first with respect to exposure, depth and parameters sampled. However, some variations were included for the purpose of making improvements (more replicate plants for analysis of epiphytes), and for reducing diving time (a reduction to 15 holdfasts). Our observations showed that the four year old kelp forest at Rogaland were very similar to the untrawled kelp forest (in contrast to Smøla), and thus holdfast fauna analysis was decided to be taken from the 3 y old forest rather than from the 4 y old as done for the Smøla study. Also holdfasts from a 6 y old forest at Smøla were analysed for fauna, due to the observed large size difference between plants from this forest and the untrawled kelp forest in this region.

Differences in kelp structure and demography and differences in epiphytes and holdfast fauna between untrawled and trawled tracks were tested by t-tests and one way ANOVAs (significance level, $a = 0.05$). Recolonization patterns were also analyzed by linear regression.

Results

Figures 2 & 3 and Table 2 show a common pattern of adult *L. hyperborea* in densities of about 10 plants per m^2 and high number of small plants in the understorey in both regions. However, considerable differences in untrawled kelp forest structure and demography between the two regions (Smøla and Rogaland) were found. The adult kelp plants at Smøla were significantly older and higher than those at Rogaland and a regional difference could also be reflected in the difference in plant densities, suggesting Smøla as most optimal for growth of *L. hyperborea*. In the following presentation, the data from trawled tracks are presented as *X* years after trawling (*X* year old forest), and the untrawled sites are given as average age of canopy plants; 7 y for Rogaland and 10 y for Smøla.

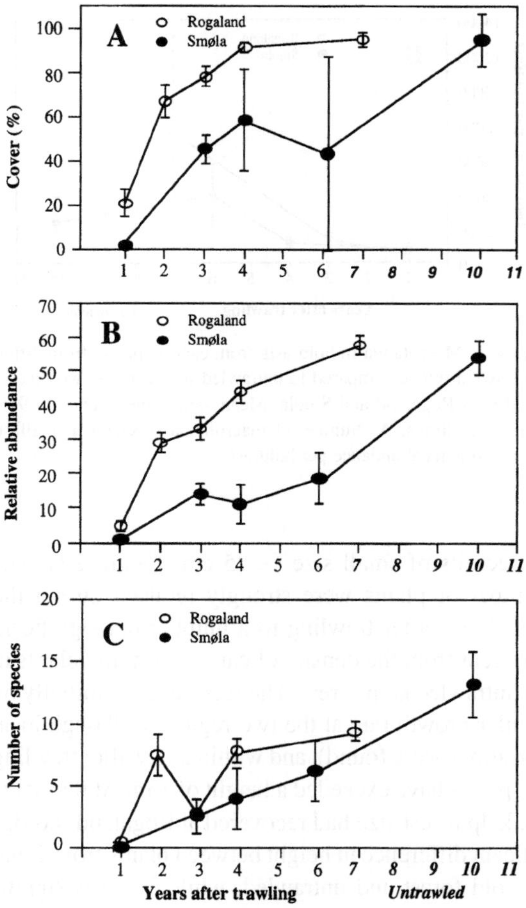

Figure 4. Occurrence of epiphytes on examined canopy plants from different trawled tracks compared to untrawled forest in the two investigated sites Rogaland and Smøla. Mean values are given with 95% confidence limits. A. Percentage cover of epiphytes, B. Relative abundance of epiphytes, C. Number of epiphytic algal species.

The pattern of restoration of the kelp forest structure is illustrated by comparing plant density and size at the different trawled areas with the untrawled sites (Figure 2). The kelp trawl removed all large plants, and at both regions the new canopy layer in the newly trawled tracks consisted of a high density (Figure 2A)

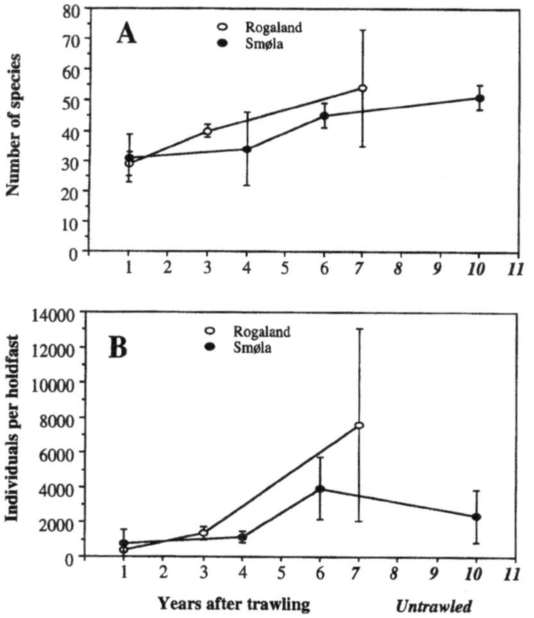

Figure 5. Macrofauna in holdfasts from canopy plants from different trawled tracks compared to untrawled forest in the two investigated sites Rogaland and Smøla. Mean values are given with 95% confidence limits. A. Number of macrofauna species per holdfast, B. Macrofauna abundance per holdfast.

of recruits of small size (< 25 cm, Figure 2B). The number of plants were strongly reduced during the first 2–3 y after trawling to a number not significant different from the density of canopy forming plants in an untrawled kelp forest. The recruits grew initially by similar growth rate at the two regions (although local variations were found), and within 2–3 y after trawling the plants have exceeded a height of 1 m. At this stage the kelp forest size had recovered at Rogaland (no significant difference in height between plants from 2 and 4 y old forest and untrawled, while 3 y was significantly higher). More than 6 y was needed for recovery of plant size at Smøla (plants from 6 y old forest was significantly smaller than plants from untrawled). The growth of the kelp holdfasts (Figure 2C) showed a similar pattern of recovery as the stipe length. Holdfast volume at Rogaland recovered after 4 y, but at Smøla the holdfasts from the 6 y old forest were still significant smaller than at untrawled.

Table 2 shows that kelp recruits established an understorey vegetation (< 25 cm in height, c.f. Figure 2B) below the new canopy as soon as one year after trawling and every year thereafter.

The age composition of the canopy forming kelp plants in the different trawled tracks (Figure 3) show

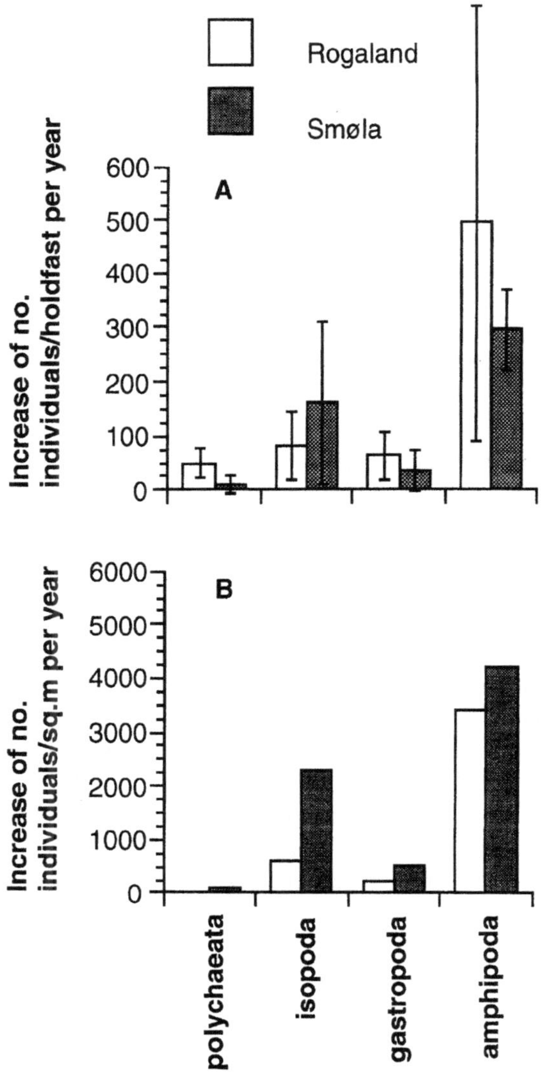

Figure 6. Colonization rate of holdfast fauna given as increase per year in number of individuals of different macrofauna groups in the two investigated sites Rogaland and Smøla. A. increase in number of individuals per holdfast per year. Mean value of slope with 95% confidence limits is given, based on linear regression analysis of fauna abundance from kelp of different years. B. estimates of increase in number of individuals per m^2 per year. (see text for further description of estimates).

that the new generations of plants have settled during more than one year prior to trawling. The age structure of this kelp forest is, however, less heterogeneous 4 y after trawling compared to the untrawled area composed of six or seven year classes. The 6 y old kelp forest at Smøla consisted of six year classes, but was dominated by 7 y old plants.

The age of kelp is important for epiphytic (total algal and faunal) cover (Figure 4A), relative abundance of epiphytes (Figure 4B), and number of epiphytic algal species (Figure 4C). The proportion of the stipes fouled by epiphytes was significantly lower in trawled tracks the first three years after trawling in both regions. Epiphytic growth can vary from encrusting algae or bryozoans to a luxuriant vegetation of voluminous red algae. The latter is characteristic on the stipes at untrawled sites, and this kind of epiphytic recovery is best illustrated by the relative abundance (Figure 4B) which shows that epiphytes needs more than four years at Rogaland and more than six years at Smøla for significant restoration. The relative rate of abundance recovery was slower at Smøla. The establishment of the different epiphytic algal species was also a slow process at Smøla compared to Rogaland. Total number of species did not differ significantly between two and four years after trawling and untrawled at Rogaland (an exception was the three year old forest that differed from the recovery trend in other ways). In contrast, the number of species gradually increased at Smøla, and still 6 y after trawling the number of epiphytic species was significantly lower than at untrawled areas. The increase at Smøla fitted a linear model (95% confidence limits: 1.1–1.5; $r^2 = 0.89$).

The macrofauna (sessile epiphytes as bryozoans was not included) found in the holdfasts are characterized by high variation within replicates both concerning number of species (taxa) per holdfast (Figure 5A) and number of individuals per holdfast (Figure 5B). In contrast to the epiphytic species, a high number of species settled one year after trawling (Figure 5A). Average number of species increased slowly by increasing age of kelp at both regions. t-tests show no significant difference in the number of species between 3 y and untrawled at Rogaland and between 6 y and untrawled at Smøla. One way ANOVA rejects the hypothesis of equal means ($p < 0.001$) for both Rogaland and Smøla. The linear regression analysis showed significant trends of increase by 5 (Rogaland) and 2.5 (Smøla) new species per year ($p < 0.0001$, $r^2 = 0.92$; $p < 0.0001$, $r^2 = 0.83$, respectively). The 95% confidence limits of the slopes are 3.8–6.5 and 1.4–2.7.

The total number of macrofauna individuals per holdfast was low during the 3 or 4 y after trawling at Rogaland and Smøla respectively. The number of individuals increased significantly approaching the untrawled (Rogaland) and the 6 y old forest (Smøla) (Figure 5B). After 6 y, no further increase in fauna abundance was found at Smøla although average holdfast volume in untrawled areas was significantly larger than holdfasts in the 6 y forest. To quantify changes in abundance of holdfast fauna during the years after trawling, we performed linear regression models of number of individuals within the four important species groups; polychaets, isopods, gastropods and amphipods (for Smøla only up to 6 y). The slopes (increase in numbers per year) with 95% confidence limits are presented in Figure 6A. All species groups, except for polychaets at Smøla, showed significant increase in number of individuals per holdfast with increasing time after trawling (the confidence limit do not cross the x-axis). Within each species group, there were no significant difference in the rate of increase between the two regions. Figure 6 also show highest mean colonization rate by amphipods.

The re-establishment of holdfast fauna abundance may be considered not as number of individuals per holdfast, but rather number of individuals per unit area of kelp forest. One year after trawling, a total of 750 individuals per holdfast and 40 plants per m^2 results in a fauna density of 30 000 per m^2. In untrawled kelp forest 5000 individuals per holdfast gives an estimate of 50 000 individuals per m^2 (10 canopy plants per m^2). Such estimates are tentative due to high variations in both plant densities and fauna abundance per holdfast (thus no error bars are given in Figure 6B). However, data on individual density of the different fauna groups may provide examples on differences in colonization patterns. The estimates on number of individuals per m^2 (density) of amphipods and isopods still showed a strong increase with kelp age at both regions, while the increase of polychaets and gastropods were less pronounced (Figure 6B). This implies that polychaets and gastropods colonize in densities close to their maximum density during the first year after trawling, while isopods and amphipods need longer time to reach maximum density.

Discussion

The results show that kelp trawling efficiently harvest all the adult canopy forming *L. hyperborea* plants, and leave a high density of recruits behind. This pool of *L. hyperborea* recruits start to grow by given improved light conditions, resulting in a new generation of plants maintaining the kelp forest. This recruitment pattern ensure persistence of a kelp forest dominated by *L. hyperborea* throughout the whole recovery (succession) period after any trawling disturbance, and not giving space for alternative kelp species (as have been found by Dayton et al., 1984; Reed & Foster, 1984). Intraspecific competition leads to reduction of the canopy plant density of the new generation approaching untrawled kelp forest size and density after 2 to 4 y. During this time, settlement of sporelings and new establishment of understorey kelp recruits ensure the potential for continuation of the kelp forest by the next trawling episode. Thus, *L. hyperborea* have the ability to resist the disturbance of repeated trawling (and probably other disturbances like storms etc.), an important property of stability (c.f. Connell & Sousa, 1983).

The age distribution of canopy plants from trawled tracks showed that the young kelp plants found after trawling were composed by more than one year class. This indicates that kelp recruits have the ability to persist as small understorey recruits for several years. Thus regrowth of kelp after trawling does not depend solely on the recruitment success in the year of trawling. The ubiquity of kelp recruits of different year classes gives this species an advantage in the succession after any trawling, regardless of season when trawling is performed. This is a further trait of the *L. hyperborea* species to maintain the dominant role in the system by a competitive advantage compared to other species.

During the first years after trawling, plant growth rate and development of kelp forest structure was found to be similar at the two regions investigated. Despite these similarities, the restoration of the kelp forest was different between the regions due to differences in maximum plant age and size. Thus, plant size recovered prior to subsequent trawling at Rogaland, but needed longer time for recovery at Smøla. Local differences were also seen, even though similar sampling sites were chosen to eliminate differences in physical factors that could introduce such differences. Three years after trawling the kelp forest at Rogaland had reached a significantly higher plant size than the four year and untrawled sites, and was composed of a high density of plants. An explanation of this may be more favorable growth conditions at this site than the neighbouring sites. The structure of the three year forest at Rogaland underline that disturbances like kelp trawling might lead to regrowth of a very dense and homogenous kelp forest.

The re-establishment of epiphytes on the kelp stipes was a slower process than the regrowth of kelp plants. Even though a high percentage of the stipes was covered by a number of species of epiphytes during 2–3 y, the relative abundance showed that epiphytes did not recover before next trawling at either of the two regions. Even when most of the stipes were fouled and most of the species were established, the abundance, or three dimensional structure of the epiphytes had not recovered. As for the kelp itself, epiphytes showed a generally slower recovery at Smøla than at Rogaland. Since age is of importance for epiphytic growth (Whittick, 1983; this study), a field subjected to repeated trawling will probably never recover in epiphytic abundance and diversity.

Another inhibiting factor for the development of epiphytic algal growth is shading (Harkin, 1981). The canopy forming plants can reduce the light penetration to the bottom by as much as 90% (Norton et al., 1977). The dense canopy of the homogenous new kelp generation in trawled tracks will inhibit light penetration more efficiently than the heterogeneous plant structure of the untrawled forests. This might be illustrated by the low number of species established in the three year forest at Rogaland, where population density was higher than at the two and four years old forest.

Evaluations of recovery of invertebrate fauna connected to the holdfasts are more difficult to make due to the high variations of fauna abundance between holdfasts of same age and size. As the kelp plants grow during the years after trawling, the holdfasts increase gradually in size (volume). This gives more room and probably also more niches to animals, and an increasing number of individuals and species per holdfast would be expected. This was found at Rogaland, and at Smøla of samples up to 6 y. The unexpected lack of further increase in fauna abundance at Smøla (mean value was lower at untrawled than at 6 y after trawling although holdfast size was about doubled) may be due to the possibility that these large holdfasts gave room for larger predatory polychaets, crustaceans and echinoderms contributing to a reduction of the numerous herbivores and detritivores. The increase in number of individuals of invertebrates per holdfast as the holdfast volume increase by age, may be of minor importance in a fauna abundance restoration context, the main objective being the restoration of fauna density per unit area. Due to the decrease in kelp plant population density with increasing number of years after trawling, estimates of recolonization per unit area showed population density of early invaders to be re-established after one year, while other species increased in density by increasing age of kelp forest. Animals with pelagic larval settlement, polychaets and gastropods, established densities approaching maximum density one year after trawling, while isopods and amphipods showed increasing density throughout the period of kelp forest restoration. The latter species recruit by brooding, and colonization into disturbed habitats might be by crawling or swimming. Due to the lack of a dispersal phase, these animals may have limited dispersal ability (e.g. Lessios et al., 1994). Differences in dispersal may be an explanation of the observed differences in colonization rates, in which case isopod and amphipod recovery might be more vulnerable if larger kelp areas are disturbed.

A relatively high number of both species and individuals of invertebrates colonized the small holdfasts one year after trawling, and a diverse community of associated flora and fauna were established subsequently trawling (5 y) at both regions. However, those species with stronger demands to habitat heterogeneity and those with highly reduced dispersal patterns contribute to a community not totally recovered by next trawling. As the kelp plants form a habitat of limited duration (maximum 10 y found at Rogaland and 14 y at Smøla), the flora and fauna of this community must be adapted to a regular colonization of new kelp plants. Thus, a kelp forest community may be recovered during a period of approximately one kelp generation.

More recent reports (Christie, 1995) have found high numbers of invertebrate species inhabiting the stipes epiphytes, and strongly increasing abundance with increasing epiphytic volumes. The reduced epiphyte abundance in the trawled tracks will thus contribute strongly to an additional retarded recovery of fauna abundance and diversity due to kelp trawling. This factor also contribute to the longer time of fauna restitution at Smøla.

Mean size and age of the kelp *L. hyperborea* varies by latitude within the area of distribution (Kain, 1971b) and within the restricted area of kelp trawling activity along the west coast of Norway (Kain, 1971a; Sjøtun et al., 1995). Our data on differences between Rogaland and Smøla support these observations. Both recovery of kelp demography and structure, as well as restoration of the flora and fauna community associated with kelp forest took longer time at Smøla where untrawled kelp forest consists of larger and older plants. The process of community change following kelp trawling at Rogaland was similar to the pattern of marine invertebrates found by Dean & Connell (1987) where species richness and abundance showed largest increase up to middle successional stages. The pattern of kelp forest community recovery at Smøla was following the model of continously increase during the succession or process of recovery (c.f. Odum, 1969).

The results from this study may be utilized in the management of kelp resources. A time interval of trawling of 5 y for both regions may be questioned, and differences in intervals between latitudinal regions may be suggested.

This study is an introductory elucidation to the understanding of patterns of community recovery of kelp forest after trawling, and thus leave a number of questions. Kelp trawling, according to our observations, fragments the kelp forest by tracks rather than totally clear larger areas. The recolonisation and recovery of kelp plants and kelp forest community reported in this study is only valid for the processes within the trawled tracks. The larger ecological consequences may depend on the dynamics formed by the proportion and distribution of trawled and untrawled kelp forest.

Acknowledgements

This study was funded by the Norwegian Directorate for Nature Management. We are grateful to Arnfinn Skadsheim, Arne Sivertsen, and the kelp trawling industry for help and information, and we thank Mohammed Abdullah for corrections to the manuscript.

References

Christie, H., 1995. Description of the kelp forest fauna at Froan, Mid Norway; variation in an exposure gradient. NINA Oppdragsmelding 368: 1–22. (in Norwegian with English summary).

Connell, J. H. & W. P. Sousa, 1983. On the evidence needed to judge ecological stability or persistence. Am. Nat. 121: 789–824.

Dayton, P. K., V. Currie, T. Gerrodette, B. D. Keller, R. Rosenthal & D. Ven Tresca, 1984. Patch dynamics and stability of some California kelp communities. Ecol. Monogr. 54: 253–289.

Dayton, P. K., M. J. Tegner, P. E. Parnell & P. B. Edwards, 1992. Temporal and spatial patterns of disturbance and recovery in a kelp forest community. Ecol. Monogr. 62: 421–445.

Dean, R. L. & J. H. Connell, 1987. Marine invertebrates in an algal succession. I. Variations in abundance and diversity with succession. J. exp. mar. Biol. Ecol. 109: 195–215.

Fredriksen, S., K. Sjøtun, T. E. Lein & J. Rueness 1995. Spore dispersal in *Laminaria hyperborea* (Laminariales, Phaeophyceae). Sarsia 80: 47–54.

Harkin, E., 1981. Fluctuations in epiphyte biomass following *Laminaria hyperborea* canopy removal. In T. Levring (ed.), Proc. Xth Int. Seaweed Symp. Göteborg, Sweden. Walter de Gruyter, Berlin and New York: 303–308.

Jones, D. J., 1971. Ecological studies on macroinvertebrate populations associated with polluted kelp forest in the North Sea. Helgoländer wiss. Meeresunters. 22: 417–441.

Kain, J. M., 1963. Aspects of the biology of *Laminaria hyperborea*. II. Age, weight and lenght. J. mar. biol. Ass. U.K. 43: 129–151.

Kain, J. M., 1971a. The biology of *Laminaria hyperborea*. VI. Some Norwegian populations. J. mar. biol. Ass. U.K. 51: 387–408.

Kain, J. M., 1971b. Synopsis of biological data on *Laminaria hyperborea*. FAO Fisheries Synopsis No. 87: 1–69.

Leinaas, H. P. & H. Christie, 1996. Effects of removing sea urchins (*Strongylocentrotus droebachiensis*): Stability of the barren state

and succession of kelp forest recovery in the east Atlantic. Oecologia 105: 524–536.
Lessios, H. A., J. R. Weinberg & V. R. Starczak, 1994. Temporal variation in populations of the marine isopod *Excirolana*: How stable are gene frequencies and morphology? Evolution 48: 594–563.
Mann, K. H., 1982. Ecology of coastal waters. A system approach. Studies in ecology Vol. 8. Blackwell scientific publications, 322 pp.
Moore, P. G., 1973. The fauna of Northeast Brittain. I. Introduction and physical environment. J. exp. mar. Biol. Ecol. 13: 97–125.
Moore, P. G., 1986. Levels of heterogeneity and the amphipod fauna of kelp holdfasts. In P. G. Moore & R. Seed (eds), Ecology of Rocky Coasts, Hodder & Stoughton, London: 274–289.
Norton, T. A., K. Hiscock & J. A. Kitching, 1977. The ecology of Lough Ine. XX. The *Laminaria* forest at Carrigathorna. J. Ecol. 65: 919–941.
Odum, E. P., 1969. The strategy of ecosystem development. Science 164: 262–270.
Oshurkov, V. V. & E. A. Ivanjushina, 1993. Effect of experimental harvesting on kelp regrowth and on the structure of the shallow-water communities of Bering Island (Commander Islands). Asian Mar. Biol. 10: 95–108.
Reed, D. C. & M. S. Foster, 1984. The effects of canopy shading on algal recruitment and growth in a giant kelp forest. Ecology 65: 937–948.
Rinde, E., H. Christie, S. Fredriksen & A. Sivertsen, 1992. Økologiske konsekvenser av taretråling: Betydning av tareskogens struktur for forekomst av hapterfauna, bunnfauna og epifytter. NINA Oppdragsmelding 127: 1–37. (in Norwegian with English summary).
Roughgarden, J., S. D. Gaines & S. W. Pacala, 1987. Supply side ecology, the role of physical transport processes. In J. H. R. Gee & P. S. Giller (eds), Organization of Communities: Past and Present. Blackwell Scientific Publications: 491–518.
Røv, N., H. Christie, S. Fredriksen, H. P. Leinaas & S. H. Lorentsen, 1990. Biologiske forundersøkelser i forbindelse med planer om taretråling i Sør-Trøndelag. NINA Oppdragsmelding 52: 1–20. (in Norwegian with English summary).
Schultze, K., K. Janke, A. Krüss & W. Weidemann, 1990. The macrofauna and macroflora associated with *Laminaria digitata* and *L. hyperborea* at the island of Helgoland (German Bight, North Sea). Helgoländer Meeresunters. 44: 39–51.
Sivertsen, K., 1991. Høsting av stortare og gjenvekst av tare etter taretråling ved Smøla, Møre- og Romsdal. Fisken og Havet 1: 1–44. (in Norwegian with English summary).
Sjøtun, K., S. Fredriksen, T. E. Lein, J. Rueness & K. Sivertsen, 1993. Population studies of *Laminaria hyperborea* from its northern range of distribution in Norway. Hydrobiologia 260/261: 215–221.
Sjøtun, K., S. Fredriksen, J. Rueness & T. E. Lein, 1995. Ecological studies of the kelp *Laminaria hyperborea* (Gunnerus) Foslie in Norway. In H. R. Skjoldal, C. Hopkins, K. E. Erikstad & H. P. Leinaas (eds), Ecology of Fjords and Coastal Waters. Elsevies science B.V.: 525–536.
Svendsen, P., 1972. Noen observasjoner over taretråling og gjenvekst av stortare, *Laminaria hyperborea*. Fiskets Gang 22: 448–460. (in Norwegian with English summary).
Sørlie, A. C. 1994. Epifyttiske alger på hapterer og stipes av *Laminaria hyperborea* (Gunn.) Foslie fra Vega i Nordland fylke. (Epiphytic algae on holdfasts and stipes of *Laminaria hyperborea* (Gunn.) Foslie from Vega, Nordland County.) Cand. scient. thesis, University of Oslo, 110 pp.
Whittick, A. 1983. Spatial and temporal distribution of dominant epiphytes on the stipe of *Laminaria hyperborea* (Gunn.) Fosl. (Phaeophyta: Laminariales) in S.E. Scotland. J. exp. mar. Biol. Ecol. 73: 1–10.

Hydrobiologia **375/376:** 59–66, 1998.
S. Baden, L. Pihl, R. Rosenberg, J.-O. Strömberg, I. Svane & P. Tiselius (eds),
Recruitment, Colonization and Physical–Chemical Forcing in Marine Biological Systems.

Pentapora fascialis (Pallas) [Cheilostomata: Ascophora] colonization of one sublittoral rocky site after sea-storm in the northwestern Mediterranean

Silvia Cocito, Francesca Ferdeghini & Sergio Sgorbini
ENEA CRAM S.Teresa, CP 316, I-19100 La Spezia, Italy

Key words: Pentapora fascialis, temperate rocky reef, storm, colonization, recruitment, Ligurian Sea

Abstract

Effects induced on the bryozoan *Pentapora fascialis* by a severe sea-storm which occurred in December 1993 at Tino Island (Ligurian Sea, NW Mediterranean) are described. Mortality, recruitment and growth of *P. fascialis* colonies have been measured directly by photographic monitoring for three and a half years. Near total mortality of the erect *P. fascialis* colonies in the photostation (11 m depth) was observed, while other benthic encrusting species seemed not to be affected. Many recruits were observed six months after the storm, but a clear relationship between recruitment and ovicells presence was not found. The percent cover increased and the number of colonies decreased with time following the storm: this was partly due to fusion between colonies. After three and a half years the cover of *P. fascialis* reached the pre-storm cover.

Introduction

Disturbance is believed not only to be a major source of temporal and spatial heterogeneity in the structure and dynamics of communities, but to be an agent of natural selection in the evolution of populations (Sousa, 1984). Consequently, the disturbance regime experienced by a community may act as a key factor in determining the life-history strategies of its components (Fong & Lirman, 1995).

Disturbance, as defined by Pickett & White (1985), may fluctuate from relatively discrete and frequent events to disruptive and episodic events. Many types of disturbances have been documented for coral reefs across a wide range of spatial and temporal scales (e.g. Pearson, 1981; Woodley et al., 1981; Karlson & Hurd, 1993; Rogers, 1993), but direct effects on temperate reef communities have been only partially explored (Harris et al., 1984; Witman, 1987; Renaud et al., 1996).

Episodic events have been shown to affect structure and function of marine benthic ecosystems strongly: understanding their effects needs long-term studies that encompass a monitoring period corresponding to the expected lifetime of the dominant organisms. The effects of a disruptive stress depend to a large extent on the characteristics of the dominant (i.e. most abundant) species, not only on their ability to withstand certain stresses, but also on their ability to recover afterwards through recolonization and regrowth (Rogers, 1993).

Pentapora fascialis is a large erect bryozoan dominating a temperate rocky reef assemblage in the Ligurian Sea (northwestern Mediterranean) (Cocito et al., 1997). During a monitoring study carried out since 1993, an exceptionally severe sea-storm, which is likely to occur less than once in 100 y, struck the area of the Gulf of La Spezia (Ligurian Sea) destructively affecting the majority of the *P. fascialis* population.

In this paper we describe the effects of such an unusual event by analysing in a single permanent station (11 m depth) how disturbance affected hard substratum assemblages and how recolonization, with particular attention to the dominant species, proceeded in the following three and a half years. Mortality, recruitment and growth of *P. fascialis* colonies were measured directly by photographic monitoring. The reproductive status of the *P. fascialis* population was assessed monthly by analysing the presence/absence

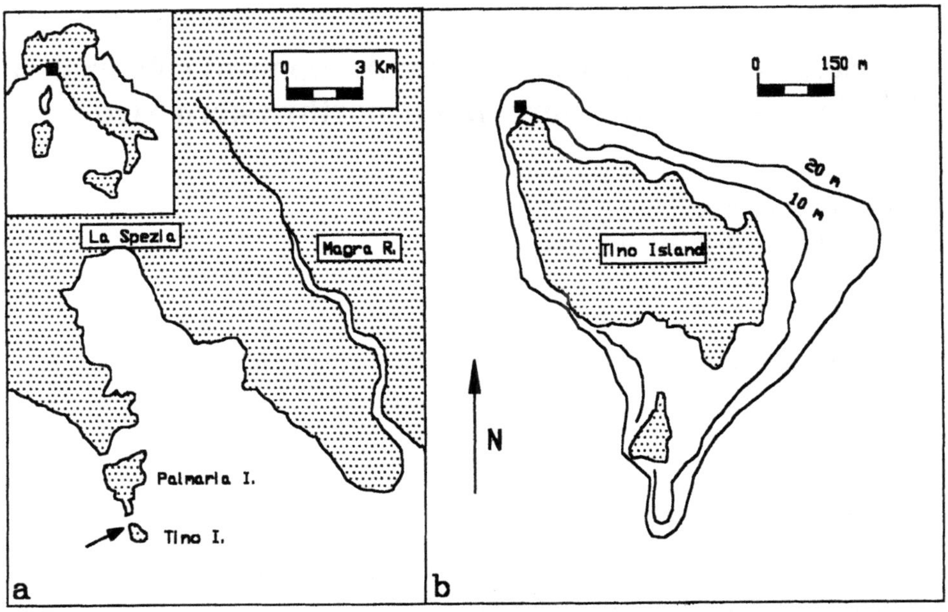

Figure 1. (a) Geographical setting of the study area (Lat. 44°01′ N; Long. 9°50′ E); (b) location of the photostation.

of ovicells, a feature of sexually mature bryozoan colonies.

Material and methods

The study area

Tino Island (44°01′ N, 9°50′ E) is located at the western border of the Gulf of La Spezia (eastern Ligurian Sea, Italy) (Figure 1a). South and southeast sides are constituted of rocky slopes of medium inclination, the western side has vertical walls, while the north vertical side is composed of rocky blocks and debris. The island is surrounded by flat, muddy bottom with average depth of about 26 m.

Hydrological and current measurements showed that the movement of water masses along the continental shelf, just off the Islands of Tino and Palmaria, a larger island to the north, is consistent with the Ligurian current, flowing towards north-west (Astraldi & Gasparini, 1992). Throughout 1993 current measurement were taken in the channel between the Islands of Tino and Palmaria through Aanderaa or electromagnetic currentometers. The area is subjected to strong currents, reaching 80 cm s^{-1} of speed.

Wave measurements registered by 'Servizio Idrografico e Mareografico Nazionale' for an eight-year period (1989–1996) during winter months recorded brief periods (1–2 d) in which significant waves reached 4 m. During winter storms waves rarely exceeded 2.5 m significant height. Moreover, they usually came from the southwest.

The station affected by the storm (Figure 1b) lies at 11 m depth facing north-west on the vertical cliff of Tino Island. The area is characterised by a hard substratum community with bryozoa (*Pentapora fascialis, Schizoporella errata*), sciaphilic algae (*Halopteris filicina, Peyssonnellia* sp., encrusting corallinaceae) and sponges (*Scopalina lophyropoda*, *Chondrosia reniformis*).

Pentapora fascialis

Pentapora fascialis (Pallas) [Cheilostomata: Ascophora] is one of the largest living bryozoans, with bright orange erect colonies: it commonly develops a mound-like form, grows to a diameter of 20 to 30 cm and has an encrusting base with convoluted bilamina sheets arising giving a honeycomb appearance (Hayward & Ryland, 1979; Sala et al., 1996). The species is particularly abundant on the northwest facing cliff of Tino Island where it forms sturdy but brittle, carbonate erect colonies of considerable size (max. diameter: 82 cm; max. height: 48 cm) (Cocito et al., 1997).

Little is known about the biology of the species: it lives on subtidal stable hard bottoms, down to 50 m or more, and exposed to strong current. It is often a predominant part of the benthos in the Mediterranean (Sala et al., 1996) and it has been found to settle on artificial substrata in the Ligurian sea (Geraci, 1974). The species is perennial (Eggleston, 1972), with a vertical growth rate up to 3.5 cm y^{-1} (Cocito & Ferdeghini, 1998).

In Bryozoa many studies concerning reproductive status have been based on analysis of presence or absence of ovicells and/or brooded embryos (Winston & Jackson, 1984; Hughes, 1990; Herrera et al., 1996). A strong relationship between ovicells presence and reproductive status has been evidenced in *Tricellaria inopinata* (Ambrogi Occhipinti, 1991). Little data is available about reproductive status of *P. fascialis*: according to Gautier (1962) it should exhibit ovicells all the year. In the Atlantic species (*P. foliacea*) an ovicell band is shown near the growing margin; another band with almost no ovicells, named 'growth check line', is produced in winter (Patzöld et al., 1987).

In Cheilostomata brooded embryos are known to develop in a non-feeding lecithotrophic larva that swims for minutes or few hours (Keough & Chernoff, 1987; McKinney & Jackson, 1989), only rarely for few days (Hughes, 1990), and to a short distance before settlement (Hayward & Ryland, 1979).

In this study, ovicell presence and distribution were assessed from different sized colonies. There were 62 *P. fascialis* colonies collected in the area from January to October 1994. Sediment was cleansed by rinsing colonies with water; subsequently, epibiotic fauna and flora were removed carefully with tweezers. Ovicell abundance was counted on standard area (25 mm^2) using a dissecting microscope in 13 ($\pm$ 6) laminae per colony. A total of 680 laminae were analyzed and ovicell abundance classes were determined. When ovicells formed a distinguishable band, its height was measured to the nearest millimeter.

Photographic monitoring

Colonisation was monitored in an area of 1 m^2 (photostation) (Hiscock, 1987) through underwater photography. No replicate samples were used. The quadrat was permanently fixed with two nails linked up with a line. Whereas monthly colonies sampling was not affected by persistent water turbidity, being visibility reduced to 1–2 m during most part of the year, photographs were taken periodically at 3–7 mo intervals from November 1993 to June 1997. A Nikonos camera with 15 mm lens and 100 ISO colour transparency film were used. Photographs were always taken at a subject-camera distance of 1 m, taking care to maintain viewpoint centrality through an optical viewfinder. Images were stored in Kodak Photo-CD (768×512 pixels; chromatic resolution of 16×10^6 colours), catalogued by month, sorted and analysed. Images have been corrected with PhotoStyler® software to restore perpendicularity. Vectorial digitisation and storage of computer-assisted drawings were produced with Autocad® software. Colonies and organisms were represented as patterned polygons. Polygons corresponding to subsequent sampling dates were superimposed to visualise colony growth. Polygon areas, which represent the total projected area on the two-dimensional substratum, were computed automatically. The complete procedure is described by Sgorbini et al. (1996): they analyse readability and distortion errors of 'useful' images and define acceptability limits of colony variations in size and growth. Moreover, they report that the resolution of the method is such to allow the recognition of colonies as small as 7 cm^2.

The occurrence of the smallest colonies or recruits, either of which was not present initially, was checked through close-up photographs. Each close-up represented an area of about 2000 cm^2 inside the quadrat; colonies as small as 1 cm^2 were recognizable.

Results

Colonisation

The monitoring study carried out since 1993 and still in progress, has involved in the same locality photostations ranging from 11 to 22 m depth. We describe the only photostation at 11 m depth affected by the storm.

Before the sea storm, in November 1993 (Table 1), the photostation (11 m depth, 70° inclination) was mainly colonized by eight *Pentapora fascialis* colonies, covering 54% of the substratum; central portions of the largest colonies showed necrosis. Encrusting sponges (*Scopalina lophyropoda*) were present on the primary substratum together with Sphacelariales (*Halopteris filicina*), encrusting corallinaceae (*Pseudolithophyllum* sp.) and the encrusting bryozoan *Schizoporella errata*; ubiquitous turf composed of brown filamentous macroalgae, small hydroids and mud covered the remaining hard substratum.

From 16 to 18 December 1993 a sea-storm struck the area of La Spezia Gulf with an unusual combination of significant mean wave height (5 m), period

Table 1. Major taxa time evolution in the photostation (1 m^2) from November 1993 to June 1997. (Algal-mud turf, seasonal hydroids and algae are not represented).

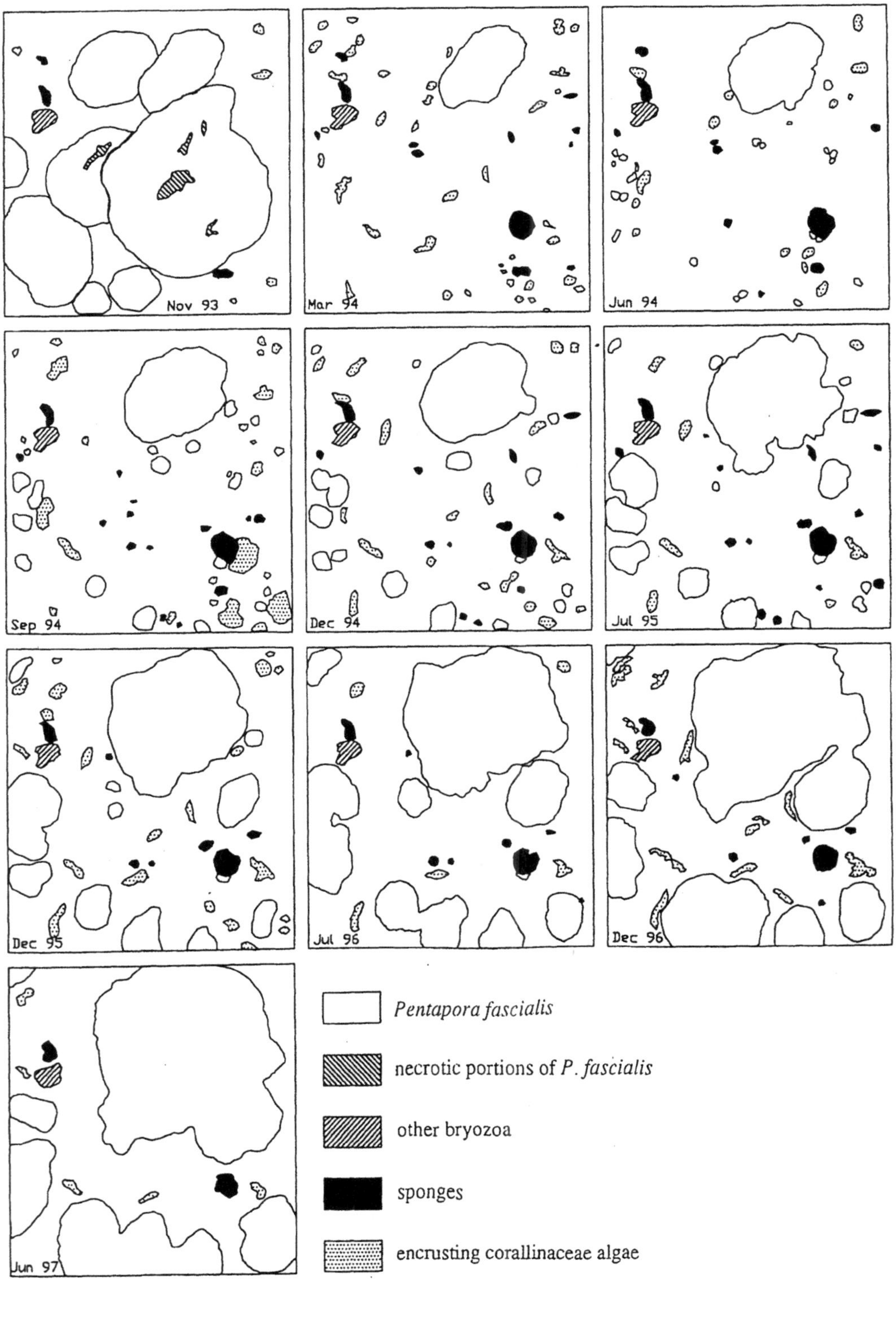

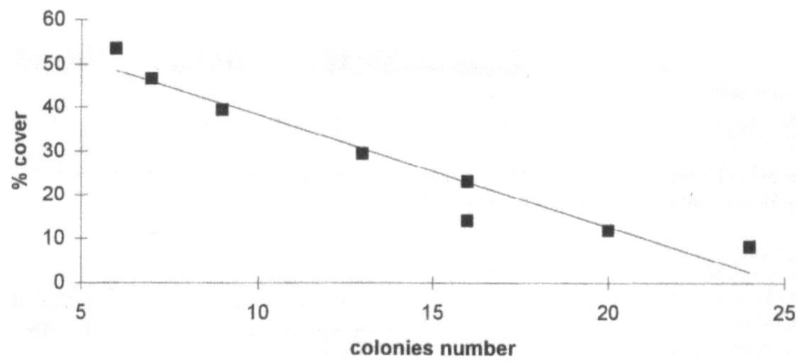

Figure 2. Pentapora fascialis percent cover versus colonies number in the photostation (1 m^2).

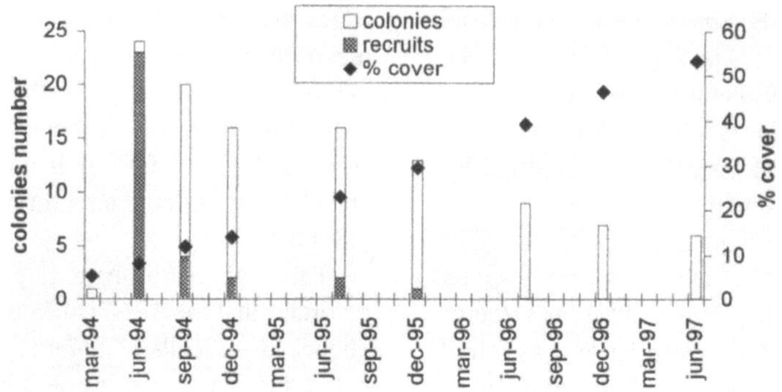

Figure 3. Relationship between *Pentapora fascialis* colonies number, recruits number and percent cover in the post-storm period.

(3 d) and direction (east). Data from the Istituto Idrografico della Marina (Genoa) (pers. com.) indicates that such a combination might occur very rarely in the eastern Ligurian Sea.

At Tino Island reef much of the surfaces facing west and northwest on the rocky, subtidal substratum down to 11 m depth were scoured severely. The spatially dominant erect forms, represented by the bryozoan *P. fascialis*, were heavily damaged, while encrusting species seemed not to be affected. Numerous remnants of *P. fascialis* colonies were seen lying mainly at the base of the cliff (Sgorbini, personal observation 1 month after the storm).

After the storm the photostation (1 m^2) was successfully photographed in March 1994. Whereas species number did not change, their abundancies varied: new specimens of the same encrusting species that were present in November 1994 became evident. Seven out of eight colonies of *P. fascialis* resulted swept away; the colony that survived was growing partially sheltered in a rock fissure and did not show damage or breakage scars on outer branching laminae. Photographs showed clearly one specimen of a massive sponge (*Cacospongia scalaris*), growing under the largest removed *P. fascialis* colony (see Table 1, March 1994). Hard substrates previously occupied by *P. fascialis* colonies were partially colonized by algal-mud turf.

During the monitoring period, a marked seasonality has been found for both algae such as Dictyotales (*Dictyota dichotoma, Padina pavonica*) spreading in summer, and hydroids (*Eudendrium racemosum, Sertularella crassicaulis* and *Campanularidae*) growing in cold months, both as epibionts and attached to the substratum.

The most evident change in the three and a half years' monitoring since the storm was the positive net growth of *P. fascialis*. Notwithstanding the sharp reduction in total cover induced by the disturbance (from 54 to 5%), its value increased quite regularly with time almost reaching the pre-storm cover value in June 1997 (53%). During the colonisation period, the percent cover increasing trend was inversely related to colonies' number ($r = 0.929, p < 0.05$) (Figure 2).

	jan	feb	mar	apr	may	jun	jul	aug	sep	oct
ovicells band height (mm)	-	n.i.	3,6	2,0	2,9	n.i.	2,6	n.i.	2,9	3,8
st. dev. (mm)	-	-	2,3	1,5	0,9	-	0,8	-	1,8	1,2

Figure 4. Pentapora fascialis: ovicell abundance classes (black rectangles: abundant, mean number 92/25 mm^2; grey rectangles: scarce, mean number 32/25 mm^2; white rectangles: absent) and ovicell band height (mm) (n.i. = not identifiable).

Recruitment

In June 1994 (Table 1), six months after the sea-storm, many brightly colored recruits of *P. fascialis* were observed: 23 recruits, almost certainly sexually generated, were found in the photostation. The size of new colonies ranged from 1.9 to 7.9 cm^2 in cover. For the recruits a first phase of planar growth or 'sheeting' was observed; afterwards, when the colony area was wider than 2 cm^2, upward growth was exhibited by raising distinguishable laminae.

The number of recruit colonies dropped from 23 in June 1994 to 4 three months later, 2 were observed in December 1994 and July 1995 and 1 in December 1995 (Figure 3). A positive correlation was found between the number of recruits and the number of colonies ($r = 0.583$, $p < 0.05$), but correlation between the number of recruits and available substratum was not significant ($r = 0.375$, $p > 0.05$).

Reduction in the number of colonies was mainly related to the interactions among growing colonies: fusion phenomena between and among colonies were observed. For example, in the zone strictly related to the presence of the survived colony of *P. fascialis,* all satellite colonies joined the biggest one through fusion mechanism. In one colony partial mortality followed by fission was observed, probably induced by persistence of algae and hydroids as epibionts or by sedimentation (see Table 1, December 1996).

About 85% of the recruits survived showed positive net growth. Two recruits overgrew *Cacospongia scalaris* but after 29 months, during which their growth was extremely reduced, they could no longer be traced.

Reproductive status

All 62 *P. fascialis* colonies collected in the area, whose diameter ranged from 2.8 to 45 cm, showed ovicell presence from February to October 1994. Ovicells were lacking only in January (Figure 4). The recruits showed ovicells, in some cases with scarce density and not forming a distinguishable band. This observation suggests that *P. fascialis* can reach reproductive ability in the early stage of colony develoment.

In January ovicells were absent, in February, June and August scarce (mean abundance 32/25 mm^2) and a clearly identifiable ovicell band near the growing edges was not present. In the other months ovicells' mean abundance was 92/25 mm^2 and each laminae exhibited a well delimited band. Statistical analysis (ANOVA Tukey's test) showed significant differences ($F = 7.904$, $p < 0.05$) in band height during March and October with the maximum values observed (3.6–3.8 mm).

The high recruitment rate (23 recruits in 1 m^2) recorded in June 1994 could be interpreted as consequence of ovicells abundance during spring months, but reduction in recruitment rates monitored in the following period was not clearly explainable.

Discussion

Pentapora fascialis, the spatially dominant species in the area affected by the storm, has several adaptations that enable it to persist in high-energy environments, including a carbonate skeleton and a mound-like shape with thick basal attachment to the substratum.

A chronic intermediate intensity disturbance regime, such as that one normally experienced by *P. fascialis* in winter months, allows the maintenance of dense adult populations at stable abundance (7 ± 1.5 colonies/m^2 at Tino Island, Cocito et al., 1997; 3.3 ± 2.5 colonies/m^2 at Medes Island, Sala et al., 1996).

On the contrary, low frequency but high intensity disturbance, such as the one experienced during the December 1993 storm at Tino Island, caused mortality of almost all colonies down to 11 m depth. On this occasion, disturbance was not size selective. A colony's vulnerability is related to its erect growth form and to degree of exposure (Woodley et al., 1981; Connell & Keough, 1985; Done, 1992). Nevertheless, several authors believe it unlikely that organisms can evolve adaptive responses to disturbance such as severe storms or hurricanes, because they occur very seldom relative to generation time (Sousa, 1984).

Observations made in the photostation at 11 m affected by the storm provide some insight on the biology of *P. fascialis*. Disturbance did change significantly species abundancies: erect species, namely *P. fascialis* colonies, were sharply reduced in number whereas encrusting (algae, sponges, bryozoa) and sheltered species seemed not to be affected. For example, one massive sponge (*C. scalaris*) was not swept away as it was growing sheltered by one of the *P. fascialis* colonies. As already underlined by Jackson (1979) growth form determines the successful in high disturbance environments: two-dimensional encrustations, more or less completely attached to the substratum, can survive storm better than mound-like forms, that exhibit erect and lateral growth and are attached to substratum along most of basal area.

Six months after the storm, 23 recruits were observed in the photostation: the bryozoa exhibited great colonising ability, probably due to ovicell production during spring months. In winter months a reduction or even a standstill of reproductive ability can be hypothesised for colonies of all sizes, as already found in the Atlantic species *P. foliacea* (Patzöld et al., 1987). Recruitment occurred with a sharply decreasing trend: a slight correlation between recruitment and colony number was found but no correlation was present between recruitment and available substratum. In a previous study, Cocito et al. (1998) did not observe recruitment of *P. fascialis* in the same area when substratum cover was more than around 40%. Larval recruitment seems to be inversely related to colony mass (Jackson & Coates, 1986).

The contribution of Cheilostomata lecithotrophic larvae to recolonization is largely unknown (Keough & Chernoff, 1987). The local position of adult colonies is known to strongly affect the spatial pattern of larval settlement (Jackson, 1985; Roberts et al., 1991), and can result in a high density of new recruits near adult colonies (Hurlbut, 1991). In this study, even though a great concentration of recruits was observed a short distance from the colony that survived the storm, its contribution to recolonization could not be identified.

Recruit growth was initially encrusting. When the colony reached an area of about 2 cm^2 upward growth occurred through differentiation of erect laminae, where adult morphology (the presence of ovicells) were observed. None of the earlier studies on *P. fascialis* referred to the above sequence of structural changes. The erect form may be interpreted as an advantage in competition for food, providing access to layers where water flow is more active, enhancing the likelihood of prey capture (McKinney & Jackson, 1989; Karlson et al., 1996; Zabala & Ballestreros, 1989).

Total cover of *P. fascialis* increased in the forty-three months quite regularly with time and was inversely related to colonies' number. Many fusion phenomena among colonies have been found, but only one fission was observed, probably by persistent epibiotic pressure or sedimentation, as observed in corals (Hughes & Jackson, 1980). In the majority of cases fusion involved only two colonies, but fusion among three colonies also occurred.

We can estimate that about three and a half years were necessary for *P. fascialis* to restore the substratum pre-storm cover (54%). Great reproductive and colonising ability, relatively fast growth rate and morphological adaptability help determine the success of *P. fascialis* as the dominant organisms.

Acknowledgements

We thank A. Peirano who participated in some diving activities and critically read the manuscript. The 'Servizio Idrografico e Mareografico Nazionale' (Rome) provided waves data. Sponges were identified by M. Pansini, hydroids by C. Morri. We are indebted to C. N. Bianchi for helpful criticism. Research on *Pentapora* fall within the scope of the project CLIMBIOCCA ('influence of CLIMatic change on the activity of BIOCostructional organisms and on their role in the CArbon cycle of the marine environment') of the Marine Environment Research Centre of ENEA (La Spezia).

References

Ambrogi Occhipinti, A., 1991. The spread of *Tricellaria inopinata* into the lagoon of Venice: an ecological hypothesis. In F. P. Bigey (ed), Bryozoa Living and Fossil, Nantes (France): 299–308.

Astraldi, M. & G. P. Gasparini, 1992. The seasonal characteristics of the circulation in the North Mediterranean Basin and their relationship with the atmospheric-climatic conditions. J. Geophys. Res. 97: 9531–9540.

Cocito, S. & F. Ferdeghini, 1998. Marcatura con colorante ed etichettatura: due metodi per misurare la crescita in briozoi calcificati. In M. Piccazzo (ed.), Atti XII Congresso Associazione Italiana Oceanologia Limnologia A.I.O.L. 2: 351–358.

Cocito, S., S. Sgorbini & C. N. Bianchi, 1997. Zonation of a suspension-feeder assemblage on a temperate rocky shoal: the influence of water current and bottom topography. In L. E. Hawkins & S. Hutchinson (eds), The Response of Marine Organisms to Their Environments, University of Southampton, UK: 183–192.

Cocito, S., S. Sgorbini & C. N. Bianchi, 1998. Aspects of the biology of the bryozoan *Pentapora fascialis* in the north-western Mediterranean. Mar. Biol. 131: 73–82.

Connell, J. H & M. J. Keough, 1985. Disturbance and patch dynamics of subtidal marine animals on hard substrata. In S. T. A. Pickett & P. S. White (eds), The Ecology of Natural Disturbance and Patch Dynamics. Academic Press Inc, London: 125–151.

Done, T. J., 1992. Effects of tropical cyclone waves on ecological and geomorphological structures on the Great Barrier Reef. Cont. Shelf Res. 12: 859–872.

Eggleston, D., 1972. Patterns of reproduction in marine Ectoprocta of the Isle of Man. J. nat. Hist. 6: 31–38.

Fong, P. & D. Lirman, 1995. Hurricanes cause population expansion of the branching coral *Acropora palmata (Scleractinia)*: wound healing and growth patterns of asexual recruits. Mar. Ecol. P.S.Z.N. I, 16: 317–335.

Gautier, Y. V., 1962. Recherchés ecologiqués sur les bryozoares chilostomes en Mediterranée occidentale. Rec. Trav. St. Mar Endoume 38: 163–168.

Geraci, S., 1974. Infralittoral bryozoa settled on artificial substrata in the Ligurian sea. Docum. Lab. Géol. Fac. Sci. Lyon, 3: 335–346.

Harris, L. G., A. W. Ebeling, D. R. Laur & R. J. Rowley, 1984. Community recover after storm damage: a case of facilitation in primary succession. Science 224: 1336–1338.

Hayward, P. J. & J. S. Ryland, 1979. British Ascophoran Bryozoans. In D. M. Kermack & R. S. K. Barnes (eds), Synopses of the British Fauna. Academic Press, London, 312 pp.

Herrera, A., J. B. C. Jackson, D. J. Highes, J. Jara & H. Ramos, 1996. Life-history variation in three coexisting cheilostome bryozoan species of the genus *Stylopoma* in Panamà. Mar. Biol. 126: 461–469.

Hiscock, K., 1987. Subtidal rock and shallow sediments using diving. In J. M. Baker & W. J. Wolf (eds), Biological Surveys of Estuaries and Coasts. Cambridge University Press, Cambridge: 198–237.

Hughes, T. P., 1990. Recruitment limitation mortality, and population regulation in open system: a case study. Ecology 7: 12–20.

Hughes, T. P. & J. B. C. Jackson, 1980. Do corals lie about their age? Some demoghraphic consequences of partial mortality, fission, and fusion. Science 209: 713–715.

Hurlbut, C. J., 1991. Larval substratum selection and postsettlement mortality as determinants of the distribution of two bryozoans. J. exp. mar. Biol. Ecol. 147: 103–119.

Jackson, J. B. C., 1979. Morphological Strategies of Sessile Animals. In G. Larwood & B. R. Rosen (eds), Biology and Systematics of Colonial Organisms. Academic Press, London and New York: 499–555.

Jackson, J. B. C., 1985. Distribution and ecology of clonal and aclonal benthic invertebrates. In J. B. C. Jackson, L. W. Buss & R. E. Cook (eds), Population Biology and Evolution of Clonal Organisms. Yale University Press, New Haven and London: 297–355.

Jackson, J. B. C. & A. G. Coates, 1986. Life cycles and evolution of clonal (modular) animals. In J. L. Harper, B. R. Rosen & J. White (eds), The Growth and Form of Modular Organisms. University Press, Cambrige: 7–22.

Karlson, R. H. & L. E. Hurd, 1993. Disturbance, coral reef communities, and changing ecological paradigms. Coral Reefs 12: 117–125.

Karlson, R. H., T. P. Hughes & S. Karlson, 1996. Density-dependent dynamics of soft coral aggregations: the significance of clonal growth and form. Ecology 77: 1592–1599.

Keough, M. J. & H. Chernoff, 1987. Dispersal and population variation in the bryozoan *Bugula neritina*. Ecology 68: 199–210.

McKinney, F. K. & J. B. C. Jackson, 1989. Bryozoan evolution. Chicago University Press, London and Chicago: 1–233.

Patzöld, J., H. Ristaed & G. Wefer, 1987. Rate of growth and longevity of a large colony of *Pentapora foliacea* (Bryozoa) recorded in their oxygen isotope profiles. Mar. Biol. 96: 535–538.

Pearson, R. G., 1981. Recovery and recolonization of coral reefs. Mar. Ecol. Prog. Ser. 4: 105–122.

Pickett, S. T. A. & P. S. White, 1985. Patch dynamics: a synthesis. In S. T. A. Pickett & P. S. White (eds), The Ecology of Natural Disturbance and Patch Dynamics. Academic Press, Orlando: 371–384.

Renaud, P. E., Jr. W. G. Ambrose, S. R. Riggs & D. A. Syster, 1996. Multi-level effects of severe storms on an offshore temperate reef system: benthic sediments, macroalgae, and implications for fisheries. Mar. Ecol. P.S.Z.N. I, 17: 383–398.

Roberts, D., D. Rittschof, E. Holm & A. R. Schmidt, 1991. Factors influencing initial larval settlement: temporal, spatial and surface molecular components. J. exp. mar. Biol. Ecol. 150: 203–211.

Rogers, C. S., 1993. Hurricanes and coral reefs: the intermediate hypothesis revisited. Coral Reefs, 12: 127–137.

Sala, E., J. Garrabou & M. Zabala, 1996. Effects of diver frequentation on Mediterranean sublittoral populations of the bryozoan *Pentapora fascialis*. Mar. Biol. 126: 451–459.

Sgorbini, S., S. Cocito & C. N. Bianchi, 1996. Underwater photography as a tool to monitor the population dynamics of a clonal organism. In G. Albertelli, A. De Maio & M. Piccazzo (eds), Atti XI Congresso Associazione Italiana Oceanologia Limnologia A.I.O.L., Genova: 819–826.

Sousa, W. P., 1984. The role of disturbance in natural communities. Ann. Rev. Ecol. Syst. 15: 353–359.

Winston, J. E. & J. B. C. Jackson, 1984. Ecology of cryptic coral reef communities. IV. Community development and life histories of encrusting cheilostome Bryozoa. J. exp. mar. Biol. Ecol. 76: 1–21.

Witman, J. D., 1987. Subtidal coexistence, storms, grazing, mutualism, and the zonation of kelps and mussels. Ecol. Monogr. 57: 167–187.

Woodley, J. D., E. A. Chornesky, P. A. Clifford, J. B. C. Jackson, L. S. Kaufman, N. Knowlton, J. C. Lang, M. P. Pearson, J. W. Porter, M. C. Rooney, K. W. Rylaarsdam, V. J. Tunnicliffe, C. M. Wahle, J. L. Wulff, A. S. G. Curtis, M. D. Dallmeyer, B. P. Jupp, M. A. R. Koehl, J. Neigel & E. M. Sides, 1981. Hurricane Allen's impact on Jamaican coral Reefs. Science 214: 749–755.

Zabala, M. & E. Ballestreros, 1989. Surface-dependent strategies and energy flux in benthic marine communities or, why corals do not exist in the Mediterranean. Scient. Mar. 53: 3–17.

Hydrobiologia **375/376:** 67–72, 1998.
S. Baden, L. Pihl, R. Rosenberg, J.-O. Strömberg, I. Svane & P. Tiselius (eds),
Recruitment, Colonization and Physical–Chemical Forcing in Marine Biological Systems.

The alternating recruitment pattern in *Ensis minor*, an exploited bivalve in the Gulf of Trieste, Italy

Donatella Del Piero & Roberto Dacaprile
Biology Department of the University of Trieste, via E. Weiss 2, 34141 Trieste, Italy

Key words: Ensis minor, fishery, recruitment, Gulf of Trieste

Abstract

E. minor is a shallow water razor clam, actively exploited by hydraulic dredges on the sandy bottom between 2.5 and 4 m depth. The fishing season lasts 6 month, from October to March. Despite of the heavy exploitation exerted in the previous years the resource exhibited dramatic recovery in 1983, 1988 and 1993 mainly due to an exceptional recruitment, that sustained the fishing activity for 2 y after the settlement. In the recruitment years the catches are normally low, but increase dramatically the following year, when little settlement occurs, probably because of space and resources competition. The last one is also reflected in the lower growth rate of the recruits when abundant settlement occurs. Two years later, when the 2+ cohort is nearly extinct and the 1+ age class is not very abundant (as stated above), the new settlement can take place and the cycle restarts. The amount of the recruitment is not predictable, but every five years very abundant age classes were observed. Apparently this process started in '83, after heavy depletion and at least two recruitment failures. The population structure was completely modified, with 0+ class absolutely dominant and the absence of the older individuals, at least in the more exploited areas, where the young-of-the-year were specially abundant. The fishery activity have been assumed to be partially involved in the maintenance of the cyclic pattern. However, by examining the data from the early seventies, preceding the full development of mechanical harvesting, it seems that this species 'naturally' adopts alternative patterns in recruitment, as other molluscs appear to do. This phenomenon is amplified by the exploitation.

Introduction

The razor clam *Ensis minor* (Chenu, 1834) is one of the most important fishery resource of the Gulf of Trieste. The species lives buried in the sandy bottom, mainly in front of the Grado and Marano lagoons and can be found down to the depth of 4 m; along the Mediterranean coast it can live down to the depth of 25 m (Van Urk, 1964). The clam generally reproduces in early springtime. According to Lebour (1938), referring to the British coasts, the planktonic larvae of genus *Ensis* spend one month in the water before they settle, but is possible that local studies could depict another pattern. The minimum marketable size of the clam is 8 cm in length and is reached in no less than one year, so the recruits settled in spring will be caught in October of the following year.

In the past this clam was harvested by hand, especially in wintertime during low tides; later, when hydraulic dredges (1970s) came into usage, harvesting has dramatically increased and *Ensis* has been heavily exploited by the local fleets. The activity is seasonal and lasts from October to March. A remarkable feature of this fishery is its alternating yield year by year, even if this pattern is not uncommon among exploited bivalves. (Tegelberg & Magoon, 1969).

These fluctuations are mainly due to variations in the recruitment strength but fishing have been reported to amplify this phenomenon (Goriup et al., 1995). In 1981 and 1982 the recruitment failed and the catches decreased, so strong limitations on the fishery were imposed. In 1984, when the exceptionally strong 1983 year class became exploitable, everything changed; the average density over the Lignano fishing ground

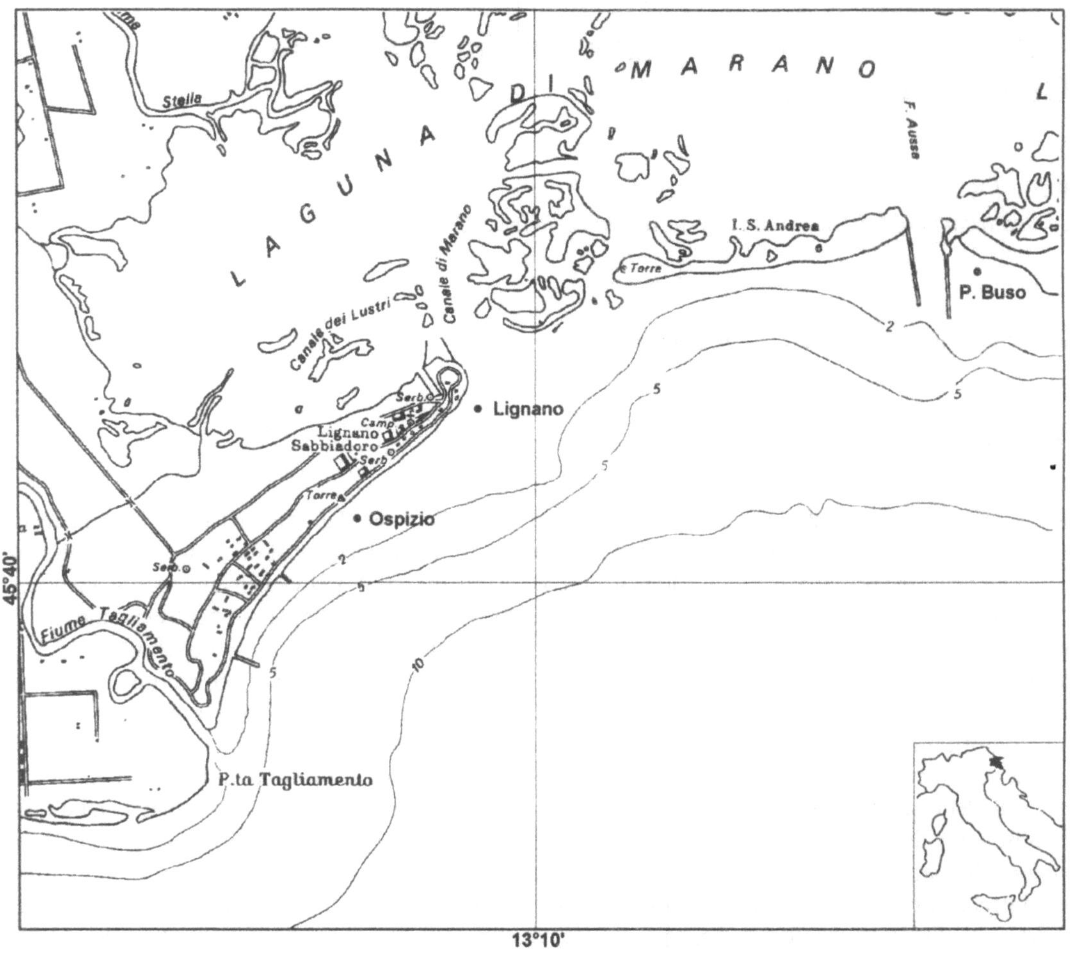

Figure 1. Sampling sites.

was estimated to be about 35 individuals m^{-2} (Del Piero, 1994) at the beginning of the fishery season. Since that time the resource seems to follow a very different pattern with quite strong recruitment every five years even if the past performance seems to be unattainable.

Materials and methods

The data presented in this paper were collected before the fishery season, or at its beginning, in 1989, 1990 and 1994 on the fishing beds usually exploited by the commercial fleet. In order to follow the growth of the newly settled cohorts, from July 1994 to March 1996 almost monthly samples were collected along a transect in front of Lignano, traditionally one of the most frequently used fishing grounds. The results concern mainly the samples collected in that area. Occasionally I will also refer to two other two sites: Ospizio and Porto Buso (Figure 1). The substrate of the sampling areas is sandy (well sorted sand) at Ospizio and Lignano, while at Porto Buso an enrichment in the fine fraction occurs.

The vessel used for sampling was a common fishing boat equipped with a commercial hydraulic dredge. Because of slight differences in the width of the fishing gear, the data on the graphs are standardised to 100 m^{-2}. The dredge is designed to catch specimens of legal size (from 8 cm upwards), but the selectivity strongly depends on the substrate and smaller clams are usually caught. In 1989 and 1990 four hauls 50 m long were done with fixed anchor;

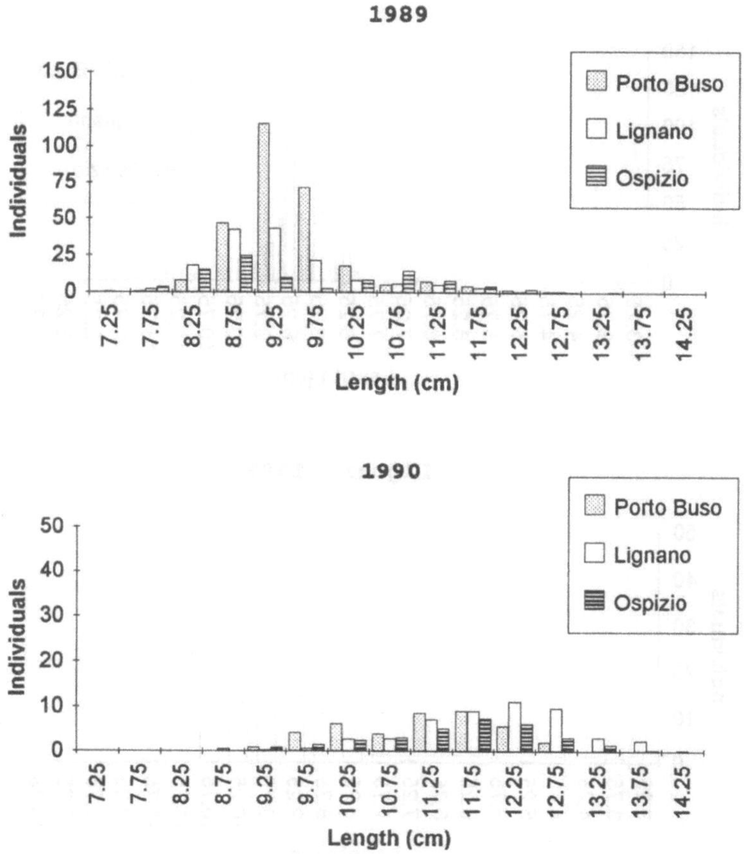

Figure 2. Frequency distribution from 1989 and 1990 samples. Data standardised to 100 m^{-2}.

from 1994 six tows (already 50 m long) were done at fixed depth between 1.5 and 4 m.

The dredge penetrates usually about 20 cm in the sediment. The molluscs collected by the gear were sorted by hand, weighted by a steelyard and packed to carry to the laboratory; occasionally subsamples were taken. In the laboratory the clams were measured with a vernier calliper to the lower mm; the frequency data were cumulated in class intervals of 5 mm.

Results

The frequency distributions obtained from samples are shown in Figures 2–3 and the Table 1 summarises the density expressed as individuals m^{-2}, as well as the total number fished, with mean length in cm and standard deviation. The age-groups were estimated from previous knowledge: it is quite easy to recognise the first two cohorts because this is a rather fast growing species, at least in the first year. Later, difficulties increase because the groups are rather clumped and it may be helpful to use automatic decomposition methods of the frequency distribution (ELEFAN or MULTIFAN) as done in previous works (Del Piero et al., 1989; Goriup et al., 1995). It is important to stress that year after year the cohorts can markedly differ in growth rate and important differences can also occur from one site to another in the same year (Del Piero et al., 1989, Goriup et al., 1995).

In 1989 at Lignano the average density was 1.51 individuals m^{-2} (Table 1) and one group was dominant, mostly belonging to the year class 1988, with the highest peak at 9.25 cm (Figure 2). At Ospizio the total density was lower (0.94 individuals m^{-2}) and the frequency distribution was clearly bimodal: the 1988 year class was recognisable at the mode of 8.75 cm and the second group, with the older age classes not easily distinguishable, was dispersed over a wider interval than in the previous case. At Porto Buso the densities were

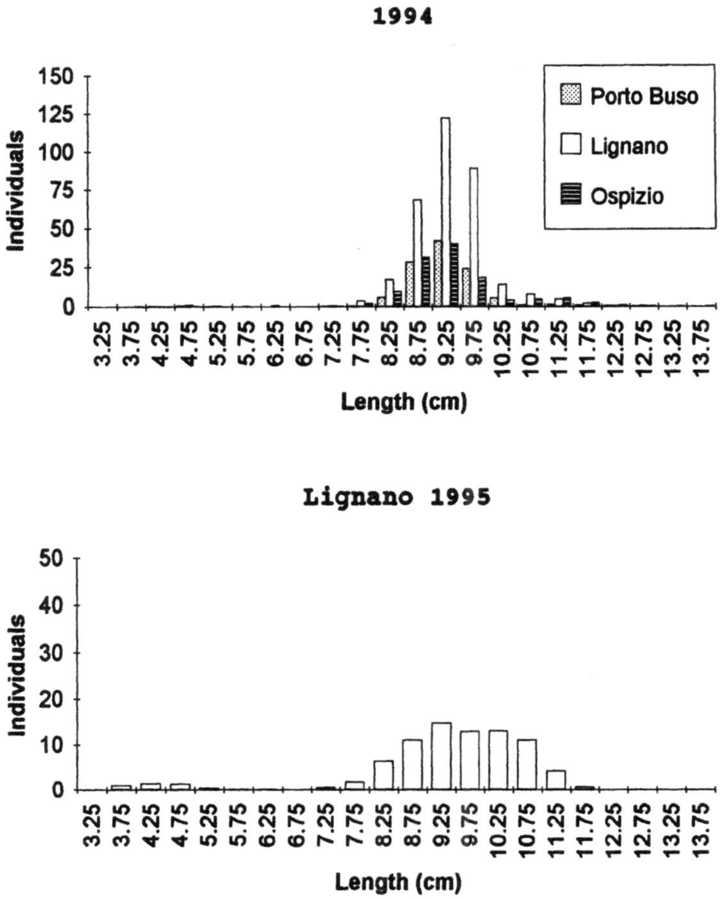

Figure 3. Frequency distribution from 1994 and 1995 samples. Standardisation as in Figure 2.

higher compared with the other sites (Table 1) and there was one peak at 9.25 cm, i.e. the 1988 y class. This situation closely resembled Lignano's one and is typical of areas subjected to intensive exploitation. In such areas only few cohorts are present normally the youngest ones (one or two years old). In less exploited areas instead, this clam can reach at least 6 y of age. In 1990 the situation was completely different: at the three stations the density was very low (Table 1 and Figure 2) and apparently no young-of-the-year settled. In this year recruitment failed in the studied areas.

The research program started again in 1994 with some differences in sampling strategies, as stated above. The sampling areas were the same as in the previous years and the results are shown in Table 1 and Figure 3. The data indicate that in 1993 finally a quite good settlement occurred.

At Ospizio and Porto Buso the densities and the length frequency distributions were similar but the structure found at Ospizio clearly differs (Figures 2–3) from 1989 when a bimodal pattern was found (Kolmogoroff-Smirnov $DN = 0.4407$, $n = 59$).

In the sampling done in October the young-of-the-year are poorly represented all over the three sites. They were sampled in larger quantity in the following months as summarised in Table 2, where mean length and standards deviation is also reported.

In October 1995 at the beginning of the fishery season (samples were taken at Lignano only) the densities are noticeably lower (0.81 individuals m^{-2}) in comparison with the 1994 results.

The average density of the recruits settled in 1995 and recorded in March 1996 is 0.19 individuals m^{-2}, quite higher than 0.10 individuals m^{-2} observed in February 1995 and referred to the 1994 settlement.

Table 1. Density in number, total number sampled, mean length (cm) and standard deviation for the three sites.

	1989	1990	1994	1995
Ospizio				
n. m^{-2}	0.94	0.32	1.19	
n. sampled	450	158	758	
mean length	9.55	11.52	9.34	
st. dev.	1.228	0.947	0.949	
Lignano				
n. m^{-2}	1.51	0.50	3.35	0.81
n. sampled	725	251	2,560	617
mean length	9.22	11.95	9.26	9.31
st. dev.	0.846	0.957	0.732	1.472
P. Buso				
n. m^{-2}	2.79	0.41	1.33	
n. sampled	1.338	204	702	
mean length	9.47	11.13	9.27	
st. dev.	0.47	0.876	0.800	

Table 2. Young-of-the-year at Lignano in 1994 and 1995: sampling month, total number caught, mean and standard deviation.

1994/5	Oct.	Nov.	Dec.	Jan.	Feb.
Total	17	34	52	61	65
Mean	5.21	5.53	5.53	5.99	5.61
St. dev.	0.714	0.636	0.622	0.439	0.528
1995/6	Oct.	Nov.	Jan.	Mar.	
Total	33	56	35	121	
Mean	4.44	5.15	5.33	5.63	
St. dev.	0.604	0.665	0.640	0.517	

Discussion

The data from 1989–1990 and 1994–1995 samples, irrespective of the quantities involved and of the modification of the structure on the Ospizio fishing grounds, are similar in pattern. In the first year there is a dominant group that can sustain the fishery, with poor settlement, followed by a noticeable decrease in the fishable stock. In this situation some appreciable settlement can occur, if the natural conditions are favourable. The low densities recorded in 1990, for example, severely limited the fishery so much that only one boat was fishing continuously during that season. In previous years at least thirty vessels were fishing only in front of Lignano for almost 4–5 months. Because no mortality occurred, except for one area occasionally subjected to dumping (Porto Buso), it's reasonable to think that the depletion, affecting all the fishing grounds and not only those mentioned in this paper, was mainly due to the fishery. So compared to the recruitment of 1983 that sustained at least two fishery seasons (Del Piero et al., 1989), the recruitment of 1988 was nearly exhausted during the '89/90 season. As regards the fishery seasons between 1991 and 1993, there are only occasional information from fishermen who reported a slight increment in the catches.

Compared with the 1989 results, the 1994 data indicate a decrease in the density at Porto Buso. The data at Lignano showed a quite good increase and that the majority of the individuals belonged to the 1993 cohort. Due to the depletion of another resource, *Chamelea gallina* (Linnaeus, 1758) and to the poor economic yield of *Callista chione* (Linnaeus, 1758), most of the fleet converted their activity into the razor clam fishery. Because of the problems involved in a reliable census of both fishing vessels and fishing activity, trust-worthy fishing effort data are not available. However, the decrease in density observed at Lignano (where monthly samples were taken) in November and in December suggests that the fishing effort was heavy with densities decreasing from 3.35 individuals m^{-2} in October to 2.42 in November, down to 1.90 in December (Biancuzzi, unpublished). The settlement of 1995 showed a lower mean length than that one found in 1994 in the same month. This gap persisted till the surveys ended (Table 2) and it does not have to be related to a slower growth rate but to a delay in the settlement time. The first recruits of 1995, in fact, were found for the first time in September with a mean length (3.68 cm) surprisingly rather close to those sampled in August 1994 (3.90 cm). Where a comparison was possible, mean length values were found to shift by one month. After checking the normality of distribution, Student's t tests applied to data were not significant for the whole season. For example the results of the last comparison, (March 1996–February 1995), are the following: mean length 5.63 ± 0.528 and 5.61 ± 0.517 respectively; $t = 0.541$, $p = 0.589$, d.o.f. = 184. The reasons of a most likely delayed settlement are due to natural causes since other colleagues, studying the plankton stages of *Engraulis* in a fixed station near Trieste (Specchi & Orlandi, unpublished), found a shift of the same range (about one month) in the appearance of eggs and larvae of anchovies.

The result of the recruitment can be affected by the quantity of adults present on the bottom at the time of settlement, e.g.: by competition for food and space. For example 61 adult individuals (1+ and more) were sampled at the depth of 2 m in May 1995 (chosen as conventional time of settlement of the young) and 66 recruits of the 1995 cohort were found in March 1996 whereas at the depth of 3 m the values were 152 and 38 respectively. Although the adult density influences the recruitment, and in this case the fishery itself could also play an important role subtracting the potentially strong competitors, on the other hand, in the years of good yield, a protracted activity could represent an important stress factor for the early settled clams. Most of the fleet can fish until the end of March when the catches are good, risking to lose part of the recruitment, while in poor-medium yield season they stop fishing in January or in the middle of February, avoiding to damage the early recruits.

The relationships between parental and recruit stocks are still under debate, but a density dependent effect can be suspected. The 1983 exceptional recruitment was widespread all over the clam beds, but undoubtedly the young density was higher in the more exploited areas where the adult density was lower and the frequency distribution extended over narrower limits than in the less exploited sites (Del Piero et al., 1989; Goriup et al., 1995).

Anyhow, as the factors influencing the amount of the recruitment are so complex, the population structure depicted by these surveys seems to be very fragile because the fishery is sustained mainly by only one cohort; therefore failures in recruitment or heavy mortality that may occur could be difficult to recover from. A more stable population structure was found at Ospizio in 1989, though associated to a quite low density, so possible fluctuations of the recruitment-parental stocks system may be dampened. High or extremely high densities (the Lignano and partly Porto Buso case) are associated with a more simple structure, lacking in older cohorts. These areas are easily and profitably exploited, but exposed to sudden collapse and subjected to the uncertainty of a recovery. The surveys show that fishery had a great part in the modification of the population structure and in the case of Lignano this reshaping seems to be irreversible.

No appreciable change of the young-of-year growth was observed in 1994 and 1995, in spite of the difference in density. On the contrary, in 1984 the mean length of the cohort settled in 1983 and sampled in October 1994 was 8.80 cm at Lignano and from 1979 to 1983 was estimated from the growth curves to be around 10 cm, depending on the year (Del Piero et al., 1989). The settlement of the 1983 follows two years of severe limitations to the fishery and the activity was stopped in February.

Conclusions

The razor clam fishery in the Gulf of Trieste in the recent years depends on the impulses given by the recruitment in 1988 and in 1993 and in this way the biennial fluctuation pattern observed in the past disappeared. The structure of the fishing beds is altered and the system appears to be adjusted on longer term fluctuations (five years in the present case). If previously it was difficult to manage this resource, now is even worse and it seems advisable to reduce the capacity of the fleet, keeping in mind that controls of catches are extremely difficult to carry out unless the fishermen are involved. The present management system based on direct control of the catch by the fishermen can help in maintaining this resource, but is necessary to supply them with scientific advice.

References

Del Piero, D., G. Orel, E. Vio, G. Brizzi & A. Zentilin, 1989. Considerazioni sulla pesca di *Ensis siliqua minor* (Chenu) nel Golfo di Trieste. Nova Thalassia 10 supp. 1: 485–492.

Del Piero, D. 1994. The clam fishery in the Gulf of Trieste. In P. G. Wells & P. J. Ricketts (eds), Coastal Zone Canada '94, 'Cooperation in the Coastal Zone', Coastal Zone Canada Association, B. I.O. Dartmouth, Nova Scotia, Canada. Conference Proceedings 4: 1645–1660.

Goriup, F., G. Brizzi, D. Del Piero, 1995. Accrescimento e struttura di popolazione di *Ensis minor* su due banchi da pesca del Golfo di Trieste. Boll. Soc. Adr. Scienze 26: 47–57.

Lebour, M. V. 1938. Notes on the breeding of some Lamellibranchs from Plymouth and their larvae. J. mar. biol. Ass. U.K. 23: 119–145.

Tegelberg, H. C., & C. D. Magoon, 1969. Growth, survival and some effects of a dense razor clam set in Washington. Proc. Natn. Shellfish. Ass. 59: 126–135.

Van Urk, R. M., 1964. The Genus *Ensis* in Europe. Basteria 28: 13–44.

Hydrobiologia **375/376:** 73–87, 1998.
S. Baden, L. Pihl, R. Rosenberg, J.-O. Strömberg, I. Svane & P. Tiselius (eds),
Recruitment, Colonization and Physical–Chemical Forcing in Marine Biological Systems.

Recolonization patterns of meiobenthic communities in brackish vegetated and unvegetated habitats after induced hypoxia/anoxia

A. Guerrini[2], M. A. Colangelo[2] & V. U. Ceccherelli[1]
[1] *Dipartimento di Biologia Evoluzionistica e Sperimentale, Università di Bologna, via Selmi 3, 40126 Bologna, Italy*
[2] *Dipartimento di Biologia, Università di Ferrara, via L. Borsari 46, 44100 Ferrara, Italy*

Key words: Meiobenthos, hypoxia/anoxia, recolonization, field experiment

Abstract

In a lagoon (Valle Smarlacca) of the 'Valli di Comacchio' complex (Northern Adriatic Sea, Italy) a phytal habitat (*Ruppia* sp.) was chosen to investigate the modifications and the following recovery of meiobenthic communities caused by induced hypoxic/anoxic disturbance. A field experiment was carried out by incubating, for three days, portions of seagrass meadow and adjacent unvegetated muddy sediment patches by means of dark benthic chambers. After these were removed, species of both nematodes and copepods showed a different degree of resistance to the induced disturbance. On the whole, on the small scale of the present experiment, meiobenthic communities appear highly resilient. As for nematodes, recolonization patterns of community recovery were differently influenced by the habitat. On bare sediment, community structure followed the natural evolution of the control habitat while, in the *Ruppia* meadow, community recovery showed more complex patterns according to the different behaviour of phytal and infaunal species, respectively. As for copepods, recolonization patterns were mainly driven by the dominant species of each habitat control. On the other hand, the heavier impact of disturbance on *Ruppia* and the slow restoration of its canopy seem to delay recolonization rates of copepod populations and differently affect the community recovery pattern.

Introduction

Shallow-water coastal and estuarine benthic communities are subject to a variety of physical and biological disturbances which vary in frequency, intensity and temporal and spatial extent. The structure of those communities (i.e. species composition and relative species abundance) and their subsequent recovery patterns, may be markedly affected by those disturbances (Turner et al., 1995).

Many early works focused on the responses of meiofaunal components of the benthic community because of their potential for faster responses to environmental changes than macrofauna (Coull & Chandler, 1992). Meiofaunal organisms might be useful biological indicators of such changes because of their short generation times. Furthermore, the meiofauna maintain a greater diversity even in environments which are subjected to considerable stresses (Gee et al., 1992).

Benthic environments of the northern Adriatic coastal lagoons have always been subjected to both natural and man-induced disturbances. In recent years, the increased eutrophication represents the prevailing anthropogenic alteration of those environments. Especially during summer, this frequently results in the occurrence of dystrophic crises which cause an organic enrichment of bottom sediments. The impact of this kind of perturbation on meiobenthic communities may depend on the habitat-specific autotrophic organisms whose decomposition rates can differently affect the kind and amount of decaying organic matter and concentrations of chemical compounds in the sediment. Moreover, microbial heterotrophic activity can be stimulated to such an extent to cause hypoxia,

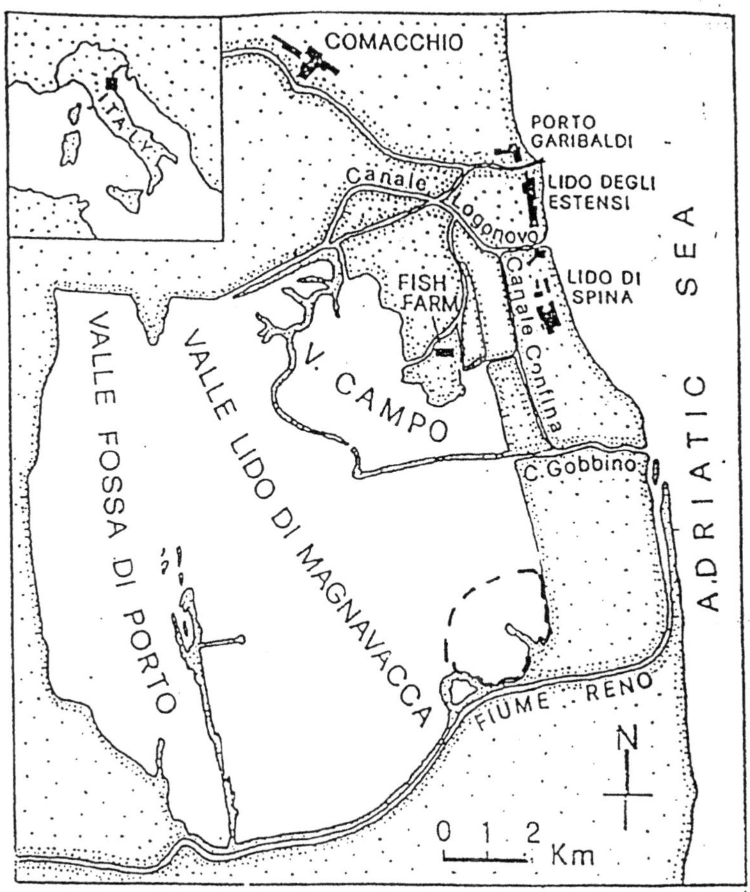

Figure 1. Map of the Valle Smarlacca (dashed perimeter) in the Valli di Comacchio complex.

or even anoxia, in the sediment and interstitial waters (Viaroli et al., 1997). These events represent the main factors of acute disturbance for biota, especially for benthic organisms.

When the primary cause of disturbance ceases, community recolonization trend depends, in part, on the resulting environmental alterations. In addition, habitat complexity (Hicks, 1986) and movement capabilities of different taxa (Palmer, 1988; Bell & Hicks, 1991) affect community recovery either towards the original structure or a different organism assemblage.

A suitable approach to study disturbance impact and recolonization processes is represented by controlled field experiments (Coull & Palmer, 1984). The aim of this study was to investigate the impact of an experimentally induced hypoxia/anoxia on macrophyte seagrass bed and on adjacent unvegetated sediment patches and the subsequent recolonization rates by nematodes and copepods of meiobenthic communities in a northern Adriatic lagoon habitat.

Materials and methods

Study area

The experiment was carried out in Valle Smarlacca, a small fish pond (about 1.9 km^2) in the Valli di Comacchio (North-Adriatic Sea, Italy), a wide complex (about 100 km^2) of shallow water impoundments connected to the Adriatic Sea *via* a network of canals (Figure 1). Valle Smarlacca is completely embanked but can receive fresh water inputs from the adjacent Reno River through man-regulated sluices. Variations of salinity (17–20‰) and depth (50–60 cm) were negligible during the experiment period.This lagoon is still characterised by a dense meadow of the aquatic phanerogam *Ruppia* sp., contrasting with the other Valli di Comacchio basins which have been experiencing an ecological catastrophe, caused by dense blooms of cyanobacteria (Sorokin et al., 1996), and are completely devoid of macrophytes now. Valle Smarlacca is regularly subjected to dystrophic crises during summer (P. Viaroli, pers.comm.), a phenomenon widely described also in other european shallow water lagoons (Castel et al., 1996; Viaroli et al., 1997). During warm summer months, the emerging *Ruppia* fronds are covered by dense tufts of epiphytic algae, the decomposition of which leads to a significant oxygen uptake as well as to sulphide production. Such serious events last several days and, under the anoxic conditions, white water can be observed together with the production of H_2S. Concentrations of sulphur and carbonates are high which impart a whitish colour to the water.

In order to compare the impact of induced hypoxia/anoxia events on meiobenthic community structures in different habitats, two experiment sites were established in this lagoon: one site in the seagrass meadow and the other one in an adjacent unvegetated muddy sediment area.

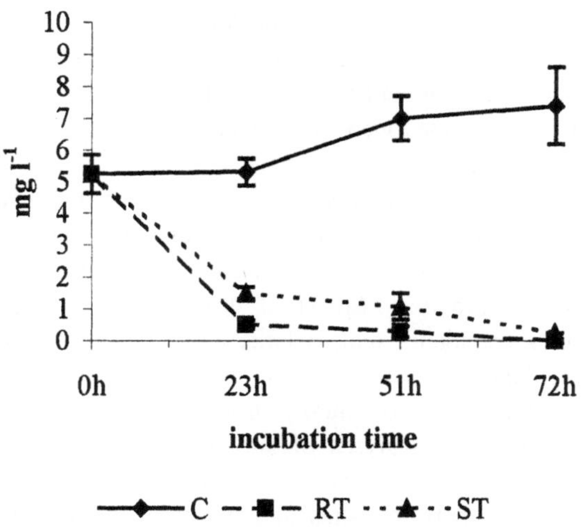

Figure 2. Trends of average oxygen concentration (± S.D.) in the water column of the controls (C), *Ruppia* dark chambers (RT) and bare sediment dark chambers (ST) respectively during the three days of incubation.

Experimental procedure

To induce anoxic disturbance, two replicate dark benthic chambers (30 cm diameter and 30 cm height) were gently forced into the bottom of both sampling sites, allowing approximately 15 cm remaining above the sediment surface. The incubation lasted 3 d. To monitor hypoxia/anoxia evolution during the incubation period, dissolved oxygen concentrations were measured by micro-Winkler procedure in water samples withdrawn by syringe from within and outside (controls) the benthic chambers every day.

At the end of incubation all the benthic chambers were completely removed to allow organism recolonization both *via* water column and by infaunal crowling from every side. The boundaries of each resulted disturbed spot were marked by 4 coloured sticks to locate spots again during the recolonization period. Similarly, 2 control areas of the same size and shape were delimited close to each disturbed spot.

Sampling was scheduled at the end of incubation period (i.e. s: start point) and on days 1, 3, 7, 14, 21, 28, 35 after incubation. At each sampling occasion, in each of the 2 experimental sites (bare and vegetated), 2 meiofaunal cores (3.14 cm^2 each) were collected from the 2 replicate incubated patches (plots A and B) and from the 2 adjacent replicate control areas (plots A and B). Every time core holes were marked to avoid re-sampling. The uppermost 5 cm sediment layer was retained, stained with Rose Bengal and fixed in 8% buffered formalin. In the laboratory, samples were sieved through both 1.00 and 0.052 mm mesh size. The fraction on 0.052 mm sieve was centrifuged three times with LUDOX-TM water solution (60%) to separate organisms from sediments. Meiobenthic organisms were counted after sorting to major taxa. At least 100 nematodes from each core were transferred to a solid watchglass containing volume mixtures of glycerol, ethanol and demineralised water (Seinhorst, 1959), collected, mounted in glycerine slides and examined microscopically. Copepods and nematodes were identified to the species, enumerated, sexed and staged.

Data handling

At each date, sampling followed a general three-way nested ANOVA design with sites (bare sediment, S, and *Ruppia* meadow, R) as fixed, treatment (control, C, and anoxia, T) as fixed and plots (replicate A and B) as random factor nested in treatment factor. Plots were considered as factor because we were interested in estimating separately the among- and within-plot (pseudoreplicate cores) components of spatial heterogeneity (Dutilleul, 1993). As dark chamber incubation differently affected the two investigated habitats (see results) two-way ANOVA design was performed also for each site separately, considering treatment as fixed factor, sampling times as random one and plots A and B as replicates, pooling data of the two pseudoreplicate cores in each plot. Time was considered an independent factor since cores were never taken from a previously cored location (Underwood & Anderson, 1994; Thrush et al.,

1996). When the results of ANOVAs were significant, differences in mean abundance among times were examined using Student-Newman-Keuls (SNK) post-hoc test. All analyses were performed using $\log_{10}(x+1)$ transformed data.

Differences in multivariate aspects of community structure among sites, control/treatment and sampling dates were analysed by non-metric multidimensional scaling ordination (MDS) of 4th root-transformed species abundance data (Clarke & Warwick, 1994), using the Bray-Curtis similarity measure. Significance of differences in community structure were tested by the following 'analyses of similarities' (ANOSIM) (Clarke, 1993): (i) a two-way nested ANOSIM for differences between sites and between control and treatment only at the start point; (ii) a two-way crossed ANOSIM for differences among combined site+treatments and among sampling times during recolonization; (iii) a two-way crossed ANOSIM for differences between control and treatment and among times for each habitat separately. After a significant global difference was detected, the same technique was employed to test pairwise differences between times in each treatment and between control/treatments at each corresponding time. The average Bray-Curtis dissimilarities between groups of samples, which were significantly different according to previous tests, were partitioned into contributions from each species using the 'similarity percentages' program (SIMPER) described in Clarke & Warwick (1994), thus pointing out the species with the greatest contribution to the differences between sample clusters.

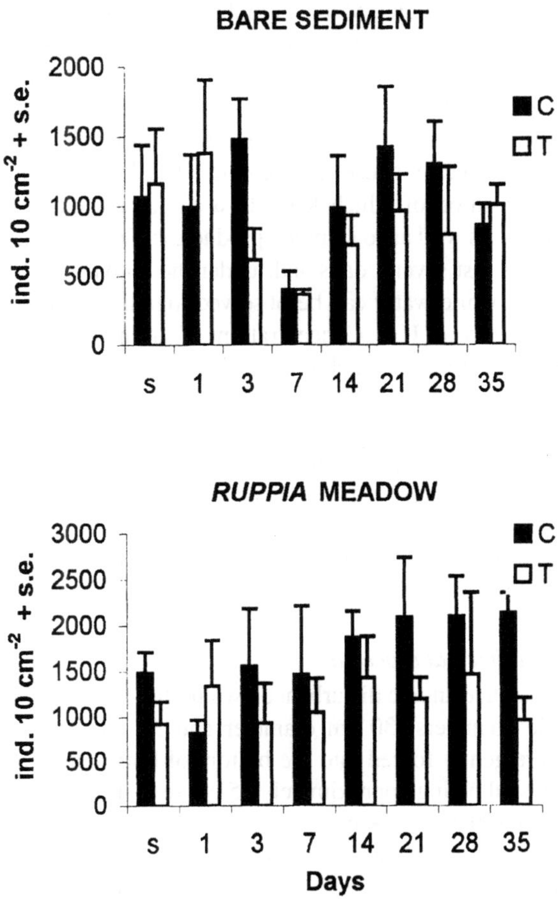

Figure 3. Average densities of total nematodes in bare sediment and in *Ruppia* meadow respectively for control (C) and treatment (T) at each sampling day (s: start, 1, 3, 7, 14, 21, 28, 35).

Results

Variations of oxygen concentrations during incubation are shown in Figure 2. Oxygen rapidly decreased within the benthic chambers of both sites within one day . At the end of the incubation oxygen was 0 mg l^{-1} inside both dark chambers with *Ruppia* and 0.2 mg l^{-1} on average inside those on bare sediment. The latter value could be considered as zero, owing to the reliability limit (0.5 mg l^{-1}) of the micro-Winkler method (Peck & Uglow, 1990). However, after the first incubation day, oxygen decline was slower in the bare sediment chambers in comparison with the phytal ones. The difference probably depended on higher respiration and decomposition rates in the phytal environment. Consequently organisms in the *Ruppia* chambers experienced anoxia longer than those in the sedimentary one.

After the removal of the chambers in the vegetated patches, the *Ruppia* leaves had not disappeared but were not thriving and looked flabby. This condition lasted about three days. After a week, however, restoration of the fronds became apparent and after two weeks the canopy looked completely restored.

A total of 36 nematodes species were recorded in the phytal habitat, 28 of them in the sedimentary one. As for copepods, out of a total of 23 species, 18 were found in the bare sediment and 21 in the *Ruppia* meadow. In both control areas nematodes were the dominant taxon, accounting for 65.7% on bare sediment and 58.5% on *Ruppia*. Copepods were the second taxon by percentage, especially in the more complex *Ruppia* habitat where they reached 31.4% of the total meiofauna compared to 15.0% in unvegetated

Table 1. *F*-values from the three-way nested ANOVA for total and major species of nematodes at the end of incubation.

Source of variation	d.f.	*F*-values			
		Total Nematodes	*Tershellingia longicaudata*	*Sabatieria punctata*	*Viscosia glabra*
Sites (S)	1	0.52	2.40	27.70*	66.08*
Treatments (TR)	1	47.79*	18.26*	0.11	0.91
Plots w TR (PLw)	2	0.01	0.03	0.87	0.43
SxTR	1	1.58	1.10	13.11	0.18
SxPLw	2	0.43	0.79	0.31	0.52
Residual	8				

* $P < 0.5$.

Table 2. Breakdown of average dissimilarity (37.94) between communities of the first (1, 3, 7 d from the start; FP) and second period of recolonization (days 14, 21, 28, 35 d; SP) into contributions from each nematode species in the bare sediment anoxic treatment. Species are ordered in decreasing contribution (up to about 70%). *X*: average abundances (± s.e.) per core (3.14 cm^2); δi: average contribution of *i*-th species to the overall dissimilarity; SD: standard deviation of δi; $\Sigma\delta i\%$: cumulative percentages of δi contributions.

Species	X_{FP}	X_{SP}	δi/SD	$\Sigma\delta i\%$
Oncholaimus skawensis	0.0 (*0.0*)	4.2 (*2.7*)	1.6	8.5
Terschellingia longicaudata	201.2 (*198.2*)	203.9 (*131.2*)	1.2	15.3
Theristus normandicus	3.0 (*2.0*)	16.5 (*7.3*)	1.5	22.0
Paralinhomoeus tenuicaudatus	2.1 (*4.1*)	2.1 (*1.5*)	1.3	28.1
Neochromadora poecilosomoides	0.1 (*0.3*)	1.3 (*0.9*)	1.3	33.9
Metalinhomoeus effilatus	3.0 (*6.4*)	1.5 (*2.0*)	1.1	39.6
Monhystera stagnalis	2.0 (*2.3*)	2.5 (*2.1*)	1.1	45.3
Dichromadora cephalata	1.1 (*2.1*)	2.0 (*1.7*)	1.1	50.9
Chromadorina sp.	0.6 (*1.2*)	1.4 (*1.1*)	1.2	56.5
Chromadorita sp.	1.4 (2.2)	1.6 (*1.6*)	1.1	61.9
Chromadora nudicapitata	0.4 (*1.1*)	1.1 (*0.8*)	1.2	67.1
Viscosia glabra	0.9 (*0.8*)	1.9 (*1.6*)	1.2	72.2

habitat (original density data per species of both taxa are available from authors).

Nematodes

The average densities of total nematodes recorded during the experiment are shown in Figure 3. At the end of the incubation, after the removal of the dark benthic chambers (start point), only the density of total nematodes and of the dominant species in the control, *Terschellingia longicaudata* De Man, 1907 was significantly lower in the disturbed patches (Table 1). On the other hand, the response to induced anoxia at the community level appeared to be more complex. Two-way nested ANOSIM reveals a significant dissimilarity only between site communities (global $R = 0.891$; $p < 0.01$). Also SIMPER analyses show more dissimilar nematode assemblages between sites than between treatments. Average dissimilarities between site samples were 51% in the control areas and 47% in the disturbed patches. The greatest contributions to dissimilarities were due to species of the Chromadoridae and Oncholaimidae families (Guerrini et al., in press). Conversely, average dissimilarities between control and treatment were 33% in the sedimentary habitat and 27% in the phytal one.

MDS ordination of all sample-points relative to the recolonization period (i.e. sites, treatments and times) for nematode assemblages emphasized that main differences of community structures were between habitats. This was supported by the two-way

Table 3. Breakdown of average dissimilarity (39.13) between communities of the first (1, 3, 7 d from the start, FP) and the second period of recolonization (days 14, 21, 28, 35 d; SP) into contributions from each nematode species in the *Ruppia* anoxic treatment. Species are ordered in decreasing contribution (up to about 70%). For symbols see Table 2.

Species	X_{FP}	X_{SP}	δi/SD	$\Sigma\delta i\%$
Chromadorita sp.	0.4 (*0.8*)	10.0 (*4.5*)	2.2	7.2
Neochromadora poecilosomoides	3.7 (*3.9*)	34.1 (*16.9*)	1.7	13.8
Oncholaimus skawensis	18.4 (*14.7*)	109.8 (*58.9*)	2.1	19.8
Viscosia glabra	8.5 (*7.6*)	46.6 (*30.1*)	1.3	25.3
Cyatholaimus gracilis	1.3 (*2.1*)	8.2 (*8.1*)	1.5	30.7
Terschellingia longicaudata	182.4 (*162.8*)	84.8 (*91.9*)	1.3	36.1
Chromadora nudicapitata	2.5 (*5.1*)	9.9 (*7.3*)	1.4	41.4
Chromadorina sp.	4.9 (5.6)	15.0 (*14.5*)	1.3	46.5
Dichromadora cephalata	0.5 (*1.1*)	7.1 (*9.4*)	1.1	51.0
Paracanthonchus heterodonthus	7.6 (*9.2*)	7.7 (*7.9*)	1.2	55.3
Oncholaimus brachycercus	0.9 (*1.9*)	2.7 (*1.7*)	1.4	59.6
Terschellingia sp.	21.8 (*18.4*)	15.1 (*13.9*)	1.1	63.2
Diplolaimella sp. a	4.5 (*4.3*)	2.8 (*2.4*)	1.2	66.9
Sabatieria punctata	81.8 (*82.5*)	56.6 (*52.4*)	1.2	70.5

crossed ANOSIM test (global $R = 0.648$; $p < 0.01$) which showed significative differences also between times (global $R = 0.221$; $p < 0.01$). Thus MDS ordinations were performed for each site separately, taking also into account that in the bare sediment benthic chambers oxygen depletion was reached later than in those of the phytal habitat (Figure 2).

No clear clusters of sample-points of control and recolonization patches respectively are recognisable for bare muddy areas (Figure 4). From day 14 onward, the treatment sample-points tend to aggregate with those of the control community. Although general two-way crossed ANOSIM test showed significant differences among times (global $R = 0.220$; $p < 0.01$), pairwise comparisons, summarised by trellis diagram (Figure 5), reveal weak dissimilarities among community structures during the experiment (e.g. many n.s. cells on Figure 5). SIMPER analysis (Table 2) shows these results were due to the constantly high abundances of *Terschellingia longicaudata*. This species did not seem to be affected at all by anoxia in this environment in comparison with the *Ruppia* meadow. Only the abundance of *Theristus normandicus* (De Man, 1890) increased during time ($F_{6,14}$: 4.64, $p < 0.01$).

At the *Ruppia* site, the MDS plot (Figure 6) shows a more distinct pattern. Two clusters can be easily distinguished, the first groups the control and treatment samples of the first week of the experiment and the second those of the subsequent weeks. Two-way crossed ANOSIM test pointed out significant differences only for time factor (global $R = 0.222$; $p < 0.01$) and pairwise comparisons (Figure 5) reveal the gradual change of the nematode community in the treatment patches since the first week. The SIMPER analysis between the sample cluster of the first week and that of the following ones (Table 3), shows increased abundances during the latter period, of Chromadoridae such as *Neochromadora poecilosomoides* (Filipjev, 1918), *Chromadorita* sp. Filipjev, 1922, and *Chromadorina* sp. Filipjev, 1918, but especially of Oncholaimidae such as *Viscosia glabra* (Bastian, 1865) ($F_{6,14}$: 3.70, $p < 0.05$) and *Oncholaimus skawensis* Ditlevsen, 1921 ($F_{6,14}$: 10.06, $p < 0.001$). These two species also exhibited significant density differences between control and treatment ($F_{1,6}$: 28.31, $p < 0.001$ and $F_{1,6}$: 16.78, $p < 0.01$ respectively). By contrast, abundances of *Terschellingia longicaudata* decreased (Table 3), although ANOVA did not yield significant effects for either factor. In the disturbed patches, densities of some rarer species, such as the strictly oxybiotic species *Anoplostoma viviparum* (Bastian, 1865) ($F_{1,6}$: 51.30, $p < 0.001$), and *Neochromadora poecilosomoides* ($F_{1,6}$: 6.03, $p < 0.01$) remained constantly lower than those in the control habitat.

Copepods

The average densities of total copepods recorded during the experiment are shown in Figure 7. The in-

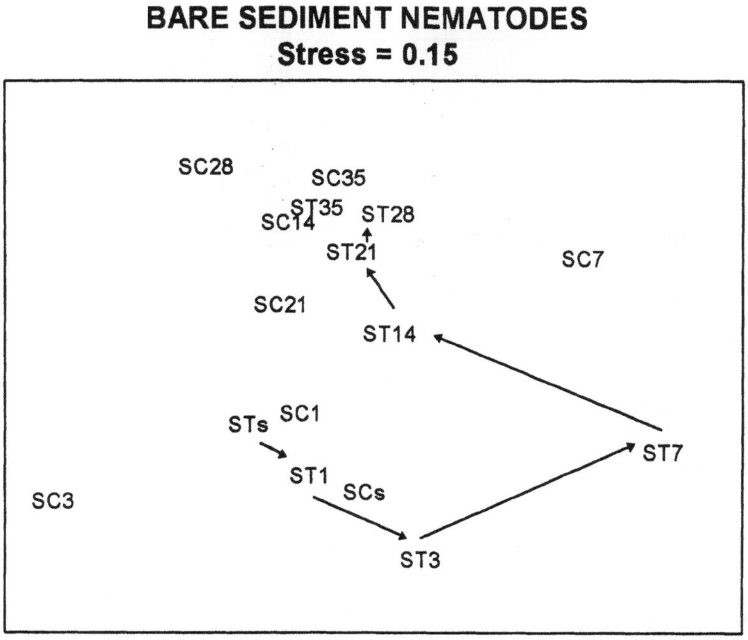

Figure 4. Two dimensional non-metric MDS ordination plot of nematode assemblages in bare sediment (S) for control (C) and treatment (T) at each sampling day (s: start, 1, 3, 7, 14, 21, 28, 35).

Table 4. *F*-values from the three-way nested ANOVA for total and major species of copepods at the end of incubation.

Source of variation	d.f.	*F*-values			
		Totals Copepods	*Canuella perplexa*	*Ectinosoma dentatum*	*Heterolaophonte strömi*
Sites (S)	1	2.99	1.18	6.94	14.07*
Treatments (TR)	1	29.34*	45.08*	122.49**	0.03
Plots w TR (PLw)	2	1.04	0.05	0.21	1.40
SxTR	1	0.07	1.63	0.14	0.01
SxPLw	2	4.30	0.35	1.80	1.47
Residual	8				

* $P < 0.5$; ** $P < 0.01$.

duced anoxia had a heavier impact on copepod than on nematode assemblages at the end of incubation, especially in the *Ruppia* habitat. ANOVA showed significant density decreases of total copepods in the disturbed patches and expecially of the dominant species *Canuella perplexa* T. & A. Scott, 1893 and *Ectinosoma dentatum* Steuer, 1940 (Table 4). Two-way nested ANOSIM revealed a significant dissimilarity between both sites (global $R = 0.427$; $p < 0.002$) and treatments (global $R = 0.385$; $p < 0.01$). SIMPER analysis (not reported here) showed that these results were due to a general abundance decrease of all species in the disturbed patches (Table 5).

As for copepod assemblages, MDS ordination of all sample-points relative to the recolonization period showed distinct community clusters. Two-way crossed ANOSIM test revealed significant differences mainly between habitats (global $R = 0.474$; $p < 0.001$) but also among sampling times (global $R = 0.335$; $p < 0.001$). To better identify the recolonization trend in each habitat, the MDS plots for each site are reported separately.

On bare sediment (Figure 8), the MDS ordination shows a quite regular evolution of the treatment date-points compared to the more crowded cluster of the control points. Two-way crossed ANOSIM

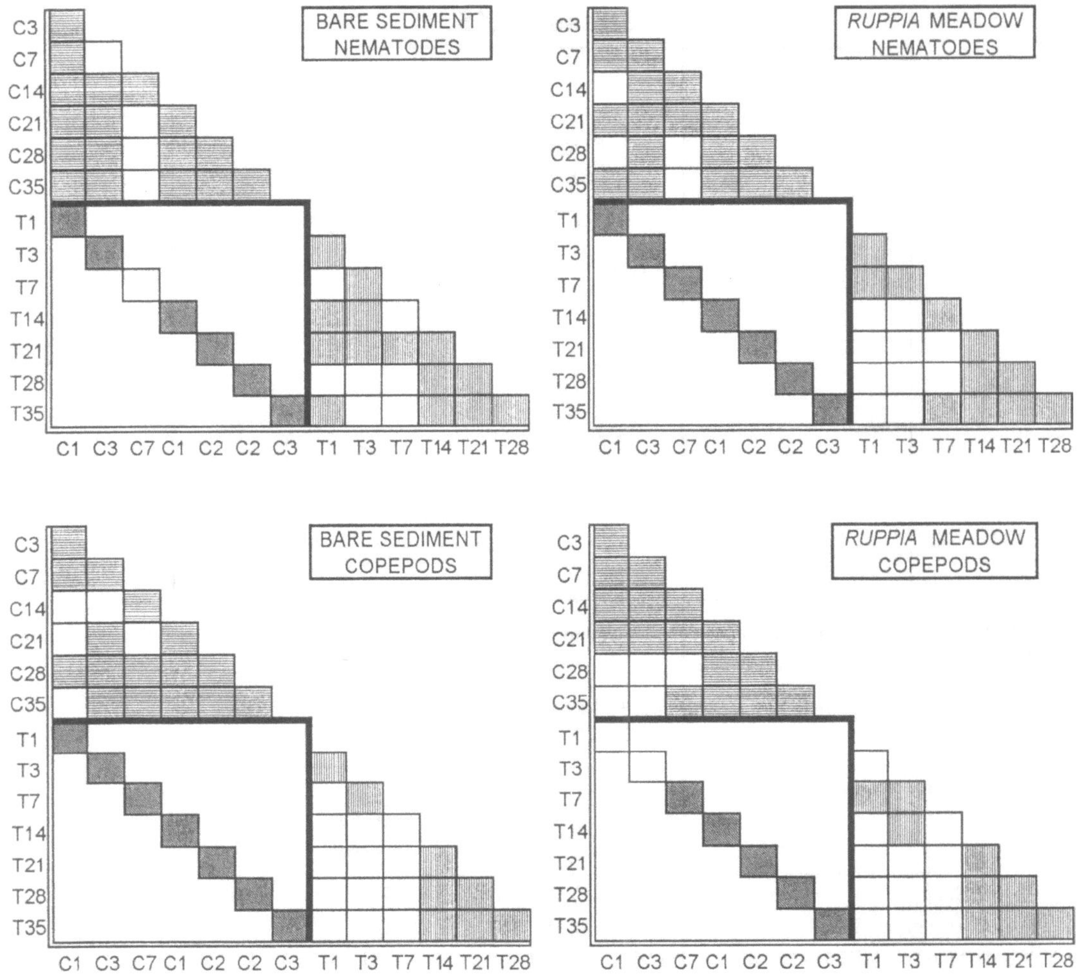

Figure 5. 'Trellis diagrams' summarizing significances of ANOSIM pairwise comparisons between sampling days for control (C) and for treatment (T), and between control and treatment at each corresponding day (1, 3, 7, 14, 21, 28, 35). Shaded areas: n.s.; blank areas: $p < 0.05$.

test indicated a significant effect of time factor only (global $R = 0.337$; $p < 0.001$). Pairwise comparisons (Figure 5) reveal that most of the differences between times are due to the evolution of community structure in the disturbed patches. Day 7 especially appears to be a discriminant date during the recovery process. This gradual structure evolution of the recolonising copepod community is even more evident in the *Ruppia* meadow. In the MDS plot (Figure 9), the date-points of the treatment describe an almost linear trajectory from the start point to the last sampling date and remain separated from the control date-point cluster. ANOSIM test revealed a significant effect of both treatment (global $R = 0.225$; $p < 0.01$) and time (global $R = 0.333$; $p < 0.001$) factors for the phytal communities. Pairwise comparisons (Figure 5) show that a clear separation between the first recolonization week (days 1–7) and the following longer period (days 14–35) also occurs in this habitat. Moreover, significant differences between community structures of the control and treatment result on the first two recolonization dates (Figure 5). Compared with bare sediment habitat, the early recolonization process in the phytal habitat appeared to be delayed.

SIMPER analysis, comparing the first week and the following four weeks for the disturbed patches in bare sediment, shows an overall recovery of all species populations (Table 6). But differences were mainly due to the high increase of abundance of the dominant species *Ectinosoma dentatum*. As for this species,

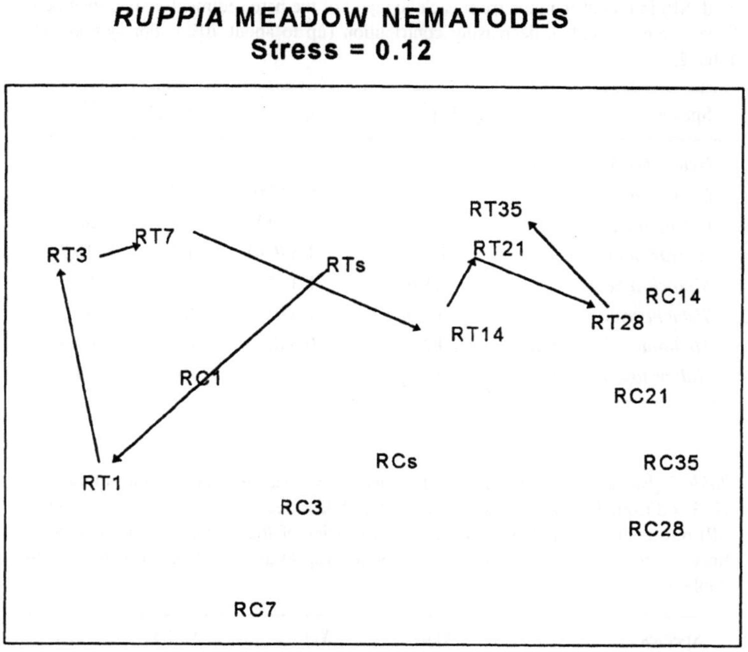

Figure 6. Two dimensional non-metric MDS ordination plot of nematode assemblages in *Ruppia* meadow (R) for control (C) and treatment (T) at each sampling day (s: start, 1, 3, 7, 14, 21, 28, 35).

Table 5. Average densities (± s.e.) per 10 cm^2 of copepod species in bare sediment (S) and *Ruppia* meadow for control (C) and treatment (T) at the end of incubation (start).

Species	SC	ST	RC	RT
Canuella perplexa	11.9 (*3.5*)	2.4 (*1.0*)	19.1 (*12.0*)	7.2 (2.5)
Ectinosoma dentatum	31.8 (*11.0*)	1.6(*0.5*)	105.7 (*31.0*)	15.9 (*4.0*)
Darcythompsonia fairliensis			0.8 (*1.0*)	0.8 (*1.0*)
Microarthriodion fallax		0.8 (*1.0*)		
Tisbe histriana	19.9 (*14.5*)		10.3 (*5.0*)	0.8 (*1.0*)
Amphiascoides brevifurca	0.8 (*1.0*)			0.8 (*1.0*)
Diosaccidae sp.				1.6 (*1.5*)
Mesochra heldti	0.8 (*1.0*)		15.9 (*11.0*)	
Nannomesochra arupinensis				0.8 (*1.0*)
Stenocaris sp.				0.8 (*1.0*)
Enhydrosoma gariene	3.2 (*1.5*)	2.4 (*1.0*)		
Enhydrosoma caeni	8.7 (2.5)	3.2 (*1.5*)	1.6 (*1.0*)	
Cletodes sp.	8.7 (7.5)	4.8 (*5.0*)	3.2 (*3.0*)	
Heterolaophonte strömi paraminuta	3.2 (2.0)	2.4 (*1.5*)	58.8 (*48.5*)	20.7 (*4.0*)
Halicyclops magniceps	1.6 (*1.0*)	0.8 (*1.0*)	19.9 (*9.0*)	2.4 (2.5)
Cyclopina gracilis	7.2 (*4.0*)		80.3 (*58.5*)	2.4 (*1.5*)
Total	97.8 (*20.0*)	18.3 (*5.5*)	315.6 (*142.0*)	54.1 (*5.5*)

Table 6. Breakdown of average dissimilarity (48.65) between communities of the first (1, 3, 7 d from the start, FP) and the second period of recolonization (14, 21, 28, 35 d; SP) into contributions from each copepod of the bare sediment anoxic treatment. Species are ordered in decreasing contribution (up to about 70%). For symbols see Table 2.

Species	X_{FP}	X_{SP}	δi/SD	$\Sigma\delta i\%$
Heterolaophonte strömi	0.9 (*0.4*)	9.2 (*1.6*)	1.7	14.9
Ectinosoma dentatum	15.7 (*4.2*)	76.1 (*7.6*)	2.0	29.5
Cyclopina gracilis	0.1 (*0.1*)	1.7 (*0.8*)	1.1	38.3
Enhydrosoma caeni	1.1 (*0.4*)	1.5 (*0.4*)	1.1	46.7
Mesochra heldti	0.1 (*0.1*)	1.3 (*0.5*)	0.9	54.0
Tisbe histriana	0.7 (*0.4*)	1.4 (*0.7*)	0.8	61.3
Michroarthridiom fallax	0.2 (*0.1*)	0.6 (*0.2*)	0.9	67.9
Halicyclops magniceps	0.2 (*0.2*)	1.1 (*0.6*)	0.8	74.5

Table 7. Breakdown of average dissimilarity (43.84) between communities of the first (1, 3, 7 d from the start; FP) and the second period of recolonization (14, 21, 28, 35 d; SP) into contributions from each copepod species of the *Ruppia* anoxic treatment. Species are ordered in decreasing contribution (up to about 70%). For symbols see Table 2.

Species	X_{FP}	X_{SP}	δi/SD	$\Sigma\delta i\%$
Cyclopina gracilis	0.9 (*0.4*)	33.3 (*4.4*)	2.2	22.1
Heterolaophonte strömi	5.9 (*1.5*)	76.5 (*7.0*)	3.3	39.0
Mesochra heldti	0.7 (*0.6*)	1.8 (*0.5*)	1.5	48.9
Canuella perplexa	0.9 (*0.4*)	2.1 (*0.6*)	1.2	57.4
Tisbe histriana	1.0 (*0.3*)	2.8 (*0.4*)	1.2	65.5
Halicyclops magniceps	1.9 (*0.8*)	2.7 (*0.8*)	1.2	73.3

two-way ANOVA only supported a significant time effect ($F_{6,14}$: 62.49, $p < 0.001$).

In the *Ruppia* meadow disturbed patches, *Ectinosoma dentatum* was not a discriminant species between the first and second recolonization phases because it was present with high abundances from day 7. Instead, the recovery pattern of the community appeared to be more determined by phytal species such as *Cyclopina gracilis* Claus, 1863 and *Heterolaophonte strömi* (Baird) *paraminuta* Noodt, 1955 (Table 7). Actually, two-way ANOVA revealed a significant time effect ($F_{6,14}$: 8.09, $p < 0.001$; and $F_{6,14}$: 13.81, $p < 0.001$ respectively) for both species. SIMPER analysis of discriminant species between control and treatment during the first two sampling dates supports the ANOSIM results about the delayed density recovery of all species in the disturbed patches (Table 8).

Discussion

Average densities of total meiofauna in the control areas were higher in the *Ruppia* meadow than in the bare sediment. This difference between seagrass and unvegetated estuarine systems is due to the enhanced habitat heterogeneity created by blades, rhizomes and roots of the same seagrasses (Trotter & Webster, 1984; Palmer, 1988). The current explanations for the greater abundances and diversity of meiofauna associated with vegetated habitats are that seagrass provides protection from predation and/or a greater abundance of food (Kern & Taghon, 1986).

Nematodes

The response of the nematode assemblages to the experimental conditions revealed that differences due to the habitats were more pronounced than those caused by induced disturbance. The nematode species pop-

Table 8. Breakdown of average dissimilarity (40.93) between the control (C) and the treatment (T) communities during the first (1, 3 d from the start) sampling dates into contributions from each copepod species of the *Ruppia* habitat. Species are ordered in decreasing contribution (up to about 70%). For symbols see Table 2.

Species	X_C	X_T	δi/SD	$\Sigma\delta i\%$
Canuella perplexa	5.6 (*1.3*)	0.7 (*0.5*)	1.8	18.1
Cyclopina gracilis	4.7 (*1.8*)	0.5 (*0.4*)	1.6	33.7
Tisbe histriana	4.9 (*1.8*)	1.0 (*0.4*)	1.3	45.6
Halicyclops magniceps	3.6 (*1.0*)	2.2 (*1.2*)	1.1	55.1
Ectinosoma dentatum	38.0 (*9.5*)	15.7 (*2.4*)	1.6	63.5
Heterolaophonte strömi	7.4 (*1.9*)	5.4 (*1.9*)	1.0	70.6

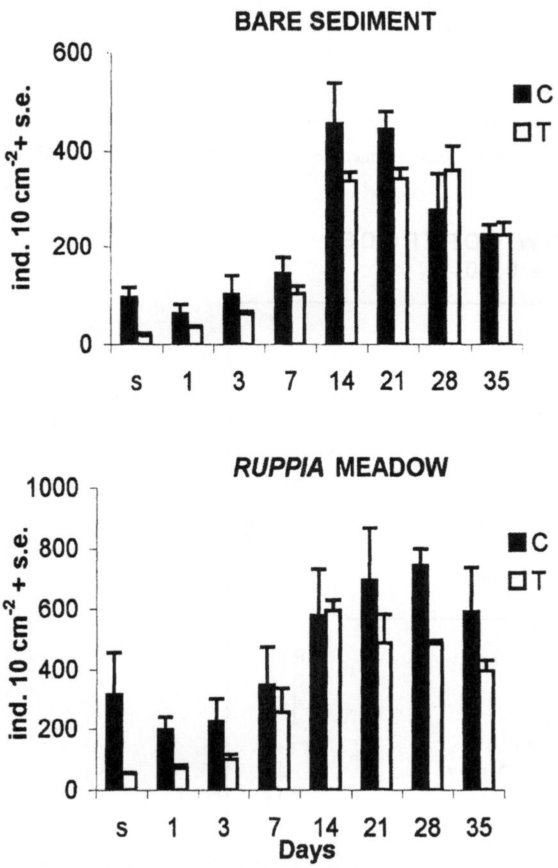

Figure 7. Average densities of total copepods in bare sediment and in *Ruppia* meadow respectively for control (C) and treatment (T) at each sampling day (s: start, 1, 3, 7, 14, 21, 28, 35).

ulations showed good resistance to the short term anoxia, mainly in the bare sediment. Previous investigations have already pointed out the general nematode tolerance to low oxygen and high sulphide concentrations (Heip et al., 1985; Meyers et al., 1987; Vopel et al., 1996). Nevertheless, in the present investigation, an abundant species such as *Terschellingia longicaudata*, which is usually considered a thiobiotic species (Nicholas et al., 1987; Austen & Widbom, 1991; Thiermann et al., 1994) was strongly affected by the anoxia in the phytal habitat. Probably, density reduction of *T. longicaudata* was due to an excess of sulphide production in that habitat owing to the amount of organic matter deriving not only from macrophyte plants but also from macroalgal epiphytes (Castel et al., 1996; Viaroli et al., 1997). The same fact could also explain the general decrease of total nematode abundances. During a further experiment (basis of a forthcoming essay) on meiofauna resistance carried out in the same sites, we also recorded variations of abiotic parameters. The dissolved sulphide released into the water column inside the dark benthic chambers was about 6 times higher in the phytal habitat than in the sedimentary one (1383 μm *vs* 243 μm) after four days of incubation. Comparably high concentrations were previously measured only within seep fields permanently devoid of meiofauna (Thiermann et al., 1994).

Unlike what was observed for copepods, a recolonization trend of nematode assemblage was not readily detectable in bare sediment, as community structure remained nearly unchanged after 3 d of hypoxia/anoxia. Only *Theristus normandicus* showed a significant density reduction during the first week, followed by a slow recovery during the successive period (after 21 d). Probably, this species needed to wait for restoration of suitable trophic conditions, perhaps increased benthic diatom abundances (Moens & Vincx, 1997).

Overall, in the phytal habitat too, nematode assemblage seemed resistant to short term anoxia, notwith-

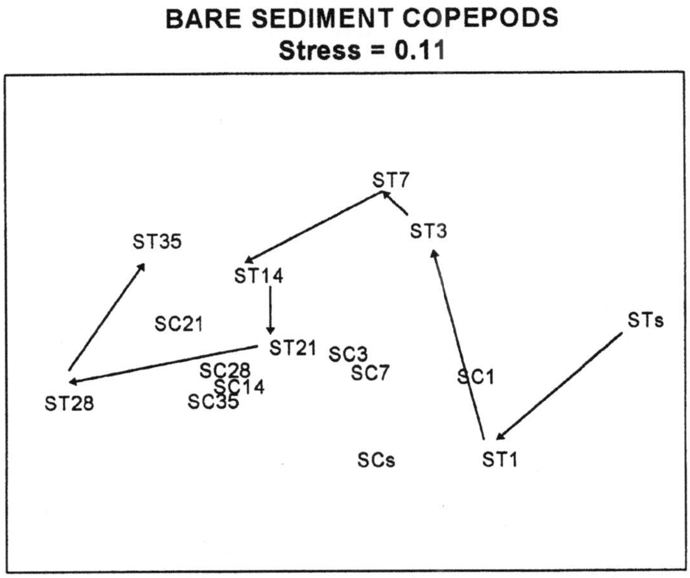

Figure 8. Two dimensional non-metric MDS ordination plot of copepod assemblages in bare sediment (S) for control (C) and treatment (T) at each sampling day (s: start, 1, 3, 7, 14, 21, 28, 35).

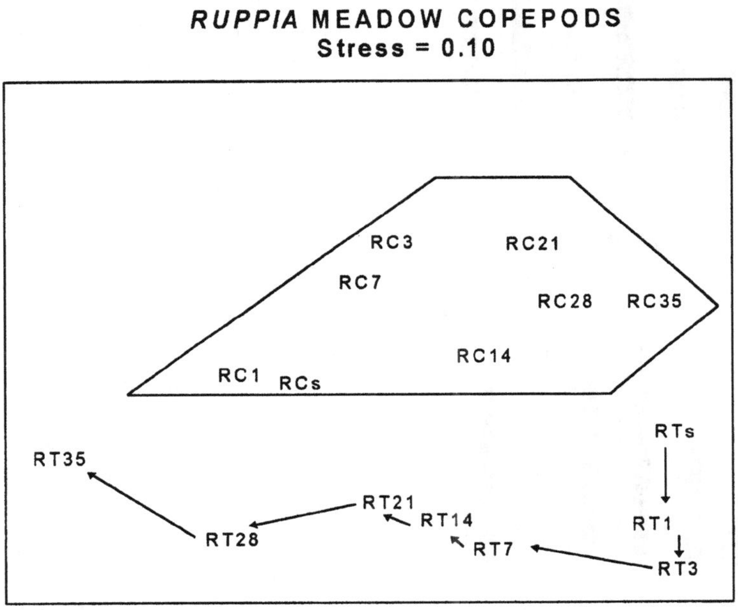

Figure 9. Two dimensional non-metric MDS ordination plot of copepod assemblages in *Ruppia* meadow (R) for control (C) and treatment (T) at each sampling day (s: start, 1, 3, 7, 14, 21, 28, 35).

standing the above mentioned decreased abundance of *Terschellingia longicaudata*. However, this species recovered quickly up to the same density of the control population one day after the end of incubation. Probably, sediment organic enrichment due to the treatment may have facilitated the settlement of this selective bacterial feeder. By contrast, during the first week of recolonization, most of Chromadoridae, epistrate feeders trophically linked to phytal habitats, and the omnivores/scavengers Oncholaimidae, *Viscosia glabra* and *Oncholaimus skawensis*, decreased in densities. This result may be explained by the reduction of trophic and 'habitat' availability. Direct field observations actually showed that, in the im-

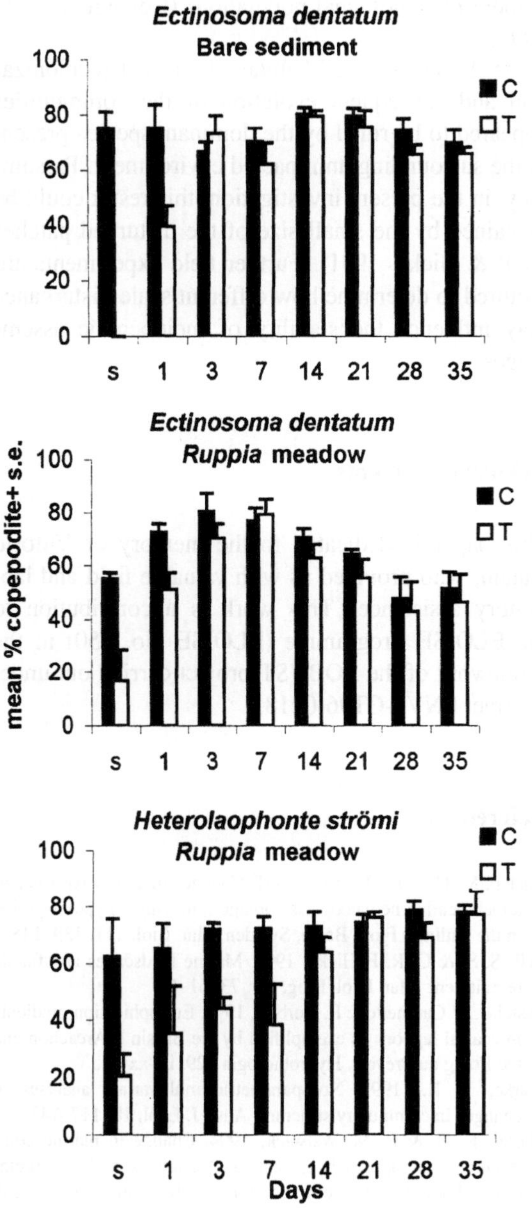

Figure 10. Average percentages per core (3.14 cm^2) of copepodites of the major species in control (C) and treatment (T) of the indicated habitats at each sampling day (s: start, 1, 3, 7, 14, 21, 28, 35).

pacted patches, decay of canopy continued after the removal of dark chambers, until day 3. Chromadoridae, *V. glabra* and *O. skawensis* could have migrated in the adjacent undisturbed areas and later recolonised the disturbed ones, after the restoration of the *Ruppia* plants and of a suitable 'pabulum' (Jensen, 1984). Nematode community structure changed again during the second phase of the recovery. Especially, the densities of Chromadoridae, *V. glabra* and *O. skawensis* were observed to markedly increase whereas those of *T. longicaudata* decreased (Table 3). Probably, in the disturbed patches, this species suffered the out-competing growth of Oncholaimidae populations.

Copepods

Expectations based on other studies (Coull & Palmer, 1984) that copepod populations would be more sensitive to disturbance than nematodes was supported by the results of this investigation. After three days of incubation, copepod densities similarly decreased (18%) in both investigated habitats. However, the overall effect of the anoxic disturbance was that almost all the species were affected, maintaining copepod assemblage from unvegetated habitat different from that of the *Ruppia* meadow (Table 5).

As a consequence of this initial situation, the main feature of copepod community response to the anoxic disturbance was a more detectable recovery pattern, clearly distinguished in two different phases: the first recolonization week (days 1–7) and the following longer period (days 14–35) of community evolution.

The first recolonization varied between the two investigated sites. In the bare sediment, total copepod density resulted not different from that of the control from day 3 onward (SNK test comparison; Figure 7). This fast initial recolonization was probably facilitated by the short distance organisms had to travel from the source of recruits (Gunnill, 1982). This appeared certainly true for the dominant species in the sedimentary habitat, *Ectinosoma dentatum*. In the control area, the population of this species was characterised by high numbers of copepodites (about 75% on average; Figure 10) which might generally be more active dispersers, *via* water column, than adults (Kern, 1990). In the second recovery phase, *E. dentatum* dominated the community of the recolonising patches mirroring density variations in the surrounding sediment.

In comparison with the simpler sedimentary habitat, recolonization processes in *Ruppia* meadow patches seemed more varied. First of all, the community recovery rate appeared to be delayed. Only after 7 d were both total copepod densities and community structure similar to the controls (SNK test comparisons; Figure 5). This slower first recolonization suggests that some restoring of the disturbed phytal habitat may be necessary for population recruitment (Bell & Hicks, 1991). During this first phase, true phytal species (e.g. *Tisbe histriana* Marcus & Por,

1961, *Mesochra heldti* Monard, 1935 and *Cyclopina gracilis*) did not settle on the disturbed patches, probably owing to the deteriorated *Ruppia* blades. But also a subcanopy-dwelling species (Hicks, 1986), such as *Heterolaophonte strömi paraminuta*, which appeared to be a faster recolonising species, confirmed this evidence. This species firstly settled in the disturbed patches mainly as adults, whereas the ratio copepodites/adults was higher and nearly constant in the control area (Figure 10). This can be due to onthogenic differences in the habitat utilisation, with adults capable of living in the sediment whereas copepodites require a thriving canopy for development because blades with associated epiphytic algae may provide more favourable nursery sites than the underlying sediments (Hall & Bell, 1993).

During the following four weeks of the experiment, when the *Ruppia* canopy recovered in the disturbed patches, almost all species, but especially the above-mentioned true phytal ones (Table 7), increased their abundances. At the same time, the whole community structure conformed to that of the control and evolved according to the natural successional pattern of the season.

On the whole, the response of meiofaunal communities to the effects of hypoxic/anoxic disturbance seems to be related both to individual requirements of the different taxa (Olafsson & Moore, 1992) and to the habitats (Bell & Hicks, 1991). Relative to copepods, nematodes apparently exhibited a greater resistance, in terms of ability to withstand perturbations. By contrast, copepod community appeared to be remarkably resilient. Copepods suffered more heavily the hypoxia/anoxia impact, but recovered faster after the disturbance (Holling, 1973). This is likely due to their higher dispersal into the water column (Palmer, 1988) which makes them (at least theoretically) capable of redistributing among disturbed and undisturbed patches.

On the other hand, intensity of the effects of the disturbance seemed to be related to the different habitats. In the macrophytal environment, the impact of anoxia was heavier, not only on copepods, but on nematodes as well. This is likely due to the increased amount of decaying organic matter, the marked deterioration of chemical compositions of sediments and to the reduction in habitat complexity. Consequently, also the recovery patterns of the communities appeared slower and more articulate in the seagrass meadow than in the bare sediment. Altered substrata must recover suitable conditions to allow the settlement of meiofaunal populations (Colangelo et al., 1996).

In the investigated habitats, both first recolonization and subsequent evolution of the communities appeared to be ruled by the dominant species present in the surrounding unimpacted environment. Presumably, in the present investigation, this result could be explained by the small size of the disturbed patches (Bell & Hicks, 1991). Further field experiments are required to determine how different scale disturbance may influence the stability of meiobenthic assemblages.

Acknowledgements

This paper is dedicated to the memory of Vittorio Gaiani, who provided us with valuable field and laboratory assistance. This work is a contribution to the ELOISE Programme (ELOISE No. 050) in the framework of the ROBUST project carried out under contract ENV4-CT96-0218.

References

Austen, M. C. & B. Widbom, 1991. Changes in a slow recovery of a meiobenthic nematode assemblage following a hypoxis period in the Gullmar Fjord Basin, Sweden. Mar. Biol. 111: 139–145.

Bell, S. S. & G. R. F. Hicks, 1991. Marine landscapes and faunal recruitment. Mar. Ecol. Prog. Ser. 73: 61–68.

Castel, J., P. Caumette & R. Herbert, 1996. Eutrophication gradients in coastal lagoons as exemplified by the Bassin d'Arcachon and the Etang du Prévost. Hydrobiologia 329: ix–xxviii.

Clarke, K. R., 1993. Non-parametric multivariate analyses of changes in community structure. Aust. J. Ecol. 18: 117–143.

Clarke, K. R. & R. M. Warwick, 1994. Change in marine communities: an approach to statistical analysis and interpretation. Plymouth Marine Laboratory, U.K. Bourne Press Ltd, Bournemouth, U.K.

Colangelo, M. A., T. Macrì & V. U. Ceccherelli, 1996. A field experiment on the effect of two types of sediment disturbance on the rate of recovery of a meiobenthic community in a eutrophicated lagoon. Hydrobiologia 329: 57–67.

Coull, B. C. & G. T. Chandler, 1992. Pollution and meiofauna: field, laboratory, and mesocosm studies. Oceanogr. Mar. Biol. annu. Rev. 30: 191–271.

Coull, B. C. & M. A. Palmer, 1984. Field experimentation in meiofaunal ecology. Hydrobiologia 118: 1–19.

Dutilleul, P., 1993. Spatial heterogeneity and design of ecological field experiments. Ecology 74: 1646–1658.

Gee, J. M., M. Austen, G. De Smet, T. Ferraro, A. McEvoy, S. Moore, D. Van Garsbeke, M. Vincx & R. M. Warwick, 1992. Soft sediment meiofauna community responses to environmental pollution gradients in the German Bight and at a drilling site off the Dutch coast. Mar. Ecol. Prog. Ser. 91: 289–302.

Guerrini, A., M. A. Colangelo & V. U. Ceccherelli. Risposta di comunità meiobentoniche a fenomeni di anossia indotta. Biol. Mar. Medit. (in press).

Gunnill, F. C., 1982. Macroalgae as habitat patch island for *Scutellidum lamellipes* (Copepoda: Harpacticoida) and *Amphithoe tea* (Amphipoda: Gammaridae). Mar. Biol. 69: 103–116.

Hall, M. O. & S. S. Bell, 1993. Meiofauna on the seagrass *Thalassia testudinum*: population characteristics of harpacticoid copepods and associations with algal epiphytes. Mar. Biol. 116: 137–146.

Heip, C., M. Vincx & G. Vranken, 1985. The ecology of marine nematodes. Oceanogr. Mar. Biol. annu. Rev. 23: 399–489.

Hicks, G. R. F., 1986. Distribution and behaviour of meiofaunal copepods inside and outside seagrass beds. Mar. Ecol. Prog. Ser. 31: 159–170.

Holling, C. S., 1973. Resilience and stability of ecological systems. Annu. Rev. Ecol. Syst. 4: 1–23.

Jensen, P., 1984. Ecology of benthic and epiphytic nematodes in brackish waters. Hydrobiologia 108: 201–217.

Kern, J. C., 1990. Active and passive aspects of meiobenthic copepod dispersal at two sites near Mustang Island, Texas. Mar. Ecol. Prog. Ser. 60: 211–223.

Kern, J. C. & G. L. Taghon, 1986. Can passive recruitment explain harpacticoid copepod distributions in relation to epibenthic structure ? J. exp. mar. Biol. Ecol. 101: 1–23.

Meyers, M. B., H. Fossing & E. N. Powell, 1987. Microdistributions of interstitial meiofauna, oxygen and sulphide gradients, and the tubes of macro-infauna. Mar. Ecol. Prog. Ser. 35: 223–241.

Moens, T. & M. Vincx, 1997. Observations on the feeding ecology of estuarine nematodes. J. mar. biol. Ass. U.K. 77: 211–227.

Nicholas, W. L., D. G. Goodchild & A. Stewart, 1987. The mineral composition of intracellular inclusions in nematodes from thiobiotic mangrove mud-flats. Nematologica 33: 167–169.

Olafsson, E. & C. G. Moore, 1992. Effects of macroepifauna on developing nematode and harpacticoid assemblages in a subtidal muddy habitat. Mar. Ecol. Prog. Ser. 84: 161–171.

Palmer, M. A., 1988. Dispersal of marine meiofauna: a review and conceptual model explaining passive transport and active emergence with implications for recruitment. Mar. Ecol. Prog. Ser. 48: 81–91.

Peck, L. S. & R. F. Uglow, 1990. Two methods for the assessment of the oxygen content of small volumes of seawater. J. exp. mar. Biol. Ecol. 141: 53–62.

Seinhorst, J. W., 1959. A rapid method for the transfer of nematodes from fixative to anhydrous glycerine. Nematologica 4: 67–69.

Sorokin, Yu. I. P. Yu. Sorokin & A. Gnes, 1996. Structure and functioning of the anthropogenically transformed Comacchio lagoonal ecosystem (Ferrara, Italy). Mar. Ecol. Prog. Ser. 133: 57–71.

Thiermann, F., R. Windoffer & O. Giere, 1994. Selected meiofauna around shallow water hydrothermal vents off Milos (Greece): ecological and ultrastructural aspects. Vie Milieu 44: 215–226.

Thrush, S. F., R. B. Whitlatch, R. D. Pridmore, J. E. Hewitt, V. J. Cummings & M. R. Wilkinson, 1996. Scale-dependent recolonization: the role of sediment stability in a dynamic sandflat habitat. Ecology 77: 2472–2487.

Trotter, D. B. & J. M. Webster, 1984. Feeding preferences and seasonality of free-living marine nematodes inhabiting the kelp *Macrocistis integrifolia*. Mar. Ecol. Prog. Ser. 14: 151–157.

Turner, S. J., S. F. Thrush, R. D. Pridmore, J. E. Hewitt, V. J. Cummings & M. Maskery, 1995. Are soft-sediment communities stable? An example from windy harbour. Mar. Ecol. Prog. Ser. 120: 219–230.

Underwood, A. J. & M. J. Anderson, 1994. Seasonal and temporal aspects of recruitment and succession in an intertidal estuarine fouling assemblage. J. mar. biol. Ass. U.K. 74: 563–584.

Viaroli, P., M. Bartoli, I. Fumagalli & G.Giordani, 1997. Relationship between benthic fluxes and macrophyte cover in a shallow brackish lagoon. Wat. Air Soil Pollut. 99: 533–540.

Vopel, K., J. Dehmlow & G. Artl, 1996. Vertical distribution of *Cletocamptus confluens* (Copepoda, Harpacticoida) in relation to oxygen and sulphide microprofiles of a brackish water sulphuretum. Mar. Ecol. Prog. Ser. 141: 129–137.

Hydrobiologia **375/376:** 89–100, 1998.
S. Baden, L. Pihl, R. Rosenberg, J.-O. Strömberg, I. Svane & P. Tiselius (eds),
Recruitment, Colonization and Physical–Chemical Forcing in Marine Biological Systems.

Active habitat selection by megalopae and juvenile shore crabs *Carcinus maenas*: a laboratory study in an annular flume

Ola Hedvall, Per-Olav Moksnes & Leif Pihl
Marine Ecology, Kristineberg Marine Research Station, Göteborg University, Sweden

Key words: Carcinus maenas, hydrodynamics, predation, passive deposition, post-settlement movements

Abstract

We studied megalopae (postlarvae) and young juveniles of the shore crab (*Carcinus maenas* L.) in laboratory experiments to examine four potentially important processes for juvenile distribution and recruitment: (1) hydrodynamic processes and passive deposition of megalopae, (2) active habitat selection of megalopae, (3) habitat specific predation rates, and (4) active habitat selection by juveniles.

In an annular flume, simulating natural current velocities in nursery areas on the Swedish west coast, we assessed the distribution of dead megalopae, live megalopae, live megalopae with predators (juvenile conspecifics and brown shrimp, *Crangon crangon*), and first instar crabs, in four simultaneously presented habitats: blue mussels (*Mytilus edulis*), eelgrass (*Zostera marina*), filamentous green algae (*Cladophora sp.* and *Chaetomorpha linum*) and bare sand. In a second experiment we studied the distribution of live megalopae between four different ephemeral macroalgae with different structural complexity (*Ulva lactuca, Enteromorpha sp., Cladophora sp.* and *Ectocarpus siliculosus*).

Dead megalopae were evenly distributed between the four habitats, whereas all other treatments showed significantly lower proportions of megalopae and juvenile crabs in the sand habitat (0–2%) compared to the structurally complex habitats (24–40%). The distribution between mussels, eelgrass and filamentous algae of live megalopae in absence of predators did not differ significantly from the hydrodynamical null hypothesis, i.e. distribution of dead megalopae. However, predation increased the proportion of megalopae significantly in the filamentous algae, providing the best refuge from predation of these habitats. First instar crabs showed a significantly different distribution compared to megalopae, with higher proportion in the algal habitat, whereas juvenile predatory crabs were found in significantly higher proportion among mussels. Megalopae selected all four different macroalgae species over open sand, but a significantly lower proportion were found in the algae with the highest structural complexity (*Ectocarpus siliculosus*; 14%) compared to the other algal species (26–30%).

These results indicate that passive deposition have little influence on the small scale (< 10 s of meters) distribution of shore crab megalopae during normal current velocities, but that active habitat selection by megalopae is the major process responsible for the non-random distribution of megalopae and juvenile shore crabs. The results further suggest that the initial distribution of megalopae between nursery habitats is quickly modified by habitat specific predation rates and size-specific movements and habitat choices by juveniles. The correlation between the habitat choice of megalopae and juvenile crabs, and the refuge value of the examined habitats suggests that habitat specific predation rates is a major selective force behind the behavior of active habitat selection in this species.

Introduction

In marine benthic organisms having complex life cycles, settlement and early juvenile stages are thought to be critical periods because of high predation mortality (Gaines & Roughgarden, 1987; Menge & Sutherland, 1987; Smith & Herrnkind, 1992; Eggleston & Armstrong, 1995; Gosselin & Qian, 1996). Young juveniles of many motile benthic organisms are often found concentrated in structurally complex habitats that are thought to provide refuge from predation (e.g. shore crabs in mussel beds: Klein-Breteler, 1976; blue crabs in seagrass beds: Heck & Orth, 1980b; spiny lobsters in macroalgae: Herrnkind & Butler, 1986; lobsters in cobblestones: Wahle & Steneck, 1991; fish in macroalgae: Carr, 1994).

Habitat specific predation rates have been demonstrated for several benthic organisms both during post-larval settlement (Barshaw et al., 1994; Dittel et al., 1996; Moksnes et al., 1997) and for juvenile stages (Wilson et al., 1987; Barshaw & Able, 1990; Smith & Herrnkind, 1992; Fernandez et al., 1993) and are often suggested as the process responsible for the non-random distribution of juveniles. However, it generally remains unclear if the non-random distribution of juveniles is caused by habitat specific predation rates, or by passive or active habitat selection of postlarvae, or by post-settlement movements of juveniles.

Recent work in both the laboratory and the field have demonstrated the importance of hydrodynamic processes in determining the distribution for benthic organisms on a small spatial scale ($\leq$ 10 s of meters) on both hard and soft bottoms (Eckman, 1983; Hannan, 1984; Butman, 1987; Butman et al., 1988; Butman, 1989; Jonsson et al., 1991; Mullineaux & Butman, 1991). Small and weakly swimming larvae can be influenced by boundary layer flow processes that may lead to a non-random distribution through passive deposition (Hannan, 1984). Flows can also influence settlement by affecting larval responses to environmental cues (Pawlik et al., 1991).

Many studies have also shown that larval settlement over very small spatial scales ($\leq$ 10 s of centimeters) is primarily an active process where larvae choose the substratum that offer greater survival during metamorphosis or for subsequent benthic stages, in response to specific biotic or physical cues (see reviews by Meadows & Campbell, 1972). Large larvae with a high swimming capacity may display habitat selection over larger scales, and many decapod postlarvae actively select structurally complex habitats over scales of meters (Botero & Atema, 1982; Herrnkind & Butler, 1986; Fernandez et al., 1993). However, most experiments with larvae of decapods have been performed in still water, and little is known about decapod larval behavior under more realistic flow regimes (but see Fernandez et al., 1994).

Migration of juveniles has just recently received attention as an important process for juvenile distribution and local population dynamics (Sogard, 1989; Palmer et al., 1996). Postlarvae and early juveniles may face a different suite of predators than larger juveniles (e.g. small benthic predators common in complex habitats, including older conspecifics). Settling postlarvae and different juvenile stages may therefore use different nursery habitats, and move between habitats to meet requirements of refuge and food, as they grow into different size-refuges from predation, resulting in one or several ontogenetic habitat shifts (Stein & Magnusson, 1976; Smith & Herrnkind, 1992; Eggleston, 1995).

To our knowledge, no attempts have been made to incorporate all of these four processes into the same study, and there is a general lack of understanding of their relative importance for distribution and recruitment of juveniles.

In the present investigation we have studied megalopae (postlarvae) and young juveniles of the shore crab (*Carcinus maenas* L.) in laboratory experiments to examine four potentially important processes for juvenile distribution and recruitment: (1) hydrodynamic processes and passive deposition of megalopae, (2) active habitat selection of megalopae, (3) habitat specific predation rates, and (4) active habitat selection of juveniles. These processes were studied in an annular flume, simulating natural current velocities, using several habitats common in nursery areas on the Swedish west coast. The results are compared with a parallel study of habitat specific predation rates, and distribution patterns of juveniles in the field, and the implication for recruitment of juvenile shore crabs is discussed.

Study organism

The shore crab is an epibenthic omnivore, common on both soft and hard intertidal and shallow subtidal substrata in north western Europe, that is an important predator on the macrofauna in the benthic community (Ropes, 1968; Klein-Breteler, 1976; Scherer & Reise, 1981; Pihl & Rosenberg, 1982; Jensen & Jensen, 1985; Pihl, 1985; Sanchez-Salazar et al., 1987). It has a complex life cycle with four pelagic zoea stages and a megalopal (postlarval) stage that settle and metamorphose into the first benthic crab stage.

On the Swedish west coast, juvenile shore crabs are found mainly in shallow (< 1 m), semi-exposed, soft sediment bays (Pihl, 1986), that show a mosaic-like distribution of small patches of habitats, dominated by blue mussel beds (*Mytilus edulis* L), shell debris, eelgrass beds (*Zostera marina* L.) and, in recent years as a result of eutrophication, mats of

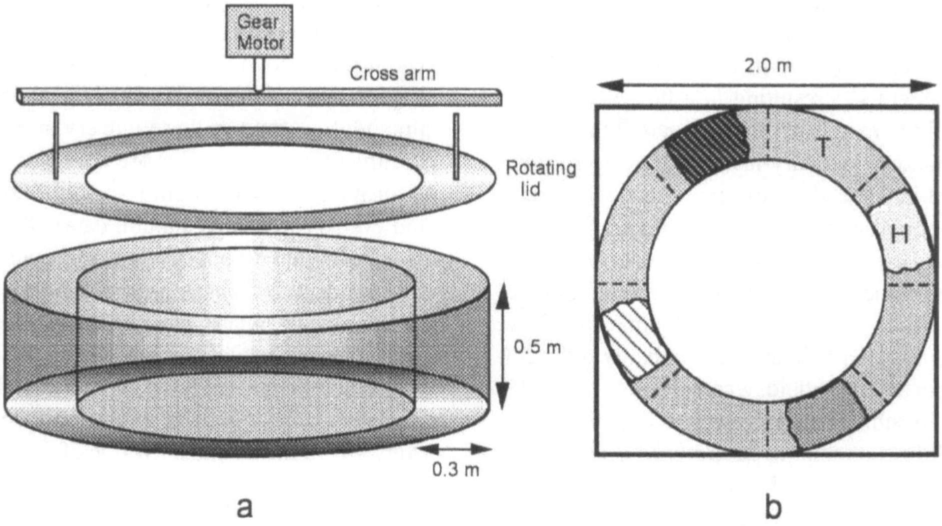

Figure 1. (a) Exploded view of the annular flume showing the cross arm above the flume that rotates the lid on the water surface. Shear stress exerted on the water by the lid causes the water circulation. (b) A schematic diagram (top view) of the annular flume showing the positions of the examined habitats. Each habitat section (H) is separated by sand covered transport sections (T).

ephemeral macroalgae with a wide range of structural complexity. Young juvenile shore crabs are not randomly distributed in these bays, but are concentrated at high densities in structurally complex habitats (over 200 juvenile crabs m^{-2}; Moksnes, unpubl. data), primarily in blue mussel beds and shell debris (Eriksson & Edlund, 1977; Pihl & Rosenberg, 1982), but also in eelgrass beds (Baden & Pihl, 1984) and in filamentous macro algae (Isaksson & Pihl, 1992), and found at much lower densities in adjacent sand or mud (< 10 m^{-2}; Moksnes, unpubl. data). Little is known about the initial distribution of megalopae, and if the non-random distribution of juveniles is set during postlarval settlement, or if it is the result of habitat specific predation rates and/or movements of juveniles.

Recent laboratory experiments have demonstrated that shore crab megalopae suffer high settling mortality in open sand (80–90%), but obtain a significant habitat refuge in mussel beds, eelgrass and filamentous green algae, the latter yielding the lowest mortality (13–14%; Moksnes et al., 1998). Further, small differences in structural complexity of ephemeral macroalgae dramatically affected predation mortality of first instar crabs, with a significant refuge obtained only in algae of medium complexity. These results indicate a potential plasticity in habitat use by megalopae and juvenile shore crab, but to evaluate the impact of these habitat specific predation rates on distribution patterns and recruitment, it is necessary to examine if megalopae settle in all of the examined habitats, or if they actively select between them.

Material and methods

Flume

All experiments were carried out in a laboratory annular flume, constructed of 4 mm transparent PVC-plastic with an outer diameter, channel width and height of 2.0, 0.30 and 0.50 m, respectively. The circumference in the middle of the channel was 5.3 m and the bottom area was 1.6 m^2 (Figure 1a). Water flow was induced by the shear stress between a rotating, circular, transparent PVC lid and the water surface. The lid was attached to a cross arm and driven by a variable electric speed engine placed above the flume.

Annular flumes driven by a rotating lids or paddles have been used in sediment transport studies (Taghon et al., 1984; Wainright, 1990; Amos et al., 1992). It is argued that the constant channel geometry and infinite flow length resulting in a fully developed benthic boundary layer is an advantage that annular flumes have over linear or oval shaped flumes (Amos et al., 1992). The disadvantage with annular flumes is the necessary trade-off between secondary circulation and cross channel current variation associated with radius-to-width ratio, and wall effects associated with the width-to-depth ratio.

In the present flume, velocity profiles measured with video techniques (Jonsson et al. 1991), in 26 cm water depth over sand bottom, with a lid speed of approximately 8 cm s^{-1} resulting in a flow velocity of approximately 3 cm s^{-1} in the center part of the flume, showed a marked boundary-layer with a logarithmic velocity profile, extending to approximately 10 cm above the bottom, and a cross-channel current variation of 33%. However, in the experiments described below, large patches of habitats were presented in the flume. Due to the constraints of the flume, the flow around the presented habitats was probably different than that expected in the field, where the flow might diverge around a habitat patch (Nowell & Jumars, 1987). This study did not attempt to describe these flows, nor to simulate natural flows and boundary layers, but only to simulate natural current velocities over living habitats and study the behavior of megalopae and juvenile crabs in flowing water.

For the present study this type of flume offered several advantages: (1) the open circular construction allowed the whole flume to be used as an experimental area, where animals, habitats and sediment could be easily removed and recovered after each trial; (2) the circular lid-driven current allowed animals to pass the test areas continuously, with similar current regimes in all areas, without disturbance and turbulence from bends, tubes, propellers or pumps; and (3) the flume takes up a small amount of space and can be built in a short time at relatively low cost.

General experimental procedures

Laboratory experiments were conducted between August and October 1996, at Kristineberg Marine Research Station, located on the west coast of Sweden.

Two experiments were carried out in order to test: (1) distribution of shore crab megalopae and juveniles, in four habitats: blue mussels, eelgrass, filamentous green algae and bare sand, as a result of passive deposition, active habitat selection and predation; and (2) active habitat selection of megalopae in four different ephemeral macroalgae with different structural complexity.

In both experiments the bottom of the flume was covered with a 25 mm even layer of sieved (750 μm) dry beach sand and filled to 26 cm water depth with filtered (750 μm) water pumped in from a nearby bay (total volume 416 l). The flow velocity above the habitats was approximately 3 cm s^{-1}, simulating normal natural current velocities in semi-exposed shallow bays on the Swedish west coast, where the tidal amplitude and surface velocities normally are below 20 cm and 4 cm s^{-1}, respectively. Mean surface current velocities during 5 weeks of continuous measurements (Coastal 3D-Acoustic Current Meter, Falmouth Sci. Inc.) in these nursery areas during the recruitment period of the shore crab was 2.2 cm s^{-1}, and increased above 10 cm s^{-1} only during unusual wind events (Moksnes, unpubl. data). The flume was divided in eight sections of equal size; four habitat sections separated by four transport sections (Figure 1b). The examined habitats were randomly allocated to the center of each habitat section where they were uniformly distributed, covering approximately a 47 cm long patch (0.14 m^2), from wall to wall, with approximately 10 cm of sand on each side of the habitat patch in each habitat section. The rest of the flume was left with bare sand leaving approximately 87 cm (0.26 m^2) of open sand between each examined habitat.

All habitats were collected in shallow bays next to the marine station, rinsed in fresh water and cleaned from epiphytes and fauna prior to use, to avoid introduction of food to the experiments.

Megalopae used in the experiments were collected daily from the surface plankton using passively fishing plankton nets placed in nursery areas, kept in 30 l aerated tanks and fed fresh blue mussel mantle prior to use. First instar juvenile crabs were raised from megalopae in 30 l, aerated tanks and fed blue mussels until molting. Predators were collected in shallow bays close to the laboratory with hand nets and kept in the laboratory in aerated circular tanks (90 l) with sediment and shells while being fed blue mussel for a minimum of 5 d before being used. Only predators actively feeding before the experiment were used, and each predator was used only once.

The trials were run for 12 h over night with constant current conditions. The flume was run for 1 h before the first animals were introduced which denoted the beginning of a trial. In all trials, 60 premolt megalopae or first instar crabs were introduced at random locations to the surface of the flume. When predators were used they were introduced 10–15 min after the larvae. Two sources of light were used in the laboratory giving 16 h L: 6 h D and 1 h of diffuse light at dusk and dawn, approximating natural fall light conditions in Sweden.

Upon termination of each trial, the rotating lid was stopped and carefully removed. Eight PVC-partitions were then pushed down, separating each habitat and transport section. Predators were first removed (if present) before habitats were separately taken out and carefully rinsed and soaked in fresh water for a minimum of one hour, whereupon the rinse water was sieved through 750 μm mesh. Filamentous macroalgae were put in alcohol with bengalrose pigment to facilitate the recovery of animals. Animals were emptied from each habitat section with a hand net, and remaining flume content, including sediment and animals from the transport section were emptied through a 750 μm mesh sieve. Recovered animals were enumerated and measured. The proportion of animals found in each habitat, excluding the transport sections, were used as the dependent variable in the statistical analysis. Replicates were randomly run in time series in each experiment. The water temperature and salinity varied between 13.2–20.1 °C and 18.5–32.2 psu, respectively during the experimental period.

Experiment 1

To examine four possible processes affecting distribution of juvenile shore crabs in the field (passive deposition of megalopae, active habitat selection of megalopae, predation on megalopae, and active habitat selection of juveniles) we tested the distribution of megalopae and juveniles between four simultaneously presented habitats: blue mussels, eelgrass, filamentous green algae and bare sand, using four different treatments: (1) 60 dead megalopae, (2) 60 live megalopae, (3) 60 live megalopae with predators, and (4) 60 live first instar crabs. The same habitats were used in all four treatments, and each treatment was replicated five times.

The four habitats consisted of (1) sand with approximately 60 live, adult (mean length 74 mm) blue mussels in a tightly packed bed with new webs of byssus threads, (2) sand with approximately 100 shoots of live eelgrass ($\approx$700 shoots m^{-2}, equivalent to natural densities in nursery areas; Baden & Pihl, 1984) uniformly planted in small bundles (i.e. 2–6 shoots), (3) sand with approximately 500 ml (17–24 g dry weight) of live, green filamentous macroalgae (branched *Cladophora* sp. and *Chaetomorpha linum* (O.F.Müll.) Kütz with stem diameter between 40–1000 μm), and (4) bare sand. In nursery areas, eelgrass shoots have a normal blade length of 15–40 cm, but in the present experiment all blades were cut to a length of less than 10 cm (mean length 81 mm) to give all habitats similar volume, and to keep the height of the habitats less than 35% of the water depth (Nowell & Jumars, 1987).

To examine if hydrodynamic conditions in the flume (simulating normal natural current velocities) and passive deposition could affect the distribution of megalopae, we tested the distribution of freeze killed megalopae. In preliminary experiments, freeze killed megalopae were found to have the same sinking velocity (1.3 cm s^{-1}) as immobilized live megalopae (ANOVA: $P > 0.05$). The treatments with dead megalopae were run for only two hours to avoid decomposition and density changes of the larvae. An even distribution was used as the extrinsic null hypothesis in the analysis.

The distribution of live megalopae was tested against both an even distribution and against the resulting mean distribution of dead megalopae, the latter representing the 'hydrodynamical null hypothesis' (Eckman, 1983).

In the predator treatment, two species of predators were used concurrently: four juvenile shore crabs (seventh to eight instars; 8–10 mm carapace width) and four brown shrimp (*Crangon crangon* L.; 33–38 mm total length). These species are the most abundant potential predators in nursery habitats and open sand habitats, respectively (Pihl, 1985), and are efficient predators on megalopae and first instar shore crabs (Moksnes et al., 1998). The distribution of each predator species was tested in a Single-Classification Test for Goodness of Fit using the pooled data from all five replicates to give sufficient accuracy to the *G*-computation (Sokal & Rohlf, 1995), using an even distribution as the extrinsic null hypothesis. The distribution of megalopae was tested with both an even distribution and the resulting mean distribution of live megalopae without predators. The latter to separate the effect of habitat choice from the effect of predators.

The distribution of first instars was examined to test if habitat selection of juveniles affect the distribution, and could, in comparison with juvenile predatory crabs, reveal a size-specific habitat preference. An even distribution, the resulting mean distribution of live megalopae, and the resulting distribution of crab predators were used as the extrinsic null hypotheses in the analyses.

All treatments were tested with *G*-statistics (Replicated Test of Goodness of Fit) where each replicate was first tested separately, followed by a test of all the data (*G*-total), the pooled data of the replicates (*G*-pooled), and a test of the heterogeneity of the replicates (*G*-heterogeneity; Sokal & Rohlf, 1995). If the null-hypothesis was found to be false in any treatment, a second analysis was performed with the sand habitat excluded to test if the structurally complex habitats differed. The p-values were adjusted accordingly to avoid type-I errors (Sokal & Rohlf, 1995).

Experiment 2

The aim of the second experiment was to test if megalopae would actively select between ephemeral macroalgae of different structural complexity (i.e. surface area per unit volume). We tested the distribution of 60 live megalopae between four simultaneously presented macroalgal habitats: (1) sand with *Ulva lactuca* L. (5–15 cm leaf width), (2) sand with *Enteromorpha sp.* (0.5–5 mm stem diameter), (3) sand with *Cladophora sp.* (40–130 μm stem diameter), and (4) sand with *Ectocarpus siliculosus* (Dillwyn) Lyngb. (10–40 μm stem diameter). Approximately 500 ml of each algal habitat was placed in each test section of the flume. Relative estimates of total surface area of each algal habitat (Moksnes et al., 1998) varied between 10 m^2 (*Ulva lactuca*) and 81 m^2 (*Ectocarpus siliculosus*). The distribution was tested with Replicated Test of Goodness of Fit ($n = 4$) both between the four different algal habitats, and between the algal habitats and the surrounding sand, using an even distribution as the extrinsic null hypothesis in both analyses.

Results

Qualitative observations

Megalopae were regularly observed in the water column during the first hour of the experiment, where they were drifting with the current while swimming upwards, or swimming against the current in the upper part of the water column, often making headway upstream in short bursts. They were observed to land on the sand and swim up in the water column again, and appeared to be able to maneuver at will. In the morning, during the last hour of the trial, no megalopae were observed swimming. Dead megalopae sank quickly to the bottom and were never observed to be moved over the bottom by the current. All observed first instar crabs remained passive in the water column while they quickly sank to the bottom (within 20 s) and the random pattern appeared not to be altered during the sinking. On the bottom most crabs started to move around on the sediment. Brown shrimp were observed to actively move and search in all habitats, but to rest only on open sand. Shore crab predators were mainly seen in the complex habitats.

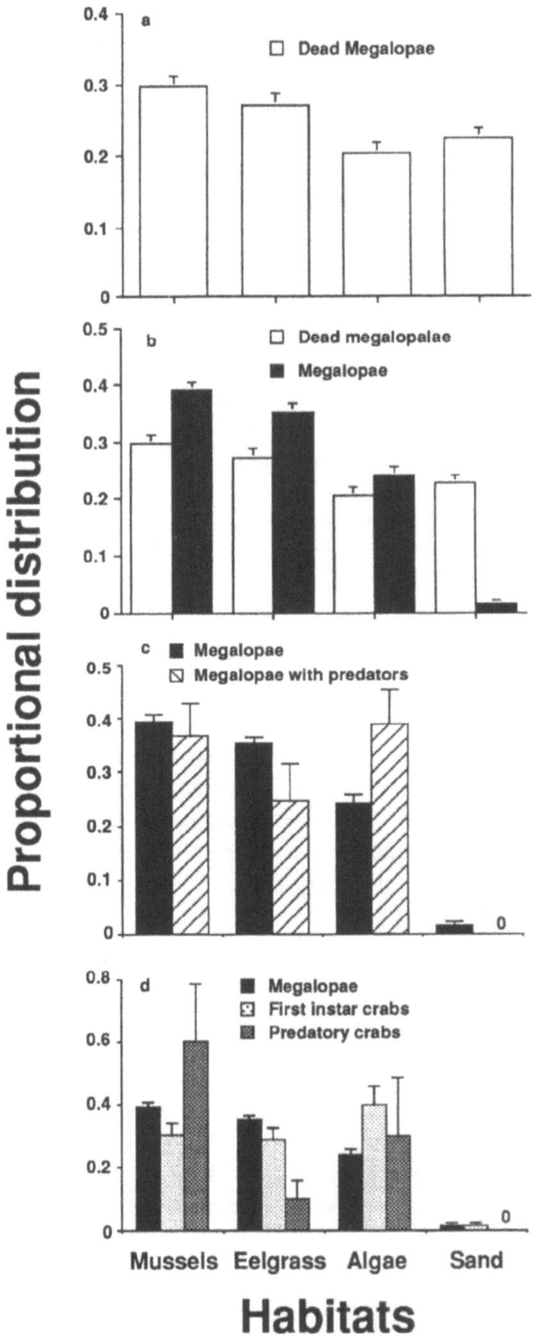

Figure 2. Mean proportional distribution (± SE) of megalopae and juvenile shore crabs in four different habitats (blue mussels, eelgrass, filamentous algae and sand) from four different treatments (dead megalopae, live megalopae, live megalopae with predators and first instar crabs; $n = 5$): (a) distribution of dead megalopae; (b) distribution of dead and live megalopae; (c) distribution of live megalopae and live megalopae with predators; and (d) distribution of live megalopae, first instar crabs, and seventh to eight instar predatory crabs. Note that the distribution of dead and live megalopae are shown in several figures where they were used as the extrinsic null hypothesis.

Table 1. Summary of *G*-statistics (Replicated Goodness of Fit Tests) from experiment 1. The distribution of each treatment was tested with different extrensic hypotheses: Ho1 = even distribution; Ho2 = distribution of dead postlarvae; Ho3 = distribution of live postlarvae; Ho4 = distribution of juvenile crab predators. Each replicate was first tested separately, follow by a test of all the data (*G*-Total), the pooled data from the replicates (*G*-Pool), and a test of the heterogeneity of the replicates (*G*-Het.). If *G*-Pool was found significant, a second analysis was performed excluding the sand habitat, using an adjusted *P*-value ($P < 0.025$).

		Replicates					*G*-Total	*G*-Pool	*G*-Het.
		1	2	3	4	5			
All habitats									
	df	3	3	3	3	3	15	3	12
Dead postlarvae									
Ho1	*G*	0.35 (ns)	0.82 (ns)	1.76 (ns)	1.40 (ns)	0.78 (ns)	5.11 (ns)	2.92 (ns)	2.19 (ns)
Live postlarvae									
Ho1	*G*	27.0***	32.6***	24.5***	33.5***	22.4***	140***	133***	6.85 (ns)
Ho2	*G*	20.7***	28.2***	20.3***	28.7***	16.7***	115***	108***	2.19 (ns)
Predators									
Ho1	*G*	24.4***	27.3***	24.9***	20.8***	35.2***	133***	107***	25.4*
Ho3	*G*	9.38*	13.4**	15.0**	7.15 (ns)	5.66 (ns)	50.6***	25.2***	25.4*
First instars									
Ho1	*G*	27.9***	30.8***	42.4***	20.3***	34.1***	156***	128***	27.7**
Ho3	*G*	19.4***	3.1 (ns)	24.5***	1.52 (ns)	4.71 (ns)	53.3***	32.7***	20.6 (ns)
Ho4	*G*	28.8***	36.2***	27.1***	29.9***	10.0*	131***	111***	20.6 (ns)
Sand excluded									
	df	2	2	2	2	2	10	2	8
Live postlarvae									
Ho1	*G*	4.76 (ns)	1.49 (ns)	1.22 (ns)	1.85 (ns)	3.53 (ns)	12.8 (ns)	10.6**	2.24 (ns)
Ho2	*G*	1.39 (ns)	0.41 (ns)	0.02 (ns)	0.42 (ns)	0.66 (ns)	2.9 (ns)	0.66 (ns)	2.24 (ns)
Predators									
Ho1	*G*	3.32 (ns)	7.84*	9.37*	3.04 (ns)	11.8**	35.4***	10.0**	25.4**
Ho3	*G*	8.24*	12.4**	14.1***	6.19 (ns)	4.40 (ns)	45.3***	19.9***	25.4**
First instars									
Ho1	*G*	10.0**	1.5 (ns)	10.7**	0.48 (ns)	3.56 (ns)	26.3*	5.85 (ns)	20.5**
Ho3	*G*	21.8***	1.58 (ns)	22.9***	3.81 (ns)	3.13 (ns)	53.2***	32.7***	20.5**
Ho4	*G*	31.9***	36.2***	27.1***	33.8***	10.0**	139***	119***	20.5**

* $P < 0.05$, ** $P < 0.01$, *** $P < 0.001$, ns $P > 0.05$ (four habitats).
* $P < 0.025$, ** $P < 0.01$, *** $P < 0.001$, ns $P > 0.025$ (sand excluded).

Experiment 1

The distribution of dead megalopae did not differ significantly from an even distribution in any replicate, with proportions varying between 21 and 30%, on average, in the four habitats (Table 1; Figure 2a). All replicates were homogenous (i.e. ratios being similar among replicates). Approximately half (52%) of all dead megalopae were found on the sand covered transport sections. Mean recovery rate of dead megalopae was 97%, the missing proportion being an estimate of loss due to handling.

The distribution of live megalopae was significantly different from both an even distribution and from the mean distribution of dead megalopae, in all homogenous replicates (Table 1; Figure 2b). Few megalopae were found on the sand habitat (1.5%) or on the transport sections (2 megalopae on average), whereas on average 24, 35 and 39% were found on the filamentous algae, eelgrass and mussel habitats, respectively. When excluding the sand habitat from the analysis, a homogenous trend of lower proportions of megalopae in the algae habitat was seen in all replicates, which was significant in the pooled data when tested against the even distribution. However, no such trend was seen when tested against the distribution of dead megalopae (Table 1). Mean metamorphosis rate of megalopae was 37% and mean recovery rate was 94%. The lower recovery rate in treatments with live megalopae compared to dead may indicate low rates

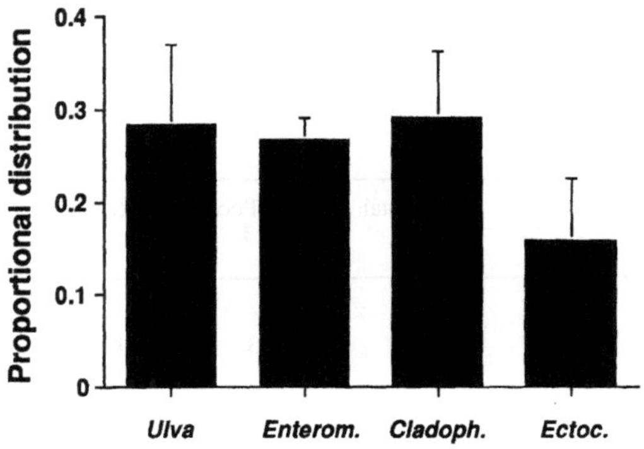

Figure 3. Mean proportional distribution (± SE) of live megalopae in four different filamentous algae (*Ulva lactuca*, *Enteromorpha sp.*, *Cladophora sp.* and *Ectocarpus siliculosus*; $n = 4$).

of cannibalism between megalopae and newly molted first instar crabs.

The distribution of live megalopae in the predator treatment was significantly different from the even distribution, also when the sand habitat was excluded, but not in all replicates (Table 1; Figure 2c). A significant difference was also found between the distribution of megalopae in the predator treatment and of the mean distribution of live megalopae without predators, even though the replicates were heterogenous (Table 1; Figure 2c). The difference in proportional distribution was caused by a reduction of mean number of megalopae in mussel, eelgrass and sand habitats (from approximately 21 to 11, 19 to 7, and 1 to 0 megalopae, with and without predators, respectively), whereas approximately the same mean number of megalopae was found in the filamentous algae (14 and 13 with and without predators, respectively). In the predator treatment, no megalopae were found on the sand habitat, and very few (1 megalopa on average) were found on the transport sections, whereas on average 39, 25 and 37% were found on the filamentous algae, eelgrass and mussel habitats, respectively. Mean metamorphosis rate of surviving megalopae was 24% and total mean recovery rate was 60%. Juvenile crab predators were not distributed evenly among all habitats ($G = 19.3$, $df = 3$, $P < 0.001$), nor among the three complex habitats ($G = 8.03$, $df = 2$, $P < 0.025$), but found mainly in the blue mussel habitat (60%) and less in the algal and eelgrass habitats (30 and 10%, respectively). No crab predators were found on the sand habitat or on the transport sections. The distribution of shrimp predators was not significantly different from an even distribution ($G = 3.20$, $df = 3$, $P > 0.05$), but showed a tendency to higher numbers on the sand then the eelgrass, mussel and algal habitats (45, 27, 18 and 9.1%, respectively). Almost half (42% on average) of all shrimp were found on the transport sections.

The distribution of first instar juvenile crabs was significantly different from the even distribution. Testing only the three complex habitats, some replicates were significantly uneven, but variation was large and the pooled data was not significant (Table 1; Figure 2d). First instar crabs also demonstrated a significantly different distribution from both live megalopae (without predators) and juvenile crab predators (Table 1). Few first instar crabs were found on the sand habitat (1.5%) or on the transport sections (4 crabs on average), whereas on average 39, 29 and 30% were found on the filamentous algae, eelgrass and mussel habitats, respectively. Mean recovery rate was 98%.

Experiment 2

Live megalopae were found in significantly higher proportion in the four macroalgal habitats (88%) compared to the open sand separating the habitats, in all replicates (Table 2). Distribution of megalopae among the four algal species was significantly different from an even distribution and showed large variation between replicates (Table 2; Figure 3). The pooled data demonstrated significantly lower proportions in the algae with the highest structural complexity (*Ectocarpus siliculosus*; 14%) compared to the other algal species (26–30%). Mean recovery rate was 95%.

Discussion

The results from the present study suggest that passive deposition have little influence on the small scale (≤ 10 s of meters) distribution of shore crab megalopae during normal current velocities, but that active habitat selection by megalopae sets the initial distribution which is modified by both subsequent predation and post-settlement movements.

Hydrodynamics and passive deposition

Under the investigated current velocity (3 cm s^{-1}), passive deposition did not cause a non-random distribution of shore crab megalopae, suggesting that this process have little impact on the distribution of megalopae and juveniles over small spatial scales (≤ 10 s of meters) in Swedish nursery areas during normal weather conditions. Limitations to this conclusion are that higher current velocities or the effect of waves

Table 2. Summary of *G*-statistics (Replicated Goodness of Fit Tests) from experiment 2. The distribution of postlarvae was tested between four species of benthic macro algae, and between the algal habitats and sand. In both analysis an even distribution was used as the extrensic null hypothesis. Each replicate was first tested separately, follow by a test of all the data (*G*-Total), the pooled data from the replicates (*G*-Pool), and a test of the heterogeneity of the replicates (*G*-Het.).

		Replicates				*G*-Total	*G*-Pool	*G*-Het.
		1	2	3	4			
Algal species	df	3	3	3	3	12	3	9
	G	13.9***	18.2***	17.4***	5.3 (ns)	54.7***	14.3**	40.4***
Alg vs sand	df	1	1	1	1	8	1	3
	G	53.0***	50.0***	45.3***	12.3***	160***	148***	12.1**

* $P < 0.05$, ** $P < 0.01$, *** $P < 0.001$, ns $P > 0.05$.

was not investigated, and that the flume did not simulate natural flows and boundary layers well. Wave induced oscillatory boundary-layers in shallow areas could create shear velocities with the capacity to resuspend megalopae (pers. com. P. R. Jonsson, 1997), and it is possible that differences in local shear velocities in the field could affect the distribution of passive megalopae, particularly in areas with larger tidal differences and higher current regimes. However, the present study demonstrate that the shore crab megalopae are not just being passively deposited. The high horizontal swimming capacity (> 3 cm s^{-1}; approximately 15 body lengths s^{-1}), and the capacity to hold on to selected substrates with its pereiopods, indicate that shore crab megalopae could avoid involuntary resuspension and redistribution also at higher flow velocities. The relative importance of hydrodynamic processes has been suggested to be related to the swimming ability of the larvae (Butman et al., 1988), and most studies that demonstrate effects of hydrodynamic processes and passive deposition have been carried out with small larvae of species with limited swimming capabilities (Eckman, 1983; Hannan, 1984; Butman, 1987; Butman et al., 1988; Butman, 1989; Jonsson et al. 1991; Mullineaux & Butman, 1991).

Active habitat selection

The ability of shore crab megalopae to maneuver and actively select settlement sites was demonstrated in the distribution of live megalopae. Approximately 95% of all megalopae were found in the three structurally complex habitats, even though these test areas only made up less than 38% of the available bottom area. The high rate of metamorphosis (37%) over the 12 h experiment indicates that most megalopae were in premolt, and thus chose a habitat for settlement. Though megalopae clearly avoided to settle on open sand habitat, they appear to be less selective between different types of structurally complex habitats, which all received a high proportion of settlers, even though filamentous algae appeared to be less preferred. When given the option of four species of ephemeral algae, megalopae again avoided open sand and settled in all four algal species, but significantly less in the alga with the highest structural complexity, *Ectocarpus siliculosus*. In recent laboratory experiments investigating habitats specific predation rates, megalopae and first instar shore crabs suffered high mortality in open sand (80–90%), but obtained a significant habitat refuge in all of the habitats investigated in the present study, except for *Ectocarpus siliculosus* and *Ulva lactuca*, which provided poor refuge under enhanced predation conditions (Moksnes et al., 1998). The folios *Ulva lactuca* was selected in high proportions in the present study, though large variation was found between replicates. These results suggest that shore crab megalopae have developed a behavior of avoiding open habitats with high predation rates, and to actively settle in a structured habitat that provides refuge from predation, but that they differentiate less between the refuge quality of habitats with different structural complexity. Active habitat selection and settlement is believed to be triggered by specific biotic or physical cues. Although the cues used by shore crab postlarva to select a settlement habitat in flowing water are unknown, the wide range of selected habitats does not suggest that it uses a positive chemical cue, but possibly a physical cue such as gradients in boundary layers, which has been suggested as an important cue for settling Dungeness crab megalopae (Fernandez et al., 1994). Positive thigmotaxis may also be involved, as suggested for settling lobster postlarvae

(Botero & Atema, 1982). The finding of active habitat selection of structurally complex habitats is supported by similar results in other decapod postlarvae from both laboratory and field studies (i.e. American lobster: Botero & Atema, 1982; spiny lobster: Herrnkind & Butler, 1986; Dungeness crab megalopae: Fernandez et al., 1993, 1994; Eggleston & Armstrong, 1995; blue crab megalopae: Morgan et al., 1996).

Predation

The distribution set by postlarval habitat selection was significantly affected by the presence of predators through an increased proportion of megalopae in the filamentous green algae. Since the distribution of the total number of predators was higher in blue mussels, and approximately even in the other three habitats, it is less likely that the change was caused by postlarval behavior, avoiding predators. The low overall recovery rate (60%) suggests instead, that the change was induced by habitat specific predation rates by brown shrimp and cannibalistic juvenile shore crabs. The mean reductions of megalopae, compared to the treatment without predators, in sand, eelgrass, blue mussels and filamentous algae (100, 55, 40 and 0%, respectively) are similar to predation rates on megalopae measured for these predators in these habitats (80–90, 19–33, 21–32 and 13–14%, respectively; Moksnes et al., 1998). We suggest that habitat specific predation rates on settling and metamorphosed megalopae cause a fast (< 12 h) change of the initial distribution of megalopae, that increases the proportion of megalopae and juvenile crabs in habitats with higher refuge value. This result is in accordance with several laboratory and field studies that have demonstrated habitat specific predation rates on decapod postlarvae and juveniles (blue crab: Wilson et al., 1987; Moksnes et al., 1997; Dungeness crab: Fernandez et al., 1993; lobsters: Barshaw & Able, 1990; Barshaw et al., 1994; mud crab: Dittel et al., 1996; spiny lobster: Smith & Herrnkind, 1992). Our result also provides support to the suggestion that settlement is a critical phase in the life cycle (Gaines & Roughgarden, 1987; Menge & Sutherland, 1987), since the shore crab megalopae suffered high mortality (37%), under simulated natural conditions, even when several types of shelter were accessible.

Post-settlement movements

First instar juvenile crabs showed a non-random distribution that was significantly different from the distribution of megalopae. They also avoided the open sand habitat, but were found in higher proportion in the filamentous algae than were megalopae. Since shore crabs are poor swimmers, particularly the young juvenile stages which have never been observed to swim by the authors, it is reasonable to assume that the redistribrution of first instar crabs was caused by movements on the sediment. These results suggest that juvenile crabs can leave the settlement habitat quickly after metamorphosis and redistribute between nursery habitats to meet requirements of shelter and food. Apart from providing the highest refuge value, filamentous green algae could possibly also constitute a food source that attracted first instar crabs. The suggestion of high motility of first instar shore crabs is supported by recent field data showing high migration between nursery habitats by the youngest juvenile shore crabs (Moksnes, 1998). The distribution of first instar crabs also differed significantly from the distribution of seventh to eighth instar predatory crabs, which were found mainly (60%) in the blue mussel habitat. This difference in distribution is consistent with field data where larger juvenile shore crabs are found in highest numbers in mussel beds and shell debris (Eriksson & Edlund, 1977; Pihl & Rosenberg, 1982) whereas the smallest instars are found also in high numbers in filamentous algae and eelgrass (Moksnes, unpubl. data), suggesting that habitat selection by juveniles is an important process for juvenile distribution. Shore crab megalopae and different juvenile stages may have different requirements of shelter and food, resulting in a size-specific movement and distribution between different nursery habitats after settlement. Larger juvenile shore crabs (3–20 mm carapace width) are very efficient predators on megalopae and first instar crabs (Moksnes et al., 1998). The use of different nursery habitats may therefore act as a spatial refuge against inter-cohort cannibalism. These suggestions of juvenile movements contrast the 'Settle and stay hypothesis' which states that colonists (in seagrass beds) are primarily planktonic settlers (Bell & Westoby 1986), but are supported by field studies showing high rates of migrating juvenile decapods into artificial seagrass beds (Sogard 1989). Size-specific ditribution between different nursery habitats have also been described for juvenile spiny lobster (Smith & Herrnkind, 1992) and Nassau grouper (Eggleston, 1995).

Implications for shore crab recruitment

The results in the present study suggest that habitat selection at settlement, habitat specific predation rates and post-settlement movements are the major processes behind the non-random distribution of juvenile shore crabs. Even though both predation and movements of juveniles affected distribution, the major part of the non-random distribution appear to be set during settlement through active habitat selection by megalopae. The correlation between the habitat choice of megalopae and juvenile crabs, and the refuge value of these habitats suggests that habitat specific predation rates is a major selective force behind the behavior of habitat selection in this species.

The ability of settling megalopae to avoid habitats with high predation rates will decrease overall mortality, at this critical stage, and increase the possibility of a coupling between larval supply and consequent recruitment of juveniles, i.e. pre-settlement regulation of the recruitment. Furthermore, active selection of nursery habitats at settlement will decrease the correlation between habitat availability and recruitment, if densities of recruits in these habitats are below carrying capacity (Whale & Steneck, 1991). Finally, the demonstrated plasticity of megalopae to select different nursery habitats and obtain refuge in them, will make the recruitment success less dependent on a single nursery habitat. Thus, these results suggest that the recruitment of juvenile shore crabs should be less affected by spatio-temporal variation in habitat distribution and availability. However, it is presently not known over what distances active habitat selection in shore crab megalopae and juveniles can concentrate settlers, or to what extent the shore crab recruitment is regulated by larval supply or post-settlement density-dependent processes. We encourage further research on the density-dependent effects on settlement, predation and juvenile migration to assess the processes behind distribution and population dynamics of the shore crab.

Acknowledgements

We sincerely thank Per Jonsson and Carl André for fruitful discussions on hydrodynamics and flume design, Jan Larsson, Gerhard Gunnarsson and Christer Gren for technical assistence during the construction of the flume, the staff at Kristineberg Marine Research Station for providing excellent facilities, and two anonymous referees for helpful comments on the manuscript. This research was made possible through grants from Swedish Council for Forestry and Agricultural Research (SFAR) and World Wide Fund For Nature (WWF) which are kindly acknowledged.

References

Amos, C. L., J. Grant, G. R. Daborn & K. Black, 1992. Sea Carousel-A Benthic, Annular Flume. Estuar. Coast. Shelf Sci. 34: 557–577.

Baden, S. P. & L. Pihl, 1984. Abundance, biomass and production of mobile epibenthic fauna in *Zostera marina* (L.) meadows, western Sweden. Ophelia 23: 65–90.

Barshaw, D. E. & K. W. Able, 1990. Tethering as a technique for assessing predation ratesin different habitats: an evaluation using juvenile lobster *Homarus americanus*. U.S. Fish. Bull. 88: 415–417.

Barshaw, D. E., K. W. Able & K. L. Heck, 1994. Salt marsh peat reef as protection for postlarval lobsters *Homarus americanus* from fish and shrimp predators: comparison with other substrates. Mar. Ecol. Prog. Ser. 106: 203–206.

Bell, J. D. & M. Westoby, 1986. Variation in seagrass high and density over a wide spatial scale: Effect on fish and decapod. J. exp. mar. Biol. Ecol. 104: 275–295.

Botero, L. & J. Atema, 1982. Behaviour and substrate selection during larval settling in the lobster *Homarus Americanus*. J. Crust. Biol. 2: 59–69.

Butman, C. A., 1987. Larval settlement of soft-sediment invertebrates: the spatial scales of pattern explained by active habitat selection and the emerging role of hydrodynamical processes. Oceanogr. Mar. Biol. annu. Rev. 25: 113–165.

Butman, C. A., 1989. Sediment-trap experiments on the importance of hydrodynamical processes in distributing settling invertebrate larvae in near-bottom waters. J. mar. biol. Ecol. 134: 37–88.

Butman, C. A., J. P. Grassle & C. M. Webb, 1988. Substrate choices made by marine larvae settling in still water and in flume flow. Nature 333: 771–773.

Carr, M. H. 1994. Effects of macroalgal dynamics on recruitment of a temperate reef fish. Ecology 75: 1320–1333.

Dittel, A., C. E. Epifanio & C. Natunewitz, 1996. Predation on mud crab megalopa, *Panopeus herbstii* H. Milne Edwards: effect of habitat complexity, predator species and postlarval densities. J. exp. Mar. Biol. Ecol. 198: 191–202.

Eckman, J. E., 1983. Hydrodynamic processes affecting benthic recruitment. Limnol. Oceanogr. 28: 241–257.

Eggleston, D. B. & D. A. Armstrong, 1995. Pre- and post-settlement determinants of estuarine Dungeness crab recruitment. Ecol. Monogr. 65: 193–216.

Eggleston, D. B., 1995. Recruitment in Nassau grouper *Epinephelus striatus*: post-settlement abundance, microhabitat features, and ontogenetic habitats shifts. Mar. Ecol. Prog. Ser. 124: 9–22.

Eriksson, S. & A-M. Edlund, 1977. On the ecological energetics of 0-group *Carcinus maenas* (L.) from shallow sandy bottom in Gullmar fjord, Sweden. J. exp. mar. biol. Ecol. 30: 233–248.

Fernandez, M., O. Iribane & D. Armstrong, 1993. Habitat selection by young of the year Dungeness crab *Cancer magister* and predation risk in intertidal habitats. Mar. Ecol. Prog. Ser. 92: 171–177.

Fernandez, M., O. Iribane & D. A. Armstrong, 1994. Swimming behaviour of dungeness crab, *Cancer magister* Dana, megalopae in still and moving water. Estuaries 17: 271–275.

Gaines, S. D. & J. Roughgarden, 1987. Fish in offshore kelp forest affect recruitment of intertidal barnacle populations. Science 235: 479–481.

Gosselin, L. A. & P. E. Qian, 1996. Early post-settlement mortality of an intertidal barnacle: a critical period for survival. Mar. Ecol. Prog. Ser. 135: 69–75.

Hannan, C. A., 1984. Planctonic larvae may act like passive particles in turbulent near-bottom flows. Limnol. Oceanogr. 29: 1108–1116.

Heck, K. L. & R. J. Orth, 1980b. Structural components of eelgrass (*Zostera marina*) meadows in the lower Chesapeake Bay-Decapod Crustacea. Estuaries 3: 289–295.

Herrnkind, W. F. & M. J. Butler IV, 1986. Factors regulating postlarval settlement and juvenile microhabitat use by spiny lobster *Panulirus argus*. Mar. Ecol. Prog. Ser. 34: 23–30.

Isaksson, I. & L. Pihl, 1992. Structural changes in benthic macrovegetation and associated epibenthic faunal communities. Neth. J. Sea Res. 30: 131–140.

Jensen, K. T. & J. N. Jensen, 1985. The importance of epibenthic predators on the density of juvenile benthic macrofauna in the Danish Wadden Sea. J. exp. mar. Biol. Ecol. 89: 157–174.

Jonsson, P. R., C. André & M. Lindegarth 1991. Swimming behavior of marine bivalve larvae in a flume boundary-layer flow: evidence for near bottom confinement. Mar. Ecol. Prog. Ser. 79: 67–76.

Klein Breteler, W. C. M., 1976. Settlement, growth and production of the shore crab, *Carcinus maenas*, on tidal flats in the Dutch Wadden Sea. Neth. J. Sea. Res. 10: 354–376.

Meadows, P. S. & J. I. Campbell, 1972. Habitat selection by aquatic invertebrates. Adv. Mar. Biol. 10: 271–382.

Menge, B. A. & J. P. Sutherland, 1987. Community regulation: variation in disturbance, competition, and predation in relation to environmental stress and recruitment. Am. Nat. 110: 351–369.

Moksnes, P-O., R. N. Lipcius, L. Pihl & J. van Montfrans, 1997. Cannibal-prey dynamicsin juveniles and postlarvae of the blue crab. J. exp. mar. Biol. Ecol. 215: 157–187.

Moksnes, P-O., L. Pihl & J. van Montfrans, 1998. Predation on postlarvae and juveniles of the shore crab *Carcinus maenas*: importance of shelter, size and cannibalism. Mar. Ecol. Prog. Ser. 166: 211–225.

Morgan, S. G., R. K. Zimmer-Faust, K. L. Heck Jr. & L. D. Coen, 1996. Population regulation of blue crabs *Callinectes sapidus* in the northern Gulf of Mexico: postlarval supply. Mar. Ecol. Prog. Ser. 133: 73–88.

Mullineaux, L. S. & C. A. Butman, 1991. Initial contact, exploration and attachment of barnacle (*Balanus amphitrite*) cyprids settling in flow. Mar. Biol. 110: 93–103.

Nowell, A. R. M. & P. A. Jumars, 1987. Flumes: Theoretical and experimentalconsiderations for simulations of benthic environments. Oceanogr. Mar. Biol. annu. Rev. 25: 91–112.

Palmer, M. A., D. Allan & C. A. Butman, 1996. Dispersal as a regional process affecting the local dynamics of marine and stream benthic invertebrates. Trends Ecol. Evol. 11: 322–326.

Pawlik, J. R., C. A. Butman & V. R. Starzak, 1991. Hydrodynamic facilitation of gregarious settlement of a reef-building tube worm. Science 251: 421–424.

Pihl, L., 1985. Food selection and consumption of mobile epibenthic fauna in shallow marine areas. Mar. Ecol. Prog. Ser. 22: 169–179.

Pihl, L., 1986. Exposure, vegetation and sediment as primary factors for mobile epibenthic fauna community structure and production in shallow marine soft bottom areas. Neth. J. Sea Res. 20: 75–84.

Pihl, L. & R. Rosenberg, 1982. Production, abundance and biomass of mobile epibenthic marine fauna in shallow waters, western Sweden. J. exp. mar. Biol. Ecol. 57: 273–301.

Ropes, J. W., 1968. The feeding habits of the green crab, *Carcinus maenas* (L.). Fish Bull 67: 183–203.

Sanchez-Salazar, M. E., C. L. Griffiths & R. Seed, 1987. The effect of size and temperature on predation of cockles *Cerastoderma edule* (L.) by the shore crab *Carcinus maenas* (L.). J. exp. mar. Biol. Ecol. 111: 181–193.

Scherer, B. & K. Reise, 1981. Significant predation on micro- and macrobenthos by the crab *Carcinus maenas* L. in the Wadden Sea. Kieler Meeresforsch. Sonderh. 5: 490–500.

Smith, K. N. & W. F. Herrnkind, 1992. Predation on early juvenile spiny lobsters *Panulirus argus* (Latreille): influence of size and shelter. J. exp. mar. Biol. Ecol. 157: 3–18.

Sogard, S. M., 1989. Colonization of artificial seagrass by fishes and decapod crustaceans: importance of proximity to natural eelgrass. J. exp. mar. Biol. Ecol. 133: 15–37.

Sokal, R. R. & F. J. Rohlf, 1995. Biometry. W. H. Freeman, New York.

Stein, R. A. & J. J. Magnusson, 1976. Behavioral response of crayfish to a fish predator. Ecology 57: 751–761.

Taghon, G. L., A. R. M. Nowell & P. A. Jumars, 1984. Transport and breakdown of fecalpellets: Biological and sedimentological consequences. Limnol. Oceanogr. 29: 64–72.

Wahle, R. A. & R. S. Steneck, 1991. Recruitment habitats and nursery grounds of the American lobster *Homarus americanus*: A demographic bottleneck? Mar. Ecol. Prog. Ser. 69: 231–243.

Wainright, S. C., 1990. Sediment-to-water fluxes of particulate material and microbes by resuspension and their contribution to the planktonic food web. Mar. Ecol. Prog. Ser. 62: 271–281.

Wilson K., K. L. Heck Jr. & K. W. Able, 1987. Juvenile blue crab, *Callinectes sapidus*, survival: an evaluation of eelgrass, *Zostera marina*, as a refuge. Fish. Bull. 85: 53–58.

Hydrobiologia **375/376:** 101–111, 1998.
S. Baden, L. Pihl, R. Rosenberg, J.-O. Strömberg, I. Svane & P. Tiselius (eds),
Recruitment, Colonization and Physical–Chemical Forcing in Marine Biological Systems.

Do barnacle larvae respond to multiple settlement cues over a range of spatial scales?

J. M. Hills[1,2,*,**], J. C. Thomason[2], J. L. Milligan[3] & M. Richardson[4]
[1] *University Marine Biological Station, Millport, Isle of Cumbrae, Ayrshire, KA28 OEG, UK*
[2] *Dept. of Marine Science and Coastal Management, University of Newcastle upon Tyne, Tyne & Wear, NE1 7RU, UK*
[3] *Gatty Marine Laboratory, University of St. Andrews, St Andrews, Fife, KY16 8LB, UK*
[4] *Harrogate Grammar School, Arthur's Avenue, Harrogate, N. Yorkshire, HG2 ODZ*
* *Corresponding author: e-mail: J.M.Hills@ncl.ac.uk*
** *Present address: Dept. of Marine Science and Coastal Management, University of Newcastle upon Tyne, Newcastle upon Tyne, Tyne & Wear, NE1 7RU, UK*

Key words: Settlement, cue, barnacle, scale, *Semibalanus balanoides*

Abstract

Numerous physical and biological factors have been identified which affect the probability of larvae settling on hard substrata. The spatial scale at which these factors operate ranges from km's to sub-mm's. The wide variety of cues that barnacle larvae respond to coupled with the subtleties of cue response to factors like surface roughness, suggests that larvae are fastidious in their choice of settlement sites and thus, (i) settlement is not rapid and, (ii) larvae carry out search behaviour to sample settlement cues. An experimental frame with settlement pits untreated or with either barnacle settlement factor, or cyprid settlement factor, or a squashed cyprid larvae were exposed for a duration of 10 minutes during the *Semibalanus balanoides* settlement season in the Clyde Sea, UK. A total of 102 of the 240 pits were settled within the 10 minutes. More settlement occurred in the chemically treated pits than the untreated pits suggesting that settlement can be both selective and rapid. Video-photography was carried out in the laboratory of the tracks of *S. balanoides* cyprids prior to settlement in pits. With untreated pits little search behaviour was identified, cyprids tended to encounter the pit and then settle. Pits treated with squashed cyprid showed a chemical cue-mediated behaviour with cyprids tending to slow down and carryout antennular crawling in the vicinity of the pit. The mean time from entering a 40 × 40 mm window around the pit and settlement was 24.9 s ($n = 11$, SE = 5.4). Within the last 1.25 s prior to settlement, cyprids settling in untreated pits moved faster than cyprids settling in CL treated pits ($P < 0.01$), with a 4 times difference between the mean speeds These data suggest that settlement can be rapid and the pre-settlement track does not necessarily display search behaviour.

Introduction

The settlement stage of sessile marine invertebrate larvae has consequences for subsequent survival and reproductive success (e.g. Connell, 1985; Walters & Wethey, 1996). Such is the importance of this critical stage of the life-cycle that considerable work has been carried out to determine the factors that affect larval settlement. Numerous factors have been identified which affect the probability, both in a positive and a negative way, of larvae settling on hard substrata. Those factors that affect settlement cover a wide range of scales from km's to sub-mm's (e.g. Raimondi, 1990; Le Tourneux & Bourget, 1988). At a large spatial scale the factors tend to be related to differential larval supply due to oceanic and tidal features (e.g. Raimondi, 1990; Pineda, 1994; Hills & Thomason, 1996; Noda & Nakao, 1996), whereas at smaller scales the factors tend to be related to physical, chemical and biological characteristics of the substratum (e.g. Chabot & Bourget, 1988; Raimondi, 1990; Keough & Raimondi, 1996; Holmes et al. 1997).

Some of the factors that have been shown to affect settlement of barnacles are summarised in Table 1. These factors that have been shown to affect settlement have been arranged in approximate order of decreasing spatial scale over which they act, from km's to < mm's. At approximately the m spatial scale there is a change from factors affecting larval supply i.e. bringing the larval to the vicinity of the settlement substrata, to those factors affecting the probability of settlement of a individual larvae in the vicinity of the surface. The change in generic processes affecting settlement rate over spatial scales have been summarised by, for example, Bourget (1988), Pineda (1994) and Hills & Thomason (1996).

In the field, substratum roughness has been shown to be the dominant influence on settlement density for the balanomorph barnacle *Semibalanus balanoides* (Wethey 1984, though see Holmes et al., 1997). Using settlement panels constructed from the same polyester resin but with different size components creating the substratum roughness, Hills & Thomason (1998a) have shown that higher settlement densities were found on surfaces with fine (<0.5 mm) and medium (<2.0 mm) scales of surface roughness compared to smoother and rougher substrata. At a small scale (<300 μm) Le Tourneux & Bourget (1988) have shown the significant preference of *S. balanoides* cyprids for sites with a high micro-heterogeneity in surface texture. Raimondi (1990) identified small pits or depressions as suitable settlement sites for cyprids of *Chthamalus anisopoma*, and when settlement was high settlement differences between sites in the Gulf of California reflected the differences in the number of suitable settlement sites.

Hills & Thomason (1996) have shown that a standard engineering index (R_a) of surface roughness and fractal dimensions could not explain variation in settlement on a range of natural substrata, as has Holmes et al. (1997) for a variety of polished natural rock surfaces. However, Hills and Thomason (1996) devised an index of the potential settlement sites (PSS) available on substrata, which is related to the size of the *S. balanoides* cyprid and the behaviour of cyprids in response to substratum roughness, which could explain much of the variation in settlement of *S. balanoides* on a range of rock types. Substratum roughness can have more subtle effects than just affecting settlement density. Hills and Thomason (1998b) have shown that substratum roughness affects the settlement pattern within a tile with a more uniform settlement density on textured surfaces compared to smooth surfaces.

It has been suggested that the settlement behaviour of *S. balanoides* occurs in three consecutive phases, 'broad exploration', 'close exploration' and, at the smallest spatial scale 'inspection' (Knight-Jones & Crisp, 1953; Crisp, 1974; Le Tourneux & Bourget, 1988), with estimated spatial scales of these behaviours of $\sim$ 1 m, $\sim$ 1 mm and < 300 μm, respectively (Le Tourneux & Bourget, 1988). At these different behavioural scales, Le Tourneux & Bourget (1988) suggested that larvae are responding to different settlement cues, with a biological cue at the broad exploration phase to avoid lethal heights in the intertidal, sites with clean surfaces for optimal attachment in the close exploration phase and the selection of heterogeneous microsites at the inspection stage. However, Connell (1961) has shown that *S. balanoides* post-settlement survival can be affected by factors such as intraspecific competition, predation, wave action and desiccation which lead to different patterns over tidal height, of settlers and adults.

The wide variety of settlement cues that larvae have been shown to respond to at a hierarchy of spatial scales, coupled with the subtleties of cue response, as identified with substratum roughness above, suggest that larvae are fastidious in their choice of settlement sites. This fastidious behaviour suggests two characteristics of larval settlement:

1. Larval settlement is not rapid as the larvae have to respond to many cues experienced on a substratum at a variety of spatial scales and make choices if they are going to swim off or settle, and if they choose to settle which exact site to select.

2. Larvae carry out a search behaviour which permits sampling of the cues within the potential settlement area.

The aims of this study were to answer the following questions:

1. If suitable settlement sites are presented in the field for a short period of time, are larvae unable to utilise those sites because of lack of time to respond to multiple cues and carry out multi-scale search behaviour or, do larvae settle rapidly and selectively to those sites?

2. Do larvae demonstrate a complex search pattern when suitable settlement sites are presented and do multiple cues affect the search pattern of the larvae?

Table 1. A non-comprehensive table of cues which affect barnacle settlement and the estimated spatial scale over which they operate

Settlement factor	Spatial scale	Species	Reference
Larval supply	km	*Chthamalus anisopoma*	Raimondi (1990)
Larval supply	km	*Semibalanus cariosus*	Noda & Nakao (1996)
Larval supply	km	*Chthamalus* spp.	Pineda (1994)
Larval supply (wind)	m/km	*Semibalanus balanoides*	Bertness et al. (1996)
Vertical zonation	m	*Balanus glandula* & *B. crenatus*	Grosberg (1982)
Larval supply	m	*S. balanoides*	Miron et al. (1995)
Larval supply	m	*S. balanoides*	Minchinton & Scheibling (1991)
Presence of *Urospora wormskjoldii*	m	*S. balanoides*	Le Tourneux & Bourget (1988)
Salinity	mm–m	*B. subalbidus*	Dineen & Hines (1994)
Rock type	cm	*S. balanoides*	Holmes et al. (1997)
Surface type	cm	*B. amphitrite, B. improvisus & B. eberneus*	O'Connor & Richardson (1994)
Texture	cm	*S. balanoides*	Hills & Thomason (1996)
Texture	cm	*Balanus* sp.	Walters & Wethey (1996)
Texture	cm	*C. anisopoma*	Raimondi (1990)
Texture	cm	*S. balanoides, B. crenatus & E. modestus*	Crisp & Barnes (1954)
Colour & illumination	cm	*Balanus* spp.	Dahlem et al. (1984)
animal extracts	cm	*S. balanoides & Elyminius modestus*	Larman & Gabbott (1975)
cracks & grooves	cm	*S. balanoides*	Wethey (1984) Chabot & Bourget (1988) Barnes (1955)
Current through crab gill	mm	*Octolasmis angulata & O. cor*	Voris et al. (1994)
clean surface	mm	*S. balanoides*	Le Tourneux & Bourget (1988)
Adult conspeciics	mm	*B. crenatus*	Miron et al. (1996)
Biofilm	mm	*various*	Keough & Raimondi (1996)
Biofilm	mm		Todd & Keough (1994)
Biofilm	mm	*B. amphritrite*	Maki et al. (1988)
Surface tension	mm	*B. c.f. variegatus*	Becker (1993)
Micro-heterogeneity	< mm	*S. balanoides*	Le Tourneux & Bourget (1988)

Methods

Settlement in response to chemical cues

On the 22nd April, 1996, a settlement frame was deployed from Keppel Pier in the Clyde Sea Scotland (Grid Reference NS177.545) during the *S. balanoides* settlement season. The frame consisted of a 1.5 m length of galvanised Dexion suspended 15 cm below two floating buoys at each end. Onto the frame were placed 6, 110 × 90 mm settlement plates made from 5 mm thick transparent polycarbonate sheet (RS Components Ltd, s/n 681–659). The plates had a 5 mm diameter single hole on two opposite sides of the plate 5 mm from the edge, and electrical tie-wraps were used to attach the plate to the frame by these holes. The plates were attached to the frame 100 mm apart so that they were parallel to the surface of the water.

Each plate had 40 pits (2 mm diameter, 2 mm deep) drilled onto the uppermost surface in regular pattern. This pit size was selected so that only one cyprid was able to settle in each pit. The pits were arranged in a 8 × 5 grid with 1 cm between pits in both rows and columns. Four treatments were used in the holes:

1. Barnacle settlement factor (BSF). This was made up following the method in Rittschof et al. (1984) except that 100 g of adult *S. balanoides* was used and BSF was made up on the morning of deployment of the experiment rather than frozen. One drop of BSF was applied to a pit using a mounting needle.

2. Cyprid settlement factor (CSF). A plankton pump was set up on Keppel Pier pumping water through a medium mesh (335 μm aperture) plankton net in a dustbin on the pier so that the water flowed out of the dustbin but the larvae were retained. The pump was left on for 24 hours pumping approximately 40 litres of water per minute. The catch in the net was strongly dominated by *S. balanoides* cyprids. The cyprids in the plankton net were removed on the morning of the experiment and used to make cyprid settlement factor. The CSF was made in a similar way to BSF on the morning of the experiment and applied using a mounting needle.

3. Cyprid larvae (CL). A single settled but unmetamorphosed cyprid larvae was removed from its settlement position with a pair of forceps and squashed in a pit on the settlement plates. The body of the cyprid was removed.

4. Untreated.

Treatments were added to the pits one hour prior to deployment in a randomised block design and allowed to air dry. Each row of eight pits was divided into two blocks of four pits, and one of each of the treatment types was added to these four pits in a randomised manner. The entire grid, thus, consisted of two blocks of four pits with a single replicate of each treatment forming a row, with a total of five rows. Each plate had a total of 10 pits of each treatment and 6 plates were used making a total of 60 pits of each of the 4 treatments.

The experiment was deployed from the pier for a total of 10 minutes. The frame was attached to the pier by a 6 m length of rope. Sea conditions were relatively rough during this experiment with an estimated force 5 S to SW winds, to which the pier is exposed. The frame experienced considerable turbulence in these conditions with the plates rotating in relation to the sea surface up to an estimated 45° and the whole frame tilting as the flotation buoys were lifted by incoming waves. After 10 minutes the experiment was taken out of the water and the number of pits with a cyprid settler was counted. Settlers were considered to be cyprids that were attached in a pit after the frame was dragged through the water back to the pier and lifted up on the pier.

Video analysis of settlement behaviour

Cyprids were filmed prior to settlement to determine the behaviour of the larva in the vicinity of the settlement site prior to settlement. A glass tank was set up (150 mm long × 50 mm wide × 100 mm deep) and filled with freshly collected Clyde sea water to within 10 mm of the top of the tank. The tank was kept in a cold room (9 °C, ±2 °C).

A set of roofing tiles (120 mm × 120 mm × 12 mm) were drilled with two holes (2 mm diameter, 2 mm deep) 20 mm apart in the centre of the tile. The pits were slightly bigger than the drill size as the slate tended to flake away around the edge of the drill bit. A set of chemical treatments were used on the slate tiles:

1. Barnacle settlement factor.
2. Cyprid larvae
3. Untreated.

Treatments were manufactured and applied in the same way as above.

Two types of trials were carried out in these experiments, with the following treatments in the two pits:

(a) CL and no treatment.
(b) CL and BSF.

Treatments were randomly arranged within each pair of pits for each trial. Each experimental comparison was carried out 9 times making a total of 18 trials.

Each trial consisted of filming the behaviour of the cyprids near the holes following immersion of the pits. A Hi8 camcorder (Canon EX1Hi) was placed approximately 50 mm away from the side of the tank and focused to give a field of view of approximately 60 mm × 40 mm. The tile was placed inside the glass tank 10 mm away from the front of the tank and the camcorder accurately focused on the tile surface. A white light source was set up using two 100 watt bulbs over the top of the tile shining at an acute angle to the front of the tile surface. A total of 150 cyprids which had been caught from Keppel Pier with a plankton pump (described above) within the last 6 hours were added to the tank using a pipette. Just prior to the addition of the cyprids the camcorder was set to record. The behaviour of the cyprids was recorded for 30 minutes following addition. The seawater was replaced between trials and a new tile and new cyprids were used for each trial.

These videos were analysed for a number of features of cyprid behaviour. Observation of the cyprids during the experiment showed that a number of cyprids might enter a pit and spend from a few seconds up to about a minute in the pit and then swim off. As these individuals had not settled in the pit they were not considered to be settlers. Settlers were considered to be individuals which remained for more than 5 minutes in the pit. No settler which remained for at least five minutes was seen to leave the pit within the remaining time to the end of the trial. Each 30 minute trial was analysed for a period of 25 minutes commencing at the addition of the cyprids to the tank. The last 5 minutes of the trial was to make sure that if a cyprid had entered a pit at 24.59 minutes of the trial there was still at least 5 minutes of tape remaining to determine if the cyprid had actually settled.

The number of pits of each treatment which were settled at the end of the 25 minutes was recorded. The time taken by each settling cyprid from the start of the trial to first entering the pit was determined. In addition, the track of each settling cyprid was recorded. A transparent acetate sheet was put over a monitor screen and an area 20 mm above and below, and 10 mm either side of the two pits was delimited making a window 40 mm horizontally and 40 mm vertically. The position of the anterior end of each settling cyprid was recorded every 0.25 of a second (i.e. every 6th frame) from when it entered this window to when it first attached to the pit. The sequential set of points over time formed the track of the cyprid prior to settlement. The x–y co-ordinates of the position of the cyprid within the 40 × 40 mm window were determined from the acetate sheet (±1 mm) The last 1.25 seconds of movement of each settling cyprid were analysed, using a macro in Minitab v. 10, with a set of standard measures of behaviour taken from (Varley et al. 1993):

(i) Mean speed. The mean speed of the cyprid (mm s^{-1}).

(ii) Mean turn angle. The angular deviation from previous direction, both clockwise and anti-clock wise were treated a positive, so values ranged from 0° (maintain same direction as previous track) to 180° (opposite direction of previous track).

(iii) Index of straightness. Calculated by dividing the beeline (or, 'as the crow flies') by the actual length of the path, thus 1 represents a straight line and values close to 0 show a highly convoluted track. Problems with the index related to values for circular tracks do not occur in this application because all settling cyprids move from the edge of the video window to a central settlement pit.

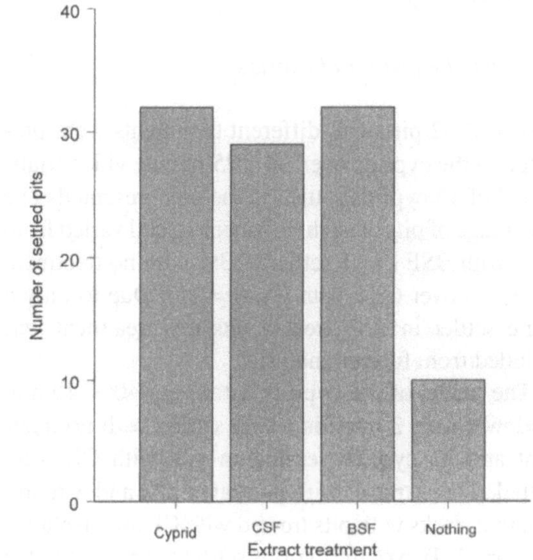

Figure 1. The number of pits settled treated with a squashed cyprid, cyprid settlement factor (CSF), barnacle settlement factor (BSF) and no chemical treatment, by cyprids after exposure of the frame for 10 minutes in April, 1996 ($\chi^2 = 13.05$, $P < 0.01$, df = 3).

Results

Settlement in response to chemical cues

A total of 102 of the 240 pits were settled within the 10 minutes of the duration of the experiment. Chi-squared analysis was used to determine if the number of pits settled between each treatment was random i.e. treatment had no effect on settlement. The chi-squared analysis showed that the number of pits settled by treatment was different to random ($P<0.01$, $\chi^2 = 13.05$, df = 3). A plot of the results by treatment (Figure 1) shows that the lowest settlement was in the untreated pits. Settlement in the three types of chemically treated pits was similar (Figure 1) with approximately 50% of the pits being settled in the 10 minutes duration of the experiment. These results suggest that both barnacle and cyprid settlement factor and squashed cyprid larvae have a similar attraction to the cyprids in the field. These results also suggest that settlement can be rapid with high amounts of settlement within a matter of minutes and that chemicals

(BSF, CSF and CL) can increase the number of settlers compared to untreated pits.

Settlement behaviour of cyprids

A total of 32 pits with different treatments were presented to the cyprids over 18 × 25 minute video trials. A total of 15 cyprids settled in the pits presented. The percentage of pits of each treatment settled varied from ~7% with BSF ($n = 1$ settler), 33% with no treatment ($n = 3$), to over 60% with CL ($n = 11$). Due to only a single settler in BSF treated pits this treatment was excluded from further analysis.

The tracks of the cyprids within the 40 × 40 mm window on the 3 cyprids which settled with no treatment and 11 cyprids settling in pits with CL were plotted. Three tracks with untreated pits and 3 representative tracks with pits treated with CL are displayed in Figure 2. Two of the cyprids settling in the untreated pits (Figure 2a & b) display tracks which are characterised by a constant upward movement. Once the pit was encountered the cyprids stopped and then settled in the pit. No searching behaviour in the area of the pit was apparent (Figure 2a & b). The third cyprid settling in the untreated pits showed the typical slow sinking and relatively fast upward swimming behaviour. The cyprid encountered the pit while sinking and then actively swam to approximately 10 mm above the pit and then moved back to the pit; the time to settlement following the initial experience of the pit was under 2 seconds.

Cyprids settling in the CL treated pits tended to show more changes in direction prior to settlement. One cyprid (Figure 2d) displayed slow sinking behaviour which was interspersed by short periods of multi-directional active swimming. Once close to the treated pit the cyprid carried out a small three-quarter turn at the rim of the pit and then settled. In the second trial the cyprid swam rapidly to within 7 mm of the pit and then proceeded to move slowly towards the pit over the surface in a fairly straight direction until it came in contact with the pit where it settled. In the next trial the cyprid settling in the CL treated pit showed a set of directional changes from 14 mm above the pit which defined a track to the pit (Figure 2f.).

The time from when the cyprid appeared within the video window to the time at which the cyprid settled in the pit was determined for the 11 settlers in the CL treated pits. The mean time to settlement was 24.9 s ($n = 11$, SE = 5.4) with a range from 2.50 s to 60.75 s. These data show that settlement was rapid from when the cyprid arrived in the video window to when it reached the pit, with the maximum time being just slightly over 1 minute.

Table 2. The mean speed, turn angle and index of straightness (and SE in brackets) for the CL treated pits ($n = 11$) and the untreated pits for the 1.25 seconds prior to settlement

Behvioural index	CL treated pits	Untreated pits
Mean speed (mm s^{-1})	4.1 (0.5)	16.4 (5.0)
Mean turn angle	47.5 (7.7)	33.6 (6.1)
Index of straightness	0.69 (0.29)	0.69 (0.09)

Mean values for the calculated behaviour indices were calculated for the 11 cyprids settling in the CL treated pits and the 3 cyprids in the untreated pits (Table 2). To test the difference between these behavioural indices the values were compared using an unbalanced, non-parametric, one-tailed Mann-Whitney tests (Minitab v. 10). Three hypotheses, generated from assessment of the graphs of the tracks (Figure 2.), were tested, namely, that the mean speed and index of straightness were greater in the tracks of the cyprid settling in the untreated compared to the CL treated pits, and that the mean turn angle was less in the untreated than the CL treated tracks.

The tracks of the cyprids settling in the untreated pits showed no greater index of straightness and no less mean turn angle than the cyprids settling in CL treated pits. However, cyprids settling in untreated pits moved faster than cyprids settling in CL treated pits ($P < 0.01$, df = 10, 2), with a 4 times difference between the mean values (Table 2.). These analyses support the descriptions of the tracks showing that the cyprids in the CL treated pits crawled over the surface in a searching pattern in the vicinity of the pit, whereas the cyprids settling in the untreated pits tend to encounter the pits by swimming.

Discussion

Rapid settlement in response to chemical cues

Work has clearly shown that cyprids respond to chemical cues such as arthropodin in BSF (e.g. Larman & Gabbot, 1975; Yule & Crisp, 1983; Rittschof et al., 1984) and the cyprid antennular secretion settlement pheromone (Walker & Yule, 1984; Yule & Walker, 1985; Clare et al., 1994). Chemically treated surfaces

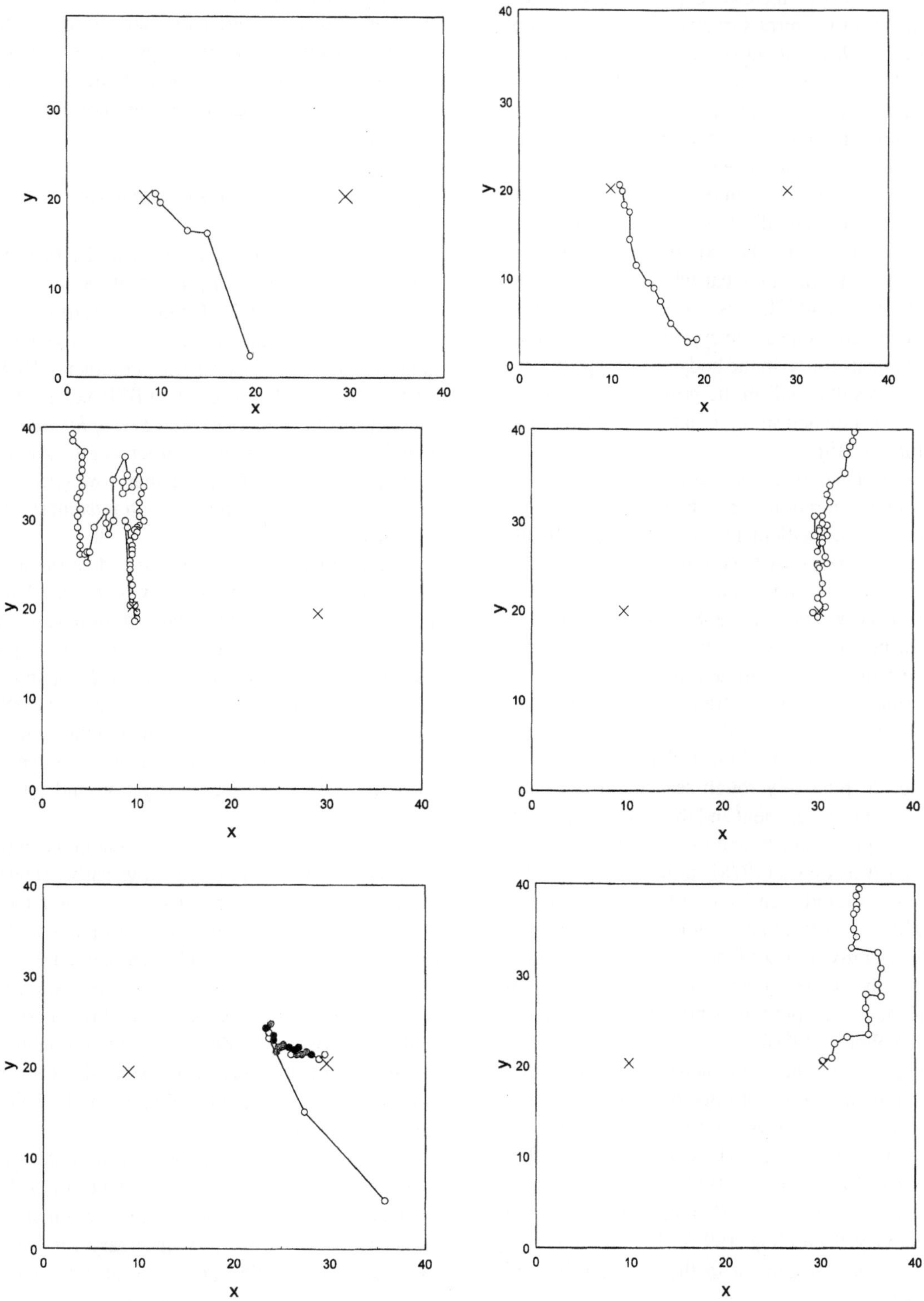

Figure 2. The tracks of the cyprid prior to settlement in the pits (marked with an X) within the 40 × 40 mm video window; each point represents 0.25 of a second. Treatment of pits: a–c, no chemical treatment; d–f squashed cyprid treatment. The shaded points in Figure 2e represent periods when the cyprid was stationary (light shading = stationary for 0.25–1 s; heavy shading > 1 s).

have been shown to increase cyprid settlement density compared to control surfaces (e.g. BSF: Larman & Gabbot, 1975; antennular secretion: Clare et al., 1994) and to change behaviour (Rittschof et al., 1984). Chemical cues emanating from adult conspecifics can play an important role as a cue for barnacle settlement in the field (Miron et al., 1996).

In the experiment presented above BSF, CSF and CL all showed a significant positive effect on settlement by the end of the experiment compared to untreated pits. It was not clear whether the attraction caused by CSF and CL was due to presence of the antennular pheromone of the cyprid, or to one or more other chemicals found in the body of the cyprid. However, these results confirm the presence of a cyprid – cyprid interaction and suggest that the attractive effect was similar to BSF.

Settlement occurred in nearly half of the pits within the 10 minute duration experiment described above. This suggests that settlement can be rapid (< 10 min). Presumably, in response to certain cues, the larvae can either carry out the three stage settlement behaviour (Le Tourneux & Bourget, 1988) rapidly, or miss out part of the behaviour, and the stronger the cue becomes the more rapid or more limited the settlement behaviour becomes. Such rapid settlement behaviour would seem to negate the possibility of fastidious searching behaviour. In light of the turbulence around the frame due to relatively rough sea conditions occurring during this experiment and the limited swimming ability of cyprids, attachment to the pit would have to occur rapidly as Lewis (1978) has suggested. If settlement was not rapid then local turbulence might limit any further attachment and would remove the cyprid from the vicinity. Keppel Pier, where this experiment took place, has a mean current speed of between 0.08 and 0.18 ms^{-1} in a predominantly unidirectional flow (Moore & Wong, 1995).

Changes in numbers of attached larvae in the field have commonly been considered in the range of a number of hours to days, and data are described at scales such as tidally (e.g. Hills & Thomason, 1996), nightly (e.g. Holmes et al., 1997), daily (e.g. Pineda, 1994 and Bertness et al., 1996), or, periodically (e.g. Minchinton & Scheibling, 1991). However, this study has shown that attachment in the field can be rapid and selective and that for some field experiments, particularly in relation to water soluble chemical cues, short experimental durations could be appropriate. In addition, short experimental durations limit confounding changes in surface characteristics which can occur rapidly (< 24 hrs; Rittschof et al., 1984), limit the consequences of gregarious settlement at higher settlement densities (Hui & Moyse, 1987; Miron et al., 1996) and permit settlement, rather than recruitment (settlement + post-settlement mortality), to be quantified.

Settlement behaviour prior to settlement

The reason for the low settlement in the BSF treated pits ($\sim 7\%$) was not clear, as the literature (e.g. Larman & Gabbot, 1975; Gabbot & Larman, 1987) and the field experiment carried out described above suggest that BSF can promote settlement. However, Rittschof et al. (1984) found that BSF had no effect on the settlement of older cultured *B. amphitrite* cyprids, though, this was due to high settlement in experiments with and without BSF, rather than low settlement in both. The reasons for the low attractiveness of BSF remain unclear.

The video trial was carried out to determine the behaviour in the cm – scale vicinity of the pit. Although this spatial scale includes the ~ 1 mm scale of the close exploration behaviour and the <300 inspection behaviour identified by Knight-Jones & Crisp (1953), Crisp (1974) and Le Tourneux & Bourget (1988), the position of the cyprid was recorded to the nearest mm in the present study and this was at too large a spatial scale to assess close exploration and inspection behaviour.

In general, the videos of settlement in the untreated pit showed that once the cyprid physically encountered the pit it stayed there rather than searching in the close vicinity of the pit. The attraction of pits as settlement sites to cyprids has been well documented (e.g. Crisp & Barnes, 1954; Raimondi, 1990). However, in this case, the pit might have had an additional attraction in that the base and sides of the pits probably had a higher micro-heterogeneity of texture (Le Tourneux & Bourget, 1988) due to the action of the drill bit while forming the holes.

Little search behaviour around the untreated pits by cyprids was identified. It appeared from the limited number of trials that ended in settlement that a direct encounter with the pit was necessary for the cyprid to settle. No searching behaviour at a distance from the pit was identified. The cyprid was thus responding to the pit as a cue at a very small distance away. It could be possible that a direct encounter was not required and that the changes in surface contour could be detected at a small distance away from the pit. Yen

& Strickler (1996) have identified a stable hydrodynamic envelope around *Euchaeta rimana*, a pelagic marine copepod, within which it can detect disturbances from predators and prey in time and space. Cyprids could possibly also be enclosed in a similar hydrodynamically stable envelope, permitting identification of surface concavities, or identification of boundary layers caused by surface roughness or conspecifics (Thomason et al., 1998) at a small distance. This hypothesis might be suitable for relatively still conditions, such as the trials considered here, but less appropriate to a wave-swept rocky shore.

When the pit was treated with CL a modified behaviour was apparent (Figure 2d–e) prior to encountering the pit. This suggests that there was some water-borne chemical attractant some distance from the pit. Weissburg & Zimmer-Faust (1991) have shown that crabs use chemo-reception to locate prey and that they 'tack' upstream in a manner similar to moths in pheromone plumes. A similar effect of a water-borne cue was not apparent in the experiments with cyprids carried out in the work presented here as the chemical cue had no effect on the turn angle of the cyprids track (Table 2). However, once the chemical was detected the cyprid moves slower in the vicinity of the pit using antennular walking over the surface. Such behaviour would tend to lead to a direct encounter with the source of the chemical which would be important in light of the limited amount of antennular secretions which permit antennular walking over the surface. Walker & Yule (1984) have estimated that the total volume of antennular secretion for a cyprid is in the order of 9425 μm^3 which permits a total of 558 antennule paces, making a track distance of 42 cm. Thus, the cyprids have limited potential for walking over a surface making directional behaviour towards a cue imperative.

Such chemical cue mediated settlement behaviour has been identified in cultured *B. amphitirite*. Rittschof et al. (1984) identified 4 distinct behaviours of cyprids in a assay tube with flowing water; swimmers, tumblers, sliders and stickers, and an intermediate group that carried out some behaviour in addition to swimming. When the assay tube was untreated a high percentage of the cyprids (60–90%) responded by swimming or intermediate behaviour. Whereas when the assay tube was coated in BSF, 35–70% of the cyprids responded by sticking (Rittschof et al., 1984). However, there is a difference between a surface coated with BSF as used by Rittschof et al. (1984) and the point source used in the experiments presented here. A coated surface would be expected to act in a plane with attraction of the surface being in a single direction (i.e. attraction towards the surface), whereas a point source would operate both towards the surface (z axis) and on the surface plane (x & y axes). However, both types of chemical cues have identified cue-mediated behavioural responses.

The distance over which the physical presence of the pit operates as a settlement cue would be small and possibly only apparent to the cyprid once it had been encountered. However, the application of chemical to the pit acts as a further attraction for the cyprid at a greater spatial scale. Thus the chemical could entrain the cyprid at a distance of a few cm and cause the cyprid to slow down and crawl over the surface using antennular walking. This effect of increasing the scale of the settlement cue could explain the results from the experiment carried out in the field. The low settlement percentage of cyprids in untreated pits was due to the limited probability of a direct encounter with the pit by a cyprid. The greater numbers of treated pits settled (by BSF, CSF and CL) could be due to entrainment of larvae at some distance away from the pit, guiding the cyprid towards the pit.

The work presented here in relating behaviour to settlement cues has taken a different approach to other workers in using actual behavioural traces of larvae to attempt to identify the reasons for variations in settlement patterns due to different characteristics of surfaces. This contrasts with other workers who have attempted to identify behaviour from within-plate (e.g. Mullineaux & Butman, 1991) and between-plate settlement patterns (e.g. Mullineaux & Garland, 1993) and relating variation in cues at different spatial scales to variation in settlement densities (e.g. Le Tourneux & Bourget, 1988). This video work presented here, although using fresh wild larvae, was carried out in a laboratory with conditions similar to slack water on a calm and sunny day. Work is presently underway by the authors filming settling cyprids in the field using underwater video-photography.

The work presented in this paper has shown that:

1. Cyprid settlement factor and squashed cyprids can be as powerful attractant to settling cyprids as barnacle settlement factor.

2. Settlement can occur over a matter of minutes in the field and even at these short time scales cyprids are selective in where they settle.

3. Video analysis showed that cyprids can approach a pit from 2 cm away and settle within a matter of seconds.

4. The track of the cyprids with no chemical treatment in the pit showed that the cyprids moved directly into the pit and settled in the pit once it was encountered. However, with squashed cyprid larvae treatment in the pits, the cyprids tended to slow down and caryout out antennular walking in the vicinity of the pit.

Acknowledgements

The camcorder used in this study was purchased through a National Environmental Research Council small grant GR9/1905. JMH was supported by the Sheina Marshall Fellowship and The Royal Society.

References

Barnes, H., 1955. Further observations on rugophilic behaviour in *Balanus balanoides* (L.). Vidensk. Medd. fra Dansk naturh. Foren. 117: 341–348.

Becker, K., 1993. Attachment strength and colonization patterns of two macrofouling species on substrata with different surface tension (in situ studies). Mar. Biol. 117: 301–309.

Bertness, M. D., S. D. Gaines & R. A. Wahle, 1996. Wind-driven settlement patterns in the acorn barnacle *Semibalanus balanoides*. Mar. Ecol. Prog. Ser. 137: 103–110.

Bourget, E., 1988. Barnacle larval settlement: The perception of cues at different spatial scales. In M. Vannini & M. Chelazzi (eds), Behavioural adaptations to the intertidal life. Plenum Press, NY.: 153–172.

Caffey, H. M., 1982. No effect of naturally-occuring rock types on settlement or survival inthe intertidal barnacle *Tesseropora rosea* (Krauss). J. exp. mar. Biol. Ecol. 63: 119–132.

Chabot, R. & E. Bourget, 1988. Influence of substratum heterogeneity and settled barnacle density on the settlement cypris larvae. Mar. Biol. 97: 45–56.

Clare, A. S., R. K. Freet & M. McClary, 1994. On the antennular secretion of the cyprid of *Balanus amphitrite amphitrite*, and its role as a settlement pheromone. J. mar. biol. Ass. U.K. 74: 243–250.

Connell, J. H., 1961. The influence of competition predtaion by *Thais lapillus* and other factors on natural populations of the barnacle *Balanus balanoides*. Ecol. Monogr. 31: 61–104.

Connell, J. H., 1985. The consequences of variation in initial settlement vs post-settlement mortality in rocky intertidal communities. J. exp. mar. Biol. Ecol. 93: 11–45.

Crisp, D. J., 1974. Factors influencing the settlement of marine invertebrate larvae. In P. T. Grant & A. M. Mackie (eds), Chemoreception in marine organisms. Academic Press, London: 177–265.

Crisp, D. J. & H. Barnes, 1954. The orientation and distribution of barnacles at settlement with particular reference to surface culture. J. anim. Ecol. 23: 142–162.

Dahlem, C., P. J. Moran & T. R. Grants, 1984. Larval settlement of marine sessile invertebrates on surfaces of different colour and position. Ocean Sci. & Eng. 9: 225–236.

Dineen, J. F. & A. H. Himes, 1994. Effects of salinity and adult extract on settlement of the oligohaline barnacle *Balanus subalbidus*. Mar. Biol. 119: 423–430.

Gabbot, P. A. & V. N. Larman, 1987. The chemical basis of gregariousness in cirripedes: A review (1953–1984). In A. J. Southward (ed.), Barnacle biology. A. A. Balkema, Rotterdam: 377–388.

Grosberg, R. K., 1982. Intertidal zonation of barnacles: the influence of planktonic zonation of larvae on vertical distribution of adults. Ecology 63: 894–899.

Hills, J. M. & J. C. Thomason, 1996. A multi-scale analysis of settlement density and pattern dynamics of the barnacle *Semibalanus balanoides*. Mar. Ecol. Prog. Ser. 138: 103–115.

Hills, J. M. & Thomason, J. C., 1998b. On the effect of size, surface area and larval behaviour on recruitment pattern and density of the barnacle (*Semibalanus balanoides*). Biofouling (in press).

Hills, J. M. & J. C. Thomason, 1998a. The effect of scales of surface roughness on the settlement of barnacle (*Semibalanus balanoides*) cyprids. Biofouling 12: 57–69.

Holmes, S. P., C. J. Sturgess & M. S. Davies, 1997. the effect of rock-type on the settlement of Balanus balanoides (L.) cyprids. Biofouling 11: 137–147.

Hui, E. & J. Moyse, 1987. Settlement patterns and competition for space. In A. J. Southward (ed.), Barnacle biology. A.A. Balkema, Rotterdam: 363–376.

Keough, M. J. & P. T. Raimondi, 1996. Responses of settling larvae to bioorganic films: Effects of large-scale variation in films. J. exp. mar. Biol. Ecol. 207: 59–78.

Knight-Jones, E. W. & D. J. Crisp, 1953. Gregariousness in barnacles in relation to the fouling of ships and to antifouling research. Nature 171: 1109–1110.

Larman, V. N. & P. A. Gabbot, 1975. Settlement of cyprid larvae of Balanus balanoides and Elminius modestus induced by extracts of adult barnacles and other marine animals. J. mar. biol. Ass. U.K. 55: 183–190.

Lewis, C. A., 1978. A review of substratum selection in free-living and symbiotic cirripeds. In F. S. Chia & M. E. Rice (eds), Settlement and metamorphosis of marine invertebrate larvae. Elsevier, NY. 207–218.

Le Tourneux, F. & E. Bourget, 1988. Importance of physical and biological settlement cues used at different sptial scales by the larvae of *Semibalanus balanoides*. Mar. Biol. 97: 57–66.

Maki, J. S., D. Rittschof, J. D. Costlow & R. Mitchell, 1988. Inhibition of attachment of larval barnacles, Balanus amphritrite, by bacterial surface films. Mar. Biol. 97: 199–206.

Minchinton, T. E. & R. E. Scheibling, 1991. The influence of larval supply and settlement on the population structure of barnacles. Ecology 72: 1867–1879.

Miron, G., B. Boudreau & E. Bourget, 1995. Use of larval supply in benthic ecology: testing correlations between larval supply and larval settlement. Mar. Ecol. Prog. Ser. 124: 301–305.

Miron, G., E. Bourget & P. Archambault, 1996. Scale of observation and distribution of adult conspecifics: their influence in assessing passive and active settlement mechanisms in the barnacle *Balanus crenatus* (Brugière). J. exp. mar. Biol. Ecol. 201: 137–158.

Moore, P. G. & Y. M. Wong, 1995. Activity and trapping characteristics in the field of *Orchomene nanus* (Krøyer) (Amphipoda: Lysianassoidea) at Millport, Scotland. J. exp. mar. Biol. Ecol 189: 143–157.

Mullineaux, L. S. & C. A. Butman, 1991. Initial contact, exploration and attachment of barnacle (*Balanus amphitrite*) cyprids settling in flow. Mar. Biol. 110: 93–103.

Mullineaux, L. S. & E. D. Garland, 1993. Larval recruitment in response to manipulated field flows. Mar. Biol. 116: 667–683.

Noda, T. & S. Nakao, 1996. Multi-scale spatial pattern of recruitment in the barnacles *Semibalanus cariosus* on the Kameda peninsula, southern Hakkaido, Japan. Hydrobiologia 324: 125–130.

O'Connor, N. J. & D. L. Richardson, 1994. Comparative attachment of barnacle cyprids (*Balanus amphitrite* Darwin 1854; *B. improvisus* Darwin 1854; & *B. eburneus* Gould 1841) to polystryene and glass substrata. J. exp. mar. Biol. Ecol. 183: 213–225.

Pineda, J., 1994. Spatial and temporal patterns in barnacle settlement rate along a southern California rocky shore. Mar. Ecol. Prog. Ser. 107: 125–138.

Raimondi, P. T., 1990. Patterns, mechanisms, consequences of variability in settlement and recruitment of an intertidal barnacle. Ecol. Monog. 60: 283–309.

Rittschof, D., E. S. Branscomb & J. D. Costlow, 1984. Settlement and behaviour in relation to flow and surface in larval barnacles, *Balanus amphitrite* Darwin. J. exp. mar. Biol. Ecol. 82: 131–146.

Thomason, J. C., J. M. Hills, A. S. Clare, A. Neville & M. Richardson, 1998. Hydrodynamic consequences of barnacle colonisation on hard substrata. Hydrobiol. (this volume).

Varley, M. J., M. J. W. Copland, S. D. Wratten & M. H. Bowie, 1993. Parasites and predators. In S. D. Wratten (ed.), Video techniques in animal ecology and behaviour. Chapman & Hall, London: 33–63.

Voris, H. K., W. B. Jefferies & S. Poovachiranon, 1994. Patterns of distribution of two barnacle species on the mangrove crab, *Scylla serrata*. Biol. Bull. 187: 346–354.

Walker, G. & A. B. Yule, 1984. Temporary adhesion of the barnacle cyprid: the existence of an antennular adhesive secretion. J. mar. biol. Ass. U.K. 64: 679–686.

Walters, L. J. & D. S. Wethey, 1996. Settlement and early post-settlement survival of sessile invertebrates on topographically complex surfaces: the importance of refuge dimensions and adult morphology. Mar. Biol. Prog. Ser. 137: 161–171.

Weissburg, M. & R. K. Zimmer-Faust, 1991. Effects of chemoreception and hydrodynamics in mediating predation. Am. Zool. 31: 103.

Wethey, D. S., 1984. Spatial pattern in barnacle settlement: day to day changes during the settlement season. J. Mar. Biol. Ass. UK 64: 687–698.

Yen, J. & J. R. Strickler, 1996. Advertisment and concealment in the plankton – what makes a copepod hydrodynamically conspicuous. Invert. Biol. 115: 191–205.

Yule, A. B. & D. J. Crisp, 1983. Adhesion of cypris larvae of the barnacle Balanus balanoides, to clean and arthropodin treated surfaces. J. mar. biol. Ass. U.K. 63: 261–271.

Yule, A. B. & G. Walker, 1985. Settlement of *Balanus balanoides*: the effect of cyprid antennular secretion. J. mar. biol. Ass. U.K. 65: 707–712.

Hydrobiologia **375/376:** 113–123, 1998.
S. Baden, L. Pihl, R. Rosenberg, J.-O. Strömberg, I. Svane & P. Tiselius (eds),
Recruitment, Colonization and Physical–Chemical Forcing in Marine Biological Systems.

Local control of recruitment in an epifaunal community and the consequences to colonization processes

Richard W. Osman[1] & Robert B. Whitlatch[2]
[1] *Academy of Natural Sciences, Estuarine Research Center, 10545 Mackall Road, St. Leonard, MD 20685, U.S.A.*
[2] *University of Connecticut, Department of Marine Science, Groton, CT 06340, U.S.A.*

Key words: predation, settlement, *Mitrella lunata*, *Anachis avara*, *Anachis lafresnayi*, *Botryllus schlosseri*, *Botrylloides diegensis*, bryozoans, subtidal

Abstract

We found that recruitment, abundance, and dominance within two subtidal epifaunal communities in southern New England, USA persist year after year over large areas of the bottom. This long-term persistence in both dominance and recruitment is not expected in such an open system with disturbances continually creating open patches for recruiting larvae whose identity and abundances change both temporally and spatially. We suggest that the persistence results from strong local control of recruitment that overrides any variability in larval production and dispersal of species from outside a site. Although local dynamics that control persistence involve all life-stages, we found that intense predation on post-settlement individuals has drastic effects. This predation alters the relative abundances of recruits, prevents the invasion of some species, and allows others to dominate. In addition, epifaunal communities are often dominated by species producing short-lived, poorly dispersed larvae. The continued local recruitment of these species at a given site can contribute to the long-term persistence of dominants already present. Based on these observations, we suggest that a system of locally reproducing, self-sustaining populations coupled with strong local environmental differences (e.g. predation on recruits) limiting the invasion of other species may better represent some subtidal benthic communities than a system with widely-dispersed larvae, recruitment dominated by production outside the community, disturbance creating continual changes in dominance, and little long-term persistence.

Introduction

Our understanding of the processes regulating marine benthic communities has been strongly influenced by a myriad of studies of rocky intertidal habitats. These communities are generally dominated by (1) a harsh physical environment which causes distinct vertical gradients in distribution (e.g. Stephenson, 1943; Lewis, 1961) or periodic disturbances of the community (e.g. Dayton, 1971; Sousa, 1979a, b, 1980), (2) strong competitive and trophic relationships controlled by one or a few key species (e.g. Connell, 1961a, b; Paine, 1966; Dayton, 1971; Underwood, 1980), and (3) recruitment which is variable and often a function of large-scale hydrographic phenomena (e.g. Shanks, 1983; Wethey, 1986; Roughgarden et al., 1988; Farrell et al., 1991; Alexander & Roughgarden, 1996). Intertidal communities conforming to this paradigm are open systems in the sense that local dynamics of disturbance and strong interactions among adults are not tightly coupled to the production and subsequent supply of new recruits. Rocky subtidal benthic communities, although generally more diverse and complex than adjacent intertidal communities, seem likely to be governed by similar processes. If, as Sole-Cava et al. (1994) have suggested, larval dispersal of subtidal species is increased over intertidal species as a consequence of their ability to survive at localities intermediate between two distant sites, then subtidal communities might be expected to be more open than their intertidal counterparts.

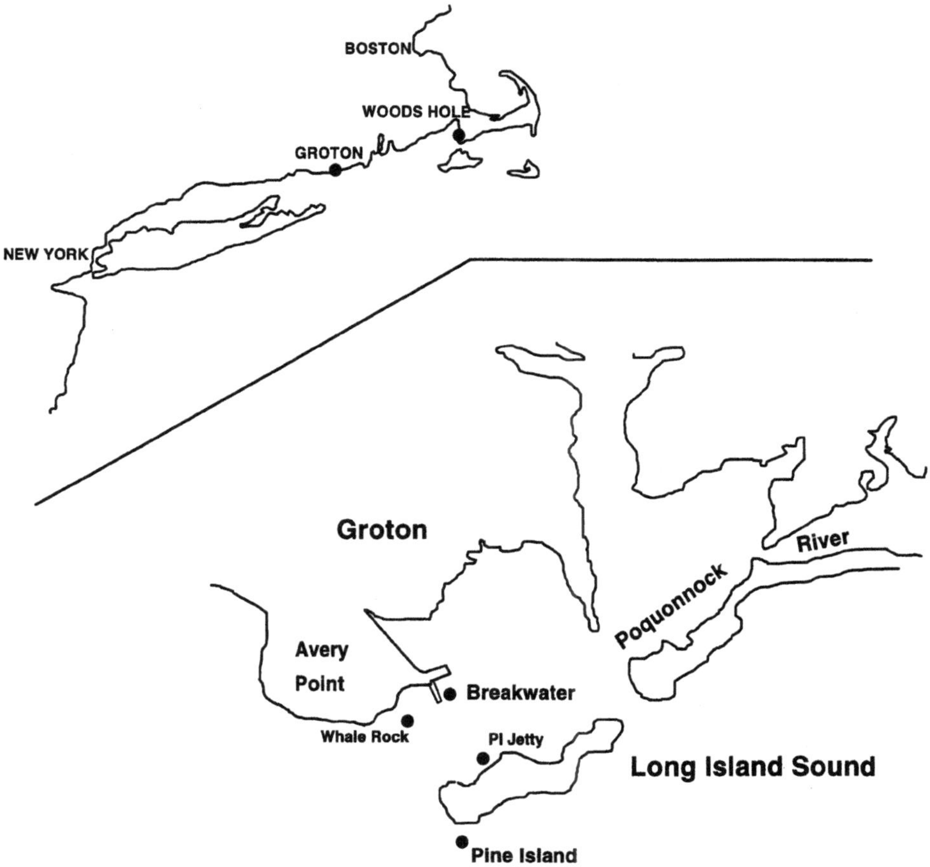

Figure 1. Map of the Groton, Connecticut, U.S.A. study sites, including the main sites at the Avery Point breakwater and Pine Island as well as the sampling sites at Whale Rock and the Pine Island Jetty.

However, there have been a significant number of empirical studies that have demonstrated or suggested that recruitment can be localized within a site or in the proximity of parental populations and that population and community dynamics can be viewed as those of a closed system. These include: (1) the limited dispersal of some algal spores (Dayton, 1973; Deysher & Norton, 1982; Dayton et al., 1984), some lecithotrophic ascidian (Olson, 1985; Grosberg, 1987; Davis & Butler, 1989; Bingham & Young, 1991; Stoner, 1992) and bryozoan (Keough, 1989) larvae, and some planktonic (Carlon & Olson, 1993) and crawling (Gerrodette, 1981) larvae of corals, (2) colony fission (Hughes & Jackson, 1980), fragmentation (Bak & Engel, 1979; Hughes & Jackson, 1980, 1985; Stoner, 1989; Lasker, 1990), and stolonal growth (Karlson et al., 1996) resulting in asexual recruitment, (3) the localized recruitment of corals on the Great Barrier Reef (Sammarco, 1983; Sammarco & Andrews, 1988), (4) the production of brooding larvae (Levin, 1984; Commito et al., 1995), benthic egg strands (Petersen & Svane, 1995), or egg masses (Bhaud & Duchene, 1996) that can localize recruitment, and (5) local hydrodynamics or entrainment of planktotrophic larvae (McShane et al., 1988; Havenhand & Svane, 1991; Bertness & Gaines, 1993; Petersen & Svane, 1995). In addition, post-settlement mortality from grazers (Vadas et al., 1992), predators (Sammarco, 1980; Sammarco & Carleton, 1981; Osman et al., 1990, 1992; Osman & Whitlatch, 1995, 1996; Carr & Hixon, 1995), and environmental factors (Vadas et al., 1992; Osman & Abbe, 1994) can also limit or prevent recruitment, regardless of the source of larvae, their distribution, or their rate of settlement. Given the breadth of phyla, life-histories, habitats, communities, and geographic regions represented by these studies, they suggest that different dynamics exist for some benthic communities which can not only result in local recruitment but also a long-term persistence in communities and populations that result from local processes dominating all aspects of species' life-histories.

We have also been finding unexpected long-term persistence and regularity in species abundance *and* recruitment within subtidal communities (Osman et al., 1992; Osman & Whitlatch, 1995, 1996). In over 10 y of study, recruitment (but not settlement) and species dominance have remained fairly constant at each field site while differing dramatically among sites. Cobble to boulder-sized substrates exposed at a site consistently developed species assemblages characteristic of the site. This occurred despite the close proximity of the sites, and that all species found at both sites could survive and reproduce at each site.

We observed these patterns in two different regions within southern New England, U.S.A. (Figure 1). In Woods Hole, MA there is a > 50-y record of dominant ascidian populations in Eel Pond (Grave, 1933; Grosberg, 1981; Osman et al., 1990, 1992) while at a nearby site in Vineyard Sound we have observed bryozoan dominance over a 25 y period (Osman et al., 1992, pers. obs.). Since 1988 we have observed similar long-term patterns of epifaunal recruitment and dominance at sites in eastern Long Island Sound.

The striking and persistent differences in species composition between nearby sites appear to be attributable to the presence or absence of predators which consume newly-settled and juvenile life-stages (Osman et al., 1992; Osman & Whitlatch, 1995, 1996). At sites where predators are present, ascidian recruitment is virtually eliminated and bryozoans dominate. At sites without abundant predators, ascidians dominate adult populations and recruitment. Predator-prey interactions involve multiple life-stages of many prey and predator species. There is no single dominant prey nor a single key predator species. Very vulnerable prey, such as solitary ascidians, are preyed on by 'micro-predators' (e.g. the snails *Mitrella lunata* and *Anachis* spp.) immediately after settlement and by larger macro-invertebrate and fish predators as 2–4 week-old juveniles. Other species such as encrusting bryozoans show little, if any, effects of the two predator guilds.

Potentially contributing to this local control of recruitment is the prevalence of sessile species that do not produce long-lived larvae capable of wide-range dispersal. For many species larval life is minutes to hours and the recruitment of these species into the local community at a site may be influenced by larval production from within the populations at the site (e.g. Olson, 1985; Davis & Butler, 1989; Stoner, 1990; Petersen & Svane, 1995). Recruitment in subtidal communities can be under intense local control as a consequence of local production of recruits and, at some sites, the combined effects of multiple guilds of predators each consuming successive life-stages of species. Whereas, persistence of some communities may result from the competitive dominance of adults (e.g. *Macrocystis* in kelp beds, Dayton, 1994), longevity (e.g. mussels, Witman, 1985; Connell, 1986), or adult predators (e.g. urchin barrens, Watanabe & Harrold, 1991), subtidal communities may also persist as a consequence of the local control of the recruitment of successive generations. A larval supply-controlled model (e.g. Roughgarden et al., 1985, 1988) with subsequent dynamics regulated by predation on and competition among adults may not be generally applicable.

Subtidal communities do remain open to larvae produced at other sites, but large numbers may be necessary to avoid being overwhelmed by post-settlement predation and local larval production. Species producing large numbers of planktotrophic larvae (e.g. the barnacles and mussels characteristic of the rocky intertidal) are thus the most probable invaders. Because recruitment of these species is likely to be uncoupled from production at a site (but see McShane et al., 1988; Todd et al., 1988; Bertness & Gaines, 1993), their persistence ultimately depends on external recruitment. Without this continuing recruitment, local dynamics should return the community to its previous state; at our sites, either ascidian or bryozoan dominance.

We have been investigating whether local dynamics involving all life-stages commonly control the distribution of species and produce long-term, *multi-generational* persistence. Specifically, we have been examining whether (1) variability among localities in locally-controlled post-settlement factors controls epifaunal recruitment resulting in site-specific differences in dominance and persistence, (2) the same local processes can prevent the invasion of new species, and (3) those species producing short-lived, poorly dispersed larvae contribute significantly to their own persistence by their continued within-site recruitment. In this paper we report some results of ongoing studies addressing these three objectives.

Methods

Studies were conducted between 1993 and 1997 at 2 field sites in eastern Long Island Sound, U.S.A., at the mouth of the Poquonnock River estuary (41E19′15″ N, 72E2′40″ W). One site was at the Avery Point breakwater and the other was near Pine Island < 1 km from the breakwater site (Figure 1). Although both sites were similar in their temperature and salinity regimes and tidal range, they were very different in several important environmental parameters. The breakwater site was in a protected area (mean tidal current $< 6\ \mathrm{cm \cdot sec^{-1}}$) with a bottom of bedrock, boulders and cobbles, and sand at 2–4 m below MLW and grading into areas of mud and eelgrass beds. Although all of the dominant species were present as adults and commonly settled onto substrates exposed at this site, the site was dominated by colonial and solitary ascidians. Except for the green crab, *Carcinus maenas*, predators were rare or absent from the breakwater site.

The Pine Island site was at 7–8 m below MLW with the bottom dominated by large boulders (1–2 m in diameter) with much of the remaining bottom covered by smaller rocks and cobbles with a few pockets of sand. The site was more exposed to wind and waves and was subjected to strong tidal currents of $\sim 60\ \mathrm{cm \cdot sec^{-1}}$ at peak flood or ebb. Unlike the breakwater site, the sessile invertebrate community was dominated by bryozoans and both small (*Mitrella lunata*, *Anachis avara*, and *Anachis lafresnayi*) and large (primarily the seastar *Asterias forbesi* and the crab *Libinia emarginata*) invertebrate predators as well as benthic feeding fish (mostly the cunner, *Tautogolabrus adspersus*) were extremely common.

From 6 July to 21 September 1994, we measured weekly recruitment of sessile species and small motile invertebrates at the two main sites and 2 other sites (Figure 1) along a transect between them. One site (PI Jetty) was on the leeward side of Pine Island at the base of a stone jetty and the other (Whale Rock) was approximately 200 m seaward of the breakwater site in an area of mixed rock and sand. Four panels (100 cm^2) were attached to a weight that held them approximately 10–20 cm above the bottom at all four sites. A 900 cm^2 piece of plastic doormat was also attached to the weight to estimate the relative abundance of small motile benthic invertebrates (L. Harris, personal communication). The panels and mat were sampled and replaced by clean substrates weekly. Upon retrieval the mat was immediately sealed in a plastic bag by divers. All invertebrates in both types of samples were counted in the laboratory with the aid of a dissecting microscope.

Experiments designed to address our objectives were conducted using 75 cm tall, 28 cm diameter PVC 'pilings' secured upright to weighted steel frames to mimic natural boulders. All pilings were in direct contact with the bottom. Depending on the experiment, one or more 100 cm^2 panels attached to each piling were used as sampling units. Our basic design consisted of 5 pilings/treatment and 1–4 panels/piling. The number of treatments varied among experiments. In recruitment experiments which examined the effects of predators on the survival of new recruits and juveniles we used 4 treatments: (1) open pilings exposed to all predators, (2) caged pilings (1 cm^2 mesh) to exclude large predators, (3) screened pilings (1 mm^2 mesh) to exclude all predators, and (4) partially screened pilings to control for artifacts of screening (see Osman & Whitlatch, 1996 for details). All cages were cylinders constructed of plastic-coated wire. Each cage was 50 cm in diameter and 85 cm tall with a sealed top and bottom. Recruits of each life-stage investigated were collected by exposing panels to larvae in the field or the laboratory, holding these substrates at the Breakwater site until a set of recruits grew to the proper life-stage, 'gardening' the panels to remove all other species or life-stages, and then deploying them on pilings at the Pine Island site. Experiments ran for 3–6 d after which panels were retrieved, photographed, and surviving individuals counted. These experiments were conducted in the summers (July through September) of 1993 and 1994 and used all four piling treatments.

An experiment examining the ability of *Mytilus* to recruit into the Pine Island bryozoan community was conducted in early July 1993 in a similar manner to other recruitment experiments with two exceptions. First, 20 panels with newly-recruited, 2–5 mm *Mytilus* on top of an underlying bryozoan community were retrieved from caged pilings that had been deployed in August 1992 and retrieved in late June 1993. Secondly, these panels were redeployed for 6 d only on 5 caged and 5 uncaged pilings at the Pine Island site. The screened and partially-screened treatments were not used. Given the short duration of the study and the use of a large cage that did not restrict the mussels' ability to move from the panels, no cage control treatments were used.

In 1994, a study was designed to examine how the age of the community or the average life-stage of indi-

Table 1. Recruitment experiments. Individual piling experiments were conducted on each of the species and the results of individual ANOVA's and Bonferonni *a posteriori* tests are shown (C = caged, O = open, S = screened, P = partially screened pilings). The range of survival and interpreted predator effects are also shown for each experiment (See also Osman & Whitlatch, 1996).

Taxon	Life-stage	Result	Survival range	Cause of mortality
Colonial Ascidians				
Botryllus schlosseri	Recruits	S > O = P = C	> 80 to < 10%	Microinvertebrate
	Juveniles	S > P = C = O	> 80 to < 10%	Microinvertebrate
	Adults	S = C = O > P	> 80 to < 40%	NONE
Botrylloides diegensis				
	Recruits	S = C = P > O	> 80 to < 20%	Fish, Macroinvertebrate
	Juveniles	S = C = P = O	> 80%	NONE
	Adults	C = O, only treatments	> 80%	NONE
Diplosoma macdonaldi	Recruits	S > P = C > O	> 80 to 0%	Microinvertebrate
Solitary Ascidians				
Ascidiella aspersa	Recruits	S > O, C & P intermediate	> 50 to 0%	Microinvertebrate
Molgula manhattensis	Recruits	C = O, only treatments	< 10 to 0%	Prob. Microinvertebrate
	Juveniles	S = P = C > O	> 80 to 0%	Fish, Macroinvertebrate
Ciona intestinalis	Recruits	C = P = S = O	< 10 to 0%	Environmental
	Juveniles	S = P = C > O	> 80 to 0%	Fish, Macroinvertebrate
Styela clava	Recruits	S = P = C = O	< 50 to 0%	Prob. Microinvertebrate
	Juveniles	S = P > C > O	> 80 to 0%	Fish, Macroinvertebrate
Bryozoans				
Cryptosula pallasiana	Juveniles	S = P = C = O	> 80%	NONE
Bugula turrita	Juveniles	S = P = C > O	> 80 to < 40%	Fish, Macroinvertebrate
Barnacles				
Balanus spp.	Recruits	S = C = P = O	> 90 to 75%	NONE
Polychaetes				
Hydroides dianthus	Recruits	S = P = O = C	> 90%	NONE
Spirorbis spp.	Recruits	S = C = P = O	> 90%	NONE
Molluscs				
Mytilus edulis	Recruits	C > O, only treatments	> 90 to 0%	Fish, Macroinvertebrate

viduals influenced the development of the community in the presence and absence of predators. On 15 July 1994, 60 clean panels were exposed at the breakwater site. After 1, 2, 3, and 4 weeks of exposure 15 of these panels were transplanted to the Pine Island site. A panel was assigned to each of 5 pilings or racks for each of 3 treatments: open pilings exposed to all predators, caged pilings exposed only to micro-predators, and suspended 2–3 m above the bottom to reduce (and usually eliminate) exposure to all predators. Given previous studies that showed no measurable effects of caging other than the exclusion of certain predators, no cage-control treatments were included in the design.

Finally, in the summer of 1997 we conducted experiments to test whether released larvae were dispersed locally. In the first of these experiments 9 pilings were deployed in a sand area adjacent the Pine Island site with a central piling and 4 pilings each at 1 m and 3 m distance. At each distance the pilings were at 90° intervals with 2 in line with the tidal flow and the other 2 perpendicular to the direction of tidal currents. The central piling of the series had 13 panels with reproductive colonies of *Botrylloides* attached to it. Recruitment of *Botrylloides* was measured on each piling using 7 clean panels attached to the remaining sites on the central piling and 20 attached to each of the 1-m and 3-m pilings. In addition 2 control pilings were placed 20 m from the main series to measure any background recruitment of the test species. We examined colonies microscopically prior to deployment to estimate the larval production for the central piling.

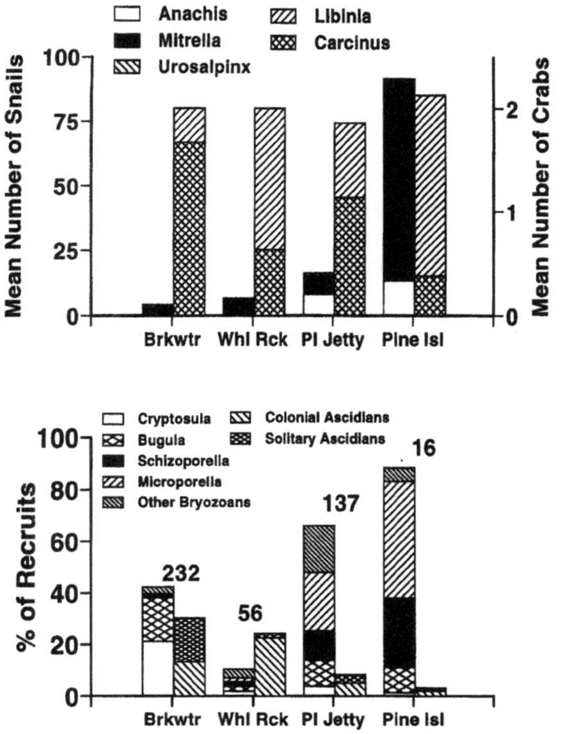

Figure 2. Mean number of small motile invertebrates (Top) collected weekly on doormat samplers and mean percent of total sessile species recruiting to panels weekly (Bottom) at the four sampling sites. Top: motile species include the gastropods *Anachis* spp. (*A. avara* and *A. lafresnayi* combined), *Mitrella lunata*, and *Urosalpinx cinerea* and new-recruits of the crabs *Libinia emarginata* and *Carcinus maenas*. Bottom: The mean cumulative recruitment of bryozoans (*Cryptosula pallasiana*, *Bugula turrita*, *Schizoporella errata*, *Microporella ciliata*, and other species) and ascidians (colonial and solitary) at each site. The numbers above each set of bars are the mean number of total recruits/wk at each site.

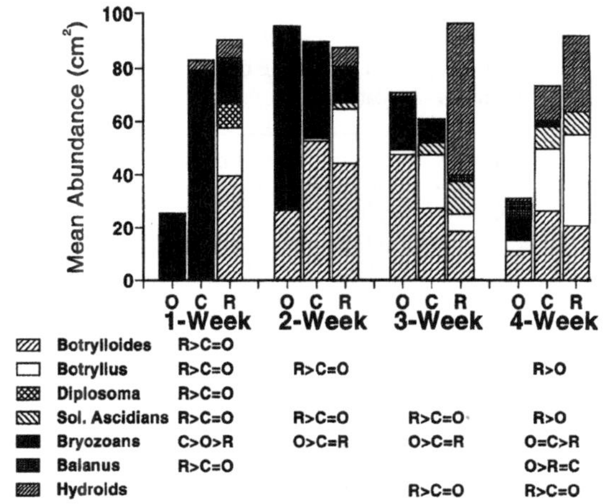

Figure 3. Comparison of the community composition among the three treatments in panels transplanted from the breakwater site to the Pine Island site. Panels transplanted after 1, 2, 3, and 4 weeks of exposure at the breakwater site are shown. The data represent mean abundance of each of the seven dominants after nine weeks. The three treatments were: open pilings (O), caged pilings (C), and suspended above the bottom on racks (R). The taxa shown are the colonial ascidians *Botrylloides diegensis*, *Botryllus schlosseri*, and *Diplosoma macdonaldi*, solitary ascidians (*Ciona intestinalis*, *Molgula manhattensis*, *Styela clava*, and *Ascidiella aspersa* combined), all bryozoans, barnacles (*Balanus* spp.), and hydroids. Significant differences for each of the taxa in each of the 4 groups of transplants are shown.

In all studies data were analyzed using analysis of variance with Bonferroni *a posteriori* tests used to compare treatment means.

Results

Our background sampling at the four sites (Figure 1) showed that the abundances of the small gastropod predators, *Mitrella lunata* ($p < 0.001$) and *Anachis* spp. ($p < 0.005$), decreased significantly from the Pine Island site to the breakwater site (Figure 2). Recruits of the crab *Libinia emarginata* but not *Carcinus* followed a similar trend but neither were significant. Recruitment of ascidians exhibited the opposite trend, significantly decreasing ($p < 0.001$) from 40% of all recruits at the Breakwater site to 1% at the Pine Island site with the Whale Rock and Pine Island Jetty sites being intermediate. Bryozoan recruitment was high at all sites except Whale Rock, but generally increased from 40% of all recruits at the breakwater to almost 90% at Pine Island. In addition the species composition of bryozoan recruits also changed with *Cryptosula pallasiana* forming a significantly greater percentage of the recruits at the breakwater ($p < 0.001$) and *Schizoporella errata* and *Microporella ciliata* dominating recruitment ($p < 0.001$) at the two Pine Island sites.

Recruitment experiments. Results of recruitment experiments summarized in Table 1 (see also Osman & Whitlatch, 1996) clearly show the effects of predators on post-settlement survival and resulting recruitment. Most new recruits of the ascidians *Botryllus schlosseri*, *Diplosoma macdonaldi*, and *Ascidiella aspersa* survived in the screened treatment but were almost eliminated in all other treatments, strongly suggesting predation by micro-predators such as *Mitrella lunata* and *Anachis* spp., but not larger invertebrates or fish. These data are in agreement with early experiments conducted in small cages with and without snails (Osman et al., 1992;, Osman & Whitlatch, 1995). Although high predator-induced mortalities of new-recruits of the solitary ascidians *Molgula manhattensis, Styela clava,* and *Ciona intestinalis* were also found in the earlier experiments, mortality was high and variable for all treatments in the piling experiments for these three species and no significant effect of treatment on recruit survival was found. However, the survival of *Styela* new-recruits was much higher in the screened treatment and no screened treatment was used in the *Molgula* experiment. Only the data for *Ciona* were inconsistent with micro-predation on recruits. Observations indicated that many new recruits of this species were poorly attached to the panel surfaces and may have been washed off panels in all treatments during periods of high tidal currents. Unlike the ascidians, barnacle and polychaete recruits suffered little mortality in any of the treatments, indicating that micro-predators had little or no effect on survival. Finally, the pattern of high survival of *Botrylloides* and *Mytilus* recruits on caged and screened pilings but not on open pilings suggests a strong effect of fish and/or macro-invertebrate predators on the recruitment of these species.

The experiments with juvenile life-stages suggest that *Botryllus* remained vulnerable to micro-predators for several weeks after settlement while *Botrylloides* escaped any measurable effect of predation. Juveniles of solitary ascidians as well as those of the bryozoan *Bugula* suffered high mortalities from either fish or macro-invertebrate predators. Juveniles of the bryozoan *Cryptosula*, on the other hand, did not appear to be affected by any of the predators. Finally, adult colonies of both *Botrylloides* and *Botryllus* were not eaten by predators.

Community development experiments. The development of the epifaunal community on panels transplanted from the breakwater site to the Pine Island site was consistent with the results of the recruitment experiments. Figure 3 contrasts the mean community composition among the treatments of the four transplant experiments near the end of the experiments. At this time most substrates had greater than 80% of their surface covered by the seven taxa represented.

Panels transplanted after 1 week of exposure at the breakwater site had a community consisting entirely of new recruits when moved. By the ninth week of the experiment, those 1-week panels placed on open and caged pilings were dominated by bryozoans with the mean abundance on caged pilings being significantly greater. Colonial ascidians were dominant on 1-week panels deployed on racks, but absent from similar panels on the caged and open pilings (Figure 3). These results indicate that ascidian recruits were removed by micro-predators on the pilings and that the community that developed was very similar to the resident bryozoan community at the site. However, panels on the racks, where micro-predation was eliminated or reduced, were dominated by colonial ascidians.

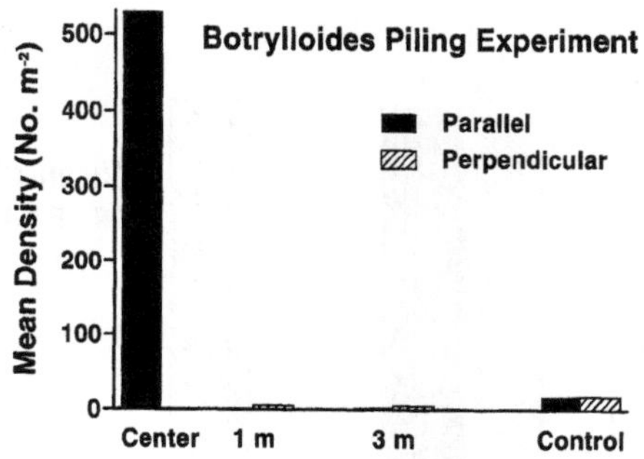

Figure 4. The distribution of *Botrylloides diegensis* recruits on pilings in the experiment measuring the local recruitment of larvae released from colonies on a central piling. Mean densities are based on 7 panels (100 cm^2) on the central piling and 20 panels on all other pilings. Data for pilings parallel and perpendicular to the direction of tidal currents are separated.

The 2-week transplants exhibited two differences from the 1-week transplants. First, *Botrylloides* shared dominance with the bryozoans on open and caged pilings resulting in it having no significant differences among the 3 treatments. Secondly, solitary ascidians were a part of the ascidian-dominated community on panels placed on racks and their abundance, although low, was significantly greater than on the two piling treatments. Again, the recruitment experiments showed that *Botrylloides* escaped the effects of predation by the time it was 1 week old. Its high abundance on 2-week panels indicates that if it can escape early predation it will become a dominant part of the community.

Finally, the results of the 3-week and 4-week transplants show that by this time some *Botryllus* had also reached a size that allowed them to escape micropredators and became a part of the dominant community on all three treatments. In addition, solitary ascidians were a part of the community on caged pilings and racks, but were still absent from open pilings. This suggests that even 4-week-old solitary ascidians may still be vulnerable to predation by fish or macroinvertebrates. It should also be noted that: (1) on some panels, particularly the 3-week panels on racks, abundant hydroids obscured encrusting species such that the abundances of these species may have been greater

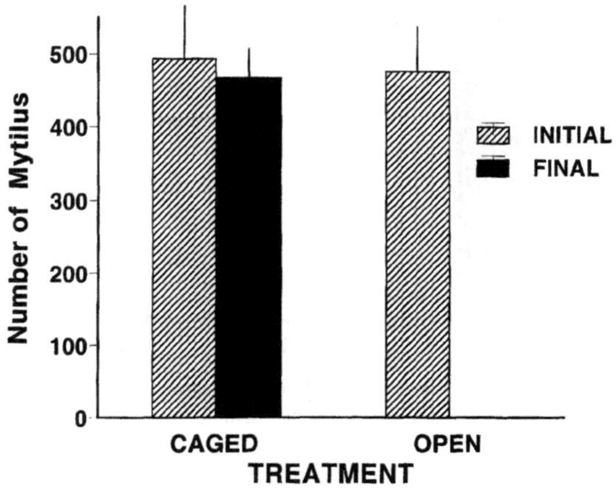

Figure 5. The mean number of *Mytilus edulis* recruits on caged and open pilings after 6 d of exposure at the Pine Island site. Initial and final means and standard errors for each treatment are shown. No *Mytilus* survived on open pilings.

than is shown and (2) the low total abundances on the 1-week open treatments actually do reflect the absence of macrofaunal cover on these panels other than encrusting bryozoans.

Local distribution experiments. At present, we only have preliminary results from one of the 1997 experiments measuring local recruitment of *Botrylloides*. When this experiment was conducted parental colonies had already begun releasing larvae and we estimated that the colonies placed on the central piling had only 1500 larvae. The experiment was run for 1 week and Figure 4 shows the distribution of larvae. Although these data suggest little, if any, recruitment of *Botrylloides* from the parent colonies at 1 and 3 m from the parents, it is clear that there was a significant recruitment of larvae from these colonies on the central piling itself. At a site with tidal current flows that can reach 60 $cm \cdot sec^{-1}$, that more than 3% of the larvae produced recruit within 1 m of the parent suggests a overwhelming effect of local recruitment. Each parental colony would easily replace itself, potentially maintaining the local population.

Invasion experiment. The *Mytilus* recruitment experiment (Table 1, Figure 5) demonstrates the ability of predators to limit or control the invasion of more widely dispersed species into this community. During the 6-d deployment of this experiment, approximately 5% of the *Mytilus* recruits on caged pilings were lost while mussels on open pilings suffered 100% mortality. Based on observations, the cause of the mortality was probably from fish such as the cunner *Tautogolabrus*, but we cannot rule out other fish, crabs, or seastars as contributing to the mortality.

In addition, since the *Mytilus* had recruited onto panels which had 70–100% cover of bryozoans, the removal of the mussels resulted in the community being returned to one dominated by bryozoans. Only small areas of the panels showed any mortality of bryozoan zooids, which would presumably be quickly regenerated. Thus, the predators quickly returned the community to one characteristic of the site.

Discussion

The results of our studies of this southern New England subtidal sessile invertebrate community strongly suggest that the dynamics of recruitment can be under intense local control. In a variety of experiments conducted with a diversity of taxa over wide spatial and temporal scales, we have found that predation of newly-settled and juvenile life-stages has its strongest influence on recruitment with: (a) the snails *Mitrella* and *Anachis* consuming new-recruits, and (b) fish and larger invertebrates preying on juvenile solitary ascidians, mussel recruits, and some bryozoan juveniles. Different species appear to escape predation at different life-stages with: (a) most bryozoans and the ascidian *Botrylloides* escaping before they are 1-wk old, (b) the colonial ascidian *Botryllus* escaping after 2–3 wk, (c) solitary ascidians escaping only as adults, and (d) mussels perhaps never completely escaping.

Community development over 2–4 months also is consistent with the results of the shorter-duration recruitment experiments with bryozoans dominating at sites with predators and ascidians dominating sites without predators. Substrates transplanted from a site without predators to a site with predators developed communities consistent with the predation experiments. The youngest communities, transplanted after 1 wk and composed of vulnerable new recruits, developed bryozoan dominance. Those transplanted after 2 wk became dominated by bryozoans and *Botrylloides*. Finally, the communities that were 3–4 wk old when transplanted became dominated by ascidians with solitary ascidians only occurring in caged treatments which excluded fish. Panels placed on racks above the bottom to remove all predators were all dominated by ascidians regardless of when they were transplanted.

The preliminary results from the dispersal experiment suggest that, for at least some species in this community, local larval settlement and resulting recruitment can be substantial. This experiment was purposely done under extremely conservative conditions in which a limited amount of substrate was available and high current velocities had the potential of quickly carrying larvae out of the local system. The high local density of recruits under these conditions suggests that at sites with more available substrate and lower current velocities local, closed-system larval settlement could have an overwhelming impact on the recruit of many species. Even in our experimental system with most *Botrylloides* larvae dispersed out of the system, local recruitment was more than sufficient to maintain a local population without any recruitment from outside the locality.

In addition, the *Mytilus* experiment suggests that mechanisms of local control, particularly predation on recruits, can prevent the invasion of this community by other species which may be more widely dispersed and also capable of settling in extremely large numbers. Many of the panels were completely covered by *Mytilus* at a density of 5 cm^{-2}, yet all individuals were quickly removed. As these mussels originally recruited only on caged pilings and not on open pilings or the surrounding boulders, this predator control of invasion was probably fairly widespread.

Finally, we have obtained consistent results from field experiments conducted at spatial scales of 18 cm^2 (Osman et al., 1992; Osman & Whitlatch, 1995) to 6600 cm^2 pilings positioned throughout a 900 m^2 study site. This consistency suggests that the effects of predators on recruitment are fairly ubiquitous over a locality and scale-independent. Most importantly, the communities that developed on experimental substrates at our field sites were dominated by the same species that have persisted at these sites over 10+ y (25–50 y if we include the Woods Hole sites) and there is no evidence of any long-term successional processes driving these communities to a different composition.

We feel that these results provide a caution against viewing regional processes as the principle controls of recruitment (Palmer et al., 1996). First, larval dispersal and settlement are not necessarily the keys to recruitment. By controlling post-settlement mortality, the community at a site can cause recruitment to be independent of changes in the supply of settling larvae. Secondly, this control may occur over much broader spatial and temporal scales than shown in earlier studies. For example, Sutherland & Karlson (1977) demonstrated that some dominant species could persist on individual substrates by inhibiting subsequent recruitment of other species. However, they and others (e.g. Osman, 1977) also showed that substrates at the same site immersed (or disturbed) at different times experienced different patterns of recruitment and developed different communities producing a mosaic of dominance among the substrates within a site. What we suggest is local control of recruitment over a spatial scale of 100's to 10 000's m^2 lasting for multiple generations (10+ y) with all substrates, regardless of size or time since last disturbed, having or developing a similar assemblage. Finally, even though the local control of recruitment can be so strong that a small group of species 'permanently' dominates a site, this does not preclude dynamic changes in relative abundance among these species on individual substrates or within patches. It is also likely that within a large local area some newly-opened patches will be invaded by different species. However, local control will inexorably return each patch within a site to the same end point. This return does not have to result from competition and predation among adults. If recruitment is tightly controlled such that only certain species filter through, then as the individuals of any invading species reach maturity and die they will be replaced by new recruits of the original dominants. Thus we hypothesize no 'multiple stable points' (Sutherland, 1974); rather we see a patch or rock with different dominants as ephemeral, lasting no more than a generation.

Our data suggest that the resident community of predators can control recruitment sufficiently to limit the colonization of available substrate to species already present within a site. It is also likely that local, within-population recruitment of many species with short-lived larval stages (e.g. bryozoans and ascidians) can contribute to the local persistence of populations and communities. In addition, these same processes should also make the locality resistant to invasion by other species. Finally, for the alternative of local control to be generally applicable to rocky subtidal communities, the local dynamics that result in persistence should operate over broad spatial and temporal scales that include communities dominated by taxa other than ascidians and bryozoans.

In conclusion, we feel that long-term persistence of benthic communities can arise as a consequence of species' life-histories that enable them to form local, self-sustaining populations with differences in dominance between sites strongly linked to environmental differences such as the intensity of predation. To some degree marine ecological systems may function as closed rather than open systems. As such, wide-scale larval dispersal may an extremely necessary process on a regional scale, but resulting regional temporal and spatial variation in the supply of larvae may be of little consequence to the local community.

Acknowledgements

We wish to thank A. Frese, M. Holt, B. Lussier, R. Malatesta, P. Mitchell, R. Stankelis, J. Siemon, E. Miles, M. Berger, and J. Street for field and laboratory assistance. This work was supported by grants from the National Science Foundation (OCE 9101815 and OCE 9123890) and the Jessie B. Cox Charitable Trust.

References

Alexander, S. E. & J. Roughgarden, 1996. Larval transport and population dynamics of intertidal barnacles: a coupled benthic/oceanic model. Ecol. Monogr. 66: 259–275.

Bak, R. P. M. & M. S. Engel, 1979. Distribution, abundance and survival of juvenile hermatypic corals (Scleractinia) and the importance of life history strategies in the parent coral community. Mar. Biol. 54: 341–352.

Bertness, M. D. & S. D. Gaines, 1993. Larval dispersal and local adaptation in acorn barnacles. Evolution 47: 316–320.

Bhaud, M. & J.-C. Duchêne, 1996. Change from planktonic to benthic development: is life cycle evolution an adaptive answer to the constraints of dispersal? Oceanol. Acta 19: 335–346.

Bingham, B. L. & C. M. Young, 1991. Larval behavior of the ascidian *Ecteinascidia turbinata* Herdman; an *in situ* experimental study of the effects of swimming on dispersal. J. exp. mar. Biol. Ecol. 145: 189–204.

Carlon, D. B. & R. R. Olson, 1993. Larval dispersal distance as an explanation for adult spatial pattern in two Caribbean reef corals. J. exp. mar. Biol. Ecol. 173: 247–263.

Carr, M. H. & M. A. Hixon, 1995. Predation effects on early post-settlement survivorship of coral-reef fishes. Mar. Ecol. Prog. Ser. 124: 31–42.

Commito, J. A., C. A. Currier, L. R. Kane, K. A. Reinsel & I. M. Ulm, 1995. Dispersal dynamics of the bivalve *Gemma gemma* in a patchy environment. Ecol. Monogr. 65: 1–20.

Connell, J. H., 1961. Effects of competition, predation by *Thais lapillus*, and other factors on natural populations of the barnacle *Balanus balanoides*. Ecol. Monogr. 31: 61–104.

Connell, J. H., 1961. The influence of interspecific competition and other factors on the distribution of the barnacle *Chthamalus stellatus*. Ecology 42: 710–723.

Connell, J. H., 1986. Variation and persistence of rocky shore populations. In Moore, P. G. & R. Seed (eds), The Ecology of Rocky Coasts, Columbia University Press, New York: 57–69.

Davis, A. R. & A. J. Butler, 1989. Direct observations of larval dispersal in the colonial ascidian *Podoclavella moluccensis* Sluiter: evidence for closed populations. J. exp. mar. Biol. Ecol. 127: 189–203.

Dayton, P. K., 1971. Competition, disturbance, and community organization: the provision and subsequent utilization of space in a rocky intertidal community. Ecol. Monogr. 41: 351–389.

Dayton, P. K., 1973. Dispersion, dispersal, and persistence of the annual intertidal alga, *Postelsia palmaeformis* Ruprecht. Ecology 54: 433–438.

Dayton, P. K., 1994. Community landscape: scale and stability in hard bottom marine communities. In Aquatic ecology: scale, pattern and process. Blackwell Scientific Publ, London: 289–332.

Dayton, P. K., V. Currie, T. Gerrodette, B. D. Keller, R. Rosenthal, & D. Ven Tresca, 1984. Patch dynamics and stability of some California kelp communities. Ecol. Monogr. 54: 253–289.

Deysher, L. & T. A. Norton, 1982. Dispersal and colonization in *Sargassum muticum* (Yendo) Fensholt. J. exp. mar. Biol. Ecol. 56: 179–195.

Farrell, T. M., D. Bracher, & J. Roughgarden, 1991. Cross-shelf transport causes recruitment to intertidal populations in central California. Limnol. Oceanogr. 36: 279–288.

Gerrodette, T., 1981. Dispersal of the solitary coral *Balanophyllia elegans* by demersal planular larvae. Ecology 62: 611–619.

Grave, B. H., 1933. Rate of growth, age at sexual maturity, and duration of life of certain sessile organisms at Woods Hole, Massachusetts. Biol. Bull. 65: 375–386.

Grosberg, R. K., 1981. Competitive ability influences habitat choice in marine invertebrates. Nature 290: 700–702.

Grosberg, R. K., 1987. Limited dispersal and proximity-dependent mating success in the colonial ascidian *Botryllus schlosseri*. Evolution 41: 372–384.

Havenhand, J. N. & I. Svane, 1991. Roles of hydrodynamics and larval behaviour in determining spatial aggregation in the tunicate *Ciona intestinalis*.

Hughes, T. P. & J. B. C. Jackson, 1980. Do corals lie about their age? Some demographic consequences of partial mortality, fission, and fusion. Science 209: 713–715.

Hughes, T. P. & J. B. C. Jackson, 1985. Population dynamics and life histories of foliaceous corals. Ecol. Monogr. 55: 141–166.

Karlson, R. H., T. P. Hughes & S. R. Karlson, 1996. Density-dependent dynamics of soft coral aggregations: the significance of clonal growth and form. Ecology 77: 1592–1599.

Keough, M. J., 1989. Dispersal of the bryozoan *Bugula neritina* and effects of adults on newly metamorphosed juveniles. Mar. Ecol. Prog. Ser. 57: 163–171.

Lasker, H. R., 1990. Clonal propagation and population dynamics of a gorgonian coral. Ecology 71: 1578–1589.

Levin, L. A., 1984. Life history and dispersal patterns in a dense infaunal polychaete assemblage: community structure and response to disturbance. Ecology 65: 1185–1200.

Lewis, J. R., 1961. The littoral zone on rocky shores – a biological or physical entity? Oikos 12: 280–301.

McShane, P. E., K. P. Black, & M. G. Smith, 1988. Recruitment processes in *Haliotis rubra* (Mollusca: Gastropoda) and regional hydrodynamics in southeastern Australia imply localized dispersal of larvae. J. exp. mar. Biol. Ecol. 124: 175–203.

Olson, R. R., 1985. The consequences of short-distance larval dispersal in a sessile marine invertebrate. Ecology 66: 30–39.

Osman, R. W., 1977. The establishment and development of a marine epifaunal community. Ecol. Monogr. 47: 37–63.

Osman, R. W. & G. R. Abbe, 1994. Post-settlement factors affecting oyster recruitment in the Chesapeake Bay, USA. In K. R. Dyer & R. J. Orth (eds), Changes in Fluxes in Estuaries: Implications from Science to Management, Olsen and Olsen Press, Denmark: 335–340.

Osman, R. W. & R. B. Whitlatch, 1995. Predation on early ontogenetic life-stages and its effect on recruitment into a marine community. Mar. Ecol. Prog. Ser. 117: 111–126.

Osman, R. W. & R. B. Whitlatch, 1996. Processes affecting newly-settled juveniles and the consequences to subsequent community development. Invert. Repro. Dev. 30: 217–225.

Osman, R. W., R. B. Whitlatch & R. J. Malatesta, 1992. Potential role of micro-predators in determining recruitment into a marine community. Mar. Ecol. Prog. Ser. 83: 35–43.

Osman, R. W., R. B. Whitlatch, R. J. Malatesta & R. N. Zajac, 1990. Onotogenetic changes in trophic relationships and their effects on recruitment. In M. Barnes & R. N. Gibson (eds), Trophic Relationships in the Marine Environment, Proc. 24th Europ. Mar. Biol. Symp., Aberdeen University Press, Scotland: 117–129.

Paine, R. T., 1966. Food web complexity and species diversity. Am. Nat. 100: 65–75.

Palmer, M. A., J. D. Allan & C. A. Butman, 1996. Dispersal as a regional process affecting the local dynamics of marine and stream benthic invertebrates. Trends Ecol. Evol. 11: 322–326.

Petersen, J. K. & I. Svane, 1995. Larval dispersal in the ascidian *Ciona intestinalis* (L.), evidence for a closed population. J. exp. mar. Biol. Ecol. 186: 89–102.

Roughgarden, J., S. Gaines, & H. Possingham, 1988. Recruitment dynamics in complex life cycles. Science 241: 1460–1466.

Roughgarden, J., Y. Iwasa, & C. Baxter, 1985. Demographic theory for an open marine population with space-limited recruitment. Ecology 66: 54–67.

Sammarco, P. W., 1980. *Diadema* and its relationship to coral spat mortality: grazing, competition, and biological disturbance. J. exp. mar. Biol. Ecol. 45: 245–272.

Sammarco, P. W., 1983. Coral recruitment across the central Great Barrier Reef: a preliminary report. In J. T. Baker, R. Carterm, P. W. Sammarco & K. Starck (eds), Proc. Great Barrier Reef Conf., James Cook Univ., Australia: 245–251.

Sammarco, P. W. & J. C. Andrews, 1988. Localized dispersal and recruitment in Great Barrier Reef Corals: the Helix experiment. Science 239: 1422–1424.

Sammarco, P. W. & J. H. Carleton, 1981. Damselfish territoriality and coral community structure: reduced grazing, coral recruitment, and effects on coral spat. Proceedings of the Fourth International Coral Reef Symposium, Manila 2: 525–535.

Shanks, A. L., 1983. Surface slicks associated with tidally forced internal waves may transport pelagic larvae of benthic invertebrates and fishes shoreward. Mar. Ecol. Prog. Ser. 13: 311–315.

Sole-Cava, A. M., J. P. Thorpe & C. D. Todd, 1994. High genetic similarity between geographically distant populations in a sea anemone with low dispersal capabilities. J. Mar. Biol. Ass. U. K. 74: 895–902.

Sousa, W. P., 1979. Disturbance in marine intertidal boulder fields: the nonequilibrium maintenance of species diversity. Ecology 60: 1225–1239.

Sousa, W. P., 1979. Experimental investigations of disturbance and ecological succession in a rocky intertidal algal community. Ecol. Monogr. 49: 227–254.

Sousa, W. P., 1980. The responses of a community to disturbance: the importance of successional age and species' life histories. Oecologia 45: 72–81.

Stephenson, T. A., 1943. The causes of the vertical and horizontal distribution of organisms between tidemarks in South Africa. Proc. linn. Soc. Lond 154: 219–232.

Stoner, D. S., 1989. Fragmentation: a mechanism for the stimulation of genet growth rates in an encrusting colonial ascidian. Bull. mar. Sci. 45: 277–287.

Stoner, D. S., 1990. Recruitment of a tropical colonial ascidian: relative importance of pre-settlement vs. post-settlement processes. Ecology 71: 1682–1690.

Stoner, D. S., 1992. Vertical distribution of a colonial ascidian on a coral reef: the roles of larval dispersal and life-history variation. Am. Nat. 139: 802–824.

Sutherland, J. P., 1974. Multiple stable points in natural communities. Am. Nat. 108: 859–873.

Sutherland, J. P. & R. H. Karlson, 1977. Development and stability of the fouling community at Beaufort, North Carolina. Ecol. Monogr. 47: 425–446.

Todd, C. D., J. N. Havenhand & J. P. Thorpe, 1988. Genetic differentiation, pelagic larval transport and gene flow between local populations of the intertidal marine mollusc *Adalaria proxima* (Alder & Hancock). Funct. Ecol. 2: 441–451.

Underwood, A. J., 1980. The effects of grazing by gastropods and physical factors on the upper limits of distribution of intertidal macroalgae. Oecologia 46: 201–213.

Vadas, R. L. Sr., S. Johnson & T. A. Norton, 1992. Recruitment and mortality of early post-settlement stages of benthic algae. Br. Phycol. J. 27: 331–351.

Watanabe, J. M. & C. Harrold, 1991. Destructive grazing by sea urchins *Stronglylocentrotus* spp. in a central California kelp forest: potential roles of recruitment, depth, and predation. Mar. Ecol. Prog. Ser. 71: 125–141.

Wethey, D. S., 1986. Local and regional variation in settlement and survival in the littoral barnacle *Semibalanus balanoides* (L.): patterns and consequences. In P. G. Moore & R. Seed (eds), The Ecology of Rocky Coasts, Columbia Univ. Press, New York: 194–202.

Witman, J. D., 1985. Refuges, biological disturbance, and rocky subtidal community structure in New England. Ecol. Monogr. 55: 421–445.

Hydrobiologia **375/376:** 125–135, 1998.
S. Baden, L. Pihl, R. Rosenberg, J.-O. Strömberg, I. Svane & P. Tiselius (eds),
Recruitment, Colonization and Physical–Chemical Forcing in Marine Biological Systems.

Recolonization dynamics in areas disturbed by bottom fishing gears

F. Pranovi[1], O. Giovanardi[2] & G. Franceschini[2]
[1]*Dipartimento Scienze Ambientali, Università di Venezia, Campo della Celestia, Castello 2737XB, 30122 Venezia, Italy*
[2]*Istituto Centrale per la Ricerca scientifica e tecnologia Applicata al Mare (ICRAM), V.le Stazione, 5, 30015 Chioggia (VE), Italy*

Key words: macrobenthos, recolonization dynamics, anthropic impact, Adriatic Sea, Lagoon of Venice

Abstract

Results of two investigations on the effects of disturbance on benthic communities in lagoon and coastal areas, caused by bottom fishing-gears ('hydraulic dredge' for clams and 'rapido', a kind of beam-trawl for soles and scallops employed in the Northern Adriatic sea), are given. Such gears, although characterised by different features and targets, have similar effects on the sea bottom: both produce deep furrows (7–13 cm for the 'rapido', up to 20 cm for the 'hydraulic dredge'), thus affecting the texture of the bottom. In 1992 ('hydraulic dredge') and in 1995 ('rapido') two different research projects were carried out; samples of benthos were collected immediately after the passage of the gears and at fortnightly-intervals, in treated and control areas. This allowed study of the modifications of the macrobenthic communities and investigation of the short and medium-term (dredge: 60 days, 'rapido': 15 days) progression of the recolonization processes in the disturbed areas. These dynamics have been analysed by giving emphasis to the species and to their time-space fluctuations. It has been found that characteristically 'non-opportunistic' species can assume an opportunistic behaviour during the initial phase of the recolonization processes of the disturbed areas.

Introduction

Results of studies on the structure of marine ecosystems, the relationships among their components and the temporal evolution of such relationships have been of increasing importance. In addition to natural disturbance, human activities have the potential to greatly modify aquatic ecosystems. Particularly, the interaction between the benthos and the abiotic component of the ecosystem has been the subject of numerous studies, both in terms of structural relationships (Sutherland, 1974; Commito, 1982; Ambrose, 1984a; Ambrose 1984b; Valderhaug & Gray, 1984) and recovery following events of anthropic disturbance (Dean & Haskin, 1964; Leppäkoski, 1975; Rosenberg, 1976; Pearson & Rosenberg, 1978; Hily, 1983; Bonsdorff et al., 1984). Chemical stresses as a consequence of pollution and dredging activities are just two examples of anthropic disturbance able to cause a drastic reduction of a benthic population, giving origin to permanent or temporary changes in the environmental abiotic features (Poiner & Kennedy, 1984; Bonvicini Pagliai et al., 1985).

Among different anthropic activities, the trawling fishery appears not only to have a direct impact on its target species, but also on benthic communities as a whole, and on the physical environment in which they live.

The effects of bottom-fishing gears on sea-bottom morphology and its communities have been assessed in many studies mainly carried out in the North Sea (Bridger, 1970; Bridger, 1972; Pickett, 1973; Reise, 1982; de Groot, 1984; ICES, 1988; Rees & Eleftheriou, 1989; BEON, 1990; Hutchings, 1990; Bergman & Hup, 1992; Jones, 1992; Hall, 1994; Kaiser & Spencer, 1996). These studies provided evidence of remarkably heavy stress on epi- and endofauna and physical disturbance on the seabed.

The Northern Adriatic Sea is a large trawlable seabed where different types of Mediterranean bottom-fishing gears are used by the commercial fleet. A concentration may cause a high fishing pressure on the bed. This has been noted in the area off Chioggia and, to a lesser extent, inside the Venice Lagoon. In the framework of two different research projects on the effects of bottom-fishing gears in the Chioggia area, our attention has been focused on the hydraulic dredge (Pranovi & Giovanardi, 1994) and the 'rapido' (a type of gear similar to a beam-trawl, only used in the Adriatic Sea) (Giovanardi et al., in prep.). The aim of this paper is to give a preliminary description of the short- to medium-term progression of the recolonization processes in two disturbed areas (one inside the Venice Lagoon and an other in the Adriatic Sea). In order to analyse the recolonization pattern of two different habitats disturbed with two different kind of fishing gears, data collected during the two research projects have been compared.

Material and methods

The two studies have been conducted in the Venetian area (Northern Adriatic Sea).

The study on the effects of the action of the hydraulic dredge was done in 1992, in a 1.5 m deep silty-sand area (sand 65%; silt 23%; sediment organic content 2.0%) located in the central basin of the Venice Lagoon (near the port of Malamocco, Figure 1). Two neighbouring zones were identified: one was regularly dredged by the commercial fishery fleet (zone 1), whereas the second had never been dredged (zone 2). The study of the effects produced by the 'rapido' was done in 1995, in a 18 m deep silty area (sand 20%; silt 78%; sediment organic content 0.5%) located in the in-shore zone (between the port of Malamocco and the Chioggia one, Figure 1). Experimental hauls were performed in three closed stations: st. 1 in a trawlable ground (treated with a single haul), st. 2 inside a two-year-old off-shore long-line mussel farm – in order to avoid interference due to the trawl-fishing activities – (treated with a single haul), st. 3, inside the farm as well (treated with 10 consecutive hauls).

Gear description

Hydraulic dredge: this gear is employed on the Northern Adriatic sandy coasts for the fishery of burrowing bivalve molluscs. The gear (Figure 2) consists of an iron cage (2.5–3 m wide) with rods having a 12 mm maximum span, several front water nozzles and two side runners. During fishing activity, the water jets (outlet pressure 1.5–3.5 bar) resuspend sediments and infaunal organisms get trapped into the cage. This fishing technique is one of the most severe in its impact on marine benthic communities, since it causes the sediment to be 'disturbed' to a 20–25 cm depth by producing about 15 cm deep furrows. Rapido: this gear, similar to a beamtrawl, consists of a rigid mouth rigged with iron teeth and a net bag where the catch is collected (Figure 2). This gear is exclusive to the Adriatic Sea and employed in the fishery of Pectinidae (*Pecten jacobaeus*, *Aequipecten opercularis*, *Chlamys* spp.) mainly on off-shore sandy bottoms and flatfish (*Solea* spp. and *Platichthys flesus*) on muddy bottoms closer to the coast-line. This type of gear produces about 5 cm deep furrows.

Sampling procedure

The experimental procedure has been set up by adapting methods used in earlier works (*e.g.* Hall et al., 1990).

Once the experimental area was identified, an experimental haul (or several, at station 3 of the 'rapido') was carried out by a commercial boat. The sampled area was immediately marked using special signs (such as buoys and posts) in order to allow its subsequent identification.

Sampling was done by SCUBA-diving with an air-lift device (hydraulic dredge: 0.5 mm mesh size bag; 'rapido': 1 mm mesh size bag) working on a 50×60 cm iron frame up to a depth of 15 cm (hydraulic dredge: three replicates; 'rapido': five replicates) within (treatment) and outside (controls) the trawled areas. The sampling procedure was later repeated (hydraulic dredge: about 20-day-intervals, 4 times in total, t0 = 3 June 1992, t1 = 24 June 1992, t2 = 15 July 1992, t3 = 5 August 1992; rapido: t0 = 5 October 1995, t1 = 20 October 1995) both within treated and control areas.

All collected specimens were kept at −15°C and then identified to the lowest possible level (species or, at least, genus). Average abundance were then calculated for each *taxon*.

Statistical analysis

Data were transformed (4th root) in order to give an equal weighting to both rare and common species.

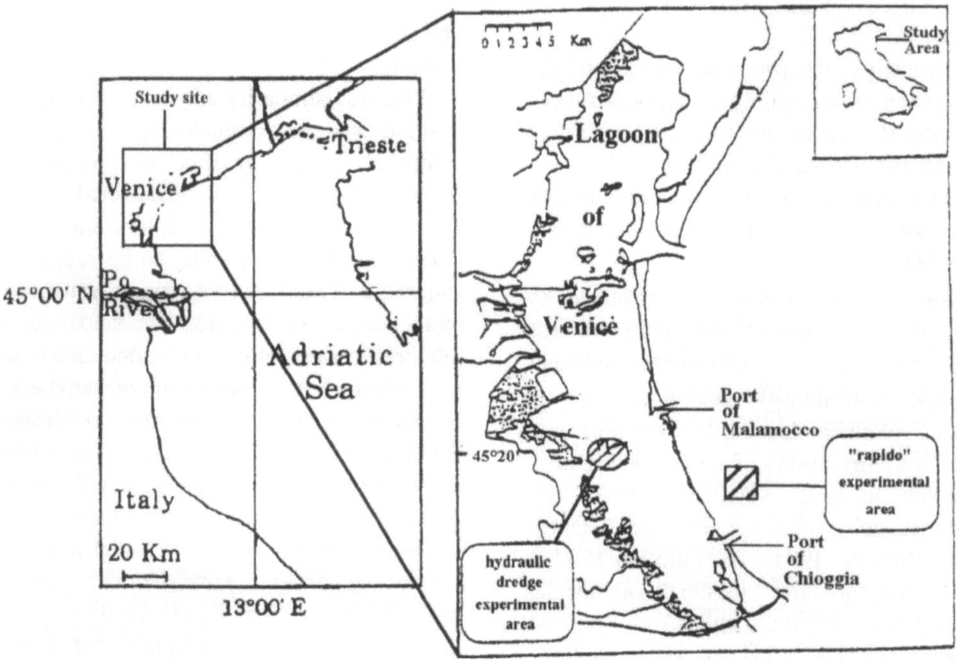

Figure 1. Location of experimental sites. The experimental area inside the Venice Lagoon, treated with the hydraulic dredge includes two adjacent zones; the experimental area treated with the 'rapido' includes two stations located inside an off-shore mussel farm and one just off it.

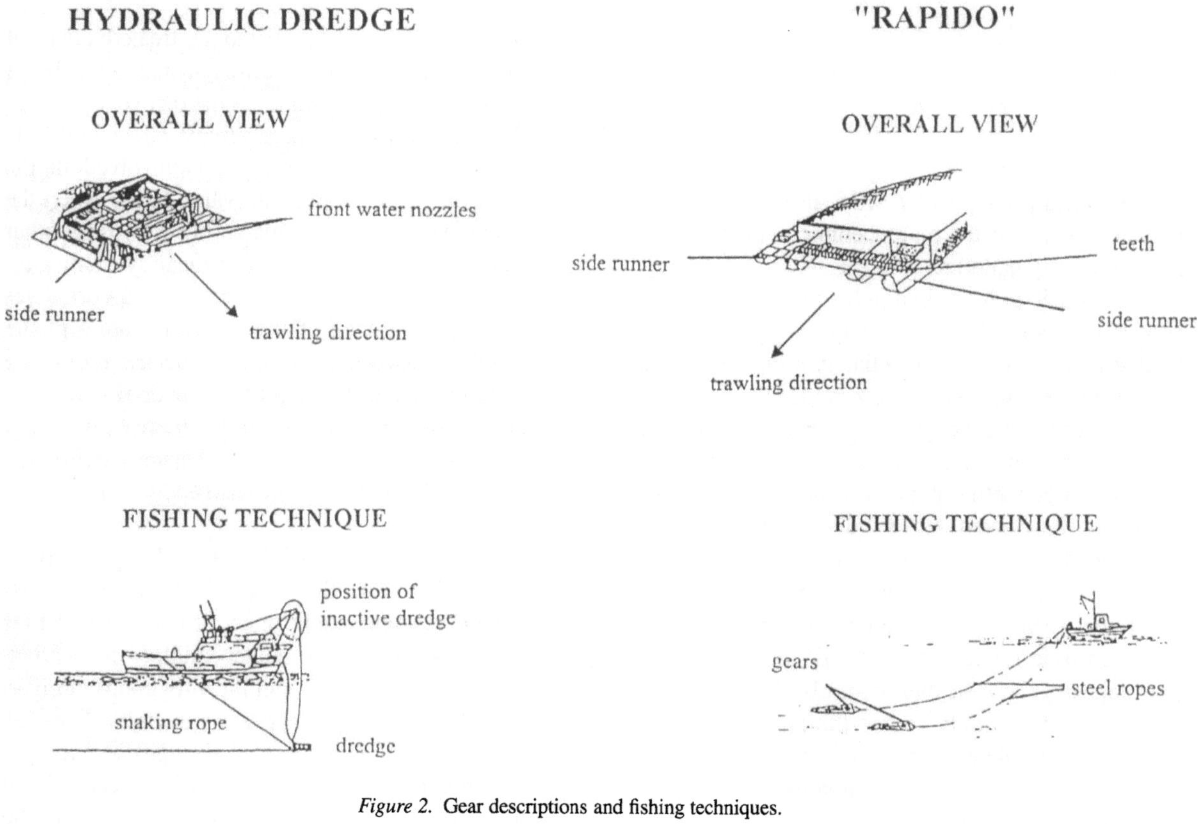

Figure 2. Gear descriptions and fishing techniques.

Bray-Curtis similarity matrix (Bray & Curtis, 1957) was calculated.

In order to analyse the evolution of the populations, a MultiDimensional Scaling – MDS – (Kruskal, 1964) was applied to compare the composition of those communities and their changes.

In order to analyse the pattern of recolonization in the areas subject to the experimental hauls two methods have been used:

- the species clustering (average group linkage) by building up a dendrogram based on Bray-Curtis similarities computed on standardised abundance;
- the average dissimilarity δ among the samples has been calculated and, then, broken down into the separate contributions of each species to δ. This was done using the SIMPER routine ('similarity percentage') within the software PRIMER (Clarke & Warwick, 1994). The ratio δ/SD(δ) has been used to measure how consistently one species contributed to δ among the samples.

The trophic structure of the sampled communities has also been described in terms of qualitative dominance, computed as a ratio of abundance of the different feeding guilds.

Results

MDS

Reordering the samples by MDS allows the description of changes in the macrobenthic populations and in the recovery patterns.

The MDS plot of hydraulic dredge samples is shown in Figure 3. Four groups are recognisable: treated area of zone 1 immediately after the haul; all other samples collected in the zone 1; all controls and treated area immediately after the haul of zone 2; all other treated areas of zone 2.

The macrobenthic populations of zones 1 and 2 (neighbouring) seem to be quite different, as the separation between controls shows.

Moreover, there is probably a different recovery ability between the two communities after the disturbance. In zone 1, except the 1t0, all other treated samples appear to be near their controls. This shows that the community rapidly begins the recovery process, resulting in an increase in similarity between treated areas and controls (after two months – t3 – the two situations are very close).

In zone 2 the group of treated areas, sampled at time t1, t2 and t3, is different from the respective controls.

The high similarity between the sample collected in the treated area immediately after the haul and its control shows the presence of a lag-time before the effects of disturbance can be detected.

MDS applied to the samples collected during the investigations carried out on the rapido (Figure 4) shows the existence of 4 groups: treated area of st. 1; treated areas of st. 2 and st. 3 at t0; the controls of the three stations at t0; the treated areas and controls at t1 (station 1 was not further repeated because it was outside the mussel farm, therefore not 'protected' from the commercial fisheries). The latter group shows a recovery process by the benthic population of st. 2 and st. 3. Moreover, there is some evidence of only slight differences between the area treated with a single passage of the gear (st. 2) and the one treated with more consecutive hauls (st. 3). The respective similarity of the samples (Figure 4) suggests that a great deal of the disturbance suffered by the benthic community is caused by the first passage of the gear.

Average dissimilarity

Hydraulic dredge: data related to treated areas of zones 1 and 2 have been separately compared in order to point out likely differences in the recovery pattern.

Zone 1: higher average diversity values (Table 1) are found when the t0 sample is involved in the comparison, with a maximum value found when the two time extremes are compared ($\delta = 67.46$). In other comparisons the difference is significantly smaller, although it never goes to 0: between t2 and t3 samples $\delta = 31.5$ (a difference of this order is found in the comparison between the samples collected at t3 in the treated and in control areas). The number of significant species follows the same general pattern. In the t0 vs. t2 comparison the significant species are virtually the same as the t0 vs. t1 one. This also applies to the t0 vs. t3 comparison.

Zone 2: the situation is found to be quite different from the one in the zone 1. The pattern is less clear and, to some extent, inexplicable. The average diversity always tends to be very high ($\delta > 58$), but, in spite of it, the species significantly contributing to δ are always between 0 and 2, as though all species contribute equally to the overall average diversity.

Rapido: the comparisons have been carried out both between the two temporal series within the

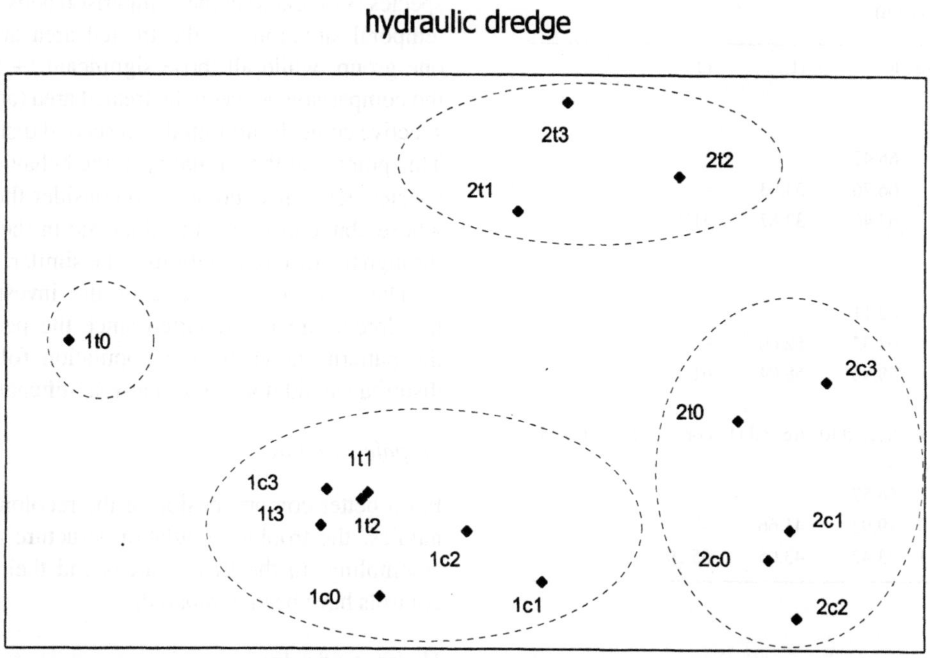

Figure 3. MDS of species abundances from samples collected in the hydraulic dredge experimental site. (1 = zone 1, fishing ground; 2 = zone 2, never trawled by bivalve fishery vessels; c = control area; t = treated area; no. = sampling date (see Materials and methods).

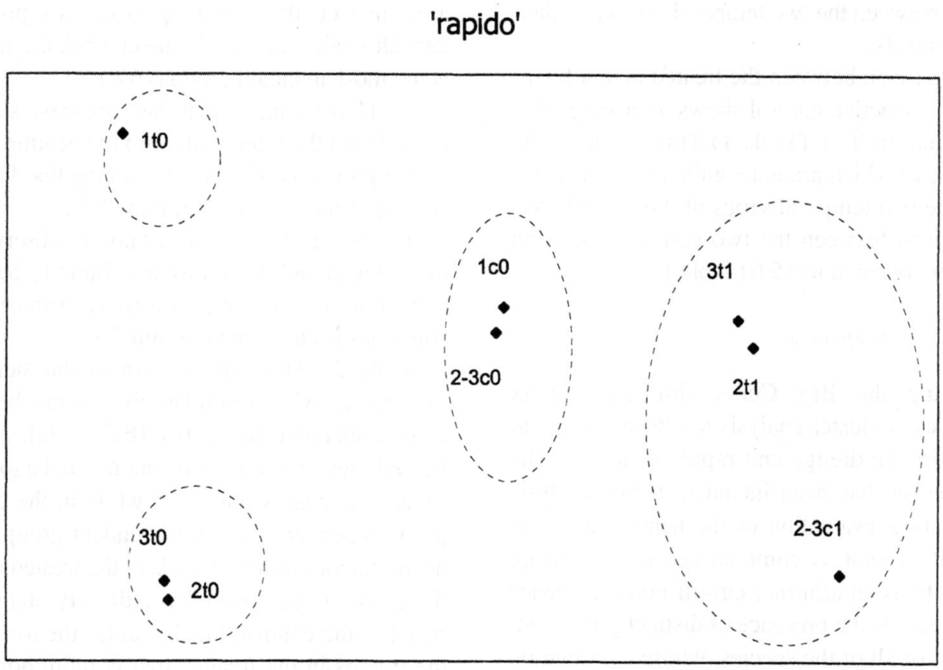

Figure 4. MDS of species abundances from samples collected in the 'rapido' experimental site. (1 = station outside the off-shore mussel farm; 2 = station inside the off-shore mussel farm treated with 1 experimental haul; 3 = station inside the off-shore mussel farm treated with 10 consecutive experimental hauls c = control area; t = treated area; no. = sampling date (see Materials and methods).

Table 1. Average diversity between samples collected in the areas treated with hydraulic dredge a) and in the 'rapido' experimental site b).

a) DREDGE	t0	t1	t2	t3
zone 1				
t0	–			
t1	66.47	–		
t2	66.76	34.63	–	
t3	67.46	32.67	31.49	–
zone 2				
t0	–			
t1	62.11	–		
t2	64.33	62.60	–	
t3	58.96	58.04	61.22	–
b) RAPIDO	treated t0	treated t1	control t0	control t1
treated t0	–			
treated t1	66.57	–		
control t0	49.92	41.66	–	
control t1	73.42	43.05	55.59	–

treated area (the data related to the area treated with 1 experimental passage and those related to the area treated with several passages have been joined together because of the homogeneity of their recovery pattern) and between the two temporal series and their respective controls.

The comparison between the treated area after 15 days and its respective control shows an average dissimilarity equal to 43.1 (Table 1) This is due to the effect of species which are not significant in the comparison of the two temporal series of the treated area. The comparison between the two controls shows an average diversity equal to 55.6 (Table 1).

Cluster analysis by species

By employing the Bray-Curtis similarity matrix among species, a cluster analysis has been applied to both the hydraulic dredge and rapido data sets. The entire species set has been included in this analysis in order to allow evaluation of the behaviour of the less frequent as well as common species. By using 40% similarity as an arbitrary cut-off point, evidence is found as regards the presence of distinct groups, including most or all of the species, whose contribution to δ is statistically significant. In the cluster of zone 1 (Figure 5) – hydraulic dredge experiment – all the significant species (∗-marked, except *Hexaplex*) are in the same group.

In the case of the rapido (Figure 6) the significant species (∗-marked) in the comparison between the two temporal situations of the treated area are found in one group, while all those significant (+-marked) in the comparison between the treated area (at t1) and respective controls are found in a second distinct group. This points out the similarity in the behaviour of such species, since it is possible to consider those species whose abundance tends to fluctuate in the same way through the different stations to be similar.

Data related to zone 2 of the investigation on the dredge are not reported since the peculiarity of the pattern shown by the population following the disturbance did not result in any significant grouping.

Trophic structures

For a better comprehension of the recolonization dynamics, the trophic population structure (abundance assembling) in the treated areas and their respective controls have been compared.

Hydraulic dredge

Zone 1. At t0 the trophic structures in the treated area and its control are similar (Figure 7a). At t1, after the beginning of the recolonization process, the control and treated area show some differences. In the treated area, in fact, the prevailing group is represented by deposit feeders (70%). In the control, the filter feeder is the most abundant group (46%).

At t2 the situation further changes, since in the treated area the filter feeder group becomes dominant, with a group distribution resembling that found in the control of the previous sampling (t1).

Finally, at t3, the distributions observed in the treated area and its control are found to be relatively similar as if a certain uniformity between the two zones has been reached (Figure 7a).

Zone 2. At t0 the pattern is the same of that of zone 1, with a similar distribution between the treated area and the control (Figure 7a); at t1 in the treated area an increase of the necrophagous and the 'others' groups is observed, while in the control deposit feeders are the most abundant group. At t2 the necrophagous group prevails in the treated area, while the group of the 'others' is still very abundant compared to the control. Finally, at t3, the distribution of the groups in the treated area is more homogeneous and similar to that of the control (Figure 7a).

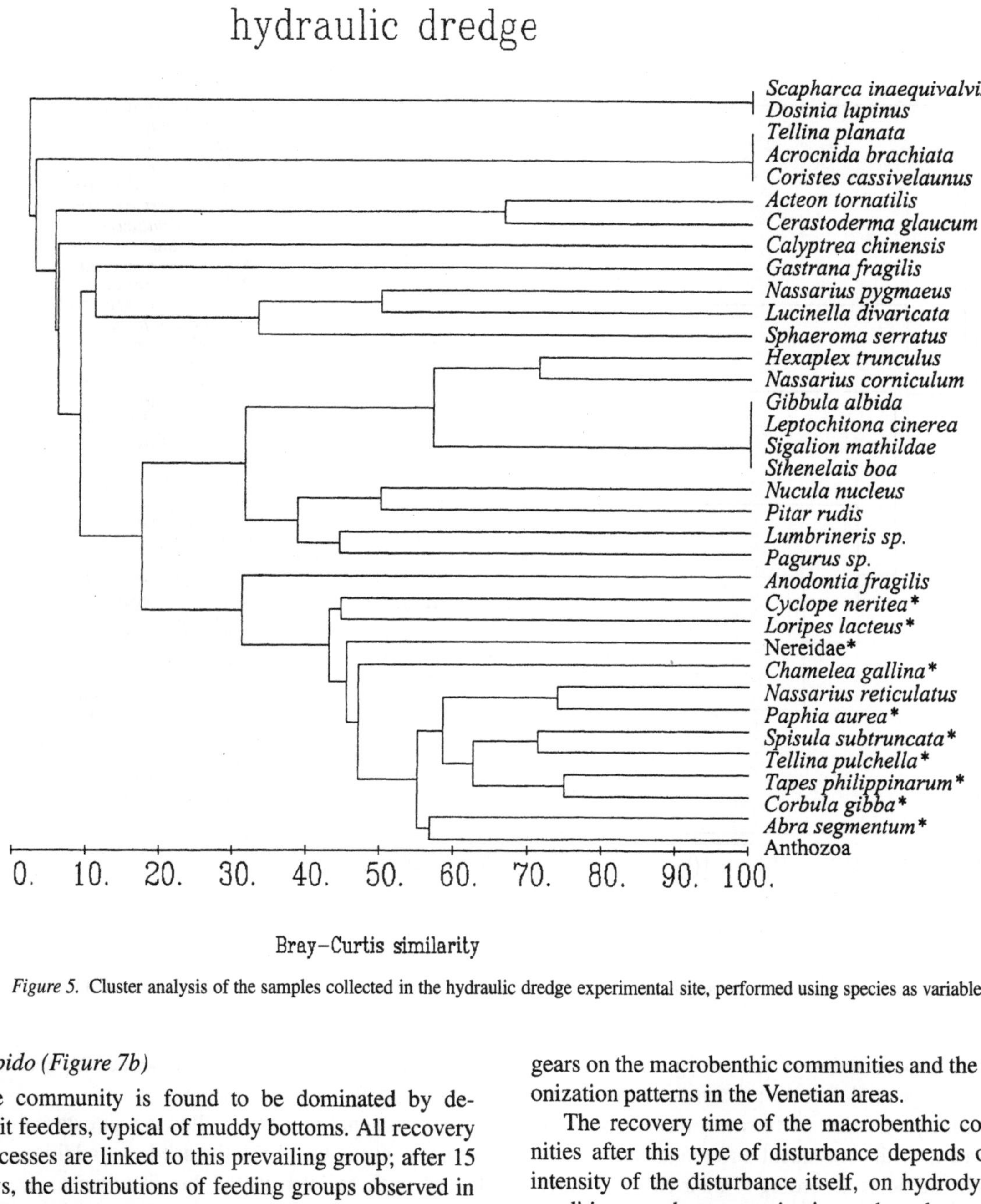

Figure 5. Cluster analysis of the samples collected in the hydraulic dredge experimental site, performed using species as variables.

Rapido (Figure 7b)

The community is found to be dominated by deposit feeders, typical of muddy bottoms. All recovery processes are linked to this prevailing group; after 15 days, the distributions of feeding groups observed in the treated area are quite similar to those observed in the controls.

Discussion and conclusions

The data collected during the different investigations allows us to draw some preliminary conclusions about the effects of disturbance produced by trawl-fishing gears on the macrobenthic communities and the recolonization patterns in the Venetian areas.

The recovery time of the macrobenthic communities after this type of disturbance depends on the intensity of the disturbance itself, on hydrodynamic conditions, on bottom grain size and on the structure of the community affected by the disturbance (Kaiser & Spencer, 1996).

It has been shown that the disturbance caused by the hydraulic dredge is more severe than the one caused by the rapido, as was expected due to the structural and working differences of the gears. This fact, associated with environmental factors, has probably produced different recovery speeds.

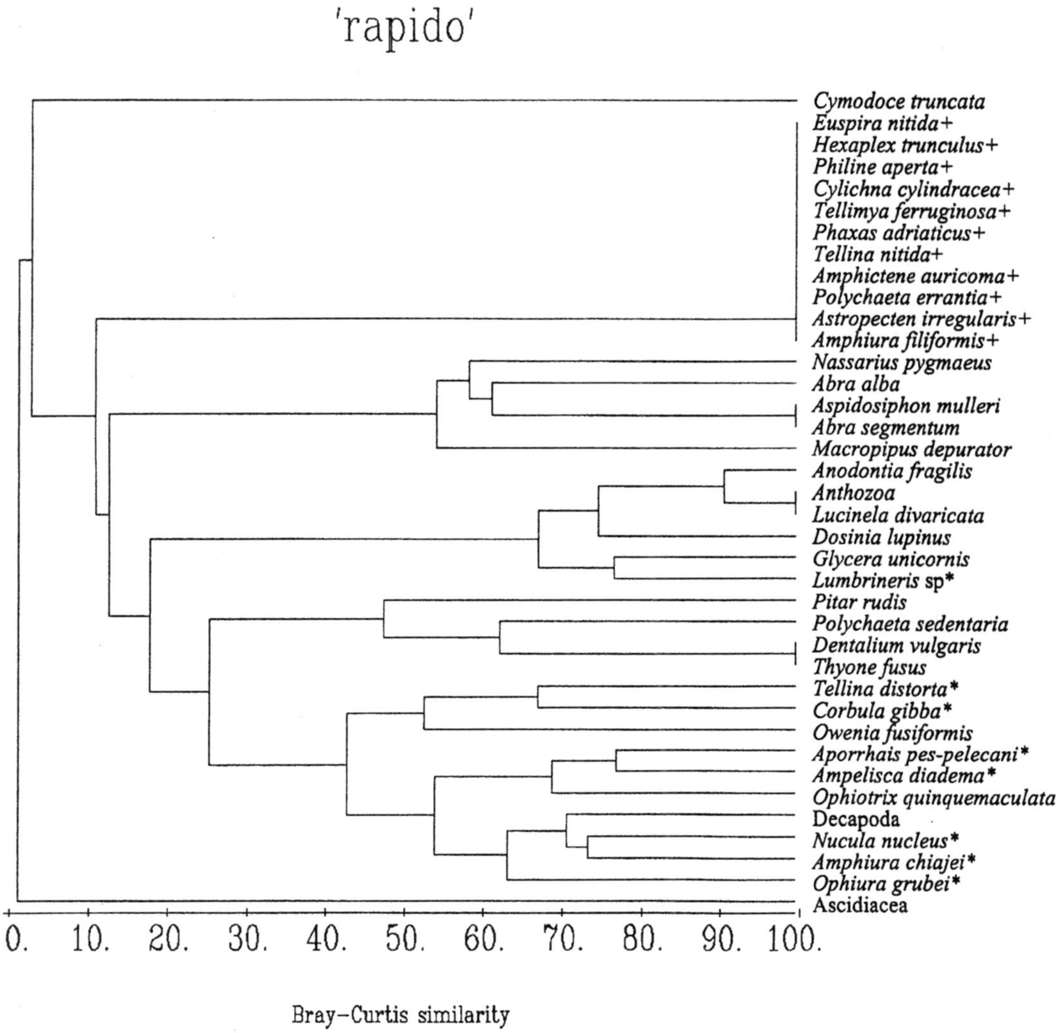

Figure 6. Cluster analysis of the samples collected in the 'rapido' experimental site, performed using species as variables.

The disturbing action of the two gears seems to have similar general features: rearrangement of the bottom with production of furrows; resuspension of the sediment, with consequent differential sedimentation of the particles according to their size ((Pranovi & Giovanardi, 1994), in the long run this could produce changes in the grain size due to the loss of finer particles, as shown in the lagoon environment); direct action on the benthic organisms with their removal or damaging.

The direct action of both gears on the benthic population causes, as shown in the analysis of the trophic groups immediately after the experimental hauls, a non-selective reduction in the abundance of all species, both those directly captured and those resuspended in the sediment plume behind the gear and transported by the water currents.

Successional processes following an external disturbance event can be influenced, in addition, by factors such as intensity of the disturbance, season (Sousa, 1980; Hawkins, 1981) and environmental features of 'disturbed' areas (Zajac & Whitlatch, 1982), and even by the community structure. In the long run, this method of fishing modifies the macrobenthic community resulting in the loss of more fragile species (Bergman & Hup, 1992; Eleftheriou & Robertson, 1992; Bergman & van Saintbrink, 1994; Kaiser & Spencer, 1996; MacDonald et al., 1996), and improving its 'recovery capacity'. This would explain the different recovery and recolonization patterns shown

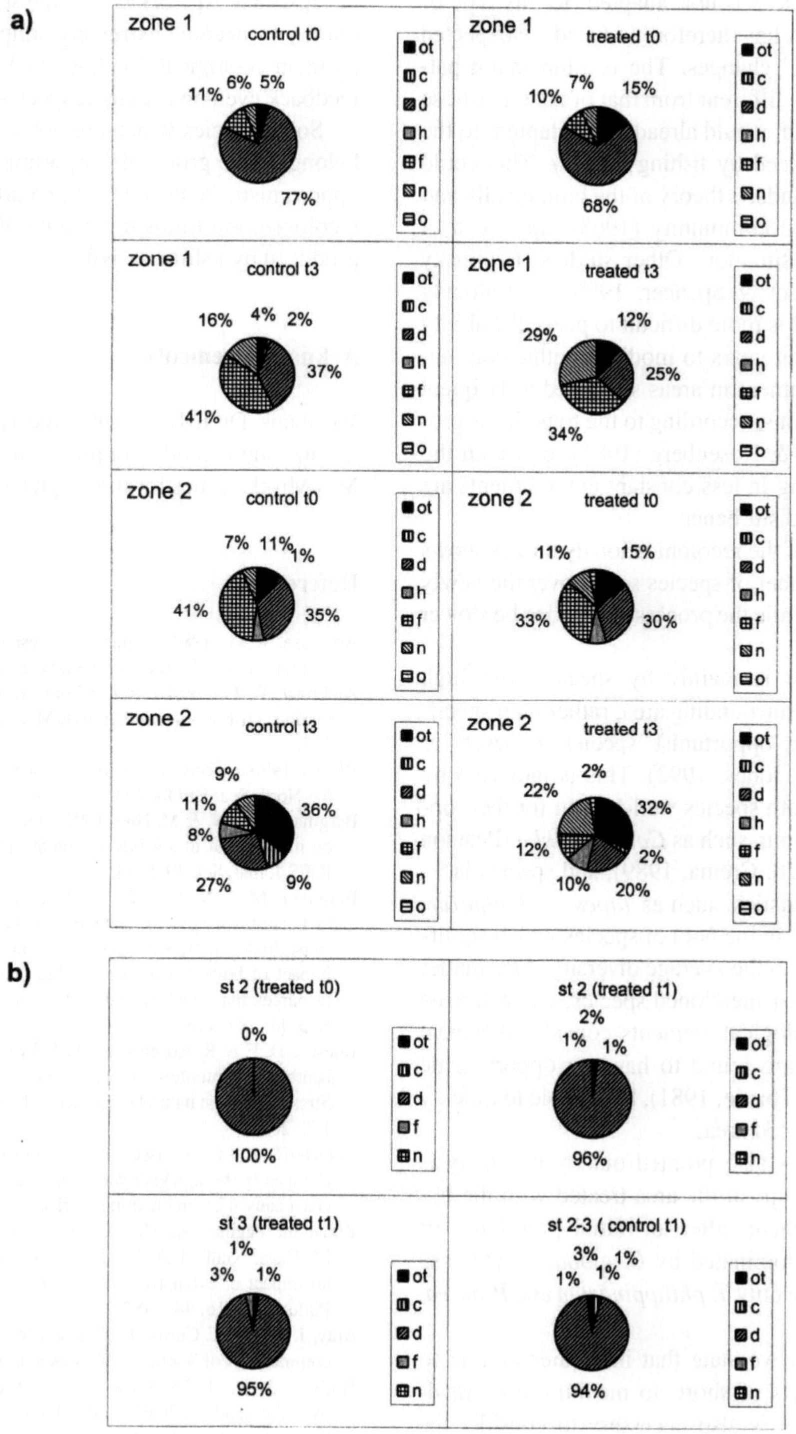

Figure 7. Trophic structure of the population in zone 1 and zone 2 of hydraulic dredge experimental site a, and in the 'rapido' experimental site b; at t=0 the stations 2 and 3 show the same trophic structure. Feeding guilds: ot = other; c = carnivorous; d = deposit feeders; f = filter feeders; h = herbivorous; n = necrophagous; o = omnivorous.

in the two areas treated with the hydraulic dredge. The community of zone 2 is not 'adapted' to this type of disturbance which has therefore induced 'unexpected and unpredictable' changes. The recolonization pattern is completely different from that of zone 1, whose population, instead, would already be 'adapted' to the disturbance produced by fishing activity. This could lend support to Sander's theory of the biologically and physically adapted community (1968), applied to a slightly different situation. Other studies (Brylinsky et al., 1994; Kaiser & Spencer, 1996) have already emphasised that it is more difficult to prove the ability of demersal fishing gears to modify benthic communities and environment in areas subjected to frequent natural perturbations, according to the hypothesis proposed by Boesch & Rosenberg (1981) for which the communities living in less constant environments are more resistant to disturbance.

The analysis of the recolonization dynamics shows that a certain number of species soon cover the newly available space, while the process is found to be slower for other species.

Recolonization is mainly by species with high abundance in the surrounding area, rather than specifically short-lived opportunist species (Boesch & Rosenberg, 1981; Jones, 1992). This is indicated by the presence of both species well-known for their opportunistic behaviour, such as *Corbula gibba* (Pearson & Rosenberg, 1978; Crema, 1989), and species lacking such characteristics, such as *Tapes philippinarum* and *Paphia aurea*, in the pool of species which significantly contribute to the average diversity. As a matter of fact, the two last-mentioned species, characterised by many morphological elements considered typical of a K strategy, are found to have an opportunistic behaviour (*sensu* Thistle, 1981), being able to quickly recolonize the treated area.

This scheme is also pointed out by the analysis of the trophic groups in the area treated with the hydraulic dredge where, after an initial prevalence of deposit feeders, dominated by *C. gibba*, an increase of filter feeders, mainly *T. philippinarum* and *P. aurea*, follows.

In conclusion, we state that in studies aiming to evaluate the effects of short- to medium-term trawl-fishing activities it is also necessary to consider the type of community the investigation is performed on. In planning management of marine demersal resources, it must be taken into consideration that benthic communities have the capacity to adapt to an anthropic 'predictable' disturbance, but that there is also a threshold beyond which even populations consisting of 'resistant' species can collapse causing the community to become extremely simplified, with loss of its main ecological functions and inevitable negative feedback even on the target species.

Some species which are not usually considered to belong to the group of 'opportunists' can assume an opportunistic behaviour within particular situations of recolonization following a state of direct disturbance produced by fishing activities.

Acknowledgements

We thank Dr I. Lombardo, the referees who helped us to significantly improve the manuscript and Mrs Mirkka Jones for the language revision.

References

Ambrose, W. G., 1984a. Influence of residents on the development of a marine soft-bottom community. J. mar. Res. 42: 633–654.

Ambrose, W. G., 1984b. Role of predatory infauna in structuring marine soft-bottom communities. Mar. Ecol. Progr. Ser. 17: 109–115.

BEON, 1990. Effects of beamtrawls fishery on the bottom fauna in the North Sea. BEON Rep. 13, 85 pp.

Bergman, M. J. N. & M. Hup, 1992. Direct effects of beamtrawling on macrofauna in a sandy sediment in the southern North Sea. ICES J. mar. Sci. 49: 5–11.

Bergman, M. J. N. & J. W. van Saintbrink, 1994. Direct effects of beam-trawling on macrofauna in sandy areas off the Dutch coast. In de Groot, S. J. & H. J. Lindeboom (eds), Environmental impact of bottom gears on benthic fauna in relation to natural resources management and protection of the North Sea, NIOZ Rep. 11: 147–178.

Boesch, D. F. & R. Rosenberg, 1981. Response to stress in marine benthic communities. In Barret, G. W. & R. Rosenberg (eds), Stress effects on natural ecosystems, J. Wiley & Sons, New York: 179–200.

Bonsdorff, E., O. Karlsson & E. Leppäkoski, 1984. Ecological changes in the brackish water environment of the Finnish West coast caused by engineering works. Ophelia 3: 33–44.

Bonvicini Pagliai, A. M., A. M. Cognetti Varriale, R. Crema, M. Curini Galletti & R. Vandini Zunarelli, 1985. Environmental impact of extensive dredging in a coastal marine area. Mar. Pollut. Bull. 16: 483–488.

Bray, J. R. & J. T. Curtis, 1957. An ordination of the upland forest communities of Southern Wisconsin. Ecol. Monogr. 27: 325–349

Bridger, J. P., 1970. Some effects of the passage of a trawl over the sea-bed. ICES, CM 1970/B:10, Gear and Behaviour Communications, 10 pp.

Bridger, J. P., 1972. Some observations on the penetration into the seabed of tickler chains on a beam trawl. ICES, CM 1972/B:7 Gear and Behaviour Communications, 9 pp.

Brylinsky, M., J. Gibson, D. C. Gordon Jr., 1994. Impacts of flounder trawls on the intertidal habitat and community of the Minas Basin, Bay of Fundy. Can. J. Fish. aquat. Sci. 51: 650–661.

Clarke, K. R., R. M. Warwick, 1994. Change in marine communities: an approach to statistical analysis and interpretation. Natural Environment Research Council, UK, 144 pp.

Commito, J., 1982. Importance of predation by infaunal polychaetes in controlling the structure of a soft bottom community in Maine, USA. Mar. Biol. 68: 77–81.

Crema, R., 1989. Aspetti della successione in comunità macrozoobentoniche marine di fondi mobili. SITE Atti 7: 315–320.

Dean, D. & H. Haskin, 1964. Benthic repopulation of the Raritan River estuary following pollution abatement. Limnol. Oceanogr. 9: 551–563.

de Groot, S. J., 1984. The impact of bottom trawling on benthic fauna of the North Sea. Ocean Management 10: 21–36.

Eleftheriou, A. & M. R. Robertson, 1992. The effects of experimental scallop dredging on the fauna and physical environment of a shallow sandy community. Neth. J. Sea Res. 30: 289–299.

Hall, S. J., D. J. Basford, N. R. Robertson, 1990. The impact of hydraulic dredging for razor clams *Ensis* sp. on infaunal community. Neth. J. Sea Res. 27: 119–125.

Hall, S. J., 1994. Physical disturbance and marine benthic communities: life in unconsolidated sediments. Oceanogr. Mar. Biol. Ann. Rev. 34: 179–239.

Hawkins, S. J., 1981. The influence of season and barnacles on the algal colonization of *Patella vulgata* exclusion areas. J. Mar. Biol. Ass. U.K. 61: 1–15.

Hily, C., 1983. Macrozoobenthic recolonization after dredging. Oceanol. Acta. Proc. 17th EMBS Brest: 113–120.

Hutchings, P., 1990. Review of the effects of trawling on macrobenthic epifaunal communities. Aust. J. Mar. Freshwat. Res. 41: 111–120.

ICES, 1988. Report of the study group on the Effects of Bottom Trawling. ICES, CM 1988/B 56, 30 pp.

Jones, J. B., 1992. Environmental impact of trawling on the seabed: a review. New Zealand J. Mar. Freshwat. Res. 26: 59–67.

Kaiser, M. J. & B. E. Spencer, 1996. The effects of beam-trawl disturbance on infaunal communities in different habitats. J. anim. Ecol. 65: 348–358.

Kruskal, J. B., 1964. Multidimensional scaling by optimizing goodness to fit a non-metric hypothesis. Psychometrika 29: 1–27.

Leppäkoski, E., 1975. Assessment of degree of pollution on the basis of macrozoobenthos in marine and brackish water environments. Acta Acad. Aboensi, Ser. B. 35: 1–90.

MacDonald, D. S., M. Little, C. Eno, K. Hiscock, 1996. Disturbance of benthic species by fishing activities: a sensitivity index. Aq. Conserv. Mar. Freshwat. Ecos. 6: 257–268.

Pearson, T. H. & R. Rosenberg, 1978. Macrobenthic succession in relation to organic enrichment and pollution of the marine environment. Oceanogr. Mar. Biol. Ann. Rev., 16: 229–311.

Pickett, J., 1973. The impact of mechanical harvesting on the Thames estuary cockle fishery. MAFF Lab. Leaflet (New Series) 29: 1–27.

Poiner, I. R., R. Kennedy, 1984. Complex pattern of changes in the macrobenthos of a large sandback following dredging. I. Community analysis. Mar. Biol. 78: 335–352.

Pranovi, F., 1994. Indagine sulle comunità bentoniche di alcuni biotopi della Laguna di Venezia. Nova Thalassia 12: 43–58.

Pranovi, F. & O. Giovanardi, 1994. The impact of hydraulic dredging for short-necked clams, *Tapes* spp., on an infaunal community in the lagoon of Venice. Sci. Mar. 58: 345–353.

Rees, H. L. & A. Eleftheriou, 1989. North Sea benthos: A review of field investigation into the biological effects of man's activities. J. Cons. int. Expl. Mer 45: 284–305.

Reise, K., 1982. Long-term changes in the macrobenthic invertebrate fauna of the Wadden Sea: are polychaetes about to take over? Neth. J. Sea Res. 16: 29–36.

Rosenberg, R., 1976. Benthic faunal dynamics during succession following pollution abatement in a Swedish estuary. Oikos 27: 414–427.

Sanders, H. L., 1968. Marine benthic diversity: a comparative study. Am. Nat. 102: 243–282.

Sousa, W. P., 1980. Experimental investigations of disturbance and ecological succession in a rocky intertidal algal community. Ecol. Monogr. 49: 227–254.

Sutherland, J. P., 1974. Multiple stable points in natural communities. Am. Nat. 108: 859–873.

Thistle, D., 1981. Natural physical disturbances and communities of marine soft bottoms. Mar. Ecol. Progr. Ser. 6: 223–228.

Valderhaug, V. A. & J. S. Gray, 1984. Stable macrofauna community structure despite fluctuating food supply in subtidal soft sediment of Oslofjord, Norway. Mar. Biol. 82: 307–322.

Zajac, R. N. & R. B. Whitlatch, 1982. Response of estuarine infauna to disturbance. I. Spatial and temporal variations of initial recolonization. Mar. Ecol. Progr. Ser. 10: 1-14.

Hydrobiologia **375/376:** 137–149, 1998.
S. Baden, L. Pihl, R. Rosenberg, J.-O. Strömberg, I. Svane & P. Tiselius (eds),
Recruitment, Colonization and Physical–Chemical Forcing in Marine Biological Systems.

Vertical migration and selective tidal stream transport in the megalopa of the crab *Carcinus maenas*

Henrique Queiroga
Departamento de Biologia, Universidade de Aveiro, Campus de Santiago, 3810 Aveiro, Portugal

Key words: recruitment, transport, larvae, vertical migration, estuaries, Portugal

Abstract

Megalopae of *Carcinus maenas* (L.) were intensively sampled in the Canal de Mira (Ria de Aveiro, Portugal) during the years of 1990 and 1991, with the use of a pump and nets, respectively. Plankton sampling was carried out along 25 h cycles conducted at fixed stations, at pre-determined depths along the water column. In 1990, the average depth of distribution of megalopae throughout the water column was affected by tidal phase: the larvae were closer to the surface during flood than ebb. Stepwise linear regression showed that a shallower distribution of the larvae was also associated with higher salinity. Other hydrological variables did not account well for the vertical distribution of megalopae. Analysis of megalopal instantaneous transport velocity indicated that their horizontal velocity depended on phase-of-tide: during ebb megalopae were transported downstream at a lower velocity than the vertically integrated water column velocity; during flood, the vertically integrated velocity of the larvae and water column were similar. Thus, a shift of vertical position during the tidal cycle controlled the transport velocity of megalopae, which was due to vertical water velocity shear current differences. During the 1991 study, significantly higher densities of megalopae were collected during flood. Main effects of phase-of-day and depth of sampling on density were not significant. However, highest densities during flood were found at mid-water, in both day phases, whereas during ebb megalopae were evenly distributed throughout the water column during the day, with low densities, or showed increasing abundances towards the bottom during the night. These observations are consistent with the hypothesis that megalopae of *C. maenas* undergo active vertical migration synchronised with the tidal cycle, which can be classified as a selective tidal stream transport mechanism to travel up estuaries. A conceptual model for the reinvasion of estuaries by *C. maenas* megalopae is proposed. According to the model, vertical movements between the bottom and the water column are controlled by exogenous factors associated with the tidal cycle.

Introduction

The portunid crab *Carcinus maenas* (L.) is the commonest European estuarine crab species. It forms a large population in the Ria de Aveiro, a bar-built estuary located on the north-western coast of Portugal (Figure 1). Queiroga et al. (1994, 1997) and Queiroga (1995, 1996) have described several aspects of the species' larval ecology. Synchronous release of first zoeae occurs during night-time ebbing tides. These larvae use selective tidal stream transport to maximise seaward transport, migrating vertically in synchrony with the tidal cycle so that they are closer to the surface during the span of the ebb. All larval stages are found in shelf waters. Shoreward advection of megalopae seems to occur during periods of relaxed northerly, upwelling-favourable, winds. Inside the estuary, megalopae are significantly more abundant during floods than during ebbs, and at night.

Migration of larvae from the sea to estuarine habitats has been recognised as a 2-step process (Boehlert & Mundy, 1988; Shanks, 1995; Forward et al., 1995): (1) cross-shelf transport of the larvae towards the shore; (2) passage through the inlet and upstream movement until an appropriate environment is found.

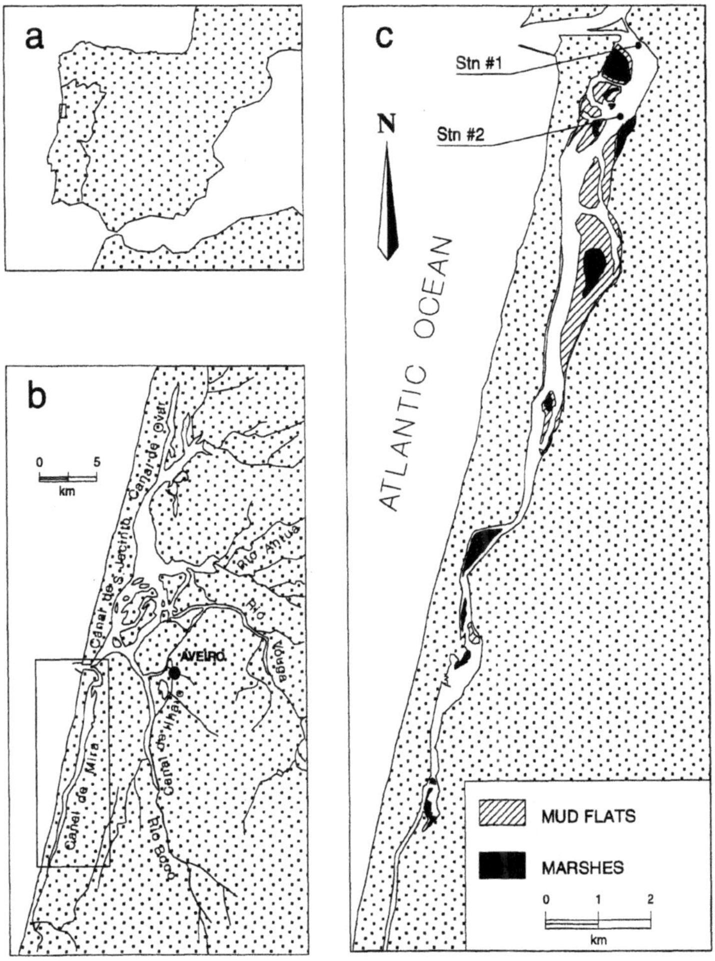

Figure 1. Study site showing the Portuguese west coast (a), Ria de Aveiro (b) and Canal de Mira with location of Stations #1 and #2 (c) at which field data was collected.

The physical factors involved in the control of these two steps differ, and thus different larval behavioural mechanisms are required in each phase.

Because planktonic larvae are unable to swim against estuarine currents, upstream movement must be accomplished through any kind of adaptation that opposes the net downstream flux of waters (Boehlert & Mundy, 1988; Epifanio, 1988). One such adaptation is selective tidal stream transport. Selective tidal stream transport is a term introduced by Harden Jones et al. (1988) to describe the migration of flat fish with the help of tidal currents. Flat fish settle on the bottom during a particular phase of the tide. When the tide turns they ascend to the water column, taking advantage of the downstream transport in the 'appropriate' direction. This mechanism involves a vertical migration rhythmic behaviour synchronised with the tidal cycle. Field investigations concerning fish larvae (Fortier & Legget, 1983; Laprise & Dodson, 1989; Rowe & Epifanio, 1994) and crab zoeae (Queiroga et al., 1997) and megalopae (Olmi, 1994) reported tidally timed vertical movements that can be classified as selective tidal stream transport mechanisms to maximise transport in estuaries, to move both up- and downstream, depending on species and larval stage. In the case of larval fish and crab zoeae the migration is entirely planktonic, and a resting-on-the-bottom phase does not occur. The larvae simply move closer to the bottom where, due to reduced water velocities caused by bottom friction, they experience slower transport velocities in the 'wrong' direction.

Previous studies on *Uca* spp. (Christy, 1982; Dittel & Epifanio, 1990; Little & Epifanio, 1991) *Callinectes sapidus* (Brookins & Epifanio, 1985; Little

& Epifanio, 1991) and *Carcinus maenas* (Queiroga et al., 1994) reported higher megalopal abundances in the water column during night-time flood tides. To explain this pattern the authors postulated that most megalopae passing through the sampling station during flood settled upstream before the turn of the tide. This behaviour would result in a net upstream transport by selective tidal stream transport. However, none of these studies described the actual process of vertical migration of the larvae. The only conclusive evidence that crab megalopae do migrate vertically in synchrony with the tide, when entering estuarine waters, is given by Olmi (1994). His study described vertical migrations of *C. sapidus* megalopae that resulted in distributions of the larvae higher in the water column during flood and at night.

In the present paper I address the temporal patterns of change of the vertical position of the megalopa of *Carcinus maenas* in an estuary, and its consequences to the transport processes of these larvae. I will show that: (i) *C. maenas* megalopae are closer to the surface during flood tide; (ii) the observed shift of vertical position according to phase-of-tide is consistent with the hypothesis that these larvae use selective tidal stream transport to travel up the estuary; (iii) the transport velocity of megalopae depends on their vertical position throughout the water column.

Materials and methods

The study site was located in the Canal de Mira (Figure 1). Tides in the Ria de Aveiro are semi-diurnal, with a mean tidal range of 2.1 m at the inlet. Circulation in the Canal de Mira can be classified as a two-layer flow with vertical mixing (Bowden, 1967; Queiroga, 1995). Two sampling experiments were carried out, hereafter referred to as the 1990 and 1991 studies. During the 1990 study plankton samples were collected with a pump (see Queiroga et al., 1994, 1997). Plankton nets were used in the 1991 study.

1990 study

Field sampling. Plankton samples were collected with a pump, fitted to a 500 mm net, during 2 sampling periods: one in the winter (February 12 to March 11) and the other in the spring (May 9 to June 7) of 1990. Each seasonal sampling period was designed as a series of 25 h sampling cycles at fixed stations. Within each 25 h sampling cycle plankton samples were collected every hour at several depths. In winter, the samples were collected in a quick succession over 1 m depth intervals within each hourly period, starting at 0.2 m. Every time the last sample in the series was collected more than 0.5 m from the bottom, an extra sample was taken at 0.2 m from the bottom. The filtered volume for a pumping time of 2 min at each depth was about 1.0 m^3. In spring, plankton samples were collected only at three levels: 0.5 m below the surface, at mid water and 0.5 m above the bottom. Pumping time at each depth was increased to 5 min, corresponding to filtered volumes of about 2.2 m^3. Although the original sampling program was conducted at three stations, very small number of megalopae were collected at two of them. Accordingly, only the sampling cycles carried at Station #2 are analysed.

Hydrological measurements were obtained every hour, at the standard 1 m depth intervals, immediately before the collection of the plankton samples. Salinity and temperature were measured during all 25 h cycles, using a Yellow Springs Instrument Model 33 SCT meter. During some of the 25 h cycles current speed and direction could also be obtained (Table 1) with the use of a Valeport BMF008 current meter. The height of the water column was always measured every hour with a sounding line. Average depth and tidal range at Station #2 were 4.7 and 2.1 m respectively.

Analysis of hydrological data. Stratification of the water column was characterised by an index of stratification, S_t, calculated as:

$$S_t = \frac{\sigma_{t\ \mathrm{surface}} - \sigma_{t\ \mathrm{bottom}}}{Z_t}, \tag{1}$$

where t is time, σ was derived from the salinity and temperature measurements according to the equation given by Fofonoff and Millard (1983) and Z_t is the instantaneous depth of the water column.

Vertical distribution of the larvae. To inspect temporal patterns of vertical position of *Carcinus maenas* megalopae, the depth specific instantaneous concentration of the larvae, C_{zt} (ind m^{-3}), was first calculated, as a function of discrete depth (z) and time (t) intervals:

$$C_{zt} = \frac{\mathrm{counts}_{zt}}{\mathrm{filtered\ volume}_{zt}}. \tag{2}$$

The instantaneous mean depth of the megalopae, $\overline{Dl_t}$, (hereafter referred to as the instantaneous depth of the megalopae), was calculated as the weighted average depth of collection:

Table 1. Data concerning the 25 h fixed station plankton sampling cycles performed in the Canal de Mira

Year	Date	Station #	Number of samples: total (with megalopae)	Number of sampling moments: total (with megalopae)	Water current data?	Salinity and temperature data?
1990 Winter	12 Feb	2	107 (14)	25 (9)	Yes	Yes
	14 Feb	2	119 (9)	25 (7)	Yes	Yes
	16 Feb	2	124 (3)	25 (3)	Yes	Yes
	18 Feb	2	105 (0)	25 (0)	Yes	Yes
	20 Feb	2	128 (1)	25 (1)	Yes	Yes
	22 Feb	2	125 (0)	25 (0)	Yes	Yes
	24 Feb[a]	2	76 (0)	13 (0)	No	Yes
	26 Feb	2	127 (0)	25 (0)	Yes	Yes
	28 Feb	2	123 (0)	25 (0)	No	Yes
	04 Mar	2	126 (1)	25 (1)	Yes	Yes
	06 Mar	2	131 (0)	25 (0)	Yes	Yes
	11 Mar	2	132 (0)	25 (0)	No	Yes
1990 Spring	09 May	2	75 (4)	25 (4)	Yes	Yes
	11 May	2	75 (11)	25 (9)	Yes	Yes
	13 May	2	75 (12)	25 (10)	Yes	Yes
	15 May	2	75 (3)	25 (2)	No	Yes
	19 May	2	75 (2)	25 (1)	Yes	Yes
	23 May	2	75 (1)	25 (1)	No	Yes
	25 May	2	75 (1)	25 (1)	No	Yes
	02 Jun	2	75 (6)	25 (5)	No	Yes
	07 Jun	2	75 (1)	25 (1)	No	Yes
1991	13 Apr	1	39 (13)	13 (6)	No	No
	22 Apr	1	39 (28)	13 (13)	No	No
	06 May	1	39 (13)	13 (10)	No	No
	14 May	1	39 (16)	13 (11)	No	No
	21 May	1	39 (12)	13 (7)	No	No
	28 May	1	39 (3)	13 (2)	No	No
	05 Jun	1	39 (1)	13 (1)	No	No
	12 Jun	1	39 (3)	13 (3)	No	No
	27 Jun	1	39 (11)	13 (6)	No	No

[a] Only the first 13 hours of the series were sampled.

$$\overline{Dl_t} = \frac{\sum_{z=l}^{n} C_{zt} \times D_{zt}}{\sum_{z=l}^{n} C_{zt}} \times \frac{1}{Z_t}, \qquad (3)$$

where D is the depth of sampling and Z is the total height of the water column. In the above equation, data are standardised by dividing the absolute mean depth of the megalopae at the moment of collection by the total height of the water column, yielding an average relative depth where surface has a value of 0 and bottom a value of 1. This transformation is convenient because it enables comparisons of megalopal position along the vertical axis independent of tidal state.

Horizontal transport. The influence of the vertical position of megalopae on their net tidal transport was assessed by first calculating the vertically integrated instantaneous current velocity $\overline{u_t}$ (m s^{-1}):

$$\overline{u_t} = \frac{\sum_{z=l}^{n} u_{zt} \times \Delta D_{zt}}{\sum_{z=l}^{n} \Delta D_{zt}}, \qquad (4)$$

where u is the longitudinal component of velocity, ΔD is the height of each stratum and $\sum_{z=l}^{n} \Delta D_{zt}$ equals Z_t, the instantaneous height of the water column. Subsequently, a vertically integrated instantaneous megalopal velocity, $\overline{u_{l_t}}$ (m s^{-1}) was calculated, and designated as the instantaneous megalopal velocity:

$$\overline{u_{l_t}} = \frac{\sum_{z=l}^{n} u_{zt} \times \Delta D_{zt} \times C_{zt}}{\sum_{z=l}^{n} \Delta D_{zt} \times C_{zt}}, \qquad (5)$$

where the different symbols have the meaning explained above. Note that if the megalopae are uniformly distributed throughout the water column, the C_{zt} terms in the above equation are all equal and cancel out to produce a value that is equal to $\overline{u_t}$ in Equation 4. The larvae are thus transported at a velocity that equals the vertically integrated current velocity. If the larvae do not distribute evenly with depth, then $\overline{u_{l_t}}$ does not equal $\overline{u_t}$. Since tidal current intensity usually increases with increasing distance from bottom due to a decrease in bottom friction, changes in vertical position result in instantaneous transport velocities of megalopae that differ from the depth integrated current velocity. The concept of larval velocity described above is similar to that reported by Rowe and Epifanio (1994), except that depth integrated larval velocity was calculated for each time interval, rather than time integrated velocity for each depth interval.

1991 study

During this study, nine 25 h plankton sampling cycles were performed at Station #1, mostly at weekly intervals, between April 13 and June 27 (Table 1). Three different nets were used, all equipped with 500 μm meshes and flowmeters: a neuston net with an opening of 0.60 × 0.30 m; a conic net with a 0.40 m diameter opening which was towed at mid-water; and a similar conical net mounted on a sledge, to sample close to the bottom. Every 2 h the three nets were deployed in a quick succession. Each tow lasted approximately 5 min. The vertical position of the neuston net was adjusted so that the immersed height of its mouth was 0.20 m on average. Assuming this, a filtered volume was calculated for the three nets, and abundances expressed in ind m^{-3}. Average filtered volumes were 35.1, 30.7 and 25.3 m^3 for the neuston, plankton and sledge net, respectively. No currents, salinity and temperature data were collected. Average depth and tidal amplitude at Station #1 were 6.5 m and 2.3 m.

Statistical analysis

For purposes of statistical analysis, samples collected during the 1990 and 1991 studies were classified using the following criteria. Tides covered by a particular 25 h fixed sampling cycle were designated by the starting date of that cycle, even when they overlapped the next calendar day. All samples collected in each sampling moment (=each 1 h period in the 1990 study; each 2 h period in the 1991 study) were classified as ebb or flood samples according to the phase of the tide, defined by high and low water slack times, during which the discrete time interval began. Similarly, the samples were classified as day time or night time, where the day phase was taken as the period between daily sunrise and sunset and the night phase as the complement of this period. All statistical analysis described bellow were made using the SPSS package.

1990 study. The significance of the effect of phases-of-tide and day, and sampling period, on the vertical position of the megalopae was analysed using a three-way analysis of variance (ANOVA), in which phase-of-tide (ebb, flood), phase-of-day (day, night) and sampling period (winter, spring), each with two levels, constituted the three factors in the analysis. Prior to analysis data were subjected to an angular transformation, which homogenised the variances in all cases. The effects of the phases of tide and day, and of the sampling period on the instantaneous megalopal velocity, were analysed by means of a three-way ANOVA as above, but the instantaneous current velocity was introduced in the analysis as a co-variable. The analysis was performed on the subset of data for which current data were obtained. No transformation was applied to the data because variances were homogeneous.

The effects of environmental variables associated with the tidal cycle on vertical position of megalopae were analysed by stepwise linear regression. The independent variables were average salinity and temperature throughout the water column for each sampling moment, vertically integrated instantaneous current velocity and height of the water. The instantaneous mean depth values were first subjected to an angular transformation. One regression was calculated for the pooled data of winter and spring.

As stated before, during the spring plankton samples were obtained only below the surface, at mid depth and above the bottom, while during the winter period had been taken at 1 m depth intervals. In order to ascertain whether the different resolution that was used to sample the water column influenced the estimates of the instantaneous vertical position and velocity of the larvae, the first 13 sampling hours of the sampling cycles performed at Station #2 on February 16, 20 and 26 and March 4 and 6 were used. These observations were selected because, due to good weather conditions, the depth of sampling could be determined very precisely. The instantaneous depth and velocity of the megalopae were first calculated, for all sampling moments, with equations 3 and 5, using the observations obtained at regular intervals. The same variables were again calculated from a subsample, now using only the samples collected at 0.2 m, at the depth level closest to mid-height of the water column, and above the bottom. The two series of values were then compared by least squares linear regression. It could be demonstrated that the two methods of sampling did not produce significantly different estimates neither of the instantaneous vertical position nor of the instantaneous velocity of the megalopae. In both cases there was a highly significant liner relation ($P < 0.001$) between the estimates obtained with sampling at surface, mid-water and bottom and at regular intervals. Moreover, the slope of the regression line and the Y-intercept never differed from 1 and 0, respectively (t-test; $P > 0.20$).

1991 study. In order to analyse the effects of depth of sampling, phase-of-tide and phase-of-day on the concentration of megalopae, a three-way ANOVA was used again. However, variances were highly heterogeneous, a feature that was impossible to correct by any of the usual transformations. Accordingly, a non-parametric ANOVA of the ranks of the concentrations, rather than of the concentrations, was employed, following the procedure described by Zar (1984, page 249). Factors in the analysis were depth level (surface, mid-water and bottom), phase-of-tide (ebb, flood) and phase-of-day (day, night).

Results

1990 study

The average concentration of the megalopa was only slightly higher in spring than in winter: 0.028 ind m^{-3} in winter versus 0.031 ind m^{-3} in spring (Queiroga et al., 1994). Average concentrations during flood and ebb were, respectively: 0.050 ind m^{-3} and 0.008 ind m^{-3} in winter, 0.051 ind m^{-3} and 0.013 ind m^{-3} in spring. Megalopae were usually concentrated closer to the surface during flood than ebb (Figure 2). Analysis of the data collected within the sampling periods of winter and spring indicated that tidal phase, phase-of-day and sampling season variably affected the average vertical position of the megalopa (Figure 3). The ANOVA (Table 2) showed that megalopae were significantly closer to the surface during flood and in spring. ($P < 0.05$ in both cases). No significant effects of phase-of-day or of the interactions were found.

An increase in average salinity (AS) caused a decrease in the relative depth (RD) of distribution of the larvae ($\sin^{-1}\sqrt{\text{RD}} = 111.4 - 1.731$ AS; $P < 0.01$; Figure 4). The other hydrological variables employed in the regression analysis did not account well for the vertical position of the megalopa. As a consequence of rainfall, decreased salinity values recorded during winter at the surface resulted in lower average salinity throughout the water column, as well as in higher stratification values than in spring (Queiroga, 1995; Queiroga et al., 1997). A scatter diagram of the relative depth of the megalopa against the stratification index (Figure 5) shows that for stratification values over 0.2, which were more frequently observed in winter, no megalopae were collected in the upper half of the water column.

Average values of the difference between the instantaneous larval velocity and the instantaneous water velocity revealed several patterns (Figure 6). Results of the ANOVA (Table 3) showed that the effect of the co-variable current velocity was highly significant ($P < 0.001$), meaning that the velocity of the larvae is related to the velocity of the water current. The results further show that the phase of the tide, and the season, have a significant influence on the transport velocity of the megalopae ($P < 0.001$ and $P < 0.01$, respectively).

1991 study

Significantly ($P < 0.05$) more megalopae were collected during flood (0.073 ind m^{-3}) than during ebb (0.030 ind m^{-3}) (Figure 7, Table 4). The main effects of depth level and phase-of-day were not significant. However, proportionally more megalopae were collected during night ebbing than in day ebbing tides, which is reflected in a significant phase-of-tide by phase-of-day interaction ($P < 0.05$). Another noticeable feature of Figure 7 is that the highest adundances during flood were found at mid-water, in both day phases, whereas during ebb the megalopae were uniformly distributed throughout the water column during the day, albeit with very low densities, or showed increasing concentrations towards the bottom during the night. The depth level by phase-of-tide interaction was not statistically significant at the 5% level, although it was very close to statistical significance ($0.06 > P > 0.05$).

Discussion

During the 1990 study, *Carcinus maenas* megalopae were 5 times more abundant in the Canal de Mira dur-

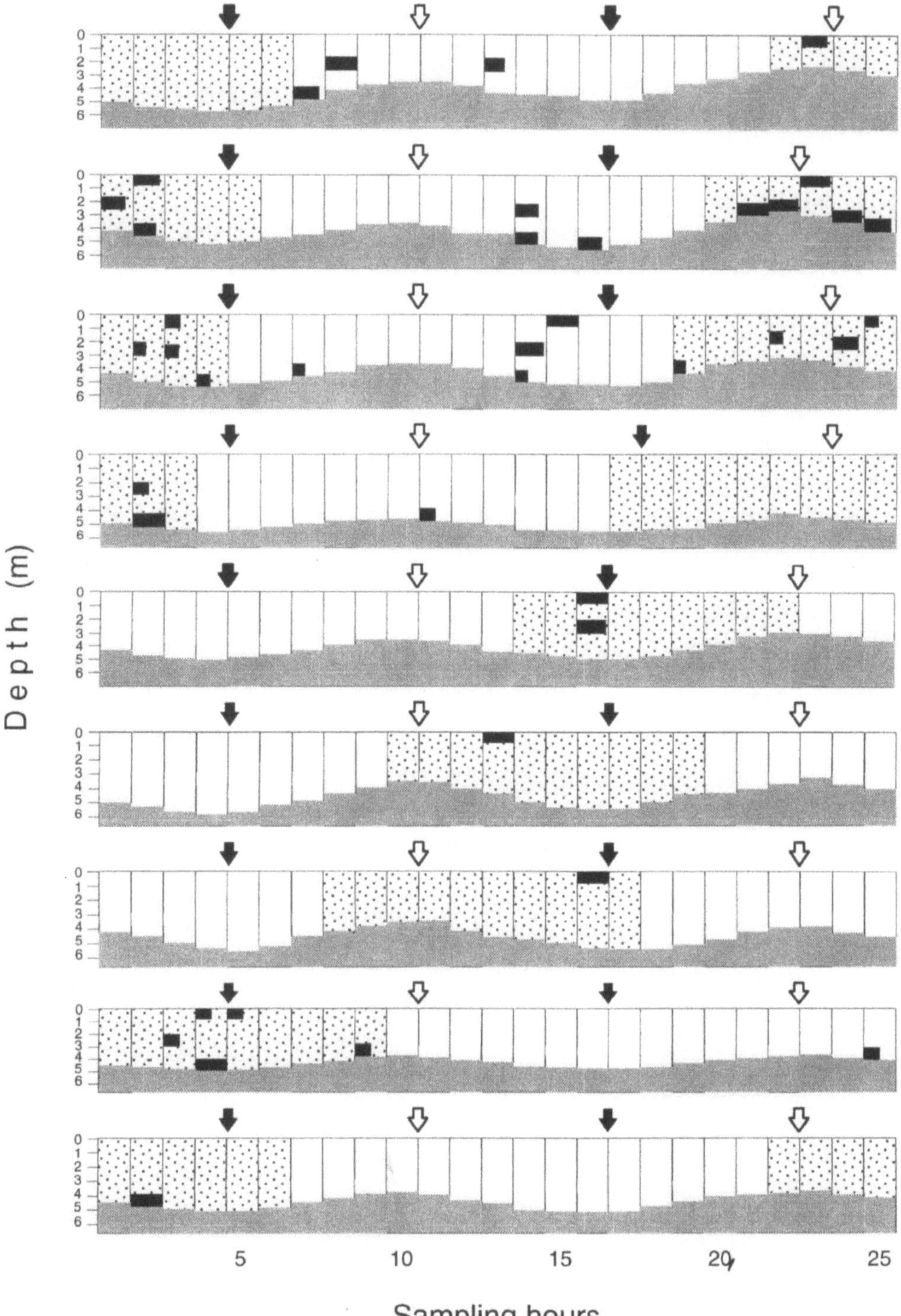

Figure 2. Carcinus maenas. Concentration of the megalopa recorded during the 1990 study, at Station #2, in each 25 h plankton sampling series, relative to depth level, tide phase and day phase (only the series where megalopae were collected are shown). Concentrations (larvae m^{-3}) are represented by horizontal bars at each depth and time interval and scaled to the maximum concentration (MC) in each 25 h series. Black arrow: high water slack; open arrow: low water slack; doted areas: night. In each 25 h series surface is set to level 0 m, while bottom, represented by the grey area in the bottom of each series, is seen rising and sinking with the tide.

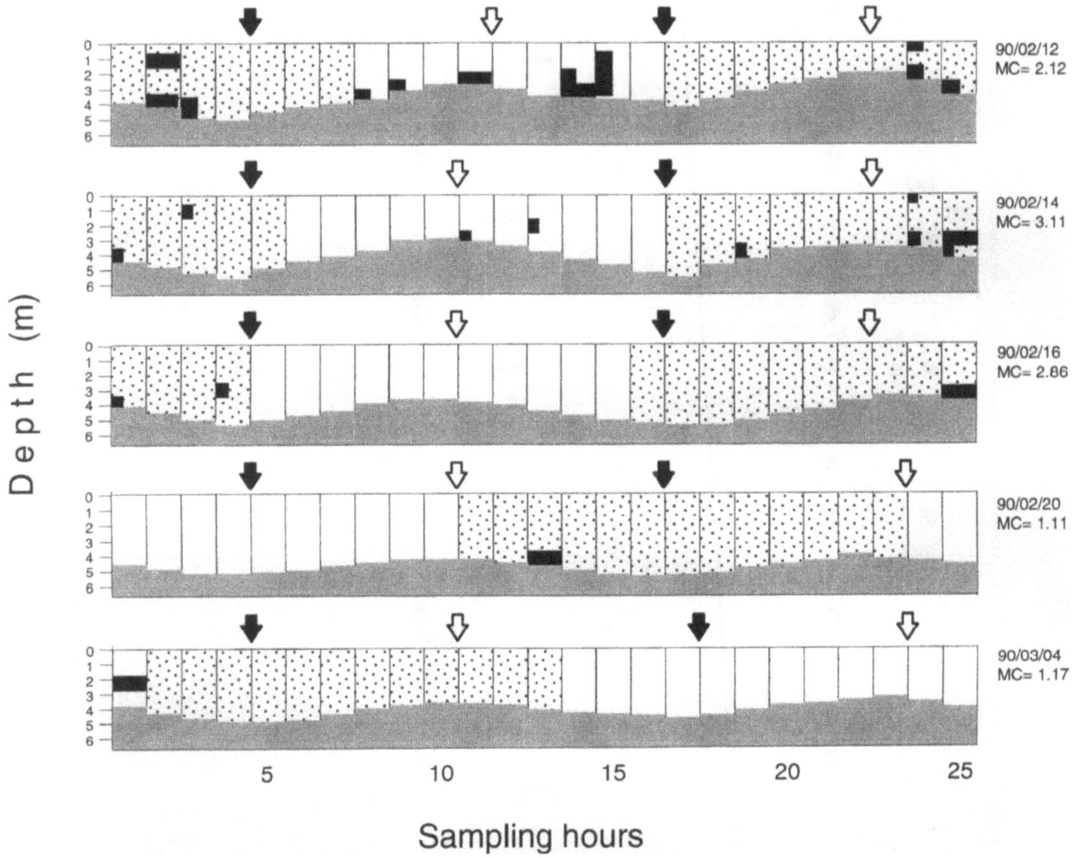

Figure 2. Continued

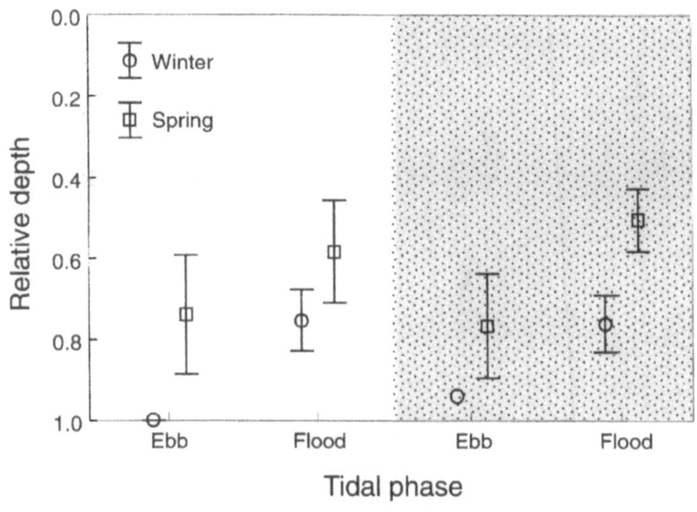

Figure 3. Carcinus maenas. Vertical position of the megalopa during the 1990 study at Station #2, according to phase-of-tide, phase-of-day and sampling period. Vertical bars represent 1 standard error. Dotted area: night.

Table 2. Three-way ANOVA of the vertical position of megalopae of *Carcinus maenas* during the 1990 study at Station #2. The null hypotheses are that, when there are megalopae in the water column, their average vertical position does not differ across treatments. df = degrees of freedom; MS = mean square; F_S = F test value; P = probability value

Source of variation	df	MS	F_s	P
Phase-of-tide	1	1960.69	4.34	$0.05 > P > 0.01$
Phase-of-day	1	586.65	1.30	$P > 0.25$
Sampling period	1	1850.15	4.10	$0.05 > P > 0.01$
Phase-of-tide by Phase-of-day	1	22.52	0.05	$P > 0.75$
Phase-of-tide by Sampling period	1	39.92	0.09	$P > 0.75$
Phase-of-day by Sampling period	1	21.76	0.05	$P > 0.75$
Phase-of-tide by Phase-of-day by Sampling period	1	225.83	0.50	$P > 0.25$
Error	47	451.59		
Total	54	555.43		

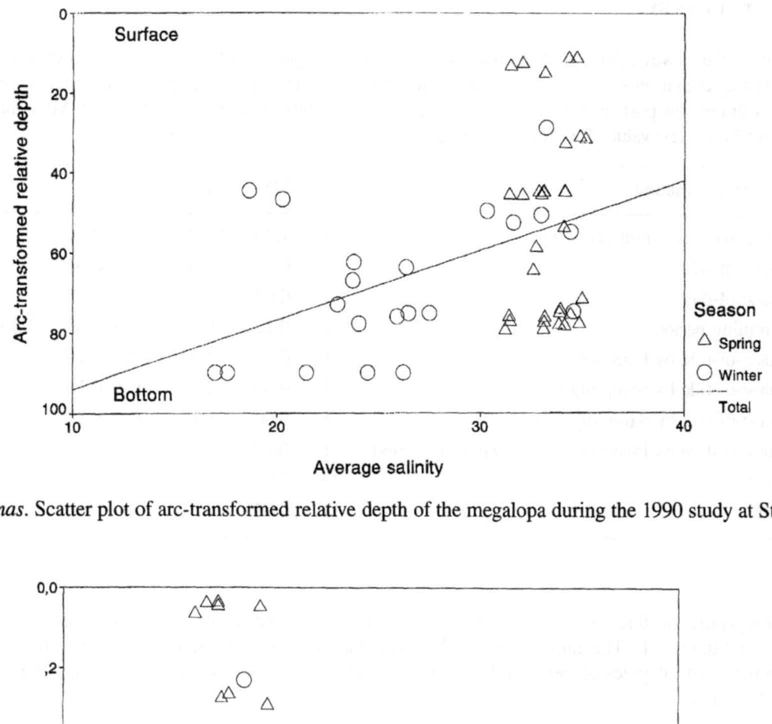

Figure 4. Carcinus maenas. Scatter plot of arc-transformed relative depth of the megalopa during the 1990 study at Station #2 as a function of salinity.

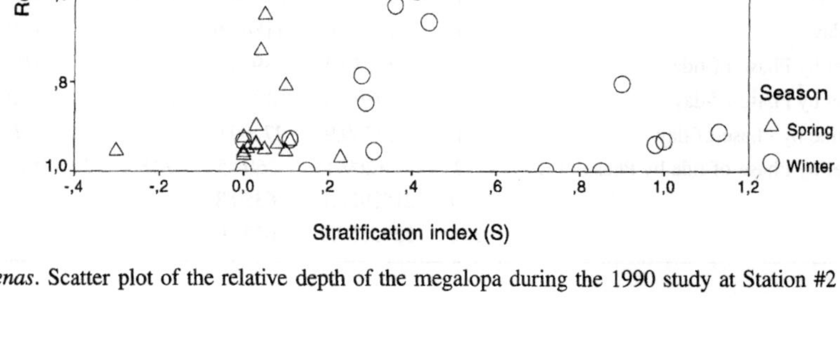

Figure 5. Carcinus maenas. Scatter plot of the relative depth of the megalopa during the 1990 study at Station #2 against the instantaneous stratification index.

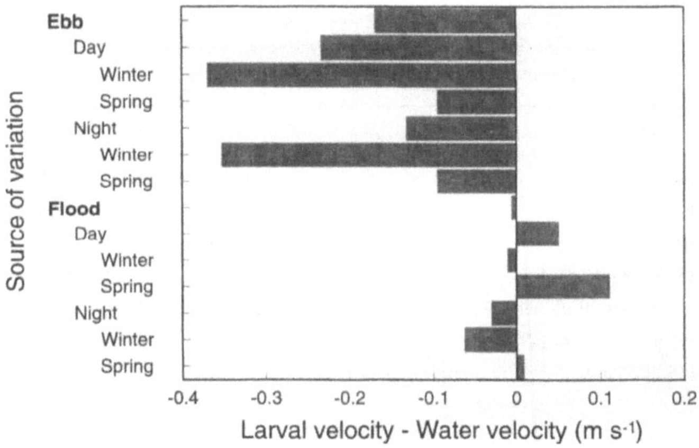

Figure 6. Difference between the instantaneous velocity of *Carcinus maenas* megalopa and the instantaneous velocity of water, recorded during the 1990 study at Station #2, by tidal state, phase-of-day and sampling period. Positive values of the difference indicate that the velocity of the larvae is higher than the current velocity.

Table 3. Three-way ANOVA of the transport velocity of megalopae of *Carcinus maenas* during the 1990 study at Station #2. The null hypotheses are that, when there are megalopae in the water column, their average transport velocity does not differ across treatments. df = degrees of freedom; MS = mean square; F_S = F test value; P = probability value

Source of variation	df	MS	F_S	P
Co-variable (Current velocity)	1	1.29	99.46	$P < 0.001$
Phase-of-tide	1	0.23	17.83	$P < 0.001$
Phase-of-day	1	0.01	0.65	$P > 0.25$
Sampling period	1	0.12	9.56	$0.01 > P > 0.001$
Phase-of-tide by Phase-of-day	1	0.01	1.15	$P > 0.25$
Phase-of-tide by Sampling period	1	0.03	2.40	$P > 0.10$
Phase-of-day by Sampling period	1	0.00	0.02	$P > 0.75$
Phase-of-tide by Phase-of-day by Sampling period	1	0.00	0.16	$P > 0.50$
Error	43	0.01		
Total	51	0.05		

Table 4. Nonparametric three-way ANOVA of the concentration of megalopae of *Carcinus maenas* during the 1991 study at Station #1. The null hypotheses are that the concentration of the megalopae does not differ across treatments. df = degrees of freedom; SS = sum of squares; MS = mean square; H = Kruskal-Wallis statistic; P = probability value

Source of variation	df	SS	MS	H	P
Depth level	2	9274.8	4637.4	1.420	$P > 0.10$
Phase-of-tide	1	29976.4	29976.4	4.590	$0.05 > P > 0.01$
Phase-of-day	1	11842.6	11842.6	1.813	$P > 0.10$
Depth level by Phase-of-tide	2	37838.4	18919.2	5.794	$P > 0.05$
Depth level by Phase-of-day	2	9805.0	4902.5	1.501	$P > 0.50$
Phase-of-tide by Phase-of-day	1	27209.9	27209.9	4.166	$0.05 > P > 0.01$
Depth level by Phase-of-tide by Phase-of-day	2	1209.1	604.5	0.185	$P > 0.75$
Error	339	2152922.1	6350.8		
Total	350	2285850.0	6531.0		

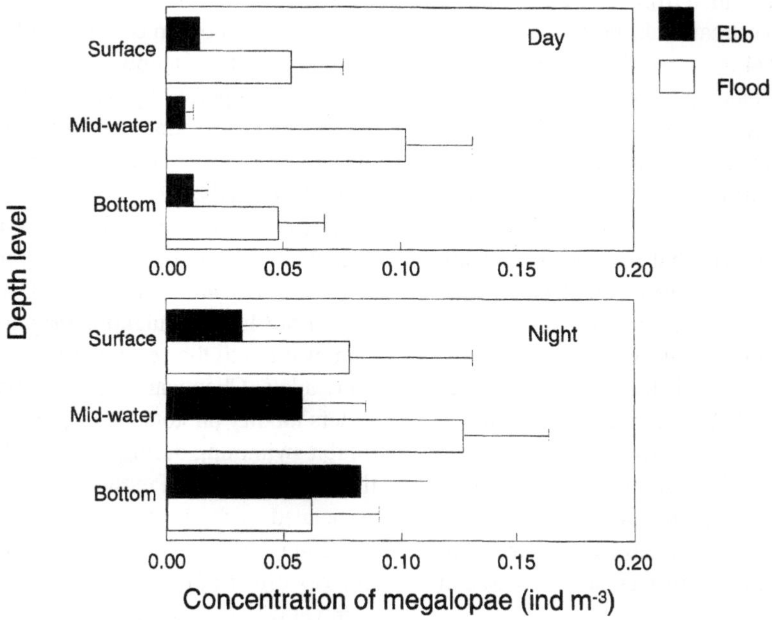

Figure 7. Carcinus maenas. Average concentration of megalopae during the 1991 study at Station #1, according to depth level, tidal phase and day phase. Thin bars represent +1 standard error.

ing the flood than ebb tide (see Queiroga et al., 1994). During the 1991 study they were 2.5 times more abundant during the rising tide. This seems to contrast with studies on *Callinectes sapidus* and *Uca* spp. that reported very low abundances during ebb (Christy, 1982; Epifanio et al., 1984; Brookins & Epifanio, 1985; Little & Epifanio, 1991; Olmi, 1994). Nevertheless, the patterns found during the present study imply a net upstream transport of *Carcinus maenas* megalopae in the Canal de Mira. Both the 1990 and 1991 studies produced evidence that megalopae were nearer the surface during flood than during ebb. Taken in conjunction, all these observations are strongly consistent with the hypothesis that *C. maenas* megalopae use selective tidal stream transport to travel up the estuarine environment.

The moment and the nature of the factors that initiate the rhythmic behaviour of crab megalopae invading the estuaries has been addressed recently. Swimming of *Callinectes sapidus* and *Uca* spp. megalopae in estuarine water can be altogether suppressed by high light intensities. This difference in photobehaviour between estuaries and coastal waters is induced by chemical cues present in the water. The absence or low densities of megalopae observed during the day in estuaries would then result from swimming inhibition by high light levels, and concurrent sinking to the bottom (Forward & Rittschof, 1994).

Differences between ebb and flood concentrations of *C. sapidus* and *Uca* spp. megalopae cannot be explained by rates of pressure change, because response thresholds obtained in the laboratory were clearly above levels experienced during flood tides in the field (Tankersley et al., 1995). However, in *C. sapidus* the ascent to the water during flood seems to be cued by increasing salinity during the rising tide (De Vries et al., 1995; Forward et al., 1995). In other cases, presence of megalopae in the water is related to rates of change in pressure and current velocity (De Vries et al., 1994). In the case of *Uca* spp. swimming to the water column seems to be caused by an endogenous activity rhythm (Tankersley et al., 1995). Freshly collected *Carcinus maenas* megalopae approaching metamorphosis, kept in the laboratory under constant conditions, showed an endogenous vertical swimming rhythm (Zeng & Naylor, 1996b). The vertical upward movement resulted from active swimming and descent from passive sinking. Surprisingly, however, the larvae moved up during the expected ebb and down during the expected flood tide, exactly like the species' first zoea did (Zeng & Naylor, 1996a). Thus, factors that trigger *C. maenas* megalopal behaviour observed in the field (this study) must be exogenous, because the endogenous rhythm described from laboratory studies would result in an downstream transport of the larvae.

Queiroga et al. (1994) reported significantly higher densities of *Carcinus maenas* megalopae during the night in the 1990 study. No difference in the depth of distribution of the larvae was detected between day phases (this report). During the 1991 study, however, phase-of-day clearly had an effect not only on the mean density of megalopae, but also on their depth of distribution. This effect depended on phase-of-tide. During the flood, night and day vertical distribution of megalopae were similar. During day ebbing tides, megalopae were uniformly distributed with depth, showing low densities; during night ebbings, concentrations increased clearly towards the bottom. This pattern could be explained assuming a light inhibition of swimming activity during the day, overrided by a reaction to an environmental variable associated with the rising tide that would cause megalopae to ascend during flood.

Temperature, current velocity and height of the water column (which is a measure of the pressure to which a megalopa resting on the bottom is subjected), did not account well for the vertical distribution of *Carcinus maenas* megalopae during the 1990 study. However, an increase in mean salinity of the water column caused a shallower distribution of the larvae. This effect suggests that a salinity increase concurrent with the rising tide might trigger swimming activity, and thus cause an ascent in the water. Previous investigations of the behavioural responses of *C. maenas* megalopae to pressure variation indicated the presence of high barokinesis, with an absolute threshold response of 10 mbar to a pressure increase. Pressure increase resulted always in upward swimming, either in darkness or in light from above, one side or bellow. When subjected to a pressure decrease of unspecified amount, megalopae sank passively (Rice, 1966; Knight-Jones and Morgan, 1966). A pressure difference of 10 mbar corresponds to about 0.10 m of water, which is much less than the water height change associated with an extreme neap tide in the Portuguese coast (ca. 1.0 m). Absolute values like the ones used in the present study, or absolute differences, in environmental variables, however, may not be adequate to describe induced behaviour of the larvae. Rates of change of environmental variables with time (e.g. Forward, 1985; 1989; Forward et al., 1989; Tankersley et al., 1995), as well as the sign of the change (e.g. Rice, 1966), which I did not test, may be of greater biological significance. So, the possibility remains that increased swimming activity by a *C. maenas* megalopa resting on the bottom could be induced by the rise of the tide, resulting in upward movement. The response to pressure reported by Rice (1966), independent of light level, is also consistent with the large numbers of megalopae collected during day and night flood tides in the 1991 study.

Thus, based on available evidence, I propose the following model for the reinvasion of estuaries by *Carcinus maenas* megalopae. Under the influence of coastal/shelf circulation, megalopae are advected shoreward until they eventually concentrate near estuarine inlets. Chemical substances present in estuarine waters modify phototactic responses of the larvae and inhibit swimming during the day. Megalopae thus settle to the bottom. Pressure and/or salinity increase associated with the rising tide then evoke upward swimming by negative geotaxis, overriding the inhibitory effect of day light levels, causing megalopae to travel with night or day flood currents into, and then up, estuaries. Decreasing salinity during ebb induces passive sinking or swimming to the bottom; this effect may be enhanced by decreasing pressure. During the course of the vertical migrations, water velocity shear current differences further control the amount of horizontal transport during each tidal phase, maximising net upstream transport.

Acknowledgements

I wish to thank Mr Fernando Morgado (Departamento de Biologia, Universidade de Aveiro) and the many students of the Departamento de Biologia who made possible the successful accomplishment of field work. I am also in debt to three anonymous referees who gave a significant contribution to the final form of the manuscript. This study received financial support from Junta Nacional de Investigação Científica e Tecnológica (Project no. 87/122), and from a scholarship granted by the Instituto Nacional de Investigação Científica.

References

Boehlert, G. W. & B. C. Mundy, 1988. Roles of behavioural and physical factors in larval and juvenile fish recruitment to estuarine nursery areas. A. Fish. Soc. Symp. 3: 51–67.

Bowden, K. F., 1967. Circulation and diffusion. In G. H. Lauff (ed.), Estuaries. American Association for the Advancement of Science. Publication No. 83, Washington, D. C.: 15–36.

Brookins, K. G. & C. E. Epifanio, 1985. Abundance of brachyuran larvae in a small coastal inlet over six consecutive tidal cycles. Estuaries 8: 60–67.

Christy, J. H., 1982. Adaptive significance of semilunar cycles of larval release in fiddler crabs (Genus *Uca*): test of an hypothesis. Biol. Bull. Mar. Biol. Lab. (Woods Hole) 163: 251–263.

De Vries, M. C., R. A. Tankersley, R. B. Forward, Jr., W. W. Kirby Smith & R. A. Luettich, Jr., 1994. Abundance of estuarine crab larvae is associated with tidal hydrological variables. Mar. Biol. 118: 403–413.

Dittel, A. I. & C. E. Epifanio, 1990. Seasonal and tidal abundance of crab larvae in a tropical mangrove system, Gulf of Nicoya, Costa Rica. Mar. Ecol. Prog. Ser. 65: 25–34.

Epifanio, C. E., 1988. Transport of invertebrate larvae between estuaries and the continental shelf. Am. Fish. Soc. Symp. 3: 104–114.

Epifanio, C. E, C. C. Valenti & A. E. Pembroke, 1984. Dispersal and recruitment of blue crab larvae in the Delaware Bay. Estuar. Coast. Shelf. Sci. 18: 1–12.

Fofonoff, N. P. & R. C. Millard, Jr, 1983. Algorithms for computation of fundamental properties of seawater. UNESCO, Paris. Technical papers in Marine Sciences No. 44.

Fortier, L. & W. C. Leggett, 1983. Vertical migrations and transport of larval fish in a partially mixed estuary. Can. J. Fish. Aquat. Sci. 40: 1543–1555.

Forward, R. B., Jr., 1985. Behavioural responses of larvae of the crab *Rhithropanopeus harrisii* (Brachyura: Xanthidae) during diel vertical migration. Mar. Biol. 90: 9–18.

Forward, R. B., Jr., 1989. Behavioural responses of crustacean larvae to rates of salinity change. Biol. Bull. 176: 229–238.

Forward, R. B., Jr. & D. Rittschof, 1994. Photoresponses of crab megalopae in offshore and estuarine waters: implications for transport. J. exp. mar. Biol. Ecol. 182: 183–192.

Forward, R. B., Jr., R. A. Tankersley, M. C. De Vries & D. Rittschof, 1995. Sensory physiology and behaviour of blue crab (*Callinectes sapidus*) postlarvae during horizontal transport. Mar. Freshwat. Behav. Physiol. 26: 233–248.

Forward, R. B., Jr., C. A. Wellins & C. U. Buswell, 1989. Behavioural responses of larvae of the crab *Neopanope sayi* to hydrostatic pressure. Mar. Ecol. Prog. Ser. 57: 297–277.

Harden Jones, F. R., M. Greer Walker & G. Arnold, 1984. Tactics of fish movement in relation to migration strategy and water circulation. In J. D. McCleave, G. P. Arnold, J. J. Dodson & W. H. Neill (eds), Mechanisms of Migration in Fishes, Plenum Press, New York: 185–207.

Knight-Jones, E. W. & B. Morgan, 1966. Response of marine animals to changes in hydrostatic pressure. Oceanogr. mar. biol. annu. Rev. 4: 267–299.

Laprise, R. & J. J. Dodson, 1989. Ontogeny and importance of tidal vertical migrations in the retention of larval smelt *Osmerus mordax* in a well-mixed estuary. Mar. Ecol. Prog. Ser. 55: 101–111.

Little, K. T. & C. E. Epifanio, 1991. Mechanism for the reinvasion of an estuary by two species of brachyuran megalopae. Mar. Ecol. Prog. Ser. 68: 235–242.

Olmi, E. J., III, 1994. Vertical migration of the blue crab *Callinectes sapidus* megalopae: implications for transport in estuaries. Mar. Ecol. Prog. Ser. 113: 39–54.

Queiroga, H., J. D. Costlow, Jr. & M. H. Moreira, 1994. Larval abundance patterns of *Carcinus maenas* (Decapoda, Brachyura) in Canal de Mira (Ria de Aveiro, Portugal). Mar. Ecol. Prog. Ser. 111: 63–72.

Queiroga, H., 1995. Processos de dispersão e recrutamento das larvas do caranguejo *Carcinus maenas* na Ria de Aveiro. Tese de Doutoramento, Universidade de Aveiro, Aveiro.

Queiroga, H., 1996. Shelf distribution and drift of the crab *Carcinus maenas* (L.) (Decapoda, Portunidae) larvae off northern Portugal in April of 1991. J. Plank. Res. 18: 1981–2000.

Queiroga, H. J., J. D. Costlow, Jr. & M. H. Moreira, 1997. Vertical migration of the crab *Carcinus maenas* first zoea: implications for tidal stream transport. Mar. Ecol. Prog. Ser. 149: 121–132.

Rice, A. L., 1966. The orientation of pressure response of some marine crustacea. Proc. Symp. on Crustacea, Mar. biol. Assn. India, Part III: 1124–1131.

Rowe, P. M. & C. E. Epifanio, 1994. Tidal stream transport of weakfish larvae in Delaware Bay, USA. Mar. Ecol. Prog. Ser. 110: 105–114

Shanks, A. L., 1995. Mechanisms for cross-shelf dispersal of larval invertebrates. In L. McEdward (ed.), Ecology of Marine Invertebrate Larvae, CRC Press, New York: 323–368.

Tankersley, R. A., L. M. McKelvey & R. B. Forward, Jr., 1995. Responses of estuarine crab megalopae to pressure, salinity and light: implications for flood tide transport. Mar. Biol. 122: 391–400.

Zar, J. E., 1984. Biostatistical analysis. 2nd edition. Prentice-Hall International, Englewood Cliffs.

Zeng, C. & E. Naylor, 1996a. Endogenous tidal rhythms of vertical migration in field collected zoea-1 larvae of the shore crab *Carcinus maenas*: implications for ebb tide offshore dispersal. Mar. Ecol. Prog. Ser. 132: 71–82.

Zeng, C. & E. Naylor, 1996b. Occurrence in coastal waters and endogeneous tidal swimming rhythms of late megalopae of the shore crab *Carcinus maenas*: implications for onshore recruitment. Mar. Ecol. Prog. Ser. 136: 69–79.

Hydrobiologia **375/376:** 151–163, 1998.
S. Baden, L. Pihl, R. Rosenberg, J.-O. Strömberg, I. Svane & P. Tiselius (eds),
Recruitment, Colonization and Physical–Chemical Forcing in Marine Biological Systems.

Temporal fluctuation in the abundance of a semelid bivalve, *Theora fragilis* (A. Adams) in Maizuru Bay, Sea of Japan

Hajime Saito, Masahiro Ueno & Isao Hayashi
Laboratory of Marine Biological Function, Graduate School of Agriculture, Kyoto University, Kyoto 606-8502, Japan
Tel.: +81-75-753-6446; fax: +81-75-753-6446; e-mail: saito@kais.kyoto-u.ac.jp

Key words: Theora fragilis, opportunistic species, Maizuru Bay, population dynamics, recruitment, sedimentation, POM

Abstract

The population dynamics of a semelid bivalve *Theora fragilis* (A. Adams) were studied in Maizuru Bay, the Sea of Japan based on data obtained through weekly or bi-weekly sampling. A sharp increase in population abundance occurred from early February to early April. However, the benthic population had almost disappeared leaving only sporadic survivors by early August. Comparison of average growth between the central and the innermost stations showed that many juveniles recruited into adult size at the latter station, where high sedimentation of particulate organic matter (POM) was observed. At the central station, the mean POM sedimentation rate was 1/3 that of the former station, and few individuals recruited into adult size. The occurrence of small juveniles was comparable at both sites. It was concluded that nutritional input is important for both recruitment into adult size and resulting fecundity.

Introduction

The semelid bivalve *Theora fragilis* (A. Adams) has been known to dominate the macrobenthic communities of eutrophic bays since the late 1930s when Miyadi and his co-workers studied the benthic fauna in various inner bays of Japan (e.g. Miyadi & Masui, 1942). Studies on the population ecology of this species were initiated in the 1970s when industrial activities caused serious pollution in coastal areas of Japan. Since then, information has been collected mainly from the western part of Japan (Mukai, 1974; Kikuchi & Tanaka, 1976; Sanukida et al., 1981; Imabayashi & Tsukuda, 1984). And, with the exception of some brief descriptions in studies on the community structure (Hayashi, 1978; Hayashi, 1983; Hayashi et al., 1992; Yokoyama & Hayashi, 1980), no intensive studies have been made in coastal regions along the Sea of Japan. The population ecology of this bivalve has not been discussed in any detail mainly due to several problems with sampling methodology, including sampling interval and sieve screen size. Since the life cycle of the bivalve is quite short, up to a few months (Kikuchi & Tanaka, *loc. cit.*), a well-designed sampling scheme is necessary to determine the population dynamics. Through frequent samplings and 0.333 mm sieve, we examined the whole population processes in Maizuru Bay – one of the major subsidiary bays of Wakasa Bay, the Sea of Japan. In this study, we focus on the relationship between the recruitment pattern to adult size and environmental conditions.

Biology of Theora fragilis

Theora fragilis is a small semelid bivalve that lives in coastal soft bottom habitats in Japan. It has also been reported to occur in Korea and Russia (Lim & Hong, 1997; Zhukova, et al., 1992), and recently has been found in Australia and New Zealand as a predominant introduced species (Poore & Rainer, 1979; Rainer, 1982; Willan, 1985; Hayward et al., 1997).

The fossil record shows that this bivalve lived in the *Turritella-Dosinia* association in Pleistocene (Nakao, 1993). *T. fragilis* is a surface deposit feeder, and uses it's siphon to collect organic matter at the sediment surface. Kikuchi & Tanaka (1976) reported that collected specimens could release gametes at any time of the year, and that individuals as small as 4 to 5 mm (shell length: SL) may spawn. This suggests that at least a portion of the population can spawn throughout the year. The clutch size of a 15mm female may reach 54 000 eggs (Imabayashi, 1991). Demersal eggs hatch in a few days, and planktonic larvae are released. It was demonstrated that the duration of the planktonic stage depends on the ambient water temperature. At 20 °C this stage lasts 9 to 13 days after fertilization, at 15 °C 26 days and at 12 °C 56 days (Imabayashi & Iwatani, 1988). The size of newly settled juveniles ranges from 0.188 to 0.245 mm (SL) suggesting that settlement can be postponed in order to search for better sediment (Imabayashi & Iwatani, *loc. cit.*). A rearing experiment showed that contact with the sediment is required for settlement (Imabayashi, 1991). Competent larvae move up and down close to the bottom and test sediment by contact (Imabayashi, *loc. cit.*). This is similar to the sediment selection behavior of *Capitella* larvae reported by Butman & Grassle (1992). Growth of benthic individuals is so fast that even the smallest individuals retained on 1 mm mesh screen can grow over 6 mm in a few months (Kikuchi & Tanaka, *loc. cit.*). Existence of the larval pool throughout the year has not been recorded, however, Imabayashi & Iwatani (1990) reported that planktonic larvae are actually present even when benthic individuals diminish from September through October. Thus larvae may be able to recruit as soon as the bottom environment becomes suitable. Tamai (1993, 1996) demonstrated that benthic adults and larvae could tolerate oxygen levels as low as 2 mg l^{-1}.

Description of sites studied

Maizuru Bay is a subsidiary of Wakasa Bay located along the Japanese coast of the Sea of Japan. The bay can be divided into three regions by its topographic shape: a narrow entrance, and the eastern and western inner parts (Figure 1). The bay is enclosed and the tidal range is quite narrow, leading to little water exchange through the entrance. As a result, the inner bay water is highly stagnant and eutrophic. In the eastern and western parts, the bottom topography is quite smooth in the central basin and has a steep coastal fringe; the maximum depth is around 20 m. Soft mud covers most areas except immediately around the islands and the submerged banks scattered throughout the bay. The bay receives freshwater discharges from several small rivers that are located in the innermost areas of both the eastern and the western parts (Figure 2). Sewage treatment plants are located at the innermost part of each. The present study covers only the eastern inner part, especially concentrating on the innermost area where *Theora fragilis* is exclusively dominant in spring. Another bivalve species, *Musculista senhousia*, and some polychaetes are also abundant. The polychaeta community of the bay mainly consists of *Paraprionospio* sp. (Form B), *Prionospio ehlersi*, *Prionospio pulchra*, *Pseudopolydora* sp., *Sigambra tentaclata*, *Notomustus* sp. and *Heteromustus* sp. etc. A venerid bivalve, *Paphia undulata*, dominates in biomass only in the innermost part adjacent to the river mouth, but its abundance is negligible.

Materials and methods

Sampling was performed weekly or bi-weekly in the eastern inner part of the bay from 22 December 1995 to 8 January 1998. To determine the temporal fluctuation pattern in abundance, the survey was carried out bi-weekly from 22 December 1995 to 25 December 1996 at only one station (Stn N). From 7 January 1997 to 8 January 1998, 17 stations were established including the previous fixed station (Figure 2). Thirteen stations were surveyed bi-weekly, and the remaining 4 stations (Stns M, N, P and Q), where the dense population of *Theora fragilis* develops every year, were surveyed at weekly intervals. The sediment was soft mud in all stations, except Stn M which contained a considerable portion of coarse particulate matter.

Sediment samples were obtained with an Ekman-Birge bottom sampler (20 cm × 20 cm × 25 cm). Three to 4 replicate samples were collected at each station. The collected sediment was washed through a 0.333 mm mesh screen and the materials retained on the screen were immediately fixed in a 10% formalin solution. The minimum size of the bivalve retained on the mesh screen is approximately 0.45 mm, which is slightly larger than the settlement size range of 0.188–0.245 mm (Imabayashi & Iwatani, 1988). In the laboratory, organisms were extracted from the washed sediment, and the bivalve specimens were preserved in 70% ethanol. Specimens were enumerated and mea-

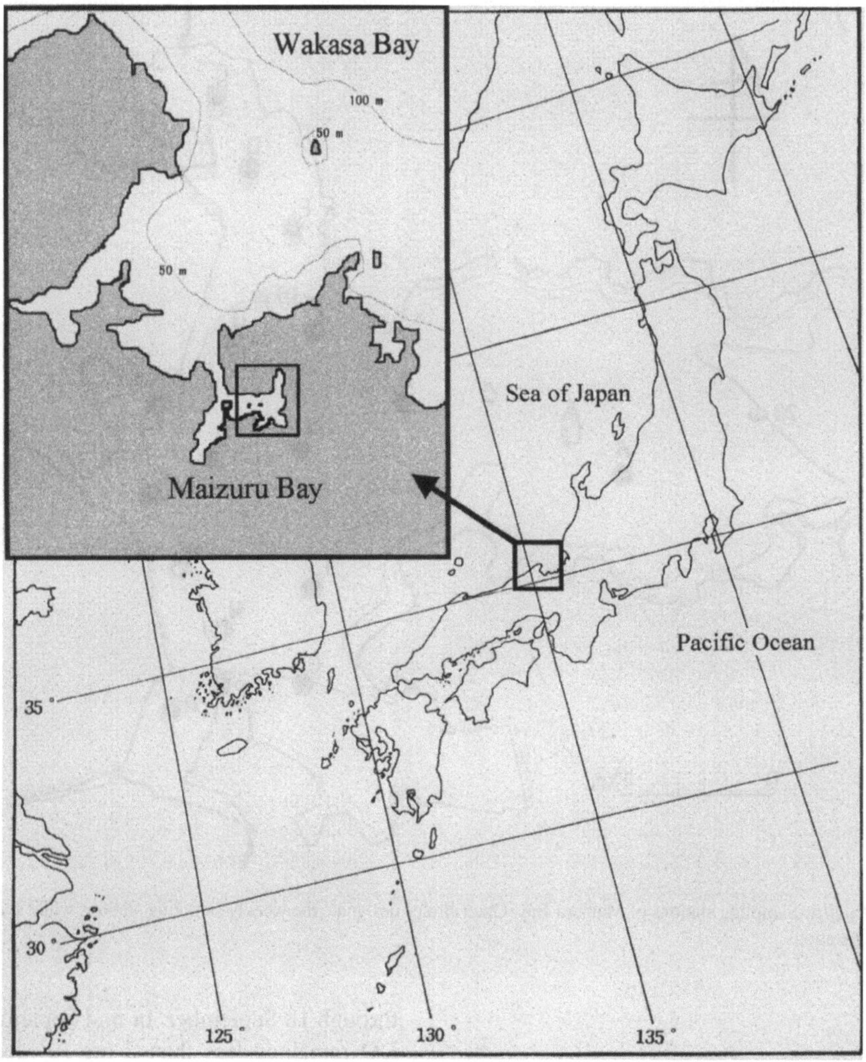

Figure 1. Theora fragilis. Location of Maizuru Bay.

sured for SL with an image analyzer (NIH image, National Institute of Health, USA).

At each survey station, abiotic data (depth, temperature, dissolved oxygen) was obtained using a CTD (Alec Electronics Co.) and a DO meter (Central Science Co.). Bottom water DO was measured at approximately 10 cm above the bottom. Surface sediment samples were collected in order to analyze total organic carbon (TOC) and total nitrogen (TN). At each sampling station, 3 core samples were taken on 30 January, 10 April, 17 July and 7 November 1997, as representatives of winter, spring, summer and autumn, respectively. Sediment traps were set approximately 1 m above the bottom at Stns A and N at 2 to 6 week intervals from 23 January to 26 November 1997 to measure the amount of particulate organic matter (POM). Three samples were obtained for each sampling station. TOC and TN were analyzed with an auto element analyzer (NA 1500, Fisons Co. Ltd.). This paper focus on data obtained from a pair of contrasting stations, Stns A and N, which represent the moderately eutrophic and the highly eutrophic stations, respectively.

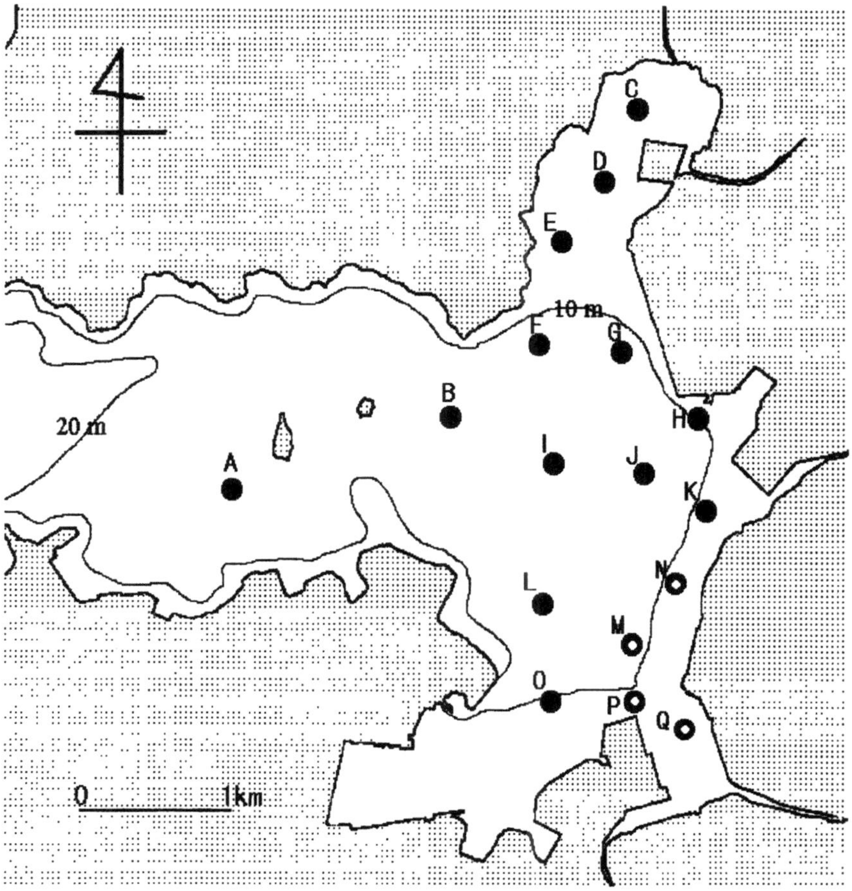

Figure 2. Theora fragilis. Sampling stations of Maizuru Bay. Open circles designate the weekly sampling stations while solid circles designate bi-weekly sampling stations.

Results

Water temperature, salinity and DO

The lowest bottom water temperature in 1997 (12 °C) was measured from 30 January through 10 April, while the highest temperature (27 °C) was observed from 29 August through 18 September (Figure 3a). On 26 September, bottom water temperature suddenly dropped to 20 °C. Bottom water salinity at Stn N usually fluctuated between 32 and 34 psu (Figure 3b). This value was significantly lower than that at Stn A (Table 1). Salinity decreased gradually at both stations from 4 July to 19 September, and suddenly increased on 26 September. Hypoxia did not occur at Stn A during the year, as bottom water DO was always higher than 4 mg l^{-1} (Figure 3c). Conversly, bottom water DO at Stn N was always less than that at Stn A, and sometimes reached the hypoxic level (less than 3 mg l^{-1}). Hypoxia was observed on 30 July, and from 12 through 18 September. In mid September, the bottom DO remained less than 2 mg l^{-1} for a few weeks. The DO condition of the bottom in the innermost area improved on 26 September.

Organic enrichment of the sediment

Figures 4a, b and c show the temporal changes in sedimentation rate for total dry weight, TOC and TN, respectively. At Stn A, organic carbon sedimentation was continuously low throughout the year (less than 1 g C m^{-2} d^{-1}). In contrast, the sedimentation rate at Stn N was always higher than that at Stn A, and several clear pulses of heavy sedimentation were discernible throughout the survey period. For example, on 6 February and 26 November, sedimentation rates were markedly high and exceeded 3 g C m^{-2} d^{-1}. Moderate sedimentation rates (2 g C m^{-2} d^{-1}) were seen on 4 July and 11 September. Since the standard deviation of 3 replicate samples was usually less than

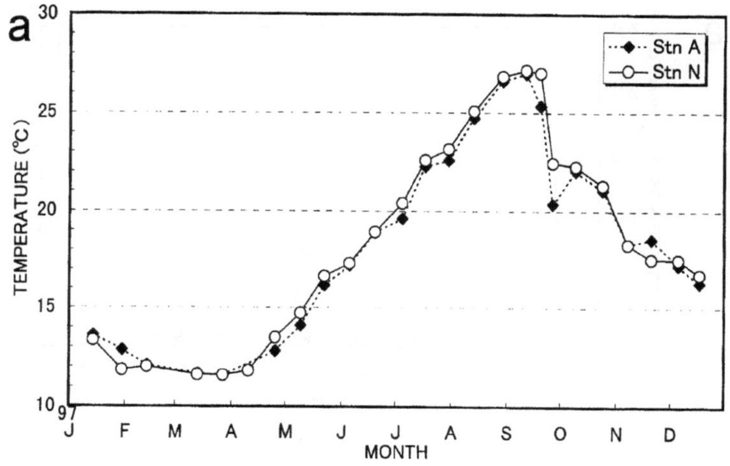

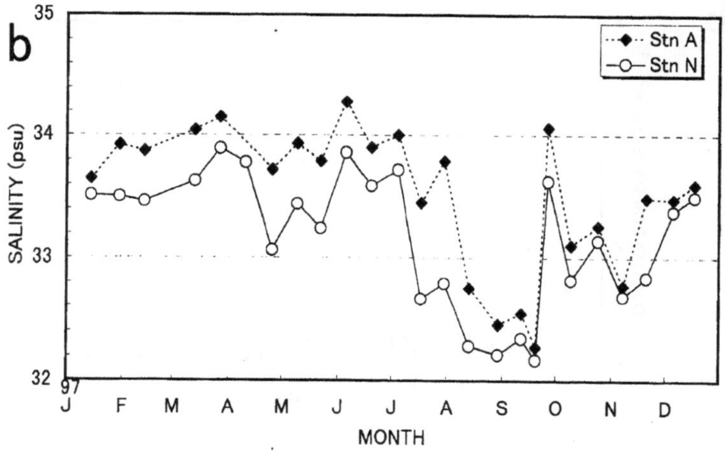

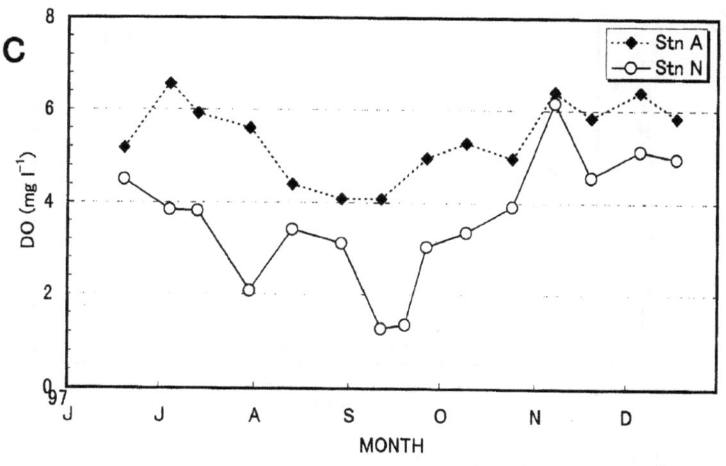

Figure 3. Theora fragilis. Temporal change of bottom water a: temperature, b: salinity, c: dissolved oxygen at Stns A & N.

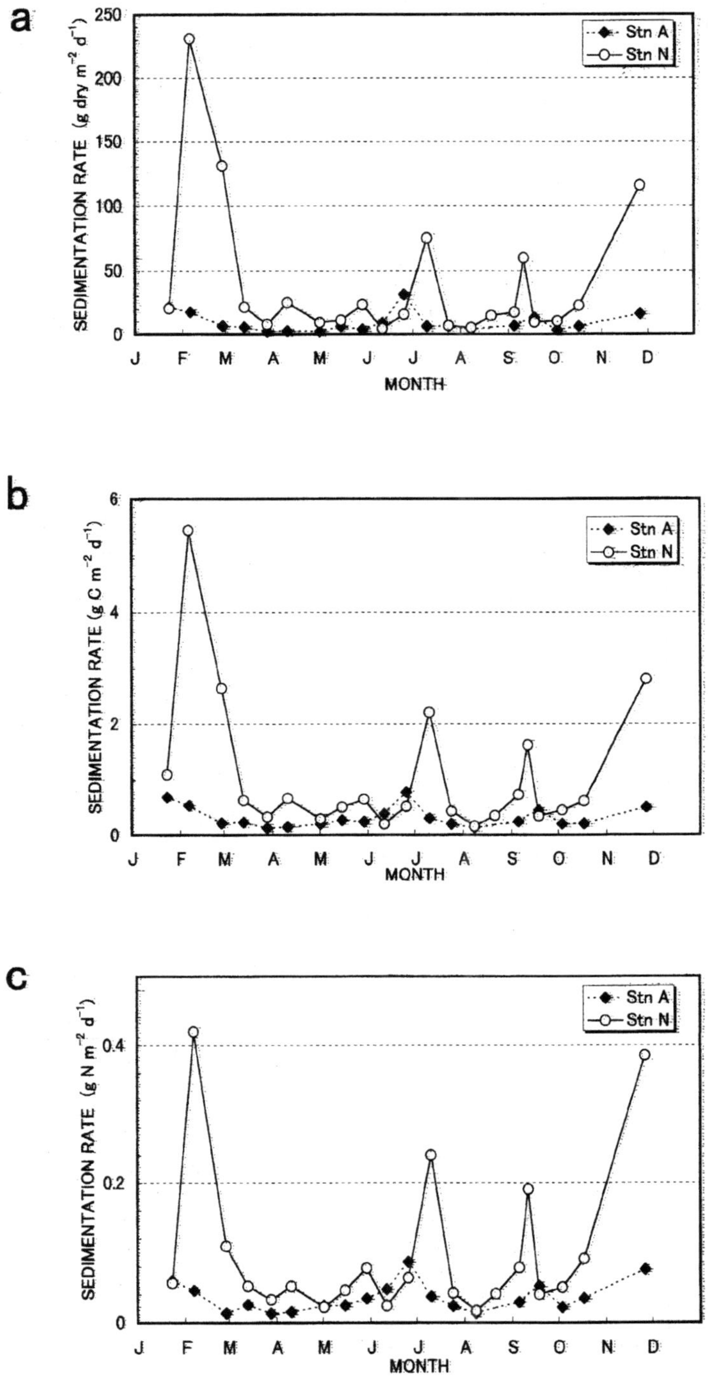

Figure 4. Theora fragilis. Temporal change of sedimentation rate at Stns A & N shown as a: dry weight, b: total organic carbon, c: total nitrogen.

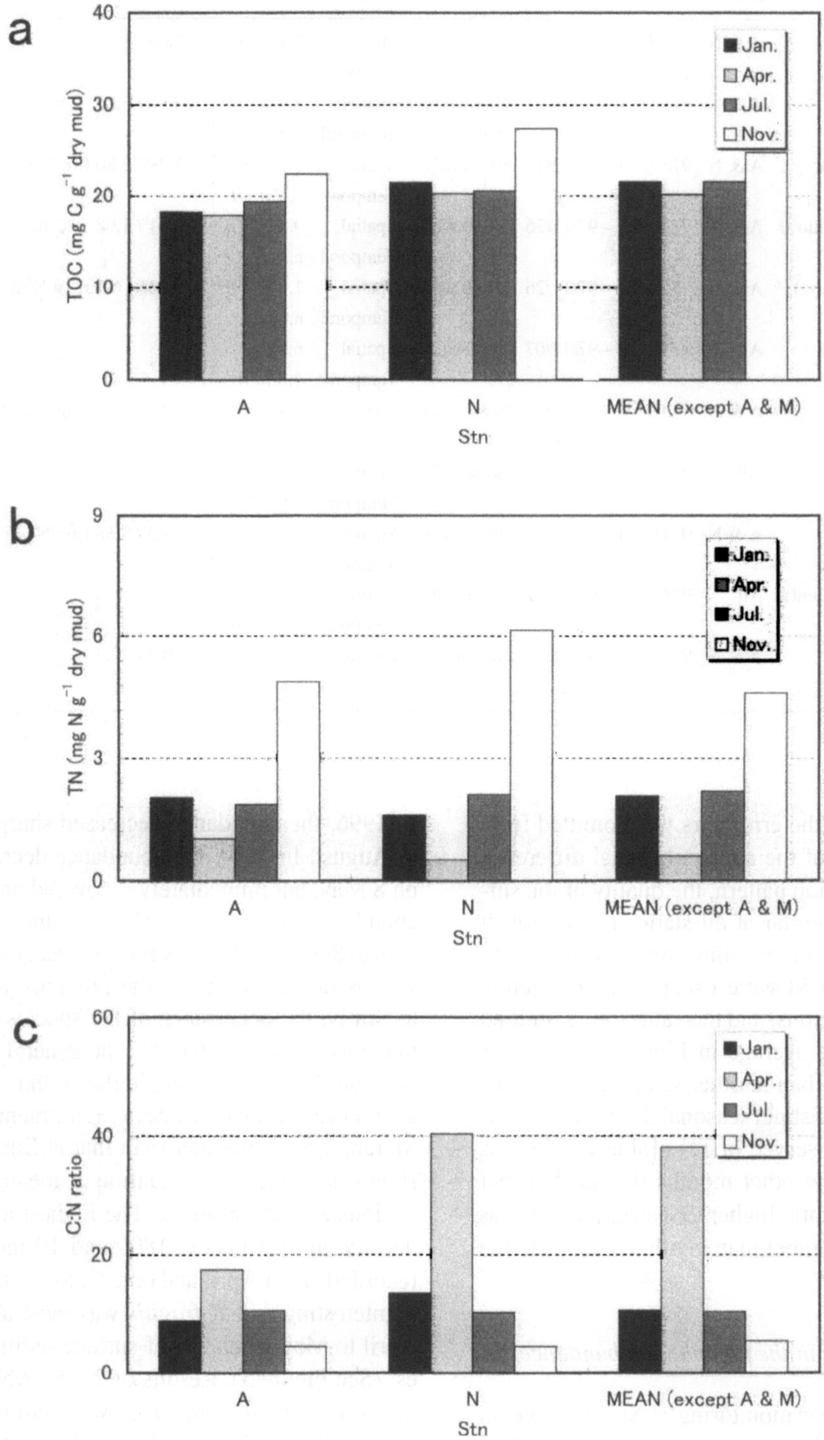

Figure 5. Theora fragilis. Comparison of surface sediment quality shown as a: total organic carbon, b: total nitrogen, c: C:N ratio.

Table 1. Theora fragilis. Two-way ANOVA of environmental variables

	Stns	Duration	Interval	Factor	p	Mean (A / N)
Temperature	A & N	97/01/14 – 97/12/17	Bi-weekly	Spatial	4.7×10^{-2}	18.4 / 18.7 (°C)
				Temporal	6.2×10^{-22}	
Salinity	A & N	97/01/14 – 97/12/17	Bi-weekly	Spatial	8.7×10^{-8}	33.5 / 33.1
				Temporal	1.5×10^{-10}	
Dissolved oxygen	A & N	97/06/19 – 97/12/17	Bi-weekly	Spatial	2.5×10^{-5}	5.39 / 3.80 (mg l^{-1})
				Temporal	7.5×10^{-3}	
TOC (Sedimentation)	A & N	97/01/23 – 97/11/26	2 to 6 week	Spatial	9.2×10^{-3}	0.31 / 1.08 (g C m^{-2} d^{-1})
				Temporal	n.s.	
TN (Sedimentation)	A & N	97/01/23 – 97/11/26	2 to 6 week	Spatial	1.0×10^{-2}	0.035 / 0.102 (g N m^{-2} d^{-1})
				Temporal	n.s.	
TOC (Sediment)	All	97/01/30 – 97/11/07	Seasonally	Spatial	n.s.	
				Temporal	4.8×10^{-3}	
	A & N	97/01/30 – 97/11/07	Seasonally	Spatial	3.4×10^{-2}	19.4 / 22.4 (mg C g^{-1} dry)
				Temporal	3.9×10^{-2}	
TN (Sediment)	All	97/01/30 – 97/11/07	Seasonally	Spatial	n.s.	
				Temporal	1.6×10^{-29}	
	A & N	97/01/30 – 97/11/07	Seasonally	Spatial	n.s.	2.45 / 2.58 (mg N g^{-1} dry)
				Temporal	1.2×10^{-2}	
C:N ratio (Sediment)	All	97/01/30 – 97/11/07	Seasonally	Spatial	n.s.	
				Temporal	4.2×10^{-6}	
	A & N	97/01/30 – 97/11/07	Seasonally	Spatial	n.s.	10.4 / 19.7
				Temporal	n.s.	

25% of the average, the error bars were omitted from the figures. In spite of the apparent spatial difference shown in sedimentation pattern, the quality of the surface sediment was similar at all stations. As a result of accidental sediment containing coarse particles, the TOC and TN at Stn M were exceptionally different from that at other stations, and the values were omitted when calculating the average in Figure 5. TOC was higher in November than in other sampling occasions (Figure 5a). A more distinct seasonal difference in sediment quality was observed in TN (Table 1). TN was lower in April than in other months (Figure 5b), but TN at Stn A was slightly higher than average. TN was much higher in November than in other seasons in the whole area.

Temporal fluctuation in the population abundance

The result of two-year monitoring at Stn N is shown in Figure 6b. It is clear that the abundance of *Theora fragilis* fluctuated seasonally. The bivalve density was estimated to be more than 3000 ind. m^{-2} from March through April. However, from August to the following January, they occurred quite sporadically. In 1996, the abundance decreased sharply from May to August. In 1997, the abundance decreased sharply on 8 May, but immediately recovered on 29 May. The abundance decreased on 12 June, and again from 25 July to 8 August. From August to December, *T. fragilis* was as rare as it was in the previous year. Contrary to Stn N, the occurrence of the species at Stn A was much lower (Figure 6a), but the general feature of the seasonal fluctuation was similar to that at Stn N. The difference was that the heavy recruitment started on 13 March, 1.5 month later than that at Stn N. Figure 6c shows the temporal fluctuation of mean density at all stations excluding Stn A. The highest and the lowest density, approximately 1800 and 10 ind. m^{-2}, were recorded on 24 April and on 9 October, respectively. It is interesting that *T. fragilis* was most abundant from April to May when TN of surface sediment was lowest (See Figure 5). Results of 2-way ANOVA showed that variations in abundance were both spatially and temporally significant (Table 2). The test of abundance including all stations showed that temporal variation was extremely significant and was superior to the significance of spatial variation. In the comparison of abundance between Stns A and N, both spatial and

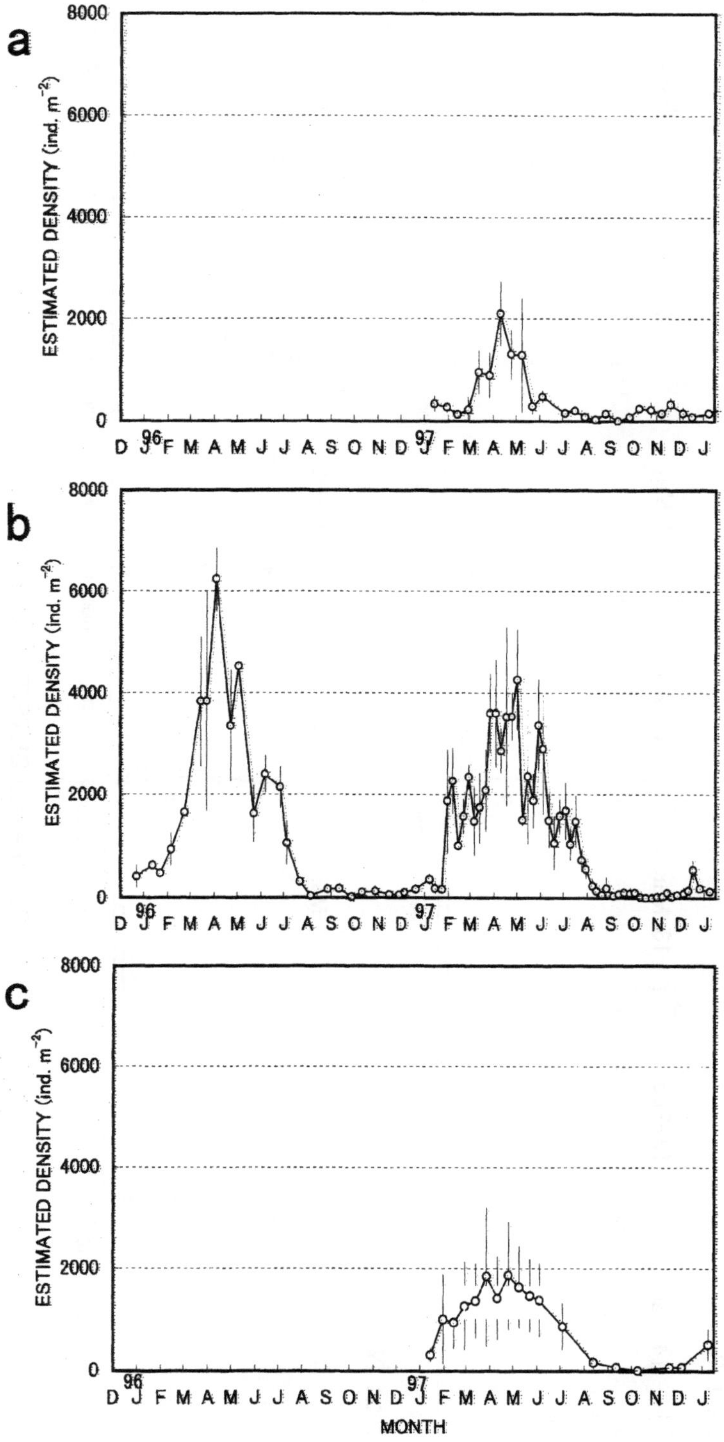

Figure 6. Theora fragilis. Temporal change of abundance. a: Stn A, b: Stn N, c: mean of all stations. Vertical bars show standard deviation.

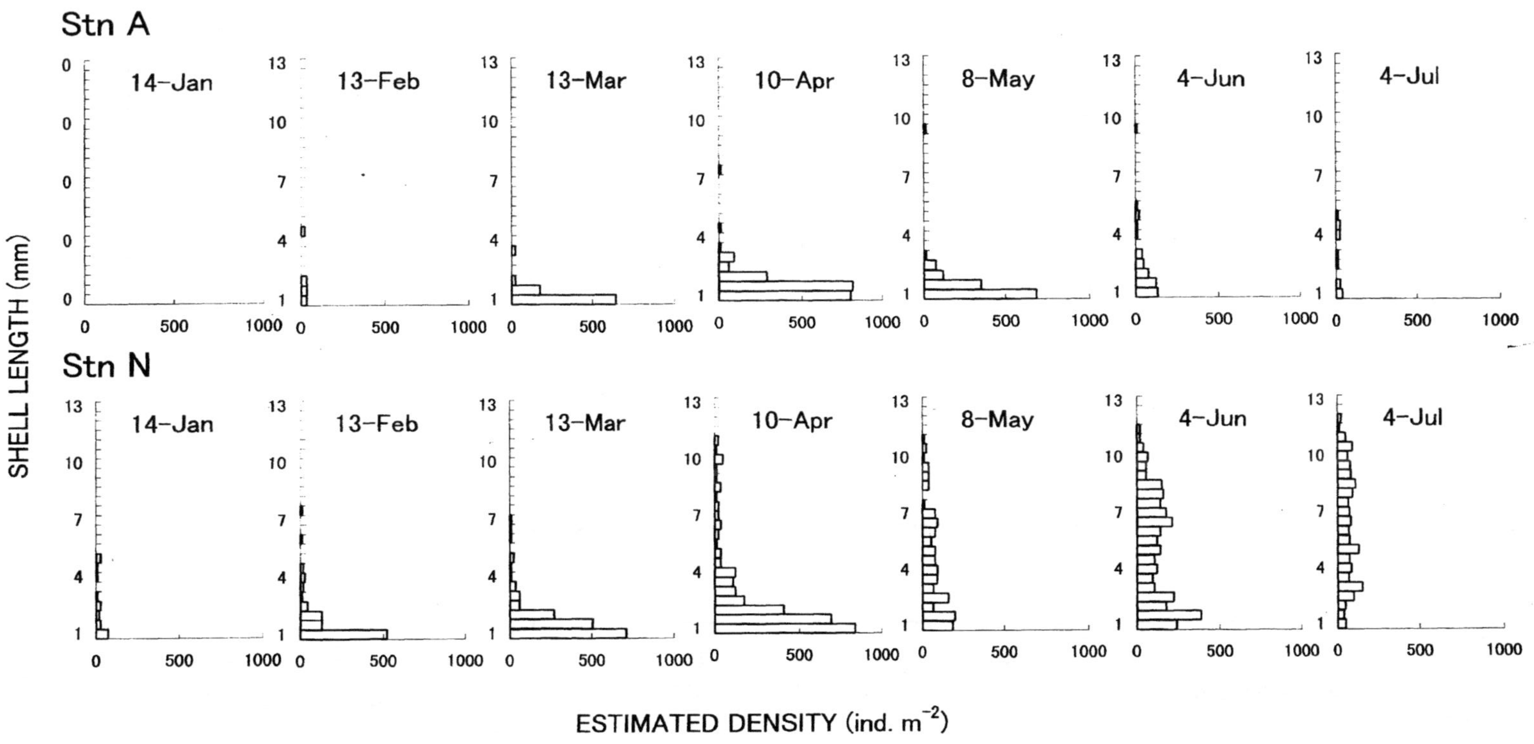

Figure 7. Theora fragilis. Temporal change of size composition at Stns A & N.

Table 2. Theora fragilis. Two-way ANOVA of abundance and biomass

	Stns	Duration	Interval	Factor	*p*	Mean (A / N)
Abundance	All	97/01/14 – 97/12/05	Monthly	Spatial	1.2×10^{-4}	
				Temporal	1.1×10^{-37}	
	A & N	97/01/14 – 98/01/08	Bi-weekly	Spatial	7.5×10^{-4}	438 / 1104 (ind. m^{-2})
				Temporal	6.8×10^{-3}	
Biomass	A & N	97/01/14 – 98/01/08	Bi-weekly	Spatial	4.3×10^{-3}	147 / 2330 (mg m^{-2})
				Temporal	n.s.	

temporal variation were significant, but the significance level of temporal variation was lower than that of spatial variation due to the long distance between the stations (about 3 km). In the comparison of estimated biomass, only spatial variation was significant because few individuals recruited into the adult size at Stn A, and the biomass was constant there all year around.

Recruitment into the adult size

The temporal change of size composition was examined at Stns A and N (Figure 7). The number of *Theora fragilis* that recruited into the adult size ($\geq$ 5 mm in SL) was quite different between the 2 stations. At Stn N, many juveniles recruited into the adult stage, and some grew over 10 mm (SL). On 14 January, 1997, total abundance was still small, and most individuals were smaller than 5 mm (SL). The number of small individuals rapidly increased to more than 500 ind. m^{-2} from February through April. A few individuals had already grown into adult size by 13 February. From 13 March to 10 April, the number of adults increased sharply, but small individuals were also abundant on 10 April. After 8 May, the number of small individuals decreased, and the major portion of the population consisted of adults. However, the occurrence of adults was sporadic at Stn A. The number of small individuals increased sharply during March, 1.5 month later than that at Stn N. Small individuals were still abundant on 8 May, and the number decreased in June. From the size composition patterns, it is quite obvious that the numbers of juveniles at the 2 stations were comparable from March through May, but the number matured at the 2 stations was markedly different.

Discussion

Since the report by Kikuchi & Tanaka (1976), *Theora fragilis* has been widely accepted as a typical opportunistic species in Japan. They pointed out that the life history of *T. fragilis* is similar to that of *Capitella* sp. in the following points: 1) larvae could be produced throughout the year, and 2) the population *in situ* tends to show mono-modal fluctuation. The definition of an opportunistic life history pattern proposed by Grassle & Grassle (1974) includes the following characteristics: 1) initial response to disturbed conditions, 2) ability to increase rapidly, 3) large population size, 4) early maturation, and 5) high mortality after settlement. *T. fragilis* exhibits several of these characteristics, including rapid population increase, large temporary population size, early maturation and high mortality, but the initial response to disturbed condition may be doubtful. Zajac & Whitlatch (1982) stated that few polychaetes in the estuarine infauna of Alewife Cove, the U.S., were regular opportunists. One of the remarkable exceptions was *Polydora ligni* which regularly displayed opportunistic responses by significantly increasing its population level in the experimentally disturbed plots above the ambient levels. In many species, initial dynamics of re-colonization may depend more on spatial and temporal variability than a particular suit of life history characteristics. In the present study, temporal variation was greater than spatial variation at all stations (Table 2). This means that the fluctuation in abundance tends to be more or less synchronous in the whole bay, although the test of abundance between Stns A and N showed that the significance levels of temporal and spatial differences were similar. In other words, the population level of this bivalve is largely regulated by the seasonal variation of the ambient condition rather than by an opportunistic habitation in disturbed areas. Taking these facts into account, we could not classify

T. fragilis as a true opportunistic species in the strict sense.

Although the general trend of the abundance of *T. fragilis* is based on temporal variation, there is a clear difference in size composition from May through July between Stns A and N (Figure 7) resulting in significant spatial difference in biomass (Table 2). The silt-clay content of sediment at Stns A and N was 98.8 and 96.9%, respectively, suggesting that the hydrodynamic characteristics of the 2 stations are similar. The only remarkable differences observed were the spatial variations of TOC and TN sedimentation (Table 1). Sedimentation was 3 times greater on average at Stn N than at Stn A (Figure 4). However, in the case of surface sediment TOC and TN, less significant differences were spatially distinguished. Greater sedimentation at Stn N than at Stn A, and no or little significant spatial difference in sediment TN and TOC does not appear to make sense, but it is possible that biological processes rapidly consumed newly settled POM. Mean biomass at Stn N was 16 times greater than that at Stn A. Mean sedimentation rate at Stn N was 3 times higher than that at Stn A. Recruitment into the adult size is more frequent at Stn N than at Stn A (Figure 7), which strongly suggests that recruitment into the adult size strictly depends on food availability.

Based on the data obtained by a 10-year monitoring program, Ueno & Sato (1993) reported that the inorganic nitrogen and phosphorus concentration and the chlorophyll content in surface water was highest in the SE innermost part. This is most likely due to the effect of land water discharge and treated sewage. Nakahara (1978) reported that there were major diatom blooms twice a year in the bay, one in early spring and the second in autumn. Although a direct relationship between diatom blooms and population dynamics of *T. fragilis* is unclear, the spatial difference of POM sedimentation between Stns A and N coincides well with the cline of chlorophyll distribution in this bay. It is more likely that active primary production in surface water supports *T. fragilis* population via benthic-pelagic coupling. Water eutrophication is definitely reflected in the recruitment pattern into the adult stage and the biomass of *T. fragilis*, and may have some effect on resulting fecundity. The species distribution is so wide that occurrence of this bivalve does not necessarily indicate pollution in the area concerned. But considering the biomass, this species could be a good long-term indicator of eutrophic conditions. Populations of *T. fragilis* seem sensitive enough to detect subtle spatial differences in average POM sedimentation.

In this study, the temporal variation of surface sediment TN was marked (Table 1). Temporal variation of TOC in sediment was small compared to that of TN (Figure 5). In classical methods of sediment analysis (e.g. ignition loss, COD), it is difficult to quantify food availability on sediment. As shown in this study, TOC in sediment was 4 to 20 times higher and more stable temporally than TN. The sharp fluctuation of TN would have been hidden and we might have concluded that sediment quality was stable throughout the year if we used classical methods of analysis. Many kinds of organic compounds are contained in sediment, and the degree of utilization differs. The present study clearly suggests the importance of protein resources for growth and recruitment into the adult size of the bivalve. Marsh & Tenore (1990) pointed out that a distinction between macro-nutrients (carbon and nitrogen) and micro-nutrients (essential polyunsaturated fatty acid and essential amino acid) is important when relating the degree of recruitment to food condition. The present findings suggested that the TN reduction in April when population abundance is high and TN elevation in November when the population size is low, is due to the high demand and turnover for protein resources by this bivalve. However, studies on micronutrients in the surface sediment might help to clarify the population dynamics of *T. fragilis* in this bay.

Acknowledgements

While this work was being done, we benefited from the support of H. Nakahara and H. Tsutsumi. The staffs of the Fisheries Research Station of Kyoto University made facilities available to accomplish this study. K. Sato, K. Fujii, M. Hosoi and H. Yamaguchi helped us on board. J. E. Harcke and N. King kindly checked our English in this paper. To all we extend our deepest thanks.

References

Butman, C. A. & J. P. Grassle, 1992. Active habitat selection by *Capitella* sp.I larvae. I. Two-choice experiments in still water and flume flows. J. mar. Res. 50: 669–715.

Grassle, J. F. & J. P. Grassle, 1974. Opportunistic life history and genetic systems in marine benthic polychaetes. J. mar. Res. 32: 253–284.

Hayashi, I., 1978. Ecological distribution of macrobenthic animals on the level bottom in Maizuru Bay. I. Spring phase of animal distribution. J. Oceanogr. Soc. Jap. 34: 24–25 (in Japanese).

Hayashi, I., 1983. Distribution of macrobenthic animals on the flat bottom in Obama Bay. Bull. Jap. Soc. Sci. Fish. 49: 1765–1775.

Hayashi, I., Y. Iguchi, T. Yoshida & T. Takahashi, 1992. Macrobenthic assemblages of Nanao Bay, Sea of Japan. Bull. Coast. Oceanogr. 30: 91–107 (in Japanese).

Hayward, B. W., A. B. Stephenson, M. Morley, J. L. Riley & H. R. Grenfell, 1997. Faunal changes in Waitemata Harbour sediments, 1930s–1990s. J. r. Soc. New Zeal. 27: 1–20.

Imabayashi, H., 1991. HINSANSOKA-FUEIYOU-KAIIKI NI OKERU SHIZUKUGAI NO CHAKUTEI KATEI [Settlement processes of *Theora lubrica* in eutrophicated hypoxic sea]. GEKKAN KAIYO 23: 656–661 (in Japanese).

Imabayashi, H. & T. Iwatani, 1988. Growth and distribution in planktonic larvae of the small bivalve *Theora lubrica* (Preliminary report). Benthos Res. 33/34: 17–24 (in Japanese).

Imabayashi, H. & T. Iwatani, 1990. Estimation of survival rate at the settlement of a bivalve *Theora lubrica*. Benthos Res. 38: 19–25 (in Japanese).

Imabayashi, H. & S. Tsukuda, 1984. A population ecology of the small bivalve *Theora lubrica* in northern Bingo-Nada (Japan). Bull. Jap. Soc. Sci. Fish. 50: 1855–1862.

Kikuchi, T. & M. Tanaka, 1976. Some aspect on the ecology of a short-lived semelid bivalve, *Theora lata* (HINDS), with special reference to its opportunistic life history. Physiol. Ecol. Jap. 17: 261–270 (in Japanese).

Lim, H. S. & J. S. Hong, 1997. Ecology of the macrozoobenthos in Chinhae Bay, Korea: 3. Community structure. J. Korean Fish. Soc. 30: 175–187.

Marsh, A. G. & K. R. Tenore, 1990. The role of nutrition in regulating the population dynamics of opportunistic, surface deposit feeders in a mesohaline community. Limnol. Oceanogr. 35: 710–724.

Miyadi, D. & T. Masui, 1942. Comparison of benthic communities of Matoya Bay and Gokasho Bay. J. Jap. Soc. Sci. Fish. 11: 111–127 (in Japanese).

Mukai, H., 1974. Ecological studies on distribution and production of some benthic animals in the coastal waters of central Inland Sea of Japan. J. Sci. Hiroshima Univ. Ser. B, Div. 1 25: 1–82.

Nakahara, H., 1978: Studies on phytoplankton in Maizuru Bay. I. Seasonal variation and vertical microdistribution. Memory of the College of Agriculture, Kyoto University 112: 49–82.

Nakao, K., 1993. Distribution of molluscan associations of the Pleistocene Toriyamahara Formation in the Miyazaki plain. Quat. Res. 32: 157–170 (in Japanese).

Poore, G. C. B. & S. A. Rainer, 1979. A 3-year study of benthos of muddy environments in Port Phillip Bay, Victoria, Australia. Estuar. coast. mar. Sci. 9: 477–498.

Rainer, S. F., 1982. Trophic structure and production in the macrobenthos of a temperate Australian estuary. Estuar. Coast. Shelf Sci. 15: 423–442.

Sanukida, S., H. Okamoto & M. Hitomi, 1981. Behavior of indicator species of marine bottom pollution. Bull. Jap. Soc. Sci. Fish. 47: 863–870 (in Japanese).

Tamai, K., 1993. Tolerance of *Theora fragilis* (Bivalvia; Semelidae) to low concentrations of dissolved oxygen. Nippon Suisan Gakkaishi 59: 615–620 (in Japanese).

Tamai, K., 1996. Temporal tolerance of larval *Theora fragilis* (Bivalvia: Semelidae) to hypoxic conditions. Fish. Sci. 62: 996–997.

Ueno, M. & K. Sato, 1993. Evaluation of water qualities in Maizuru Bay, based on Oceanographic survey by Fisheries Research Station of Kyoto University. Report of Fisheries Research Station, Kyoto University 2: 46–57 (in Japanese).

Willan, R. C., 1985. Successful establishment of the Asian mussel *Musculista senhousia* in New Zealand. Records of the Auckland Institute and Museum 22: 85–96.

Yokoyama, H. & I. Hayashi, 1980. Zonation and species diversity of smaller macobenthos in the western part of Wakasa Bay (the Sea of Tango). J. Oceanogr. Soc. Jap. 36: 46–58.

Zajac, R. N. & R. B. Whitlatch, 1982. Responses of estuarine infauna to disturbance. I. Spatial and temporal variation of initial recolonization. Mar. Ecol. Prog. Ser. 10: 1–14.

Zhukova, N., V. I. Kharlamenko, V. I. Svetashev & I. A. Rodionov, 1992. Fatty acid as smarkers of bacterial symbionts of marine bivalve molluscs. J. exp. mar. Biol. Ecol. 162: 253–263.

Hydrobiologia **375/376:** 165–176, 1998.
S. Baden, L. Pihl, R. Rosenberg, J.-O. Strömberg, I. Svane & P. Tiselius (eds),
Recruitment, Colonization and Physical–Chemical Forcing in Marine Biological Systems.

Do hydrodynamic factors affect the recruitment of marine invertebrates in a macrotidal area?

The case study of Pectinaria koreni (Polychaeta) in the Bay of Seine (English Channel)

E. Thiébaut[1], Y. Lagadeuc[2], F. Olivier[3], J. C. Dauvin[2] & C. Retière[3]
[1] *Laboratoire d'Océanographie Biologique et d'Ecologie du Plancton Marin, URA CNRS 2077, Université Pierre et Marie Curie – Paris VI, Bâtiment A, Case 6, 4 Place Jussieu, 75252 Paris cedex 05, France*
[2] *Station Marine, Université de Lille I, B.P. 80, 62 930 Wimereux, France*
[3] *Muséum National d'Histoire Naturelle, Laboratoire Maritime de Dinard, URA CNRS 699, 17 Avenue Georges V, BP 28, 35801 Dinard cedex, France*

Key words: larval dispersal, postlarval drifting, recruitment, *Pectinaria koreni*, English Channel

Abstract

For marine benthic invertebrates exhibiting complex life cycles, changes in populations' distribution and abundance are governed by a large variety of physical, chemical and biological processes. From field observations in the Bay of Seine and laboratory experiments conducted since 1987 on the polychaete *Pectinaria koreni*, the present study highlights the relative importance of hydrodynamical and biological factors which affect individuals within both the planktonic and benthic phases at different scales of space and time in a macrotidal area. *Pectinaria koreni* is one of the main macrofaunal component of the *Abra alba* muddy fine sand community of the eastern Bay of Seine. Despite a highly advective and diffusive environment, a relative larval retention near adult population was reported due to some local hydrodynamics features (e.g. tidal residual circulation, Seine river plume front) and the interaction between the vertical current structure and the larval vertical migration. Although larval retention could be disrupted by wind induced currents, multiple spawning events over the reproductive period increase the likelihood that at least one larval cohort ensures a high recruitment during the life-span. Following a massive settlement whatever the sediment grain size, the newly settled larvae exhibited a high immediate decrease of their densities as a result of postlarval mortality and migration. Postlarval drifting was induced by a combination of physical factors (i.e. tidal currents and swell) and postlarval behaviour in response to sediment texture and adult/settlers interactions. According to the hydrodynamics of the bay, this process may generate a postlarval transport from offshore bottoms to coastal suitable habitats and counteract the demographic effects of larval dispersal. A conceptual model of factors governing the recruitment and population maintenance of *Pectinaria koreni* is proposed and discussed in comparison with results obtained on another polychaete, *Owenia fusiformis*, in the same area.

Introduction

In marine coastal zones, the majority of benthic invertebrates exhibit a complex life cycle including a planktonic larval phase and two bottom-dwelling juvenile and adult phases. For such species, the recruitment, defined as the supply of early juveniles to established adult populations, includes both pre- and post-settlement processes (Connell, 1985). It shows large variations on a variety of temporal and spatial scales so that the success or the failure of the recruitment can regulate ultimately populations' distribution and abundance (Rougharden et al., 1988; Gaines & Bertness, 1992). The successive processes which governed the recruitment are mainly:

(1) egg fertilization and offspring production which is controlled by both organisms characteristics (e.g. degree of aggregation of conspecific adults) and the external physical environment (e.g. current velocity) (Levitan et al., 1992);
(2) larval dispersal which is determined by local circulation features (i.e. eddy-diffusion and advection)

and larval attributes (e.g. duration of planktonic larval life and vertical migrations) (Scheltema, 1986; Hill, 1991);

(3) metamorphosis and larval settlement which depends on both substratum encouter rate related to the near-bottom concentration of competent larvae and bottom boundary-layer flows, and the substratum acceptance or rejection (Butman, 1987);

(4) post-settlement events including juvenile mortality (Olafsson et al., 1994; Gosselin & Qian, 1997) and juvenile migration which results from passive resuspension and/or active dispersal behaviour (Armonies, 1994).

Thus, the recruitment depends on a large variety of hydrodynamical and biological properties as well as their interactions which affect the probability that an individual of any stage make the transition to the next one (Eckman, 1996).

Within the English Channel, characterized by intense tidal currents average velocity of which is about 1 m s^{-1} (Salomon, 1991), the macrotidal hydrodynamical regime can greatly affect the achievement of benthic marine invertebrates life cycle. While currents may advecte and disperse eggs, larvae and post-larvae, they also determine adult benthic communities distribution related to superficial sediment composition, so that characteristic species of fine sand communities form isolated insular populations which are confined in bays and estuaries (Cabioch et al., 1982).

Since 1987, field observations in the eastern Bay of Seine and laboratory experiments were conducted on a target-species, the polychaete *Pectinaria koreni*, to determine which factors govern the recruitment of marine invertebrates in a macrotidal area. The aim of the present study is to highlight the relative importance of hydrodynamical and biological processes which affect individuals within both the planktonic and benthic phases. A conceptual model of factors controlling the recruitment and population maintenance of *Pectinaria koreni* is proposed and discussed in comparison with results obtained on other benthic invertebrates in the same area.

Study area

The Bay of Seine is a quadrilateral embayment, ca. 65 by 140 km wide and 5 to 30 m deep, located along the French coasts of the eastern English Channel. The major freshwater inputs into the bay are due to the Seine River, discharge of which varies seasonally from a maximum of up to 2000 m^3 s^{-1} in winter to a minimum of 100–200 m^3 s^{-1} in summer. Maximum tidal current velocities decrease from 1.8 to 0.5 m s^{-1} from the north-western part to the eastern part of the bay (Chabert d'Hières, 1986). In the latter area, tidal flows are asymmetrical so that flood currents are stronger but shorter in duration than ebb currents.

Residual circulation depends on 3 factors: (1) tide, (2) horizontal density gradients due to the Seine freshwater input and (3) meteorological factors, especially wind (Le Hir et al., 1986) (Figure 1). Tide induces a general drift of water masses to the east (up to 6–7 cm s^{-1}) in the northern part of the bay and to the west along the coasts of Calvados. It also generates anticyclonic gyres which can locally increase the residence time of particles. The gyre off the Cap d'Antifer favours the Seine freshwater flow in a north-westerly direction. Because of horizontal density gradients, the eastern part of the bay acts as a partially well-mixed estuary with a 2-layer circulation. The diluted surface layer exhibits a net seaward residual flow which reaches 5 cm s^{-1} while the deeper saline layer is characterised by a net landward residual flow which varies between 2 and 5 cm s^{-1}. At coastal shallower depths, wind action causes an increase of residual current velocity which can reach 30 cm s^{-1} during strong winds (> 15 m s^{-1}).

In the eastern part of the bay, benthic communities are distributed in response to hydrodynamic and sedimentary gradients, so that four different communities are distinguished from the coast to the open sea: a muddy *Macoma balthica* community, a muddy fine sand *Abra alba-Pectinaria koreni* community, a medium sand *Ophelia borealis* community and a sandy gravel *Glycymeris glycymeris* community (Cabioch & Gentil, 1975).

The target species: Pectinaria koreni

The tubicolous polychaete *Pectinaria koreni,* which behaves as a sub-surface deposit feeder, is one of the most common species of the *Abra alba* muddy-fine sand community (Thiébaut et al., 1997). Its densities show strong seasonal variations with a minimum in winter (i.e. few hundred ind. m^{-2}) and a maximum in summer (i.e. up to 3000 ind. m^{-2}) after the recruitment (Gentil et al., 1986). Between 1986 and 1991, the adult population exhibited weak fluctuations in terms of abundance and distribution so that highest densities were regularly reported off the Seine estuary, off Deauville and Cabourg, and in a coastal zone from the

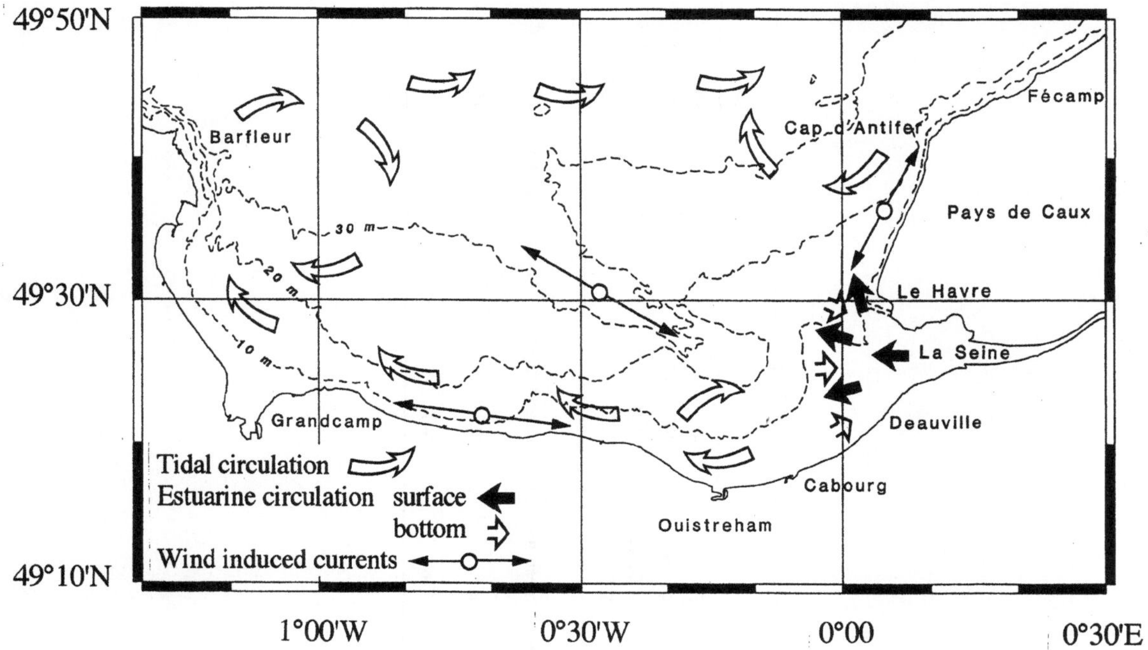

Figure 1. Schematic representation of the residual circulation in the Bay of Seine (modified from Le Hir et al., 1986).

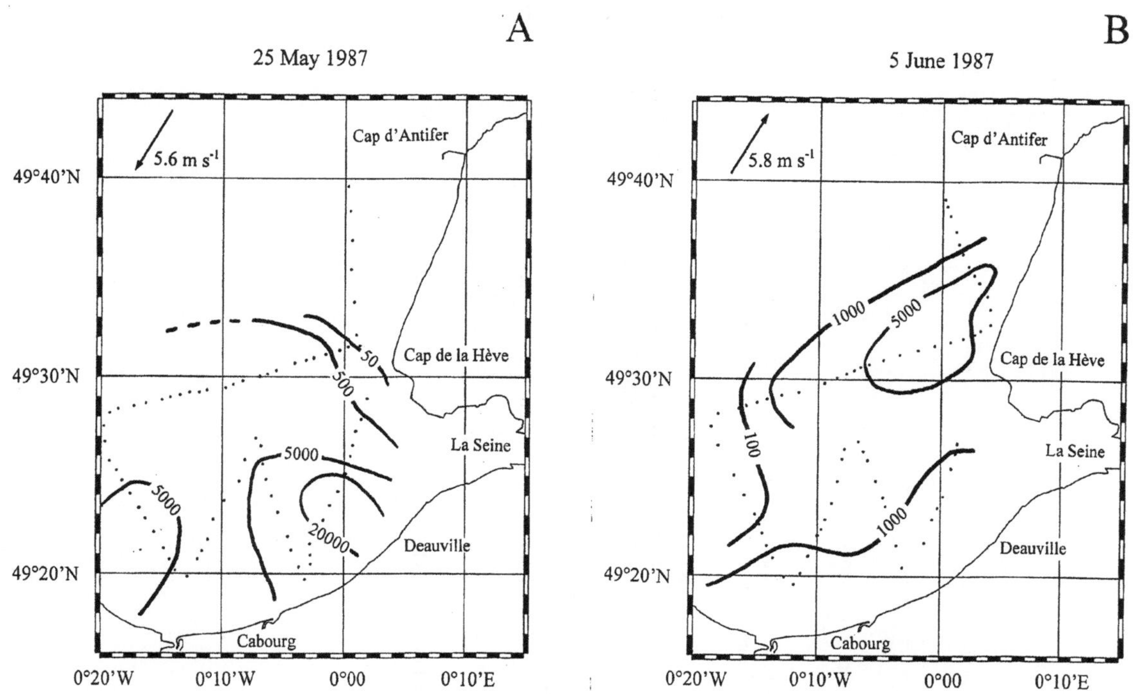

Figure 2. Schematic representation of *Pectinaria koreni* larval distribution at 3-m depth in relation to wind direction and intensity. (A) Distribution of a 10 d old larval cohort on 25 May 1987 under the influence of northern and north-eastern wind. (B) Distribution of a 10 d old larval cohort on 5 June 1987 under the influence of western and south-western wind (from Lagadeuc, 1992b). Larval densities are expressed as ind. m^{-3}. Dots indicate sampling stations. Arrows indicate wind direction. The averaged wind velocity over the 10 d preceding the sampling is given.

Table 1. Mean densities of *Pectinaria koreni* settlers at differents stations of the eastern Bay of Seine sampled on 9–12 June 1987 during the peak-settling (from Lambert, 1991). Sediments characteristics were defined from samples collected in February 1988.

Station	R1	R2	E	A	B	R5	R6	R7	R8
Latitude	49°25.08′ N	49°26.28′ N	49°27.13′ N	49°27.20′ N	49°27.30′ N	49°28.18′ N	49°29.05′ N	49°30.00′ N	49°30.96′ N
Longitude	0°07.08′ E	0°03.20′ E	0°02.95′ E	0°01.50′ E	0°00.20′ E	0°02.82′ W	0°04.38′ W	0°07.09′ W	0°08.97′ W
Depth (m)	1.5	2	8	10	9	9.5	12	14	17.5
Sediment characteristics									
substrate type	mud	mud	mud	sandy mud	muddy fine sand	clean fine sand	clean fine sand	clean med. sand	clean med. sand
% $< 63\ \mu m$	–	4	16	34	2	0	0	–	0
% $\geq 500\ \mu m$	–	0	8	2	10	4	6	–	45
Settlers density (ind m^{-2})	1118	529	11301	28409	16481	176	647	2060	0

Cap de La Hève to the Cap d'Antifer (Thiébaut et al., 1997).

Pectinaria koreni is an univoltine species with a short life-span of 12–18 months (Elkaïm & Irlinger, 1987). It has two main spawning periods in March–April and May–June and exhibits a high fecundity ranged from 10 000 to 12 000 ovocytes per female (Elkaïm & Irlinger, 1987; Irlinger et al., 1991). Following an external fertilization, the pelagic larval development which was estimated to be about 11–13 d from *in situ* observations, included two trochophore stages, three metatrochophore stages and one aulophore stage (Lagadeuc & Retière, 1993). The larval life span of the successive trochophore and metatrochophore stages are 1, 2–3, 4–5, 2 and 1 d respectively; the planktonic period of the aulophore stage is undetermined.

Larval dispersal

Horizontal transport of *Pectinaria koreni* larvae is mainly governed by tidal residual circulation and wind-induced currents so that larvae can be advected up to 40 nautical miles from the adult population (Lagadeuc, 1990). In accordance with the tidal residual circulation, a general drift of larvae occurs principally in westerly and north-westerly directions, and to a lesser extent, in north-easterly direction (Lagadeuc, 1992b). Although such a transport can generate a sizeable loss of larvae outside the eastern Bay of Seine, the low intensity of tidal residual currents off the Seine estuary is sufficient to promote larval retention near the adult population. Moreover, in the north-easterly direction, the frontal structure of the plume acts as a physical boundary which limits offshore export of larvae (Thiébaut, 1996).

Episodic wind events can modify the predictive pattern of tidal dispersal of larvae and disrupt larval retention (Lagadeuc, 1992b). During *Pectinaria koreni* reproductive period, from March to July, main wind directions are north/north-east and west/south-west (Table 2). Larval distribution and transport in relation to wind-induced circulation are given for two different 10 d old larval cohorts sampled in May and June 1987 (Figure 2). On 25 May, larval population was transported westward under the influence of north-easterly winds: maximal larval densities (> 2000 ind. m^{-3}) were located off Deauville and off Ouistreham while larval densities in the north of the Cap de La Hève were low (< 20 ind. m^{-3}) (Figure 2A). Conversely, on 5 June, larvae exhibited maximal densities off the Seine estuary and off the Cap de La Hève (> 5000 ind. m^{-3}) and, were dispersed northward in response to westerly and south-westerly winds (Figure 2B).

Larval vertical distribution is related to water column stratification (Lagadeuc, 1992a). When vertical stratification is strong, larval vertical distribution varies according to larval development stages. Trochophore stages are mainly reported in surface waters at the level of the halocline while metatrochophore stages are sampled below the halocline (Figure 3B). The aulophore stage is confined within near-bottom waters. Conversely, when the stratification is low in response to an increase of vertical turbulence, all larval stages are distributed evenly over the whole water column (Figure 3A).

Metamorphosis and settlement

The metamorphosis, defined as an irreversible set of anatomical and physiological changes of individuals (Butman, 1987), begins in the water column by the transformation of metatrochophore into aulophore larvae, i.e. competent larval stage (Lambert et al., 1996). It is a gradual process which continues over about two weeks after the settlement by the purchase of all morphological structures of juveniles. Although *Pectinaria koreni* aulophores have lost their typically larval morphological characteristics, they do not exhibit all the organs of the juveniles even as buds and are not well adapted to benthic life when they reach the bottom.

Under still water conditions, *Pectinaria koreni* postlarvae settle preferentially on muds and sandy muds rather than on muddy fine sands or clean fine sands, and are able to actively select a particular substrate by crawling at the sediment surface on a microspatial scale (i.e. rate of displacement: 0.36 to 1.90 cm d^{-1}) (Desroy et al., 1997). Although this mechanism may be involved in the microscale distribution of postlarvae, field observations at the scale of the eastern Bay of Seine reveal that larvae may settle over a large edaphic spectrum, from muds to clean medium sands, regardless of substrate suitability for adults (Table 1).

Following the settlement, postlarvae exhibit a high mortality rate ranging from 41% to 100% over two weeks (Lambert, 1991; Lambert et al., 1996). This large decrease in postlarval densities may be explained by (1) metamorphosis as a critical period for survival, (2) post-settlement processes like predation and (3) postlarval dispersal.

Table 2. Variations of climatic characteristics (i.e. wind regime and Seine freshwater discharge) from 1967 to 1996 in the eastern Bay of Seine during the *Pectinaria koreni* reproductive period. Wind characteristics (i.e. 8 values per day) were measured at the signal station of the Cap de la Hève (Meteofrance data). Daily Seine freshwater discharge was measured at a lock-gate located 150 km upstream of Le Havre Harbour (data from the Navigation Department of Rouen Harbour).

		March	April	May	June	July
Frequency of main wind direction (%)	N-NE	29.2	41.7	37.9	38.4	34.7
	W-SW	31.8	23.2	25.3	29.6	32.3
Monthly wind speed (m s^{-1})	Average	6.5	5.9	5.4	5.1	5.1
	Min.	4.3	4.3	3.9	4.1	3.3
	Max.	9.1	8.8	7.6	6.5	6.7
Nb of days with wind speed ≥ 7 m s^{-1}	Average	12.1	9.3	7.0	5.7	5.4
	Min.	1	2	0	1	0
	Max.	20	21	15	12	12
Seine freshwater discharge (m^3. s^{-1})	Average	625	586	436	332	265
	Min.	312	256	178	125	121
	Max.	1480	1280	1110	673	517

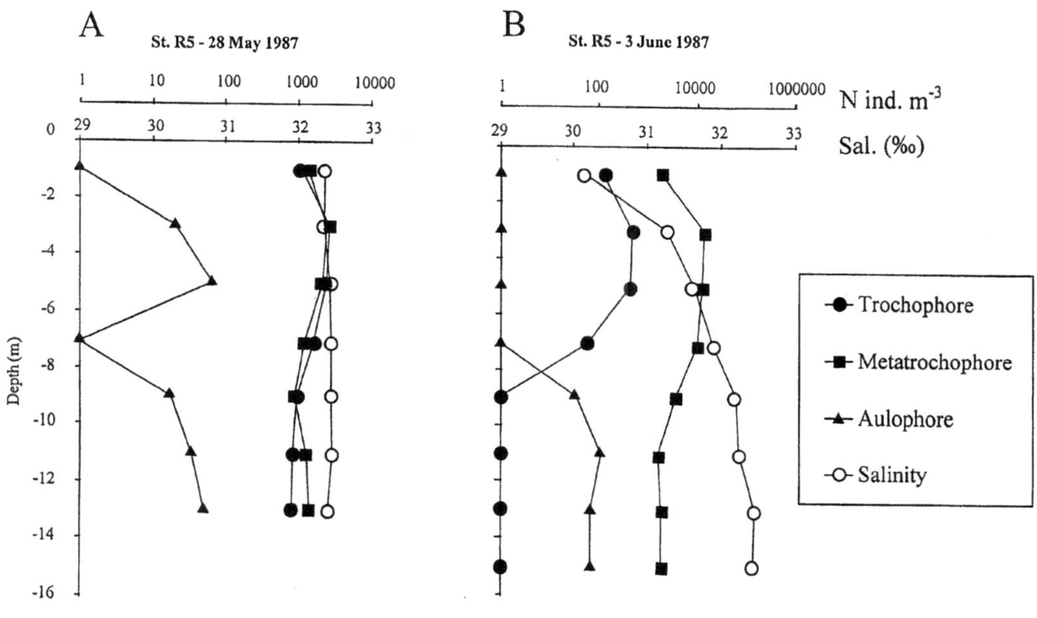

Figure 3. Vertical distribution of *Pectinaria koreni* larval stages at one station off the Seine estuary (49°28.18′ N–0°02.82′ W) in relation to water column stratification. (A) Vertical homogeneity (28 May 1987). (B) Vertical stratification (3 June 1987). (from Lagadeuc, 1992a).

Postlarval dispersal

Postlarval dispersal results from combined action of hydrodynamical factors in the bottom boundary layer and postlarval behaviour. Flume experiments have shown that postlarvae and juveniles can actively leave the substrate by the secretion of a long mucous thread which increases the hydrodynamic drag on the organisms and enables them to be resuspended in the water column (Lambert, 1991; Olivier et al., 1996b; Desroy et al., 1997). Such a behaviour is induced by unfavourable ecological conditions like the occurrence of an unsuitable substrate or high densities of conspecific adults (Olivier et al., 1996b). Adults effects on postlarval dispersal result either from direct action on postlarvae or from sediment disturbance generated by bioturbation.

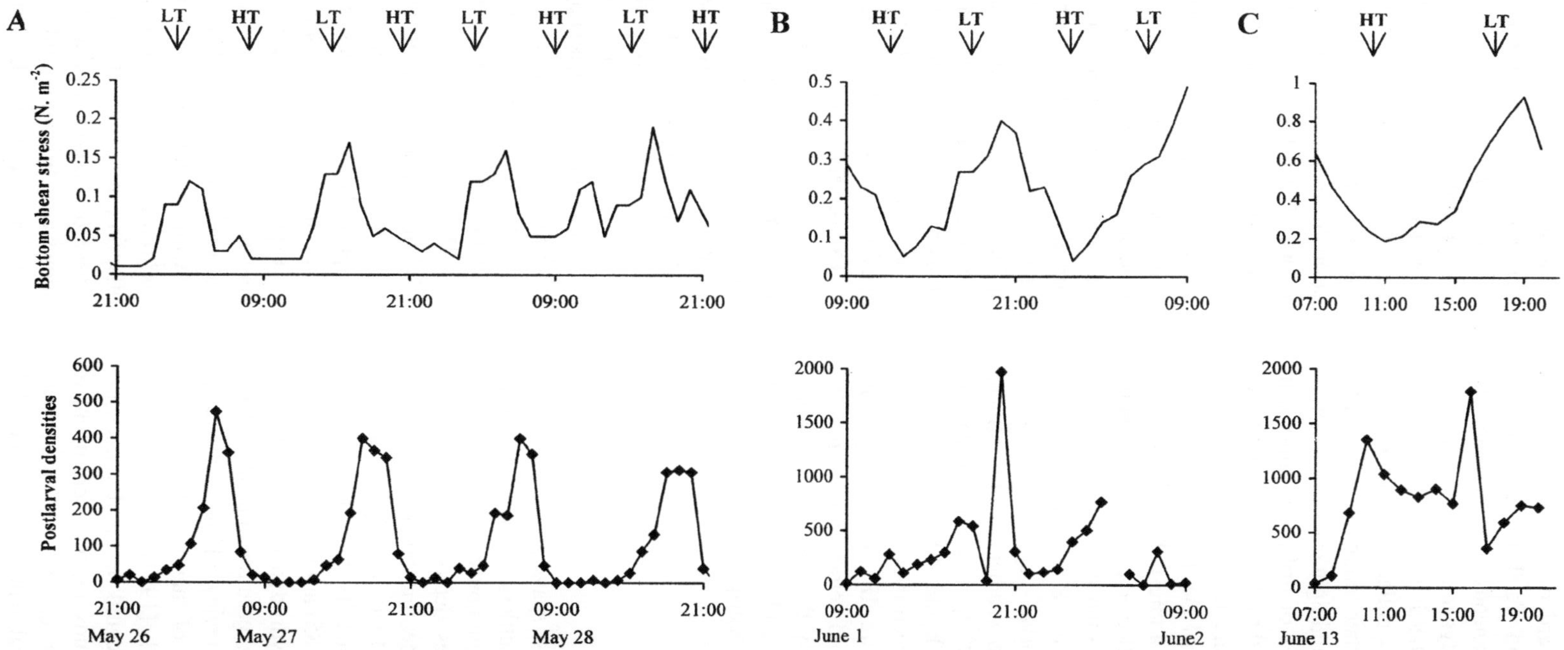

Figure 4. Temporal evolution of *Pectinaria koreni* drifting postlarvae in relation to semi-diurnal tidal cycle and bottom shear stress at one station off the Seine estuary (49°26.60′ N–0°01.30′ E). (A) Depth-averaged postlarval densities (ind. m^{-2}) from 26 to 28 May 1992 (from Thiébaut et al., 1996). (B) Near-bottom postlarval densities (ind. 400 m^{-3}) from 1 to 2 June 1992 (from Olivier et al., 1996a). (C) Near-bottom postlarval densities (ind. 400 m^{-3}) on 13 June 1992 (from Olivier et al., 1996a).

From different time-series carried out in May-June 1992 at one station seaward of the Seine estuary, Olivier et al. (1996a) and Thiébaut et al. (1996) demonstrated that instantaneous tidal currents and swell are the main hydrodynamical factors involved in postlarval dispersal in the field (Figure 4). Observations with pump samples collected over the whole water column in calm weather conditions (mean wind velocity of ~ 5.2 m s^{-1}) highlight cyclic fluctuations of depth-averaged densities of drifting postlarvae in relation to semi-diurnal tidal cycle (Figure 4A). Maximum densities (i.e. > 100 ind. m^{-2}) were consistently observed at the beginning of the flood when the bottom shear stress exceeded a threshold value of 0.1 N m^{-2}. Furthermore, data collected using a suprabenthic sledge which samples the water column from 0.10 to 1.45 m above the sea-bed reveal that the importance of tidal currents on postlarval dispersal depends on weather conditions. In calm weather conditions (mean wind velocity of 2.1 m s^{-1}), maximum density of drifting postlarvae (i.e. 2000 ind. 400 m^{-3}) was also reported during the flood even if cyclic variations are not clearly apparent (Figure 4B). Conversely, with moderate winds (mean wind velocity of 5.8 m s^{-1}), swell increases significantly the bottom shear stress, which exceeded 0.2 N m^{-2} and reached 1 N m^{-2}, and induces a near permanent resuspension of postlarvae (Figure 4C).

Discussion

Over the past two decades, spawning, larval dispersal, settlement and early settlers mortality of marine invertebrates have been commonly studied to understand how these processes determine the structure and abundance of adult benthic populations. However, most studies have focused only on one life stage and have emphasized its role. For instance, Gaines & Rougharden (1985) argued that larval supply in combination with hydrodynamical factors play a crucial role in generating recruitment patterns. Conversely, from a synthetic review of major processes regulating the recruitment of marine soft-bottom invertebrates, Olafsson et al. (1994) have shown that post-settlement mortality could be the dominant determinant. The present study on the recruitment of *Pectinaria koreni* in the eastern Bay of Seine reveals that the spatial and temporal maintenance of the adult population could not be explained by one or few key-factors but is a result of a combination of a large variety of favourable and repressive factors which include biological and hydrodynamical properties acting on different life stages (Figure 5).

Pectinaria koreni larval dispersal is caused by its biological properties and by local hydrodynamic features which could favour either larval retention near the parental population or larval export to open sea (Figure 5). In relation to the two-layer estuarine circulation, the ontogenic vertical migration of larvae could be one of the dominant factor that ensures their retention off the Seine estuary. Younger larvae, mainly located in the diluted surface layer are carried downstream whereas older larvae present in the deeper saline layer are advected upstream to the area of adult population. As the influence of Seine freshwater was estimated to extend around about 15 nautical miles from the mouth of the estuary (Thiébaut et al., 1992), the ontogenic vertical migration is efficient in terms of larval retention only if larvae can benefit of the two-layer circulation and are not transported too far offshore. The low tidal residual circulation and the Seine River plume front may ensure that high larval densities remain in the vicinity of the Seine estuary.

Conversely, wind effects induce a large variability in larval dispersal and may disrupt larval retention at the proximity of adult population resulting of the processes cited above:

(1) When wind velocity exceeds 7 m s^{-1}, ontogenic vertical migration is not reported and larvae are evenly distributed over the whole water column so that no differential transport occurs for the different larval stages (Lagadeuc, 1992a);
(2) Moderate north-easterly winds are able to break up the Seine River plume front and generate a dilution in surface of the plume waters and larvae (Thiébaut, 1996);
(3) Wind-induced currents modify over the few days of larval life duration the direction of larval transport; they also increase residual currents velocity in the coastal zones and consequently the larval export to offshore waters (Lagadeuc, 1992b).

As wind events cause an unpredictable hydrodynamical environment at a short temporal scale, multiple spawning events over the 4 months of *Pectinaria koreni* reproductive period could prevent the loss of all offspring in response to a catastrophic event and increase the likelihood that at least one larval cohort allows a significant recruitment each year. Due to the variations of climatic characteristics in the eastern Bay of Seine from March to July, the relative importance of wind events on larval dispersal and recruitment should

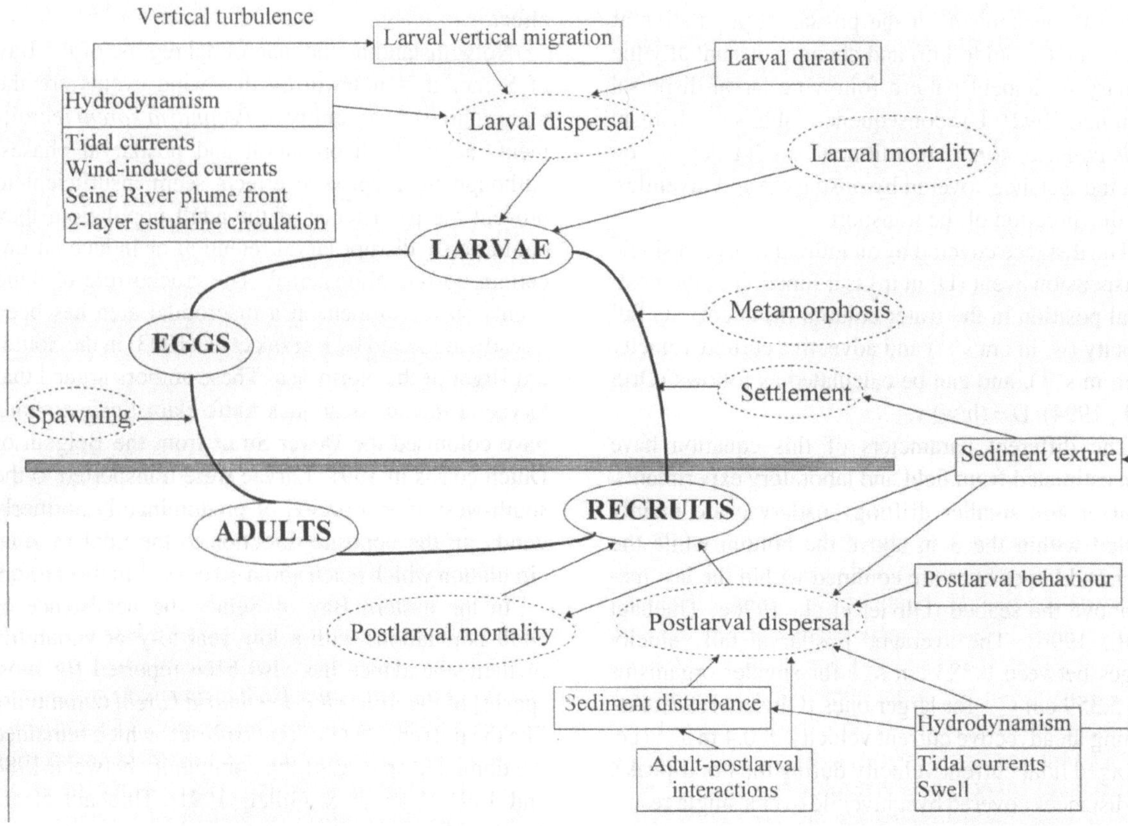

Figure 5. Conceptual model of biological and hydrodynamical factors governing *Pectinaria koreni* recruitment in the eastern Bay of Seine. Solid oval: life-cycle stages. Dashed oval: processes governing the recruitment. Solid rectangle: hydrodynamical and biological factors involved in the recruitment.

vary with time. Thus, higher monthly wind speed and higher number of days with a wind speed ≥ 7 m s^{-1} in March–April should decrease the probability of larval retention at the beginning of the reproductive period, even if stronger Seine freshwater discharge at this time could strengthen the two-layer estuarine circulation (Table 2).

Pectinaria koreni postlarval drifting depends on both active behaviour (i.e. the secretion of a mucous thread) in response to an unfavourable biotic and abiotic environment and the bottom shear stress generated by tidal currents and swell (Figure 5). According to the size of drifted postlarvae and juveniles reported from field studies, which varied between 0.9 and 62.5 mm (Lambert et al., 1996; Olivier et al., 1996a), it appears that postlarvae and juveniles retain the ability to be transported by thread drifting over few months, independent of the adaptations of organisms to the benthic life. While crawling at the sediment surface induces a postlarval motion at microscale (i.e. cm), thread drift-

Table 3. Distance covered by a drifted postlarva over a single resuspension event for different size classes and heights above the sea-bed. Postlarval fall velocity has been measured experimentally by Olivier (1997). Horizontal current velocity is assumed to be constant and equal to 0.4 m s^{-1}.

Size classes (mm)	Height above the seabed (cm)	Fall velocity (cm s^{-1})	Covered distance (m)
1–2	300	0.383–0.506	313–237
	200		209–158
	100		104–79
2–4	150	0.880–1.904	68–31.5
	100		45.5–21.0
	50		22.7–10.5
4–10	100	4.190–5.359	9.5–7.5
	50		4.8–3.7
	25		2.4–1.9

ing in the water column may represent a significant dispersal mechanism for the postlarval population at mesoscale (i.e. m to km) and change substantially the primary settlement pattern following larval dispersal (Günther, 1992). The consequences of this mechanism on *Pectinaria koreni* recruitment should depend on both the distance covered by postlarvae and juveniles, and the direction of the transport.

The distance covered by an individual over a single resuspension event (D, in m) is a function of the postlarval position in the water column (h, in cm), its fall velocity (w, in cm s^{-1}) and advective current velocity (v, in m s^{-1}), and can be calculated as follows (Orth et al., 1994): $D = (h/w)\ v$.

The different parameters of this equation have been estimated from field and laboratory experiments. Younger and smaller drifting postlarvae are mainly located within the 3 m above the bottom while the older and larger ones are confined within the last meter above the seabed (Olivier et al., 1996a; Thiébaut et al., 1996). The averaged postlarval fall velocity ranges between 0.383 cm s^{-1} for smaller organisms and 5.359 cm s^{-1} for larger ones (Olivier, 1997). Assuming an advective current velocity of 0.4 m s^{-1} (i.e. averaged tidal current velocity during the flood-peak), the distances covered by a juvenile over a single resuspension event are estimated for different size classes (Table 3). These calculations highlight that spatial scales of postlarval drifting vary greatly from few hundreds meters for smaller postlarvae to few meters for larger juveniles. Thus, just young postlarvae may be subject to a consequent transport by thread drifting over successive resuspension events at the scale of the eastern Bay of Seine. For old ones, this mechanism may only be involved in the exploration of prospective habitats or in response to disturbance at local scale.

The direction of the transport changes according to the main hydrodynamical factors involved in postlarval drifting. In calm weather conditions, the relationship between postlarval resuspension and semi-diurnal tidal currents would favour a regular and landward transport during the flood and could lead over several tidal cycles to a migration of young postlarvae from unsuitable offshore bottoms towards coastal muddy fine sands, on a distance of several km. Such a migration could act as a feed-back mechanism which partly counteracts the demographic effects of larval dispersal on the recruitment. Conversely, the swell related to moderate or strong winds induces a near permanent resuspension of postlarvae independently of the tidal cycle and generates an episodic transport which sometimes strengthens or sometimes is opposed to the tidal effect.

Notwithstanding the macrotidal regime of the Bay of Seine, it is noteworthy that wind events are the main source of variability of *Pectinaria koreni* recruitment, acting both on larval and postlarval phases. Although their episodic effects seem insufficient to prevent the persistence of the adult population, they could partly disrupt larval retention or induce an uncertain postlarval dispersal. The crucial role of wind events on recruitment in a macrotidal area has been already suggested by Luczak et al. (1993) in the Southern Bight of the North Sea. These authors argued that larvae of the american jack knife clam *Ensis directus* have colonised the Dover Strait from the Belgian or Dutch coasts in 1991. Larvae were transported to the south-west after a period of predominantly northerly winds, in the opposite direction to the tidal residual circulation which reach about 10 cm s^{-1} in this region.

In the eastern Bay of Seine, the persistence of adult populations with a low year-to-year variability of their abundance has also been reported for most species of the *Abra alba-Pectinaria koreni* community like the polychaete *Owenia fusiformis* which remained the dominant species of the community between 1986 and 1991 (Dauvin & Gillet, 1991; Thiébaut et al., 1997). The comparison of biological properties between *Pectinaria koreni* and *Owenia fusiformis* reveals how different processes involved in the recruitment may ensure the maintenance of adult populations in the same hydrodynamical environment (Table 4). Although wind can greatly modify the scheme of residual circulation, its frequent changes in speed and direction vanish partly its effects on long time scales in comparison with the permanent tide action (Salomon, 1991). Therefore, its consequences on larval dispersal should depend on larval life duration so that larval transport is principally governed by wind for *Pectinaria koreni* and by tidal residual circulation for *Owenia fusiformis*. The dominant role of tide for the latter species would limit the likelihood of a total export of larval population to offshore unsuitable substrates (Thiébaut et al., 1994). Nevertheless, the analysis of *Owenia fusiformis* population structure between 1986 and 1988 have shown significant year-to-year variability in juvenile abundance so that yearly high recruitment alternated with yearly moderate or low recruitment (Dauvin & Gillet, 1991). Thus, the long life-span and the multivoltine reproduction of *Owenia fusiformis* may be considered as analogous to the multiple spawning events per year of *Pectinaria koreni* to

Table 4. Comparison of main biological properties of *Pectinaria koreni* and *Owenia fusiformis* larvae, postlarvae and adults, and their consequences on the recruitment in relation to hydrodynamical factors. [1] Ménard et al. (1989); [2] Gentil et al. (1990); [3] Thiébaut et al. (1992); [4] Dauvin (1992).

Biological properties	*Pectinaria koreni*	*Owenia fusiformis*	Consequences on the recruitment
Life-span	12–18 months	3–4 y[1]	Multiple spawning events over one or several years for *P. koreni* and *O. fusiformis* respectively increase the probability that at least one larval cohort generates a high recruitment during the life-span.
Reproduction	univoltine	multivoltine[2]	
Nb of spawning events/year	high (> 3–4)	low (~ 1–2) [2]	
Larval life duration	11–13 d	~ 1 month [3]	As the effects of wind-induced currents on the residual circulation decreases with time, dispersal of *P. koreni* larvae is more sensitive to wind events than dispersal of *O. fusiformis* larvae.
Larval vertical migration	ontogenic	ontogenic[3]	For both species, interactions between ontogenic vertical migration of larvae and two-layer estuarine circulation could favour larval retention off the Seine estuary.
Larval habitat selection	no	no[4]	At the scale of eastern Bay of Seine, primary settlement of both the species is related to larval supply and results of a passive deposition; it is independent of sediment grain size.
Postlarval behaviour	postlarval drifting	no postlarval drifting[4]	While *O. fusiformis* settlement is near definitive, behaviour of *P. koreni* postlarvae induces a postlarval drifting which may change primary settlement pattern and generate a net landward transport.

increase the probability that at least one larval cohort permit a high recruitment during the life-span of both the species.

Acknowledgements

This study forms a part of the contribution of the 'GDR Manche' framework to the French Program on the Determinism of the Recruitment (PNDR-GLOBEC France). It was supported by the CNRS and IFREMER.

References

Armonies, W., 1994. Drifting meio- and macrobenthic invertebrates on tidal flats in Königshafen: a review. Helgoländer Meeresunters. 48: 299–320.

Butman, C. A., 1987. Larval settlement of soft-sediment invertebrates: the spatial scales of pattern explained by active habitat selection and the emerging role of hydrodynamical processes. Oceanogr. Mar. Biol. annu. Rev. 25: 113–165.

Cabioch, L. & F. Gentil, 1975. Distribution des peuplements benthiques dans la partie orientale de la Baie de Seine. C. r. Acad. Sci., Paris 280D: 571–574.

Cabioch, L., J. C. Dauvin, C. Retière, V. Rivain & D. Archambault, 1982. Evolution de peuplements benthiques des fonds sédimentaires de la région de Roscoff, perturbés par les hydrocarbures de l'Amoco-Cadiz. Neth. J. Sea Res. 16: 491–501.

Chabert d'Hières, G., 1986. La circulation des eaux en Baie de Seine. Actes de Colloques IFREMER 4: 15–23.

Connell, J. H., 1985. The consequences of variation in initial settlement vs. post-settlement mortality in rocky intertidal communities. J. exp. mar. Biol. Ecol. 93: 11–45.

Dauvin, J. C. & P. Gillet, 1991. Spatio-temporal variability of the demographic structure of *Owenia fusiformis* Delle Chiaje (Annelida: Polychaeta) from the Bay of Seine (eastern English Channel). J. exp. mar. Biol. Ecol. 152: 105–122.

Dauvin, J. C., 1992. Cinétique du recrutement et croissance des juvéniles d'*Owenia fusiformis* Delle Chiaje en baie de Seine (Manche orientale). Oceanol. Acta 15: 187–196.

Desroy, N., F. Olivier & C. Retière, 1997. Effects of individual behaviors, inter-individual interactions with adult *Pectinaria koreni* and *Owenia fusiformis* (Annelida, Polychaeta), and hydrodynamism on *Pectinaria koreni* recruitment. Bull. mar. Sci. 60: 547–558.

Eckman, J. E., 1996. Closing the larval loop: linking larval ecology to the population dynamics of marine benthic invertebrates. J. exp. mar. Biol. Ecol. 200: 207–237.

Elkaïm, B. & J. P. Irlinger, 1987. Contribution à l'étude de la dynamique des populations de *Pectinaria koreni* Malmgren (Polychète) en Baie de Seine orientale. J. exp. mar. Biol. Ecol. 107: 171–197.

Gaines, S. D. & J. Rougharden, 1985. Larval settlement rate: a leading determinant of structure in an ecological community of the marine intertidal zone. Proc. natn. Acad. Sci. U.S.A. 82: 3707–3711.

Gaines, S. D. & M. D. Bertness, 1992. Dispersal of juveniles and variable recruitment in sessile marine species. Nature 360: 579–580.

Gentil, F., J. P. Irlinger, B. Elkaïm & F. Proniewski, 1986. Premières données sur la dynamique du peuplement macrobenthique des sables fins envasés à *Abra alba* de la Baie de Seine orientale. Actes de Colloques IFREMER 4: 409–419.

Gentil, F., J. C. Dauvin & F. Ménard, 1990. Reproductive biology of the polychaete *Owenia fusiformis* Delle Chiaje in the Bay of Seine (eastern English Channel). J. exp. mar. Biol. Ecol. 142: 13–23.

Gosselin, L. A. & P. Y. Qian, 1997. Juvenile mortality in benthic marine invertebrates. Mar. Ecol. Prog. Ser. 146: 265–282.

Günther, C. P., 1992. Dispersal of intertidal invertebrates: a strategy to react to disturbances of different scales. Neth. J. Sea Res. 30: 45–56.

Hill, A. E., 1991. Vertical migration in tidal currents. Mar. Ecol. Prog. Ser. 75: 39–54.

Irlinger, J. P., F. Gentil & V. Quintino, 1991. Reproductive biology of the polychaete *Pectinaria koreni* (Malmgren) in the Bay of Seine (English Channel). Ophelia Suppl. 5: 343–350.

Lagadeuc, Y., 1990. Processus hydrodynamiques. Dispersion larvaire et recrutement en régime mégatidal. Exemple de *Pectinaria koreni* (Annélide polychète) en baie de Seine (Manche). Thèse de doctorat, Université de Lille I, 279 pp.

Lagadeuc, Y., 1992a. Répartition verticale des larves de *Pectinaria koreni* en Baie de Seine orientale: influence sur le transport et le recrutement. Oceanol. Acta 15: 109–118.

Lagadeuc, Y., 1992b. Transport larvaire en Manche, exemple de *Pectinaria koreni* en Baie de Seine. Oceanol. Acta 15: 383–395.

Lagadeuc, Y. & C. Retière, 1993. Critères d'identification rapide des stades de développement des larves de *Pectinaria koreni* (Malmgren) (Annélide Polychète) de la Baie de Seine (Manche). Vie Milieu 43: 217–224.

Lambert, R., 1991. Recrutement d'espèces benthiques à larves pélagiques en régime mégatidal. Cas de *Pectinaria koreni* (Malmgren), annélide polychète. Thèse de doctorat, Université de Rennes I, 176 pp.

Lambert, R., C. Retière & Y. Lagadeuc, 1996. Metamorphosis of *Pectinaria koreni* (Annelida: Polychaeta) and recruitment of an isolated population in the English Channel. J. mar. biol. Ass. U.K. 76: 23–36.

Le Hir, P., J. C. Salomon, C. Le Provost, G. Chabert d'Hières & J. L. Mauvais, 1986. Approche de la circulation résiduelle en Baie de Seine. Actes de Colloques IFREMER 4: 63–72.

Levitan, D. R., M. A. Sewell & F. S. Chia, 1992. How distribution and abundance influence fertilization success in the sea urchin *Strongylocentrotus franciscanus*? Ecology 73: 248–254.

Luczak, C., J. M. Dewarumez & K. Essink, 1993. First record of the american jack knife clam *Ensis directus* on the French coast of the North Sea. J. mar. biol. Ass. U.K. 73: 233–235.

Ménard, F., F. Gentil & J. C. Dauvin, 1989. Population dynamics and secondary production of *Owenia fusiformis* Delle Chiaje (Polychaete) from the Bay of Seine (eastern English Channel). J. exp. mar. Biol. Ecol. 133: 151–167.

Olafsson, E. B., C. H. Peterson & W. G. Ambrose Jr, 1994. Does recruitment limitation structure populations and communities of macro-invertebrates in marine soft sediments: the relative significance of pre- and post-settlement processes. Oceanogr. Mar. Biol. annu. Rev. 32: 65–109.

Olivier, F., 1997. Régulation des populations de recrues en régime mégatidal: effet du couplage physique-biologie en couche limite benthique. Thèse de doctorat, Université de Rennes I, 135 pp.

Olivier, F., C. Vallet, J. C. Dauvin & C. Retière, 1996a. Drifting in post-larvae and juveniles in an *Abra alba* (Wood) community of the eastern part of the Bay of Seine (English Channel). J. exp. mar. Biol. Ecol. 199: 89–109.

Olivier, F., N. Desroy & C. Retière, 1996b. Habitat selection and adult-recruit interactions in *Pectinaria koreni* (Malmgren) (Annelida: Polychaeta) post-larval populations: results of flume experiments. J. Sea Res. 36: 217–226.

Orth, R. J., M. Luckenbach & K. A. Moore, 1994. Seed dispersal in a marine macrophyte: implications for colonization and restoration. Ecology 75: 1927–1939.

Rougharden, J., S. Gaines & H. Possingham, 1988. Recruitment dynamics in complex life cycles. Science 241: 1460–1466.

Salomon, J. C., 1991. Hydrodynamic action on benthic macrofauna in tidal coastal zones. ICES mar. Sci. Symp. 192: 15–23.

Scheltema, R. S., 1986. On dispersal and planktonic larvae of benthic invertebrates: an eclectic overview and summary of problems. Bull. mar. Sci. 39: 290–322.

Thiébaut, E., J. C. Dauvin & Y. Lagadeuc, 1992. Transport of *Owenia fusiformis* larvae (Annelida: Polychaeta) in the Bay of Seine (eastern English Channel). Mar. Ecol. Prog. Ser. 80: 29–39.

Thiébaut, E., J. C. Dauvin & Y. Lagadeuc, 1994. Horizontal distribution and retention of *Owenia fusiformis* larvae (Annelida: Polychaeta) in the Bay of Seine. J. mar. biol. Ass. U.K. 74: 129–142.

Thiébaut, E., 1996. Distribution of *Pectinaria koreni* larvae (Annelida: Polychaeta) in relation to the Seine river plume front (eastern English Channel). Estuar. coast. Shelf Sci. 43: 383–397.

Thiébaut, E., J. C. Dauvin & Z. Wang, 1996. Tidal transport of *Pectinaria koreni* postlarvae (Annelida: Polychaeta) in the Bay of Seine (eastern English Channel). Mar. Ecol. Prog. Ser. 138: 63–70.

Thiébaut, E., L. Cabioch, J. C. Dauvin, C. Retière & F. Gentil, 1997. Spatio-temporal persistence of the *Abra alba-Pectinaria koreni* muddy-fine sand community of the eastern Bay of Seine (English Channel). J. mar. biol. Ass. U.K. 77: 1165–1185.

Hydrobiologia **375/376:** 177–189, 1998.
S. Baden, L. Pihl, R. Rosenberg, J.-O. Strömberg, I. Svane & P. Tiselius (eds),
Recruitment, Colonization and Physical–Chemical Forcing in Marine Biological Systems.

Effects of green algal mats on infaunal colonization of a New England mud flat – long-lasting but highly localized effects

Martin Thiel* & Les Watling
Darling Marine Center, University of Maine, Walpole, ME 04573, USA
* *Address for future correspondence: Martin Thiel, Smithsonian Marine Station, 5612 Old Dixie Highway, Fort Pierce, FL 34946, USA; email: thiel@sms.hboi.edu*

Key words: algal mats, soft-bottom infauna, community change, mud flats, reproduction

Abstract

The response of benthic infauna to green algal mats is commonly examined in either small-scale short-term experiments or large-scale long-term data sets that are averaged over large areas. In this study, we used a small-scale approach to study long-term effects caused by the recurring appearance of green algal mats in the early 1990's on a New England mud flat. Algal mats frequently covered the inner parts of the mud flat, where algae persisted for 6 months between July and December, and were incorporated into the sediment after the growth season. These inner parts of the mud flat differed from the outer parts, where green algae never occurred. Between 1979 and 1996, infaunal numbers increased ten-fold at the 37 stations in the inner part, but not at the 19 stations in the outer part of the mud flat. Detritus-feeding annelids primarily contributed to the numerical increase of infaunal colonizers; grazing gastropods and amphipods, and suspension- and filter-feeding bivalves showed no change. The recurring development and deposition of green algae in the inner part of the mud flat resulted in a localized yet persistent change in the infaunal community. The results suggest that it is necessary to follow the fate of algal mats after the growth season. We propose that effects on infaunal colonizers are most severe and long-lasting where decaying algal mats finally become incorporated into the sediment. Brood protecting annelids are most likely to benefit from detrital material provided by moderately dense algal mats.

Introduction

Growth of dense green algal mats has been reported from intertidal and subtidal soft-bottoms from many parts of the world. Dense algal mats can have a strong impact on macrofauna organisms (Nicholls et al., 1981; Soulsby et al., 1982; Breber, 1985; Buttermore, 1977; Hull, 1987; Raffaelli et al., 1989, 1991; Everett, 1994; Peterson et al., 1994), resulting in death of some species while others may thrive. In many areas where mass development of green algae has been recorded, simultaneous changes of soft-bottom communities have been observed (Reise, 1982; Reise et al., 1989; Raffaelli et al., 1991). Most of these studies are based on qualitative samples (Reise et al., 1989) or selected stations that were sampled in years before and years after the mass occurrence of green algal mats (e.g. Nicholls et al., 1981; Raffaelli et al., 1991). Several experimental studies have shown that algal mats initially have a detrimental effect on macrofaunal numbers, but may at a later stage or after the removal of the algal cover result in increased numbers (although reduced diversity) due to rapid colonization by so-called opportunist species (e.g. Everett, 1994; Norkko & Bonsdorff, 1996a, b). Although these studies show that substantial changes occurred in macrofauna numbers, they usually do not allow for a detailed resolution on a larger spatial scale.

Green algal mats clearly have an effect on macrofauna communities, but it is not known how far-reaching these effects are, i.e., whether they also affect macrofauna in neighboring sediments not usually covered by the algae. Experiments by Thrush (1986) have shown that macrofauna in the immediate vicinity (0.5 m) can be affected by dense macroalgal mats, yet it is not known whether areas further away from the

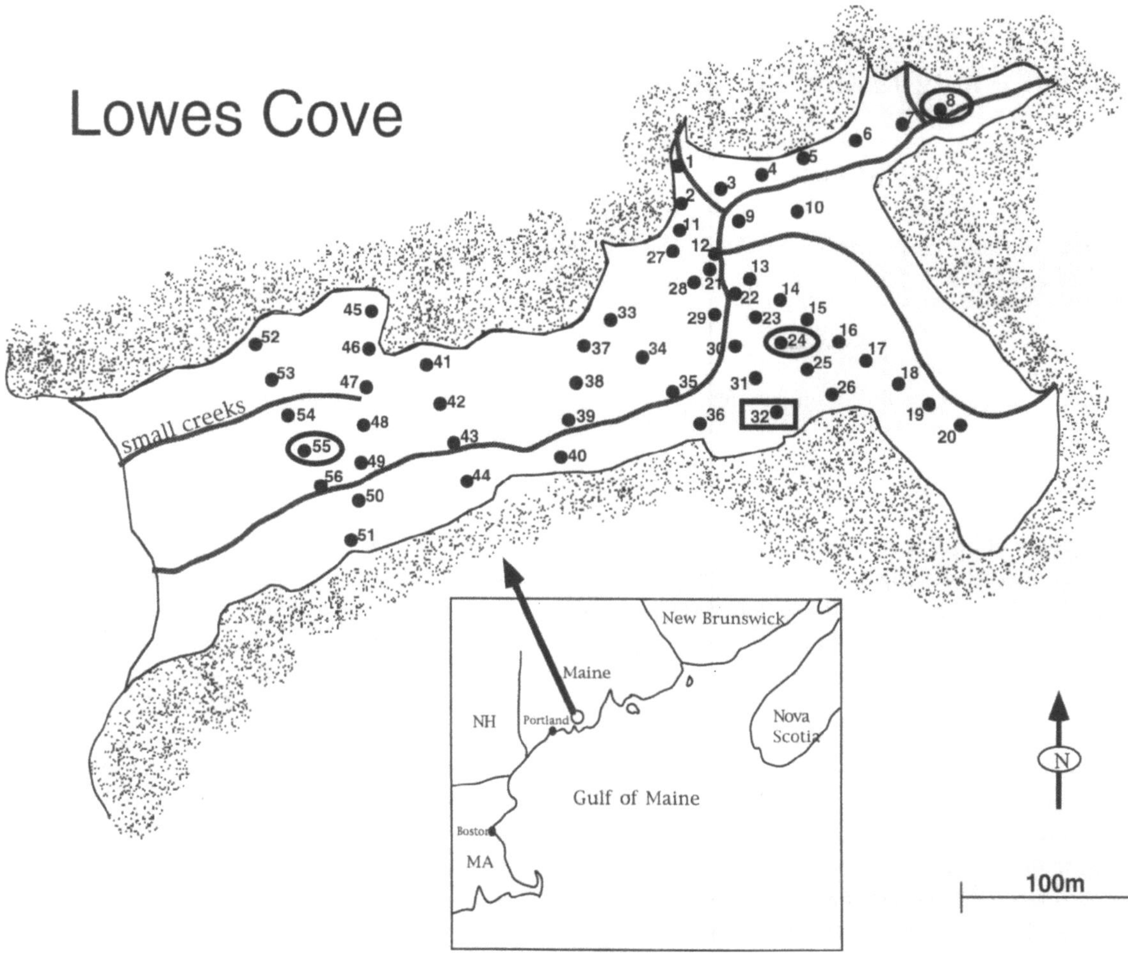

Figure 1. Study area Lowes Cove in the Damariscotta Estuary, Maine; macrofauna samples were taken at 56 stations in June 1979 and June 1996; biomass of green algae was determined every month at stations 8, 24 and 55 in 1996; additional macrofauna data were evaluated for station 32.

algal mats are also affected. Since green algal mats usually redevelop each year in the same areas of a mud flat, but never or rarely cover other parts of a mud flat (e.g. Reise, 1983; Reise & Siebert, 1994) two scenarios are possible. In the first scenario, macrofauna changes under green algal mats may have some larger scale effect on neighboring areas not usually covered by algae. In the second scenario, macrofauna changes only occur on mud flat areas immediately affected by green algal mats, but macrofauna in areas adjacent to the mats are not influenced. In this study we used a small scale sampling approach that allows for a detailed spatial resolution to examine which of the two scenarios is more likely to occur. Available information on the life-history characteristics of the most common macrofauna species is used to examine the reasons for the observed pattern in macrofauna changes.

Material and methods

Study site

Lowes Cove is a sheltered cove in the Damariscotta Estuary, Maine, USA. The mud flat is surrounded on three sides by wooded areas (see Figure 1). Freshwater discharge into the Damariscotta Estuary is on the order of 1–3 m^3 s^{-1} (McAlice, 1977; cited in Mayer et al., 1996), resulting in salinities of 30 at the study site. Maximum nitrate levels in the estuarine waters range

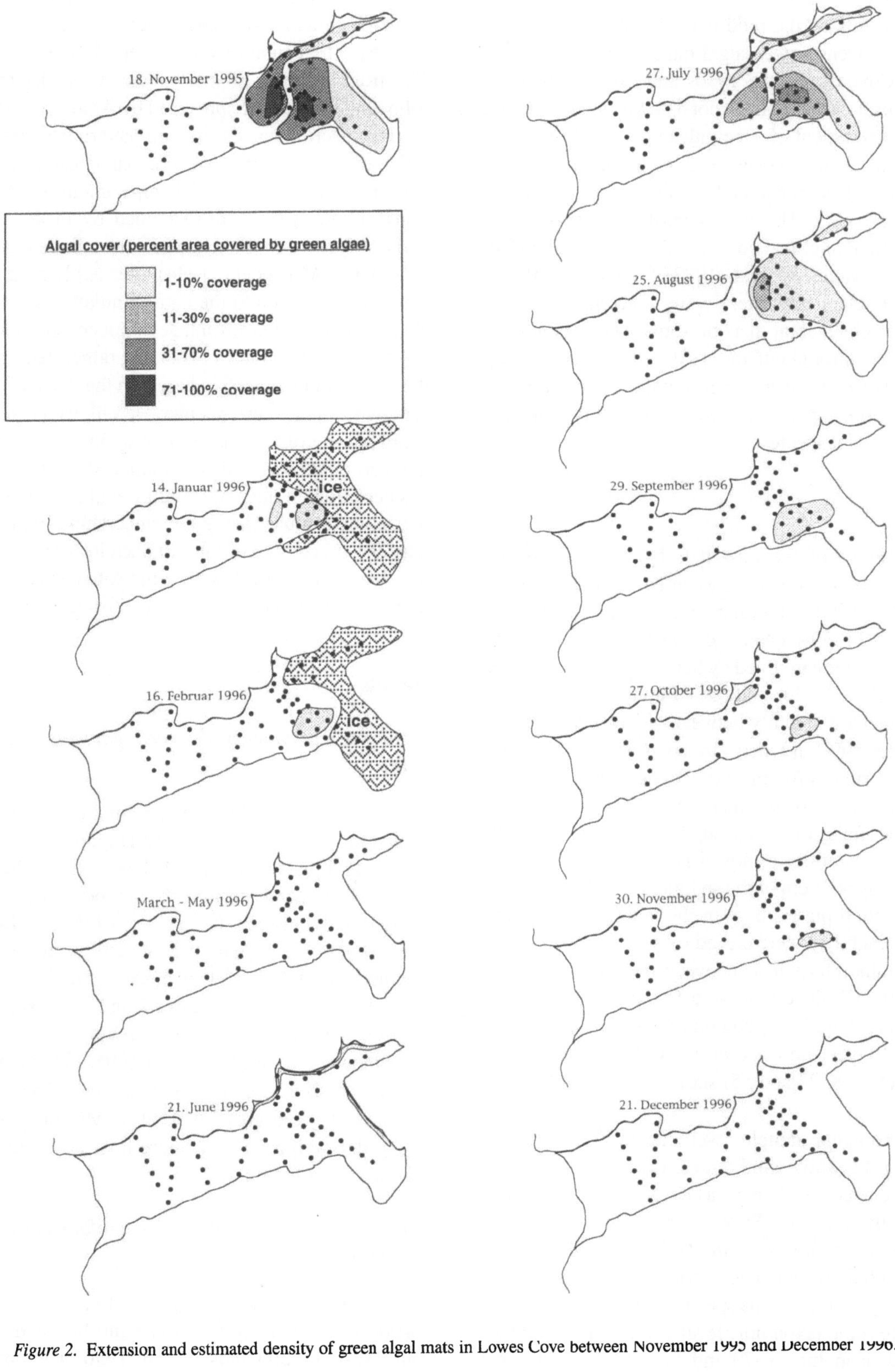

Figure 2. Extension and estimated density of green algal mats in Lowes Cove between November 1995 and December 1996.

between 6 and 12 μm NO_3-N (McAlice, 1993; Mayer et al., 1996).

Sediments on the mud flat are clayey silt (McAlice 1993). Dense green algal mats did not develop in Lowes Cove during the 1970s and early 1980s (personal observations). It is not known when the first mass development of green algae occurred in Lowes Cove, but at least since 1993 green algal mats developed each summer and persisted throughout the fall (pers. obs.). The first scientific observations of dense green algal mats on mud flats in Maine are reported by Vadas & Beal (1987) for the mid 1980s. However, clam harvesters were concerned with the effects of green algal mats on some mud flats in eastern Maine throughout the 1970s (Arbuckle, 1982). In Lowes Cove the mats are composed primarily of *Enteromorpha* species (*E. prolifera* usually being the most abundant species).

Sampling

The coverage of the mud flats by green algae was qualitatively monitored between November 1995 and December 1996 according to a scale (1–10% areal coverage; 11–30% coverage; 31–70% coverage; 71–100% coverage) provided by Schories (1995). At three stations (Nrs. 8, 24, and 55 – Figure 1) green algae were collected monthly between November 1995 and December 1996 and their biomass (dry weight) determined. A ring (176 cm^2 surface area and 1 cm high) was placed onto the sediment. A knife was used to cut the algae along the inner edge of the ring, the algal layer was carefully removed and placed in a plastic bag. The upper 1 cm of the sediment was then scooped into a second plastic bag. In the laboratory, the contents of each bag were sieved over a 1 mm-sieve. All macrofauna intertwined between the algal layer was removed. All algae that were found in the sediment layer were sorted out, combined with the algae from the algal layer and dried at 70 °C for 24 h. At each station (Nrs. 8, 24 and 55) six replicate samples were taken.

Macrofaunal samples were taken at 56 stations covering the whole mud flat area (Figure 1). At each station, one core (78 cm^2 surface area) has been taken to a depth of 12 cm. Each sample was sieved over a 500-μm mesh, preserved in 4%-formalin, then stored in 70%-ethanol, and later sorted under a dissecting microscope. Macrofauna species were identified to the lowest feasible taxonomic level and counted. Samples were taken in June 1979 and June 1996. Additional data from one selected station (Nr. 32, data from Thiel, 1997; unpubl. data) were used to reveal whether the observed infaunal changes under the mat are a persistent phenomenon or only reflect a singular event. The trophic levels of macrofauna were determined following the example provided by Ambrose (1993).

The macrofauna distribution pattern between June 1979 and June 1996 were compared using a fuzzy cluster analysis. The relationships among the 56 stations in each year were examined by means of the divisive fuzzy c-means clustering technique using the program FCM (Podani, 1994) after applying a cube-root transformation to the data (Findlay et al., 1995). This method has the advantage of not constraining any sample to an individual cluster, but rather determining the degree of membership based on the distance of the sample from the cluster center. Herein, we present the results of the fuzzy cluster analysis in a geographical manner, rather than a list of membership functions as has been used in the past (Findlay et al., 1995). On the maps, any sample having a membership affinity of at least 0.1 (maximum affinity to one cluster can be 1.0) to a second or third cluster is indicated as sharing its membership among two or more clusters.

Results

Extension and biomass of green algal mats in Lowes Cove

Green algal mats only developed in the eastern, most sheltered, parts of Lowes Cove (Figure 2). In 1995, dense algal mats that had developed during the summer of 1995 persisted until November but gradually disappeared between November 1995 and January 1996 (Figure 2). Green algae started to redevelop on the mud flats in June 1996 and reached maximum coverage in August 1996 (Figures 2 and 3). Coverage by green algae was comparatively low in 1996 (compare November 1995 and November 1996) but some algal patches persisted until late fall (Figure 4). The average biomass of green algae in August 1996 ranged from < 1 g dry weight m^{-2} to more than 200 g dw m^{-2} (Figure 3 and Figure 4).

Abundance and distribution of macrofauna in 1979 and 1996

Macrofauna densities ranged between 6 700 and 74 000 individuals m^{-2} in 1979 and between 10 100 and 237 000 individuals m^{-2} in 1996. The densities

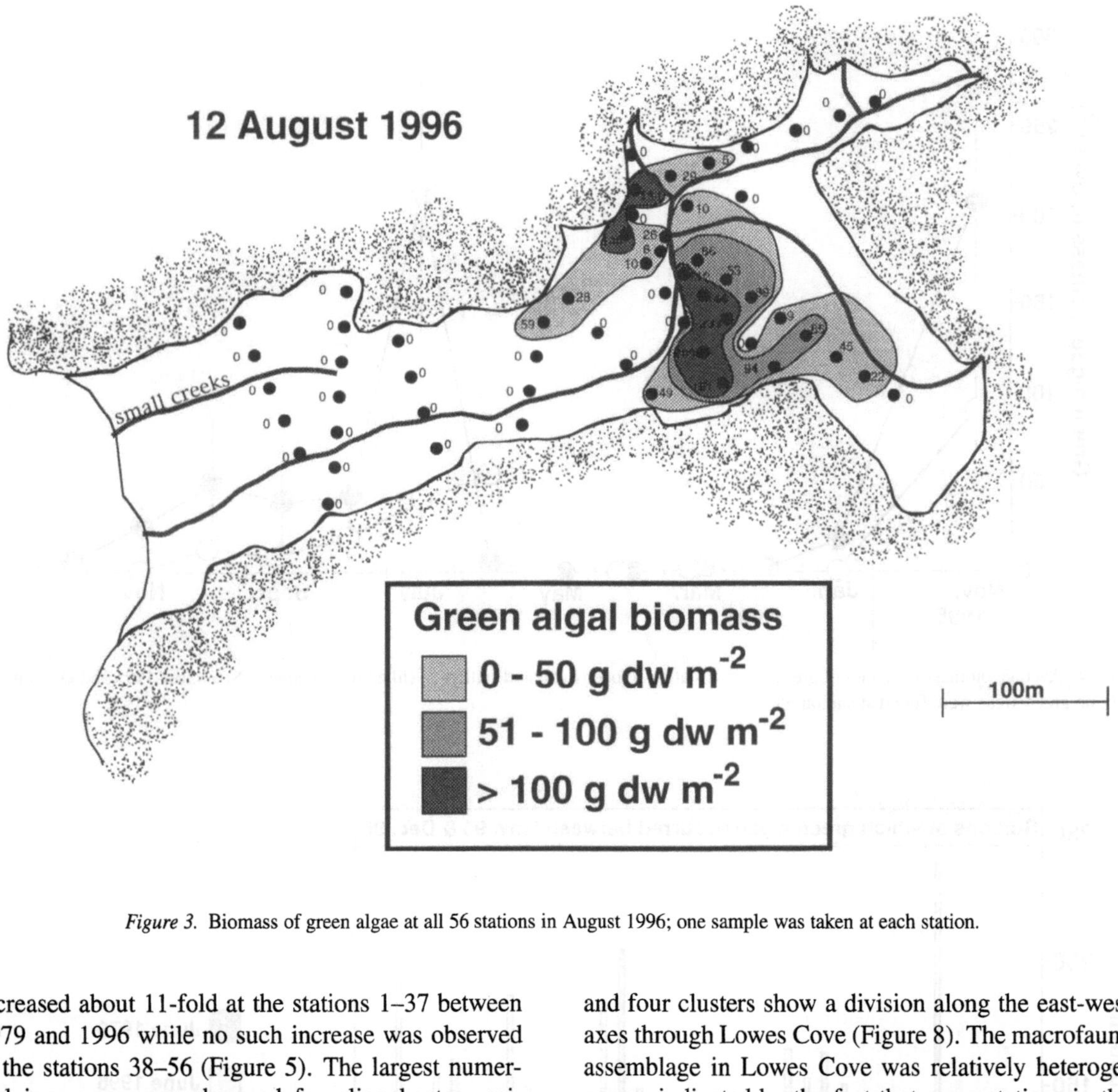

Figure 3. Biomass of green algae at all 56 stations in August 1996; one sample was taken at each station.

increased about 11-fold at the stations 1–37 between 1979 and 1996 while no such increase was observed at the stations 38–56 (Figure 5). The largest numerical increases were observed for oligochaetes, spionid polychaetes, *Capitella* sp. and *Leitoscoloplos* sp., while few changes occurred in bivalves, *Hydrobia* sp., *Corophium volutator* and the polychaetes *Nereis* sp., *Heteromastus filiformis* and *Mediomastus ambiseta* (Figure 6a, b). At station 32 which is frequently covered by green algae (see Figure 2 and Figure 3), a persistent nearly 3-fold increase in macrofauna numbers between the late 1970s and the early 1990s occurred (Figure 7). This increase in abundance of macrofauna can primarily be attributed to subsurface-living deposit-feeders (Figure 7).

The results of the fuzzy cluster analysis show that the station grouping pattern between June 1979 and June 1996 has distinctly changed. In June 1979, the results for the clustering partition in two, three and four clusters show a division along the east-west axes through Lowes Cove (Figure 8). The macrofauna assemblage in Lowes Cove was relatively heterogeneous, indicated by the fact that some stations in the eastern-most parts had substantial membership in the same clusters as some of the western-most stations. In June 1996, the clusters split along a north-south axis at about the middle of Lowes Cove (Figure 8). The eastern cluster(s) closely overlap with the areas that were frequently covered by green algae (compare Figure 2 and Figure 3). The partition to three clusters reveals a cluster that closely overlaps with the distribution center of green algae in Lowes Cove, while partitioning into 4 clusters splits off a '*Corophium*' cluster on the northern edge of Lowes Cove from the remainder of the western cluster (Figure 8).

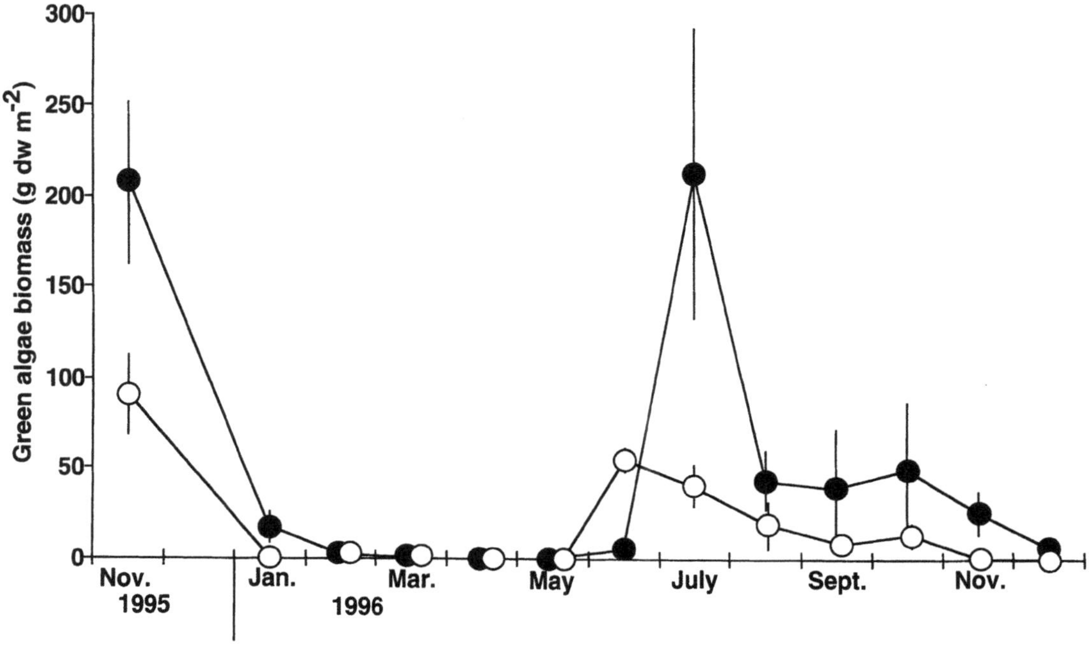

Figure 4. Average biomass (± 1 s.e.) of green algae at station 8 (open dots) and station 24 (filled dots) between November 1995 and December 1996; no green algae were found at station 55.

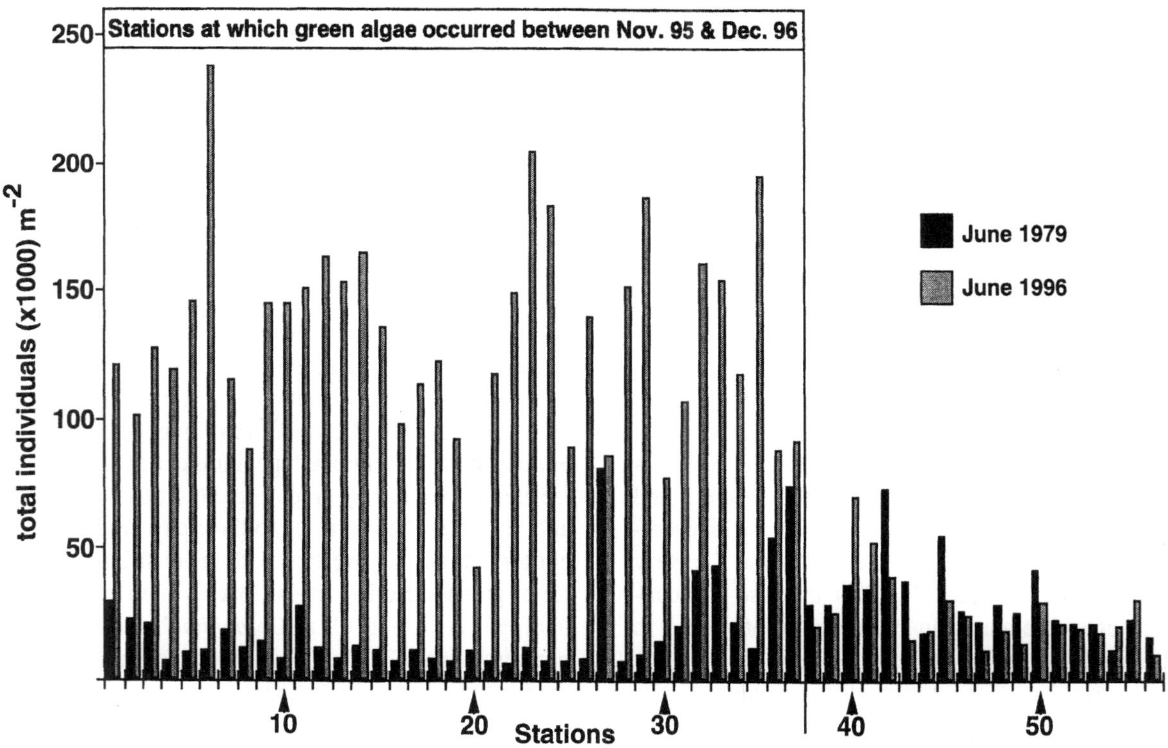

Figure 5. Total macrofauna abundance (individuals m^{-2}) at all 56 stations in Lowes Cove in 1979 and 1996.

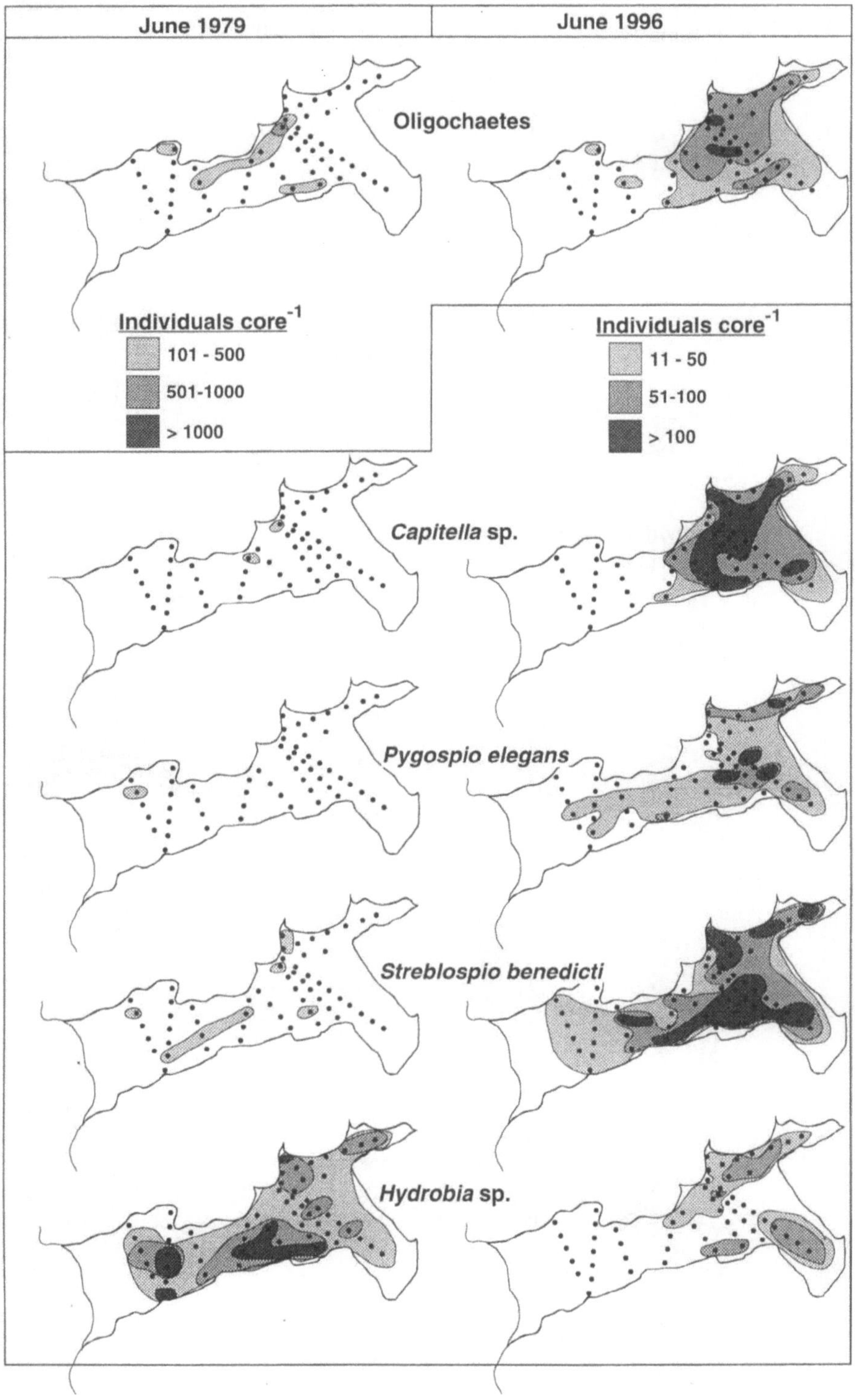

Figure 6a. Abundance and distribution of major macrofauna organisms (individuals core^{-1}) in 1979 and 1996.

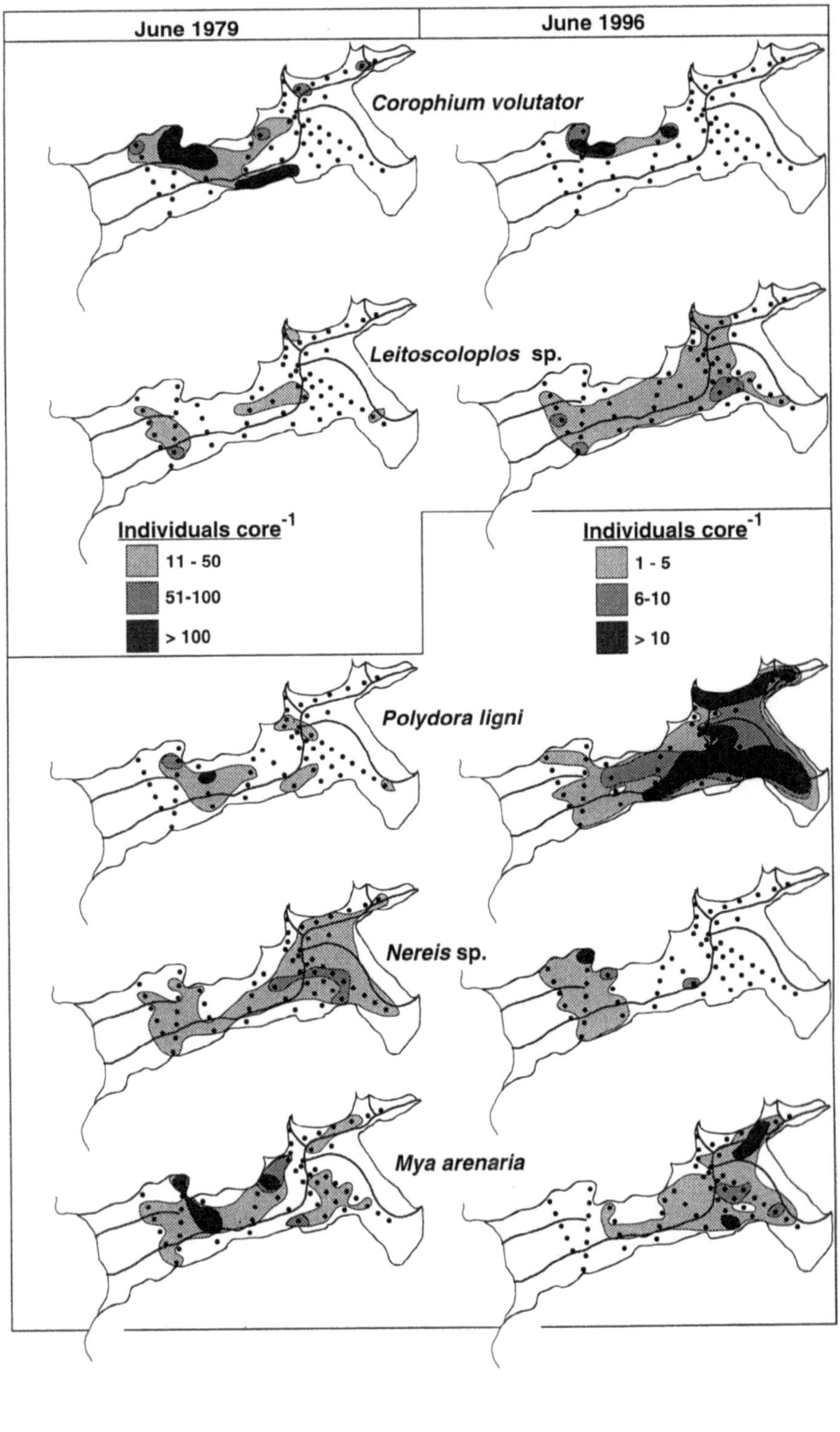

Figure 6b. For legend see p. 183.

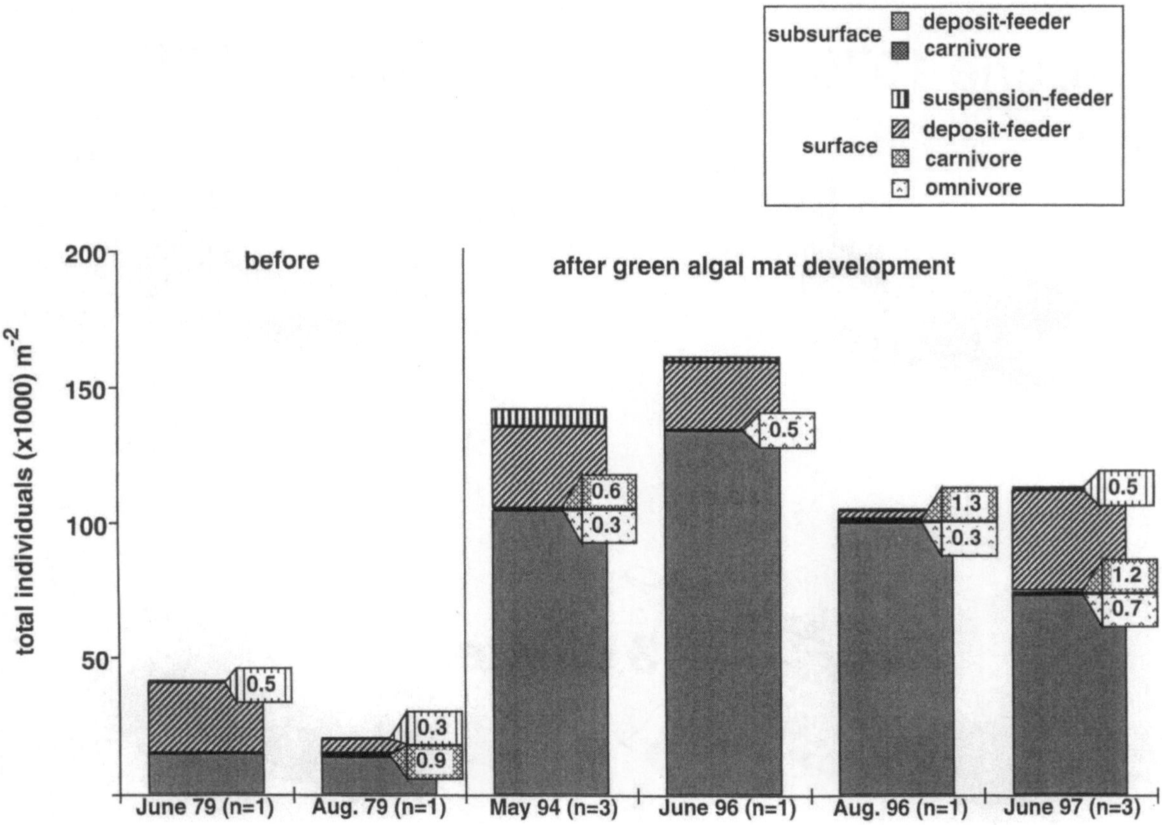

Figure 7. Total macrofauna abundance (individuals [$\times$ 1000] m^{-2}) at station 32 in Lowes Cove in June 1979, August 1979, May 1994, June 1996, August 1996 and June 1997; at each station $n = 1$–3 samples were taken; small boxes show numbers for minor functional groups.

Discussion

During the early 1990s green algal mats frequently covered the eastern, sheltered, parts of Lowes Cove which may be considered a relatively typical New England mud flat. Between the late 1970s and the early 1990s a strong increase in macrofauna abundance occurred in parts of the mud flat that are usually covered by green algae but not in other parts. This increase is particularly due to subsurface-dwelling, deposit-feeding, annelids.

Extension, biomass and persistence of green algal mats in Lowes Cove

Lowes Cove is a relatively sheltered mud flat, green algae develop autochthonously, and remain on the mud flats until well after the growth season. Green algal mats persist throughout the fall and during the winter they are finally incorporated into the sediments by macrofaunal activities (see Figure 4). Similarly, for Chesapeake Bay Kemp et al. (1994) reported that macrophytes persist longer at the site of origin in a sheltered cove than in an open embayment. This prolonged persistence of algal mats in one location is not typical for mud flats in the European Wadden Sea or subtidal waters in the Irish Sea and the Baltic Sea. In these areas, strong currents or storms may dislodge algal mats within days or weeks after first deposition (e.g. Thrush, 1986; Reise & Siebert, 1994; Norkko & Bonsdorff, 1996b). The dislodged algae then accumulate and are redeposited at other locations (Bonsdorff, 1992).

Algal biomasses found in Lowes Cove are relatively moderate compared to those found in other study areas or used in experimental studies. For example, Bonsdorff (1992) found algal biomasses of 150–830 g dw m^{-2}, Everett (1994) found an average algal biomass of ≈ 300 g dw m^{-2}, Hull (1987) applied in the most dense experimental treatment approximately 300 g dw m^{-2} (conversion factor fw to dw: 10), and

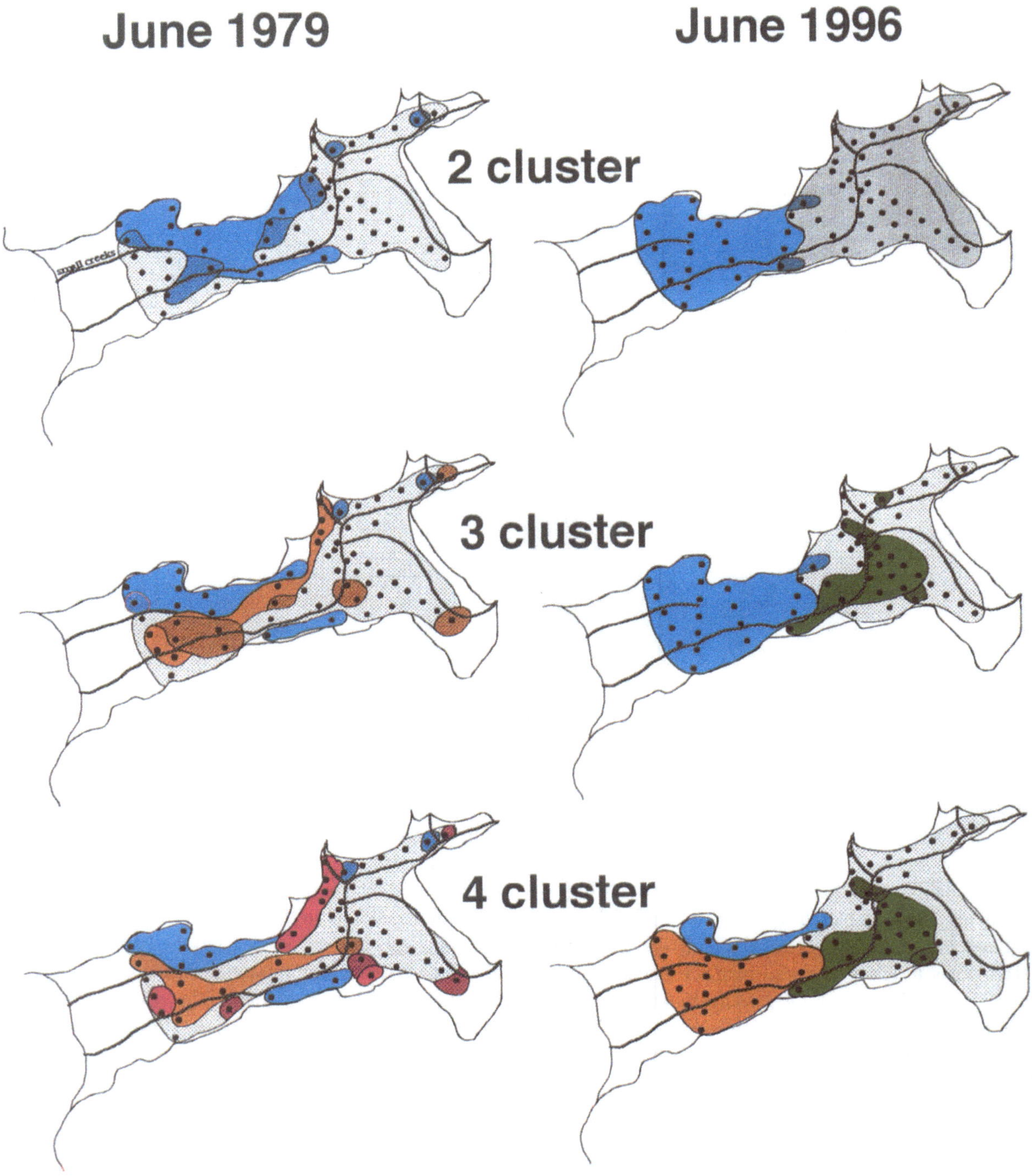

Figure 8. Results of fuzzy cluster analysis for $n = 56$ samples from June 1979 and $n = 56$ samples from June 1996; presented are partitions in two, three, and four cluster; cluster are indicated by different colours and shading; samples that have affinities of ≥ 0.1 to more than one cluster are indicated by overlapping colours and shading; only one sample (station 52 in June 1979, 3 cluster partition) had affinities to three clusters.

Table 1. Abundance changes and reproductive traits of major macrofauna species in Lowes Cove, Maine.

	Sum June 1979	Sum June 1996	Change 79–96	> 5-fold increase	Larval stage	Duration
Mya arenaria	144	133	n		p	l
Macoma balthica	251	101	–		p	l
Gemma gemma	0	159	+		p	l
Hydrobia sp.	3413	1262	–		b	
Corophium volutator	1578	753	–		b	
Oligochaetes	2853	24827	+	+	b	
Heteromastus filiformis	42	24827	+		p	m
Mediomastus ambiseta	534	124	–		p?	?
Capitella capitata	67	4705	+	+	b/p	m
Streblospio benedicti	268	5246	+	+	b/p	m
Pygospio elegans	63	1831	+	+	b/p	
Polydora ligni	42	630	+	+	b/p	
Leitoscoloplos sp.	376	1194	+		b/p	m
Tharyx sp.	109	51	–		p	m
Eteone sp.	50	262	+	+	b/p	m
Exogone hebes	11	231	+	+	b?	
Nereis sp.	121	57	–		p	l
Nemertean (white)	0	136	+		b	

Changes: *n* = no change; – = decrease; + = increase.
Larval stage: p = planktonic; b = benthic; b/p = both reported.
Duration of pelagic phase: l = long; m = moderate.

Norkko & Bonsdorff (1996a, b) applied artificial algal mats of ≈440 g dw m^{-2}, amounts that were rarely reached in Lowes Cove (see Figures 3 and 4). Despite their relatively large areal extension and prolonged persistence, algal mats in Lowes Cove appear to have moderate biomass. They do not reach biomasses for which strong detrimental effects on macrofauna have been reported (Hull, 1987; Bonsdorff, 1992; Everett, 1994; Norkko & Bonsdorff, 1996a, b).

Macrofauna changes in Lowes Cove between 1979 and 1996

Few species decreased in abundances between June 1979 and June 1996 (e.g. *Mediomastus ambiseta*, *Nereis* spp., *Hydrobia* sp.), but these decreases do not show a distinct pattern. Some species decreased relatively evenly throughout the mud flat (*M. ambiseta*) while others disappeared only from the eastern parts (*Nereis* spp.) or from the western parts (*Hydrobia* sp.). The data obtained for snails *Hydrobia* sp. need to be interpreted with caution because these snails completely vanished from Lowes Cove during the 1980s for unknown reasons (own observations). While no consistent spatial pattern was recognizable for species that decreased in abundance, a distinct pattern could be recognized for species that increased in abundance. Several macrofauna species increased in abundance but these increases are largely limited to the 37 stations that were at least temporarily covered by green algal mats in 1996 (Table 1). The distinct pattern as revealed by the fuzzy cluster analysis for June 1996 strongly indicates that the observed macrofauna response is an immediate result of the occurrence of the green algal mats. While in June 1979 several diverse infaunal associations were irregularly distributed over the mud flats, the situation had drastically changed in June 1996 when a clear separation is recognizable between the stations affected by green algal mats and those stations that were unaffected. It thus appears that green algae have a strong impact on infaunal colonization, and those effects may overshadow effects of e.g. predation, adult-larval interactions and sediment-disturbance.

The increase in macrofauna abundance is primarily due to subsurface-dwelling, deposit-feeding, annelids that are tolerant to, or even benefit from, enrichment of sediments with organic matter (oligochaetes, spionid polychaetes, *Capitella* sp.). This dominance shift towards burrowing detritivores under algal mats was also observed by Norkko & Bonsdorff (1996b). The species responsible for this shift are those most likely to benefit from organic detritus in form of decaying green algal parts because they are able to tolerate hypoxia commonly associated with these algal mats. They are also relatively mobile within the sediment and may move higher up in the sediment in response to decreasing oxygen levels.

Is life history responsible for highly localized effects?

When discussing the colonization pattern of marine soft-bottoms, consideration of the life history of potential colonizers is important (Thrush & Townsend, 1986; Shull, 1997). In defaunated sediments, the first colonizers usually are species with planktonic larval stages. These initial colonizers are generally poor competitors and usually disappear at a later successional stage when larger organisms that are competitively superior reach adult size (e.g. Watling, 1975; Pearson & Rosenberg, 1978). Temperate intertidal mud flats are inhabited by macrofauna species with highly diverse reproductive strategies. Larvae of some species have long-lasting (months to weeks) planktonic larval stages (e.g. large bivalves, large polychaetes and other worms) while others spend very short time periods in the water column (days to hours – e.g. some polychaetes and enteropneusts) or do not enter the water column at all (e.g. small bivalves and gastropods, small polychaetes, oligochaetes). The latter species with brood protection are particularly common in intertidal habitats compared to other habitats (Clutton-Brook, 1991). They may be most tolerant to moderate changes in an intertidal environment since they protect their offspring during the most sensitive early larval stages. While many planktonic larvae may not be capable, or are too substrate-selective (see Butman, 1987), to colonize disturbed areas, benthic larvae that enjoy parental protection may be more prone and competent to colonize moderately disturbed areas. Indeed, most of the species that increased in abundance in Lowes Cove between 1979 and 1996 provide some form of brood protection (see Table 1). Small crawl-away larvae that do not undergo a planktonic phase emerge from oligochaete egg-cocoons, from maternal burrows (*Capitella* sp. – see e.g. Rasmussen, 1973; Pearson & Pearson, 1991; Baoling et al., 1991) or from maternal tubes (*Pygospio elegans* – Rasmussen, 1973; *Streblospio benedicti* – Dean, 1965; Grassle & Grassle, 1974). Brood protecting species may be able to convert algal detritus immediately into offspring and would thus benefit most from moderate algal coverage (for a detailed discussion see also Grassle & Grassle, 1974). Pearson & Pearson (1991) note that brood protecting *Capitella* species are favored in environments with rich local food resources (such as algal detritus). The suggestion that brood protecting species are favored in moderately enriched environments is supported by the findings of Norkko & Bonsdorff (1996b) who recorded high abundances of oligochaete egg cocoons within 10 d after experimental establishment of algal mats. The ability of crawl-away larvae to recruit immediately into the parental habitat without a costly detour through the water column may primarily be responsible for the strong abundance increase of the subsurface-dwelling, deposit-feeding, annelids in Lowes Cove. This 'neighbourhood recruitment' may also be responsible for the highly localized macrofauna changes observed in Lowes Cove. Macrofauna species with planktonic larvae (e.g. bivalves *Macoma balthica* and *Mya arenaria*, and nereid polychaetes) may not be able to benefit from moderate algal coverage because recruits are too sensitive to potential hypoxia or successful settlement is adversely affected by green algal mats (Ólafsson, 1988; Bonsdorff, 1992).

Acknowledgements

Our very special thanks go to Heidi Ryder for her inexhaustible patience in sorting samples that contained more oligochaetes than we ever wished for. The comments from two anonymous reviewers helped improve the original manuscripts. This study was supported by the University of Maine/University of New Hampshire Sea Grant Office in the 1970s and by the Gulf of Maine Foundation in 1996. Further financial support was received from the Graduate School and the School of Marine Sciences both at the University of Maine.

References

Ambrose, W. G., Jr., 1993. Effects of predation and disturbance by ophiuroids on soft-bottom community structure in Oslofjord: results of a mesocosm study. Mar. Ecol. Prog. Ser. 97: 225–236.

Arbuckle, J. K., 1982. Effects of disturbance on the macrofauna of a mudflat community in eastern Maine. Unpubl. M.Sc.-thesis, University of Maine, Orono: 44 pp.

Baoling, W., Q. Pei-Yuan & Z. Sonling, 1991. Morphology, reproduction, ecology and allozyme electrophoresis of three *Capitella* sibling species in Qingdao (Polychaeta: Capitellidae). Ophelia Suppl. 5: 391–400.

Bonsdorff, E., 1992. Drifting algae and zoobenthos – effects on settling and community structure. Neth. J. Sea Res. 30: 57–62.

Breber, P., 1985. On growing of the carpet-shell clam (*Tapes decussatus* (L.)): Two years' experience in Venice lagoon. Aquaculture 44: 51–56.

Butman, C. A., 1987. Larval settlement of soft-sediment invertebrates: the spatial scales of patterns explained by active habitat selection and the emerging role of hydrodynamical processes. Oceanogr. Mar. Biol. annu. Rev. 25: 113–165.

Buttermore, R. E., 1977. Eutrophication of an Impounded estuarine lagoon. Mar. Poll. Bull. 8: 13–15.

Clutton-Brock, T. H., 1991. The evolution of parental care. Princeton University Press, New Jersey, 1–352.

Dean, D., 1965. On the reproduction and larval development of *Streblospio benedicti* Webster. Biol. Bull. 128: 67–76.

Everett, R. A., 1994. Macroalgae in marine soft-sediment communities: effects on benthic faunal assemblages. J. exp. mar. Biol. Ecol. 175: 253–274.

Findlay, R. H., L. Watling & L. M. Mayer, 1995. Environmental impact of salmon net-pen culture on marine benthic communities in Maine: a case study. Estuaries 18: 145–179.

Grassle, J. F. & J. P. Grassle, 1974. Opportunistic life histories and genetic systems in marine benthic polychaetes. J. mar. Res. 32: 253–284.

Hull, S. C., 1987. Macroalgal mats and species abundance: a field experiment. Estuar. coast. Shelf Sci. 25: 519–532.

Kemp, W. M., W. R. Boynton & A. J. Hermann, 1994. Simulations models of an estuarine macrophyte ecosystem. In B. C. Patten, Jørgensen S. E. & S. I. Auerbach (eds), Complex Ecology, the Part-Whole Relation in Ecosystems. Prentice Hall, Englewood Cliffs, N.J.: 262–278.

Mayer, L. M., D. W. Townsend, N. R. Pettigrew, T. C. Loder, M. W. Wong, D. Kistner-Morris, A. K. Laursen, A. D. Schoudel, C. Conairis, J. Brown & C. Newell, 1996. The Kennebec, Sheepscot and Damariscotta River Estuaries: Seasonal Oceanographic Data. University of Maine, Department of Oceanography Technical Report No. 9601: 1–110.

McAlice, B. J., 1993. Environmental characteristics of the Damariscotta River Estuary, Maine. Darling Marine Center, Spec. Publ. No. 1: 1–119.

Nicholls, D. J., C. R. Tubbs & F. N. Haynes, 1981. The effect of green algal mats on intertidal macrobenthic communities and their predators. Kieler Meeresforsch., Sonderh. 5: 511–520.

Norkko, A. & E. Bonsdorff, 1996a. Rapid zoobenthic community responses to accumulations of drifting algae. Mar. Ecol. Prog. Ser. 131: 143–157.

Norkko, A. & E. Bonsdorff, 1996b. Population responses of coastal zoobenthos to stress induced by drifting algal mats. Mar. Ecol. Prog. Ser. 140: 141–151.

Ólafsson, E. B., 1988. Inhibition of larval settlement to a soft bottom benthic community by drifting algal mats: an experimental test. Mar. Biol. 97: 571–574.

Pearson, M. & T. H. Pearson, 1991. Variation in populations of *Capitella capitata* (Fabricius, 1780) (Polychaeta) from the West Coast of Scotland. Ophelia Suppl. 5: 363–370.

Pearson, T. H. & R. Rosenberg, 1978. Macrobenthic succession in relation to organic enrichment and pollution of the marine environment. Oceanogr. Mar. Biol. annu. Rev. 16: 229–311.

Peterson, C. H., E. A. Irlandi & R. Black, 1994. The crash in suspension-feeding bivalve populations (*Katelysia* spp.) in Princess Royal Harbour: an unexpected consequence of eutrophication. J. exp. mar. Biol. Ecol. 176: 39–52.

Podani, J., 1994. Multivariate data analysis in ecology and systematics – a methodological guide to the SYN-TAX 5.0 package. SPB Academic Publishing, The Hague.

Raffaelli, D., S. Hull & H. Milne, 1989. Long-term changes in nutrients, weed mats and shorebirds in an estuarine system. Cah. Biol. mar. 30: 259–270.

Raffaelli, D., J. Lima, S. Hull & S. Pont, 1991. Interactions between the amphipods *Corophium volutator* and macroalgal mats on estuarine mudflats. J. mar. biol. Ass. U.K. 71: 899–908.

Rasmussen, E., 1973. Systematics and ecology of the Isefjord marine fauna (Denmark). Ophelia 11: 1–482.

Reise, K., 1982. Long-term changes in the macrobenthic invertebrate fauna of the Wadden Sea: are polychaetes about to take over? Neth. J. Sea Res. 16: 29–36.

Reise, K., 1983. Sewage, green algal mats anchored by lugworms, and the effects on Turbellaria and small Polychaeta. Helgoländer. Meeresunters. 36: 151–162.

Reise, K., E. Herre & M. Sturm, 1989. Historical changes in the benthos of the Wadden Sea around the island of Sylt in the North Sea. Helgoländer. Meeresunters. 43: 417–433.

Reise, K. & I. Siebert, 1994. Mass occurrence of green algae in the German Wadden Sea. Dt. hydrogr. Z., Suppl. 1: 171–180.

Schories, D., 1995. Populationsökologie und Massenentwicklung von *Enteromorpha* spp. (Chlorophyta) im Sylter Wattenmeer. Bericht aus dem Institut für Meereskunde an der Christian-Albrechts-Universität Kiel 271, 145 pp.

Shull, D. H., 1997. Mechanisms of infaunal polychaete dispersal and colonization in an intertidal sandflat. J. mar. Res. 55: 153–179.

Soulsby, P. G., D. Lowthion & M. Houston, 1982. Effects of macroalgal mats on the ecology of intertidal mudflats. Mar. Poll. Bull. 13: 162–166.

Thiel, M., 1997. Extended parental care in estuarine amphipods. Unpubl. PhD-thesis, University of Maine, 188 pp.

Thrush, S. F., 1986. The sublittoral macrobenthic community structure of an Irish sea-lough: effect of decomposing accumulations of seaweed. J. exp. mar. Biol. Ecol. 96: 199–212.

Thrush, S. F. & C. R. Townsend, 1986. The sublittoral macrobenthic community composition of Lough Hyne, Ireland. Estuar. coast. Shelf Sci. 23: 551–574.

Vadas, R. L. & B. Beal, 1987. Green algal ropes: a novel estuarine phenomenon in the Gulf of Maine. Estuaries 10: 171–176.

Watling, L., 1975. Analysis of structural variations in a shallow estuarine deposit-feeding community. J. exp. mar. Biol. Ecol. 19: 275–313.

Hydrobiologia **375/376:** 191–201, 1998.
S. Baden, L. Pihl, R. Rosenberg, J.-O. Strömberg, I. Svane & P. Tiselius (eds),
Recruitment, Colonization and Physical–Chemical Forcing in Marine Biological Systems.

Hydrodynamic consequences of barnacle colonization

J. C. Thomason[1], J. M. Hills[2,*], A. S. Clare[3], A. Neville[4] & M. Richardson[5]
[1] *Department of Marine Sciences and Coastal Management, University of Newcastle upon Tyne, Newcastle upon Tyne, Tyne & Wear, NE1 7RU, UK*
[2] *University Marine Biological Station Millport, Isle of Cumbrae, Ayrshire, KA28 OEG, UK*
[3] *Marine Biological Association, The Laboratory, Citadel Hill, Plymouth, PL1 2PB, UK*
[4] *Department of Mechanical & Chemical Engineering, Heriot-Watt University, Riccarton, Edinburgh, EH14 4AS, UK*
[5] *Harrogate Grammar School, Arthur's Avenue, Harrogate, N. Yorkshire, HG2 ODZ*
* *Present address: Department of Marine Sciences and Coastal Management, University of Newcastle upon Tyne, Newcastle upon Tyne, Tyne & Wear, NE1 7RU, UK.*

Key words: hydrodynamics, barnacle, drag, boundary layer, optimum flow

Abstract

The aim of this project was to study the hydrodynamic consequences of substratum colonization by barnacles. Replicas of individual barnacles and barnacle colonies, and living colonies were studied in a 5 m seawater flume using tracer dyes, macro video-photography and image digitization and in a small flume fitted with differential pressure sensors. Using replicas, colonization densities were manipulated to determine the dynamics of complex barnacle populations. These techniques enabled the quantification of drag, boundary layers and their interrelationships with mean flow velocity and population density.

These data show that mean stream velocity affected the size and magnitude of the boundary layer and that at higher velocities the flow over the height at which the cirri were operating was decreased proportionally greater than at low mean stream velocities. The effect on flow velocity over the cirri with increased boundary layer at higher mean stream velocities was assessed and an optimum mean stream velocity for flow over cirri identified. Density of barnacles was shown to have an effect on the flow characteristics over the population, with flow changing from independent flow at low densities, through interactive flow to skimming flow at higher densities. Drag was also shown to be affected by barnacle density. The economic impact of barnacle fouling on the world maritime fleet is estimated and discussed.

Introduction

Research on barnacles and associated hydrodynamics has been largely concerned with the governance of settlement processes, at scales ranging from oceanic to surface shear gradients (e.g., Alexander & Roughgarden, 1996; Bertness et al., 1996; Crisp, 1955; Eckman, 1996; Mullineaux & Butman, 1991; Noda & Nakao, 1996; Rittschof et al., 1984; Shanks, 1986; Walters & Wethey, 1991; Walton-Smith, 1946; Wethey et al., 1988) or the effects of flow on feeding processes (e.g., Anderson, 1981; Southward, 1955; Trager et al., 1994; Trager, 1988; Trager et al., 1992; Trager et al., 1990). Likewise, there are numerous studies of the effects of fouling on hull speed and fuel costs for ships and the problems of fouling on static offshore structures such as oil rigs (e.g. Bowden & Davison, 1974; Carlton, 1994; Freeman, 1977; Gucinski et al., 1984; Harderlie, 1984; Lewthwaite et al., 1985; Pidgeon, 1993). However there has been little work on the cumulative impact of community barnacle development on surface hydrodynamics. Likewise there has been little work to compare the hydrodynamics of feeding and non-feeding barnacles and the hydrodynamic relationships between upstream individuals and their downstream neighbours. Therefore the aims of this study were to: (1) determine the effects of colony size and growth on surface hydrodynamics; (2) compare the effects of feeding and non-feeding barnacles on the boundary layer; (3) to examine the hydrodynamic effects of barnacles on neighbouring downstream conspecifics; and (4) to predict optimal flow rates for feeding. The study utilised two common fouling species of balanomorph barnacle, the boreo-arctic *Semibalanus balanoides* and the circum-tropical *Balanus amphitrite*.

Materials and methods

Settlement and replication of Semibalanus balanoides

S. balanoides cyprids were settled onto cast polyester resin tiles deployed from Keppel Pier adjacent to University Marine Biological Station Millport, Clyde Sea, UK (Grid Reference NS 177.545) during the spring cyprid settlement season. By varying the texture of the tiles it was possible to settle cyprids at different densities, smooth tiles having the lowest densities (ca. 0.2 barnacles cm^{-2}) and finely textured tiles having the greatest (ca. 3 cyprids cm^{-2}). Details of tile manufacture and the effects of texture on the settlement density of *S. balanoides* are described in Hills & Thomason (1997) and Hills et al. (1997). The tiles were deployed from Keppel Pier until the barnacles had reached adulthood (approximately 9 months). Impressions were made of the colonies using polyvinylsiloxane dental impression material and replicas made using Devcon 2-tonne epoxy resin following the technique of Marrs (1995). This technique is highly accurate with an accuracy of $< 1\ \mu$m. The tiles with low density settlement (smooth) produced colonies with widely spaced large conical individuals, whereas the high density tiles (fine texture) produced colonies with hummocks of tightly packed, flute-shaped individuals. These replicas were used to study the hydrodynamics of a large individual in relation to others some distance apart, and the hydrodynamics of a complex aggregated colony. They were also used to take friction loss measurements (see below).

Additionally, a single large (1.5 cm basal diameter) specimen of *S. balanoides* was impressioned and replicated repeatedly. By temporarily gluing these replicas to acrylic tiles with petroleum jelly, it was possible to manipulate experimentally the density and pattern of barnacle 'colonies'. These experimental colonies were made of rows of barnacles, with 3 barnacles in each row and colonies with 3, 5, 7, and 9 rows were examined. The rows were either aligned or offset from one another. Each barnacle was spaced equidistant (1.5 cm) from its neighbours.

Culture and replication of Balanus amphitrite

B. amphitrite used in flume experiments were cultured at the Marine Biological Association, Plymouth, U.K. Nauplii released from broodstock were raised to the settlement stage cypris larva at 25 °C, on a 16 h: 8 h light : dark cycle, in filtered (~0.7 fm), Eddystone (English Channel) seawater (FESW), to which antibiotics had been added. The nauplii were fed with *Skeletonema costatum* and reached the cyprid stage in 4 d. Cyprids were filtered from the batch culture and stored in FESW at 6 °C for 3 d. They were then presented with roughened (sanded with P120 grit wet and dry paper) acrylic panels and kept at 25 °C until settlement had occurred. The juvenile barnacles were raised at 25 °C in daily changes of 0.45 fm FESW and fed *S. costatum* until large enough to filter *Artemia* sp. When the barnacles had reached approximately 2 mm, they were transferred to 20 °C and grown on for polyvinylsiloxane impressions to be made. Impressions were taken at the age of three, seven and eleven months. Mean basal diameters were 0.45 cm, 0.79 cm, and 1.2 cm, respectively. Further details of *B. amphitrite* culture can be found in Rittschof et al. (1984) and Rittschof et al. (1992).

Large flume studies

A large flume at Plymouth Marine Laboratory, UK was used in conjunction with dye tracer and video techniques to visualise and quantify flows and boundary layers. This flume is 5 m long, 1 m wide, and 0.5 m deep. Water flow is controlled by a variable speed hydraulic pump. The flume was filled to a depth of 0.3 m with Eddystone seawater and there were 4 m from the inflow tubes to the experimental area. Synchronous recordings of experiments were made using two Hi8 video cameras, one positioned vertically above the subject, the other laterally. A macro lens was used on one camera to obtain close-up footage of individuals and of small groups of barnacles. Dye streams were introduced from 1 ml syringes containing non-toxic water soluble food dyes. The syringe needles were inserted and glued (cyanoacrylate adhesive) into catheter tubing (1 mm OD). Six syringes were used simultaneously, the distal ends of the catheters were attached 1.5 cm apart to a fine stainless steel tube (1.5 mm

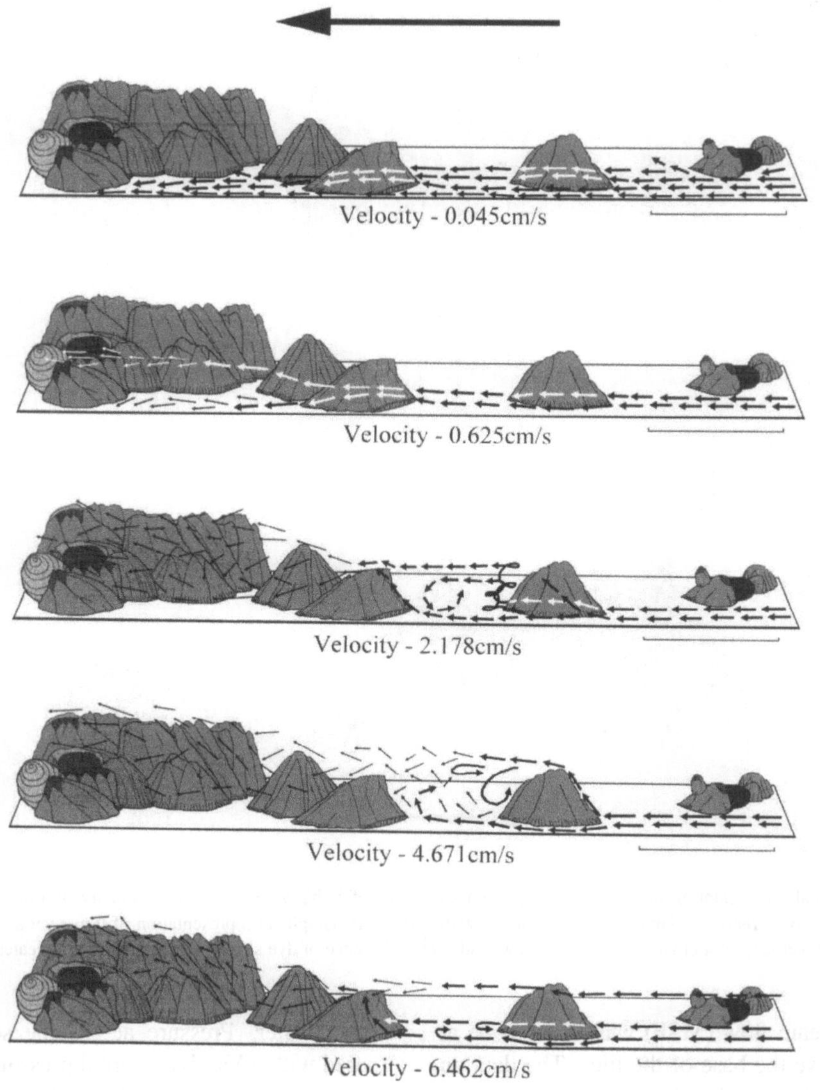

Figure 1. Graphical representation of flow over a replica colony of *S. balanoides* at flow rates from 0.045–6.462 cm s^{-1}. Width of arrows indicates coherence of dye stream and large arrow at top indicates flow direction. Scale bars = 2 cm.

OD). This enabled six continuous dye streams to be produced at the same time at different heights (or at different lateral positions) above the specimen, with different dye colours aiding identification. Flow was uniform in the flume and various flow rates, ranging from 0.045 cm s^{-1} to 6.5 cm s^{-1} were used. Replicas and colonised acrylic tiles were placed in a 1 × 0.2 m acrylic surround of the same thickness and with a long tapered leading edge. This obviated any problems of turbulence generation by the leading edge of the cast or tile. Images were either captured directly from tape using a frame grabber or drawn onto acetates placed on the monitor screen and scanned using a high resolution Nikon Scantouch 210 flatbed scanner. Sequences were analysed, individual dye streams being followed across the field of view of the camera, using image analysis software (either SIS analySIS or UTHSCA Image Tool) to obtain qualitative and quantitative assessments of the hydrodynamics. Graphical output was obtained using Adobe Photoshop and MathSoft Axum, and modelling conducted using Minitab.

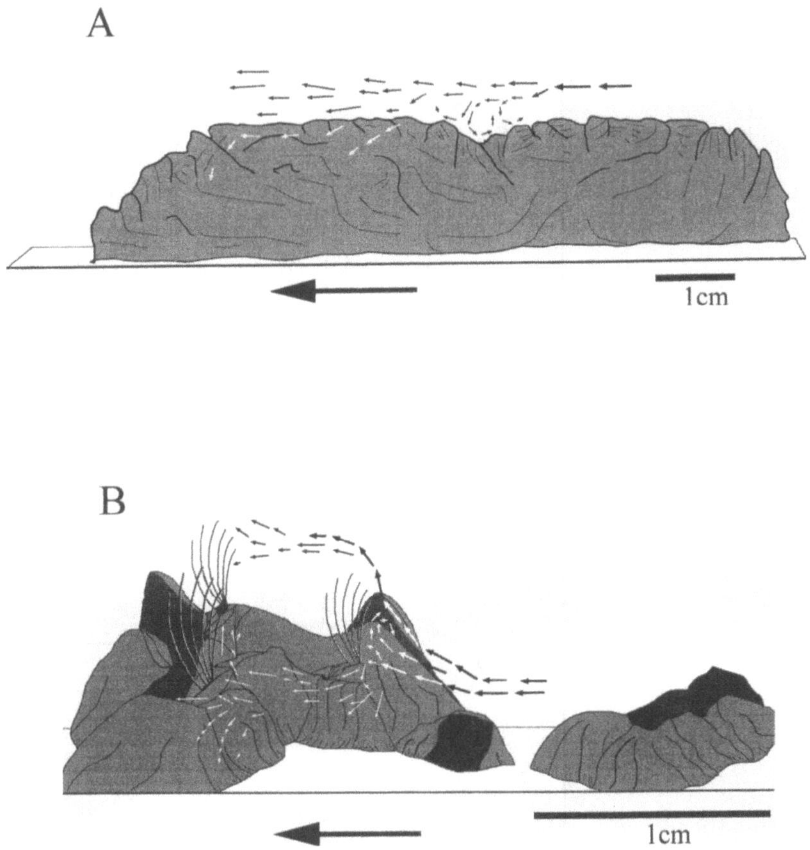

Figure 2. (A) Graphical representation of flow over a high density colony of *S. balanoides* at a flow velocity of 4 cm s^{-1}. Width of arrows indicates coherence of dye stream and large arrow indicates flow direction. (B) Graphical representation of flow over a small group of feeding *B. amphitrite* at a flow velocity of 4 cm s^{-1}. Width of arrows indicates coherence of dye stream and large arrow indicates flow direction.

The mean stream flow (MSF) was taken to be at a point 6 cm above the base of the tiles. The data on flow velocities at different heights above subjects were determined by assuming an exponential curve for the boundary layer (see Vogel, 1994) and by fitting the raw data to the curve using least-squares linear regression. Boundary layers were then derived from these equations and back-transformed to give curvilinear data. These were expressed in terms of percentage MSF.

Small flume studies

The purpose of this part of the study was to accurately quantify the hydrodynamic response of experimental and natural colonies of barnacles. A small (50 × 50 × 1000 mm) closed-loop flow rig was used to measure the frictional characteristics across a 50 × 50 mm removable section to be measured. The rig used a variable speed pump and was filled with fresh water. Pressure across the section was measured with a Validyne digital pressure transducer. The theory was based on the Darcy formula with modifications made for the square section such that the wetted perimeter and cross sectional area were considered. The frictional head loss was determined from the reduced Darcy equation in conjunction with the Energy equation such that:

$$h_f = \frac{fl}{m}\frac{u^2}{2g} = \frac{\delta p}{\rho g}, \tag{1}$$

where, f is the friction factor, l is the section length, m is the ratio of the cross sectional area to contact perimeter, u is the free stream velocity (approximated as MSF), g is the gravitational constant, p is the pressure, ρ is the density, and h_f is the frictional head loss. Further details and the rationale for this equation can be found in Douglas et al. (1995). The tests were

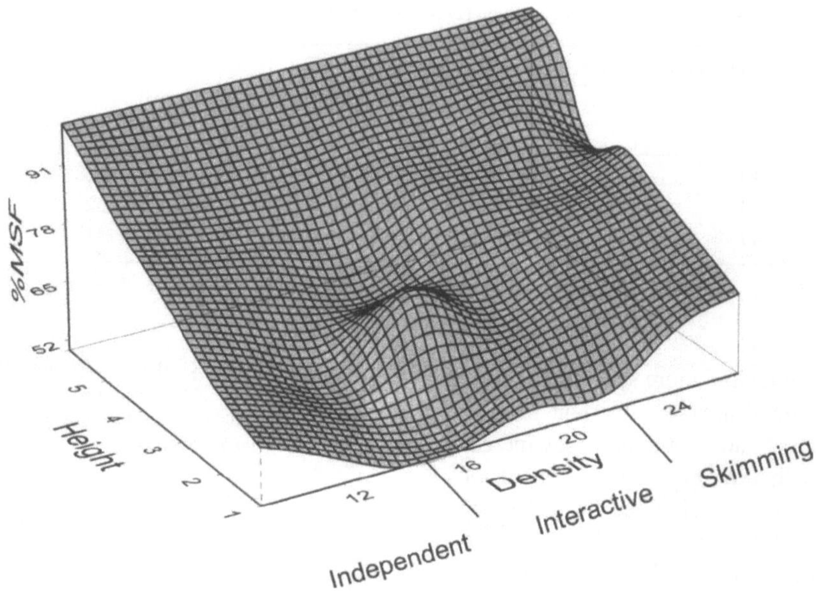

Figure 3. Three dimensional plot of barnacle density, height above base (cm) and percentage mean stream flow (%MSF) for experimental colonies. Indicated are the divisions between independent, interactive and skimming flow.

conducted in a turbulent flow regime with Re varying between 4000–17500. Two sets of experiments were conducted: arrangement of single barnacles in rows (offset or aligned) at different flow rates; and measurements made of frictional losses for replicas of dense and non-aggregated colonies. In terms of the power required to maintain flow, barnacles induce an additional power requirement which can be quantified by:

$$P = \rho g h_f Q, \tag{2}$$

where P is the power required in Watts, and Q is the volumetric flow rate (Douglas et al., 1995).

Results

Neighbour effects and the consequences of colony form and density

Dye streams were interpreted both in profile and in plan view to give a picture of the flows around a single large conical barnacle and of its effect on the flow to its downstream neighbours. An example of the profile analysis is provided in Figure 1. At very low flow rates (0.05–0.63 cm s^{-1}) the dye stream was unperturbed, remaining very coherent within the boundary layer. As the flow rate increased (2.28–4.67 cm s^{-1}) the dye stream was distorted increasingly by the upstream barnacles and helical von Karman vortices were generated, these disappearing as the flow rate increased (4.67–6.46 cm s^{-1}) and skimming flow became apparent. With the appearance of vortices, downstream flow to the neighbouring barnacles became turbulent.

The flow over complex hummock-forming colonies was examined to determine the possible consequences for barnacles living between hummocks (Figure 2A). At speeds ca. > 2 cm s^{-1}, it was found that the generation of vortices within the boundary layer by the upstream hummocks enabled barnacles within troughs to have access to significant amount of water flow across the colony.

The relationship between barnacle density, and MSF at flow rates between 9–12 cm s^{-1} for experimental colonies is summarised in Figure 3. At low barnacle density the boundary layer effects are independent with each barnacle individually affecting the hydrodynamic characteristics. At higher densities, the flow becomes interactive with small groups of barnacles having characteristic flows. Full skimming flow was identified at high densities with less disruption of MSF than was apparent with the other flow types.

The data for frictional head loss and power requirement are given in Tables 1 and 2. From these data it can be concluded that the number of barna-

Table 1. Frictional head loss (h_f) determined from equation (1) and Power requirements (P) determined from equation (2) for experimental colonies.

Flow rate (cm s^{-1})	Barnacle arrangement					
	None		Aligned		Staggered	
	h_f	P	h_f	P	h_f	P
10	0.0007	0.0017	0.0014	0.0034	0.0028	0.0068
20	0.0015	0.0073	0.0077	0.0037	0.0126	0.062
40	0.0025	0.027	0.0182	0.178	0.0253	0.248

cles affects the frictional loss more severely than their arrangement, the frontal area is important in determining frictional loss, and frictional losses increase disproportionally with flow rate. These conclusions are consistent with the data obtained from the large flume experiments, in particular the observations on changes from independent through to skimming flow.

Table 2. Frictional head loss (h_f) determined from equation (1) and Power requirements (P) determined from equation (2) for replicas of *S. balanoides* colonies.

Colony density	h_f	P
Low (smooth tile)	0.0146	0.0143
Medium (coarse tile)	0.0189	0.0185
High (medium tile)	0.0224	0.219
Very High (fine tile)	0.0288	0.282

The effect of feeding on the boundary layer

The effect of feeding by a single barnacle on its down stream neighbours was also investigated. At low flow rates (< ca. 4 cm s^{-1}) cirral beating disturbed the boundary layer and flow around downstream barnacles became turbulent. At higher flow rates where beating of the cirri ceased and they were permanently extended, boundary layers around the upstream barnacle's cirral basket prevented flow between its cirri, though disturbance to the flow caused significant turbulence enabling some downstream individuals to filter water effectively (Figure 2b). The effect of feeding by *B. amphitrite* at flow rates between 8–25 cm s^{-1} is presented in Figure 4. Feeding at slow flow rates, as expected, increased the boundary layer thickness (55% MSF at 1 cm compared with 65% MSF for non-feeding barnacles), this effect becoming more pronounced at a medium flow rate (70% MSF at 1 cm compared with 85% MSF). At fast flow rates there was no significant difference between boundary layers of barnacles feeding and non-feeding.

Prediction of optimal flow rates for feeding

To predict the optimal flow rates for barnacle feeding, it was assumed that the cirral basket for a 1 cm basal diameter individual extends to 1 cm above the substratum. This ratio was based on personal unpublished field observations of this species. By determining the percentage of MSF at 1 cm for velocities ranging from 5–13 cm s^{-1}, a linear equation was derived:

$$\begin{aligned}&\%\text{MSF at } 1\text{ cm} = 119.9 - 0.66\text{MSF}\\&(r^2 = 0.51, P < 0.01)\end{aligned} \quad (3)$$

From this equation predictions were calculated to obtain flow at 1 cm and these were plotted against MSF (Figure 5). From this deterministic model it is clear that the optimal flow rate (to maximise hydrodynamic particle supply to the cirri), as a function of MSF and boundary layer effects, is between 8–10 cm s^{-1}.

The effect of barnacle growth the boundary layer

The effect of growth on boundary layers by *B. amphitrite* at flow rates between 9–12 cm s^{-1} are presented in Figure 6. The small (0.45 cm diameter, 3 months old) barnacles produced a thicker boundary layer than the larger 7 (0.79 cm) and 11 (1.2 cm) months old) barnacles: 40% MSF at 1 cm above the substratum compared with 70% MSF, respectively.

Discussion

Water flowing over sessile organisms is essential for the removal of waste and eggs or larvae, and for the delivery of oxygen and food (Okamura, 1985; Trager

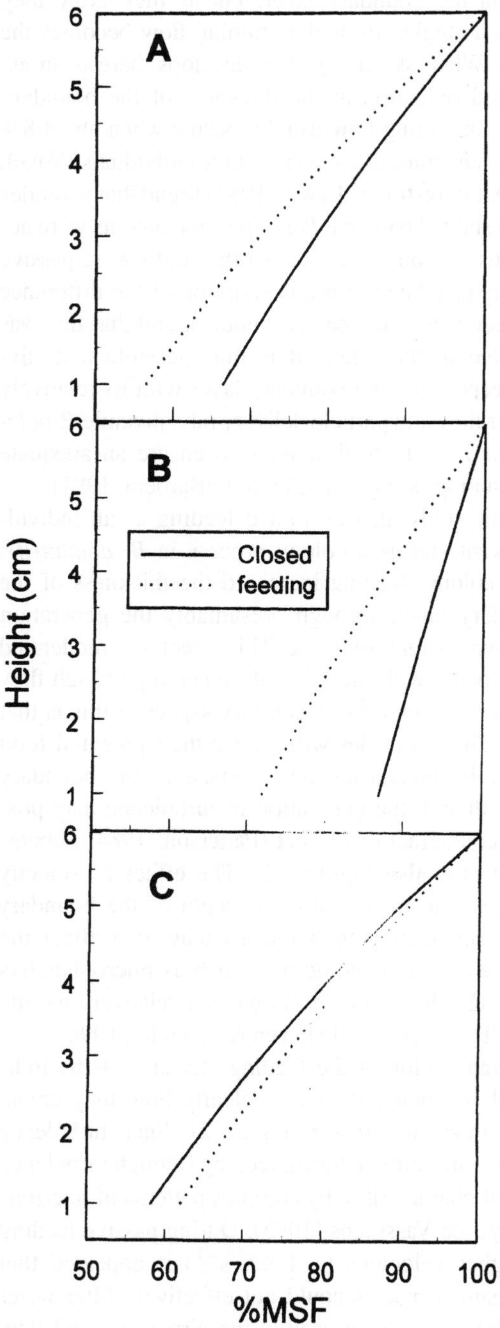

Figure 4. Plot of the effect of feeding by *B. amphitrite* on boundary layers at three flow velocities. (A) mean flow = 9.7 cm s^{-1}, (B) mean flow = 11.9 cm s^{-1} (C) mean flow = 19.9 cm s^{-1}. The percentage of mean stream flow (%MSF) is plotted against height above the colony base.

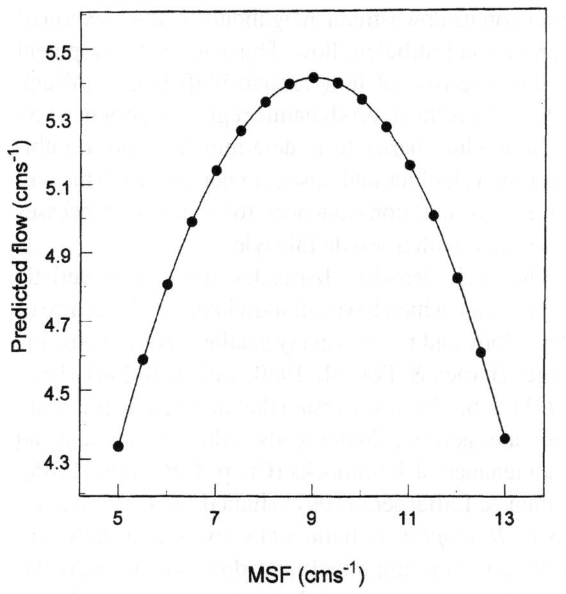

Figure 5. Plot of predicted flow (cm s^{-1}) at 1cm above a barnacle against mean stream flow (MSF, cm s^{-1}). Optimal flow is between 8 and 10 cm s^{-1}.

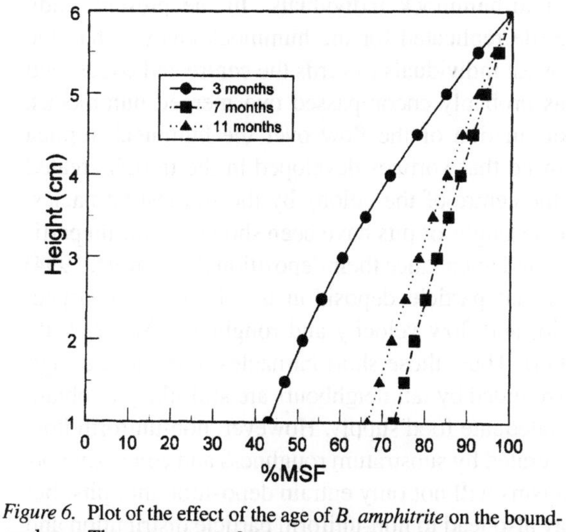

Figure 6. Plot of the effect of the age of *B. amphitrite* on the boundary layer. The percentage of mean stream flow (%MSF) is plotted against height above the colony base.

et al., 1990; Vogel, 1994). The environment in which barnacles live can be hydrodynamically very variable, with flow ranging from essentially zero in calm atidal conditions to 14 m s^{-1} within large waves, and this variability can determine the morphology, size, survivorship and distribution of species (Denny, 1994). The qualitative results presented herein demonstrate how flow characteristics alter over a colony and how an upstream individual can have a large hydrodynamic effect on its downstream neighbours by the creation of vortices and turbulent flow. This pattern development is characteristic of flow around bluff bodies (Vogel, 1994). Thus the hydrodynamic regime experienced by an individual barnacle is determined by both ambient flow velocities and upstream neighbours. This will have important consequences for transport processes associated with a sessile lifestyle.

At high densities barnacles form characteristic hummocks, which have tall individuals in the centre of the colony and progressively smaller ones towards the edge (Barnes & Powell, 1950; Pullen & LaBarbera, 1991). It has been suggested that differential prey capture rates across colonies leads to the development and maintenance of hummocks (Crisp & Bourget, 1985). Pullen & LaBarbera (1991) studied the hydrodynamics of *B. amphitrite* hummocks, examining flows for upstream, mid-hummock and downstream individuals. They demonstrated that a hummock acts like a large bluff body (see Vogel, 1994), and that velocities were slower (ca. $\times$ 4) and prey capture rates lower (ca. $\times$ 5) around downstream individuals than upstream or mid-hummock individuals. In the present study the tile replicated for the hummock analysis had the shortest individuals towards the centre and edges, and thus probably encompassed two merged hummocks. Examination of the flow over the hummock replica showed that vortices developed in the trough created at the centre of the colony by the shortest barnacles. Such troughs or pits have been shown to entrain particles and to enhance their deposition (Snelgrove, 1994) and that particle deposition is related to pit aspect ratio, and flow velocity and roughness (Yager et al., 1993). Thus, those short barnacles living in a trough surrounded by tall neighbours are still likely to obtain an adequate food supply. However, non-uniform flow generated by substratum roughness and colonizing organisms will not only entrain deposition into pits, but may also lead to non-uniform particle distribution and may therefore determine the size of prey captured by barnacles in different regions of a colony (Abelson & Loya, 1995; Abelson et al., 1993). Thus it may be that food quality, rather than quantity is important in determining the formation and maintenance of barnacle hummocks (c.f. Crisp & Bourget, 1985; Pullen & LaBarbera, 1991). The consequences of being a short barnacle surrounded by tall neighbours includes effects on fecundity and mortality processes as well as prey capture.

The hydrodynamic consequences of barnacle growth have been poorly researched to date. This study has shown how at small sizes, barnacles tend to act as individual elements, thus having a large effect on the boundary layer, but as they grow they act as a single unit and skimming flow becomes the norm. When skimming flow develops there is an associated reduction in the thickness of the boundary layer. Skimming flow usually occurs when about 8% of the substratum is covered with individuals (Vogel, 1994). Interestingly, Lewis (1981) found that juveniles of the stalked barnacle *Pollicipes polymerous* were active suspension feeders whilst the adults were passive feeders (see below for a description of the difference between active and passive modes), and that this was reflected in their diet. It is thus possible that, living deeper within a boundary layer with its relatively slower flow and particle delivery rate, juvenile *P. polymerous* have to feed actively to ensure an adequate food supply (see also Pullen & LaBarbera, 1991).

This study also examined feeding as an individual event and as a colony process in *B. amphitrite*. As a colony, feeding increased the thickness of the boundary layer, through presumably the generation of downstream turbulence. This effect was moderated by velocity, with the effect disappearing at high flow rates (> 14 cm s^{-1}). A contrary aspect of this is that by feeding barnacles will reduce their potential food supply by increasing the thickness of the boundary layer, though the generation of turbulence may possibly counteract this effect (Patterson, 1984; Sebens, 1997a, see also Figure 2B). The effect of velocity changes (mean local flow or depth of the boundary layer) on feeding in barnacles may also affect the capture of non-avoiding prey such as microplankters (Hunt & Alexander, 1991) but not relatively mobile prey (Trager et al., 1994; Yen & Strickler, 1996).

Examination of the feeding mechanisms by individual *B. amphitrite* shows clearly how they create downstream turbulence (Figure 2). Such turbulence will be intermittent during feeding (though turbulence is itself characterised by complex patterns of intermittencey; see Vassilicos (1995). During passive feeding at higher velocities (>4 cm s^{-1}) it appeared that upstream barnacles could not effectively filter water as a boundary layer around the cirri prevented flow between individual cirri. However, turbulence generated by the upstream individual enabled downstream neighbours in the colony to filter effectively during passive suspension feeding episodes. These observations are contradictory to the data presented by Trager et al. (1990) who studied feeding in *S. balanoides*. This variance between these two studies is probably a consequence of species-specific differences in cirral morphology (see Anderson, 1981). Thus there are identifiable beneficial aspects to feeding in a colony.

Barnacles have three modes of feeding: normal beat, fast beat and extension (Crisp & Southward, 1961). The first two modes are often termed active feeding, the latter mode passive feeding (Pullen & LaBarbera, 1991; Trager et al., 1990). Conversely most suspension feeders are restricted to just one mode (Jørgensen, 1966). In some species of barnacle microplankton capture is augment by pumping water through the mouth parts (Anderson, 1981; Hunt & Alexander, 1991). The switch from active to passive feeding is triggered by ambient flow rate and the exact point is species specific. For instance, for *S. balanoides* it is at 3 cm s^{-1} (Trager et al., 1990), and for *Tetraclita ramosa* it is between 100–140 cm s^{-1} (Hunt & Alexander, 1991). The switch from active to passive mode also involves the orientation of the cirral basket into the current. In the present study the switch point for *B. amphitrite* was at a velocity of ca. 4 cm s^{-1}. Pullen (1991) found the switch point to be at 2–3 cm s^{-1} for the same species. Such a variance is not unexpected as ambient hydrodynamic regimes where the animals live appear to modify the switch point (Southward, 1955). The consequences for a colony may be that upstream or tall barnacles passively feed whilst downstream or short neighbours are forced to actively feed (see Pullen & LaBarbera, 1991). Similar changes in feeding behaviour under varying hydrodynamic conditions are also found in sea anemones (Sebens, 1987).

We predicted an optimal flow for prey capture in *S. balanoides* of 9 cm s^{-1}. This is not related to the feeding mechanisms employed, nor indicative of the switch point velocity, but is a measure of optimal particle supply as a function of boundary layer and velocity effects. Flow at Keppel Pier is predominately uni-directional (225°) and the velocity ranges from 10–18 cm s^{-1}, with the flow chiefly between 10–14 cm s^{-1} (Wong & Moore, 1996). Extreme hydrodynamic conditions constrain species distributions (Denny, 1994) and the optimum flow for feeding in a variety of non-barnacle species has been shown to be that experienced at sites where the species are most common (Hunt & Alexander, 1991; Leonard et al., 1988; Patterson, 1984; Sebens, 1997b). Thus there would be appear to be a conflict between our laboratory data and field observations. It is likely under these flow conditions that *S. balanoides* is feeding passively most of the time and that given the densities present at the site (see Hills & Thomason, 1996) they exist most of the time in a skimming flow regime. However as they thrive at Keppel pier, reaching a large size (1–2 cm diameter) and sexual maturity within 9 months the field conditions would seem optimal (i.e. there is an adequate food supply despite skimming flow over the colonies). Further work is therefore required to resolve the conflict between laboratory and field observations.

This study has shown how barnacle density has complex effects on community hydrodynamics. Density alters the flow characteristics of the substratum with low density colonies acting as a series of individuals, medium density colonies producing interacting flow regimes and high density colonies generating skimming flows (see Vogel, 1994). These characteristic flows alter the boundary layer characteristics (Figure 6) and will ultimately have an impact on the particle capture rates of colony members. These data are supported by the results from the small flume experiments, whereby density increases resulted in a non-proportional increase in frictional head loss, from 0.0146 to 0.0288 for low and high density colonies, respectively. Similar results were obtained for experimental colonies. These data also show that the arrangement of barnacles in a colony does not affect the frictional loss as much as frontal area. Kimura (1985), in a study of net and pipe fouling by an unnamed species of balanomorph barnacle, found that drag coefficient was the same for Reynolds numbers ranging from 2000–40 000. The nets and pipes studied were heavily fouled and the constancy of the drag coefficient indicates that the colonisers were creating large boundary layers. This is in agreement with the present study. The small flume data also show how, economically, low levels of barnacle fouling could have a severe impact on the maritime industry. The world commercial fleet currently uses some 700 mt of fuel oil at an annual cost of \$ 2.16 billion (@ \$ 21 per barrel, June 1997). Assuming form (hull-shape) friction remains constant, then based on the power requirement data, skin friction losses due to slight to severe barnacle fouling would increase the fuel costs to \$ 2.19–\$ 2.77 billion, respectively. Likewise, heavy fouling on non-powered craft such as yachts would result in a decrease in speed by up to 28%.

In conclusion, we have demonstrated how the hydrodynamic regime around barnacles and the substratum they have colonised alters both with flow velocity, age/size of the barnacles, feeding/non-feeding, colony density and position of an individual within a colony.

Acknowledgements

This study was funded by NERC grant GR9/01905. JMH was supported by the Royal Society and Sheina Marshall Fellowship, and ASC by grant-in-aid from the Marine Biological Association. The authors are grateful to Robert Young for technical assistance.

References

Abelson, A. & Y. Loya, 1995. Cross-scale patterns of particulate food acquisition in marine benthic environments. Am. Nat. 145: 848–854.

Abelson, A., T. Miloh & Y. Loya, 1993. Flow patterns induced by substrata and body morphologies of benthic organisms, and their roles in determining availability of food particles. Limnol. Oceanogr. 38: 1116–1124.

Alexander, S. E. & J. Roughgarden, 1996. Larval transport and population dynamics of intertidal barnacles: a coupled benthic/oceanic model. Ecol. Monogr. 66: 259–275.

Anderson, D. T., 1981. Cirral activity and feeding in the barnacle *Balanus perforatus* Bruguiere (Balanidae), with comments on the evolution of feeding mechanisms in thoracican cirripedes. Phil. Trans. r. Soc., Lond. Ser. B 291: 411–449.

Barnes, H. & H. T. Powell, 1950. The development, general morphology and subsequent elimination of barnacle populations, *Balanus crenatus* and *B. balanoides*, after heavy initial settlement. J. anim. Ecol. 19: 175–179.

Bertness, M. D., S. D. Gaines & R. A. Wahle, 1996. Wind-driven settlement patterns in the acorn barnacle *Semibalanus balanoides*. Mar. Ecol. Prog. Ser. 137: 103–110.

Bowden, B. S. & N. J. Davison, 1974. Resistance increments due to hull roughness associated with form factor extrapolation methods. N. M. I. Ship, TM 3800.

Carlton, J. S., 1994. Marine propellers and propulsion. Butterworths Heinemann, Oxford.

Crisp, D. J., 1955. The behaviour of barnacle cyprids in relation to water movement over a surface. J. exp. Biol. 32: 569–590.

Crisp, D. J. & E. Bourget, 1985. Growth in barnacles. Adv. mar. Biol. 22: 199–244.

Crisp, D. J. & A. J. Southward, 1961. Different types of cirral activity of barnacles. Phil. Trans. r. Soc., Lond. Ser. B 243: 271–308.

Denny, M. W., 1994. Extreme drag forces and the survival of wind- and water-swept organisms. J. exp. Biol. 194: 97–115.

Douglas, J. F., J. M. Gasiorek & J. A. Swaffield, 1995. Fluid Mechanics. Longman, London, 819 pp.

Eckman, J. E., 1996. Closing the larval loop: linking larval ecology to the population dynamics of marine benthic invertebrates. J. exp. mar. Biol. Ecol. 200: 207–237.

Freeman, J. H., 1977. The marine fouling of fixed offshore installations. Department of Energy, London, ISSN 0309-8184.

Gucinski, H., R. E. Baier, A. E. Meyer, M. S. Fornalik & R. W. King, 1984. Surface microlayer properties affecting drag phenomena in seawater. Sixth int. Conf. mar. Corr. Foul. 585–604.

Harderlie, E. C., 1984. A brief overview of the effects of macrofouling. In J. D. Costlow & R. C. Tipper (ed.), Marine Biodeterioration: An Interdisciplinary Study. US Naval Research Institute, Annapolis: 163–166.

Hills, J. M. & J. C. Thomason, 1996. A multi-scale analysis of settlement density and pattern dynamics of the barnacle *Semibalanus balanoides*. Mar. Ecol. Prog. Ser. 138: 103–115.

Hills, J. M. & J. C. Thomason, 1997. The effect of scales of surface roughness on the settlement of barnacle (*Semibalanus balanoides*) cyprids. Biofouling 12: 57–69.

Hills, J. M., J. C. Thomason & J. Muhl, 1997. A high-precision method for the manufacture of complex three-dimensional surfaces. Biofouling (submitted).

Hunt, M. J. & C. G. Alexander, 1991. Feeding mechanisms in the barnacle *Tetraclita squamosa* (Bruguiere). J. exp. mar. Biol. Ecol. 154: 1–28.

Jørgensen, C. B., 1966. Biology of Suspension Feeding. Pergamon Press, Oxford.

Kimura, H., 1985. Hydrodynamic resistance of pipes and nets fouled with the acorn barnacle. Bull. jap. Soc. sci. Fish. 51: 549–555.

Leonard, A. B., J. R. Strickler & N. D. Holland, 1988. Effects of current speed on filtration during suspension feeding in *Oligometra serripena* (Echinodermata: Crinoidea). Mar. Biol. 9: 111–125.

Lewis, C. A., 1981. Juvenile to adult shift in feeding strategies in the pedunculate barnacle *Pollicipes polymerus* (Sowerby) (Cirripedia: Lepadomorpha). Crustaceana 41: 14–20.

Lewthwaite, J. C., A. F. Molland & K. W. Thomas, 1985. An investigation into the variation of ship skin frictional resistance with fouling. Trans. r. Inst. nav. Arch. 127: 269–284.

Marrs, S., J. C. Thomason, M. Cowling & T. Hodgekiess, 1995. A replica method for the study of marine biofilms. J. mar. biol. Ass. U.K. 75: 759–762.

Mullineaux, L. S. & C. A. Butman, 1991. Initial contact, exploration and attachment of barnacle (*Balanus amphitrite*) cyprids settling in flow. Mar. Biol. 110: 93–103.

Noda, T. & S. Nakao, 1996. Multi-scale spatial pattern of recruitment in the barnacles *Semibalanus cariosus* at fishing ports on the Kameda Peninsula, southern Hokkaido, Japan. Hydrobiologia 324: 125–130.

Okamura, B., 1985. The effects of ambient flow velocity, colony size, and upstream colonies on the feeding success of Bryozoa. J. exp. mar. Biol. Ecol. 83: 69–80.

Patterson, M. R., 1984. Patterns of whole colony prey capture in the octocoral *Alcyonium siderium*. Biol. Bull. 167: 613–629.

Pidgeon, J. D., 1993. Critical review of current and future marine antifouling coatings. Lloyd's register of shipping engineering services, London, 93/TIPEE/4787.

Pullen, J. & M. LaBarbera, 1991. Modes of feeding in aggregations of barnacles and the shape of aggregations. Biol. Bull. 181: 442–452.

Rittschof, D., E. S. Branscomb & J. D. Costlow, 1984. Settlement and behaviour in relation to flow and surface in larval barnacles, *Balanus amphitrite* Darwin. J. exp. mar. Biol. Ecol. 82: 131–146.

Rittschof, D., A. S. Clare, D. J. Gerhart, S. A. Mary & J. Bonaventura, 1992. Barnacle *in vitro* assays for biologically active substances: toxicity and settlement inhibition assays using mass cultured *Balanus amphitrite amphitrite* Darwin. Biofouling 6: 115–122.

Sebens, K. P., 1987. Coelenterata. In F. J. Vernberg & T. J. Pandian (ed.), Animal Energetics. Vol. 1. Academic Press, New York: 55–120.

Sebens, K. P., 1997a. Adaptive responses to water flow: morphology, energetics, and distribution of reef corals. Proc. 8th int. Coral Reef Symp. (in press).

Sebens, K. P., 1997b. Effects of water flow and branch spacing on particle capture by the reef coral *Madracis mirabilis* (Duchassaing and Michelloti). J. exp. mar. Biol. Ecol. 211: 1–28.

Shanks, A. L., 1986. Tidal periodicity in the daily settlement of intertidal barnacle larvae and an hypothesized mechanism for the cross-shelf transport of cyprids. Biol. Bull. 170: 429–440.

Snelgrove, P. V. R., 1994. Hydrodynamic enhancement of invertebrate larval settlement in microdepositional environments: colonizing tray experiments in a muddy habitat. J. exp. mar. Biol. Ecol. 176: 149–166.

Southward, A. J., 1955. Feeding of barnacles. Nature 175: 1124–1125.

Trager, G., Y. Achituv & A. Genin, 1994. Effects of prey escape ability, flow speed, and predator feeding mode on zooplankton capture by barnacles. Mar. Biol. 120: 251–259.

Trager, G. C., 1988. Can barnacles count to 10 – suspension feeding in variable flow and the role of learning. Am. Zool. 28: A 155–A 155.

Trager, G. C., D. Coughlin, A. Genin, Y. Achituv & A. Gangopadhyay, 1992. Foraging to the rhythm of ocean waves – porcelain crabs and barnacles synchronize feeding motions with flow oscillations. J. exp. mar. Biol. Ecol. 164: 73–86.

Trager, G. C., J. S. Hwang & J. R. Strickler, 1990. Barnacle suspension-feeding in variable flow. Mar. Biol. 105: 117–127.

Vassilicos, J. C., 1995. Turbulence and intermittency. Nature 374: 408–409.

Vogel, S., 1994. Life in moving fluids. The physical biology of flow, Princeton University Press, Princeton, N.J, 467 pp.

Walters, L. J. & D. S. Wethey, 1991. Settlement, refuges, and adult body form in colonial marine-invertebrates – a field experiment. Biol. Bull. 180: 112–118.

Walton-Smith, F. G., 1946. Effect of water currents upon attachment and growth of barnacles. Biol. Bull. 90: 51–70.

Wethey, D. S., M. W. Luckenbach & C. A. Kelly, 1988. Larval settlement in barnacles: influence of water flow. In M.-F. Thompson, R. Sarojini & R. Nagabhushanam (ed.), Marine Biodeterioration: Advanced Techniques Applicable to the Indian Ocean. Oxford & IBH Publishing, New Dehli: 499–511.

Wong, Y. M. & P. G. Moore, 1996. Observations on the activity and life history of the scavenging isopod, *Natatolana borelealis* Lilljeborg (Isopoda: Cirolanidae) from Loch Fyne, Scotland. Estuar. coast. Shelf Sci. 42: 247–262.

Yager, P. L., A. R. M. Nowell & P. A. Jumars, 1993. Enhanced deposition to pits: a local food source for benthos. J. mar. Res. 51: 209–236.

Yen, J. & J. R. Strickler, 1996. Advertisement and concealment in the plankton – what makes a copepod hydrodynamically conspicuous? Invert. Biol. 115: 191–205.

Hydrobiologia **375/376:** 203–216, 1998.
S. Baden, L. Pihl, R. Rosenberg, J.-O. Strömberg, I. Svane & P. Tiselius (eds),
Recruitment, Colonization and Physical–Chemical Forcing in Marine Biological Systems.

The influence of epilithic microbial films on the settlement of *Semibalanus balanoides* cyprids – a comparison between laboratory and field experiments

R. C. Thompson[1,*], T. A. Norton[2] & S. J. Hawkins[3]
[1] *Department of Marine Sciences and Coastal Management, University of Newcastle, Dove Marine Laboratory, Cullercoats, North Shields, Tyne & Wear, NE30 4PZ, UK*
[2] *Port Erin Marine Laboratory, School of Biological sciences, University of Liverpool, Port Erin, Isle of Man, IM9 6JA, UK*
[3] *Division of Biodiversity and Ecology, School of Biological Sciences, University of Southampton, SO16 7PX, UK*
[*] *Corresponding author: Present address: Division of Biodiversity and Ecology, School of Biological Sciences, University of Southampton, Bassett Crescent East, Southampton, SO16 7PX, UK.*
Tel. +44 1703 594405; e-mail: rct@soton.ac.uk

Key words: barnacle, biofilm, cypris, recruitment, fouling, microalgae, succession, rocky shore, intertidal

Abstract

Barnacle cypris larvae show considerable exploratory behaviour prior to habitat selection. The influence of natural epilithic microbial fouling organisms on the settlement of *Semibalanus balanoides* cyprids (Crustacea: Cirripedia) was examined using laboratory and field based investigations. In choice chambers, cues from microbial films were important; cyprids preferred surfaces with a mature microbial film to either unfilmed surfaces or those with a developing film. Cyprids also discriminated between filmed rocks from different tidal heights, preferentially selecting those from the mid-shore which is their usual zone. Filmed surfaces which had previously been colonised by barnacles were selected in preference to unfilmed surfaces, but the presence of an adult barnacle did not enhance settlement on either filmed or unfilmed surfaces. However, laboratory experiments were not consistent with settlement in the field which was predominantly influenced by the proximity of conspecifics and by traces of previous barnacle colonisation. These factors increased settlement, and seemed to over-rule cues from microbiota within the film. Difficulties in the application of laboratory based studies to settlement in the natural environment are discussed.

Introduction

Microbial films form on hard surfaces immersed in water (Wahl, 1989). In shallow waters they are mainly comprised of diatoms and cyanobacteria, and play a key role in the ecology of benthic communities. Such films provide a food resource for grazers such as molluscs and fishes (Underwood, 1984a; Hawkins et al., 1989; Edgar & Shaw, 1995). In addition, microbial films form an interface between a substratum and the water column and in this respect they can influence settlement and succession by benthic invertebrates and macroalgae (Wahl, 1989).

Most benthic organisms have a dispersive stage in their life cycle which ends when the larva or propagule encounters a surface and settles, before developing into a sessile adult (for reviews see Santelices, 1990; Norton, 1992 for algae and; Pawlik, 1992; Rodriguez et al., 1993 for invertebrates). Microbial films have been shown to promote settlement of a variety of invertebrates (e.g. Polychaetes: Knight-Jones, 1951; Wilson, 1955; Pawlik & Butman, 1993, Gastropods: Scheltema, 1961, Bivalves, Tritar et al., 1992, Bryozoans: Mihm et al., 1981; Brancato & Woollacott, 1982, Echinoderms: Cameron & Hinegardner, 1974; Johnson et al., 1991, and Ascidians: Keough & Raimondi, 1995) including barnacle larvae (Le Tourneux & Bourget, 1988; Keough & Raimondi, 1995). However, there are also some accounts of microbial films reducing settlement (e.g. bryozoans Mihm et al., 1981 and barnacles Maki et al., 1992).

Most studies examining factors affecting settlement have been conducted in laboratory conditions, for example using choice chambers. This approach has helped to focus on specific questions relating to particular settlement cues, but cannot mimic the complexity of the natural environment. In the field settlement is influenced by a range of physico-chemical and biological factors such as tidal currents, surface topography and the presence of conspecifics. In addition behavioural cues may be less important in a highly turbulent environment. Many studies have focused on interactions between a single species from the film, for example using cultures of bacteria (Dillon et al., 1989; Maki et al., 1992; Tritar et al., 1992) or diatoms (Huang & Boney, 1985) and a settling organism, but the influence of a particular species may be reversed by the presence of other species (Meadows & Williams, 1963; Huang & Boney, 1985). In addition studies have frequently been conducted using micro-organisms grown on artificial surfaces such as plastic dishes or glass microscope slides (e.g. Todd & Keough, 1994). Yet, the surface properties of artificial surfaces differ from those of natural substrata and this may considerably influence both settlement and recruitment (Mihm et al., 1981; Wahl, 1989; Henschel et al., 1990 for examples of little difference between surfaces; but see also Becker, 1993; Anderson & Underwood, 1994). Relatively few authors have assessed the influence of microbial films on natural rock surfaces (but see Strathmann et al., 1981; Le Tourneux & Bourget, 1988) and the relative contribution of microbial films to settlement and colonisation in natural conditions is not clear. In the present study we examine the influence of natural epilithic biofilms on the settlement of marine invertebrates using larvae of the intertidal barnacle *Semibalanus balanoides* as a model.

Barnacle cypris larvae exhibit considerable habitat selection behaviour prior to settlement, they have been extensively studied in both laboratory and field conditions and are believed to respond to various settlement cues at a range of spatial scales (Crisp & Meadows, 1963; Crisp, 1974; Crisp, 1984; Bourget, 1988; Neal & Yule, 1994, Clare, 1995; Walker, 1995). Laboratory experiments have shown that cyprids settle preferentially on surfaces which are covered by

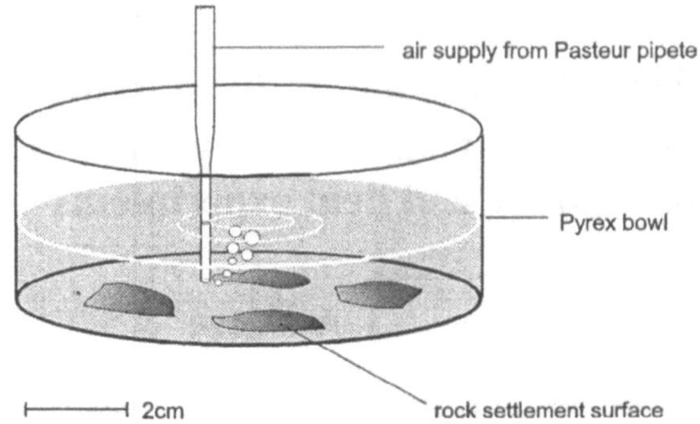

Figure 1. Choice chamber with air supply and four settlement surfaces.

microbial films (Crisp & Meadows, 1963) and that they are able to distinguish between films of different ages and films grown in different conditions (Neal & Yule, 1994; Wieczorek et al., 1995). Cyprids may therefore be able to use cues in the microbial film to aid habitat selection. Where environmental conditions are favourable the micro-organisms within these films can reproduce rapidly, and it has been suggested that cyprids may gain 'up to date' information about conditions at a particular position on the shore (Neal & Yule, 1994; Wieczorek et al., 1995) or determine shore level (Strathmann et al., 1981; Raimondi, 1988b; Johnson & Strathmann, 1989) from cues within the film.

Cyprids can also distinguish between different types of rock (Holland et al., 1984; Raimondi, 1988a; Holmes et al., 1997) and usually settle preferentially on rugose surfaces (Yule & Walker, 1984; Bourget, 1988). However, the presence of conspecifics or their remnants provides an exceptionally strong inducement to settle (Knight Jones, 1953; Wethey, 1984; Bourget, 1988; Raimondi, 1988b; Crisp, 1990; Raimondi, 1991) and the relative contribution of microbial cues compared to those from conspecifics remains unclear.

In this study a combination of laboratory and field experiments were used to demonstrate the influence of intertidal epilithic biofilms on barnacle settlement and to determine the relative importance of these films compared to better known, settlement cues.

Methods

Study sites

Samples of rock for laboratory choice chambers were collected from the mid-tide level (~3. 4 m above LAT) on a moderately exposed shore at Port St Mary, Isle of Man (4°44′ W., 54°4′ N.) (see: Southward,

Table 1. Summary of barnacle settlement on surfaces with differing microbial films in choice chamber. Twenty replicate chambers, each containing 20 cyprids, were used in all experiments except for that of 10/6/94 where 9 cyprids were used in each chamber. * $p < 0.05$.

Date	Substrate choices (origin of rock sample in settlement chamber)	Matrix of Tukey-Type multiple comparison test				
20/5/95	newly exposed unfilmed rock surface (chiselled & burned)	UF	–			
	poorly developed film, not previously colonised surface (prepared as (F) but allowed 2 weeks to acquire microbial film)	PF	ns	–		
	filmed, not previously colonised (surface chiselled & burned, then allowed 4 months to acquire microbial film)	F	*	*	–	
			UF	PF	F	
10/6/94	newly exposed unfilmed rock surface	UF	–			
	moderately exposed lower shore	ELS	ns	–		
	moderately exposed mid shore	EMS	*	*	–	
	moderately exposed upper shore	EUS	ns	ns	*	–
			UF	ELS	EMS	EUS
2/5/95	newly exposed unfilmed rock surface	UF	–			
	as (F) above	F	ns	–		
	filmed & previously colonised by adult barnacles (adult barnacles removed by scraping then allowed 4 months to acquire microbial)	FPC	*	ns	–	
			UF	F	FPC	
10/5/95	newly exposed unfilmed rock surface	UF	–			
	as (UF) but with an adult barnacle transplanted on to rock surface	UFB	ns	–		
	as (F) above	F	*	ns	–	
	as (F) but with an adult barnacle transplanted on to rock surface	FB	*	ns	ns	–
			UF	UFB	F	FB

1953, for site descriptions). Field experiments were prepared in areas of dense barnacle cover at the same site.

Choice chambers

These experiments were conducted at the Port Erin Marine Laboratory, Isle of Man during June 1994 and May 1995, when barnacle cyprids could easily be collected using a plankton net trawled back and forth along the side of a jetty nearby. Plankton samples from these trawls were viewed in Bogorov trays with the aid of a dissecting microscope. *Semibalanus balanoides* cyprids were removed individually using a Pasteur pipette, transferred to a beaker of filtered seawater and kept at 10°C until they were introduced into choice chambers, normally within 24 h of collection.

Choice chambers were situated in a controlled temperature room maintained at similar conditions to those in the sea at that time of year (10°C, 16 h light 8 h dark). Twenty chambers were used for each experiment except on one occasion when availability of cyprids reduced replication to nine (Table 1). Each chamber consisted of a 300 ml Pyrex basin which had been washed and sterilised in an autoclave. Chambers were filled with 200 ml of filtered (0.22 μm) seawater and aerated with a steady stream of bubbles delivered from a centrally positioned Pasteur pipette connected to an air supply (Figure 1). Water movement generated by bubbles rising from the air supply circulated cyprids around the chambers maximising the likelihood of them encountering each of the surfaces offered and may also have promoted settlement (Crisp, 1955).

Settlement substrata were rocks chips, each approximately 25 mm in diameter and 5–10 mm thick. These were removed, using a hammer and chisel, from randomly located positions at selected tidal heights on the shore. Samples were not colonised by barnacles, cyprids or other macrobiota and had no visible remnants of a previous barnacle matrix on their surface. As far as possible rock chips were used on the day of collection, but when necessary they were stored four a maximum of 3 d at 4°C. With the exception of surfaces which had been manipulated experimentally it was assumed that the rocks supported a microbial film typical of the part of the shore from which they had been collected. Four choice chamber experiments were conducted in all. Depending upon the experiment, three or four different types of rock surface were introduced into each tank and placed face upwards using a pair of sterile tweezers. Care was taken not to cross-contaminate samples during collection and handling. Samples were positioned around the edges of the tanks and were separated by a gap of about one centimetre (Figure 1). In order to ensure an even spatial distribution of each rock type the orientation of samples was rotated in a clockwise direction as each successive chamber was prepared.

Approximately 20 cyprids were introduced into each tank using a Pasteur pipette. This number was selected to provide sufficient data from the number of tanks used, but not to be so large as to cause problems of cyprid-cyprid interactions (see Rittschof et al., 1992; Clare et al., 1994; Wieczorek et al., 1995). Settlement was recorded daily until all cyprids had either settled or become inactive.

Filmed, partially filmed and unfilmed rocks were used to determine whether cyprids could differentiate between filmed and unfilmed surfaces. These were obtained by chiselling away the top few millimetres of bedrock from several areas ($\sim 15 \times 15$ cm) of the mid-shore during both January and April 1995. Rock chips were removed from these areas for use in the choice chamber experiments during May 1995. At this time the chiselled areas had been accessible to colonising organisms for 12 weeks (January clearance) and 2 weeks (April clearance) providing 'filmed' and 'partially filmed' surfaces, respectively. 'Unfilmed rock' was prepared by removing the top surface of the bedrock, as described previously, and then immediately removing rock chips from these newly exposed areas so that none of the surface had previously been exposed to seawater.

Further trials were conducted to establish whether cyprids could distinguish between microbial communities from different shore levels. Rock samples were collected from the mid-shore ($\sim 50\%$ cover of adult *Semibalanus balanoides*, 3.4 m above lowest astronomical tide [LAT]), whilst upper (4.4 m above LAT) and lower (2.0 m above LAT) shore samples were collected on the extreme edges of this zone ($> 5\%$ coverage of adult barnacles). One sample from each level together with a sample of unfilmed rock (prepared as above) was placed in each chamber.

Additional experiments were used to determine the principal cue(s) which enabled cyprids to detect rocks that had originated from the mid-shore. In the first of these experiments unfilmed surfaces, filmed surfaces without traces of adult barnacles, and filmed surfaces with traces of adult barnacles were used. With the exception of the unfilmed surfaces (prepared as described above), these were prepared on the mid-shore during January 1995, allowing a natural microbial community to develop for three months before barnacle settlement commenced. Filmed surfaces without previous traces of barnacles were produced by chiselling away the top of the rock surface. Whilst, filmed surfaces with traces of adult barnacles were created by gently scraping barnacles from areas of dense barnacle matrix, on the mid-shore, using a paint scraper. Rock samples of each film type were chiselled from the shore during June 1995 and placed in each of the choice chambers (Table 1).

A subsequent experiment investigated the relative importance of cues from filmed surfaces compared to those provided by conspecific adult barnacles. Here, adult barnacles were carefully removed from intertidal rock using a sharp scalpel. This procedure was easiest with solitary barnacles living on a smooth rock surface. One barnacle was fixed onto a smooth area of each test substrata by gluing the outer edge of the rostrum to the rock with small spots of super-glue (cyanoacrylate). The rocks were placed into filtered seawater for 24 h whilst the glue cured. Rocks on which the barnacles had survived transplantation (i.e. observed to be feeding normally) were then used in the choice chambers together with one rock of each type with glue, but no barnacle as a control.

Field experiments

Three areas of shore were selected which had exceptionally dense barnacle cover ($\sim 95\%$). Typically, these were at the edge of horizontal ledges that were exposed to tidal currents. Treatments were prepared at two distances from the adult barnacle population. This was achieved by clearing circular areas of either 30 cm or 50 cm diameter from the barnacle matrix using a paint scraper. Care was taken to choose positions which had a relatively smooth underlying rock surface, and would be surrounded by at least 20 cm of matrix after clearance.

Within each circular clearance area three treatments were prepared during January 1995. These surfaces were identical to those described in the choice chamber experiments above ('unfilmed', 'filmed' and 'filmed with barnacle remnants'). However the

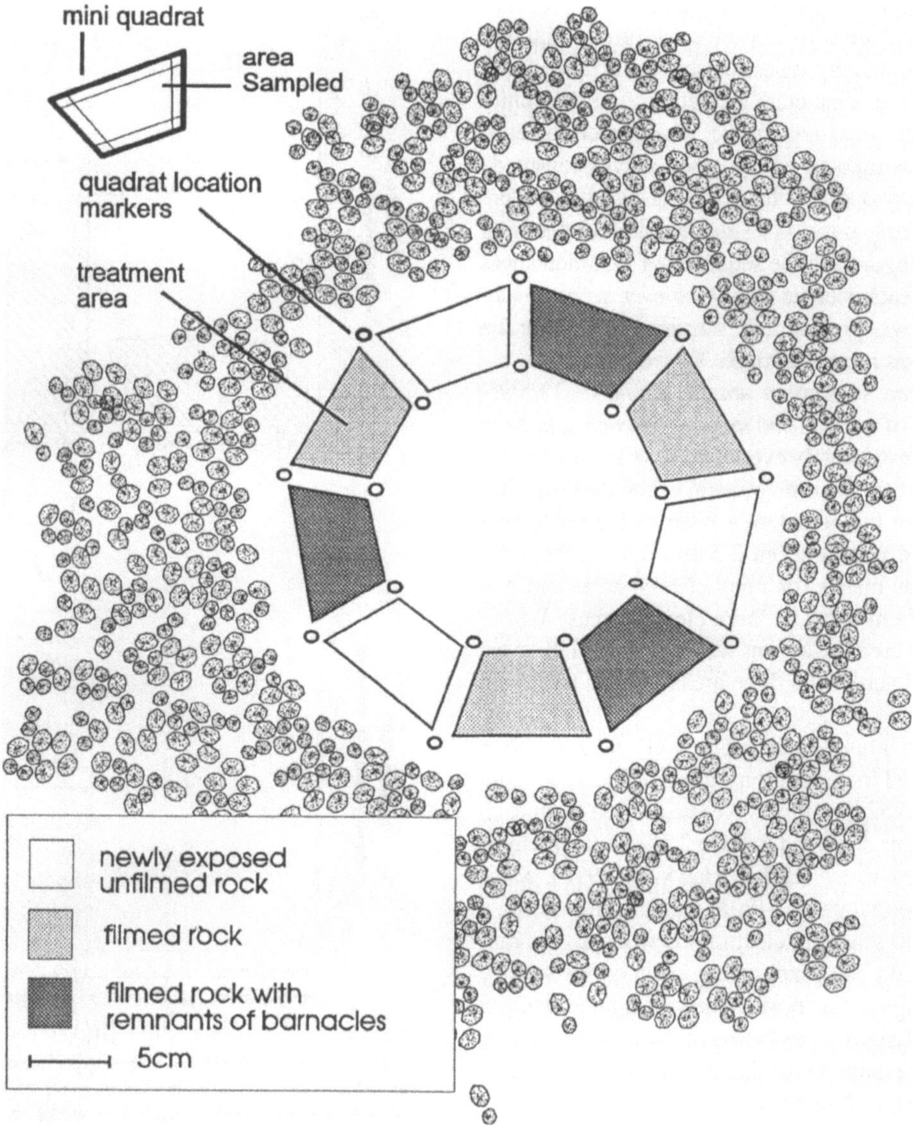

Figure 2. Example of settlement area with differing microbial films at mid-tide level, Port St Mary, Isle of Man. Experimental treatments were positioned within circular areas which had been cleared so that surrounding barnacles were at least 3.5 cm away. Identical treatments were also prepared 13.5 cm away from surrounding barnacles (see text).

method of preparation was refined slightly. Unfilmed surfaces and filmed surfaces were prepared by removing the upper 1 mm of rock using a sharp wood chisel. The newly exposed surface was then heated with a blow lamp to denature all traces of arthropodin, a substance released by adult barnacles which promotes cyprid settlement (Knight Jones, 1953; Crisp & Meadows, 1963; Yule & Crisp, 1983). During this procedure care was taken to shield the biofilm on the surrounding rock from thermal damage. Areas of the scraped barnacle matrix, formed when the circular clearance areas were prepared, were used as filmed surfaces previously colonised by barnacles. Treatment areas were allowed to become fouled by microbiota during the next three months before the barnacle settlement season commenced. At the start of settlement in April 1995, the surface of unfilmed treatments was again removed using a wood chisel leaving newly exposed rock.

Three duplicates of each treatment were prepared in each cleared area. To achieve a uniform spatial distribution of treatment types, these were arranged in a circle (25 cm in diameter) located around the centre of the cleared areas (Figure 2). The corners of each treatment was marked with a small depression made using a battery operated hand drill. These holes were later used to help position a small quadrat for counting settlement (Figure 2). The sequence of treatments was the same in each cleared area. However, as each successive area was prepared, the starting position for the treatments was rotated through 30° from north.

During the settlement season (25/4/95–23/5/95) the numbers of cyprids and metamorphosed spat were recorded approximately every two days from a central 12.5 cm^2 portion the mini quadrat of the quadrat (Figure 2). Hence the actual area from which settlement was recorded was between 3.5 and 6.5 cm from the adult barnacle matrix for small cleared areas and was 13.5 to 16.5 cm away in large cleared areas. Within each cleared area settlement was determined by averaging counts across the three areas of each treatment type.

Direct counts of microbiota were made on rock chips collected from additional areas of each treatment type. Six rock chip samples were collected from each surface type air dried and fixed in 2.5% gluteraldehyde in filtered sea water (Hill & Hawkins, 1990). Samples were then viewed at 480× magnification using a Philips XL 30 scanning electron microscope. Six randomly located photographs were taken on the surface of each sample. These were later examined using a binocular microscope (64× magnification) to provide reasonable resolution for identification with a large field of view (see Patterson et al., 1986).

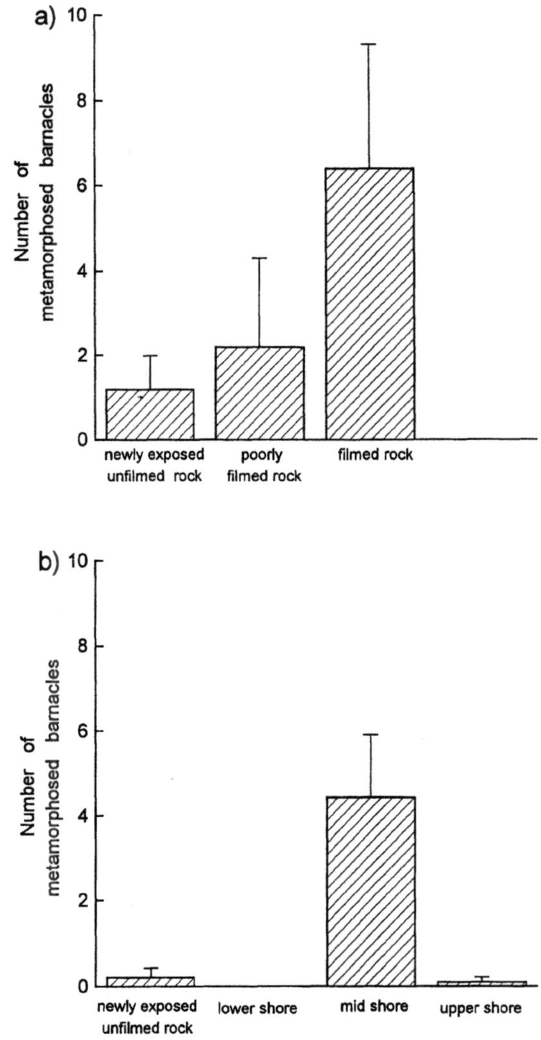

Figure 3. Number (mean ± 1 SE) of cyprids plus metamorphosed barnacles on rock surfaces in choice chambers a) newly exposed unfilmed surface, poorly filmed (two weeks to acquire a microbial film) and filmed surfaces (three months to acquire a microbial film) b) microbial films on rock surfaces collected at differing tidal heights at Port St Mary, Isle of Man.

Data analyses

Counts of cyprid settlement from choice chambers showed strong heteroscedasticity which could not be removed by transformation and so data were analysed using Friedman's test in Minitab (v. 10.1) on an IBM personal computer, followed by Tukey-Type multiple comparisons (Zar, 1984, p. 200)

Data from field experiments conformed to Cochran's test for homogeneity of variance (Winer et al., 1971) after square root transformation and were analysed using the GLM command in Minitab. 'Distance' and 'surface type' were both treated as fixed effect factors in the analyses. Plots of residuals were examined after each ANOVA to check that error terms were normally distributed. Comparisons between levels within treatment factors were selected a priori and determined using Bonferroni comparisons (Maxwell and Delaney, 1990, p. 190).

Cyprid settlement data were analysed from days when settlement was maximal, as choices made on these occasions would have a large impact on overall recruitment to the adult population and were also most likely to reveal differences between treatments. Analyses of counts for metamorphosed spat were made towards the end of the settlement season. In essence, abundance at this time represented the effect of the treatments on overall settlement for the season. As treatment plots were not independent with time settlement was not compared between sampling dates.

Results

Choice chambers

Cyprids settled on rock surfaces in all chambers with significant differences in abundance between treatments in each experiment (Table 1). Cyprids selected surfaces with a developed biofilm in preference to surfaces with a poorly developed film or an unfilmed surface (Figure 3a, Table 1) and settled preferentially on rocks from the mid-shore (Figure 3b, Table 1). None of the rocks used were colonised by barnacles prior to exposure in the choice chambers. However, rocks from the mid-shore were more likely to have previously been colonised than those from the upper or lower shore. Therefore it remained uncertain whether cyprids were responding to traces of adult barnacles on the rock surfaces or to some aspect of the microbial film which was typical of the mid-shore.

Subsequent work showed that settlement was significantly greater on surfaces which had previously been colonised by barnacles than those which had not (Figure 4a, and Table 1). Settlement was not enhanced by addition of an adult conspecific to either filmed but not previously colonised, or unfilmed surfaces. However, as before settlement was significantly greater on filmed surfaces than on unfilmed surfaces (Figure 4b, Table 1).

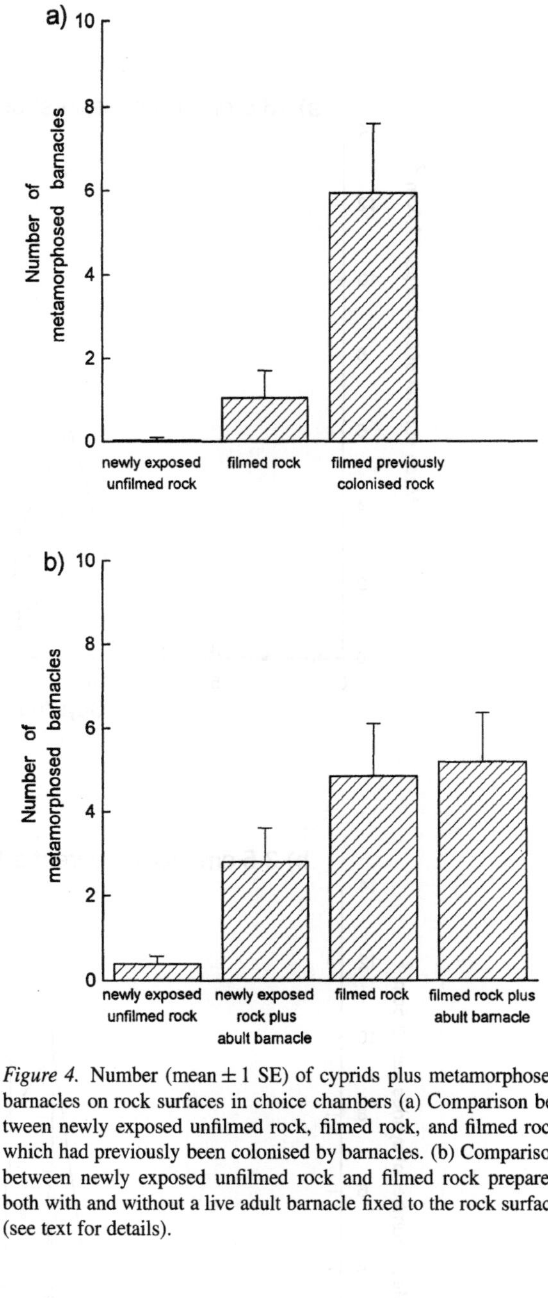

Figure 4. Number (mean ± 1 SE) of cyprids plus metamorphosed barnacles on rock surfaces in choice chambers (a) Comparison between newly exposed unfilmed rock, filmed rock, and filmed rock which had previously been colonised by barnacles. (b) Comparison between newly exposed unfilmed rock and filmed rock prepared both with and without a live adult barnacle fixed to the rock surface (see text for details).

Field experiments

Settlement of cyprids varied between days (Figure 5). However, the same preference for surface types was maintained throughout the settlement period. Areas close to the adult barnacle population received greater settlement than those which were further away. Settlement was greatest on filmed surfaces which had previously been colonised by barnacles and was least on unfilmed surfaces.

Analysis of variance revealed significant differences in the abundance of cyprids between treatments on both day 11, 6/5/95 and day 13, 8/5/95 (Table 2). On both occasions settlement was greater in treatments closer to the adult population, but this difference was only significant on day 13. Variation in settlement between film types was most evident on day 13 with significant differences between each of the three treatments. Settlement was greatest on previously colonised surfaces and least on unfilmed ones. These two film types were also significantly different on day 11. Clearly, settling cyprids could differentiate between the three surfaces offered, but the relative contribution of either 'films' or 'distance from adults' remained unclear because of high settlement variability.

Substrata preferences observed for cyprids were also reflected in the abundance of metamorphosed

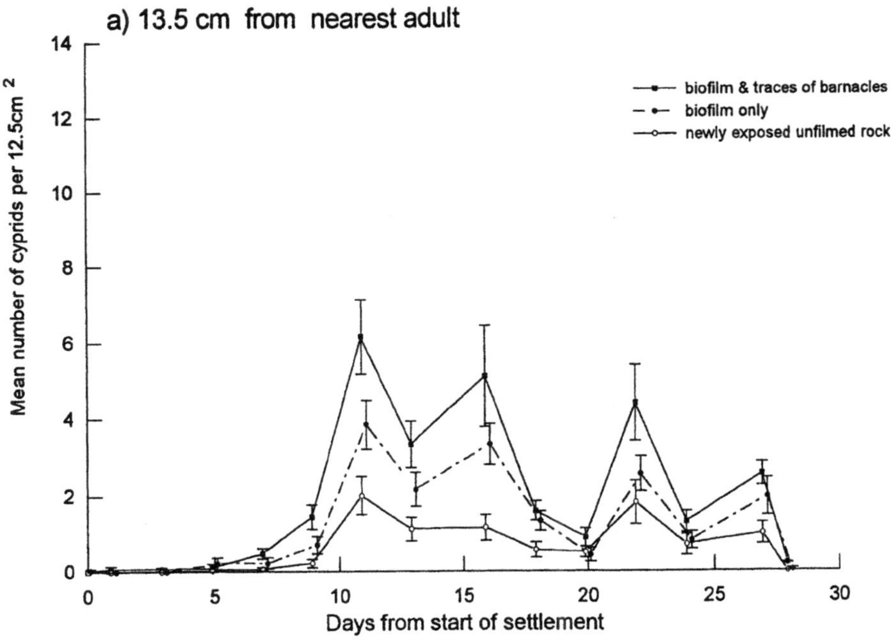

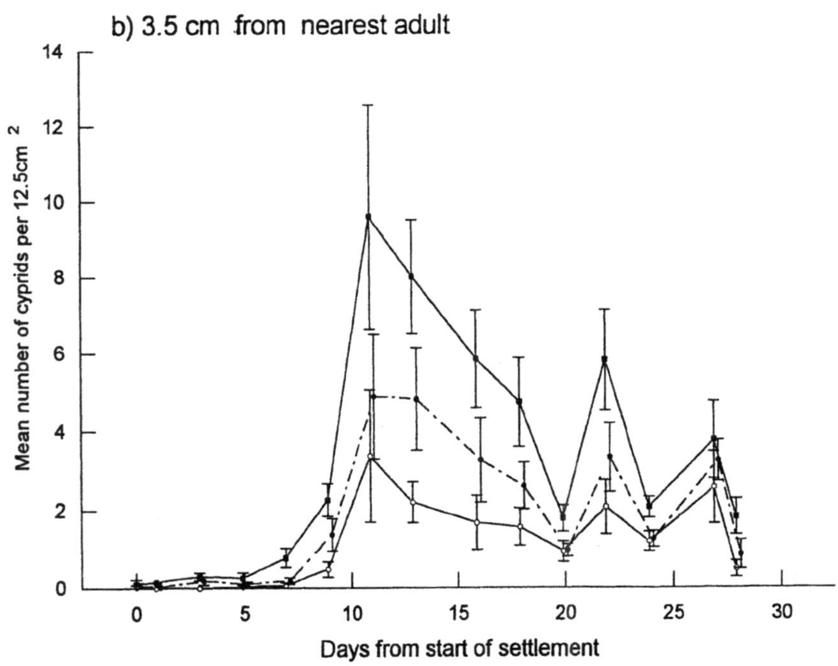

Figure 5. Number (mean ± 1 SE) of barnacle cyprids in treatment areas of differing microbial films, at different distances from the surrounding barnacle population at mid-tide level, Port St Mary, Isle of Man during April and May 1995. Data points have been offset slightly for clarity.

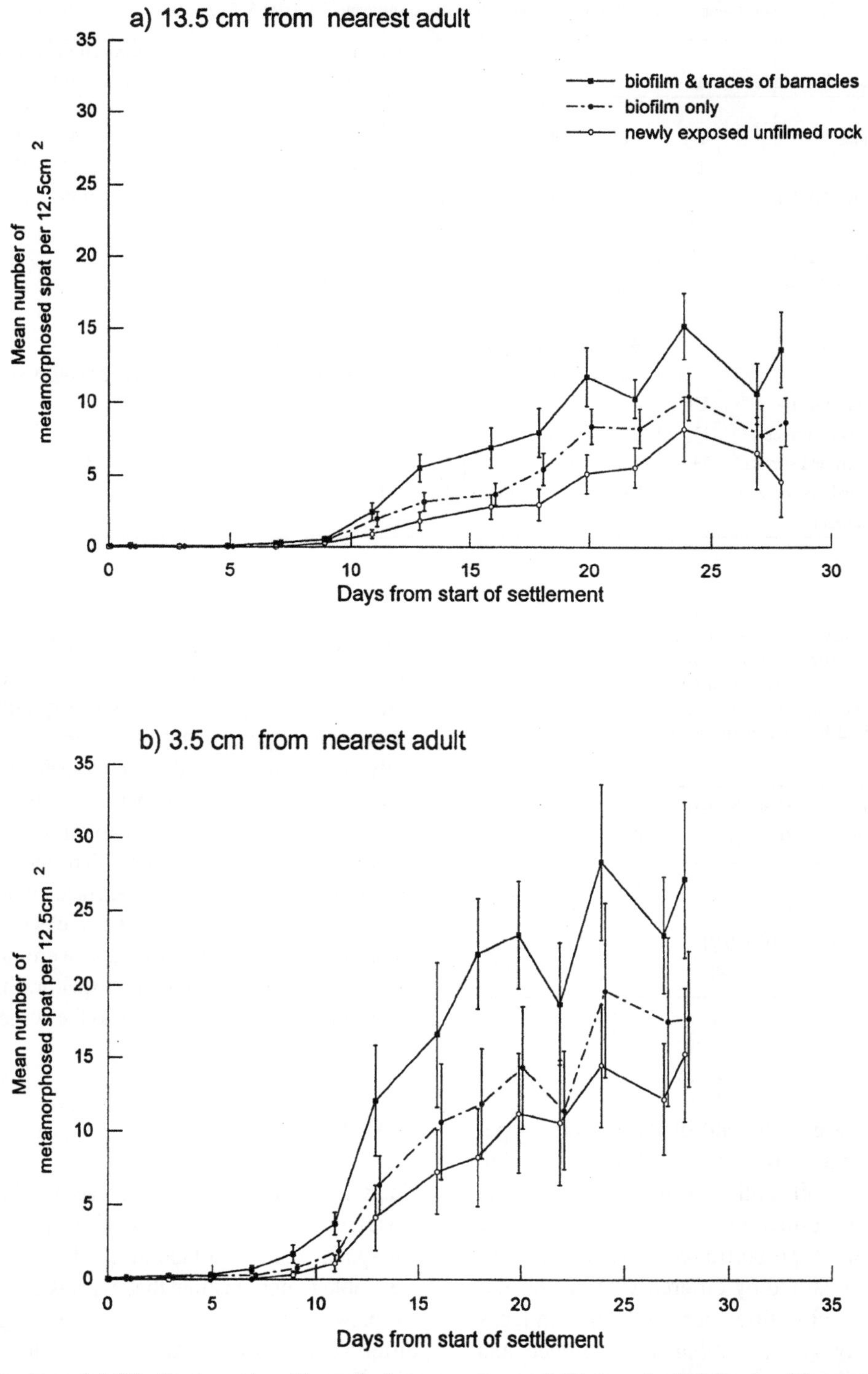

Figure 6. Number (mean ± 1 SE) of metamorphosed barnacles in treatment areas of differing microbial films, at different distances from the surrounding barnacle population at mid-tide level, Port St Mary, Isle of Man during April and May 1995. Data points have been offset slightly for clarity.

Table 2. Analysis of variance for factors influencing settlement of *Semibalanus balanoides* cyprids at the peak of settlement and the abundance of metamorphosed spat present towards the end of the settlement season on the mid-shore, Port St Mary, Isle of Man. Data were square root transformed. 'Distance' (from adult barnacles) and 'film' (type of microbial film present) were treated as a fixed factors, ns = not significant, * $p < 0.05$, ** $p < 0.01$, *** $p < 0.001$.

		Cyprid settlement						**Metamorphosed spat**		
Source of variation	df	Day 11 (6/5/95)			Day 13 (8/5/95)			Day 22 (19/5/95)		
		MS	*F*	*p*	MS	*F*	*p*	MS	*F*	*p*
Distance	1	1.06	1.20	ns	6.78	16.68	***	15.99	9.50	**
Film	2	7.02	7.96	***	5.44	13.37	***	8.76	5.21	**
Distance * Film	2	0.09	0.10	ns	0.31	0.77	ns	0.32	0.19	ns
Error	48	0.88			0.41			1.68		
Cochran's test:										
C_{crit}, P0.05 = 0.38		C = 0.32			C = 0.30			C = 0.35		
		Bonferroni analysis of differences between film types								
		Day 11 (6/5/95)			Day 13 (8/55/95)			Day 22 (19/5/95)		
UF = unfilmed	UF	–			–			–		
F = filmed surface	F	ns	–		*	–		ns	–	
FPC = filmed surface	FPC	*	ns	–	*	*	–	*	ns	–
Previously colonized by barnacles		UF	F	FPC	UF	F	FPC	UF	F	FPC

Table 3. Abundance of microbiota on rock surfaces (mean ± 1 SE) prepared in differing ways UF = unfilmed, F = filmed and not previously colonised, PF = poorly developed film and not previously colonised surface, FPC = filmed and previously colonised by adult barnacles (see Table 1 for details of preparation).

	Diatoms no. per mm^2	Cyanobacteria % cover	Crustose coralline algae % cover
UF	0	0	0
PF	2.84 ± 2.07	0	0
F	2.84 ± 1.33	2.03 ± 0.71	7.5 ± 5.14
FPC	0.71 ± 0.71	1.0 ± 0.49	4.0 ± 2.01

spat recorded towards the end of the settlement period (19/5/95, Figure 6). Surfaces close to the adult population had significantly greater settlement than those which were further away (Table 2). This factor accounted for greater proportion of the variation in settlement than that caused by differences between film types. There were also differences between film types with significantly greater settlement on filmed and previously colonised surfaces compared to unfilmed surfaces. Filmed surfaces which had not previously been colonised had intermediate levels of settlement.

Microbiota were evident on many of the settlement surfaces when viewed using a scanning electron microscope (Table 3). The data were not compared statistically owing to the small sample size and considerable heteroscedasticity. However, some qualitative differences were apparent: unfilmed surfaces, not surprisingly, were uncolonised, filmed surfaces and filmed surfaces with remnants of barnacles, which had been exposed to seawater for several months, had a more diverse fouling community, with diatoms, cyanobacteria and crustose coralline algae. Surfaces which had been exposed to seawater for just two weeks (used in one of the choice chamber experiments) were not colonised by either cyanobacteria or crustose coralline algae but had a similar cover of diatoms to the filmed surfaces.

Discussion

Habitat selection by cyprid larvae was influenced by a variety of cues which probably operate at different spatial scales. In laboratory choice chambers we were able to demonstrate that cyprids could differentiate between rock surfaces which had originated from differing tidal levels. At a scale of less than 10cm (the diameter of the choice chambers) settlement was enhanced by a mature microbial film, whilst the presence of an adult barnacle had little effect. However, experiments in the field showed that settlement was considerably enhanced by conspecific adults which were 3 cm or more away, and by traces of previous barnacle colonisation at a scale of less than 3 cm.

In laboratory choice chambers cyprids distinguished between rock surfaces from differing levels on the shore, settling preferentially on those from the mid-shore. Similar results have been obtained in field conditions by Strathman et al. (1981) who transplanted panels which had been fouled on the lower shore, to the upper shore, and showed that *Balanus glandula* could be induced to settle above its normal range. They concluded that cyprids were able to differentiate between some feature of the micro-flora which varied between shore levels. Raimondi (1988b) demonstrated that cyprids were attracted to surfaces coated with extracts from other shore dwellers typical of the tidal height where the adult barnacle population was most abundant. The cause of the enhanced settlement on mid-shore rocks in the present study is unclear. It may have resulted from a greater abundance of barnacle remnants on rock surfaces at that tidal level, or possibly from chemical exudates released by neighbouring barnacles which had become absorbed into the microbial film on the mid-shore (e.g. Crisp & Meadows, 1963). Alternatively some characteristic of the mid-shore microbial community may have provided a stimuli which enhanced settlement.

Several researchers have suggested that barnacle cyprids might use cues from microbial films to locate the most appropriate microhabitat or shore level at which to settle (e.g. Le Tourneux & Bourget, 1988; Wieczorek et al., 1995). The abundance of both diatoms and cyanobacteria in the intertidal zone does vary with tidal height (Underwood, 1984b; Dye & White, 1991) and similar variations were evident at Port St Mary (RCT – unpublished data). However, the optimal tidal level for micro-organisms within these films varies according to the climatic conditions experienced whilst the tide is out. Some species being found higher on the shore during cool humid weather than during warmer periods (Aleem, 1950, RCT unpublished data). Consequently it seems more likely that macrobiota such as conspecifics or macroalgae which have a relatively stable vertical zonation pattern would offer a more reliable cue to settlement (e.g. Strathmann et al., 1981).

Laboratory experiments conducted during the present study confirm that filmed rock surfaces were selected in preference to poorly filmed and unfilmed surfaces. Rocks with mature films had a diverse microbiota with diatoms, cyanobacteria and crustose coralline algae, whilst poorly developed films were only colonised by diatoms. Possibly these differences in the microbial community provided a cue which enhanced settlement on surfaces with mature films. An excellent study by Wieczorek et al. (1995) also demonstrated that settlement was greater on older films. Whilst other researchers have reported the opposite trend (Maki et al., 1990; Neal & Yule, 1994; Keough & Raimondi, 1995). These observations at first appear contradictory however, a range of different film types (single species and multi-species) and several barnacle species (*Balanus amphitrite, Balanus perforatus, Elminius modestus, Balanus variegatus*) were used by the different researchers. Clearly, differing outcomes might be expected from such a diverse array of experiments, and we feel it may be inappropriate to generalise about the influence of microbial films on the settlement of marine invertebrates (see Keough & Raimondi, 1995, 1996).

In laboratory experiments where traces of previous barnacle colonisation were not present (10/5/95 and 20/5/95), settlement was greater on filmed surfaces than on unfilmed surfaces. However, when such traces were present, it was more difficult to detect significant differences between filmed and unfilmed surfaces, in either the laboratory or the field (3 out of 4 comparisons were not significant). Here, settlement was significantly greater on filmed surfaces which had previously been colonised by adult barnacles than on unfilmed surfaces. Filmed surfaces (which had not been colonised by barnacles) received levels of settlement which were intermediate, but not different, to that on filmed surfaces which had previously been colonised by barnacles and unfilmed surfaces. Microscopic examination did not reveal any major differences in the abundance of diatoms, cyanobacteria and crustose coralline algae between filmed surfaces and filmed surfaces which had previously been colonised by adult barnacles. Consequently, it appears that when traces of previous colonisation are present on a settlement surface these may offer a more important cue to settlement than that offered by the microbial film alone.

A note of caution must be made here as there are considerable difficulties in resolving the microbiota within fouling communities. Scanning electron microscopy as used here only resolves organisms at the surface of the film (MacLulich, 1986; Hill & Hawkins, 1990). Additionally, there is considerable natural spatial variability in the distribution of both diatoms and cyanobacteria, and relative to their size (20 μm) comparatively large areas of substratum must be sampled in order to accurately describe the microbial community present. In the this study we examined 6 rock fragments of each film type giving a total area of 1.4 mm^2. Barnacle cyprids are able to sample numerous separate areas of substratum, to a maximum of around 638 paces of 50 μm diameter (Walker & Yule, 1984), prior to settlement. In addition cyprids have putative chemoreceptors in their carapace (Jensen et al. 1995) which may be used to sample the boundary layer immediately above a substratum. Consequently, the total area of rock surface which we viewed may not be representative of that sampled by a cyprid prior to settlement.

In the field, the presence of conspecific adults was an important factor promoting barnacle settlement, but in choice chambers addition of an adult barnacle to filmed surfaces did not enhance settlement. Numerous previous studies have shown that cyprids frequently settle shortly after encountering an adult barnacle (Knight Jones, 1953; Crisp, 1974; Crisp, 1984; Wethey, 1984; Bourget, 1988; Crisp, 1990). The results of the present study indicate that the proximity of conspecifics may be unimportant at a scale of a few centimetres (within the confines of a choice chamber) and that at this level of habitat selection the presence of a microbial film provides a greater stimuli. At an even finer scale of resolution, Le Tourneaux & Bourget (1988) suggests that bare uncolonised areas of substratum are selected for final settlement.

Our study raises an interesting issue as to the applicability of laboratory versus field-based experiments on larval settlement (see comments by Bourget, 1988; Pearce & Scheibling, 1991). For instance, in the laboratory we were able to demonstrate that cyprids selected filmed (not previously colonised surfaces) in preference to unfilmed surfaces. However, it was not possible to clearly demonstrate this effect in the field where the presence of conspecifics and traces of previous barnacle colonisation provided such a strong cue to settlement that the influence of the film itself was overruled. Consequently, whilst experimental conditions may be easier to control in the laboratory than in the field, the outcome of these studies only indicate possible rather than actual cues and do not necessarily provide answers which are relevant to settlement in the natural environment.

References

Aleem, A. A., 1950. Distribution and ecology of British marine littoral diatoms. J. Ecol. 38: 75–106.

Anderson, M. J. & A. J. Underwood 1994. Effects of substratum on the recruitment and development of an intertidal estuarine fouling assemblage. J. exp. mar. Biol. Ecol. 184: 217–236.

Becker, K., 1993. Attachment strength and colonisation patterns of two macrofouling species on substrata with different surface tension (*in situ* studies). Mar. Biol. 117: 301–309.

Bourget, E., 1988. Barnacle larval settlement: the perception of cues at different spatial scales. In G. Chelotti & M. Vennini (eds), Behavioural Adaptations to Intertidal Life. NATO ASI Series. Plenum Press: 155–173.

Brancato, M. S. & R. M. Woollacott, 1982. Effect of microbial films on settlement of bryozoan larvae (*Bugula simplex, B. stolonifera* and *B. turrita*). Mar. Biol. 71: 51–56.

Cameron, R. A. & R. T. Hinegardner, 1974. Initiation of metamorphosis in laboratory cultured sea urchins. Biol. Bull. Mar. Biol. Lab., Woods Hole, Mass. 146: 335–342.

Clare, A. S., R. K. Freet, & M. jr. McClary, 1994. On the antennular secretion of the cyprid of *Balanus amphitrite amphitrite* settlement pheromone. J. mar. biol. Ass. U.K. 74: 243–250.

Clare, A. S., 1995. Chemical signals in barnacles: Old problems, new approaches. In Schram, F. R. & Hoeg, J. T. (eds), New frontiers in barnacle evolution. Crustacean Issues 10. A. A., Balkema, Rotterdam: 49–67.

Crisp, D. J., 1955. The behaviour of barnacle cyprids in relation to water movements over a surface. J. exp. Biol. 32: 569–590.

Crisp, D. J., 1974. Factors influencing the settlement of marine invertebrate larvae. In P. T. Grant & A. M. Mackie (eds), Chemoreception in Marine Organisms. Academic Press, London: 177-265.

Crisp, D. J., 1984. An overview of research on marine invertebrate larvae, 1940–1980. In: J. D. Costlow & R. C. Tipper (eds), Marine Biodeterioration: an interdisclipinary study. Naval Institute Press, Annapolis, Maryland: 103–126.

Crisp, D. J., 1990. Field experiments on the settlement, orientation and habitat choice of *Chthamalus fragilis* (Darwin). Biofouling 2: 131–136.

Crisp, D. J. & P. S. Meadows, 1963. Adsorbed layers: the stimulus to settlement in barnacles. Proc. r. Soc. London, Series B, 158: 364–387.

Dillon, P. S., J. S. Maki & R. Mitchel, 1989. Adhesion of *Enteromorpha* swarmers to microbial films. Microb. Ecol. 17: 39–47.

Dye, A. H. & R. A. White, 1991. Intertidal microalgal production and molluscan herbivory in relation to season and elevation on two rocky shores on the east coast of Southern Africa. South Afr. J. mar. Sci. 11: 483–489.

Edgar, G. J. & C. Shaw, 1995. The production and trophic ecology of shallow-water fish assemblages in southern Australia: II. Diets of fishes and trophic relationships between fishes and benthos at Western Port, Victoria. J. exp. mar. Biol. Ecol. 194: 83–106.

Hawkins, S. J., D. C. Watson, A. S. Hill, S. P. Harding, M. A. Kyriakides, S. Hutchinson & T. A. Norton, 1989. A comparison of feeding mechanisms in microphagous, herbivorous, intertidal, prosobranchs in relation to resource partitioning. J. Mollusc. Stud. 55: 151–165.

Henschel, J. R., G. M. Branch & P. A. Cook, 1990. The colonisation of artificial substrata by marine sessile organisms in False Bay. 2. Substratal material. South Afr. J. mar. Sci. 9: 299–307.

Hill, A. S. & S. J. Hawkins, 1990. An investigation of methods for sampling microbial films on rocky shores. J. mar. biol. Ass. U.K. 70: 77–88.

Holland, D. C., D. J. Crisp, R. Huxley & J. Sisson, 1984. Influence of oil shale on intertidal organisms: Effect of oil shale extract on settlement of the barnacle *Balanus balanoides* (L.). J. exp. mar. Biol. Ecol. 75: 245–255.

Holmes, S. P., C. J. Sturgess & M. S. Davies, 1997. The effect of rock-type on the settlement of *Balanus balanoides* (L.) cyprids. Biofouling 11: 137–147.

Huang, R. & Boney, A. D., 1985. Individual and combined interactions between littoral diatoms and sporelings of red algae. J. exp. mar. Biol. Ecol. 85: 101–111.

Jensen, P. G., J. Moyse, J. T. Hoeg & H. Al-Yahya, 1994. Comparative SEM studies on lattice organs: putative sensory structures on the carapace of larvae from Ascothoracida and Cirripedia (Crustacea Maxillopoda Thecostraca). Acta zool. 75: 125–142.

Johnson, C. R., D. C. Sutton, R. R. Olson & R. Giddins, 1991. Settlement of crown-of-thorns starfish: Role of bacteria on surfaces of coralline algae and a hypothesis for deep water recruitment. Mar. Ecol. Progr. Ser. 71: 143–162.

Johnson, L. E. & R. R. Strathmann, 1989. Settling barnacle larvae avoid substrata previously occupied by a mobile predator. J. exp. mar. Biol. Ecol. 128: 87–103.

Keough, M. J. & P. T. Raimondi, 1995. Responses of settling invertebrate larvae to bioorganic films: Effects of different types of films. J. exp. mar. Biol. Ecol. 185: 235–253.

Keough, M. J. & P. T. Raimondi, 1996. Responses of settling invertebrate larvae to bioorganic films: Effects of large scale variation in films. J. exp. mar. Biol. Ecol. 207: 59–78.

Knight Jones, E. W., 1953. Laboratory experiments on gregariousness during settling in *Balanus balanoides* and other barnacles. J. exp. Biol. 30: 584–598.

Knight-Jones, E. W., 1951. Gregariousness and some other aspects of settling behaviour of *Spirobis*. J. mar. biol. Ass. U.K. 30: 201–222.

Le Tourneux, F. & E. Bourget, 1988. Importance of physical and biological settlement cues used at different spatial scales by the larvae of *Semibalanus balanoides*. Mar. Biol. 97: 57–66.

MacLulich, J. H., 1986. Experimental evaluation of methods for sampling and assaying intertidal epilithic microalgae. Mar. Ecol. Progr. Ser. 34: 275–280.

Maki, J. S., D. Rittschof & R. Mitchell, 1992. Inhibition of larval barnacle attachment to bacterial films: An investigation of physical properties. Microb. Ecol. 23: 97–106.

Maki, J. S., D. Rittschof, M. O. Samuelsson, U. Szewzyk, A. B. Yule, S. Kjelleberg, J. D. Costlow & R. Mitchell, 1990. Effect of marine bacteria and their exopolymers on the attachment of barnacle cypris larvae. Bull. mar. Sci. 46: 499–511.

Maxwell, S. E. & H. D. Delaney, 1990. Designing experiments and analysing data. Wadsworth Publishing Company. Belmont, California: 901.

Meadows, P. S. & G. B. Williams, 1963. Settlement of *Spirobis borealis* Daudin larvae on surfaces bearing micro-organisms. Nature 198: 610–611.

Mihm, J. W., Banta, W. C., & Loeb, G. I., 1981. Effects of adsorbed organic and primary fouling films on bryozoan settlement. J. exp. mar. Biol. Ecol. 54: 167–179.

Neal, A. L. & A. B. Yule, 1994. The tenacity of *Elminius modestus* and *Balanus perforatus* cyprids to bacterial films grown under different shear regimes. J. mar. biol. Ass. U.K. 74: 251–257.

Norton, T. A., 1992. Dispersal by macroalgae. Brit. Phycol. J. 27: 293–301.

Patterson, D. M., R. M. Crawford & C. Little, 1986. The structure of benthic diatom assemblages: a preliminary account of the use and evaluation of low-temperature scanning electron microscopy. J. exp. mar. Biol. Ecol. 96: 279–289.

Pawlik, J. R., 1992. Chemical ecology of the settlement of benthic marine invertebrates. Oceanogr. Mar. Biol. annu. Rev. 30: 273–335.

Pawlik, J. R. & A. C. Butman, A. C., 1993. Settlement of a marine tube worm as a function of current velocity: Interacting effects of hydrodynamics and behaviour. Limnol. Oceanogr. 38: 1730–1740.

Pearce, C. M. & R. E. Scheibling, 1991. Effect of macroalgae, microbial films, and conspecifics on the induction of metamorphosis of the green sea urchin *Strongylocentrotus droebachiensis* (Mueller). J. exp. mar. Biol. Ecol. 147: 147–162.

Raimondi, P. T., 1988a. Rock type affects settlement, recruitment, and zonation of the barnacle *Chthamalus anisopoma* Pilsbury. J. exp. mar. Biol. Ecol. 123: 253–267.

Raimondi, P. T., 1988b. Settlement cues and determination of the vertical limit of an intertidal barnacle. Ecology 69: 400–407.

Raimondi, P. T., 1991. Settlement behaviour of *Chthalamus anisopoma* larvae largely determines the adult distribution. Oecologia 85: 349–360.

Rittschof, D., A. S. Clare, D. J. Gerhart, M. Sister Avelin & J. Bonaventura, 1992. Barnacle *in vitro* assays for biologically active substances: Toxicity and settlement inhibition assays using mass cultured *Balanus amphitrite amphitrite* Darwin. In L. Evans (ed.), Aspects of Current Research in the US Navy Biofouling Program: 115–122.

Rodriguez, S. R., F. P. Ojeda & N. C. Inestrosa, 1993. Settlement of benthic marine invertebrates. Mar. Ecol. Progr. Ser. 97: 193–207.

Santelices, B., 1990. Patterns of reproduction, dispersal and recruitment in seaweeds. Oceanogr. Mar. Biol. annu. Rev. 28: 177–276.

Scheltema, R. S., 1961. Metamorphosis of the veliger larvae of *Nassarius obsoletus* (Gastropoda) in response to bottom sediment. Biol. Bull. Mar. Biol. Lab. Woods Hole, Mass., 120: 92–109.

Southward, A. J., 1953. The ecology of some rocky shores in the south of the Isle of Man. Proc. Transact. Liverpool Biol. Soc. 59: 1–50.

Strathmann, R. R., E. S. Branscomb & K. Vedder, 1981. Fatal errors in set as a cost of dispersal and the influence of intertidal flora on set of barnacles. Oecologia 48: 13–18.

Todd, C. D. & M. J. Keough, 1994. Larval settlement in hard substratum epifaunal assemblages: A manipulative field study of the effects of substratum filming and the presence of incumbents. J. exp. mar. Biol. Ecol. 181: 159–187.

Tritar, S., D. Prieur & R. Weiner, 1992. Effects of bacterial films on the settlement of the oysters, *Crassostera gigas* (Thunberg, 1793) and *Ostrea edulis*, Linnaeus, 1750 and the Scallop *Pecten maximus* (Linnaeus, 1758). J. Shellfish Res. 11: 235–330.

Underwood, A. J., 1984a. Microalgal food and the growth of the intertidal gastropods *Nerita atramentosa* Reeve and *Bembicium nanum* (Lamarck) at four heights on a shore. J. exp. mar. Biol. Ecol. 79: 277–291.

Underwood, A. J., 1984b. The vertical distribution and seasonal abundance of intertidal microalgae on a rocky shore in New South Wales. J. exp. mar. Biol. Ecol. 78: 199–220.

Wahl, M., 1989. Marine epibiosis. I. Fouling and antifouling: Some basic aspects. Mar. Ecol. Progr. Ser. 58: 1–2.

Walker, G. & A. B. Yule, 1984. Temporary adhesion of the barnacle cyprid: The existence of an antennular adhesive secretion. J. mar. Biol. Ass. U.K. 64: 679–686.

Walker, G., 1995. Larval settlement: Historical and future perspectives. In Schram, F. R. & J. T. Hoeg (eds), New Frontiers in Barnacle Evolution. Crustacean Issues 10. A. A., Balkema, Rotterdam: 69–85.

Wethey, D. S., 1984. Spatial pattern in barnacle settlement: Day to day changes during the settlement season. J. mar. Biol. Ass. U.K., 64: 687–698.

Wieczorek, S. K., A. S. Clare & C. D. Todd, 1995. Inhibitory and facilitatory effects of microbial films on settlement of *Balanus amphitrite amphitrite* larvae. Mar. Ecol. Progr. Ser. 119: 1–3.

Wilson, D. P., 1955. The role of micro-organisms in the settlement of *Ophelia bicornis* Savigny. J. mar. Biol. Ass. U.K., 34: 531–543.

Winer, B. J., D. R. Brown & K. M. Michels, 1971. Statistical Principles in Experimental Design. McGraw-Hill, New York.

Yule, A. B. & D. J. Crisp, 1983. Adhesion of cypris larvae of the barnacle, *Balanus balanoides*, to clean and arthropodin-treated surfaces. J. mar. Biol. Ass. U.K. 63: 261–271.

Yule, A. B. & G. Walker, 1984. The temporary adhesion of barnacle cyprids: Effects of some differing surface characteristics. J. mar. Biol. Ass. U.K. 64: 429–439.

Zar, J. H., 1984. Biostatistical Analysis. Prentice-Hall Inc., New Jersey, USA. 718 pp.

Hydrobiologia **375/376:** 217–226, 1998.
S. Baden, L. Pihl, R. Rosenberg, J.-O. Strömberg, I. Svane & P. Tiselius (eds),
Recruitment, Colonization and Physical–Chemical Forcing in Marine Biological Systems.

Scale-dependent benthic recolonization dynamics: life stage-based dispersal and demographic consequences

R. B. Whitlatch[1,*], A. M. Lohrer[1], S. F. Thrush[2], R. D. Pridmore[2], J. E. Hewitt[2], V. J. Cummings[2] & R. N. Zajac[3]
[1] *Department of Marine Sciences University of Connecticut Groton, CT 06340, USA*
[2] *National Institute of Water and Atmospheric Research Box 11-115 Hamilton, New Zealand*
[3] *Department of Biology and Environmental Science The University of New Haven West Haven, CT 06516, USA*
* *Corresponding author: email: whitlatc@uconnvm.uconn.edu; Fax: +860-405-9153; Voice: +860-405-9154*

Abstract

Soft-sediment recruitment dynamics are dependent upon two sources of colonists; larvae transported from the water column and post-settlement movement of juvenile and/or adult life-stages across the seabed. Differences in the relative dispersal ability of the different life-stages into disturbed patches of habitat should vary predictably with the spatial scale of the disturbance. Smaller patches with a greater edge:surface area ratio should be more influenced by the post-settlement colonist pool than larger patches possessing a smaller edge:surface area ratio. A life stage-based recolonization model, using a *Polydora cornuta* life table, has been developed to describe how differences in the immigration rates of larvae, juveniles and adults can influence within-patch recovery times. Model results indicate that immigration of adult stages into disturbed patches has the least influence on patch recovery time. In contrast, post-settlement juvenile stages generally has a pronounced effect on patch population dynamics. Experimental evidence of scale-dependent migration of different life-stages to disturbed patches is also presented for the spionid polychaete *Boccardia syrtis*. Future research on scale-dependent recolonization dynamics in soft-sediment habitats should focus on acquisition of within-patch demographic data in order to more fully understand the importance of post-settlement life-stage movement in regulating population and community dynamics.

Introduction

Studies on macroinfaunal recolonization dynamics of soft-sediment habitats have a relatively long and well developed history in the field of marine benthic ecology (e.g. reviews of Pearson & Rosenberg, 1978; Thistle, 1981; Hall et al., 1994). Considerable attention has been directed to examination of the recovery process following natural and human-induced disturbances and a variety of conceptual models have been developed which assess the role disturbance plays on the population and community dynamics of benthic habitats. Most infaunal succession models place primary focus on the broad-scale nature of the physical environment and how it sets the boundary conditions for the recovery process (Zajac & Whitlatch, 1985; Zajac et al., this issue). These models primarily view recolonization processes and successional dynamics based on species' adaptations to the abiotic environment, and examine how subsequent changes in physical conditions influence the nature of the recovery process (Pearson & Rosenberg, 1978; Rhoads & Boyer, 1982). While the primary focus of these models has been placed on describing the nature of the recovery process, few (e.g. Johnson, 1970, 1971; Van Blaricom, 1982; Zajac & Whitlatch, 1991; Gunther, 1992) have explicitly examined the interactive roles of recolonization processes and patch dynamics and how these factors may vary with spatial scale of the disturbance (see Thrush et al., 1996, for a review of the marine soft-sediment habitat literature).

It is now widely recognized that ecological patterns and processes operating at one spatial scale may not necessarily be similar to those found at another spatial scale (e.g. Dayton and Oliver 1980; Levin & Huggett, 1990; Gascon & Travis, 1992). Here we present the

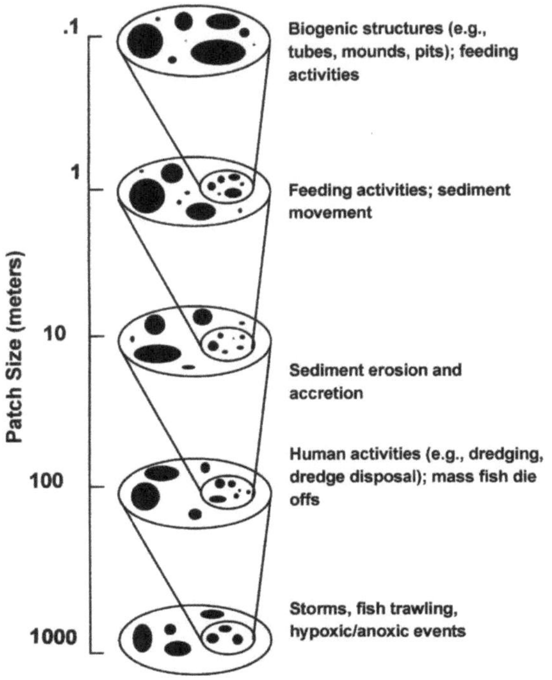

Figure 1. A hierarchial patch structure typical of soft-sediment habitats (modified from Kotliar & Wiens, 1990) indicating the relative spatial extent over which various natural and human-induced processes are interacting in soft-sediment marine habitats.

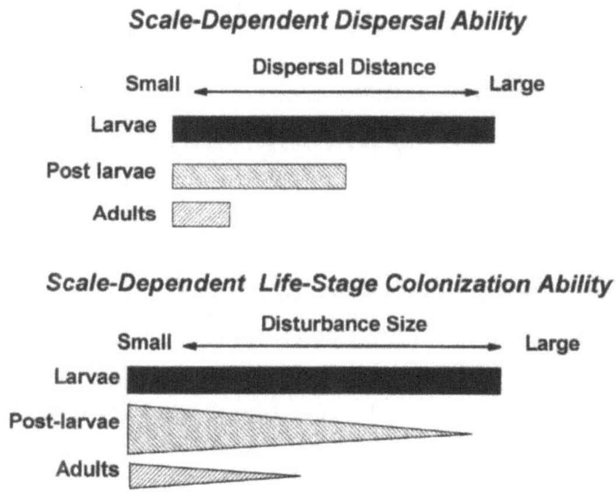

Figure 2. A conceptual model of scale-dependent life-stage dispersal of benthic macroinfaunal species and its effect on the recolonization dynamics of disturbed patches of sediment (modified from Gunther, 1992). Because of differences in dispersal ability, the relative abundance of different benthic invertebrate life-stages are predicted to vary with the spatial scale of disturbance.

general framework of a patch-based, soft-sediment recolonization model and demonstrate how it can be used to explore potential population-level patch dynamics relative to the spatial scale of disturbance and demographic consequences relative to patch recovery. We also present results of a field experiment explicitly designed to study the effects of the scale of disturbance on soft-sediment macroinfaunal recolonization dynamics in an intertidal sandflat habitat.

Background and current perspectives

Natural and human-induced disturbances are recognized as important structuring forces in affecting soft-sediment population and community dynamics. These disturbances frequently vary in spatial extent (Figure 1). Small-scale disturbances at spatial scales of less than 1 m^2 are typically generated by the foraging movements of benthic predators (e.g. fish and crustacean pits) or through disruption of the sediment fabric by the feeding activities of infauna (e.g. biogenic mounds and depressions). Meso-scale disturbances (meters to decimeters) are usually caused by the disruption of the sediment by natural physical processes (e.g. sediment erosion and accretion) or through the actions of humans (e.g. dredging, trawling). A limited number of biotic agents (e.g. marine mammal foraging, mass mortalities of fish) can also disturb sediments at this scale. Large-scale disturbances (decimeters to kilometers) are generally caused by physical processes (e.g. storms, anoxic events) or large-scale human-related activities (e.g. petrochemical spills, fish trawling).

In soft-sediment habitats, populations and communities are patchily distributed over a range of spatial scales (Thrush, 1991) and each patch often represents a different stage of recovery from a given disturbance (Johnson, 1973; Hall et al., 1994). The prevailing view of soft-sediment successional dynamics is that some species (‘opportunistic’ or ‘indicator’ species) respond quickly to recently disturbed habitat patches, while other species are more constrained in their population responses and recolonize disturbed patches at much slower rates (e.g. Grassle & Grassle, 1974; McCall, 1977). It is important, however, to recognize that disturbances often generate variable conditions which can lead to multifaceted species’ population responses (Zajac & Whitlatch, 1991). Opportunistic species responses may not always occur following a disturbance (e.g. Zajac & Whitlatch, 1982; Bonsdorff, 1990; Thrush et al., 1996) and the mode of response to disturbed habitat patches is frequently dependent upon habitat conditions, species’ life histories and species mobility patterns (e.g. Commito et al., 1995; Shull, 1997).

Soft-sediment recolonization modes and recovery dynamics have been hypothesized to be related to disturbance size (Smith & Brumsickle, 1989; Gunther, 1992; Hall et al., 1994). The dispersal ability of many types of benthic invertebrates can vary with life-stage (Figure 2). While it is recognized that the planktonic larval stages of many benthic invertebrates have the potential of wide dispersal, there is an increasing awareness regarding the importance of smaller-scale, post-settlement dispersal of juvenile and adult life-stages contributing to recolonization of disturbed soft-sediment habitats (Gunther, 1992, for review). One of the important consequences of life-stage differences in dispersal ability is the prediction that small-scale disturbances should contain a wider representation of different life-stages of a given species than those inhabiting large-scale disturbances (Figure 2). Differences in dispersal ability should result in disturbance size-dependent recovery times, variations in the population structure of organisms residing in different patch sizes, and subsequent differences in within-patch population parameters (e.g. population growth rate). While there are an increasing number of studies examining modes of dispersal and colonization of infaunal organisms following a disturbance (e.g. Emerson & Grant, 1992; Commito et al., 1995; Shull, 1997), relatively few have explicitly examined the presence of scale-dependent differences in life-stage recolonization to disturbed habitat patches (e.g. Smith & Brumsickle, 1989; Thrush et al., 1996). Results do indicate the presence of species- and habitat-specific differences in dispersal ability and mode of recolonization (Thrush et al., 1996; Shull, 1997).

Studies on soft-sediment patch dynamics have primarily focused on recolonization responses of benthic organisms in relation to relatively small-scale disturbances. These studies reveal that the specific characteristics of a disturbance can play an important role in affecting the dynamics of the population/community recolonization process. Different functional groups of organisms often respond differently to the specific type of disturbed patch, location of the patch, and the time the patch was disturbed (e.g. Zajac & Whitlatch, 1982; Savidge & Taghon, 1998; Snelgrove, 1994). Few have actually examined recolonization responses at the demographic level or whether scale-dependent recruitment actually leads to differences in between-patch population dynamics. Interpreting the effects of disturbance from population abundance data alone can

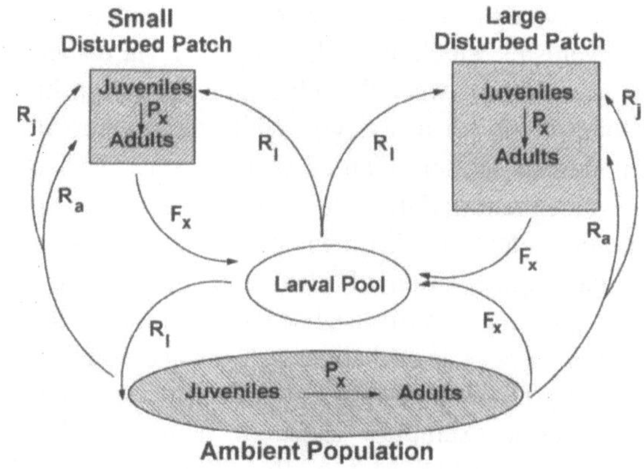

Figure 3. A life stage-based recolonization metapopulation model. Populations of organisms inhabiting undisturbed (ambient) and disturbed patches of sediment consist of two life stages (juvenile and adults). Larvae produced by adults inhabiting ambient and recovery sub-populations (F_x) are mixed together in a common larval pool. Colonization into disturbed patches is from two sources; a larval pool (R_l) and the migration of juvenile (R_j) and adult (R_a) life-stages from the ambient population. Within each sub-population, P_x = probability of survival from one life-stage to the next.

be both misleading and inappropriate. For example, Zajac & Whitlatch's (1989) study of the responses of the polychaete *Nephtys incisa* to a meso-scale anthropogenic disturbance (dredge material disposal) found that while there were no significant differences in the abundance of the polychaete between a reference (ambient) population and ones inhabiting the center and periphery of the disturbed area, demographic characteristics (e.g. population growth rate) indicated the latter populations had been negatively affected by the disturbance. The only modeling attempt examining demographic aspects of soft-sediment patch dynamics concluded that larval colonization was of primary importance during the initial stages of the successional process (Zajac & Whitlatch, 1991). The authors noted post-settlement lateral movement of juvenile and adult stages may potentially be critical in affecting rapid within-patch population increases; however, artifacts associated with sample collection and processing often precluded an adequate estimation of post-settlement stage recruitment.

Benthic invertebrate life-stage recolonization dynamics: a metapopulation approach

A metapopulation model has been constructed to illustrate how the size of a disturbed patch of habitat can result in among-patch variability in population dynamics. The metapopulation approach permits description of the dynamics of spatially fragmented populations of organisms which are linked together via dispersal of individuals between the sub-populations (Hanski & Gilpin, 1991, for a review). While the metapopulation concept is more than 25 y old (Levins, 1970), it has gained renewed interest in the ecological and conservation literature. With increasing human-induced habitat fragmentation of plant and animal populations, greater emphasis has been placed on describing the dynamics of local habitat patches and conditions necessary to maintain regional population persistence (e.g. marine examples: Zajac & Whitlatch, 1991; Quinn et al., 1993; Man et al., 1995; Whitlatch & Osman, 1997). For example, permanent or transient physical or biological differences among patches often results in local variation in mortality and reproductive rates of organisms. As a result, some patches may act as 'source' sub-populations whose reproduction maintains less favorably situated 'sink' populations (Pulliam, 1988; Pulliam & Danielson, 1991). Because of the patchiness of benthic species assemblages, populations should best be viewed as aggregations of interconnected sub-populations inhabiting a particular region. It is important, therefore, to understand the balance between local control of the dynamics of sub-populations and any broader-scale control resulting from the connectivity of all the sub-populations within a given region.

Figure 3 depicts a graphical representation of a metapopulation demographic model which is designed to assess the relative importance of patch-size on life-stage dependent modes of infaunal patch recovery. The model permits assessment of population characteristics within a given patch and allows comparisons between patches. The model is composed of spatially distinct sub-populations of organisms residing in different patch sizes. The patches are embedded within an ambient (undisturbed) population of organisms. Recolonization of each patch is dependent upon two colonist sources from the ambient population: (a) a supply of individuals recruiting from a common larval pool, and (b) a supply of juvenile and adult life-stages migrating from the ambient population to the sub-populations via lateral transport across the seabed. Populations of organisms inhabiting the different patches and ambient sediment are broadly classified as pre-reproductive (juveniles) and reproductive (adult) life-stages. The nodes in the model are connected by arrows depicting the probabilities of colonization (R_x) from the different life stages, and within sub-population survivorship (P_x) into the next mode. Within-patch and ambient population larval production both contribute to the common larval pool. Recolonization of post-settlement life stages to the disturbed patches is only from individuals residing in the ambient population by assuming the patches are very small relative to the ambient population and that they are sufficiently separated to minimize between-patch colonization.

The transition from our conceptual model to a mathematical one was facilitated by the use of STELLA II© software. Modelling in STELLA is a graphical programming process. First, variables that influence the system's behavior are classified and arranged in a graphical way using icons. Then, the mathematical relationships between variables are explicitly stated. The basic design of our life-stage model is similar to the three-stage insect model developed by Hannon and Ruth (1997).

Once constructed, the model was parameterized using life stage-specific survivorship and fecundity data for *Polydora cornuta* given in Zajac (1991). Average values from only the spring and spring cohorts were used. While the *Polydora* population normally consisted of six to eight separate life-stages (larvae and two juvenile and three to five adult stages; see Zajac, 1991), for simplicity the life-history matrix was collapsed into three categories: larvae, juveniles and adults. Larvae remained in the water column for two weeks, settled and developed into juveniles over a four week period. Six weeks after release of larvae, individuals matured into adults and began reproducing. Adults had the potential of remaining in the population for approximately eight additional weeks until their eventual death.

Intrinsic rates of population increase were calculated and were faithfully reproduced using matrix algebra in MATLAB© software. Our model included a discrete logistic equation (and thus a carrying capacity) for each life-stage. The life-stages eventually reached stable densities at which they would remain unless perturbed. Two 'patches' were modelled; an ambient patch and a defaunated patch. The defaunated patch was embedded within the much larger ambient patch and all parameters were the same in the two patches (with the exception of area and initial worm density). The ambient patch was allowed to stabilize before any immigration was allowed into the defaunated patch. Once inhabited, the previously defaunated patch did contribute to the larval pool. An initial rate of immigration for each of the three life-stages was set prior to every simulation. Rates of immigration declined asymptotically from initial rates as densities in the defaunated patch approached those of the ambient population. A truly defaunated patch has a net gain of immigrants but as the patch fills up, immigration and emigration begin to balance out.

The model permits examination of how variability in the two colonist sources can influence within-patch population dynamics and how these may change with the size of the disturbed patch. For instance, assuming similar conditions in each of the different sizes of disturbed habitat: (1) the density of colonists vertically advected from the larval pool should be independent of the size of a disturbed habitat, while (2) colonization into disturbed patches through lateral advection of post-settlement life-stages should be dependent upon surface:edge ratios of a given patch. Larger patches with lower edge-to-surface area ratios should have proportionately lower densities of recruiting post-larval life-stages than smaller patches possessing higher edge-to-surface area ratios. We predict, therefore, that within-patch population size-structure and dynamics should be dependent upon patch size. Smaller patches should recover to ambient population characteristics (e.g. stable stage-distribution, population density) faster than sub-populations inhabiting larger patches. Within-patch population dynamics should also vary between different patch sizes. Patch-size differences in population size structure should contribute to between-patch variability in subpopulation growth rates.

We simulated the effects of patch size by altering immigration rates of different life-stages and examining the time needed for the patch to recover to ambient conditions. Recovery time was defined when the patch reached a stable stage-distribution similar to the ambient population. We feel this currency of patch recovery provides a more realistic description of within-patch population dynamics as opposed to comparing population densities between populations inhabiting ambient sediments and disturbed patches (e.g. Zajac & Whitlatch, 1989). While data are currently unavailable on scale-dependent differences in life-stage recruitment for *Polydora cornuta*, Shull (1997) reported that for small-scale disturbances (experimentally defaunated sediment: 100.2 cm^2 surface area), recruitment rates of post-settlement life-stages (juveniles and adults) were about four times greater than larvae in the first 10 d of a 20 d experimental deployment. Since differences in colonization rates by different polychaete life-stages will vary with season and habitat, we elected to

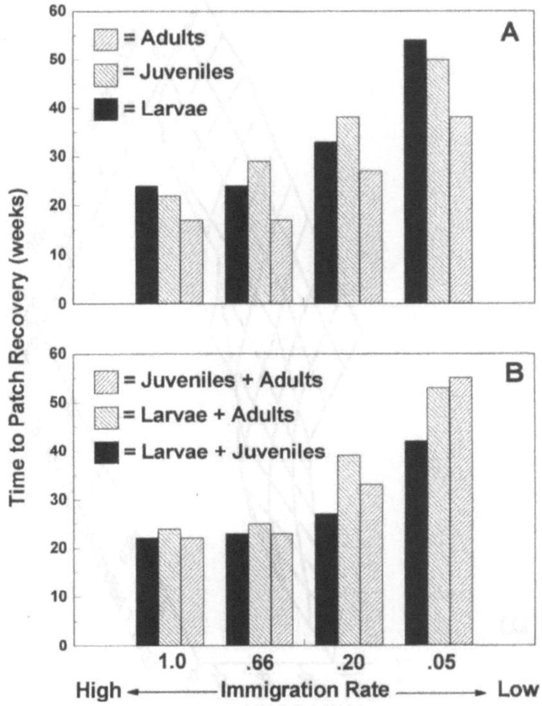

Figure 4. A = life stage-based recolonization model simulations showing the relationship of single *Polydora cornuta* life-stage recolonization rate on the time to recovery of populations in disturbed patches of habitat. B = Model simulations showing the effect of two-stage recolonization rates on the recovery time of disturbed patches. Recovery time = time (wks) for the patch to mimic the ambient population stable-stage distribution (see text for further details).

model dispersal over a range of values varying from 0.05 to 1.0.

Model results indicate that when only one life-stage was allowed to immigrate into the disturbed patches, adults contributed to the fastest patch recovery times (Figure 4a). It should be noted that with only adults immigrating into the patches, the stable stage-distribution has a density of no larvae and juveniles. Larvae, in contrast, had the slowest effect on recovery because they affect juveniles and adults; the stable stage-distribution has non-zero densities for all three life-stages. Due to this feature, coupled with the fact that the adult life-stage does not achieve densities greater than zero until six weeks after larvae begin recruiting, a longer time is required for the population to stop fluctuating and reach a stable stage-distribution. Juvenile life-stage immigration had an intermediate recovery time because this pool affected only one other life-stage (adults). In all three

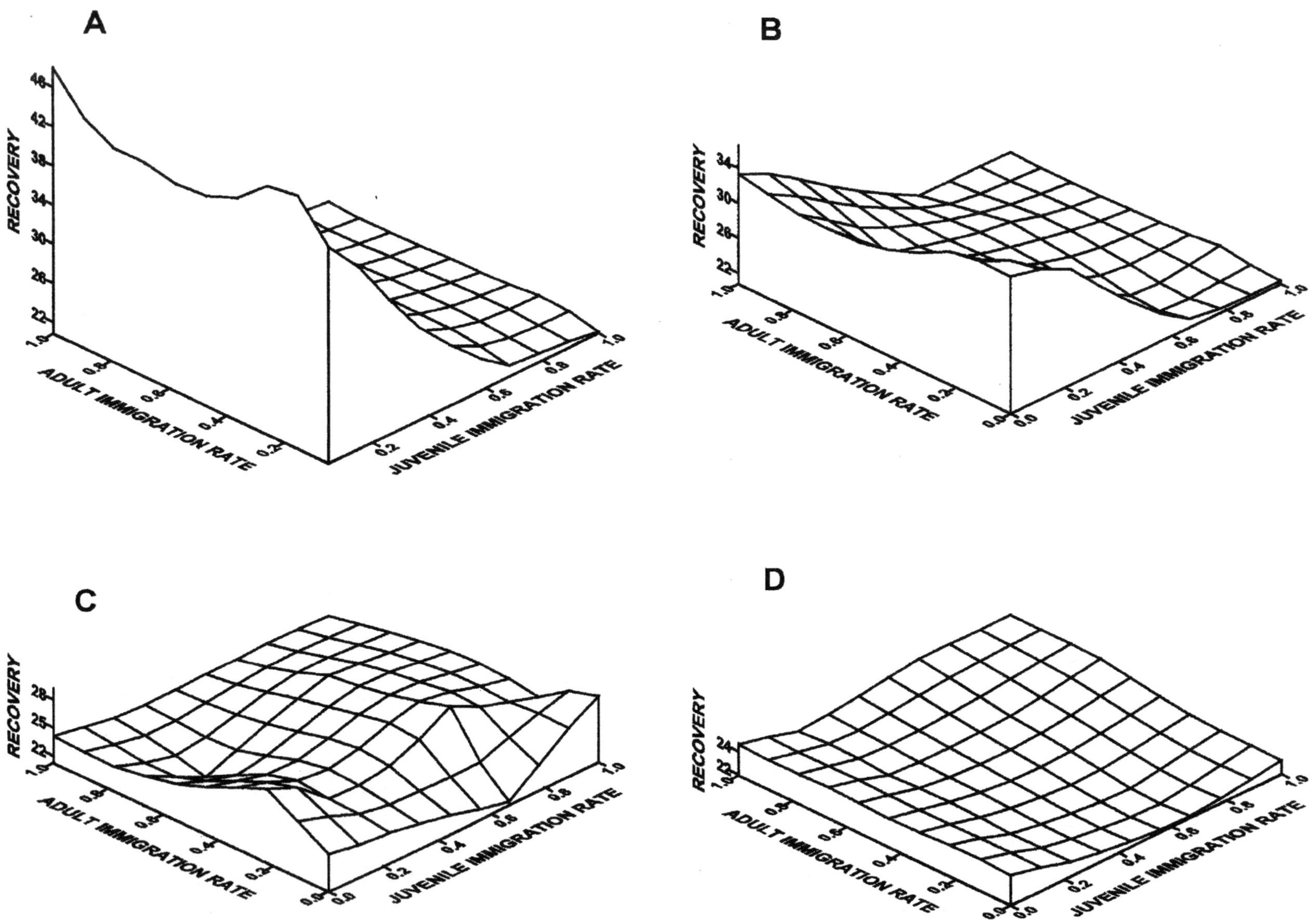

Figure 5. Model simulations showing the effect of three-stage recolonization rates on the recovery of disturbed patches. A = larval recruitment rate = 0.05; B = larval recruitment rate = 0.2; C = larval recruitment rate = 0.66; D = larval recruitment rate = 1.00. Recovery time = time (wks) for the patch to mimic the ambient population stable-stage distribution.

cases, increased immigration rates resulted in shorter patch recovery times. When two life-stages were allowed to recruit into disturbed patches, recovery was lengthened relative to single life-stage migration (Figure 4b). This result is somewhat of an artifact since it takes more time for two life-stages to simultaneously reach a stable stage-distribution than it does for one life-stage. The two life-stage model simulations can, however, be compared with each other (Figure 4b). In all cases, when two life-stages were recruiting, an increase in the immigration rate led to a reduced patch recovery time. In contrast to results in Figure 4a, juvenile life-stage recruitment tended to have a larger effect on patch recovery time when compared to adult migration.

When all three polychaete life-stages were allowed to recruit into the disturbed patches, model results indicate that larval inputs are very important to patch recovery time (Figure 5). This is best illustrated by comparing model simulations where larval input was low (Figure 5a) to simulations in which larval recruitment rate was high (Figure 5d). When larval recruitment is relatively low, increases in juvenile life-stage immigration had a greater effect on patch recovery time than increases in adult life-stage immigration (Figure 5a, 5b). However, when juvenile and adult recolonization rates were at their maximum values, recovery time was not necessarily the fastest (Figure 5c and 5d). Consistent with Figure 2b, life-stage effects on patch recovery time occurred in the following order of importance: larval recolonization > juvenile recolonization > adult recolonization.

Model results are generally consistent with the earlier metapopulation model of Zajac & Whitlatch (1991) which demonstrated changes in relative importance of larval versus adult colonization of open patches on the seasonal contributions to overall population maintenance of *Polydora cornuta*. The present model does, however, indicate that juvenile migration can improve patch recovery time even when larval immigration was set at its maximum rate. Further work is needed to assess the importance of differences in the timing and temporal variability of inputs from the two recruitment sources.

Patch-size effects on recolonization of a polychaete: an experimental test

A field experiment was conducted to examine the effects of disturbance size on the recolonization dynamics of an intertidal sandflat infaunal assemblage in Manakau Harbour, New Zealand. Details of the experimental design and community-level results are given in Thrush et al. (1996). During the study, the spionid polychaete *Boccardia syrtis* was the most abundant species recolonizing the disturbed plots (25–50% of the organisms sampled). While we currently do not possess a complete understanding of the demographic properties of *Boccardia* (e.g. a stage-specific survivorship and fecundity schedule), data were collected to assess whether general population characteristics varied with respect to size of the disturbed plots.

Different-sized patches (0.203 m^2; 0.81 m^2; 3.24 m^2) were created by smothering the sediment; after one month the sediment was completely devoid of macroinfauna. Core samples were collected from each of the plots and surrounding ambient sediments over a 9 month period. *Boccardia* collected in the samples were counted and measured (width of the 5th setiger) using a dissecting microscope and digitizing pad. Individuals were classified into two general size categories: 'juveniles' (<0.25 mm) and 'adults' (>0.25 mm) based on observations and previous work (Thrush et al., 1996). Rates of post-settlement *Boccardia* movement during the early phase of recolonization were determined by deploying bed load traps (e.g. Commito et al., 1995) placed in the different plot sizes and in surrounding ambient sediments 10 and 23 d following initiation of sampling. Traps were retrieved after 12 h and contents were examined under a dissecting microscope.

Temporal changes in the abundance of *Boccardia* in disturbed plots indicated the lack of an initial rapid increase in the first phase of recolonization. Also, no significant differences in recolonization abundance of *Boccarida* occurred between plot sizes over the course of the experimental period (235 d). Temporal changes in density of populations residing in the ambient sediments made at least as important a contribution to the recovery of the experimental plots as did within-plot population dynamics (Thrush et al., 1996). Despite the lack of plot-size differences in *Boccardia* population abundance, scale-dependent life-stage recruitment was evident in the early phase of recolonization process (Figure 6a). During the first 10 d of the experiment, small patches had a significantly greater percentage of adults than medium and large disturbances. By day 23, however, no significant differences in the percentage of adults were found in the three plot sizes.

Bedload traps revealed very low numbers of post-settlement stages were being transported into the dis-

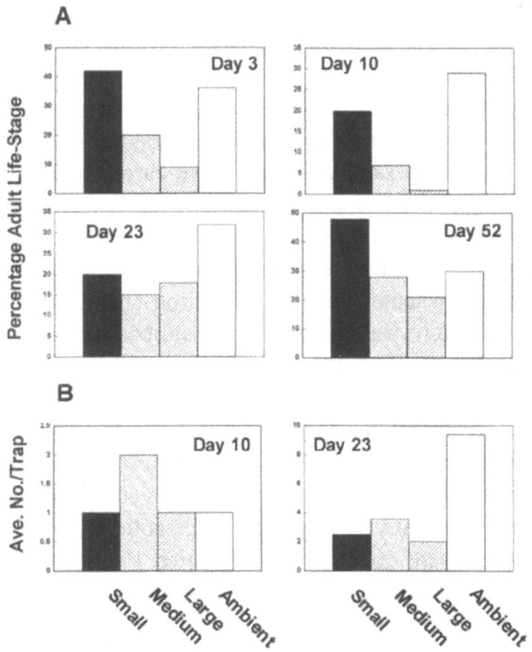

Figure 6. A = the relationship of disturbed plot size to the frequency of *Boccardia syrtis* adult:juvenile life-stages for four different sampling periods following initial sediment defaunation (see text). Significant differences were found among the three plot sizes on Day 3 and Day 10, with the smallest size having higher proportions of adults than medium and largest plots. B = average numbers of *Boccardia syrtis* found in bedload traps which were deployed in different plot sizes on two separate 12 h periods. No significant differences in the numbers of *Boccardia* were recorded between the experimental plots and ambient sediments for Day 10. No differences were detected for the three plot sizes on Day 23 while ambient sediments had significantly more organisms in the trap samples.

turbed plots at the time of the deployments and no evidence of scale-related differences in the disturbed plots were found (Figure 6b). Bedload transport of post-settlement stages is common within the harbour (Commito et al., 1995) and the low numbers of organisms found in the bedload traps likely is a consequence of quiescent wind/wave conditions at the time of deployment.

Despite the lack of significant bedload transport of organisms into the plots, results are consistent with the prediction that larger disturbances should have fewer adult stages than smaller disturbances. Comparisons of juvenile and adult stage-structure in the experimental plots with surrounding sediments revealed that small- and medium-sized plots appeared to recover (in terms of life-stage structure) in about 20 d while large plots recovered in approximately 50 d. Results, therefore, are generally consistent with model simulations which demonstrated the effect of disturbance size on within-patch population recovery time.

Conclusions

Soft-sediment recovery dynamics are frequently reliant upon two different sources of colonists; larvae transported from the water column and post-settlement life-stages (juveniles and adults) laterally advected across the seabed. Larval and post-settlement recolonization rates have been shown to vary with season, habitat and species' life history characteristics (e.g. Zajac & Whitlatch, 1982; Thrush et al., 1996; Shull, 1997). In addition, the relative contribution of the two recolonization sources should vary with size of disturbed habitat. Smaller patches with a greater edge:surface area ratio should be more influenced by the post-settlement recruit pool than larger patches with a smaller edge:surface area ratio. Our life stage-based recolonization model points to several potential types of scale-related differences in demographic characteristics during recolonization of disturbed patches. First, patch recovery rates can be greatly influenced by post-settlement transport, particularly juvenile stages, to disturbed patches. While recruitment regulation is largely considered to be controlled by larval inputs in hard substrate habitats containing sessile invertebrate species (e.g. Roughgarden et al., 1985; 1988), caution should be exercised in generalizing this conclusion to soft-sediment systems which possess two different, and often highly independent, sources of colonists. Secondly, while lateral migration of adult life-stages to small-scale disturbances may contribute to local rapid increases in population density (e.g. Levin, 1984), these life-stages may not necessarily contribute to within-patch population dynamics when compared to other life-stages (particularly late stages or post-reproductive individuals).

Our experimental studies revealed evidence of scale-dependent recolonization of different polychaete life-stages, though the effects appeared restricted to the initial phase of the recolonization process. While post-settlement transport of juvenile and adult life-stages to disturbed patches is likely to be common in a wide variety of habitats, data on the flux (both immigration and emigration) of this post-settlement colonist pool are needed to fully understand the role this process plays in patch recovery and stability as well as regional population dynamics. Future research on scale-dependent recolonization dynamics in soft-sediment habitats should focus on acquisition of within-patch demographic data (e.g. stage-specific growth and survivorship) in order to more fully understand the processes responsible for the regulation of soft-sediment population and community dynamics.

Acknowledgements

We are grateful to R. Ford, M. Taylor, L. Van Dam and S. Hasler who assisted with field and laboratory work in New Zealand and C. Fuller and D. Shull who helped with field work in the United States. J. R. Wood-Rodriguez provided help with Figures 1 and 3. Comments from R. Osman and two anonymous reviewers on a previous version of the manuscript are greatly appreciated. Field work was funded by FRST CO 1211 and the U.S. National Science Foundation, and model development was supported by a grant from the Jessie B. Cox Charitable Trust.

References

Bonsdorff, E., 1980. Macrobenthic recolonization of a dredged brackish water bay in S.W. Finland. Ophelia (suppl.) 1: 145–155.

Commito, J. A., S. F. Thrush, R. D. Pridmore, J. E. Hewitt & V. J. Cummings, 1995. Dispersal dynamics in a wind driven benthic system. Limnol. Oceanogr. 40: 1513–1518.

Cummings, V. J., R. D. Pridmore, S. F. Thrush & S. F. Thrush, 1995. Post-settlement movement by intertidal benthic macroinvertebrates: Do common New Zealand species drift in the water column? New Zealand J. mar. Freshwat. Res. 29: 59–67.

Dayton, P. K. & J. S. Oliver, 1980. An evaluation of experimental analysis of population and community patterns in marine environments. In K. R. Tenore & B. C. Coull (eds), Marine Benthic Dynamics. University of South Carolina Press, (SC): 93–120.

Emerson, C. W. & J. Grant, 1992. The control of soft-shell clam (*Mya arenaria*) recruitment on intertidal sandflats by bedload sediment transport. Limnol. Oceanogr. 36: 1288–1300.

Gascon, C. & J. Travis, 1992. Does the spatial scale of experimentation matter? A test with tadpoles and dragonflies. Ecology 73: 2237–2243.

Grassle, J. F. & J. P. Grassle, 1974. Opportunistic life histories and genetic systems in marine benthic polychaetes. J. mar. Res. 32: 253–284.

Gunther, C. 1992. Dispersal of intertidal invertebrates: a strategy to react to disturbances of different scales? Neth. J. Sea Res. 30: 45–56.

Hall, S. J., D. Raffaelli & S. F. Thrush, 1994. Patchiness and disturbance in shallow water benthic assemblages. In A. G. Hildrew, P. S. Giller & D. Raffaelli (eds), Aquatic Ecology: Scale, Pattern and Processes. Blackwell Scientific, Oxford, England: 333–375.

Hannon, B. & M. Ruth, 1997. Modeling Dynamic Biological Systems. Springer-Verlag, New York: 399 pp.

Hanski, I. & M. Gilpin, 1991. Metapopulation dynamics: brief history and conceptual domain. Biol. J. Linnean Soc. 42: 3–16.

Johnson, R. G., 1970. Variation in diversity within benthic marine communities. Am. Nat. 104: 285–300.

Johnson, R. G., 1973. Conceptual models of benthic marine communities. In T. J. M. Schopf (ed), Models in Paleobiology. Freeman, Cooper (CA): 148–159.

Kotliar, N. B. & J. A. Wiens, 1990. Multiple scales of patchiness and patch structure: a hierarchial framework for the study of heterogeneity. Oikos 59: 253–260.

Levin, L. A., 1984. Life history and dispersal patterns in a dense infaunal polychaete assemblage: community structure and response to disturbance. Ecology 65: 1185–1200.

Levin, L. A. & D. V. Huggett, 1990. Implications of alternative reproductive modes for seasonality and demography in an estuarine polychaete. Ecology 71: 2191–2208.

Levins, R., 1970. Extinction. In M. Gertenhaber (ed), Some Mathematical Problems in Biology. Am. Math. Soc. (RI): 77–107.

McCall, P. L., 1977. Community patterns and adaptive strategies of the infaunal benthos of Long Island Sound. J. mar. Res. 35: 221–266.

Man, A., R. Law & N. V. C. Polunin, 1995. Role of marine reserves in recruitment to reef fisheries: a metapopulation model. Biol. Conserv. 71: 197–204.

Pearson, T. H. & R. Rosenberg, 1978. Macrobenthic succession in relation to organic enrichment and pollution of the marine environment. Oceanogr. Mar. Biol. Ann. Rev. 16: 229–311.

Pulliam, H. R., 1988. Sources, sinks and population regulation. Am. Nat. 132: 652–661.

Pulliam, H. R. & B. J. Danielson, 1991. Sources, sinks and habitat selection: a landscape perspective on population dynamics. Am. Nat. 137: S50–S66.

Quinn, J. F., S. R. Wing & L. W. Botsford, 1993. Harvest refugia in marine invertebrate fisheries: models and applications to the red sea urchin, *Strongylocentrotus franciscansus*. Am. Zool. 33: 537–550.

Rhoads, D. C. & L. F. Boyer, 1982. The effects of marine benthos on physical properties of sediments: A successional perspective. In P. L. McCall & M. J. S. Tevesz (eds), Animal-Sediment Relations. Plenum Publishing (NY): 3–52.

Roughgarden, J., Y. Iwasa & C. Baxter, 1985. Demographic theory for an open marine population with space-limited recruitment. Ecology 66: 54–67.

Roughgarden, J., S. Gaines & H. Possingham, 1988. Recruitment dynamics in complex life cycles. Science 241: 1460–1466.

Savage, W. B. & G. L. Taghon, 1988. Passive and active components of colonization following two types of disturbance on an intertidal sandflat. J. exp. mar. Biol. Ecol. 115: 137–155.

Shull, D. H., 1997. Mechanisms of infaunal polychaete dispersal and colonization in an intertidal sandflat. J. mar. Res. 55: 153–179.

Smith, C. R & S. J. Brumsickle, 1989. The effect of patch size and substrate isolation on colonization modes and rate in an intertidal sediment. Limnol. Oceanogr. 34: 1263–1277.

Snelgrove, P. V. R., 1994. Hydrodynamic enhancement of invertebrate larval settlement in microdepositional environments: colonization tray experiments in a muddy habitat. J. exp. mar. Biol. Ecol. 176: 149–166.

Thistle, D., 1981. Natural physical disturbances and the communities of marine soft bottoms. Mar. Ecol. Progr. Ser. 6: 223–228.

Thrush, S. F., 1991. Spatial patterns in soft-sediment communities. Trends Ecol. Evol. 6: 75–79.

Thrush, S. F., R. B. Whitlatch, R. D. Pridmore, J. E. Hewitt, V. J. Cummings & M. R. Wilkinson, 1996. Scale-dependent recolonization: the role of sediment stability in a dynamic sandflat habitat. Ecology 77: 2472–2487.

Van Blaricom, G. R., 1982. Experimental analysis of structural regulation in a marine sand community exposed to oceanic swell. Ecol. Monogr. 52: 283–305.

Whitlatch, R. B. & R. W. Osman, 1997. Oyster reefs as metapopulations: approaches for restoring and managing spatially fragmented habitats. In M. Luckenbach, R. Mann & J. Wesson (eds), Oyster Reef Habitat Restoration: A Synopsis and Synthesis of Approaches. Virginia Institute of Marine Sciences Press, Gloucester (VA): in press.

Zajac, R. N. & R. B. Whitlatch, 1982. Responses of estuarine infauna to disturbance. I. Spatial and temporal variation of initial recolonization. Mar. Ecol. Prog. Ser. 10: 1–14.

Zajac, R. N. & R. B. Whitlatch, 1985. A hierarchial approach to modeling soft-bottom successional dynamics. In P. Gibbs (ed), Proceedings of the 19th European Marine Biological Symposium. Cambridge University Press, Cambridge, England: 265–276.

Zajac, R. N. & R. B. Whitlatch, 1989. Natural and disturbance-induced demographic variation in an infaunal polychaete, *Nepthys incisa*. Mar. Ecol. Progr. Ser. 57: 89–102.

Zajac, R. N. & R. B. Whitlatch, 1991. Demographic aspects of marine, soft sediment patch dynamics. Am. Zool. 31: 808–820.

Zajac, R. N., 1991. Population ecology of *Polydora ligni* (Polychaeta: Spionidae). II. Seasonal demographic variation and its potential impact on life history evolution. Mar. Ecol. Progr. Ser. 77: 207–220.

Hydrobiologia **375/376:** 227–240, 1998.
S. Baden, L. Pihl, R. Rosenberg, J.-O. Strömberg, I. Svane & P. Tiselius (eds),
Recruitment, Colonization and Physical–Chemical Forcing in Marine Biological Systems.

Recolonization and succession in soft-sediment infaunal communities: the spatial scale of controlling factors

Roman N. Zajac[1,*], Robert B. Whitlatch[2] & Simon F. Thrush[3]
[1] *Department of Biology and Environmental Science The University of New Haven West Haven, CT 06516, U.S.A.*
[2] *Department of Marine Sciences The University of Connecticut Groton, CT 06340, U.S.A.*
[3] *National Institute of Water and Atmospheric Research Box 11-115 Hamilton, New Zealand*
* *Corresponding Author: e-mail: zajacrn@charger.newhaven.edu*

Key words: disturbance, infauna, marine, recolonization, soft-sediment, succession

Abstract

Succession in marine soft-sediment habitats has been studied extensively and several general models of successional dynamics have been developed. However, few researchers have addressed how successional dynamics may change over different spatial scales. Here we extend a model that focuses on the factors that control recolonization and succession. These factors comprise three levels of a hierarchy which include environmental conditions, life history and population processes and biotic interactions. Using this hierarchical framework, we consider the spatial scales at which different factors operate, and argue that the relative mix and intensity of factors controlling succession change at different spatial scales. As a result, successional dynamics may vary considerably as the spatial scale of disturbance increases. At small scales, factors at each level of the hierarchy are important. The greater potential for biotic interactions at this scale may be particularly critical. At meso- to large scales, population processes and environmental conditions have the most influence on successional dynamics. Due to these differences, responses to small-scale ($\lesssim 1$ m^2) as well as large-scale ($\gtrsim 1$ hectare) disturbances may be quite variable. Within this range ($\gtrsim 1$ m^2 $\lesssim 1$ hectare), short- and long-term responses to disturbance may be relatively more predictable and conform to current models of succession in soft-sediment habitats.

Introduction

Disturbance is recognized as an important factor structuring marine and estuarine soft-sediment communities (Grassle & Sanders, 1973; Johnson, 1973; McCall, 1977; Woodin, 1981; Thistle, 1981). Because of its very nature, the physical fabric of soft-sediment habitats is relatively easily disturbed by a variety of processes over a broad range of spatial and temporal scales (Figure 1). It is not surprising then that recolonization and succession in soft-sediments have been studied fairly extensively (see Rhoads & Boyer, 1982; Probert, 1984; Thrush, 1991; Hall et al., 1994 for reviews, and references therein), and several models have been proposed that attempt to generalize these dynamics (discussed in more detail below). Moreover, this information plays a critical role in documenting impacts and assessing ecological responses to human induced disturbance events.

Over the past several years, many ecologists have turned their attention to elucidating how ecological phenomena vary at different spatial and temporal scales (e.g. Dayton & Tegner, 1984; Schneider & Piatt, 1986; Wiens, 1989; Levin, 1992; Angel, 1994), including the dynamics of benthic communities (e.g. Barry & Dayton, 1991; Thrush, 1991; Dayton, 1994; Hall et al., 1994). In this paper, we focus on the dynamics of recolonization and succession in soft-sediment habitats at different spatial scales. There are several components of spatial scale that can be considered including spatial grain, or patch size, and spatial extent, or the area or distance over which specific phenomena occur. More specifically, we address two questions regarding these components of spatial scale.

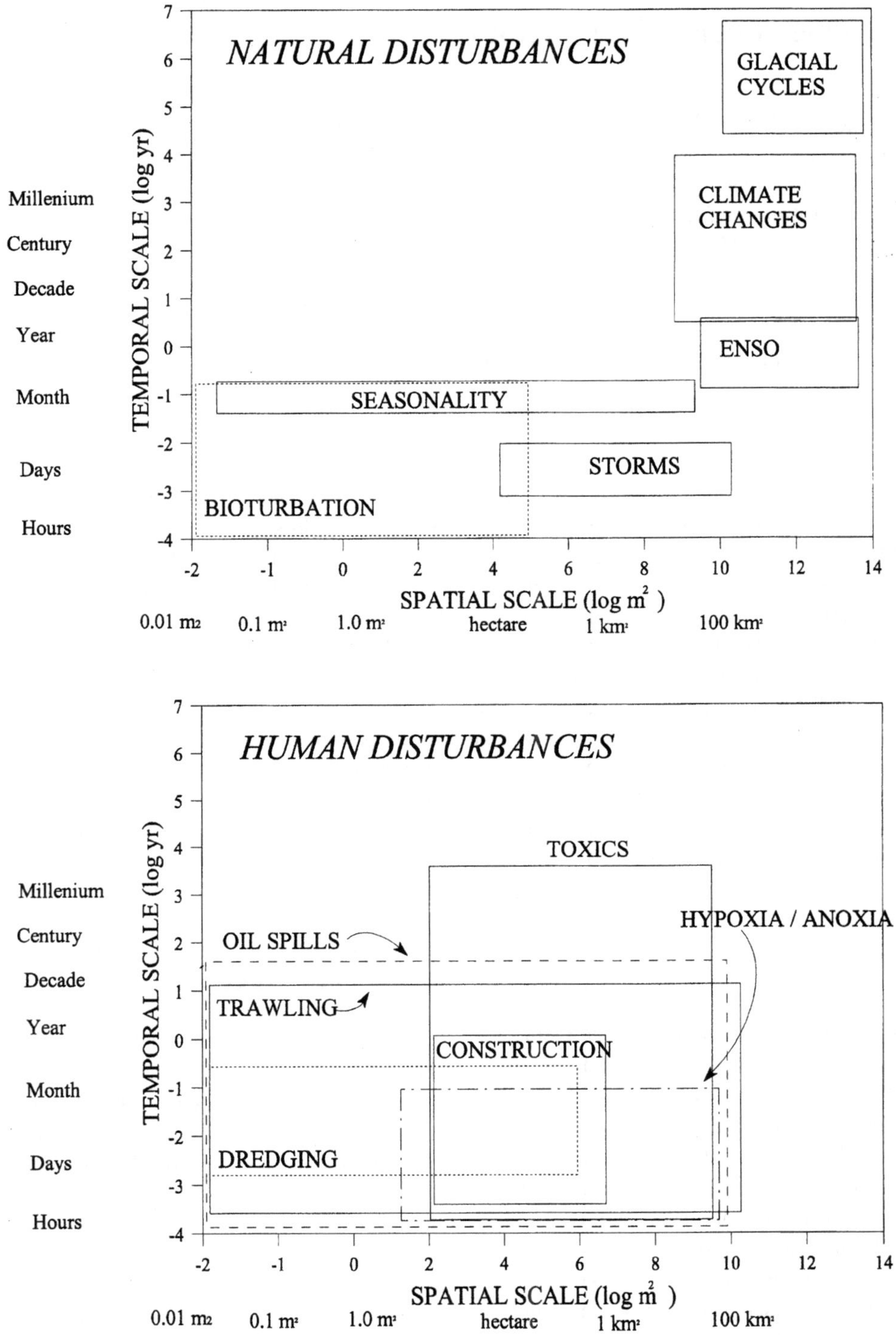

Figure 1. Conceptual depiction of the spatial and temporal scales of different types of natural (top) and human (bottom) disturbances which impact soft-sediment habitats.

Table 1. Ecological characteristics of Stage I and Stage III type species. The characteristics of Stage II species are intermediate between Stages I and III. Table adapted from McCall (1977).

Stage I	*Stage III*
1. Opportunistic Species	1. Equilibrium Species
a. Many reproductions per year	a. Few reproductions per year
b. High recruitment	b. Low recruitment
c. Rapid development	c. Slow development
d. Early colonizers	d. Late colonizers
e. High death rate	e. Low death rate
2. Small	2. Large
3. Sedentary	3. Mobile
4. Deposit Feeders (mostly surface feeders)	4. Deposit and Suspension Feeders
5. Brood Protection; Lecithotrophic Larvae	5. No Brood Protection; Planktotrophic Larvae

First, do factors controlling recolonization and succession in soft-sediment habitats operate across similar spatial scales (i.e. their spatial extent)? Second, how might any differences in the operational scale of the controlling factors affect patterns of recolonization and succession relative to the scale of disturbance (i.e. the size of the disturbance patch)?

We feel that these are important questions for several reasons. Experimental studies of recolonization and succession in soft-sediment habitats have been conducted primarily at small scales (on the order of < 5 m^2), and only a small number have considered spatial scale explicitly (Smith & Brumsickle, 1989; Ruth et al., 1994; Thrush et al., 1996). The extent to which we can use this information to predict the pace and pattern of succession at larger scales will be dependent on our understanding of how the controlling factors studied in these experiments scale up. This is especially critical in trying to elucidate the effects of large-scale human disturbances such as dredge disposal and trawling. In addition, current recolonization and succession models do not explicitly consider changes in spatial scales. By considering the scales over which controlling factors operate we hope to extend current models so that testable predictions can be developed for a variety of situations when disturbance impacts soft-sediment habitats.

Consideration of spatial scale in soft-sediment succession models

Several models of recolonization and succession in soft-sediment communities have been developed based on both experimental and observational studies. Two of the most commonly cited are those of Pearson & Rosenberg (1978) and Rhoads et al. (1978). These models predict a specific sequence of successional stages following disturbance (Figure 2). The stages are comprised of species with particular life modes and life histories that they are differentially adapted to conditions within each successional stage (Table 1). As given, neither of these models explicitly consider the spatial scales over which we might expect the predicted community changes to occur. However, examination of the data used to develop the models, and information from similar studies, provides some insights as to the robustness of the models relative to changes in the scale of disturbance.

Rhoads et al. (1978) developed their model based on small-scale (0.1 m^2 defaunated boxes) experimental disturbances (McCall, 1977) and on monitoring of community responses following dredge spoil disposal which disturbed an area approximately 29 000 m^2. There were differences in the timing and extent of recolonization among the two sites. For example, in the experimental boxes, there was a very rapid (10 d) and high influx of opportunistic species, whereas in the disposal area the influx of opportunistic species was not as great and occurred over an extended period of time (200 d). These differences were attributed to several factors, including location differences (shallow vs. deep for the experimental and disposal site, respectively) affecting the potential supply of larval colonists, year to year variability in larval availability, potential toxic leachate effects on larval settlement at the dumpsite and differences in methodology (e.g. the use of different sieve sizes). Although there were differences, groups of species with particular characteristics representing different stages were found, forming the basis for the model (Figure 2, Table 1). Thus, we may hypothesize that there were similar factors controlling recolonization and succession following these disturbances, but that these were operating differentially with respect to the size and location of the disturbance to cause the varying responses.

Based on studies of macrofaunal responses to organic pollution, Pearson & Rosenberg (1978) developed a model of benthic succession that shares many of the same elements as that proposed by Rhoads et al.

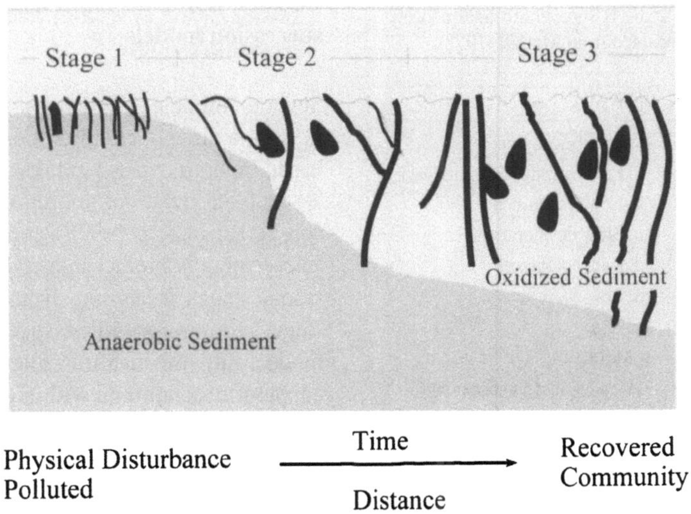

Figure 2. Model of soft-bottom community responses to disturbance / pollution gradients. Stage 1 represents early successional communities following disturbance or communities in polluted habitats. Stage 3 communities represent successional endpoints or communities in non-polluted habitats. Stage 2 are intermediate in both scenarios. See text for more details. The figure is redrawn from Rhoads et al. (1978). The characteristics of species dominating particular stages are given in Table 1.

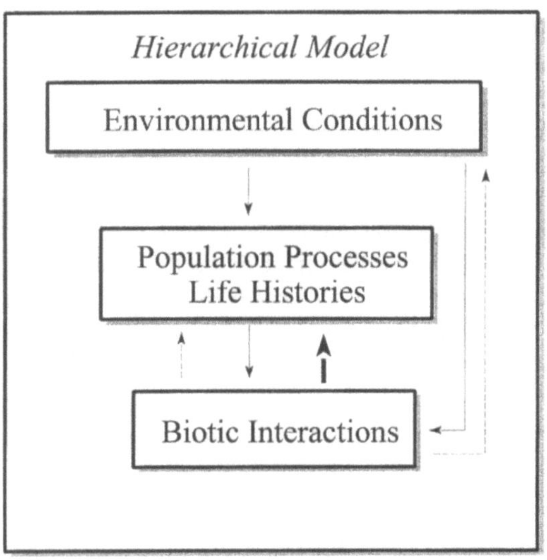

Figure 3. Hierarchy of factors controlling successional processes as proposed by Zajac and Whitlatch (1985). The thin solid lines indicate the flow of information on ecological time scales. Although information flows primarily from higher levels to lower ones, the dotted arrow lines reflect the possible upward flow of information (e.g. biotic interactions may alter sediment characteristics). Information can also flow on evolutionary time scales (as suggested by the thick arrow).

(1978). A critical aspect with respect to the themes being discussed here is that the model is based on studies of primarily large impacted areas, including portions of lochs and fjords. This suggests that the factors controlling successional dynamics following large-scale pollution abatement are similar to those operating in the small-scale disturbances such as conducted by McCall (1977) and in meso-scale disturbances such as dredge spoil disposal (Rhoads et al., 1978).

Taken together, these studies support the notion that successional dynamics are fairly similar over different spatial scales as well as the critical factors controlling these dynamics. However, results from other studies indicate that successional dynamics do not always conform to the predictions of these models and can be quite variable (e.g. Kaplan et al., 1975; Van Blaricom, 1982; Zajac & Whitlatch, 1982a, 1982b; Elmgren et al., 1983; Smith & Brumsickle, 1989; Thrush et al., 1991, 1996). It is still not clear then to what extent successional dynamics change over different spatial scales and in different situations (i.e. the type of disturbance and habitat), and what are the critical factors controlling succession at different spatial scales.

The scales at which factors controlling recolonization and succession operate

Zajac & Whitlatch (1985) suggested that the factors controlling soft-sediment succession can be viewed as a hierarchy comprised of three interacting levels:

Table 2. Examples of how factors controlling recolonization and succession change over different spatial scales. This list is not exhaustive and is meant to illustrate scale differences among the hierarchical levels shown in Figure 3. Larger dots indicate that a factor will likely be more important at the given scale than at others. Open positions in the table indicate that the factor does not control succession at the given scale. Questions marks indicate uncertainty as to the whether the factor operates at the given scale or not. See text for details.

Factors	Spatial scale		
	Small ($\lesssim 1$ m^2)	Meso (1 m^2–1 hectare)	Large ($\gtrsim 1$ hectare)
Environmental conditions			
Temperature	•	●	●
Salinity	•	●	●
Sediment type	•	●	●
Food resources	●	●	●
Hydrodynamics			
* Larval transport	●	●	●
* Sedimentary and food resource effects	●	●	●
* Interactions with boundary layer affecting larval settlement	●		
Life history and population processes			
Life span		●	●
Generation time		●	●
Population growth rate		●	●
Reproductive mode[1]	●	●	●
Larval development mode[2]	●	●	●
Active adult motility	●	•	
Feeding mode	●		
Biotic interactions			
Tolerance	•	●	●
Facilitation	●	?	
Inhibition	●	?	
Predation	●	•	

[1] Semelparity versus iteroparity.
[2] Direct development versus planktonic stage.

environmental conditions, species life histories and biotic interactions (Figure 3). The environmental level is comprised of factors such as temperature, salinity, sediment characteristics, hydrodynamics and food levels. The life history level is comprised of species traits such as life span, generation time, reproductive periodicity and output, mode of larval development, feeding mode and level of motility. The biotic interaction level comprises all the possible types of interactions that may occur, including facilitation, tolerance and inhibition (Connell & Slatyer, 1977) predation, and others (e.g. Woodin, 1983). These levels are quasi-independent, but to some extent each successive level sets the boundary conditions for the next level depending on circumstances (see examples in Zajac & Whitlatch, 1985). Also, we do not intend to suggest that natural processes are exactly structured in a hierarchical manner. Our model is meant to be used as way of exploring the interactions among different sets of factors and processes and their relative influence on infaunal succession under different circumstances.

Although a general association of environmental factors, population processes and biotic interactions with characteristic spatial and temporal scales can be inferred (e.g. Hall et al., 1994), the model did not address the question of scale explicitly. In this section, we elaborate on the model by assessing the spatial extent over which factors at each level of the hierarchy may operate relative to successional dynamics. After doing, so we address the question of which spatial scales each of the factors are likely to exert their most influence on recolonization and succession relative to the scale of the disturbance.

Environmental factors

It is well known that physical and chemical environmental factors, such as temperature and salinity, respectively, can vary over many spatial scales (Table 2). For example, in intertidal habitats, pits (on the order of < 1 m^2) dug by crabs and other foraging organisms can collect water during low tide and temperature and salinity in the pits may vary from exposed surrounding sediments. Such small-scale differences may affect local recolonization patterns, but it is easier to envision differences in these factors influencing responses to disturbance at larger scales. For example, salinity gradients in estuaries result in differential distribution of organisms (e.g. Boesch, 1977) and in turn alter the suite of species that may colonize following a disturbance (Zajac & Whitlatch, 1982a). Other physical and chemical factors are also likely to have their most pronounced effect at meso- to large scales, although certain chemical constituents released by infauna may have effects primarily over small scales (e.g. Woodin, 1991).

Infauna have various degrees of sediment type fidelity (Gray, 1974; Rhoads, 1974), and it is reasonable to assume that differences in sediment type will affect recolonization and succession by controlling which species settle is a disturbed patch with particular sediment characteristics. Unfortunately, there is surprisingly little experimental work that has been done to test sediment effects on recolonization and succession. Zajac & Whitlatch (1982a, 1982b) showed that there were few differences in initial recolonization among small-scale (0.229 m^2) sand and mud experimental disturbances for estuarine communities, although differences were more apparent in longer-term succession plots. At large scales, changing sediment characteristics within a disturbed area, specifically increasing deposition of fine sediment particles, appeared to be responsible for successional changes after severe flooding in a subtropical estuary (Moverley et al., 1986).

In contrast, other physical factors may have significant influences on succession over a range of spatial extents. Key among these is hydrodynamics, which can affect larval settlement and transport at several scales (e.g. Eckman, 1983; Butman, 1987; Palmer, 1988; Günther, 1992; Keough & Black, 1996). Hydrodynamics also affect the distribution of food resources which may have a critical role in shaping successional dynamics in soft-sediment habitats (Thistle, 1981).

Life history and population processes

The roles life history and population attributes play in shaping recolonization and succession in soft sediments have received much attention (e.g. Grassle & Grassle, 1974; McCall, 1977; Pearson & Rosenberg, 1978; Rhoads et al., 1978; Levin, 1984; Zajac & Whitlatch, 1991) and form much of the basis of the models discussed above (Figure 2, Table 1). For this level of the hierarchy, we propose that characteristics most closely associated with an individual will be more critical at small scales, whereas those which integrate population demography and species life history operate over larger spatial extents. The former include, for example, feeding mode and adult motility. The latter group includes, for example, population growth rate (as a function of population patterns in survivorship and fecundity), life span, generation time and the meta-population characteristics of species in the region. In addition to these factors, there are two critical attributes, reproductive mode and larval development mode, which operate and are likely critical to species responses to disturbance across all spatial scales.

As far as is known, most adult infauna have limited active motility (e.g. Günther, 1992; and references therein). Tubiculous forms are generally sedentary but can relocate over distances on the order of centimeters. Errant species move to a greater extent but most of these movements are over small scales. Some species, however, can actively enter the water column as adults and be transported by currents over distances of meters to kilometers (e.g. Dean, 1978; Dauer et al., 1982). Apart from active movements, storms (e.g. Dobbs & Vozarik, 1983) and near bottom flows (Emerson & Grant, 1992; Committo et al., 1995; Shull, 1997) can passively suspend adult infauna into the water column, causing them to be transported also over large distances.

Average life span, generation time and population growth rate set the pace of population processes. Differences in life span will expose individuals in populations to different suites of temporal changes, whereas differences in generation time will have the same affect but in the context of different cohorts of individuals. Generation time can also have significant effects on population growth rate. Population growth rate establishes the potential for populations to maintain themselves in particular habitats and to respond to various environmental conditions (Zajac & Whitlatch, 1991). We suggest that the greatest variability in these population characteristics likely occurs at meso- to large spatial scales. However, few studies have ad-

Table 3. Hypothesized scales of dispersal and relative differences in exposure to varying environments and period of exposure of species with different combinations of reproductive and larval development modes. See text for further discussion.

	Reproductive mode		
	Monotelic	Polytelic	Semi-continuous breeders
Larval development mode			
Planktonic			
Dispersal	Multi-scale	Multi-scale	Multscale
Environmental exposure	Wide array	Wide array	Wide array
Exposure period	Shorter	Shorter	Extended
Benthic			
Dispersal	Small to mesoscales	Small to mesoscales	Small to mesoscales
Environmental exposure	Local	Local	Local
Exposure period	Shorter	Shorter	Extended

dressed variation in population processes of benthic fauna across a range of spatial scales, so it is difficult to assess the extent to which this supposition is accurate. Zajac & Whitlatch (1988, 1989) found populations of the polychaete *Nephtys incisa* exhibited significant variation in individual growth, size-structure and fecundity among sites 200 to 3 km apart, but that despite these differences the populations formed a relatively coherent demographic unit having similar population growth rates. Alon & Stancyk (1982) found significant variation in life history traits among two populations of grass shrimp (*Palaeomenetes pugio*) among estuaries with different salinity regimes approximately 15 km apart. In contrast, Olive & Morgan (1984) found that populations of several *Nephtys* species in widely separated beach areas in the British Isles had similar age-structures, and suggested that they may have similar population dynamics which are keyed into periodic recruitment failures. Although our focus here is on spatial variation, it should be noted that there can be significant short-term temporal differences in the demographic characteristics of local infaunal populations (e.g. Levin & Hugget, 1990; Zajac, 1991) which may be of greater magnitude than spatial differences.

Infauna exhibit a variety of reproductive and larval development modes, and their population dynamics may vary considerably over different spatial scales depending on the combination of these two traits. Infaunal species can divided into three broad groups based on their reproductive modes: monotelic, polytelic and semi-continuous breeders (Olive & Clark, 1978). Monotelic species reproduce once and die, polytelic species usually have one major reproductive event per year over their lifetime, and semi-continuous breeders reproduce on a continuous basis upon reaching sexual maturity that may be interrupted by, for example, seasonal changes in environmental conditions. Larval development can likewise be split into two general groups, planktonic and benthonic, but there are many species which brood larvae to some advanced stage and then release them into the plankton for various periods of time (e.g. Olive & Clark, 1978).

Although a detailed discussion is not feasible here, Table 3 presents general combinations of these traits and possible key impacts on dispersal and the degree and extent to which populations will be exposed to different environmental conditions. In general, these two traits operate over the full array of spatial scales relevant to soft-sediment environments, but the range of scales for species with particular traits are different. The greatest difference occurs between species with semi-continuous reproduction relative to other combinations of these traits.

Planktonic development puts larvae up into the water column where they can be carried by currents over a wide range of distances. Thus, monotelic and polytelic species with planktonic development, may disperse over a similar range of scales and be exposed to a wide array of environmental conditions. The major difference here is that the larval cohort in any one particular year is comprised from the reproductive efforts of several adult cohorts for polytelic species, and just by one cohort of adults reaching sexual maturity

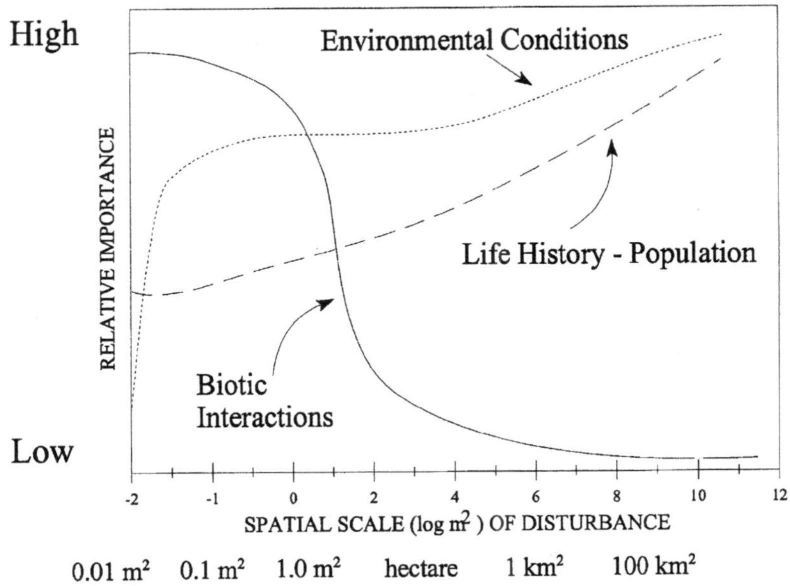

Figure 4. Predicted relative levels of importance of each level of the hierarchy (Figure 3) in controlling successional dynamics over different scales of disturbance.

for monotelic species. In both cases, release of larvae usually takes place over a short period of time once during a year. This reduces the exposure period and, depending on the duration of larval development, will affect the array of environments the larvae are exposed to. The degree of exposure of larvae to different environments (locally and at larger scales) will determine the probability of larvae encountering a disturbed patch and the chances of colonization, relative to the timing of competency and settlement. In comparison, semi-continuous breeders with planktonic larvae will also be dispersed broadly over a wide array of habitats, but this will occur over an extended period of time.

Benthic development likely leads to dispersal over relatively smaller distances, although dispersal may be somewhat greater under high flow conditions. Thus, individuals of species with benthic development will be exposed to more local environments. We see these conditions applying to both semelparous and iteroparous species with benthic development. Exposure periods are also shorter given the short-term periodicity of reproduction noted above. The situation will be similar for semi-continuous breeders except that the exposure period will be extended.

Biotic interactions

We consider biotic interactions within the context of three broad groups which were suggested as mechanisms for successional change by Connell & Slatyer (1977). Facilitation comprises interactions in which one group of organisms enhances the establishment of another. Inhibition results in groups of organisms preventing or significantly reducing the establishment of another group. This may occur via competition for resources such as food and/or space. Tolerance represents the null situation in which organisms establish populations in an area depending primarily on their ability to live within prevailing environmental conditions. To this set of interactions we also add predation.

Given the relatively small size of most infauna and their limited motility, it is generally assumed that direct biotic interactions occur over small distances. Experimental studies of competition in soft sediments have primarily been conducted at small scales using cores of manipulated densities and cages (see Wilson, 1990 for a review). Based on this body of work there is ample evidence that competitive interactions can shape infaunal communities at these scales by inhibiting settlement and establishment of individuals. Studies which have focused on recolonization and succession, using similar approaches have also shown that inhibition and facilitation can shape responses to small scale disturbances (e.g. Gallagher et al., 1983; Whitlatch & Zajac, 1985).

The extent to which biotic interactions extend beyond the ambit of direct individual-to-individual interactions depends on the size and nature of patches formed by particular species, their density within the patch and the nature of the interactions. For example, Thrush et al. (1996) found that sediment stabilization by tube-building polychaetes was an important factor structuring sandflat communities. Recovery in large disturbance patches (3.24 m^2) in this community was slower than in smaller patches (0.203 m^2) because of greater sediment instability in the larger patches and the lack of facilitation provided by an established tube bed. Most studies of biotic interactions in soft-sediments provide measures of population densities, however, few provide information on the characteristics and sizes of patches where the interactions take place. To assess how biotic interactions might extend over larger scales it is important to obtain this type of information (e.g. McArdle & Blackwell, 1989; Thrush et al., 1989), and then assess interactions amongst such patches (e.g. Peterson & Black, 1987).

Predatory interactions among infauna may inhibit the establishment of individuals and shape community structure (Ambrose, 1991). Predation on infauna by epifauna, can also be an important structuring force in infaunal communities (e.g. Virnstein, 1977; Woodin, 1981; Knieb 1985; Riese, 1985 and references therein). As with inhibition and facilitation, it is fairly well known that predatory interactions take place at small scales in soft-sediments. This may be particularly the case with respect to interactions among infauna. With respect to epifauna, predatory interactions can occur over multiple scales depending on the foraging area and abundance of the predator. For example, individual epibenthic predators such as shrimp, crabs and fish forage locally, and their effects are easily quantified by assessing reduction of infauna in foraging pits. At high densities their populations may significantly affect large areas of soft-sediment habitats. Examples of large-scale predatory effects on soft-sediment populations include foraging by overwintering birds on mudflats (Wilson, 1988, 1989) and whale feeding on soft-sediment infauna (Oliver & Slattery, 1985).

How the scale of controlling factors may affect recolonization and succession

Based on the above discussion, we hypothesize that the relative importance of each level of the hierarchy in controlling recolonization and succession changes as the spatial scale of the disturbance changes. In general, environmental factors exert control over multiple scales, life history and population processes are critical at primarily meso- to large scales, whilst biotic interactions have the most effect at small scales (Figure 4). These differences may be significant in shaping the dynamics of recolonization and succession relative to the size, and possibly type, of disturbance.

Disturbances affecting infaunal communities occur at a variety of scales (Figure 1) (Hall et al., 1994). Yet as noted above, models predicting successional dynamics in soft-sediments do not explicitly consider spatial scale. Can the models developed by Rhoads et al. (1978) and Pearson & Rosenberg (1978) be augmented by incorporating spatial scale? We do so by considering several components of recolonization and succession, including the predictability of the various stages of succession, the rate of recovery and the relative abundances of different types of species during succession (i.e. opportunistic vs. non-opportunistic species) in relation to the spatial scale (extent) of controlling factors and the scale (size) of disturbance.

The models shown in Figure 2 predict a successional sequence comprising an initial recolonization by opportunists (Stage I) followed by a progressive re-establishment of climax type species (Stage III). However, the predictability of these stages and the relative abundance of different species types during succession may change as the scale of disturbance changes (Figure 5). [The nature of the disturbance, i.e. polluted vs. non-polluted, may also be a critical factor as discussed below.] At small spatial scales (< 0.1 m^2), particular factors at the life history – population processes level of the hierarchy will be primarily in play, causing the predictability of any particular successional stages to be relatively low and the relative variation in the mix of species type to be high. Specifically, adult motility plays a key role here, as well as the availability of competent larvae for settlement. Since the disturbance scale is small, the relative probability of various species colonizing the disturbance patch is similar, depending on the characteristics of the community set by prevailing environmental factors. Key among these may be small-scale flow patterns, whereas factors such as salinity, temperature and sediment type have little effect on influencing succession.

As the scale of disturbance increases to around 1 m^2, we hypothesize an increase in the relative predictability of successional stages and that the variation in the relative abundance of non-opportunistic species

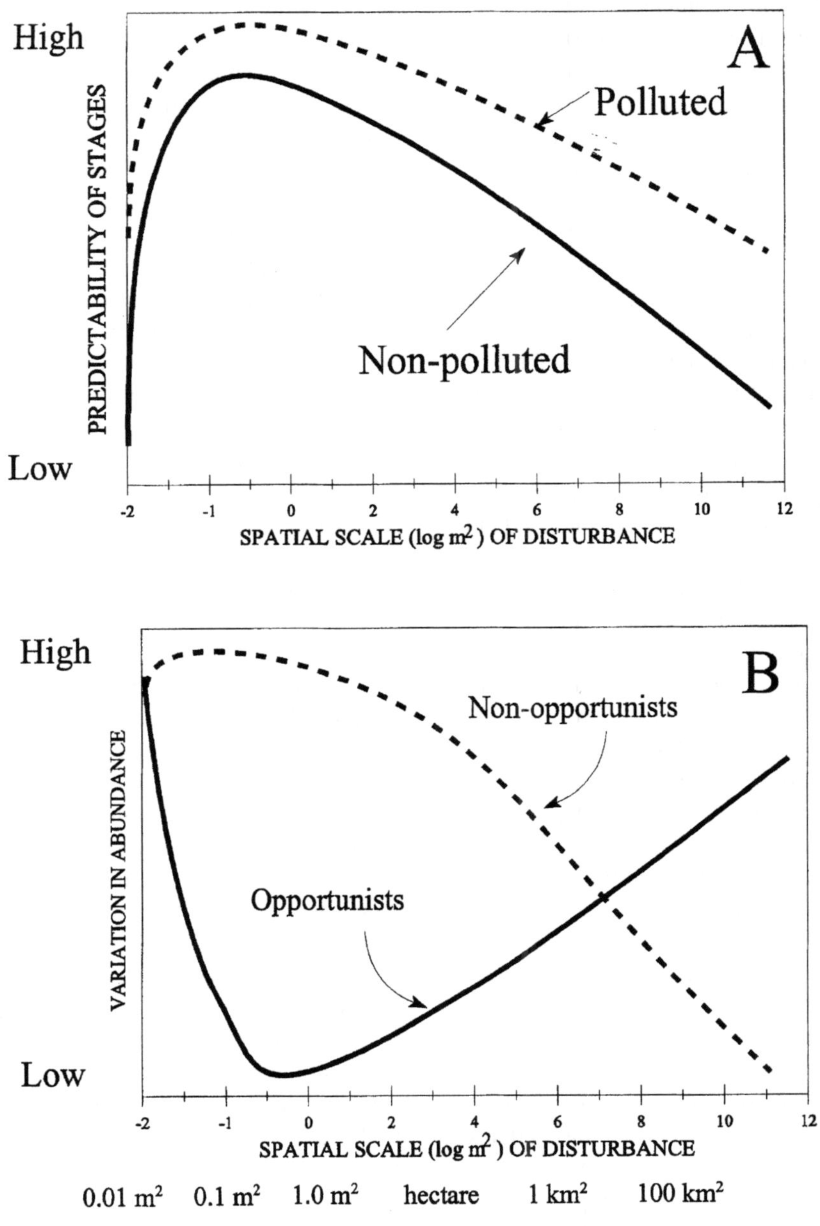

Figure 5. Predicted changes in A) the relative predictability of successional stages shown in Figure 2 relative to the spatial scale of disturbance, and B) the variation in the abundance of opportunistic (Stage I like) and non-opportunistic (Stage III like) species relative to the spatial scale of disturbance. In 5A, the curve noted non-polluted refers to natural disturbances or human disturbances that do not involve potentially toxic materials; polluted refers to human disturbances that involve potentially toxic materials including high levels of organic loading.

will decline. We base this prediction on the results of experimental work presented in support of the model (e.g. McCall, 1977) and the interaction of certain hierarchical factors. As disturbances approach this scale, the ambit of active adult motility is exceeded on the most part, and recolonization is shaped by the availability of settling infauna and the hydrodynamics of the area that can deliver both larvae via the water column and any adults that may be moved by bedload transport. This results in the settlement of primarily small, opportunistic-type species, setting up the potential for successional sequences as shown in Figure 2.

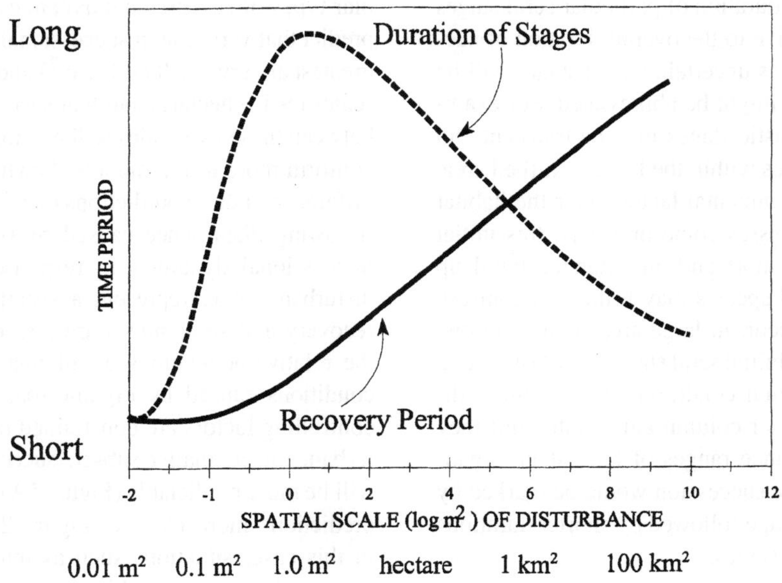

Figure 6. Predicted changes in the duration of successional recovery (solid line) and the duration of particular stages (dotted line), relative to the spatial scale of disturbance.

At spatial scales from approximately 1 m^2 to > a hectare, we hypothesize a decline in the predictability of successional stages, an increase in the variability of opportunistic responses and a decline in the variation of colonization by non-opportunistic species (Figure 5). At these spatial scales, many more environmental factors begin influencing succession as large-scale differences in, for example, sediment type and flow affect the distribution of source populations providing colonists to initiate recolonization, and their population dynamics to advance successional changes. Differences in demographic characteristics and population growth rates among source populations and within the disturbed area itself are also likely to be greater. Adult motility has little effect, and successional dynamics will depend on variable transport of larvae of a greater mixture of species.

In addition to shifts in successional characteristics at different spatial scales, we also suggest that there are changes in the duration of overall recovery and of particular stages (Figure 6). For very small disturbances (< 0.1 m^2), recovery may be relatively fast, on the order of days to weeks, because of the potential for a rapid influx of adult, juvenile and larval colonists from surrounding sediments and the water column. Under these circumstances, if there are any discernable successional stages, they will occur over very short time periods. As the area of disturbance increases to approximately 1.0 m^2, we suggest that the duration of recovery will increase, but the extent to which this occurs may depend on potential biotic interactions that are prevalent at this scale. Inhibition type interactions may increase particular successional periods as later arriving colonists cannot establish populations, whereas facilitation may shorten the recovery process by enhancing the settlement of later stage species. The absence of a critical interaction at this and larger scales may also extend the recovery period, as suggested by results presented by Thrush et al. (1996). At these scales, the duration of particular stages are also likely to be long relative to the overall duration of the succession. Disturbance areas are small enough so that they can be potentially colonized in high numbers by opportunistic species, but large enough to perhaps be beyond the scale of effective adult migration.

As the spatial scale of disturbance increases, the duration of successional recovery should increase (Figure 6). This may be a function of simply that the area undergoing recovery increasingly approaches, or eventually exceeds, the potential for source populations to supply colonists that can establish populations over short periods of time (months to a year). Succession will be dependent on factors that operate over large scales, including the mix of demographic characteristics of potential source populations and large-scale transport processes (Table 2). At these scales, we also predict that the duration of particular seral stages may be shorter relative to the overall duration of succession. However, it is uncertain which stages will be extended and which might be abbreviated. For example, initial opportunistic stages may be transient and occur in small patches within the large disturbed area. As large-scale environmental factors alter the habitat and population processes come into play, this initial stage may be very short and an extended build up of later successional species may follow. In contrast, the opposite may occur, in large areas that were disturbed by pollution. Initial seral stages may last several years as environmental conditions both in the sediments and in the water column ameliorate until they are within the tolerance ranges of later stage organisms. In this case, the succession would be marked by an extended initial stage followed by shorter durations of later successional stages.

Conclusions

Much of our understanding of factors controlling recolonization and succession is based on experimental studies conducted at spatial scales of generally less than 1 m^2 (Hall et al., 1994). These have necessarily focused on a small portion of the suite of factors comprising the hierarchy shown in Figure 3. Our knowledge of disturbance, recolonization and succession relative to meso- and large scale controlling factors is rudimentary and observational / correlational in nature. In this paper we have strived to fashion a framework based on available information so that the study of soft-sediment successional processes might be effectively approached at these larger scales.

Based on our interpretations of the information available it appears that current succession models (Figure 2) are applicable to some small-scale natural disturbances and to large-scale cases of disturbance by organic pollution. Using the basic aspects of these models and the hierarchical framework, we suggest that successional dynamics may change considerably as the spatial scale increases. The mix and intensity of controlling factors changes at different spatial scales (Figure 4). At small scales, factors within all three levels of the hierarchy are important. The greater potential for biotic interactions at this scale may be particularly critical. At meso- to large scales, population processes and environmental conditions have the most influence on successional dynamics. Because of these differences, there may be spatial thresholds for particular types of successional dynamics. Specifically, we predict that variation in successional dynamics will be greatest at very small (< 0.1 m^2) and at meso- to large scales (> 1.0 hectare) and that successional dynamics between these two scales will be more predictable and conform more to the models shown in Figure 2. One difference in these predictions may involve succession following disturbance caused by organic pollutants. Successional dynamics in response to this type of disturbance may represent a specific case in which recovery and seral stages are governed primarily by the relative adaptations of infauna to environmental conditions caused by organic loading. Because the controlling factors are constrained by the type of disturbance to a smaller subset, successional dynamics will be more predictable (Figure 5A), following model predictions more closely (Figure 2). However even in this case, situations such as intermittent hypoxia may result in more variable successional dynamics depending on the timing, degree and duration of these events.

A potentially interesting aspect of changes in successional dynamics across different spatial scales is the relationship to the biodiversity of a particular region. In a species-rich region there may be a more diverse set of life histories and demographic dynamics displayed by resident species than in less diverse regions. If successional processes are keyed to the variation present in factors at different spatial scales, then we might predict that soft-sediment successional dynamics will be more variable as regional species richness increases. Unfortunately to our knowledge there are no studies that have addressed this possibility.

An important question facing benthic ecologists and those charged with managing and protecting sea floor habitats is, To what extent can we use small-scale experiments to predict larger scale responses? Our predictions suggest that this may not be possible because the relative mix of factors controlling successional process at different spatial scales may be fundamentally different. To gain a better understanding of successional processes in soft-sediments it will be necessary to push the spatial experimental envelop by simulating disturbances of various types at larger scales. This may be especially important and indeed practical for certain kinds of disturbances such as trawling. For situations when it is not practical, e.g. periodic hypoxia / anoxia, monsoons and large-scale dredging, specific sets of predictions and specific sampling designs should be formulated that address multi-scale processes in order to tease apart the relative importance of environmental factors, population and life history processes and biotic interactions in controlling successional dynamics.

References

Alon, N. C. & S. E. Stancyk, 1982. Variation in life history patterns of the grass shrimp *Palaemonetes pugio* in two South Carolina estuarine systems. Mar. Biol. 68: 265–276.

Ambrose, W. G., 1991. Are infaunal predators important in structuring marine soft-bottom communities? Am. Zool. 31: 849–860.

Angel, M. V., 1994. Spatial distribution of marine organisms: patterns and processes. In P. J. Edwards, R. M. May & N. R. Webb (eds), Large-Scale Ecology and Conservation Biology. Blackwell Scientific Publications, Oxford: 59–109.

Barry, J. P. & P. K. Dayton, 1991. Physical heterogeneity and the organization of marine communities. In Kolasa, J. & S.T.A. Pickett (eds), Ecological Heterogeneity. Springer-Verlag, New York: 270–320.

Boesch, D. F., 1977. A new look at the zonation of benthos along the estuarine gradient. In B. C. Coull (ed), Ecology of the Marine Benthos. University of South Carolina Press, Columbia (SC): 245–266.

Butman, C. A., 1987. Larval settlement of soft-sediment invertebrates: The spatial scales explained by active habitat selection and the emerging role of hydrodynamical processes. Oceanogr. mar. Biol. Ann. Rev. 25: 113–165.

Committo, J. A., C. A. Currier, L. R. Kane, K. A. Reinsel & I. M. Ulm, 1995. Dispersal dynamics of the bivalve *Gemma gemma* in a patchy environment. Ecol. Monogr. 65: 1–20.

Connell, J. H. & R. O. Slatyer, 1977. Mechanisms of succession in natural communities and their role in community stability and organization. Am. Nat. 111: 1119–1144.

Dauer, D. M., R. M. Ewing, J. W. Sourbeer, W. T. Harlon & T. L. Stokes, 1982. Nocturnal movements of the macrobenthos of the Lafayette River, Virginia. Int. Rev. ges. Hydrobiol. 67: 761–775.

Dayton, P. K., 1994. Community landscape: scale and stability in hard bottom marine communities. In P. S. Giller, A. G. Hildrew & D. G. Rafaelli (eds), Aquatic Ecology: Scale, Pattern, and Process. Blackwell Scientific Publications, London: 289–332.

Dayton, P. K. & M. J. Tegner, 1984. The importance of scale in community ecology: A kelp forest example with terrestrial analogs. In P. W. Price, C. N. Slobodchikoff & W. S. Gaud (eds), A New Ecology: Novel Approaches to Interactive Systems. John Wiley & Sons, New York: 457–483.

Dean, D., 1978. Migration of the sandworm *Nereis virens* during winter nights. Mar. Biol. 45: 165–173.

Dobbs, F. C. & J. M. Vozarik, 1983. Immediate effects of a storm on coastal infauna. Mar. Ecol. Prog. Ser. 11: 273–279.

Eckman, J. E., 1983. Hydrodynamic processes affecting benthic recruitment. Limnol. Oceanogr. 28: 241–257.

Elmgren, R., S. Hansson, U. Larsson, B. Sundelin, & P. D. Boehm, 1983. The 'Tsesis' oilspill: acute and long-term impact on the benthos, Mar. Biol. 73: 51–65.

Emmerson, C. W. & J. Grant, 1992. The control of soft-shell clam (*Mya arenaria*) recruitment on intertidal sandflats by bedload sediment transport. Limnol. Oceanogr. 36: 1288–1300.

Gallagher, E. D., P. A. Jumars & D. D. Trueblood, 1983. Facilitation of soft-bottom succession by tube builders. Ecology 64: 1200–1216.

Grassle, J. F. & J. P. Grassle, 1974. Opportunistic life histories and genetic systems in marine benthic polychaetes. J. mar. Res. 32: 253–284.

Grassle, J. F. & H. L. Sanders, 1973. Life histories and the role of disturbance. Deep Sea Res. 20: 643–659.

Gray, J. S., 1974. Animal sediment relationships. Oceanogr. Mar. Biol. Ann. Rev. 12: 223–261.

Günther, C. P., 1992. Dispersal of intertidal invertebrates: A strategy to react to disturbances of different scales? Neth. J. Sea Res. 30: 45–56.

Hall, S. J., D. Raffaelli & S. F. Thrush, 1994. Patchiness and disturbance in shallow water benthic assemblages. In P. S. Giller, A. G. Hildrew & D.G. Rafaelli (eds), Aquatic Ecology: Scale, Pattern, and Process. Blackwell Scientific Publications, London: 333–375.

Johnson, R. G., 1973. Conceptual models of benthic communities. In T. J. M. Schopf (ed), Models in Paleobiology. Freeman Cooper and Co., San Francisco: 148–159.

Kaplan, E. H., J. R. Welker, M. G. Kraus, & S. McCourt, (1975) Some factors affecting the colonization of a dredged channel, Mar. Biol. 32: 193–204.

Keough, M. J. & K. P. Black, 1996. Predicting the scale of marine impacts; Understanding planktonic links between populations. In R. J. Schmitt & C. W. Osenberg (eds), Detecting Ecological Impacts – Concepts and Applications in Coastal Habitats. Academic Press, San Diego: 199–234.

Kneib, R. T., 1985. Predation and disturbance by grass shrimp, *Palaemonetes pugio* Holthuis, in soft-substratum benthic invertebrate assemblages. J. exp. mar. Biol. Ecol. 93: 91–102.

Levin, L. A., 1984. Life history and dispersal patterns in a dense infaunal polychaete assemblage: Community structure and response to disturbance. Ecology 65: 1185–1200.

Levin, L. A. & D. V. Huggett, 1990. Implications of alternative reproductive modes for seasonality and demography in an estuarine polychaete. Ecology 71: 2191–2208.

Levin, S. A., 1992. The problem of pattern and scale in ecology. Ecology 73: 1943–1983.

McArdle, B. H. & R. G. Blackewell, 1989. Measurement of density variability in the bivalve *Chione stutchburyi* using spatial autocorrelation. Mar. Ecol. Progr. Ser. 32: 245–252.

McCall, P. L., 1977. Community patterns and adaptive strategies of the infaunal benthos of Long Island Sound. J. mar. Res. 35: 221–226.

Moverley, J. H., P. Saenger & M. A. Curtis, 1986. Patterns of polychaete recolonization in Queensland subtropical estuaries following severe flooding. Hydrobiologia 134: 227–235.

Olive, P. J. W. & R. B. Clark, 1978. The physiology of reproduction. In P. J. Mill (ed), Physiology of Annelids. Academic Press, London: 271–368.

Olive, P. J. W. & P. J. Morgan, 1984. A survey of the age structure of beach populations of *Nephtys* spp. in the British Isles: the basis of population fluctuations. In J. Boutler (ed), Proceedings of the 17th European Marine Biological Symposium. Oceanol. Acta special volume: 141–145.

Oliver, J. S. & P. N. Slattery, 1985. Destruction and opportunity on the sea floor: effects of graywhale feeding. Ecology 66: 1965–1975.

Palmer, M. A., 1988. Dispersal of marine meiofauna: a review and conceptual model explaining passive transport and active emergence with implications for recruitment. Mar. Ecol. Prog. Ser. 48: 81–91.

Pearson, T. H. & R. Rosenberg. 1978. Macrobenthic succession in relation to organic enrichment and pollution of the marine environment. Oceanogr. Mar. Biol. Ann. Rev. 16: 229–311.

Peterson, C. H. & R. Black, 1987. Resource depletion by active suspension feeders on tidalflats: influence of local density and tidal elevation. Limnol. Oceanogr. 32: 143–166.

Probert, P. K., 1984. Disturbance, sediment stability and trophic structure of soft-bottom communities. J. mar. Res. 42: 893–921.

Reise, K. 1985. Tidal Flat Ecology: An Experimental Approach to Species Interactions. Springer Verlag, Berlin, 191 pp.

Rhoads, D. C., 1974. Organism-sediment relations on the muddy sea floor. Oceanogr. Biol. Ann. Rev. 12: 263–300.

Rhoads, D. C. & L. F. Boyer, 1982. The effects of marine benthos on physical properties of sediments: A successional perspective. In P. L. McCall & M. J. S. Tevesz (eds), Animal-Sediment Relations. Plenum Publishers, New York: 3–52.

Rhoads, D. C., P. L. McCall & J. Y. Yingst (1978). Production and disturbance on the estuarine seafloor. Am. Sci. 66: 577–586.

Ruth, B. F., D. A. Flemer & C. M. Bundrick, 1994. Recolonization of esturaine sediments by macroinvertebrates: Does microcosm size matter? Estuaries 17: 606–613.

Schneider, D. C. & J. F. Piatt, 1986. Scale-dependent correlation of seabirds with schooling fish in a coastal ecosystem. Mar. Ecol. Progr. Ser. 32: 237–246.

Shull, D. H., 1997. Mechanisms of infaunal polychaete dispersal and colonization in an intertidal sandflat. J. mar. Res. 55: 153–179.

Smith, C. R. & S. J. Brumsickle. 1989. The effect of patch size and substrate isolation on colonization modes and rate in an intertidal sediment. Limnol. Oceanogr. 34: 1263–1277.

Thistle, D., 1981. Natural physical disturbances and communities of marine soft-bottoms. Mar. Ecol. Prog. Ser. 6: 223–228.

Thrush, S. F., 1991. Spatial patterns in soft-bottom communities. Trends in Ecol. Evol. 6: 75–79.

Thrush, S. F., J. E. Hewitt & R. D. Pridmore, 1989. Patterns in the spatial arrangements of polychaetes and bivalves in intertidal sandflats. Mar. Biol. 102: 529–535.

Thrush, S. F., R. D. Pridmore, J. E. Hewitt & V.J. Cummings, 1991. Impact of ray feeding disturbances on sandflat macrobenthos: do communities dominated by polychaetes or shellfish respond differently? Mar. Ecol. Progr. Ser. 69: 245–252.

Thrush, S. F., R. B. Whitlatch, R. D. Pridmore, J. E. Hewitt, V. J. Cummings & M. R. Wilkinson, 1996. Scale-dependent recolonization: The role of sediment stability in a dynamic sandflat habitat. Ecology 77: 2471–2487.

VanBlaricom, G. R., 1982. Experimental analyses of structural regulation in a marine sand community exposed to oceanic swell. Ecol. Monogr. 52: 283–305.

Virnstein, R. W., 1977. The importance of predation by crabs and fishes on benthic infauna in Chesapeake Bay. Ecology 58: 1199–1217.

Whitlatch, R. B. & R. N. Zajac. 1985, Biotic interactions among estuarine infaunal opportunistic species. Mar. Ecol. Prog. Ser. 21: 299–311.

Wiens, J. A., 1989. Spatial scaling in ecology. Funct. Ecol. 3: 385–397.

Wilson, W. H., 1988. Shifting zones in a Bay of Fundy soft-sediment community: patterns and processes. Ophelia 29: 227–245.

Wilson, W. H., 1989. Predation and the mediation of intraspecific competition in an infaunal community in the Bay of Fundy. J. exp. mar. Biol. Ecol. 132: 221–245.

Wilson, W. H., 1990. Competition and predation in marine soft-sediment communities. Ann. Rev. Ecol. Syst. 21: 221–241.

Woodin, S. A., 1981. Disturbance and community structure in a shallow water sand flat. Ecology 62: 1052–1066.

Woodin, S. A., 1983. Biotic interactions in recent marine sedimentary environments. In M. J. S. Tevesz & P. L. McCall (eds), Biotic Interactions in Recent and Fossil Benthic Communities. Plenum Press, New York: 3–38.

Woodin, S. A., 1991. Recruitment of infauna: positive or negative cues? Am. Zool. 31: 797–807.

Zajac, R. N., 1991. Population ecology of *Polydora ligni* (Polychaeta, Spionidae). II. Seasonal demographic variation and its potential impact on life history evolution. Mar. Ecol. Progr. Ser. 77: 207–220.

Zajac, R. N. & R. B. Whitlatch, 1982a. Responses of estuarine infauna to disturbance. I. Spatial and temporal variation of initial recolonization. Mar. Ecol. Prog. Ser. 10: 1–14.

Zajac, R. N. & R. B. Whitlatch, 1982b. Responses of estuarine infauna to disturbance. II. Spatial and temporal variation of succession. Mar. Ecol. Prog. Ser. 14: 15–27.

Zajac, R. N. & R. B. Whitlatch. 1985. A hierarchical approach to modeling soft-bottom successional dynamics. In Gibbs, P. E. (ed), Proceedings of the 19th European Marine Biological Symposium. Cambridge University Press, Cambridge: 265–276.

Zajac, R. N. & R. B. Whitlatch, 1988. The population ecology of *Nephtys incisa* in central Long Island Sound prior to and following disturbance. Estuaries 11: 117–133.

Zajac, R. N. & R. B. Whitlatch, 1989. Natural and disturbance induced demographic variation in an infaunal polychaete, *Nephtys incisa*. Mar. Ecol. Progr. Ser. 57: 89–102.

Zajac, R. N. & R. B. Whitlatch, 1991. Demographic aspects of marine, soft sediment patch dynamics. Am. Zool. 31: 808–820.

Hydrobiologia **375/376:** 241–254, 1998.
S. Baden, L. Pihl, R. Rosenberg, J.-O. Strömberg, I. Svane & P. Tiselius (eds), Recruitment, Colonization and Physical–Chemical Forcing in Marine Biological Systems.

Physiological flexibility; a necessity for life in anoxic and sulphidic habitats

Lars Hagerman
Marine Biological Laboratory, Copenhagen University, Strandpromenaden 5, Dk-3000 Helsingør, Denmark
E-mail: mbllh@inet.uni2.dk

Key words: tolerance, aerobic/anaerobic metabolism, blood pigment

Abstract

Shallow marine areas and brackish water are often characterized by extensive short- or long-term fluctuations in physical parameters: temperature, salinity, oxygen. Sulphide can accumulate as a consequence of oxygen deficiency.

Diel fluctuations in oxygen can in non-exposed shallow waters vary from 0 to 300% saturation, at the same time temperature can increase to 30 °C while salinity can decrease or increase. These types of short-term fluctuations impose a great demand on the immediate response of the organism: the organism must be able to maintain 'normal' physiological processes and behaviour as long as possible to minimize, for instance, the risk of predation by more tolerant species. Apparently competing species can on a small scale be separated in space by different abilities to tolerate diel or local fluctuations in environmental conditions. Short-term adaptations are often different from those used under long-term fluctuations which often cover larger areas.

Energetics are most favourable with aerobic metabolism. The maintenance of a constant aerobic metabolism over as wide an oxygen range as possible is influenced by various ex- and in-trinsic factors but must be considered a primary adaptation to decreasing oxygen tensions in active benthic animals. Active animals that often are exposed to large variations in oxygen tensions have developed the best behaviour to increase the availability of oxygen and thus to regulate their aerobic metabolic rate despite an often low tolerance of severe hypoxia or anoxia. High temperatures decrease the ability to regulate and might force an anaerobic metabolism. Besides behavioural and ventilatory/circulatory adaptations improved haemolymph pigment characteristics are important in supplying oxygen under hypoxia. These adaptations are well developed in eurythermal and euryhaline shallow water and rock-pool crustaceans but best developed in sediment-living organisms which besides hypoxia can also be influenced by hypercapnia and sulphide. The highly toxic sulphide complicates the pattern. Active epibenthic organisms can simply avoid sulphide by moving away. Endobenthic or inactive organisms must try to exclude it temporarily, or adapt to the presence of sulphide. Burrow-constructing benthic invertebrates are dependent on their ability to ventilate and thus oxygenate at least parts of their burrows to oxidize the sulphide diffusing into the burrow water. During hypoxia in the overlying water or during ventilatory pauses the oxygen in the burrows will not be sufficient and the animal will be exposed to sulphide.

Animals digging and crawling into the sediment more frequently and over longer time are more continuously exposed to severe hypoxic, anoxic and sulphidic conditions and have therefore developed strategies permitting survival under long anoxic and high sulphide exposure. Long-term exposure involves adaptations to decrease metabolic expenditure during aerobic metabolism and to use anaerobic pathways with a miminum drain on energy stores, both as a protection and to prolong survival. Benthic animals from a wide range of phyla have developed different strategies resulting in survival for many weeks under adverse environmental conditions.

Introduction

Fluctuations in environmental parameters will always be more or less stressful to marine and brackish organisms. The extent of the stress depends on the physiological adaptability of the organism which in turn determines the tolerances of an environmental factor or, in reality, a combination of several environmental factors. It is the organisms most exposed to variations, fluctuations and extremes of environmental factors that are also those with best developed physiological and biochemical adaptive potentials and best tolerances. This ought to be a natural fact, but when studying literature it is often described as a surprise that certain organisms are found where they live as, for instance, salinity or oxygen tension was low.

Variations in these environmental factors can be short-term or long-term, fluctuations within hours, as for tidal variations, fluctuations with a 24 h periodicity, seasonal or even variations from year to year. It is of course clear that periodicity as well as amplitude of the variations are of utmost importance to the organisms. Fluctuations with the greatest amplitude occur in shallow waters such as rock-pools or shallow fjords, fluctuations or extremes with the longest duration can occur in greater water volumes such as for instance the Baltic.

Salinity does not vary very much except in tidal estuaries and sometimes in rock-pools. In tidal estuaries it can vary during low tide due to freshwater overflowing the tidal areas, even though the fluctuations in the sediment can show a considerable delay and have less amplitude than the overlying water (Kinne, 1966). Fluctuations in rock-pools are caused by evaporation or rainfall and long-term calm weather can increase or reduce the salinity considerably (Ganning, 1971a; Morris & Taylor, 1983). In brackish areas where salinity already is low, for instance in the Baltic, the freshwater component of the fauna can be important, so in the case of a marine organism decreasing salinity can be stressful, but for a limnic organism it is the increasing salinity that can be disasterous (Ganning, 1971b). In both cases, osmo/ionregulatory capacities, low permeability and regulation via the excretory organs determines if the organism can survive or not. Conditions in rockpools can be extreme not only with regard to salinity, temperature can vary to over swimming-pool temperatures in summer and bottom-freezing in winter (Ganning, 1971a; Morris & Taylor, 1983). There is also a great heterogeneity between pools and also within the individual pool. Rockpools and to a certain extent estuaries are probably the only biotopes where salinity shows pronounced short-term rhytmicity. It is clear then that only euryhaline organisms will be found here. Examples of such organisms with wide salinity tolerances are the crustaceans *Palaemon adspersus*, *Palaemonetes varians* and *Praunus flexuosus* (McLusky, 1979; Hagerman & Uglow, 1983; Nielsen & Hagerman, 1998). One should note that the tolerances are not the same in summer and winter, and also that the changes are not the same from species to species. These species are excellent osmoregulators. It should also be noted that osmoregulation means many things; not only anisosmotic extracellular regulation as in these crustaceans, but also isosmotic intracellular regulation, that is regulation via organic substances as amino acids. Euryhaline molluscs stick to this latter type of regulation (Lange, 1978). Extreme salinities often coincide with temperature extremes in the rockpools and this makes the environment even more stressful, especially at high temperatures as the metabolic demand both of organisms and of the whole ecosystem will increase tremendously.

Oxygen varies from a considerable over-saturation during the day to severe hypoxia at night (Ganning, 1971a; Truchot & Duhamel-Jouve, 1980; Nielsen & Hagerman, 1998). Photosynthetic activities as well as respiration will influence the CO_2 content and thus the pH. The wide fluctuations in oxygen and associated factors in rockpools are quite substantial. Water PO_2 can fall at night to zero within hours, total carbon dioxide and PCO_2 increases, the latter up to 1–3 torr. During daytime the water PO_2 can increase to 400–600 torr, pH can increase to above 10 and total carbon dioxide decreases and PCO_2 can fall as low as 0.0001 torr (Truchot & Duhamel-Jouve, 1980). The main processes responsible for such variations are biological, i.e. photosynthesis and respiration. Hypoxic nocturnal conditions in rockpools are always associated with hypercapnic waters, and conversely hyperoxic always with hypocapnic. In addition to this, substantial amounts of sulphide, ammonia and other compounds can accumulate at nights.

The changes seen in rockpools are also seen in shallow areas where water movement is low, for instance in the Danish fjords, in the inner part of the Baltic archipelagos and in lagoon-like areas (Muus, 1967; Nielsen & Hagerman, 1998). These areas all show drastic short-time fluctuations, although often with another periodicity, and have a diverse fauna characterised by organisms able to adjust their behaviour or physiology to the extremes which often prevail just for a couple of hours.

Behaviour, tolerance

Behavioural adaptations to avoid hypoxia or otherwise adverse conditions are straightforward. The simplest is to move away, horizontally or vertically. Closer to the water surface PO_2 is often a little higher and a typical response is to seek the surface. One species, the shrimp *Palaemonetes varians* can even lay in the water surface or jump out of the water and adhere to *Phragmites* to increase the availability of oxygen (Hagerman & Uglow, 1984). The shore crab, *Carcinus*, may also partly emerge from hypoxic water (Hill et al., 1991) Species with a low tolerance, for example the brown

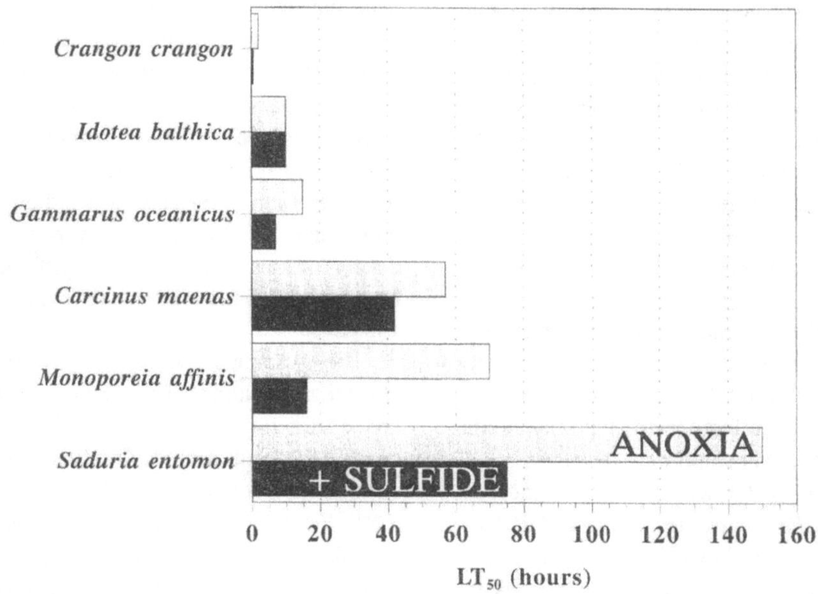

Figure 1. Tolerance of anoxia and of hypoxia + sulphide in some Baltic crustaceans. From Dries & Theede, 1973; Hagerman & Vismann, 1993; Vismann, 1996; Sandberg, Vismann & Hagerman, 1998.

shrimp *Crangon crangon* (Dries & Theede, 1974) and most fishes (Pihl et al., 1991) simply migrate to more oxygenated waters. Otherwise it is just a question of tolerance, how long do the adverse conditions prevail? Tolerance in itself is a question of physiological capacity and adaptability. The tolerance varies from species to species and the classical study of Dries & Theede (1974) gives us a clear picture of the tolerances of many Baltic invertebrates; the lower the activity and the longer the possible lifespan the better is the tolerance of anoxia. For example, *Crangon crangon* has an LT_{50} in anoxia of only 2 h while the bivalve *Arctica islandica* has an LT_{50} of 2000 h (Dries & Theede, 1974). Tolerance is further complicated if sulphide is also involved (which occurs in nature if severe hypoxia or anoxia prevails). Sulphide shortens tolerance due to its toxic effects. In the non-tolerant *Crangon*, survival is halved when sulphide (5 μM) is present under hypoxia compared to anoxia (Figure 1) (Hagerman & Vismann, 1995). The small Baltic amphipod *Monoporeia affinis* has an anoxic mortality of less than 1% h^{-1}, this increases to 6% h^{-1} in 50 μM sulphide (Sandberg et al., 1998). The Baltic isopod *Saduria entomon* has a high tolerance of anoxia (LT_{50} 150 h), this tolerance is reduced by 50% in hypoxia (20–30% sat.) and 150 μM sulphide (Figure 1) (Hagerman & Vismann, 1993). Another example is the mudshrimps *Calocaris macandreae* and *Callianassa subterranea* where survival is less in 1 mm sulphide under near normoxic conditions than under anoxic conditions (Johns et al., 1997). It should be noted that in the latter cases the tolerances are at least day(s), in the former the tolerances were minutes. The high tolerance of the mudshrimps is due to the fact that they can be exposed to sulphide in their burrows or when digging, while *Crangon* rarely is exposed to sulphide and avoids it by violent escape reactions.

Spatial distribution

Species which are closely phylogenetically and/or ecologically related, almost 'co-existing' species have developed small physiological differences with regard to, for instance, tolerances of hypoxia/anoxia and sulphide which separates them spatially and sometimes also temporally. The marine and brackish water polychaetes *Neanthes (Nereis) virens* and *Hediste (Nereis) diversicolor* can in some areas be found together. Both are benthic species, living in burrowed tubes. Long term survival under severe hypoxia (10% saturation) showed *H. diversicolor* to be more tolerant than *N. virens*. Under hypoxia and sulphide incubation (175 μM, pH 8.0–8.2) the difference was more evident. Short-time exposure to high sulphide concentrations under anoxia showed that the acute toxicity of sulphide in *N. virens* was twice that of *H. diversicolor* (Vismann, 1990). Short time exposure reflects situations when worms are penetrating anoxic sediment when constructing new tubes, the former experiment reflects conditions which can prevail especially during summertime. Coexistence of these two species is possible in a wide range of environmental conditions. In nature the species are separated by interference competition: aggression and territorial behaviour. Sulphide is often not homogeneously distributed and *H. diversicolor* is often found in the more sulphide-rich patches in zones where the two species are overlapping. In other words, the higher sulphide detoxification capacity in *H. diversicolor* allows it to stay in patches where the sulphide is too high for *N. virens* (Vismann, 1990). It is clear, that a high tolerance to hypoxia/anoxia correlates with tolerance to sulphide.

Another example is the seasonal coexistence of two shrimp species *Palaemon adspersus* and *Palaemonetes varians*. In winter they occur together but are spread out at depths of 2–3 m in some Danish fjords, in summertime they migrate to shallow water where the density is higher, but now they are spatially separated (Hagerman & Uglow, 1983). *P. adspersus* is bigger and more aggressive than *P. varians*. If held together at high densities in an aquarium, *Palaemonetes* will swim close to the surface, while *Palaemon* will rest on the bottom. Both species are euryhaline, but *Palaemonetes* can be regarded as holeuryhaline (Hagerman & Uglow, 1983) and also more tolerant of high temperatures. Comparing their tolerances of severe hypoxia, *Palaemonetes* show a much better tolerance (Nielsen & Hagerman, 1998). This difference in tolerance of low salinity, high temperature and low oxygen allows *Palaemonetes* to live in the innermost part of the fjord where the environmental variations are more extreme. The spatial differentiation between the two species in summer can be within 1–2 m.

The infaunal brittlestar *Amphiura filiformis* and the epifaunal *Ophiura albida* live in and on the same substrate. In general, tolerance of hypoxia and sulphide is closely related to the *in situ* levels, It can therefore be expected that the infaunal *A. filiformis* is more tolerant of these factors than the epibenthic *O. albida*. And so it is, *A. filiformis* has a significantly higher tolerance of both hypoxia and sulphide than *O. albida*. In oxygen saturations less than 1% they had LT_{50} values of 7.5 and 2.5 d respectively and the presence of even very small concentrations of sulphide (pH 8.1) decreased the survival significantly, strongly suggesting sulphide to be the key or limiting factor for the survival of both species (Vistisen & Vismann, 1997). During exposure to oxygen of less than 10%, all *A. filiformis* emerged from the sediment, a normal reaction for all infaunal species. On the sediment they often lifted the disc up into the water column and survived for many days like this. *O. albida*, on the other hand, showed increased activity (escape response) lasting for 2 to 4 d after which they turned to inactivity with the disc lifted. At oxygen saturations below 1% they quickly became moribound. Under anoxia, even very low sulphide concentrations (0–20 μM total sulphide) are important for survival of the brittle stars (Vistisen & Vismann, 1997).

The examples used show how organisms living close together and sharing certain resources like food, are separated in space due to different tolerances to a key factor, often sulphide or oxygen or to other factors such as temperature. Over the years we have studied the Baltic isopod *Saduria entomon* (Hagerman & Szaniawska, 1988, 1990, 1991, 1992, 1994; Hagerman & Vismann, 1993, 1997; Vismann & Hagerman 1993, 1996). Even if this species is extremely euryhaline and euryoxic etc., it is rather stenothermal and this prevents it from penetrating into very shallow water in summer and also from migrating out into the Øresund and the Kattegat area (but here competiton from crabs etc is also much greater).

Aerobic respiration in hypoxia

Short-term adaptations to low oxygen always involve keeping an aerobic metabolism as long as possible, both with regard to time and to as low an oxygen tension as possible. It is often just a question of hours, either a low-tide period or just a few hours at night before the oxygen increases again, as mentioned earlier. Many species with a low tolerance survive overnight hypoxia or even anoxia. A mobile, active species like the brown shrimp *Crangon crangon* has a tolerance of severe hypoxia/anoxia which is very temperature and salinity dependent, LT_{50} in 10% oxygen saturation can be so low as 1.5 h in a salinity of 5 psu but increases to 6 h at 20 psu (Hagerman & Szaniawska, 1986).

The usual way to increase oxygen availability is to increase the ventilation (and the irrigation), that is the flow of water over the permeable areas primarily the gills. In polychaetes this means increased ventilatory movements, in crustaceans an increased scaphognathite and/or uropod/pleopod beat rate. Heart rate seems to remain relatively the same during hypoxia, although such things as cardiac arrests do not occur (Grieshaber et al., 1994). Stroke volume can sometimes increase. Many species can thus regulate

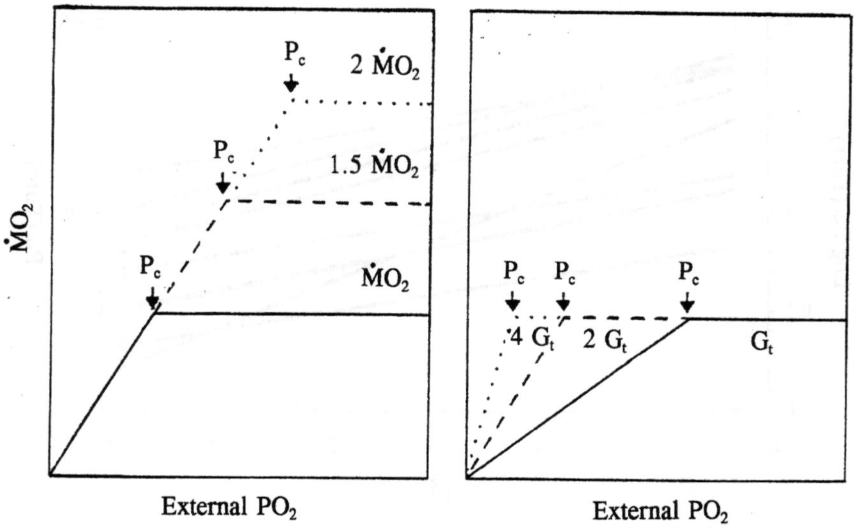

Figure 2. Changes in oxygen consumption with oxygen tension of the water. A. Changes in P_c due to changes in factors like temperature, salinity and variations in activity level (incl. laboratory stress). B. Changes due to improved oxygen characteristics of the blood pigment and improved internal circulation. MO_2: aerobic respiration, P_c: critical oxygen tension, G_t: gas conductance. $MO_2 = G_t$ (external PO_2 – internal PO_2). Redrawn from Herreid, 1980.

their oxygen consumption rate, their aerobic respiration, over a wide range of oxygen tension. They are, with a not specially good terminology called oxygen 'regulators'. Other species are 'conformers', their respiration decreases when environmental oxygen tension decreases. It can, however, be discussed if this latter type really exist among invertebrates tolerant of hypoxia. The point where O_2 consumption starts to fall is called the P_{crit} or P_c and this 'point' is dependent on both environmental conditions and the physiological state of the animal. P_c increases with temperature, it depends on salinity, on time of day, on activity, on size in some species, on amount of respiratory pigment and of its oxygen binding characteristics, on circulation and respiration and in crustaceans also on moult stage (Figure 2) (Herreid, 1980). Finally, and extremely important: it depends on laboratory stress. In general the lower the P_c the better the tolerance to anaerobic conditions. We can look a little historically at where P_c has been placed in some common crustaceans. For instance *Homarus* was originally found to be an oxygen conformer (Thomas, 1954). This experiment was done with fettered animals in a closed chamber. Later, the same species was found to be a relatively good regulator using animals in an open respiratory system and only partially fettered. Later again, the same species was found to regulate oxygen uptake down to ca. 25% saturation, now with lobsters allowed to become quiescent but freely mobile (Butler et al., 1978). Similar examples can be found with *Crangon*, *Palaemon*, *Saduria* and many other species. So, as our knowledge of the biology, ecology and physiology of a species increases the P_c will move more and more to the left. It can, however, not go to zero even if it can be rather close like in *Saduria* (Hagerman & Szaniawska, 1988).

The functional role of the blood pigment

To increase the availability of oxygen is only part of the adaptation to hypoxia, the oxygen must enter the body via diffusion and be transported to the metabolizing tissues where it is needed. This transport takes place either with the oxygen physically dissolved in the haemolymph or bound to a blood pigment, normally haemoglobin or haemocyanin.

The physically dissolved oxygen is a function of oxygen in the environment, diffusion barriers over the gills/body surface and the activity level of the animal. The physically dissolved oxygen tension has been measured in the pericardium of *Saduria entomon* with a microelectrode implanted directly in the pericardium. Resting animals show small, regular variation in postbranchial oxygen tensions, sudden bursts of activity causes an immediate reduction in oxygen tension and periods of no ventilation cause anoxic haemolymph within a few minutes (Vismann & Hagerman, 1993). Only circa 30% of the gill area was effective in oxygen transfer but this area can possibly

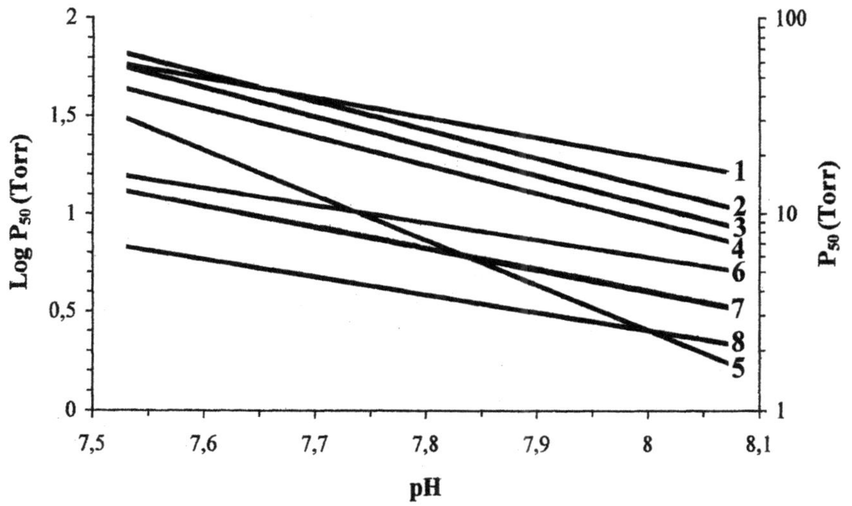

Figure 3. The relation between P_{50} (log P_{50}) and pH for haemocyanin from a number of crustaceans: 1: *Palaemon serratus* (10 °C) (Taylor & Spicer, 1989), 2: *Palaemon adspersus* (15 °C) (Weber & Hagerman, 1981), 3: *Crangon crangon* (15 °C) (Hagerman & Vismann, 1995), 4: *Palaemonetes varians* (Nielsen & Hagerman, 1998), 5: *Saduria entomon* (10°C) (Hagerman & Vismann, 1997), 6: *Nephrops norvegicus* (10 °C) (Johns, 1996), 7: *Calocaris macandreae* (10 °C) (Johns, 1996), 8: *Natatolana borealis* (10 °C) (Taylor & Moore, 1995).

be increased by a more active ventilation. This study clearly showed that the amount of physically dissolved oxygen depends totally on the momentary behaviour of the animal and that it is the immediately available oxygen source.

The changes of the blood pigment during hypoxia are primarily an increase in the concentration of pigment, a synthesis resulting in an increase in the total amount of oxygen that can be carried. This synthesis during hypoxia has been shown again and again and is certainly valid for all organisms. The synthesis takes place, however, only if the animal can be active and feeding (Hagerman, 1986). The blood pigment oxygen affinity, P_{50}, is dependent on temperature, on blood pH which in turn depends on acid base balance, CO_2 content, lactic acid concentration etc (Truchot, 1992). It is natural to think that species normally exposed to hypoxia also have an efficient blood pigment with a low P_{50}, i.e. a high affinity for oxygen (the sigmoid curve is to the left), a pronounced Bohr shift and a high cooperativity. This is true for a variety of crustaceans when compared at similar temperatures and *in vivo* pH. The Bohr effect, the change in logP_{50}/change in pH, shows that with a normal Bohr effect O_2 affinity decreases with decreasing pH. In many species moderate hypoxia is accompanied by blood alkalosis (washing CO_2 out of the body due to increased ventilation, this changes acid base balance and results in increased pH) (Truchot, 1987) and this moves the affinity curves to the left, i.e. it adaptively increases O_2 affinity.

It has also been observed in several crustaceans that exposure to hypoxia can enhance oxygen affinity (De Fur et al., 1990; Truchot, 1992). This can be caused by increased levels of urate and lactate (Lallier et al., 1987; Lallier & Truchot, 1989) where the urate comes from ammonia which changes to urate production during hypoxia and the lactate results from a partial anaerobic metabolism. Other modulating factors can also be involved (Bridges et al., 1997). The specific effect of lactate ions counteracts the Bohr effect. During, for example, exercise (functional) anaerobiosis, the accumulation of H^+ ions associated with lactate will decrease the haemolymph pH and thus counteract the increase caused by hyperventilation while lactate ions *per se* will improve the affinity. The affinity can, however, also change without modulating factors by long-term adaptation causing changes in sub-unit structure (Mangum, 1990; De Fur et al., 1990). An example of this is again the brackish shrimp *Palaemonetes varians* in which oxygen affinity increased from a P_{50} of 18 Torr to 10 Torr at pH 7.80 after 4 d adaptation at 24 °C and 10 Torr hypoxia (Nielsen & Hagerman, 1998). This is at an oxygen tension where the metabolism in this species is still fully aerobic. Some species, like *Saduria entomon* already have such a low P_{50} (Figure 3) that a further change due to hypoxia is more or less meaningless.

Instead the blood pigment is permanently adapted to hypoxic conditions (Hagerman & Vismann, 1997). The hypoxia-induced respiratory alkalosis will also *in vivo* further increase the haemocyanin oxygen affinity.

Burrow constructing animals like *Nephrops norvegicus* and especially mud-shrimps (*Upogebia deltaura* and *Calocaris macandreae*) live in burrows that for their oxygenation is totally dependent on the activity of the individual animal. The burrows are often deep and complex. Even under normoxic conditions in the overlying water there is a progressive hypoxia with depth inside the burrow. The pH profile in the burrow is similar to that of oxygen with a lower pH in the deeper parts, reflecting the hypercapnic conditions in these regions (Astall et al., 1997). *Upogebia* seems to ventilate the burrow with its pleopods with the same rate regardless of PO_2, but the time spent irrigating increased with decreasing PO_2 with a maximum irrigation around 35 torr and the scaphognathite and heart continued to beat slowly in *Upogebia* even after 10 h of anoxia (Astall et al., 1997). When the mud-shrimps stop ventilating the PO_2 of the water steadily declines, but just body movements or irrigation results in rapid increases. With regard to oxygen consumption, heart and scaphognathite rates the mud-shrimps behave as 'normal' crustaceans with a good adaptation to hypoxia but not to the same extent as burying crustaceans. The haemocyanin of mud-shrimps has been shown to have an oxygen affinity that is higher than for non-burrowing crustaceans, but not so high as for some digging crustaceans like *Saduria* (Figure 3). The P_{50} for two *Upogebia* species at an *in vivo* pH of 7.8 were 6.3 and 7.5 torr (Astall, 1993) while for *Saduria* at pH 7.8 the P_{50} was 3.2 torr (Hagerman & Vismann, 1997). The isopod *Natatolana borealis* can maintain its aerobic respiration to an oxygen tension of 15–25 torr (2.0–3.3 kPa) but has a low haemocyanin oxygen carrying capacity with a high affinity (2.99 torr, pH 7.8, 10 °C Taylor & Moore, 1995). As mentioned earlier for decapods as well as amphipods or isopods, the high oxygen affinity may be an important adaptation to prolonged hypoxic periods provided an aerobic metabolism is possible.

One can both speculate on and calculate the role of the blood pigment. The O2 capacity of haemocyanin in crustaceans is low (1–3 vol%) compared to that of haemoglobin in vertebrates (6–15 vol%), it is only around double what can be carried in physical solution (Mangum, 1980). One reason for the high individual variability of haemocyanin is that it serves not only as an oxygen carrier but also an organic reserve. Haemolymph protein and haemocyanin concentrations show a steady decrease in starving animals (Hagerman, 1983, 1986). It is therefore not a general rule that exposure to hypoxia induces more blood pigment or increases the affinity for oxygen. It depends very much on whether a blood pigment is actually required. Is the blood pigment necessary at all? Recent investigations on the small Baltic amphipod *Monoporeia affinis* which is often exposed to hypoxia show a pigment where oxygen affinity is low, 4 kPa (30 torr) at pH 7.5, 10 °C, the Bohr factor is low and the cooperativity is low, in comparison with for example decapods (Hagerman et al., 1997). All this points to a very modest role of the blood pigment and a 'simple' structure. The amphipods are very small, they are rather permeable and much of the gas exchange in amphipods is extrabranchial. This reduces the need for an O_2 transporting pigment; the physically dissolved oxygen is enough remembering that this species is a cold water species, that its metabolic rate is low and oxygen availability even at hypoxia is better than at higher temperatures. With the relatively high O_2 carrying capacity of the haemolymph the pigment may, however, still be used as an emergency O_2 reserve during short term stress as exposure to short term hypoxia or during bursts of activity, i.e. functional hypoxia (Hagerman et al., 1997).

That animals can live without a blood pigment is seen in some Antarctic fish and also a population of *Cancer pagurus* has been found without blood pigment (even if this later case must be regarded as an isolated phenomenon) (Bottoms, 1977). It has also been found that hyperiid amphipods do not possess haemocyanin even though they are active swimmers capable of diel vertical migrations and they are relatively large compared with other amphipods (Spicer & Morritt, 1995).

Adaptive strategies to sulphide exposure

Burrow-living organisms face the added problem that they must get rid of excretory products like ammonia which accumulate in the burrow both as a result of excretion from the animal but also from microbial activity in the burrow wall. Ammonia concentrations can be more than 10-fold those of the overlying water (Astall et al., 1997). Ventilation of the burrows is necessary not only for respiratory purposes and to flush out excretory products but also for oxidising sulphide diffusing from the sediment. During hypoxia there might not be enough oxygen to oxidise the sulphide and the animal then has the option to exclude the sulphide from entering the body or, if this is impossible to 'detoxify' it to harmless thiosulphate, sulphite or sulphur (Vismann, 1991b; Völkel & Grieshaber, 1995). These latter cases still need some oxygen. If these adaptations are not possible, the animal has to protect itself by turning to an anaerobic metabolism. A certain tolerance of sulphide is necessary not only due to exposure to periodic hypoxia, but also when digging into sediment the organism cannot avoid being exposed to high concentrations of sulphide.

A primary protection to exposure to sulphide is thus to try to escape. If this is not possible the animal may be able to prevent sulphide from entering the body. This can be done by closing the valves in bivalves. Exclosure is in all animals only a short-term protection, if oxygen and ions can enter the body via diffusion, then sulphide can also enter. The diffusion of sulphide into the body is pH dependent and it is the internal pH that decides how much of the sulphide will be in the form of H_2S, the most poisonous component as shown for *Crangon* by Vismann (1996). So, protection by preventing diffusion is more or less impossible. Some sulphide can be trapped in the integument by binding to iron, for example in the priapulid *Halicryptus spinulosus* (Oeschger & Vetter, 1992). After a couple of hours, however, this mechanism does not work any longer and sulphide enters the body. But if sulphide in the sediment is patchily distributed, even this trapping can be quite efficient for short-term exposure. Once in the body the sulphide must, however, be detoxified (oxidised) to more harmless substances. Most of the sulphide entering the tissues is oxidised to thiosulphate and some to sulphite and in this way the animal is able to prevent sulphide concentrations in the body increasing to lethal concentrations (see e.g. Vismann, 1991a; Johns et al., 1997; Völkel & Grieshaber, 1994; Jahn et al., 1996; Hagerman & Vismann, 1993). The production of thiosulphate is energetically the most favourable since 1 mole of sulphide is removed for every 1.5 moles of oxygen. The oxidation to sulphite and sulphate requires 3 and 4 moles of oxygen per mole of sulphide (Johns et al., 1997). The oxidation takes place both in the hepatopancreas and in the muscle tissues. It seems to vary a little from organism to organism but generally the hepatopancreas seems to be important. It has been shown that the mud-shrimp *Calocaris macandreae* can use aerobic metabolism when exposed to up to 25 μM sulphide under severe hypoxia before switching to anaerobic metabolism (Johns et al., 1997). The increase in anaerobic metabolites, in this case lactate, was coincident with a decrease in thiosulphate accumulation. It was also evident that at normoxic sulphide exposure the change to an anaerobic metabolism took place at higher sulphide exposure. Similar phenomen have been shown for the burying isopod *Saduria entomon* (Vismann, 1991a; Hagerman & Vismann, 1993) and the data show an oxygen-dependent detoxification mechanism. It is also clear that when no oxygen is available, the sulphide oxidation mechanism in the mitochondria does not work (Völkel & Grieshaber, 1996). The mitochondrial oxidation of sulphide has been shown in a number of animals for instance in *Halicryptus spinulosus* (Oeschger & Vetter, 1992),the polychaetes *Heteromastus filiformis* (Oeschger & Vismann, 1994) and *Arenicola marina* (Völkel & Grieshaber, 1994) and the crustacean *Saduria entomon* (Vismann, 1991a). The mitochondrial sulphide oxidation works only at low micromolar sulphide concentrations and, of course, in the presence of oxygen. At higher sulphide concentrations a complete inhibition of the respiratory chain occurs. Thus, we can see that an anaerobic metabolism is initiated at higher sulphide concentrations even if oxygen is present at concentrations that would normally permit an aerobic metabolism. The anaerobic metabolism is used both as a defence against high sulphide concentrations and to anoxic conditions.

Anaerobic metabolism under long-time hypoxia/anoxia

All behavioural changes and physiological adaptations dealt with so far, with the exception of the tolerance and some of the hydrogen sulphide problems, are thus only valid if oxygen is also available, even if it is only in small amounts. If absolutely no oxygen is present or only hypoxia and sulphide, the animal has to turn to anaerobic metabolism and then quite different strategies take over depending on whether the anoxic conditions prevail for a longer or for a shorter time.

Facultative anaerobiosis is a general phenomenon among benthic invertebrates but the metabolic pathways used in anaerobiosis can vary from group to group, from species to species and sometimes also within the species or individual depending on external and internal circumstances (for a detailed recent review on anaerobic metabolic pathways, see Grieshaber et al., 1994). Anaerobic metabolism can be either a functional (exercise) or an environmental anaerobiosis, the latter caused by anoxia or by high sulphide concentrations. Active species such as crustaceans and some errant polychaetes use normal glycolysis with lactate (lactic acid) as the anaerobic end-product resulting from both exercise and environmental anaerobiosis. The tolerance to anoxia is closely connected to anaerobic capacity and the pathway used. Low tolerance, that is short survival, is always associated with glycolysis with lactate as end-product. The accumulated lactate depends on the amount of substrate, mostly glycogen, and the rate at which the glycogen store is depleted. For all invertebrates it is an advantage to keep an aerobic metabolism as long as possible since this gives most ATP (38) per mole glucosyl/glycogen oxidized (Hochachka & Somero, 1984). To use anaerobic metabolism with the same energy output would mean an enormous drain on the energetic stores and survival would be very short. It is therefore important also for invertebrates with lactate as the end-product during exposure to environmental anoxia to change behaviour and to decrease activity to a minimum (i.e. resting, inactivity) to make sure that glycogen will last as long as possible. The yield of ATP in glycogen – lactate fermentation is only 3 moles per mole glucosyl/glycogen oxidized (compared to 38!). So during anoxia the glucosyl/glycogen consumption rate would need to rise 12–13 times if demand for ATP remained unchanged. The organism can therefore, to survive as long as possible, either have more glycogen in the body, or utilize other more efficient fermentations, or reduce its metabolic rate more than being just inactive. It seems that all three mechanisms are valid depending on the organism and the circumstances. Animals with a high tolerance of anoxia generally have a high glycogen level but this can still account for only a small part of the total tolerance time. Substrates other than glycogen are sometimes present, e.g. the different phosphagens (phosporylated guanidinium compounds). It is just one step to transphosphorylate phosphagen to ATP so that energy can immediately become available during the start of severe hypoxia (de Zwaan & Putzer, 1985). This also means, however, that the phosphagen stores are quickly depleted. The phosphagen levels vary in different species but mobile (active) animals have higher contents than inactive species. This also indicates that the phosphagens are used mostly in functional anaerobiosis such as escape reactions when the abdominal muscle in crustaceans is used (Onnen & Zebe, 1983).

The degradation of the amino acid L-aspartate with a synthesis of L-alanine (the aspartate-alanine shuttle) under anaerobiosis has also been shown to occur in a number of marine organisms such as *Scoloplos armiger* (Schöttler & Grieshaber, 1988) and *Cardium edule* (Meinardus-Hager & Gäde, 1981). The use of aspartate can function if the concentration of pyruvate increases from glycogen breakdown and if the enzyme malate dehydrogenase restores NAD^+ for oxidation of glyceraldehyde-3-phosphate. Pyruvate can therefore accept the amino group instead of serving as an H^+ sink. Alanine would be the result of all this and is probably an end product of anaerobiosis in anoxia tolerant marine animals having aspartate as a substrate for example in molluscs and some polychaetes (de Zwaan & Putzer, 1985; Grieshaber et al., 1994). The use of aspartate is, like glycogen to lactate in crustaceans, under short-term severe hypoxia/anoxia in the groups mentioned. The rest of the aspartate which is not degraded to alanine, i. e. the carbon-containing part, is converted to succinate (Zebe, 1975). It is, via malate, dehydrated to fumarate and reduced to succinate in the mitochondria.

Glycolysis follows the normal cytoplasmic Embden-Meyerhof-Parnas route during severe hypoxia/anoxia until the formation of PEP, phosphoenolpyruvate. Here PEP-carboxykinase catalyses carboxylation of PEP to oxaloacetate. Oxaloacetate comes initially from aspartate (if the organism had any) but in longer lasting anaerobiosis it comes from glycogen. The oxaloacetate is reduced to malate and the rest of the anaerobic metabolism takes place in the mitochondria. The important step here is a dehydration catalyzed by fumarase to fumarate. The enzyme succinate dehydrogenase reduces fumarate to succinate giving another two moles ATP per glycosul unit that is fermented (Holwerda & de Zwaan, 1979). The reaction will only work if the PO_2 within the mitocondria is below 0.2 torr which is the P_{50} of the cytochrome c oxidase. Below this, the respiratory chain will be limited by oxygen and succinate will increase. Succinate can therefore be a good proof of a mitochondrial anaerobic metabolism. For animals surviving long periods of anoxia it is better than lactate or opines (from phosphagens) as these latter will have depleted their substrate more rapidly. In other words, animals producing lactate do not normally produce succinate at all or at the same time.

Malate is not only metabolized to succinate, but also to other substances like pyruvate, alanine, acetate and propionate. In *Mytilus* de Zwaan et al. (1981) showed that 0.61 mol succinate and 0.11 mol acetate and 0.08 mol propionate was formed from 1 mol malate. The anaerobic metabolism with succi-

Table 1. Summary of anaerobic pathways and their respective ATP yield. After Hochachka & Somero (1984) and de Zwaan & Putzer (1985).

Glycosyl U + $6O_2$ + 39ADP → $6CO_2$ + $6HO_2$ + **39ATP**
Glycosyl U +3ADP → 2 lactate + **3ATP**
Glycosyl U + 2ASP + 4.71ADP → 1.71 succinate + $1.14CO_2$ + 2ALA + **4.71ATP**
Glycosyl U + 2ASP + 6.43ADP → 1.71 propionate + $2.8CO_2$ + 2ALA + **6.43ATP**
Glycosyl U + $0.86CO_2$ + 4.71ADP → 1.71 succinate + **4.71ATP**
Glycosyl U + 6.43ADP → 1.71 propionate + $0.86CO_2$ + **6.43ATP**

nate as the end-product is known from a large number of organisms, e.g. *Halicryptus spinulosus* (Oeschger, 1990), *Arenicola marina* (Schöttler et al., 1984a) and *Arctica islandica* (Oeschger, 1990). *Hediste (Nereis) diversicolor* is an example of a flexible species which can resort to both glycolysis with lactate and glycolysis with succinate as the end-product (Schöttler et al., 1984b), probably due to different behavioural strategy during the period of anoxia. The further production of volatile fatty acids, mainly acetate and propionate, is known in many molluscs and annelids such as *Cerastoderma edule* (Gäde, 1975), *Arenicola marina* (Zebe, 1975; Pörtner et al., 1979) *Nephtys hombergi* (Schöttler, 1982) and some species of *Nereis* (Schöttler, 1979), *Halicryptus spinolosus* (Oeschger, 1990) etc. The fatty acids are excreted into the surrounding water. It should be emphasized again that the production of fatty acids takes place during prolonged anoxia, i. e. only in the very tolerant species. This pathway is thus of utmost adaptational importance for survival under long-time anoxia.

The energy output when using different pathways under anaerobiosis compared to aerobic metabolism has been mentioned several times. To go to succinate gives a total of 4.71 ATP, to go further to propionate gives a total of 6.43 ATP, meaning that the substrates will last longer with non-lactate pathways (Table 1).

Some normally lactate-producing species have developed alternatives if they are exposed to anoxia or hydrogen sulphide combined with hypoxia for a long period. The flexibility in *Hediste (Nereis) diversicolor* (Vismann, 1990) and the possibility in this species to produce both lactate and succinate depending on external circumstances (Schöttler et al., 1984b) has already been mentioned. Another interesting case is the brackish water isopod *Saduria entomon*. This species normally produces lactate as in all crustaceans. It has been noted, however, that some individuals under long anoxia have low levels of lactate, the concentration remaining at almost normoxic levels. This led us to study in more detail the anoxic behaviour of *Saduria*. Individuals were frequently observed to be active, swimming and crawling under anoxia, whilst some remained inactive the whole time. It turned out that active animals accumulated lactate (and mortality was high), those that were inactive did not and had a low mortality. A search for other end-products revealed no succinate or propionate but a clear accumulation of alanine, obviously a strategy for the long time anoxic survival which is unique among crustaceans (Hagerman & Szaniawska, 1990). The same pathway was found when *Saduria* was exposed to sulphide and hypoxia (Hagerman & Vismann, 1993). There is an explanation for the accumulation of alanine under anoxia. The glycolytic flux can proceed if pyruvate is transformed to alanine by the action of glutamate-pyruvate transaminase (GPT). Glutamate can be regenerated from ketoglutarate, ammonia and NADH in the glutamate dehydrogenase reaction (Carlsson & Gäde, 1986). The enzymes catalysing these reactions have been shown to be present in relatively high activities in *Limulus*. It is also possible that the increase in alanine level, in turn inhibits pyruvate kinase and thus causes a decline in metabolic rate.

This last phenomenon, a decline in metabolic rate, is also a necessary adaptation during exposure to severe hypoxia, anoxia or sulphide. The development of respiratory techniques, combined with direct calorimetric studies has made it possible to study the importance of both aerobic and anaerobic metabolism in relation to metabolites and total energy expenditure. Metabolic depression under hypoxia/anoxia is a great advantage for survival. Metabolic processes run slower and substrates or fuels are conserved and the negative influences of metabolites on the internal milieu are decreased. Important enzymes in this regulation of the flow of glycolysis are phosphofructokinase (PFK) and pyruvatekinase but more mech-

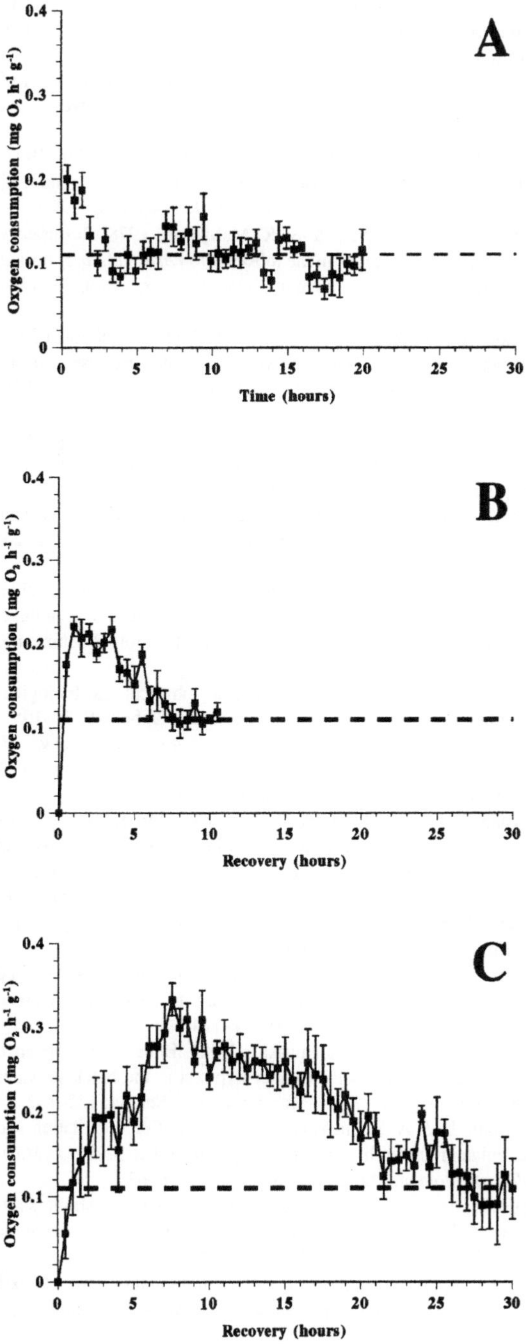

Figure 4. Normoxic respiration (A) and respiration during recovery from anoxia (B) and from hypoxia (5 torr) and sulphide (41 μM) (C) exposure in *Saduria entomon*. From Vismann & Hagerman (1996).

anisms are necessary (Storey, 1988). Some of these mechanisms as well as the role of PFK have been questioned in recent years (Biethinger et al., 1991), but a metabolic depression during long-term anoxia is a fact. *Mytilus edulis* is a good example of a species with anoxic metabolic depression. De Zwaan & Wijsmann (1976) showed that the glycogen stores of this bivalve would only be sufficient for normoxic ATP provision by anaerobic fermentation for 3 d, but by undergoing a metabolic decrease *Mytilus* can use its stores under anoxic conditions for weeks. The priapulid *Halicryptus spinulosus*, the bivalves *Astarte borealis* and *Arctica islandica* have been shown to reduce their metabolic rate to 1% of the aerobic rate (Oeschger, 1990; Oeschger et al., 1992). The isopod *Saduria entomon* is another example, it reduces its metabolism after 40–60 h in anoxia to 5–7% of the normoxic level thus making survival possible for ca 150 h in anoxia, unique for a crustacean in boreal regions (Hagerman & Szaniawska, 1990; Normant et al., 1998).

Recovery

To be alive after exposure to anoxia or sulphide is not necessarily the same as surviving. Animals can recover or be moribund. Processes occurring after exposure, during the recovery phase, are therefore also important. Animals that are exposed to such severe hypoxia that they turn to an anaerobic metabolism and accumulate metabolites have to restore their fuel resources and reoxidize at least part of the metabolites when aerobic metabolism is again possible. This is seen as an oxygen debt, illustrated here by the oxygen debt during recovery after 8 h exposure to severe hypoxia (0.1 kPa) in *Saduria entomon* (Vismann & Hagerman, 1996). This oxygen debt, of varying size, is a general phenomenon seen after severe hypoxia/anoxia in marine organisms (Herreid, 1980).

Sulphide has as demonstrated in *Crangon*, *Monoporeia*, *Calocaris* and others, an effect on the survival even if the animal relies on an anaerobic metabolism. The oxygen consumption of *Saduria entomon* after exposure to hypoxia and sulfide is much higher, 5 times the oxygen debt under hypoxia and the normoxic level was reached after 28 h recovery (Figure 4) compared to 8 h after hypoxia alone (Vismann & Hagerman, 1996). Around half of the oxygen used after hypoxia is used for removing lactate. After exposure to hypoxia and sulphide approximately 20% of the oxygen debt was used for lactate removal.

The metabolic shutdown, or decrease, in *Saduria entomon* is 70% in hypoxia and 95% in long-term anoxia but only 38% in hypoxia with sulphide. This agrees well with the increased flow of energetic reserves (and thus a shorter survival) when sulphide is present compared to hypoxia/anoxia alone.

The part of the oxygen debt related to removal of reduced sulphur is as little as 0.04%, i.e. totally negligible. In reality it is even lower as most of the thiosulphate is excreted (Grieshaber, pers. comm.). Oeschger & Pedersen (1994) also noted in the bivalve *Scrobicularia plana* that oxidation of sulphide during recovery was of no significance for the oxygen debt.

So, the remaining 80% of the oxygen debt is caused by other phenomena. Among these might be an increased activity (but this was not seen after sulphide exposure), restoring ionic levels in haemolymph and cells, protein synthesis, regeneration of ATP stores etc. All these are processes which have arrested during the sulphide exposure (Vismann & Hagerman, 1996).

Conclusion

To be able to survive adverse environmental conditions the organism can 'choose' between two strategies: either a short-time survival where the drain on energy stores is high and the animal can remain relatively active, or a long-time survival combining larger energetic stores with a decreased metabolic rate. Some very flexible organisms can combine both strategies and these organisms are those that also show best tolerances to hypoxia, anoxia and hydrogen sulphide. Sediment-living organisms have greater adaptive potential to hypoxia/anoxia and sulphide than shallow-water epibenthic species.

Acknowledgements

I would like to thank Dr Bent Vismann for cooperation and for many fruitful discussions during the preparation of this manuscript.

References

Astall, C. M., 1993. Comparative physiological ecology of some mud-burrowing shrimps (Crustacea: Decapoda: Thalassinidea). Ph. D. thesis, University of Glasgow, Scotland.

Astall, C. M., A. C. Taylor & R. J. A. Atkinson, 1997. Behavioural and physiological implications of a burrow-dwelling lifestyle for two species of Upogebiid mudshrimps (Crustacea: Thalassinidea). Estuar. coast. Shelf Sci. 44: 155–168.

Biethinger, M., R. Hoffmann & H. W. Hofer, 1991. Phosphofructokinase from mollusc muscle is activated by phosphorylation. Arch. Biochem. Biophys. 287: 263–267.

Bottoms, A. B., 1977. Some aspects of the respiratory physiology of *Cancer pagurus* L. (Crustacea, Decapoda) in Scottish west coast waters. Ph. D. thesis, University of Stirling, Scotland.

Bridges, C. R., V. Hupperts, A. A. Eshky & A. C. Taylor, 1997. Haemocyanin oxygen transport in *Ocypode* spp.: modulation of oxygen affinity? J. mar. biol. Ass. U.K. 77: 145–158.

Butler, P. J., E. W. Taylor & B. R. McMahon, 1978. Respiratory and circulatory changes in the lobster (*Homarus vulgaris*) during long term exposure to moderate hypoxia. J. exp. Biol. 73: 131–146.

Carlsson, K. H. & G. Gäde, 1986. Metabolic adaptation of the horseshoe crab, *Limulus polyphemus*, during exercise and environmental hypoxia and subsequent recovery. Biol. Bull. 171: 217–235.

deFur, P. L., C. P. Mangum & J. E. Reese, 1990. Respiratory responses of the Blue Crab *Callinectes sapidus* to long-term hypoxia. Biol. Bull. 178: 46–54.

Dries, R. R. & H. Theede, 1974. Sauerstoffmangelresistenz mariner Bodenevertebraten aus der Westlichen Ostsee. Mar. Biol. 25: 327–333.

Gäde, G., 1975. Anaerobic metabolism of the common cockle, *Cardium edule*. 1. The utilization of glycogen and accumulation of multiple end products. Arch. int. Physiol. Biochim. 83: 879–886.

Ganning, B., 1971a. Studies on chemical, physical and biological conditions in Swedish rockpool ecosystem. Ophelia 9: 51–105.

Ganning, B., 1971b. On the ecology of *Heterocypris salinus*, *H. incongruens* and *Cypridopsis aculeata* (Crustacea: Ostracoda) from Baltic brackish-water rockpools. Mar. Biol. 8: 271–279.

Grieshaber, M. K., I. Hardewig, U. Kreutzer & H. O. Pörtner, 1994. Physiological and metabolic responses to hypoxia in invertebrates. Rev. Physiol. Biochem. Pharmacol. 125: 43–147.

Hagerman, L., 1983. Haemocyanin concentration of juvenile lobsters (*Homarus gammarus*) in relation to moulting cycle and feeding conditions. Mar. Biol. 77: 11–17.

Hagerman, L., 1986. Haemocyanin concentration in the shrimp *Crangon crangon* (L.) after exposure to moderate hypoxia. Comp. Biochem. Physiol. 85A: 721–724.

Hagerman, L. & A. Szaniawska, 1986. Behaviour, tolerance and anaerobic metabolism under hypoxia in the brackish-water shrimp *Crangon crangon*. Mar. Ecol. Prog. Ser. 34: 125–132.

Hagerman, L. & A. Szaniawska, 1988. Respiration, ventilation and circulation under hypoxia in the glacial relict *Saduria (Mesidotea) entomon*. Mar. Ecol. Prog. Ser. 47: 55–63.

Hagerman, L. & A. Szaniawska, 1990. Anaerobic metabolic strategy of the glacial relict isopod*Saduria (Mesidotea) entomon*. Mar. Ecol. Prog. Ser. 59: 91–96.

Hagerman, L. & A. Szaniawska, 1991. Ion regulation under anoxia in the brackish water isopod *Saduria (Mesidotea) entomon* Ophelia 33: 97–104.

Hagerman, L. & A. Szaniawska, 1992. *Saduria entomon*: ecophysiological adaptations for survival in the Baltic. In E. Bjørnestad, L. Hagerman & K. Jensen (eds), Proc. 12th Baltic Marine Biologists Symp. Olsen & Olsen, Fredensborg: 71–76.

Hagerman, L. & A. Szaniawska, 1994. Haemolymph nitrogen compounds and ammonia efflux rates under anoxia in the brackish water isopod *Saduria entomon*. Mar. Ecol. Prog. Ser. 103: 285–289.

Hagerman, L. & R. F. Uglow, 1983. The influence of temperature on the osmoregulation of the brackish-water shrimp *Palaemonetes varians* Leach. Ophelia 22: 229–236.

Hagerman, L. & R. F. Uglow, 1984. The influence of hypoxia on the blood regulation of the brackish water shrimp *Palaemonetes varians* Leach. J. exp. mar. Biol. Ecol. 76: 157–165.

Hagerman, L., E. Sandberg & B. Vismann, 1997. Oxygen binding properties of haemolymph from the benthic amphipod *Monoporeia affinis* (Lindström) from the Baltic. Mar. Biol. 130: 209–212.

Hagerman, L. & B. Vismann, 1993. Anaerobic metabolism, hypoxia and hydrogen sulphide in the brackish water isopod *Saduria entomon* (L.). Ophelia 38: 1–11.

Hagerman, L. & B. Vismann, 1995. Anaerobic metabolism in the shrimp *Crangon crangon* exposed to hypoxia, anoxia and hydrogen sulfide. Mar. Biol. 123: 235–240.

Hagerman, L. & B. Vismann, 1997. Oxygen binding characteristics of haemocyanin in the Baltic isopod *Saduria entomon*. Ophelia 46: 141–151.

Herreid, C. F., 1980. Hypoxia in invertebrates. Comp. Biochem. Physiol. 67A: 311–320.

Hill, A. D., A. C. Taylor & R. H. C. Strang, 1991. Physiological and metabolic responses of the shore crab *Carcinus maenas* (L.) during environmental anoxia and subsequent recovery. J. exp. mar. Biol. Ecol. 150: 31–50.

Hochachka, P. W. & G. N. Somero, 1984. Biochemical adaptation. Princeton University Press, Princeton N.J., 181 pp.

Holwerda, D. A. & A. de Zwaan, 1979. Fumarate reductase of *Mytilus edulis* L. Mar. Biol. Lett. 1: 33–40.

Jahn, A., I. Gamenick & H. Theede, 1996. Physiological adaptations of *Cyprideis torosa* (Crustacea; Ostracoda) to hydrogen sulphide. Mar. Ecol. Prog. Ser. 142: 215–223.

Johns, A. R., 1996. Sulphide metabolism in burrowing marine crustacea. Ph.D. thesis, University of Glasgow.

Johns, A. R., A. C. Taylor, R. J. A. Atkinson & M. K. Grieshaber, 1997. Sulphide metabolism in thalassinidean crustacea. J. mar. biol. Ass. U.K. 77: 127–144.

Kinne, O., 1966. Physiological aspects of animal life in estuaries with special reference to salinity. Neth. J. Sea Res. 3: 222–244.

Lallier, F., F. Boitel & J. P. Truchot, 1987. The effect of ambient oxygen and temperature on haemolymph L-lactate and urate concentrations in the shore crab *Carcinus maenas*. Comp. Biochem. Physiol. 86A: 255–260.

Lallier, F. & J. P. Truchot, 1989. Modulation of haemocyanin oxygen-affinity by L-lactate and urate in the prawn *Penaeus japonicus*. J. Exp. Biol. 147: 133–146.

Lange, R., 1978. Isosmotic intracellular regulation and euryhalinity in marine bivalves. J. exp. mar. Biol. Ecol. 5: 170–179.

Mangum, C. P., 1980. Respiratory function of the hemocyanins. Am. Zool. 20: 19–38.

Mangum, C. P., 1990. Inducible O_2 carriers in the crustaceans. In O. Kinne, E. Kinne-Saffran & K. W. Beyenbach (eds), Comparative Physiology. Karger, Basel: 92–103.

McLusky, D., 1979. Some effects of salinity and temperature on the osmotic and ionic regulation of *Praunus flexuosus* (Crustacea; Mysidacea) from Isefjord. Ophelia 18: 191–203.

Meinardus-Hager, G. & G. Gäde, 1981. Anaerobic metabolism of the common cockle, *Cardium edule* –IV Time dependent changes of metabolites in the foot and gill tissue induced by anoxia and electrical stimulation. Comp. Biochem. Physiol. 70B: 271–277.

Morris, S. & A. C. Taylor, 1983. Diurnal and seasonal variation in physico-chemical conditions within intertidal rock pools. Estuar. coast. Shelf Sci. 17: 339–355.

Muus, B., 1967. The fauna of Danish estuaries and lagoons. Meddr.Danm. Fisk. – Havunders. 5: 1–316.

Nielsen, A. & L. Hagerman, 1998. The influence of short term hypoxia on metabolism and haemocyanin oxygen transport in two palaemonid prawns. Mar. Ecol. Prog. Ser. 167: 177–183.

Normant, M., A. Szaniawska & G. Graf, 1998. Heat production in *Saduria entomon* (Isopoda) from the Gulf of Gdansk during an experimental exposure to anoxic conditions. Mar. Biol. 131: 269–274.

Oeschger, R., 1990. Long-term anaerobiosis in sublittoral marine invertebrates from the western Baltic Sea: *Halicryptus spinulosus* (Priapulida), *Astarte borealis* and *Arctica islandica* (Bivalvia). Mar. Ecol. Prog. Ser. 59: 133–143.

Oeschger, R. & T. F. Petersen, 1994. Influence of anoxia and hydrogen sulphide on the energy metabolism of *Scrobicularia plana* (da Costa)(Bivalvia). J. exp. mar. Biol. Ecol. 184: 255–268.

Oeschger, R., H. Peper, G. Graf & H. Theede, 1992. Metabolic responses of *Halicryptus spinulosus* (Priapulida) to reduced oxygen levels and anoxia. J. exp. mar. Biol. Ecol. 162: 229–241.

Oeschger, R. & R. D. Vetter, 1992. Sulfide detoxification and tolerance in *Halicryptus spinulosus* (Priapulida): a multiple strategy. Mar. Ecol. Prog. Ser. 86: 167–179.

Oeschger, R. & B. Vismann, 1994. Sulphide tolerance in *Heteromastus filiformis* (Polychaeta): mitochondrial adaptations. Ophelia 40: 147–158.

Onnen, T. & E. Zebe, 1983. Energy metabolism in the tail muscles of the shrimp *Crangon crangon* during work and subsequent recovery. Comp. Biochem. Physiol. 74A: 833–838.

Pihl, L., S. P. Baden & R. J. Diaz, 1991. Effects of periodic hypoxia on distribution of demersal fish and crustaceans. Mar. Biol. 108: 349–360.

Pörtner, H. O., B. Surholt & M. K. Grieshaber, 1979. Recovery from anaerobiosis of the lugworm *Arenicola marina* L.: changes of metabolite concentrations in the body-wall musculature. J. Comp. Physiol. 133: 227–231.

Sandberg, E., B. Vismann & L. Hagerman, 1998. Tolerance of the Baltic amphipod *Monoporeia affinis* to hypoxia, anoxia and hydrogen sulfide. Ophelia (in press).

Schöttler, U., 1979. On the anaerobic metabolism of three species of *Nereis* (Annelida). Mar. Ecol. Prog. Ser. 1: 249–254.

Schöttler, U., 1982. An investigation on the anaerobic metabolism of *Nephtys hombergii* (Annelida: Polychaeta). Mar. Biol. 71: 265–269.

Schöttler, U., G. Wienhausen & J. Westermann, 1984a. Anaerobic metabolism in the lugworm *Arenicola marina* L.: the transition from aerobic to anaerobic metabolism. Comp. Biochem. Physiol. 79B: 93–103.

Schöttler, U., B. Surholt & E. Zebe, 1984b. Anaerobic metabolism in *Arenicola marina* and *Nereis diversicolor* during low tide. Mar. Biol. 81: 69–73.

Schöttler, U. & M. K. Grieshaber, 1988. Adaptation of the polychaete worm *Scoloplos armiger* to hypoxic conditions. Mar. Biol. 99: 215–222.

Spicer, J. I. & D. Morritt, 1995. Oxygen carriage by the haemolymph of hyperiid amphipods. J. mar. biol. Ass. U.K. 75: 997–998.

Storey, K. B., 1988. Suspended animation: the molecular basis of metabolic depression. Can. J. Zool. 66: 124–132.

Taylor, A. C. & P. G. Moore, 1995. The burrows and physiological adaptations to a burrowing lifestyle of *Natatolana borealis* (Isopoda: Cirolanidae). Mar. Biol. 123: 805–814.

Taylor, A. C. & J. I. Spicer, 1989. Interspecific comparison of the respiratory response to declining oxygen tension and the oxygen transporting properties of the blood of some Palaemonid prawns (Crustacea: *Palaemonidae*). Mar. Behav. Physiol. 14: 81–91.

Thomas, H. J., 1954. The oxygen uptake by the lobster (*Homarus vulgaris* Edw.). J. Exp. Biol. 31: 228–251.

Truchot, J. P., 1987. Comparative aspects of extracellular acid-base balance. Springer, Berlin Heidelberg New York, 248 pp.

Truchot, J. P., 1992. Respiratory function of arthropod hemocyanins. Adv. Comp. Envir. Phys. 13: 377–410.

Truchot, J. P. & A. Duhamel-Jouve, 1980. Oxygen and carbon dioxide in the marine intertidal environment: diurnal and tidal changes in rockpools. Respir. Physiology 39: 241–254.

Vismann, B., 1990. Sulfide detoxification and tolerance in *Nereis (Hediste) diversicolor* and *Nereis (Neanthes) virens* (Annelida: Polychaeta). Mar. Ecol. Prog. Ser. 59: 229–238.

Vismann, B., 1991a. Physiology of sulfide detoxification in the isopod *Saduria (Mesidotea) entomon*. Mar. Ecol. Prog. Ser. 76: 283–293.

Vismann, B., 1991b. Sulfide tolerance: physiological mechanisms and ecological implications. Ophelia 34: 1–27.

Vismann, B., 1996. Sulfide species and total sulfide toxicity in the shrimp *Crangon crangon*. J. exp. Mar. Biol. Ecol. 204: 141–154.

Vismann, B. & L. Hagerman, 1993. Activity and ventilation-induced variations of postbranchial oxygen tensions in the isopod *Saduria entomon*. Mar. Biol. 116: 631–635.

Vismann, B. & L. Hagerman, 1996. Recovery from hypoxia with and without sulfide in *Saduria entomon*: oxygen debt, reduced sulfur and anaerobic metabolites. Mar. Ecol. Prog. Ser. 143: 131–139.

Vistisen, B. & B. Vismann, 1997. Tolerance to low oxygen and sulfide in *Amphiura filiformis* and *Ophiura albida* (Echinodermata: Ophiuroidea). Mar. Biol. 128: 241–246.

Völkel, S. & M. K. Grieshaber, 1994. Oxygen dependent sulfide detoxification in the lugworm*Arenicola marina*. Mar. Biol. 118: 137–147.

Völkel, S. & M. K. Grieshaber, 1995. Sulfide tolerance in marine invertebrates. Adv. Comp. Envir. Physiol. 22: 233–257.

Völkel, S. & M. K. Grieshaber, 1996. Mitochondrial sulphide oxidation in *Arenicola marina*. Evidence for alternative electron pathways. Europ. J. Biochem. 235: 231–237.

Weber, R. E. & L. Hagerman, 1981. Oxygen and carbon dioxide transporting qualities of hemocyanin in the hemolymph of a natant decapod *Palaemon adspersus*. J. Comp. Physiol. 145: 21–27.

Zebe, E., 1975. *In vivo* Untersuchungen über den Glucose-Abbau bei *Arenicola marina* (Annelida: Polychaeta). J. Comp. Physiol. 101: 133–145.

de Zwaan, A., D. A. Holwerda & P. R. Veenhof, 1981. Anaerobic malate metabolism in mitochondria of the sea mussel *Mytilus edulis* L. Mar. Biol. Lett. 2: 131–140.

de Zwaan, A. & V. Putzer, 1985. Metabolic adaptations of intertidal invertebrates to environmental hypoxia (a comparison of environmental anoxia to exercise anoxia). Ed. Laverack, M. S., Symp. Soc. Exp. Biol. 39: 33–62.

de Zwaan, A. & T. C. M. Wijsmann, 1976. Anaerobic metabolism in bivalvia (mollusca). Characteristics of anaerobic metabolism. Comp. Biochem. Physiol. 54B: 313–324.

Hydrobiologia **375/376**: 255–264, 1998.
S. Baden, L. Pihl, R. Rosenberg, J.-O. Strömberg, I. Svane & P. Tiselius (eds), Recruitment, Colonization and Physical–Chemical Forcing in Marine Biological Systems.

Manganese in the haemolymph and tissues of the Norway lobster, *Nephrops norvegicus* (L.), along the Swedish west coast, 1993–1995

Susanne P. Eriksson & Susanne P. Baden
Göteborg University, Institute of Marine Ecology, Kristineberg Marine Research Station, S-45034 Fiskebäckskil, Sweden

Key words: manganese, Crustacea, sediment, hypoxia, biomarker, Faroe Islands

Abstract

Spatial and temporal differences in manganese levels in Norway lobsters, *Nephrops norvegicus*, were compared with the concentrations of manganese in their environment. Animals were collected twice yearly (spring and autumn) from seven stations along the Swedish west coast and from one site in the Faroe Islands, during 1993–94, and analysed for manganese tissue concentration and content. Animals were also collected from the Swedish stations in the autumn of 1995 and compared with animals from a stressful environment, frequently exposed to hypoxia.

There were large spatial differences and the animals collected in the Faroe Islands contained (in most tissues) one order of magnitude less manganese than the animals collected along the Swedish west coast. The manganese level of the haemolymph correlated most closely with the manganese concentration the animal was exposed to in the field. The manganese concentration of the female gonad tissue did however not differ with space nor time and remained stable around 5.1 μg Mn g^{-1} dw tissue throughout the investigation.

Animals taken from an area with known repeated hypoxia in the bottom water, had high levels of manganese in especially the gills. Their mean manganese concentration was over 20 times higher (1560 μg Mn g^{-1} dw tissue) than the manganese concentration in animals from the other Swedish stations. They also had more than threefold the amount of manganese in the brain, giving a mean concentration of 193 μg Mn g^{-1} dw tissue.

Introduction

Manganese (Mn) is an abundant metal in most marine soft sediments. The Mn cycle is intercalated with the oxygen and organic carbon cycles as an intermediate oxidant/reductant (Aller, 1994). Primary chemical factors controlling sedimentary Mn cycling are the oxygen content of the overlying water, the penetration of oxygen into the sediments and benthic organic carbon supply (Lynn & Bonatti, 1965; Grill, 1978; Balzer, 1982; Sundby et al., 1986; Hunt & Kelly, 1988). Manganese is released from the sediment in its divalent ionic form (Mn^{2+}) when top sediment layers become reduced (Balzer, 1982; Kremling, 1983; Sundby et al., 1986; Kremling et al., 1997). Its release occurs by diffusion processes as a result of a steep Mn^{2+} concentration gradient across the sediment pore water and bottom water interface (Emerson et al., 1983; Balzer, 1982; Kremling, 1983; Jung et al., 1996). During normoxic conditions in the bottom water, the sediment pore water may contain dissolved Mn^{2+} concentrations of 0.16–19.0 mg l^{-1} (Canfield et al., 1993; Aller, 1994; Magnusson et al., 1996), whereas bottom water concentrations normally are around 0.18–16.5 μg l^{-1} (Laslett & Balls, 1995; Hall et al., 1996). During hypoxia, < 16% oxygen saturation, the Mn^{2+} of the bottom water can increase to concentrations of 1.50 mg l^{-1} (Balzer, 1982). Bioturbation promotes oxidation-reduction reactions in hypoxic sediments by increasing the total area of oxic-anoxic boundaries and thereby the surface available for diffusive exchange (Aller, 1988; Aller, 1994). Remobilised Mn has also been seen to reprecipitate and concentrate near burrow walls (Aller, 1988).

Table 1. Physical and chemical data from the eight stations investigated along the Swedish west coast and the single station in the Faroe Islands. Sediment variables were only determined for the eight stations sampled in Sweden in 1995. (n.d. = not determined; [a] data from Magnusson et al., 1996).

Stn.	Name	Depth (m)	Gear	Sediment type	Sediment mean (max) (μg Mn g^{-1} dw)	Porewater mean (max) (μg Mn ml^{-1})	% dw. organic content
F	Faroe Islands	60	Creel	n.d.	n.d.	0.18[a]	n.d.
1	Koster	39	Creel	sandy-clay	192 (199)	0.23 (0.72)	4.4
2	Leran	61	Trawl	clay	320 (449)	3.48 (7.17)	7.4
3	Sörgrundet	60	Trawl	sandy-clay	295 (392)	5.29 (12.91)	6.0
4	Gråskär	40	Creel	clay	402 (440)	0.81 (1.56)	12.1
5	Tovan	40	Creel	sandy-clay	162 (197)	0.31 (1.17)	2.7
6	Stora Pölsan	52	Trawl	clay	405 (487)	1.82 (4.22)	10.1
7	Klåback	50	Trawl	clay	633 (970)	4.65 (9.08)	11.2
8	Laholm Bay	37	Trawl	sandy-clay	357 (498)	1.05 (3.06)	8.4

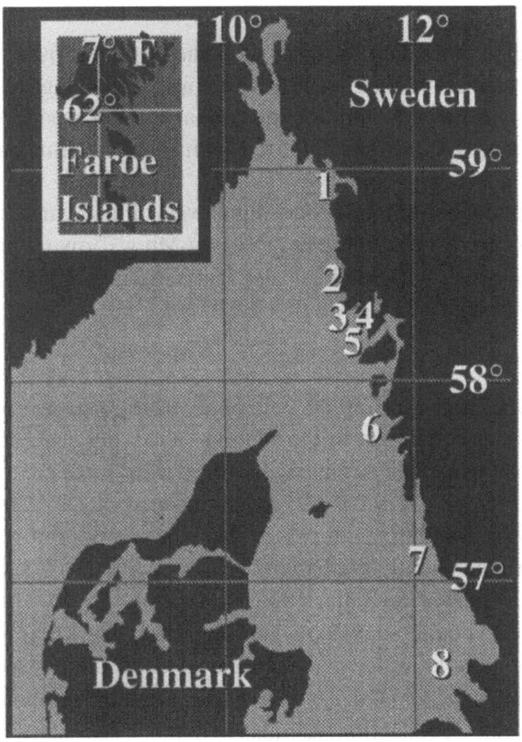

Figure 1. Sampling stations along the Swedish west coast and in the Faroe Islands (F). Station 1–8 from north to south; Koster, Leran, Sörgrundet, Gråskären, Tovan, Stora Pölsan, Klåback and the Laholm Bay.

Marine invertebrates are frequently used as indicators of heavy metal pollution in the aquatic environment because of their ability to concentrate metals from the surrounding medium and to integrate temporal variations in heavy metal concentrations. Manganese becomes bioavailable when it is reduced to Mn^{2+} and can readily be taken up by benthic fauna (Luoma, 1983). Manganese is an essential trace metal involved in many metabolic functions of both plants and animals (Johnson & Nielsen, 1990). It is in aquatic crustaceans mainly found in the calcified parts of the animals where it is thought to cause little damage (Bryan & Ward, 1965; Alikhan, 1989; Steenkamp et al., 1994; Baden et al., 1995). Manganese in excess affects the nervous system (Chemeris et al., 1982; Beirao, 1989), but little is known about other possible effects of elevated Mn concentrations in the inner soft tissues. High Mn concentrations have been found in the crabs *Callinectes sapidus* and *Potamonautes warreni* collected in metal polluted areas (Weinstein et al., 1992; Steenkamp et al., 1994) and in Norway lobsters (*Nephrops norvegicus*) following autumnal hypoxia in the SE Kattegat (Sweden) (Baden et al., 1994).

The decapod *Nephrops norvegicus* lives on soft sediments where it digs elaborate tunnels. Entry of dissolved Mn occurs mainly via gill transport (Bryan & Ward, 1965; Baden et al., 1995), but food (and hereby ingested sediment) has been suggested to be the major uptake route during normal conditions with low concentrations of Mn in the surrounding medium (Bryan & Ward, 1965). Net accumulation of Mn in *N. norvegicus* occurs when the animals are exposed to dissolved

Mn^{2+} concentrations exceeding 1.75 mg l^{-1} (Baden et al., 1995). Excretion of Mn mainly takes place across the gills, but a small amount also passes across the antennal glands (out) with the urine. Laboratory experiments with labelled ^{54}Mn have shown that 55–70% (for most tissues) and 85% (for haemolypmh) of the Mn taken up is lost after five days of excretion (Baden et al., 1995).

The aim of this study is to investigate spatial (site) and temporal (season and year) differences of manganese (Mn) levels in the Norway lobster, *Nephrops norvegicus,* and to compare these with the concentrations of manganese in their environment. Since manganese in invertebrates has been suggested as a possible biomarker for oxygen depletion in eutrophic areas (Förlin et al., 1996), we have also included results with animals from an area frequently exposed to autumnal hypoxia.

Material and methods

Sampling in 1993–1994

Seven known commercial fishing grounds along the Swedish west coast (Stn. 1–7, from north to south) and one fishing location in the Faroe Islands were chosen for the investigation (Figure 1). Average depths ranged from 39 to 61 m between sites and all locations had a cohesive soft sediment type typical for *Nephrops norvegicus* grounds (Table 1). There had been no reported abnormalities in the catches (indicating severe oxygen depletion) from these areas during the last five years prior to the investigation.

Animal collection and preparation
Norway lobsters, *Nephrops norvegicus,* (size range from 30–66 mm carapace length) were caught either by trawl or with creels whichever was typical for fishing in each area (Table 1). Animals were collected twice yearly, in spring (May) and autumn (October). Twelve intermoult stage individuals (C_4) were sampled on every occasion from each station. All lobsters were weighed and haemolymph (HL) samples were taken before the animals were frozen. The animals were dissected and separated into six different tissues; exoskeleton (EXO), gills (GL), gonad (GON), hepatopancreas (HP), muscle (MUS) and rest of body parts (RBP) which included brain, stomach, antennal glands, eyes and intestine.

The manganese (Mn) concentration was measured in the exoskeleton, gills and haemolymph of all of the animals collected ($n = 12$ from each station and sampling date). The Mn concentration was also measured in female gonads (F.GON). To maintain an even number of replicates in the statistical analysis 3 female gonads were randomly selected from each sampling occasion and site.

The Mn concentration was also measured in the male gonads, HP, MUS and RBP of six individuals from each station and sampling date. Total body content of Mn and the condition of the animals was then estimated in these animals. The total Mn was calculated from the Mn concentration of each tissue and the animals tissue proportions. The wet weight (ww), dry weight (dw) and ww/dw ratio of dissected tissues from the whole animal was calculated and standardised to a lobster of 100 g wet weight. Differences in tissue proportions, due to the animals sex or size, were small and could be ignored (unpublished data). Haemolymph weight (volume) was estimated as being the difference between the animals total wet weight and the sum of all dissected tissues (EXO, GL, GON, HP, MUS and RBP). The condition of the *N. norvegicus* was evaluated using the dw and ww/dw ratio of the hepatopancreas and muscle tissue.

Sampling in the autumn of 1995

One *N. norvegicus* location in the Kattegat (Stn. 8, depth 37 m) (see Figure 1 and Table 1), known for autumnal oxygen depletion in the bottom water, was added for comparison to the other Swedish locations (Stn. 1–7) in 1995. Animal collection and preparation was conducted in the same manner as in 1993–1994 except that the brain of each individual was separated from the rest of the body parts during dissection. Manganese concentrations were measured in the exoskeleton, gills, haemolymph and brain (BR) of all animals.

Physical and chemical sampling from the Swedish sites in 1995

Physical and chemical data were collected from the eight Swedish locations (Stn. 1–8) in the autumn of 1995, at the same time as the lobsters were caught. Oxygen concentration was measured in the bottom water (15–20 cm above the sediment) using the Winkler technique (SIS, 1975) and salinity and oxygen was measured at 0.5 m intervals throughout the water column to identify the depth of the pycnocline. The bottom water was considered to be hypoxic when oxygen saturation was 20% or lower. This is the level

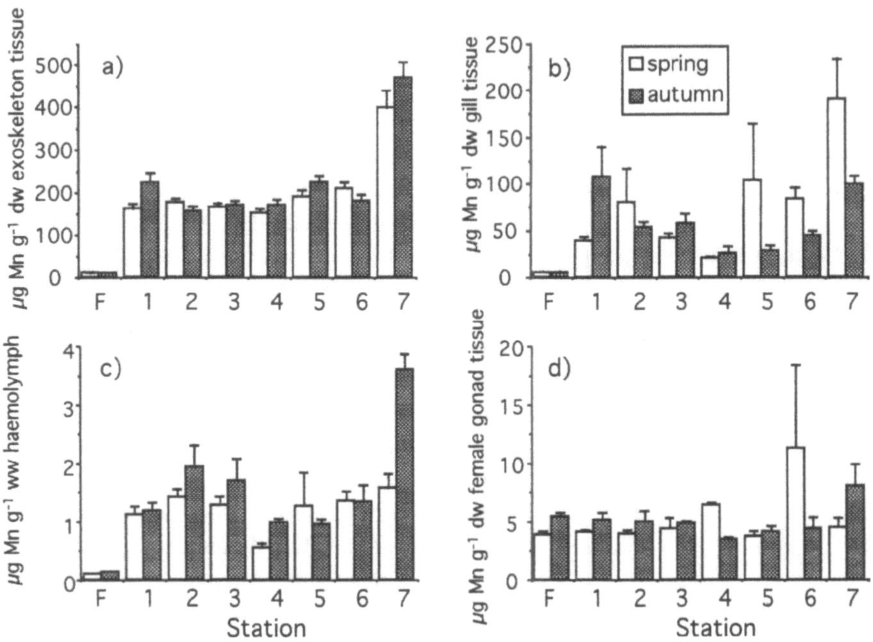

Figure 2. The manganese concentration (mean μg Mn g^{-1} $\pm$ SE) of the a) exoskeleton ($n = 24$), b) gills ($n = 24$), c) haemolymph ($n = 24$) and d) female gonad ($n = 6$) from *Nephrops norvegicus* collected at one location in the Faroe Islands (F) and seven different locations along the west coast of Sweden (Station 1–7, from north to south) in the spring and autumn of 1993 and 1994.

where the oxygen saturation of the water starts to affect both adult *N. norvegicus* and the Mn leakage from the sediment (Baden et al., 1990; Balzer, 1982). Sediment was collected using a multiple corer to give intact sediment profiles. Only the top 13 cm sediment of two best cores from each site were used for further analysis. One of the selected cores was immediately sliced into 1 cm sections and each section centrifuged separately at 2000 rpm for 30 min. Pore water content (% volume) was estimated and the water saved to measure its Mn concentration. The centrifuged sediment fractions were freeze dried and samples were taken from these to measure total Mn concentration of the sediment. In the second core, the Mn concentration of the pore water was measured using the agarose gel technique (Mortimer et al., 1998). Solutes (in this case, dissolved Mn^{2+}) in the pore water are extracted using a diffusive equilibrium between a permanent gel phase and the pore water. After the sample was hauled on deck, a grid filled with agarose gel (saturated with distilled water) was instantly driven down through the sediment core and left undisturbed for a minimum of six hours. The grid was then removed and each gel piece (the grid compartments being sections of 2 mm) was equalised in distilled water before being analysed for Mn^{2+}.

Metal analysis in animal tissues and sediment

The concentration of Mn was determined in haemolymph and tissues of *N. norvegicus* and in bottom water, pore water and sediment by atomic absorption spectrophotometry flame and graphite furnace (GBC 932 AA). Samples were checked against LUTS-1 (non-defatted lobster hepatopancreas) and BCSS-1 (sediment) reference material from the National Research Council in Canada (< 3% deviation from certified values). Prior to metal analysis the freeze dried samples (< 0.2 g tissue or sediment) and haemolymph ($\approx$ 1 ml) were dissolved in 1–2 ml concentrated nitric acid (65% pro analysis) and digested in a CEM microwave extraction system (MES 1000). The samples and standards were all diluted with distilled water to a matrix concentration of 20% nitric acid.

Statistical analysis

Norway lobsters collected in 1993–94 were tested for differences in Mn concentration. The animals, haemolymph and tissue (EXO, GL and F.GON) con-

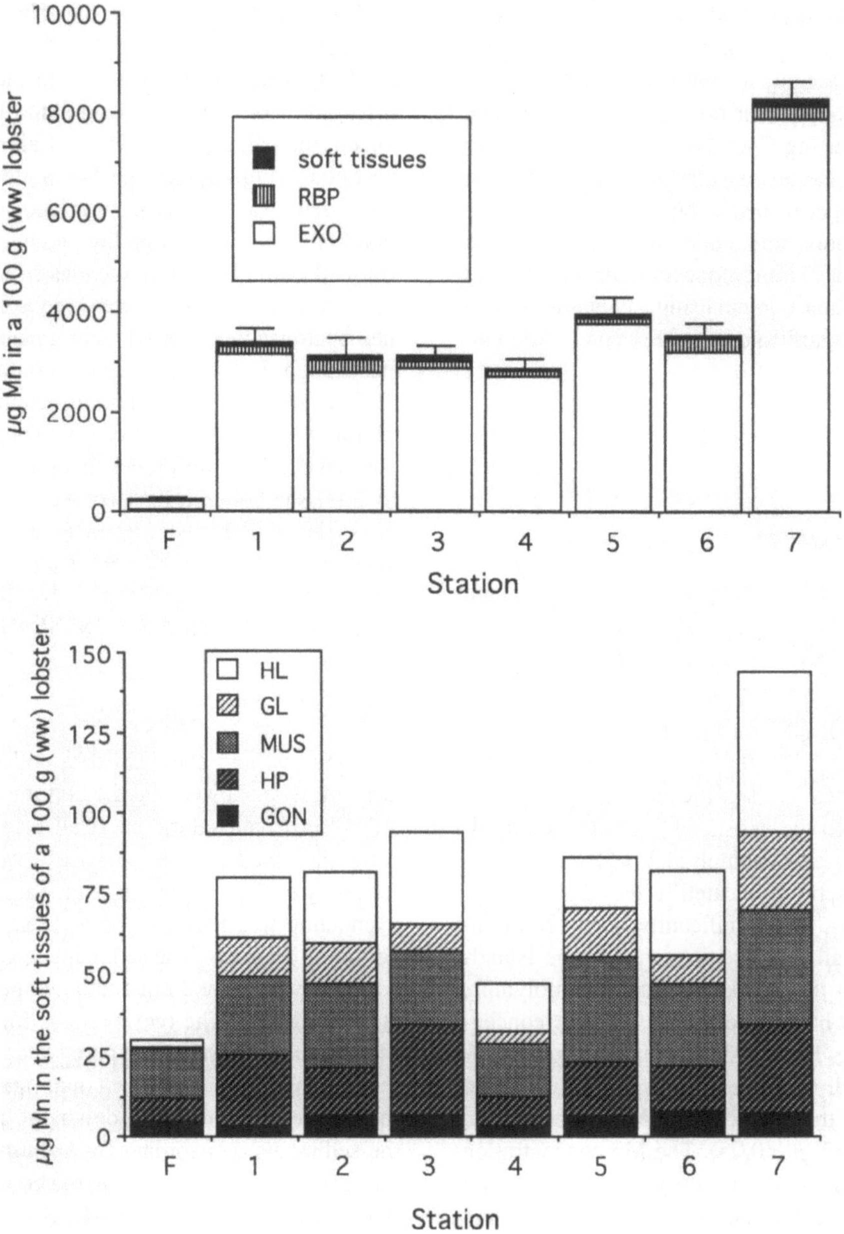

Figure 3. The total body content of manganese in *Nephrops norvegicus* collected at one location in the Faroe Islands (F) and seven different locations along the west coast of Sweden (Station 1–7, from north to south) in 1993–94. All tissue values are standardised to a 100 g lobster (wet weight) and $n = 24$. (a) The animals total body content (mean μg Mn $\pm$ SE) where tissues have been separated into soft tissues (haemolymph, gills, muscle, hepatopancreas and gonad), rest of body parts (RBP) and exoskeleton (EXO). (b) The soft tissues separated into individual tissues; haemolymph (HL), gills (GL), muscle (MUS), hepatopancreas (HP) and gonad (GON).

centrations were checked against site ($n = 8$), season ($n = 2$) and year ($n = 2$) using 3-way-ANOVAs followed by Scheffe's a posteriori analysis. Two-tailed unpaired t-tests were used when comparing the seasonal difference in animal condition (using hepatopancreas and muscle weight) and the Mn content in female and male gonads. Tests for homogeneity of variance were carried out using Cochran's test, and whenever this test showed heterogeneity, a $\log_{10}$ transformation of the data was performed. The mean Mn concentration of the pore water and sediment was tested against the animals tissue concentration of Mn and the sediments organic content using a linear regression model. Level of significance was set as $p < 0.05$ in all instances.

Results

Results from 1993–1994

The manganese (Mn) concentration in the exoskeleton, gills, haemolymph and female gonad from animals collected in 1993–94 are shown in Figures 2a–d. There were significant differences in the exoskeleton, gills and haemolymph concentrations between sites (ANOVA. df 7,352. $p < 0.05$). The station sampled in the Faroe Islands had significantly lower Mn concentrations in both the exoskeleton (compared to all Swedish stations) and haemolymph (all but Stn. 4–5). Station 7, which was located furthest south of the Swedish locations, had significantly higher Mn concentrations than all other stations (the Faroe Islands included) in both the exoskeleton and haemolymph. This station also had higher gill tissue Mn concentrations than the Faroe Islands and Station 4. A significant overall increase from spring to autumn was detected for the haemolymph Mn concentration (ANOVA. df 1,352. $p < 0.05$). The Mn concentration in the female gonad showed no spatial or temporal differences (Figure 2d). No differences were found between the two years investigated and no interactions could be prooved between sites, season and year for any of the four tissues investigated.

The condition of the *Nephrops norvegicus* caught in 1993–94 showed that the dry weight of the hepatopancreas and muscle increased significantly from spring to autumn whereas the ww/dw ratio decreased. The dry weight of the hepatopancreas increased from 1.54 (SD 0.46) to 1.82 (SD 0.65) and the muscle from 7.86 (SD 0.80) to 8.30 (SD 1.20). The ww/dw ratio decreased from 2.90 (SD 0.62) to 2.39 (SD 0.55) in the hepatopancreas and from 4.72 (SD 0.51) to 4.37 (0.51) in the muscle tissue.

The total Mn content in the lobsters (average of all sampling occasions in 1993–94) are shown in Figure 3a–b. The animals collected in the Faroe Islands had at least one order of magnitude lower total Mn content than all stations (Nr. 1–7) along the Swedish west coast (Figure 3a). The Mn content of each separate tissue was consistent in relation to the animal's total body content, so that the percentage of the total Mn load found in each tissue was similar for animals from all stations. There were three exceptions to this; the hepatopancreas, muscle and gonad tissues of animals caught in the Faroe Islands contained 7, 10 and 12 times more Mn than animals caught in Sweden in relation to total Mn body content (Figure 3b). There was no significant difference in total Mn content of the gonad tissue between the two sexes. This was mostly due to the reproductive organs large difference in size since the male gonad on average contained a Mn concentration of 33.2 μg Mn g^{-1} dw (SE 2.4) and the female only 5.1 μg Mn g^{-1} dw (SE 0.6).

Results from 1995

The physical and chemical data collected in the autumn of 1995 showed that the bottom water contained 50–80% oxygen saturation at all locations (Stn. 1–8 along the Swedish west coast) and were thus not affected by oxygen depletion at the sampling time. The pycnocline was found to be at least 10 m from the bottom in all areas. The sediments contained 20–35% water by volume at 3 cm depth and between 10–25% at 10 cm depth. The two methods for measuring the Mn^{2+} concentration of the pore water showed similar results (Figure 4). The Mn^{2+} maxima (indicating the redoxcline) were found at depths of 2.2–9.1 cm into the sediments, concordant for bioturbated sediments during normoxic conditions in the bottom water (Canfield et al., 1993; Aller, 1994). Small differences in results between the two techniques were probably due to local variations, since the analyses were taken from two different cores and thus not totally identical. Since the resolution of the gel grid is higher, these data were used in the statistical analysis.

The Mn concentration measured in the animals collected in 1995 are shown in Table 2. The regression analysis showed that there was a significant positive correlation between the animals' haemolymph concentration and that of the porewater mean (Regression

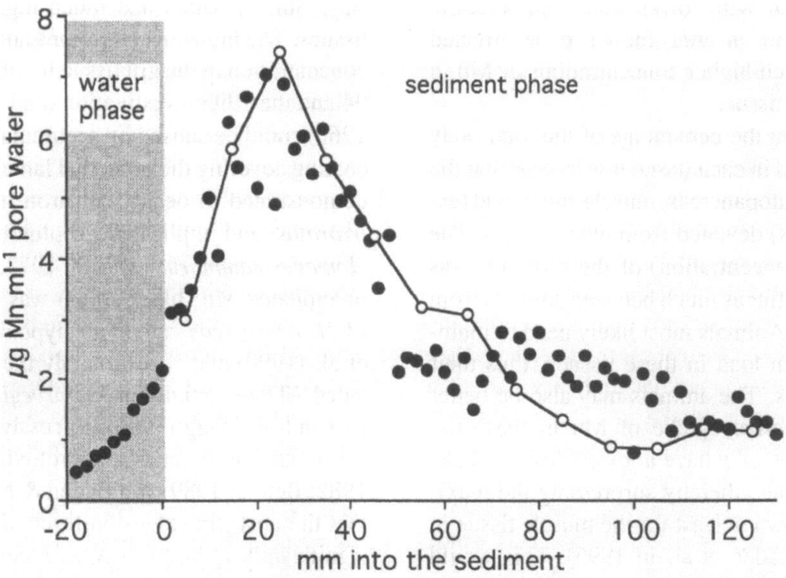

Figure 4. An example of the manganese concentration (μg Mn ml^{-1}) from pore water collected by centrifuging sediment (white dots) or by agarose gel diffusion (black dots). The reading are derived from two separate samples taken at the same site and time.

Table 2. The manganese concentration (mean μg Mn g^{-1} tissue ± SE, n = 12) in *Nephrops norvegicus* collected at eight different locations along the Swedish west cost (Stn. 1–8, from north to south) in the autumn of 1995.

Stn.	Exoskeleton (μg Mn g^{-1} dw)	Gills (μg Mn g^{-1} dw)	Haemolymph (μg Mn g^{-1} ww)	Brain (μg Mn g^{-1} dw)
1	193.2 (29.9)	25.2 (6.4)	1.12 (0.20)	22.8 (2.6)
2	223.4 (28.8)	58.2 (15.0)	3.31 (0.74)	37.7 (4.9)
3	213.9 (16.0)	54.4 (11.0)	5.27 (1.17)	49.0 (5.3)
4	182.5 (27.2)	37.1 (4.9)	1.54 (0.20)	24.1 (3.0)
5	254.6 (9.7)	31.7 (3.4)	3.44 (1.67)	25.0 (3.8)
6	208.1 (12.0)	51.4 (5.8)	2.83 (0.50)	33.0 (4.0)
7	435.7 (35.5)	72.6 (11.5)	7.35 (2.33)	59.7 (6.0)
8	304.4 (29.9)	1560 (910)	4.30 (1.38)	192.9 (77.5)

df 1,7. $p < 0.05$, $R^2 = 0.62$). The additional location, Station 8 (located furthest south, in the Kattegat) had very high Mn concentrations in the gill and brain (> 20 and > 3 times respectively), compared to all other stations. Positive correlations between the Mn concentration of the gills and the porewater mean (Regression df 1,6. $p < 0.05$, $R^2 = 0.70$), the brain and the porewater mean (Regression df 1,5. $p < 0.05$, $R^2 = 0.87$) and the gills and the sediment mean (Regression df 1,5. $p < 0.05$, $R^2 = 0.64$) were found, but only when the high values from Stn 8 were excluded from the analyses. No correlation was found between the mean Mn concentration of the porewater and that of the sediment. There was however a positive correlation between the Mn concentration of the sediment and its organic content (Regression df 1,6. $p < 0.05$, $R^2 = 0.74$).

Discussion

The manganese (Mn) level in the Norway lobster, *Nephrops norvegicus*, showed both spatial and temporal differences in various body tissues. The animals Mn concentrations were lower in areas with little Mn in the sediment and with lower water temperature. Animals collected in an area known to be affected by hypoxia had much higher concentrations of Mn in particularly the gill tissue.

When comparing the percentage of the total body content of Mn found in each tissue it was clear that the Mn level of the hepatopancreas, muscle and gonad (especially the females) deviated from other tissues. The Mn content (and concentration) of these tissues was lower and did not differ as much between animals from different locations. Animals most likely need to maintain an essential Mn load in these tissues, thus their minimum Mn levels. The animals may also be better at controlling excessive uptake of Mn in these tissues, or these tissues may have a lower/slower uptake rate than other tissues, thereby surpressing the maximum Mn level (shown at least for the muscle tissue in *N. norvegicus* by Baden et al. in 1995). This would explain the stable Mn loads in the hepatopancreas, muscle and gonad. In addition, animals in the Faroe Islands may depend more on food for their essential uptake of Mn whereas the animals from the Swedish locations might also be exposed to Mn by the surrounding water. Metals taken up by crustaceans from solution tend to accumulate mainly in the gills (and exoskeleton), whereas metals taken up from their diet appear to be accumulated mainly in the hepatopancreas (and muscle) (Brouwer et al., 1995; Canli & Furness, 1995).

The fact that the investigation from 1995 showed a strong relationship between the Mn concentration of the haemolymph and that of the sediment porewater also indicates that the Swedish animals are exposed to bioavailable Mn from the surrounding water. It also partly explains the spatial differences found in total Mn load (and concentration) in the exoskeleton, gills and haemolymph of the *N. norvegicus* collected in 1993–94. The station in the Faroe Islands deviated from the Swedish locations by having an order of magnitude lower Mn load and concentration in these tissues. Interestingly, most of the animals from the Swedish locations contained similar levels of Mn with the exception of Klåback (Stn. 7), the station furthest to the south during 1993–94. Lobsters from this location had significantly higher Mn values in the exoskeleton and haemolymph when compared with lobsters from the other Swedish stations.

The Mn concentration of the gills and brain from animals collected in 1995 correlated well with the porewater concentration, with the exception of the most southerly location in the Kattegat (the Laholm Bay, Stn. 8) which had much higher values in both tissues. The high concentrations and variability of Mn concentration in the gill tissue found in both the 1993–94 and the 1995 investigation (and by Bryan & Ward, 1965) could be caused by accumulation in the mucous coating covering the branchial lamellae. This has been demonstrated to occur with iron in the crab *Cancer irroratus*, and implicated for plutonium in the lobster *Homarus gammarus* (Martin, 1973; Swift, 1992). Reprecipitated Mn (black spots) was found on the gills of *N. norvegicus* caught in hypoxic areas by Baden et al. (1990) and was probably the cause for the elevated ATPase activity in *N. norvegicus* gills measured by Canli & Stagg (1996). Excessive concentration of Mn is known to act as a neurotoxin (Chemeris et al., 1982; Beirao, 1989) and Baden & Neil (1998) showed that the post-flip extension force of the abdomen was significantly reduced in *N. norvegicus* exposed for 3 weeks to 10 mg dissolved Mn l^{-1}. These animals had a mean brain concentration of 250 μg g^{-1} dw. In this study the animals caught in the southern Kattegat (an area often exposed to autumnal hypoxia) contained on average 192.9 μg Mn g^{-1} dw (SE 77.5). Earlier investigations (unpublished data) have indicated that Mn accumulated in the brain comes not only from the haemolymph, but may also come directly from the medium via the antennules. These are the main chemosensory organs of crustaceans, and are described in detail by Johansson et al. (1996). The high values from the animals in the southern Kattegat may therefore be an indicator of excess Mn exposure. The Kattegat has experienced repeated autumnal hypoxia in recent decades over extensive areas (Rosenberg et al., 1992; Hansen et al., 1995) and whilst the oxygen tension was normal at the time of sampling in this study, this does not discount the possibility of fluctuating conditions giving rise to intermittent Mn leakage from the sediment.

The Mn concentration of the exoskeleton did not correlate with the sediment or the porewater in the Swedish stations. Some other unknown factor must therefore influence the concentration of the exoskeleton. This may be because the amount of Mn incorporated into the exoskeleton depends primarily on the Mn concentration the animal is exposed to during the calcification process; stages A and B. Manganese readily replaces calcium in its carbonate form (Nassrallah-Aboukais et al., 1996), and about 98% of the Mn content in marine aquatic lobsters is incorporated in the calcium carbonates of the animals exoskeleton (Bryan & Ward, 1965; Baden et al., 1995;

this study). It has been suggested that lobsters can also use moulting to dispose of excess metals (Bryan & Ward, 1965; Ward, 1966; Swift, 1992). Since *N. norvegicus* rarely eat more than a few legs of its cast exuviae (pers. obs.), as opposed to *Homarus* spp. which eat most of their shell after moulting (Bryan & Ward, 1965; Aiken, 1980), moulting could serve as an important regulator of the Mn content in *N. norvegicus.*

Manganese release from the sediment to the water column leads to depletion of Mn in anoxic marine sediments (Calvert & Pedersen, 1993; Jung et al., 1996). The net Mn flux is also accelerated by organic enrichment and temperature increase (Hunt, 1983; Hunt & Kelly, 1988; Aller, 1994) and this may explain the increase in the lobsters haemolymph Mn concentration from spring to autumn. No correlation was found between the total Mn concentration of the sediment and the concentration of the dissolved (and bioavailable) Mn^{2+} found in the pore water. One must thus investigate how much of the total Mn found in the sediment at each station that can become bioavailable, before making conclusions about animal concentrations in relation to total sediment concentrations. The oxides and hydroxides of Mn act as efficient scavengers of other metals (Glasby, 1984; Petersen et al., 1995), and phosphate (Yao et al., 1996). When Mn remobilisation occurs in the sediment, these elements are simultaneously released into the water column with Mn^{2+}. MnO_2 has also been shown to help to transform hazardous compounds such as PCP into less toxic compounds (Petrie et al., 1995). Elevated Mn leakage from the sediment, for whatever reason, is thereby also an indicator of other elements becoming more bioavailable to benthic living organisms.

In conclusion, the manganese (Mn) concentration of the Norway lobster *Nephrops norvegicus* may be used to monitor bioavailable manganese in the field. The result indicates that the relationship between Mn load of the inner soft tissues (hepatopancreas, muscle and gonad) compared to the exoskeleton, gills and haemolymph might be useful to differentiate between manganese exposure from food and the surrounding water. The manganese level of the haemolymph correlates most closely with the manganese concentration the animal is exposed to in the field. We have also found some indications that the lobsters accumulate high levels of manganese in the gills and brain when exposed to manganese in stressful habitats. The concentration of the exoskeleton might prove to be an important long-term indicator, but more studies are required into this aspect.

Acknowledgements

We are greatful to Drs Steve Arnott, Lars Hagerman and one referee for valuable comments on the manuscript. Several local fishermen and temporary assistants are thanked for help in collection and dissection of animals. We would also like to acknowledge Dr Stefan Hulth for lending the gel-grid equipment and Kaldbak Marinebiological Station for providing facilities when sampling in the Faroe Islands. Financial support was given by the Swedish Environmental Protection Agency Grant. No. 12404b (SPE) and by the Swedish Council for Forestry and Agricultural Research Grant No. 40.0259/96 (SPB).

References

Aiken, D. E., 1980. Molting and growth. In J. S. Cobb & B. F. Phillips (eds), The Biology and Management of Lobsters. Academic Press, New York: 91–163.

Alikhan, M. A., 1989. Magnesium and manganese regulation during moult cycle in *Porcellio spinicomis* Say (Porcellionidae, Isopoda). Bull. envir. Contam. Toxicol. 42: 699–706.

Aller, R. C., 1988. Benthic fauna and biogeochemical processes in marine sediments: The role of burrowing structures. In T. H. Blackburn & J. Sørensen (eds), Nitrogen cycling in Coastal Marine Environments. John Wiley & Sons Ltd, Chichester: 301–338.

Aller, R. C., 1994. The sedimentary Mn cycle in Long Island Sound: Its role as intermediate oxidant and the influence of bioturbation, O_2, and C_{org} flux on diagenetic reaction balances. J. mar. Res. 52: 259–295.

Baden, S. P. & D. M. Neil, 1998. Accumulation of manganese in the haemolymph, nerve and muscle tissue of *Nephrops norvegicus* (L.) and its effect on neuromuscular performance. Comp. Biochem. Physiol. 119A: 351–360.

Baden, S. P., M. H. Depledge & L. Hagerman, 1994. Glycogen depletion and altered copper and manganese handling in *Nephrops norvegicus* following starvation and exposure to hypoxia. Mar. Ecol. Prog. Ser. 103: 65–72.

Baden, S. P., S. P. Eriksson & J. M. Weeks, 1995. Uptake, accumulation and regulation of manganese during experimental hypoxia and normoxia by the Decapod *Nephrops norvegicus* (L.). Mar. Pollut. Bull. 31: 93–102.

Baden, S. P., L.-O. Loo, L. Pihl & R. Rosenberg, 1990. Effects of Eutrophication on Benthic Communities Including Fish – Swedish West Coast. Ambio 19: 113–122.

Balzer, W., 1982. On the distribution of iron and manganese at the sediment/water interface: thermodynamic versus kinetic control. Geochim. Cosmochim. Acta. 46: 1153–1161.

Beirao, P. S., 1989. Sodium- and calcium-dependent mechanisms in the action of the secretory epithelium of a clam mantle. J. exp. Biol. 145: 395–402.

Brouwer, M., J. Enghild, T. Hoexum-Brouwer, I. Thogersen & A. Truncali, 1995. Primary structure and tissue-specific expression of blue crab (*Callinectes sapidus*) metallothionein isoforms. Biochem. J. 311: 617–622.

Bryan, G. W. & E. Ward, 1965. The absorption and loss of radioactive and non-radioactive manganese by the lobster, *Homarus vulgaris*. J. mar. biol. Ass. U.K. 45: 65–95.

Calvert, S. E. & T. F. Pedersen, 1993. Geochemistry of recent oxic and anoxic marine sediments: Implications for the geological record. Mar. Geol. 113: 67–88.

Canfield, D. E., B. B. Jørgensen, H. Fossing, R. Glud, J. Gundersen, N. B. Ramsing, B. Thamdrup, J. W. Hansen, L. P. Nielsen & P. O. J. Hall, 1993. Pathways of organic carbon oxidation in three continental margin sediments. Mar. Geol. 113: 27–40.

Canli, M. & R. W. Furness, 1995. Mercury and cadmium uptake from seawater and from food by the Norway lobster *Nephrops norvegicus*. Envir. Toxicol. Chem. 14: 819–828.

Canli, M. & R. M. Stagg, 1996. The effects of *in vivo* exposure to cadmium, copper and zinc on the activities of gill ATPase in the Norway lobster, *Nephrops norvegicus*. Arch. envir. Contam. Toxicol. 31: 494–501.

Chemeris, N. K., V. N. Kazachenko, A. N. Kislov & A. L. Kurchikov, 1982. Inhibition of acetylcholine responses by intracellular calcium in *Lymnaea stagnalis* neurones. J. Physiol. 323: 1–19.

Emerson, S., L. Jacobs & B. Tebo, 1983. The behaviour of trace metals in marine anoxic waters. In C. S. Wong, E. Boyle, K. W. Bruland, J. D. Burton & E. D. Goldberg (eds), Trace metals in sea water. Plenum Press, New York: 579–608.

Förlin, L., S. P. Baden, S. P. Eriksson, Å. Granmo, E. Lindesjöö, K. Magnusson, R. Ekelund, A. Esselin & J. Sturve, 1996. Effects of contaminants in Roundnose grenadier (*Coryphaenoides rupestris*) and Norway lobster (*Nephrops norvegicus*) and contaminant levels in mussels (*Mytilus edulis*) in the Skagerrak and Kattegat compared to the Faroe Islands. J. Sea Res. 35: 209–222.

Glasby, G. P., 1984. Manganese in the marine environment. Oceanogr. Mar. Biol. annu. Rev. 22: 169–194.

Grill, E. V., 1978. The effect of sediment-water exchange on manganese deposition and nodule growth in Jervis Inlet, British Columbia. Geochim. Cosmochim. Acta 42: 485–494.

Hall, I. R., D. J. Hydes, P. J. Statham & J. Overnell, 1996. Dissolved and particulate trace metals in a Scottish Sea Loch: an example of a pristine environment? Mar. Pollut. Bull. 32: 846–854.

Hansen, I. S., G. Aertebjerg & K. Richardson, 1995. A scenario analysis of effects of reduced nitrogen input on oxygen conditions in the Kattegat and the Belt Sea. Ophelia 42: 75–93.

Hunt, C. D., 1983. Variability in the benthic Mn flux in coastal marine ecosystems resulting from temperature and primary production. Limnol. Oceanogr. 28: 913–923.

Hunt, C. D. & J. R. Kelly, 1988. Manganese cycling in coastal regions: Response to eutrophication. Estuar. coast. Shelf Sci. 26: 527–558.

Johansson, K. U. I., L. Gefors, R. Wallen & E. Hallberg, 1996. Structure and distribution patterns of aestetascs and male-specific sensilla in *Lophogaster typicus* (Mysidacea). J. Crust. Biol. 16: 45–53.

Johnson, P. E. & F. H. Nielsen, 1990. Copper, manganese, cobalt and magnesium. In A. M. Pearson & T. R. Dutson (eds), Meat and Health, Advances in Meat Research. Elsevier Applied Science, London: 275–299.

Jung, H.-S., C.-B. Lee, Y.-G. Cho & J. K. Kang, 1996. A mechanism for the enrichment of Cu and depletion of Mn in anoxic marine sediments, Banweol intertidal flat, Korea. Mar. Pollut. Bull. 32: 782–787.

Kremling, K., 1983. The behaviour of Zn, Cd, Cu, Ni, Co, Fe, and Mn in anoxic Baltic waters. Mar. Chem. 13: 87–108.

Kremling, K., J. J. S. Tokos, L. Brügmann & H.-P. Hansen, 1997. Variability of dissolved and particulate trace metals in the Kiel and Mecklenburg Bights of the Baltic Sea, 1990–1992. Mar. Pollut. Bull. 34: 112–122.

Laslett, R. E. & P. W. Balls, 1995. The behaviour of dissolved Mn, Ni and Zn in the Forth, an industrialised, partially mixed estuary. Mar. Chem. 48: 311–328.

Luoma, S. N., 1983. Bioavailability of trace metals to aquatic organisms – a review. Sci. Tot. envir. 28: 1–22.

Lynn, D. C. & E. Bonatti, 1965. Mobility of manganese in diagenesis of deep-sea sediments. Mar. Geol. 3: 457–474.

Magnusson, K., R. Ekelund, G. Dave, Å. Granmo, L. Förlin, L. Wennberg, M.-O. Samuelsson, M. Berggren & E. Brorström-Lundén, 1996. Contamination and correlation with toxicity of sediment samples from the Skagerrak and Kattegat. J. Sea Res. 35: 223–234.

Martin, J. L.-M., 1973. Iron metabolism in *Cancer irroratus* (Crustacea: Decapoda) during the intermoult cycle with special reference to iron in the gills. Comp. Biochem. Physiol. 46A: 123–129.

Mortimer, R. J. G., M. D. Krom, P. O. J. Hall, S. Hulth & H. Ståhl, 1998. Use of gel probes for the determination of high resolution solute distributions in marine and estuarine pore waters. Mar. Chem. in press.

Nassrallah-Aboukais, N., A. Boughriet, J. C. Fischer, M. Wartel, H. R. Langelin & A. Aboukais, 1996. Electron paramagnetic resonance (EPR) study of Cu^{2+} and Mn^{2+} ions interacting as probes with calcium carbonate during the transformation of vaterite into cubic calcite. J. Chem. Soc., Faraday Trans. 92: 3211–3216.

Petersen, W., K. Wallman, L. Pinglin, F. Schroeder & H.-D. Knauth, 1995. Exchange of trace elements at the sediment-water interface during early diagenesis processes. Mar. Freshwat. Res. 46: 19–26.

Petrie, R. A., J. E. McLean & R. C. Sims, 1995. Treatment of pentachlorophenol with manganese oxide addition to biotic and abiotic sediments. Haz. Waste Haz. Mat. 12: 271–282.

Rosenberg, R., L.-O. Loo & P. Möller, 1992. Hypoxia, salinity and temperature as structuring factors for marine benthic communites in a eutrophic area. Neth. J. Sea Res. 30: 121–129.

SIS (The Standardizing committe' of Sweden), 1975. Titrimetrisk bestämning av halten löst oxygen hos vatten. Swedish standard, SIS 028114: 1–6.

Steenkamp, V. E., H. H. du Preez, H. J. Schoonbee & P. H. van Eeden, 1994. Bioaccumulation of manganese in selected tissues of the freshwater crab, *Potamonautes warreni* (Calman), from industrial and mine-polluted freshwater ecosystems. Hydrobiologia 288: 137–150.

Sundby, B., L. G. Anderson, P. O. J. Hall, Å. Iverfeldt, M. M. Rutgers van der Loeff & S. F. G. Westerlund, 1986. The effect of oxygen on release and uptake of cobalt, manganese, iron and phosphate at the sediment-water interface. Geochim. Cosmochim. Acta 50: 1281–1288.

Swift, D. J., 1992. The accumulation of Plutonium by the European Lobster (*Homarus gammarus* L.). J. envir. Radioact. 6: 1–24.

Ward, E. E., 1966. Uptake of plutonium by the lobster *Homarus vulgaris*. Nature 209: 625–626.

Weinstein, J. E., T. L. West & J. T. Bray, 1992. Shell disease and metal content of blue crabs, *Callinectes sapidus*, from the Albemarle-Pamlico estuarine system, North Carolina. Arch. envir. Contam. Toxicol. 23: 355–362.

Yao, W. & F. J. Millero, 1996. Adsorption of phosphate on manganese dioxide in seawater. Envir. Sci. Technol. 30: 536–541.

Hydrobiologia **375/376:** 265–285, 1998.
S. Baden, L. Pihl, R. Rosenberg, J.-O. Strömberg, I. Svane & P. Tiselius (eds),
Recruitment, Colonization and Physical–Chemical Forcing in Marine Biological Systems.

Do physical and chemical factors structure the macrobenthic community at a continental slope in the NE Atlantic?

Els Flach[1] & Laurenz Thomsen[2]
[1] *NIOO-CEMO, P.O. Box 140, 4400 AC Yerseke, The Netherlands*
[2] *GEOMAR, Wischhofstrasse 1–3, 24148 Kiel, Germany*

Key words: Macrobenthos, community structure, continental slope, flow velocity, C/N, NE Atlantic

Abstract

Macrofauna density, biomass and community structure together with several characteristics of the sediment and flow velocity were estimated in May 1994 and August 1995 at seven stations ranging from 208 m to 4470 m water depth along the OMEX-transect in the Goban Spur area (NE Atlantic). In 1994 four additional stations were sampled at a parallel transect about 40 km SSE of the OMEX-transect. In 1995 two additional transects were sampled, one in the Porcupine Seabight ~ 100 km NNW and one along the slope at ~ 3500 m water depth situated ~ 200 km SSE of the OMEX-transect.

An overall trend in decrease in density and biomass with increasing water depth was found, but no depth related pattern in mean individual weight could be observed. Mean individual weight, however, did show a negative relationship with flow velocities. Correspondence-analyses and single linkage clustering of the community structure showed three more or less depth related clusters, representing a shelf community, an upper-slope and a lower-slope community. These clusters coincided with differences in grain-size,% organic C and total N within the sediment and differences in flow velocities. However, some of the stations at similar depths were not clustered together. Grain-size did not differ at stations with similar depth, but the% of C and N and flow velocities could differ markedly. Stations at similar depth, but with different physical and/or chemical conditions showed differences in density, biomass, mean individual weight and in macrobenthic community structure. More filter-feeding taxa were observed at stations with higher flow velocities, whereas more subsurface deposit-feeders were found at stations with higher sedimentation rates.

Thus, besides the effects of water depth on macrobenthic community structure, other physical and chemical factors (such as flow velocities and organic matter supply) can be important structuring factors as well.

Introduction

This study is part of the Ocean Margin EXchange (OMEX) programme of the EU. The general aim of OMEX is to study the physical, chemical and biological processes at ocean margins – the shelf break and the continental slope – that determine the transport of material from the shelf to the deep sea. The aim of the subproject Benthic Biology within OMEX is to evaluate the role of biological processes in the benthic boundary layer and in the sediments in the cycling of particulate organic material. As part of this subproject we studied the macrobenthic community structure along four transects at a continental margin in the NE Atlantic in relation to sediment composition and flow velocities.

Macrofaunal densities and biomass are generally known to decrease with increasing water depth, which is generally assumed to be correlated to decrease in food availability (Sibuet et al., 1989). However, remarkable differences in densities at similar depths are found in different geographical regions (Jumars & Gallagher, 1982). Biomass values also show large variations at similar depths, when comparing values from different regions (Rowe, 1983). Schaff et al. (1992) and DeMaster et al. (1994) found large differences in macrofauna densities at three sites at 850 m, 150–180 km apart, on the North Carolina slope. These differences could be related to differences in organic carbon deposition rate. Diaz et al. (1994) also conclude that high rates of accumulation of organic-rich sediments lead to high standing stocks of benthos at the continental slope of Cape Hatteras. At the shelf of the northern Beringan Chukchi Sea, Grebmeier et al. (1988) found a relationship between production in the water column and the C/N ratio at the sediment surface and the benthic biomass. They conclude that the quality and quantity of organic carbon deposited to the benthos directly influence benthic biomass.

Besides depth related patterns in density and biomass, also depth related zonation patterns in community structure are generally observed. Carney et al. (1983) state that the fauna of the bottom oceans changes in species composition with increasing depths and that this change is so pronounced that several taxa (even above the species level) have restricted and circumscribed vertical ranges. Rex (1981) found that the most pronounced change in faunal composition with depth occurs at the shelf-slope transition, that rates of species replacement are more gradual below the shelf-slope break, but that the pattern and overall rate vary considerably between faunal groups. However, the causes of species replacement are still poorly understood and probably include a wide variety of physical and biotic factors (Rex, 1981). Besides differences in community structure along the depth gradient also differences in community structure at similar depths were observed (Schaff et al. 1992, Blake & Grassle, 1994). Again, these differences coincided with differences in organic carbon supply. Gooday & Turley (1990) suggest that the patchy distribution of much of the organic material deposited on the ocean-floor probably plays a major role in structuring deep-sea benthic ecosystems.

Billett et al. (1983) reported seasonal pulses of detrital material to bathyal and abyssal depths in the Porcupine Seabight in the NE Atlantic. This material appears to be derived directly from the surface primary production and to sink rapidly to the deep-sea floor (Billett et al., 1983). Once on the sea bed, the detrital material may be moved over the sediment surface by bottom currents. When currents exceed about 7 cm s^{-1} (at 1 m above the bottom), the material can be resuspended (Lampitt, 1985). Lampitt (1985) speculates that because resuspension of detritus will make the material more readily available to suspension feeders, quite subtle changes in the near-bottom current regime may have substantial effects on the structure of benthic and benthopelagic communities. Some species can switch from deposit-feeding to passive suspension feeding as the flux of suspended particles increases (Taghon & Greene, 1992), which can be advantageous in an environment with fluctuating flow velocities. Creutzberg et al. (1984) observed along transects in the southern North Sea a succession of feeding types, which could be related to bottom characteristics and hydrographic conditions. In a trench in the Skagerrak, Rosenberg (1995) also found a correlation between faunal distribution and richness and food availability through advective near-bottom processes on the slope and accumulation at the bottom of the trench.

In the present study we try to correlate benthic macrofauna density, biomass, mean individual weight and community structure with sediment composition and hydrodynamic conditions along bathymetric gradients and at similar depths some hundreds of km apart to test the hypothesis that 'Structural characteristics of benthic communities can give important indirect but time integrated information about the quantity and quality of organic material arriving at the sea floor'.

Material and methods

Study site

Seven stations (A, I, B, II, F, III, E) were sampled along the OMEX-transect in spring (May 1994) and summer (August 1995). The OMEX-transect is situated at the Goban Spur area in the NE Atlantic. Samples were taken from the continental shelf of the Celtic Sea and along the continental slope down to the Porcupine Abyssal Plain, at water depths ranging from 208 to 4470 m. Some additional stations were sampled along an about 40 km SSE situated transect in May 1994 (D86-8/12), and along a transect in the Porcupine Seabight (P1-4) situated about 100 km NNW and one transect along the slope (A2-4) at $\sim$ 3500 m water depth situated about 200 km SSE of the OMEX-transect in August 1995 (Figure 1). In Table 1 the sampling dates, positions and water depths are given for all stations.

Macrofauna

Macrofauna samples were taken with the circular boxcorer of the Netherlands Institute for Sea Research (NIOZ). Different numbers of boxcore-samples of different sizes are available. Boxcores with a diameter of 30 cm (mainly from the shallowest stations) and 50 cm

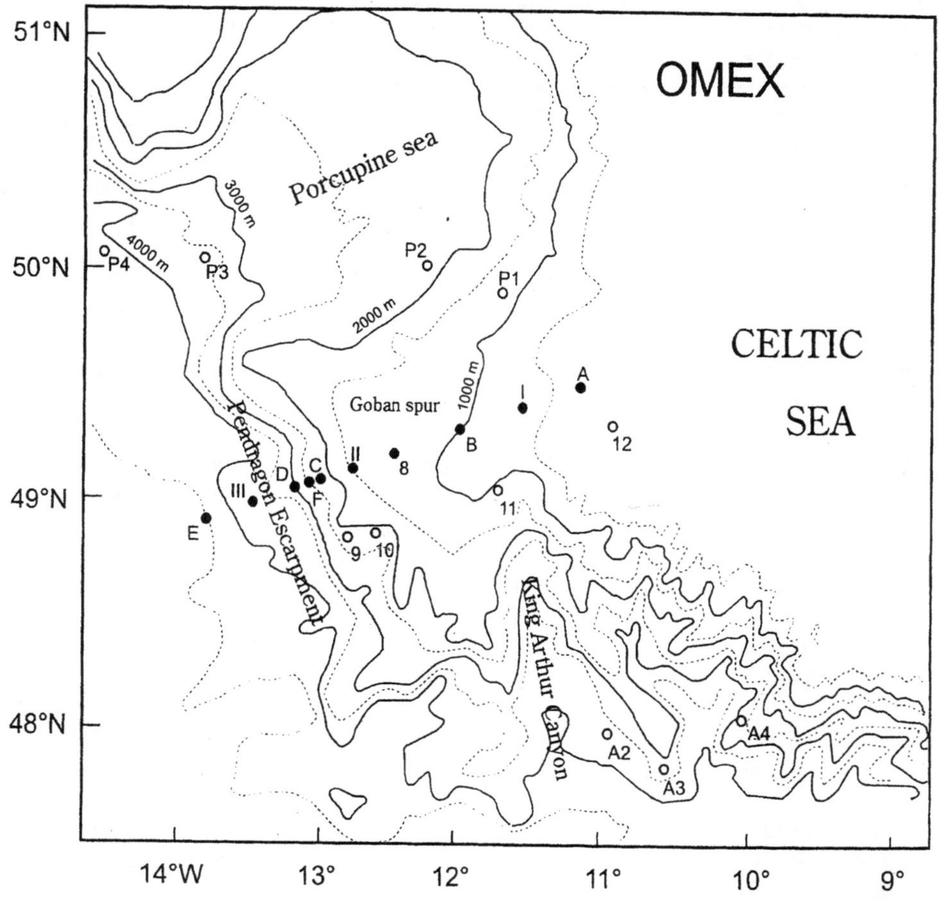

Figure 1. Sampling stations along the OMEX-transect (black dots, A to E) at the Goban Spur sampled in May 1994 and August 1995, along the extra transect sampled in May 1994 (open dots, (D86-) 9 to 12) and along the two extra transects sampled in August 1995 (open dots, Porcupine P1 to P4 and Along-slope A2 to A4).

were used (Table 1). Some subsamples were taken out of some of the boxes for other purposes, resulting in different sample-sizes (Table 1). For this reason mean densities and biomass per m^2 are given.

Boxcore-samples were sieved on a 0.5 mm mesh width, stored in 4% buffered formaldehyde, stained with Rose Bengal and sorted under a stereo microscope (10x). Macrofauna was sorted into major taxonomic groups (phyla, order, class). These groups were used to estimate differences in community structure between the stations using the correspondence-analysis from the BioDiversity-program of SAMS/Natural History Museum London and the single linkage tree diagram of the Statistica Program. Biomass was estimated as wet weight per major taxon after drying the animals a few seconds on absorbent paper. Weighing was done with 0.1 mg accuracy. Because of the small size of most individuals no attempt was made to puncture shells of bivalves to drain them of water. Biomass values were converted into organic C-content per major taxon using the conversion factors given by Rowe (1983). Meiofaunal taxa (e.g. Nematoda, Harpacticoida) were excluded from all analyses. Regression-lines (log_{10}-linear) were computed for densities, biomass and mean individual weight against depth, using the single boxcore-samples. Differences/similarities between stations were tested using ANOVA. Means per station with standard deviation were calculated and tested with the Statistica Program (Basic Statistics).

Abiotic factors

Small subcores of 10 cm diameter were taken out of the boxcores for sediment analysis. In 1994 samples were sliced in different sediment layers to a depth of 15 cm for grain-size analysis. Because very homoge-

Table 1. Sampling stations for macrofauna along the OMEX-transect (st. A–E), the parallel transect of 1994 (st. D86-8/12), the transect in the Porcupine Seabight (st. P1–P4) and the along slope transect of 1995 (st. A2–A4). Ø = diameter of boxcore. O = total area sampled

Station	Sampling date	Position N	Position W	Depth (m)	Macrofauna		
					n boxcores	Ø (cm)	O (cm^2)
A	23-5-'94	49°29.7′	11°08.4′	208	3	30	2060.6
	18-8-'95	49°28.52′	11°12.45′	231	3	30	2120.7
I	23-5-'94	49°24.9′	11°31.4′	670	2	30	1413.7
	19-8-'95	49°24.70′	11°31.86′	693	2	50	3927.0
B	24-5-'94	49°22.4′	11°45.1′	1034	2	50	3927.0
	20-8-'95	49°22.00′	11°47.99′	1021	3	50	4339.4
II	26-5-'94	49°11.3′	12°49.7′	1425	2	50	3927.0
	21-8-'95	49°11.19′	12°49.17′	1457	3	50	4417.9
F	28-5-'94	49°09.5′	13° 05.3'	2182	2	50	3769.9
	22-8-'95	49°09.06′	13°05.38′	2256	3	50	4339.4
III	30-5-'94	49°05.2′	13°25.9′	3673	2 + 1	50 + 30	4633.8
	23-8-'95	49°03.99′	13°25.84′	3648	3	50	4417.9
E	1-6-'94	49°02.3′	13°42.2′	4460	2 + 1	50 + 30	4633.8
	24-8-'95	49°02.29′	13°42.20′	4470	3	50	4339.4
D86-8	4-6-'94	49°14.0′	12°30.0′	1140	3	50	5890.5
D86-9	5-6-'94	48°50.5′	12°45.0′	2190	3	50	5654.9
D86-10	7-6-'94	48°51.9′	12°38.8′	2290	2	50	3927.0
D86-11	9-6-'94	49°04.5′	11°44.5′	1120	3	50	5890.5
D86-12	9-6-'94	49°22.0′	11°00.0′	185	1	30	706.9
P1	27-8-'95	49°55.7	11° 47.8	1441	2	50	2375.9
P2	28-8-'95	49°56.9′	12°23.8′	2210	2	50	2375.9
P3	29-8-'95	50°02.7′	13°58.3′	3670	2	50	2375.9
P4	26-8-'95	50°05.3′	14°40.2′	4115	2	50	2375.9
A2	1-9-'95	47°51.2′	10°54.5′	3880	2	50	2375.9
A3	2-9-'95	47°46.9′	10°27.2′	3341	2	50	2375.9
A4	3-9-'95	48°01.9′	10°11.8′	3824	2	50	2375.9

neous sediments were found (Flach & Heip, 1996a), only one mixed sample per station was analysed in 1995. Particle size of the sediments was estimated using a Malvern Particle Sizer 3600 EC. In August 1995 at all stations porosity was estimated in 1 cm layers to a depth of 15 cm. Also for carbon and nitrogen analysis small subcores were sliced in different sediment layers in 1994 (Flach & Heip, 1996a). In 1995 only the upper 1 cm of the sediment, sliced in layers of 2.5 mm, and one deeper layer (9–10 cm) were analysed. These samples were immediately frozen at –25 °C on board and analysed with a Carlo Erba type NA-1500 elemental analyzer according to Nieuwenhuize et al. (1994). Carbon was partitioned in inorganic and organic fractions by acidification with 25% HCl in situ within silver sample cups.

Flow velocity in the benthic boundary layer (BBL) was measured with a thermistor flow meter at 30 cm

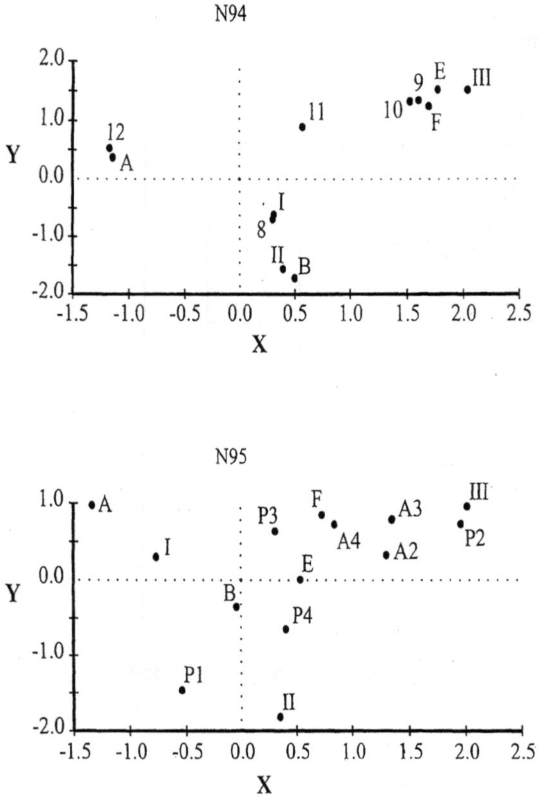

Figure 2. Results of the correspondence-analyses of the Biodiversity Program of (a) the stations sampled in May 1994 (OMEX-stations A to E and extra stations (D86-) 8 to 12) and of (b) the stations sampled in August 1995 (OMEX-stations A to E, Porcupine-stations P1 to P4 and Along-slope stations A2 to A4).

height above bottom (ADM Instruments, Germany) fixed to the BIOPROBE bottom water sampler (Thomsen et al., 1994). A normal launch involved lowering the instrument system to the sea floor using a one conductor cable (10 mm diameter, 200 g m^{-1} in water). One buoy (Benthos, 25 kg net buoyancy) was tied to the wire cable 30 m above the instrument. Flow velocity, light transmission in the BBL, compass direction (position) and battery capacity of BIOPROBE were monitored via the one conductor cable.

Results

Macrofauna 1994

Figure 2a shows the result of the correspondence-analysis of the macrofaunal community structure at the stations sampled in May 1994. This figure shows three clear clusters of stations: the two shelf stations A and D86-12 at 208 and 185 m water depths form one cluster, the upper-slope stations I, B, II and D86-8 at depths between 670 and 1425 m water depth form a cluster and the lower slope stations F, III, E and D86-9 and 10 form the third cluster. Station D86-11 at 1120 m is placed apart, although closest clustered to the upper-slope cluster at similar depths, which is situated between the shallow- and deep-cluster. This result suggests that the community structure of the macrofauna is related to water depth and that there are different communities at shelf, upper-slope and lower-slope. The tree diagram (Figure 3a) shows a more or less similar pattern, except that station D86-8 (1140 m) is clustered within the lower-slope community and that station D86-11 is not clustered far apart from the other upper-slope stations but clustered together with station II.

Total macrofauna densities decreased with increasing water depth ($\log_{10}$n m^{-2}, $R = -0.94$, $p < 0.001$, $n = 28$; Flach & Heip, 1996a). Figure 4 shows the numerical densities of the major taxonomic groups at the 12 stations. The two shelf stations A and D86-12 had the highest densities and were mainly dominated by polychaetes. The upper-slope stations (~ 600–1500 m) showed significant ($p < 0.05$) differences in total densities. The three stations at ~ 1100 m (B, D86-8 and D86-11) were also significantly ($p < 0.05$) different in total densities, but total density at station D86-11 (1120 m) was not significant different from station II at 1425 m. However, the upper slope stations (except station D86-11) showed a similar relative distribution of the major groups. At these stations high numbers of small ophiuroids were found, whereas at the shelf stations high numbers of small echinoids were found (Figure 5a). The three stations around 2200 m water depths showed remarkable similarities in total densities ($p = 0.45$) and relative abundance of the major groups (Figure 4). The two deepest stations also had similar densities ($p = 0.38$) but about three times lower than the ~ 2200 m stations ($p < 0.01$). However, the relative abundance of the major groups were more or less similar at these lower-slope stations, characterized by very low numbers of echinoderms and relatively high numbers of crustaceans. Whithin the group of crustaceans isopods were mainly dominant at the lower-slope stations, decapods were only found at the two shelf stations, whereas at the upper-slope stations amphipods, tanaids, isopods, cumaceans and ostracods, were all present at more or less similar densities, except station I at 670 m, which had high numbers of tanaids (Figure 5b). Molluscs counted for

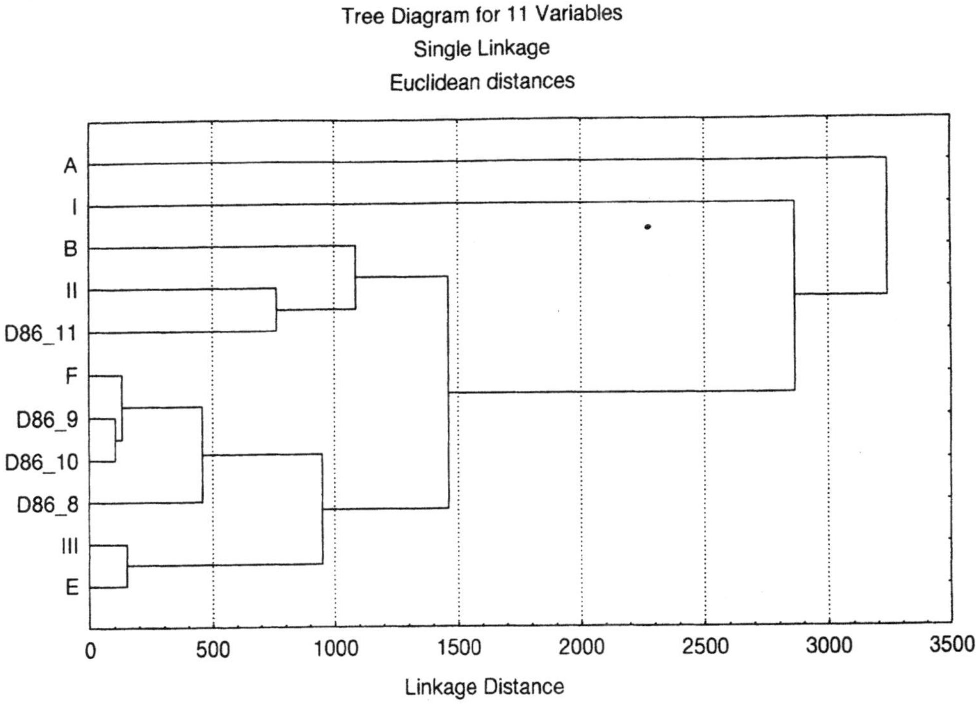

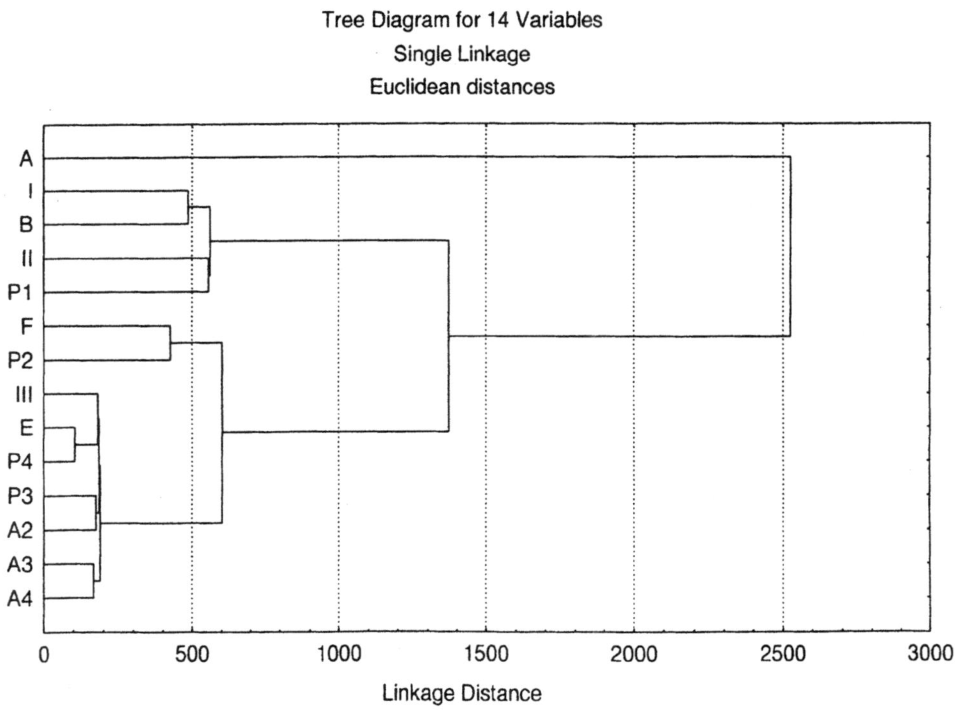

Figure 3. Single linkage tree-diagrams of the Statistica Program of (a) the stations sampled in May 1994 (OMEX-stations A to E and extra stations (D86-) 8 to 12) and of (b) the stations sampled in August 1995 (OMEX-stations A to E, Porcupine-stations P1 to P4 and Along-slope stations A2 to A4).

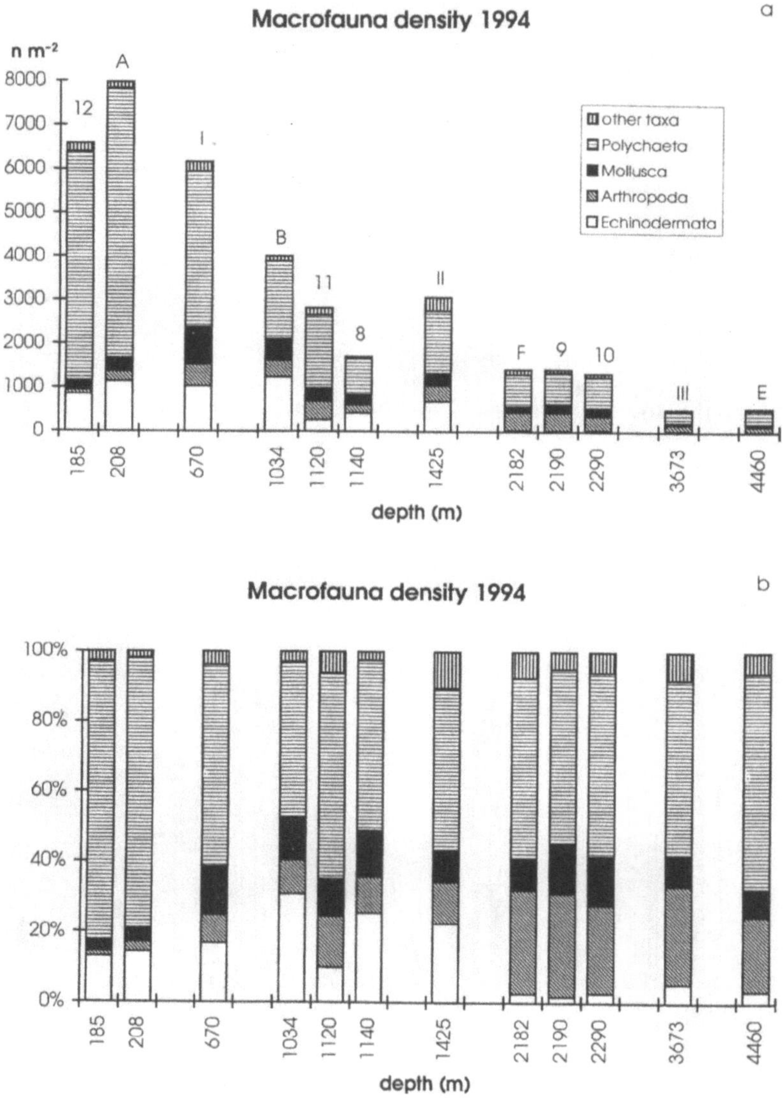

Figure 4. Absolute (n m^{-2}) and relative abundance of the major taxa of the stations sampled in May 1994 (OMEX-stations A to E and extra stations (D86-) 8 to 12).

~ 10% of the macrofauna at all stations except the two shelf stations, where very few molluscs were found. Gastropods were nearly absent at the lower-slope stations, whereas relatively high numbers of scaphopods were found there (Figure 5c). Although relatively low numbers of other small taxa were found, the high numbers of filter-feeding taxa (Tunicata, Porifera and Hydro- and Bryozoa (= Ectoprocta)) were remarkable at station II (1425 m). These taxa were also found at the stations B and 11 at 1034 and 1120 m. At station I (670 m) sipunculids dominated, which were also found at relatively high numbers numbers at the ~ 1100 m and ~ 2200 m stations. Very large sipunculids were found at station B and 10, and large seapens at station B. These large specimens were excluded from the biomass data (see Flach & Heip, 1996a). Biomass also decreased with increasing water depth, when these large individuals were excluded ($\log_{10}$ g C m^{-2}, $R = -0.81$, $p < 0.001$, $n = 28$; Flach & Heip, 1996a). Biomass was high and not significantly different at the three shallow OMEX-stations A (208 m), I (670 m) and B (1034 m). However, at the other shelf

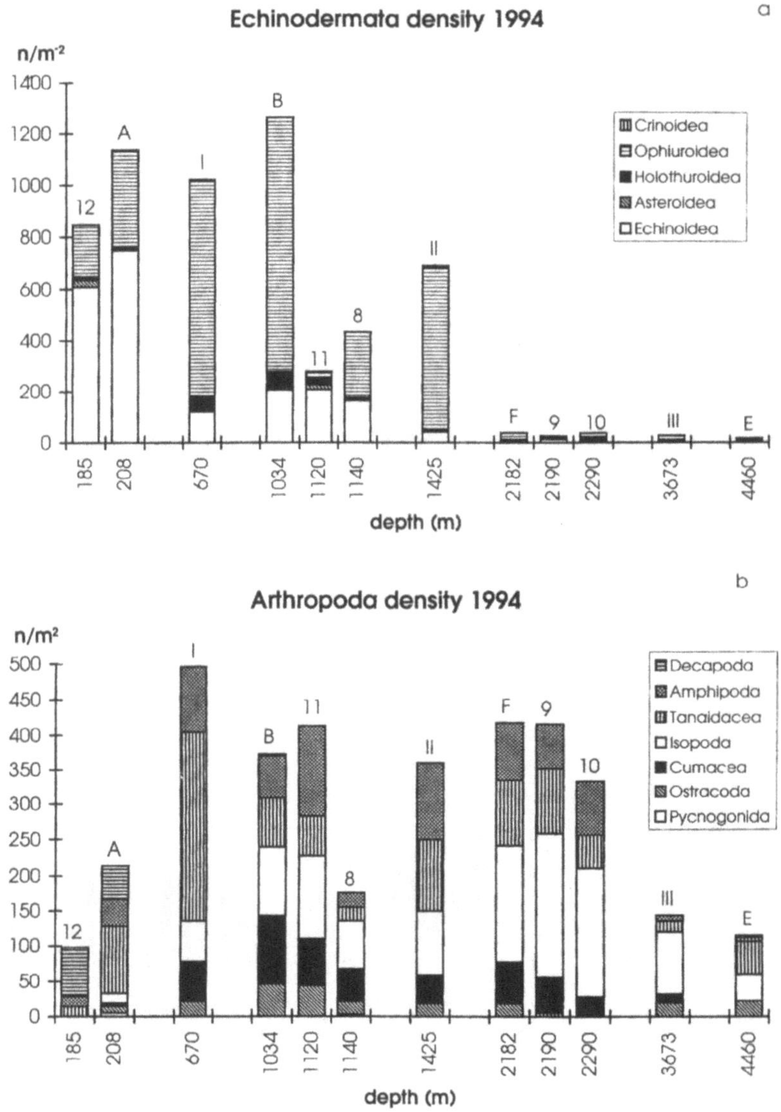

Figure 5. Densities (n m^{-2}) of the taxa used for the correspondence-analysis and tree diagram of the stations sampled in May 1994 (OMEX-stations A to E and extra stations (D86-) 8 to 12). Sipu = Sipunculida, Nemer = Nemertini, Oligo = Oligochaeta, Tuni = Tunicata, Pori = Porifera and ... zoa = Hydrozoa + Anthozoa + Bryozoa + Scyphozoa.

station D86-12 (185 m) biomass was much lower (but only 1 sample). At the other ~ 1100 m stations, 11 and 8, biomass was significantly ($p < 0.01$) lower than at station B (Figure 6a). Mean individual weight did not decrease with increasing water depth, but showed even an insignificant increase ($\log_{10}$ mg C, $R = 0.32$, $p = 0.09$, $n = 28$). At station B at 1034 m the highest mean individual weight was found, even without the two large individuals (Figure 6b). Mean individual weight was significantly higher at this station at 1034 m than at the shelf station A ($p = 0.003$) and than at the other two stations at ~ 1100 m ($p < 0.05$). Station II at 1425 m water depth showed a relatively low biomass and low mean individual weight. The mean individual weight was significantly higher ($p = 0.04$) at the deep station III at 3673 m than at station II at 1425 m. The three stations around 2200 m water depth were very similar again, and did not show any significant differences neither in mean individual weight nor in biomass, when the two large sipunculids

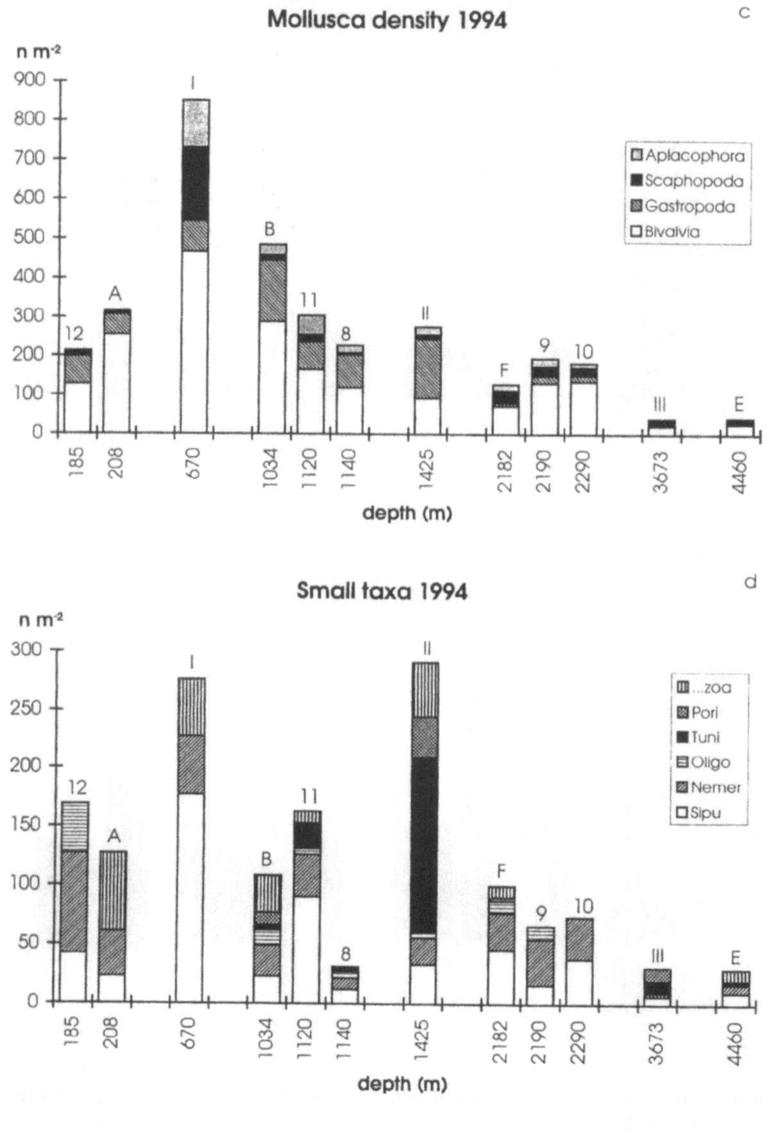

Figure 5. Continued

from station 10 were excluded. However, when they were included biomass and mean individual weight were significantly ($p < 0.05$) higher at station 10 than at the other two ~ 2200 m stations.

Macrofauna 1995

The correspondence-analysis of the stations from August/September 1995 (Figure 2b) did not show the same kind of clusters as found in 1994 with partly the same stations (Figure 2a). The shelf station A was again placed far away from all other stations, with station I at 693 m as its closest neighbour. The two stations at ~ 1450 m (P1 and II) were more or less placed together with station B at 1021 m, representing the upper-slope community. The two stations at ~ 2200 m were placed far apart, with the OMEX station F grouped close together with the along-slope station A4 at 3824 m and the Porcupine station P2 placed in the neighbourhood of station OMEX-III at 3648 m. The other station at ~ 3600 m station P3 was placed far from OMEX-III and closer to station A4. The other two along slope stations A3 at 3341 m and A2 at 3880 m were clustered close together. The

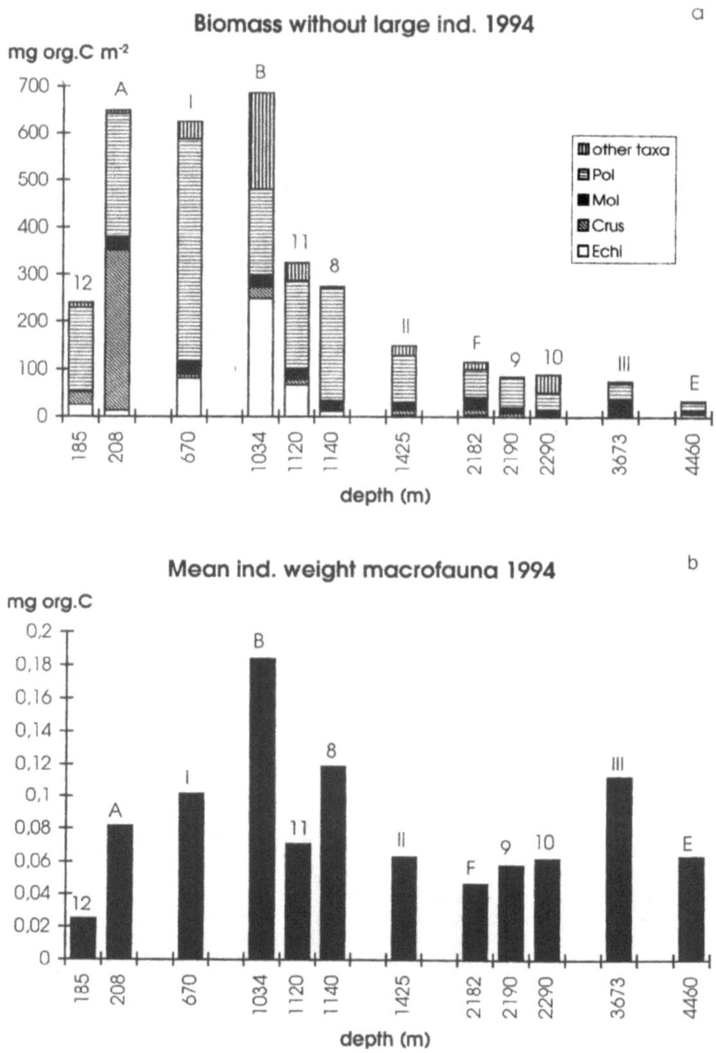

Figure 6. Biomass (mg organic C m^{-2}) and mean individual weight (mg organic C) excluding some extremely large individuals (see Flach & Heip, 1996a) of the stations sampled in May 1994 (OMEX-stations A to E and extra stations (D86-) 8 to 12).

correspondence-analysis thus showed that although some depth related pattern could be observed, stations at similar depth can have different community structures. This, however, could not be related to geographical regions as stations from the same transect were placed far apart (Figure 2b). The tree diagram (Figure 3b) also placed the shelf station A far apart and grouped the upper-slope stations together. The ~ 2200 m stations F and P2, however, were grouped together at some distance from all other lower-slope stations. The two deepest stations (> 4000 m) showed the highest similarity and also the other deep stations P3 and A2 (3670 and 3880 m) and the stations A3 and A4 (3341 and 3824 m, resp.) were grouped closely together.

Figure 7 shows the density of the macrofauna in 1995 and the relative abundances of the major taxonomic groups. Macrofauna densities again decreased with increasing depth ($\log_{10} n\ m^{-2}$, $R = -0.94$, $p < 0.001$, $n = 34$). The shelf station A had the highest density and was mainly dominated by the polychaetes. The upper-slope stations I, B, P1 and II were not significantly different ($p > 0.4$) in total densities, but station I had higher numbers of polychaetes and less crustaceans than the other stations. The two stations at ~ 2200 m had more or less similar densities

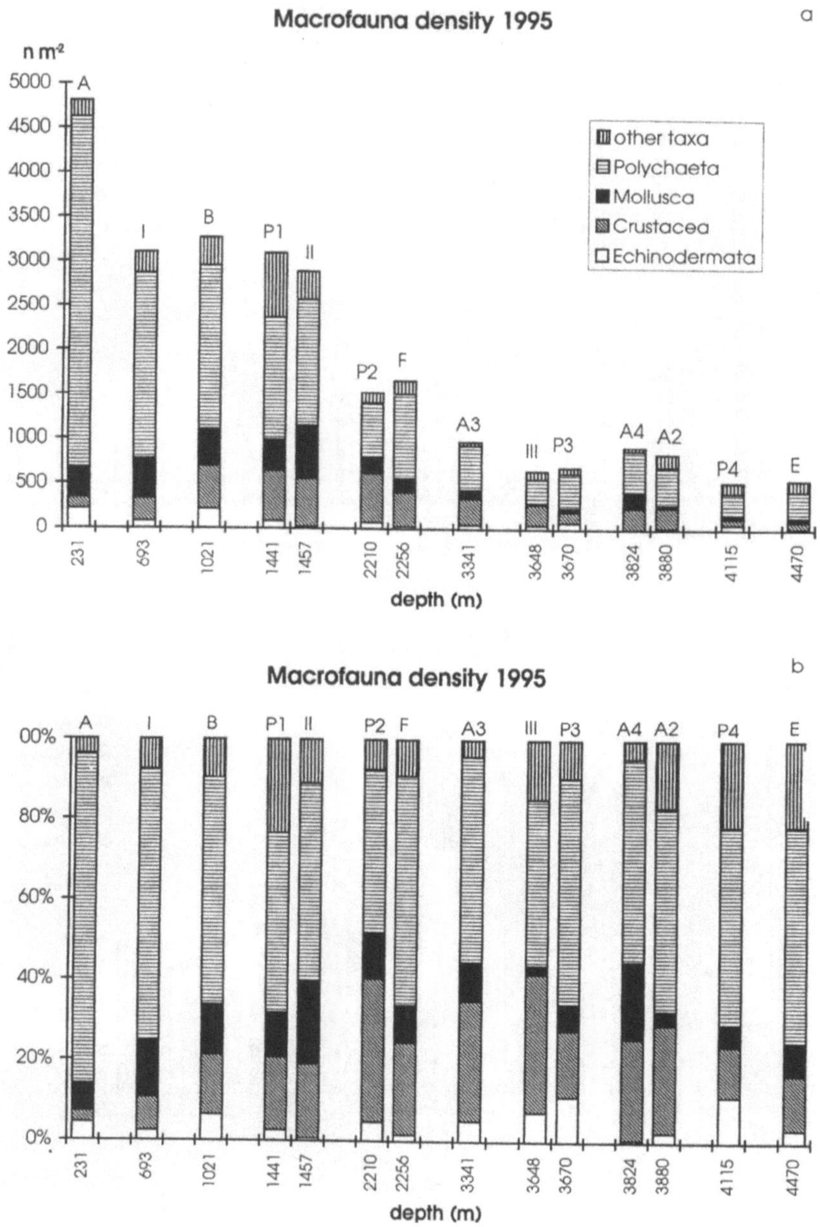

Figure 7. Absolute (n m^{-2}) and relative abundance of the major taxa of the stations sampled in August 1995 (OMEX-stations A to E, Porcupine-stations P1 to P4 and Along-slope stations A2 to A4).

($p = 0.43$), but the OMEX station F had higher numbers of polychaetes and less crustaceans than the Porcupine station P2. The same was found for the two stations at $\sim$ 3600 m, but this time the Porcupine station P3 had higher numbers of polychaetes and less crustaceans than the OMEX-III station. The three stations of the along slope transect showed significantly ($p < 0.05$) higher densities compared to the OMEX- and Porcupine-transect at these depths, with relatively low numbers of polychaetes and high numbers of crustaceans. Within the echinoderms a remarkable difference between the OMEX and the Porcupine stations could be observed with higher numbers of ophiuroids in the Porcupine Seabight and higher numbers of

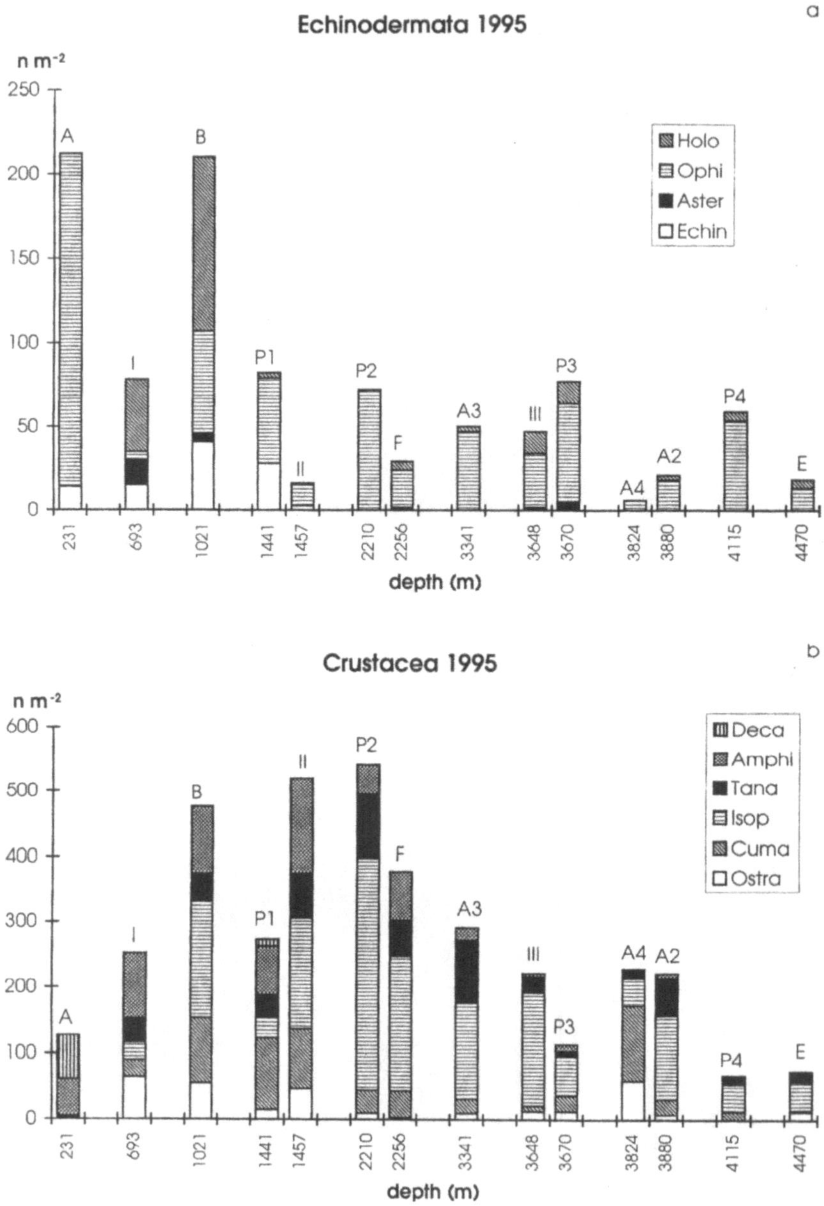

Figure 8. Densities (n m^{-2}) of the taxa used for the correspondence-analysis and tree diagram of the stations sampled in August 1995 (OMEX-stations A to E, Porcupine-stations P1 to P4 and Along-slope stations A2 to A4). Sipu = Sipunculida, Nemer = Nemertini, Oligo = Oligochaeta, Tuni = Tunicata, Pori = Porifera and . . . zoa = Hydrozoa + Anthozoa + Bryozoa + Scyphozoa.

holothurians at the Goban Spur (Figure 8a). Figure 8b shows the differences between the stations of the crustaceans. Besides the already mentioned differences in number at stations at similar depths, the totally different community of the shelf station A and the high number of cumaceans instead of isopods at the deep station A4 are remarkable. The molluscs showed extremely high numbers of the gastropod *Benthonella tenella*) at station OMEX-II and high numbers of bivalves at station A4 (Figure 8c). High numbers of filter-feeding taxa as Porifera, Tunicata, Hydrozoa and Bryozoa (= Ectoprocta) were found at the upper-slope station B, P1 and II and relatively also at the deep stations A2, P4 and E (Figure 8d). Sipunculids were

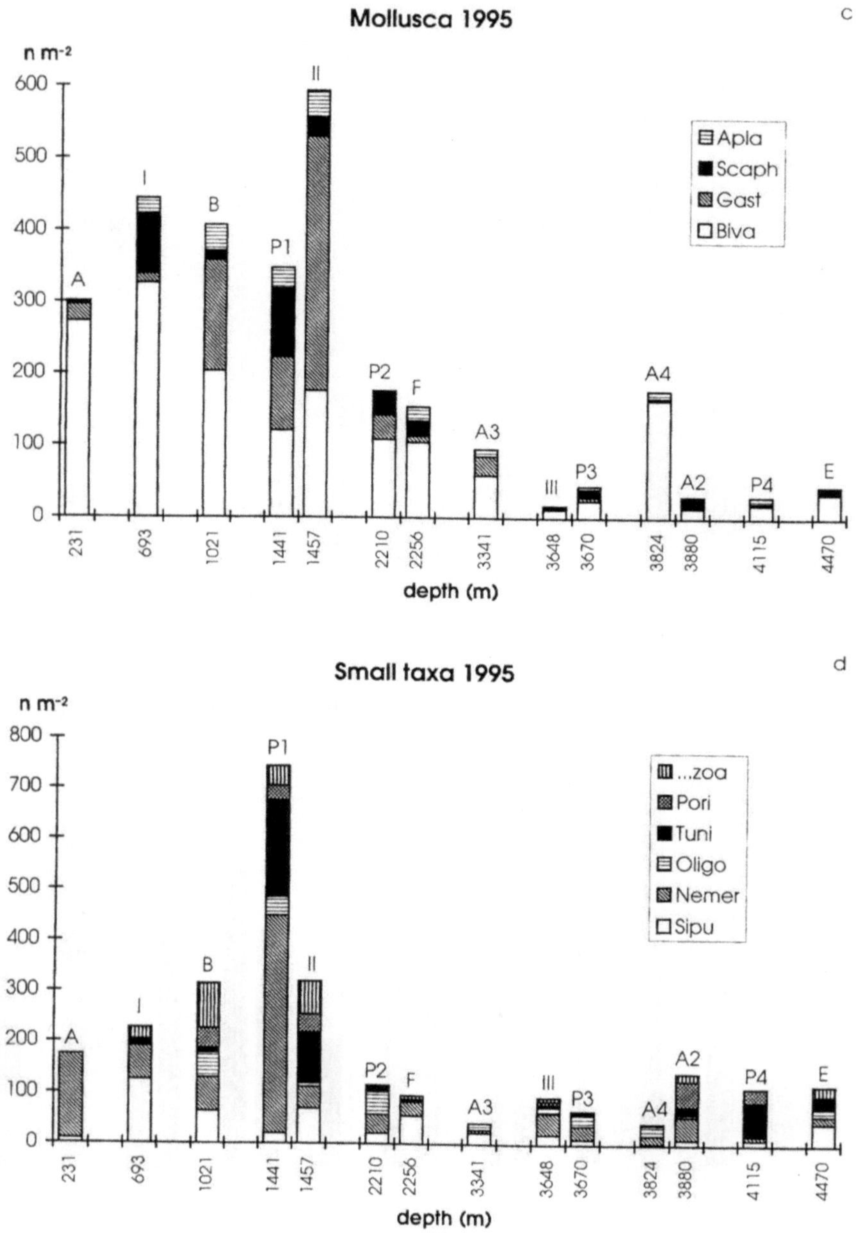

Figure 8. Continued.

found at all stations, but at station B and P2 extremely large ones were found. These sipunculids weighed more than all other specimens together. Figure 9a therefore shows the biomass without these extreme large individuals. The overall trend was a decrease in biomass with increasing water depth ($\log_{10}$ g C m^{-2}, $R = -0.87$, $p < 0.001$, $n = 34$), but the three shallowest stations did not show significant differences in biomass again. At mid-slope depths the OMEX-stations had significantly ($p < 0.05$) higher biomass than the Porcupine stations (but the large sipunculid at P2 was excluded) and at greater depths the Porcupine stations had higher biomass ($p = 0.07$). The along-slope station A3 showed a relatively low biomass and A4 a high biomass. This pattern is reflected in the mean individual weight of the specimens (Figure 9b). The upper-slope stations I and B had significantly higher mean individual weights than the shelf stations A ($p <$

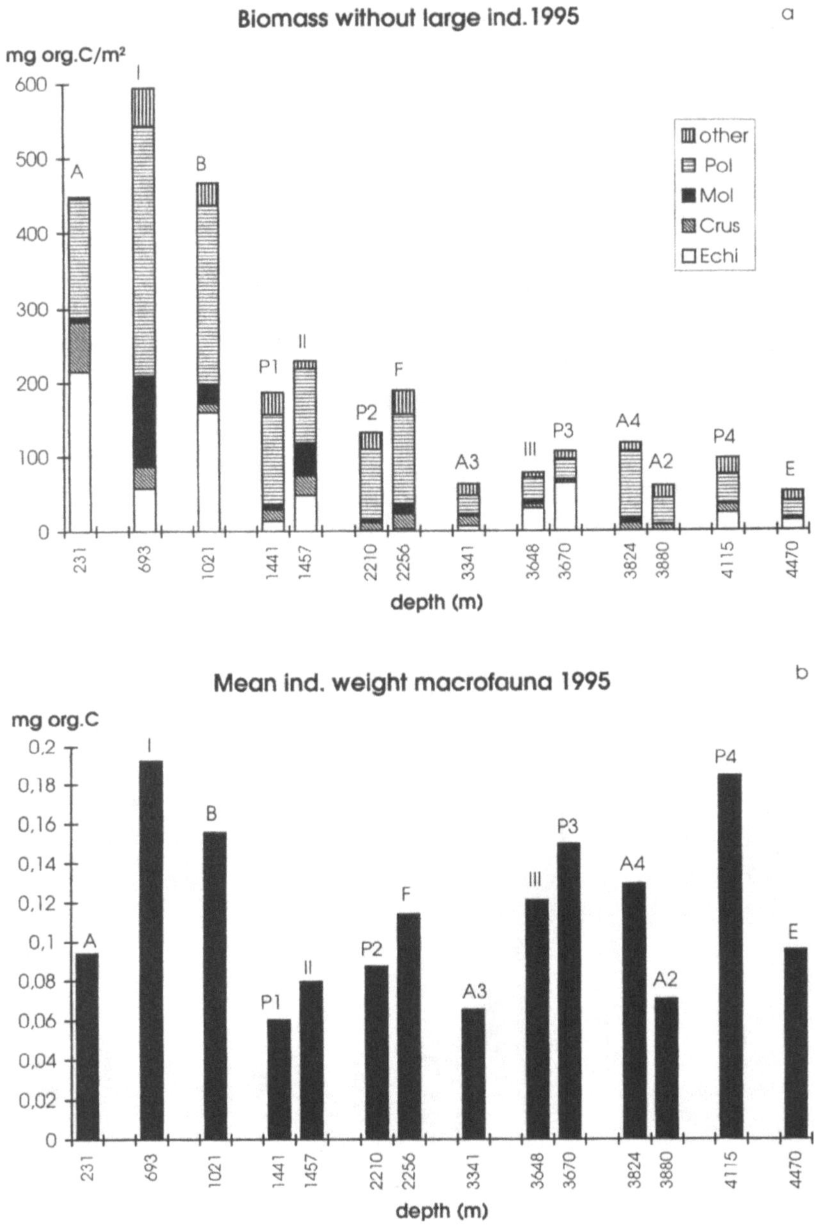

Figure 9. Biomass (mg organic C m^{-2}) and mean individual weight (mg organic C) excluding some extremely large individuals of the stations sampled in August 1995 (OMEX-stations A to E, Porcupine-stations P1 to P4 and Along-slope stations A2 to A4).

0.05). The deep stations P3 and P4 and also the stations A4 and III had relatively high mean individual weights, significantly ($p < 0.05$) higher than the shallower stations P1 and II at $\sim$ 1400 m and the along slope stations A2 and A3. The deepest Porcupine station P4 had also a significant ($p < 0.05$) higher mean individual weight than the deepest OMEX station E. So, again no depth related pattern in mean individual weight could be observed ($\log_{10}$ mg org. C, $R = 0.12$, $p = 0.48$, $n = 34$).

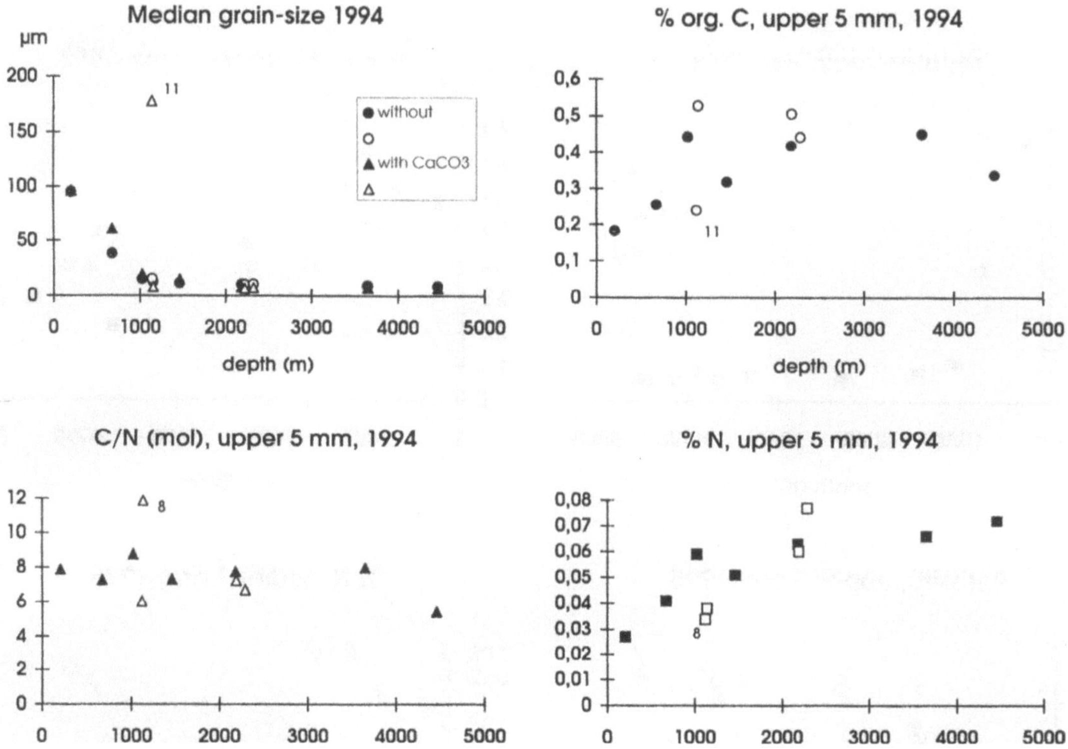

Figure 10. Percentage (%) of total nitrogen (N) and organic carbon (C) and the C/N ratio (mol) in the upper 5 mm of the sediment and the median grain-size (μm) of the upper 5 cm of the sediment (with and without calciumcarbonate $CaCO_3$) at the OMEX-stations (black symbols) and the extra stations (open symbols) in May 1994.

Abiotic factors

In 1994, the sediment composition of the two ~ 200-m stations was very different from all other stations having a sandy sediment with low% $CaCO_3$, and low% of org. C and total N (Figure 10). The median grain-size at the stations at ~ 2200 m and the two deepest stations was similar, also the% org. C and N were more or less similar at these five stations (Figure 10). The upper-slope stations showed some differences, especially station D86-11 was exceptional. At this station a very high median grain-size was found, due to a high abundance of dead solid foram shells. The sediment at this station also contained relatively low% of total N and organic C. Station D86-8, had a high% of org. C, but a low% of N resulting in a high C/N ratio. Whereas the OMEX station B at about the same depth had both high% of org. C and N. At mid-slope depths high flow velocities were found, whereas at the deep stations low flow velocities occurred (Figure 12).

Sediment grain-size and porosity showed a clear depth related pattern in 1995 and all stations deeper than ~ 2000 m showed similar sediments (Figure 11). However, the% org. C and total N showed differences at similar depths (Figure 11). The station at 1400 m in the Porcupine Seabight had much lower values than the OMEX-station at that depth. The% org. C and N at the two deep Porcupine stations and the three along slope stations were comparable to the OMEX-station at 2200 m, and thus much higher than at similar depth at the OMEX transect. The flow velocities at the OMEX-stations and the Porcupine stations at similar depths showed more or less similar values (Figure 12). The along slope station A4 had more or less similar flow velocity as the OMEX and Porcupine station at ~ 3500 m (2–3 cm s^{-1}), whereas the flow velocities at the two along slope stations A2 and A3 were much higher (Figure 12).

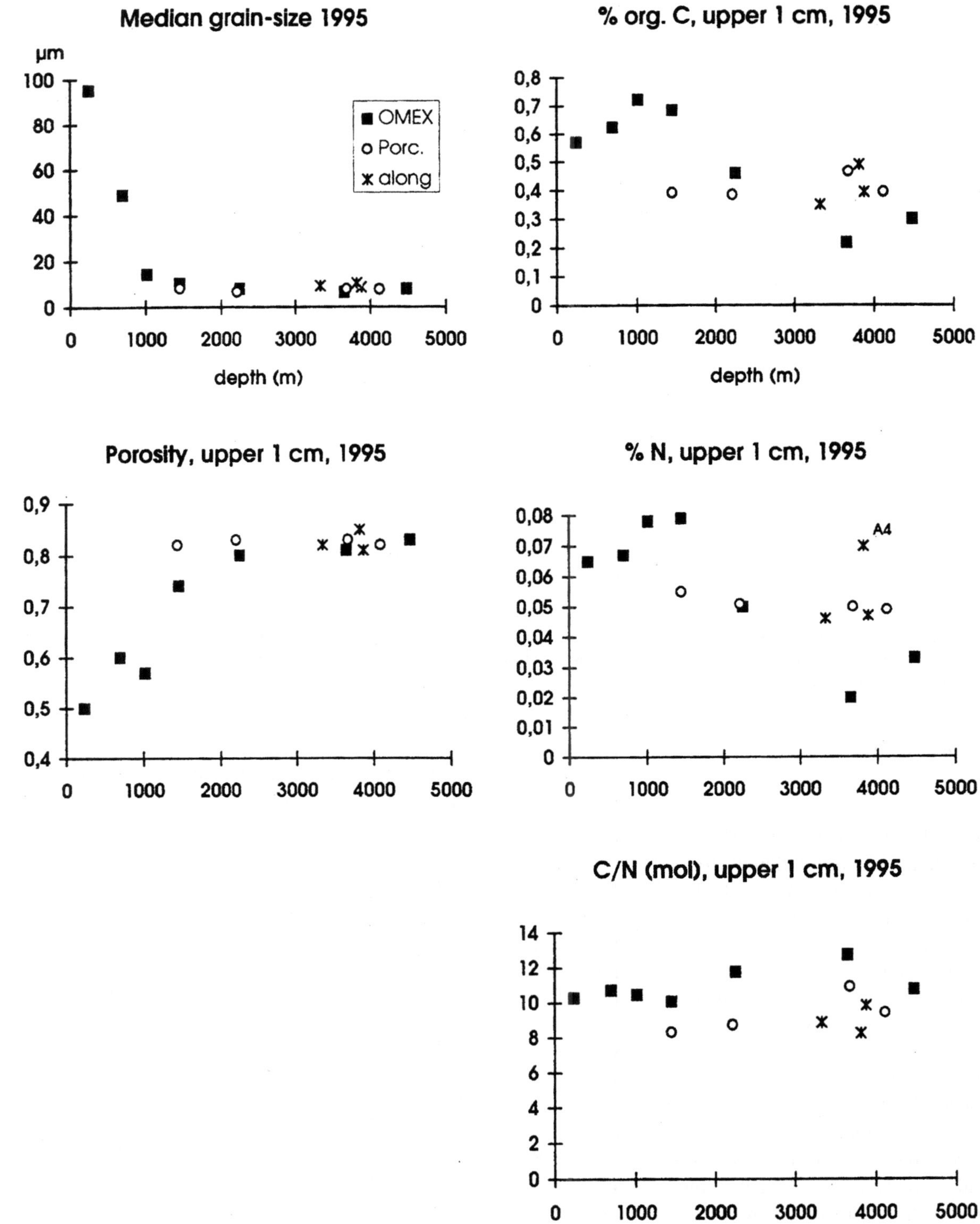

Figure 11. Percentage (%) of total nitrogen (N) and organic carbon (C), the C/N ratio (mol) and the median grain-size (μm) and porosity in the upper 1 cm of the sediment at the OMEX, Porcupine and Along-slope stations in August 1995.

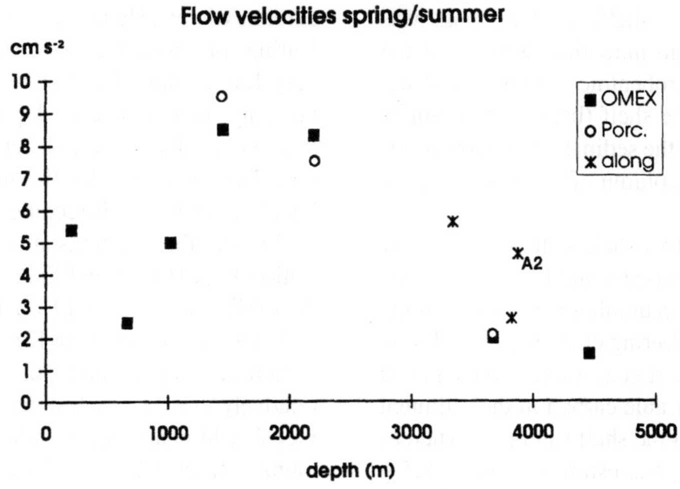

Figure 12. Flow velocities (cm s^{-1}) at 30 cm above the bottom at the OMEX, Porcupine and Along-slope stations during spring/summer.

Discussion

An overall trend in decrease in density and biomass with increasing water depth was found, but no depth related pattern in mean individual weight could be observed. Mean individual weight, however, did show a significant negative relationship with flow velocities ($R = -0.61$, $p = 0.025$, $n = 13$) and a positive, but non significant ($R = 0.33$, $p = 0.24$, $n = 14$) relationship with the% of organic carbon within the sediment. Although the% of organic carbon within the sediment poorly reflects the food availability for the fauna, it is still related to the rain of carbon and gives in combination with the sediment accumulation rate some indication of the carbon flux (Sibuet et al., 1989). Some stations at similar depth showed significant differences in density, biomass and mean individual weight. Using the individual boxcores for the correspondence-analyses, similar patterns were found, indicating that the community structure in the boxcore-samples per station were similar and thus gave a good estimation of the community structure at that site (Figure 13). The main conclusion from the correspondence-analyses and tree diagrams was that the macrobenthic community could be divided in three groups representing a shelf (break), an upper-slope and a lower-slope community.

Shelf community

At the two shelf stations A and D86-12 the macrobenthic community was totally different from all other stations. The macrobenthic community was dominated by small polychaetes (mainly opportunistic, interface-feeding Spionidae and patches of filter-feeding Serpulidae and Sabellidae). Characteristic for the shelf fauna were also the high numbers of Echinoidea (especially Spatangidae) and Decapoda. Among the Mollusca, filter-feeding Cardiacea were found, which were not found at the slope stations. Of the whole macrobenthic community ~ 65% were filter- and interface-feeders (Flach et al., in press). Although the highest den-

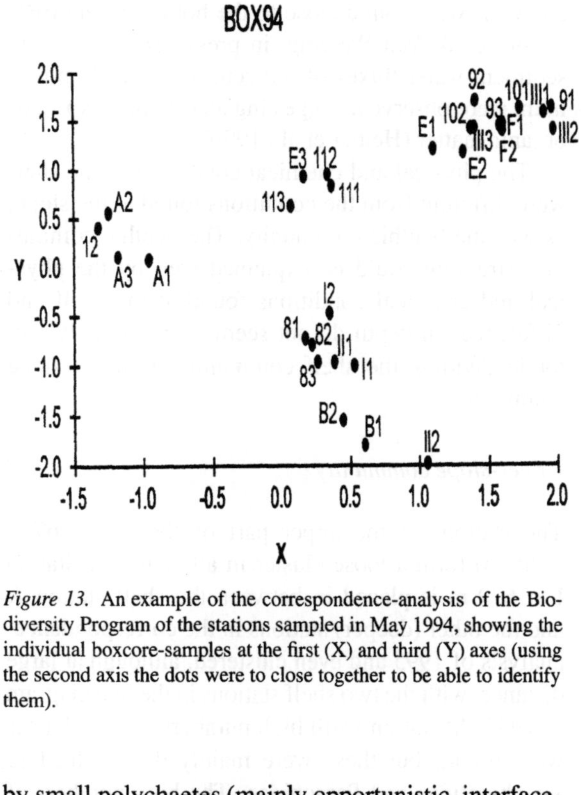

Figure 13. An example of the correspondence-analysis of the Biodiversity Program of the stations sampled in May 1994, showing the individual boxcore-samples at the first (X) and third (Y) axes (using the second axis the dots were to close together to be able to identify them).

sities were found on the shelf, total biomass was not high due to low mean individual weights of the macrobenthos. The macrobenthic community structure thus suggests that the shelf (break) environment is relatively instable, that the sediment has a low nutritive value and the water column offers the main food source.

The sediment at the two shelf stations was sandy with shell remains, low porosity and low% of organic C and total N. Sediment accumulation rates were high, ~ 30 cm per 7 ka (Van Weering et al., in press). Using a diagenetic model Soetaert et al. (in press) estimated the total load of mineralisable carbon in the sediment to be only 4.5 g C m^{-2} at the shelf station A, whereas at the deep station III is was estimated to be 38.5 g C m^{-2}. Flow velocities were intermediate, ~ 5.5 cm s^{-1} during spring/summer and ~ 15 cm s^{-1} during autumn/winter, and high concentrations of TPM, POC and CPE were found close to the bottom in the BBL (Thomsen & Van Weering, in press). Relatively low sediment-water fluxes of oxygen, nitrate and ammonium were observed, suggesting an off-shelf export of organic matter (Helder et al., 1996).

The physical and chemical conditions at the shelf were different from the conditions found at the slope, as was the benthic community. The benthic community structure could be explained well by the physical and chemical conditions found at the shelf and differences in depth do not seem to be the only factor in dividing the shelf community from the slope community.

Upper-slope community

The stations at the upper part of the slope (670–1465 m) form a loose cluster in all analyses. Station I at 670 m is placed in between the shelf station A and the other (deeper) stations in the correspondence-analysis of 1995 and even clustered, although at large distance, with the two shelf stations in the tree diagram of 1994. At station I still high numbers of polychaetes were found, but these were mainly deposit-feeding Ampharetidae and Paraonidae. The high numbers of bivalves and sipunculids at this station, of which most could be characterized as interface-feeders (Flach et al., in press), were also remarkable.

The sediment composition showed intermediate values between station A and B, both in grain-size and in% org. C and N. The sediment accumulation rate was also intermediate between A and B (Van Weering et al., in press), as were the flow velocities in autumn/winter (Thomsen & Van Weering, in press). During spring/summer, however, flow velocities were very low at this site (Figure 12). Station I is thus not only intermediate in depth, but also in physical and chemical factors and benthic community structure. The low flow velocities in spring/summer favour interface- and deposit-feeding species.

In 1994 three stations at ~ 1100 m were sampled, station B (1034 m) and D86-8 (1140 m) situated on the OMEX-transect and D86-11 (1120 m) situated ~ 35 km SE at the top of the King Arthur Canyon. These stations, although at about the same depth and situated relatively close to each other, differed in all aspects (density, biomass, mean individual weight and community structure). D86-11 was placed far apart from all other upper-slope stations in the correspondence-analysis. At this station very low numbers of small ophiuroids were found (Figure 5a), which were very abundant at all other upper-slope stations. Polychaetes were relatively abundant at this station, mainly Ampharetidae (surface deposit-feeders) living in relatively solid sand tubes. Other abundant taxa were amphipods and sipunculids, whereas also filter-feeding taxa as Tunicata and Hydrozoa were found at this station. In the tree diagram station 11 was grouped together with station OMEX-II at which very high numbers of these filter-feeding tunicates and hydroids were found. Figure 10 shows that the sediment characteristics of this two stations at ~ 1100 m were also different. Station 11 had a very high median grain-size, which was due to high numbers of solid foram shells. The% of org. C and total N within the sediment was low at this station, as was the sediment accumulation rate (only ~ 10 cm per 7 ka, Van Weering et al., in press). In the tree diagram station 11 was grouped together with station OMEX-II and both stations had in common that they are exposed to ... a) enhanced vertical fluxes under low flow conditions (OMEX-II in summer and D86-11 all over the year from the canyon walls) which would support surface deposit feeding and ... b) enhanced horizontal fluxes and predominant filter feeding during times of high flow conditions (OMEX-II in winter and D86-11 at the King Arthur Canyon, which is exposed to high energetic internal tides (Huthnance, 1995)). In this tree diagram station D86-8 was, although apart, grouped between the lower-slope stations. At this station 8 total abundance was low and few animals were found especially in the upper 1 cm layer, resulting in a vertical distribution pattern that indeed was very similar to the lower-slope stations (Flach & Heip, 1996a). Within the polychaetes relatively high numbers of subsurface deposit-feeders were found, which were also characteristic of the lower-slope fauna. The sediment accumulation rate at station 8 was high (~ 22 cm per 7 ka, Van Weering et al., in press). The % of org. C within the sediment was high at this station, but the % of N very low, resulting in a high C/N ratio. At this station 8 thus sedimentation of refractory material seems to occur, which is also more characteristic for the lower part of the slope.

Station 8 was geographically situated between the OMEX-stations B and II. At these two stations ~ 80% of the macrofauna was found in the upper 1 cm of the sediment (Flach & Heip, 1996a). Station B was characterised by a very high total biomass, due to a high mean individual weight. At this station the highest% org. C and N were found in the sediment and Soetaert et al. (1995) conclude that the carbon arriving at this station must be very reactive, with a mean degradability of 0.0037 d^{-1}, and therefore consisting mainly of fresh material. Net sediment accumulation rates were not very high (~ 16 cm per 7 ka, Van Weering et al., in press) and resuspension of material seems to occur during autumn/winter as relatively high flow velocities were measured (Thomsen & Van Weering, in press). Feeding conditions thus seem to be very good at this station, especially for interface-feeders which were found in relatively high numbers at this station (Flach et al., in press). At station II high numbers of filter-feeders were found, not only in the macrofauna but also in the megafauna and Foraminifera (Flach et al., 1997). In 1995, station II was clustered together with the Porcupine station P1 at a similar depth (~ 1450 m). This station P1 was also characterised by high numbers of filter-feeding taxa. At both stations high flow velocities were observed (Figure 12). Both stations also had low mean individual weights, at station P1 even the lowest value measured. The % org. C and N within the sediment were also relatively low at this station P1, but at station II they were high (Figure 11). Sediment accumulation rates at station II, however, were very low resulting in an accumulation of only ~ 5 mgC cm^{-2} ka^{-1} (Van Weering et al., in press). Physical factors thus favour a benthic community of small filter-feeding animals at both ~ 1450 m sites.

Lower-slope community

The lower-slope community was clearly distinct from the upper-slope community. On the whole, the lower-slope community was dominated by deposit-feeders (Flach et al., in press) and relatively more fauna was found in deeper sediment layers (Flach & Heip, 1996a). Duineveld et al. (1997) observed an uncoupling in food supply between the upper- and lower-slope. The upper slope seemed mainly to be fuelled by springbloom material, whereas the major food pulse to the lower slope comes from the summer offshore bloom. In August 1995, Duineveld et al. (1997) observed a mucous layer of fresh phytodetritus on top of the sediment at all stations deeper than 3000 m. Differences in seasonal variations within the macrofauna was also observed between the upper- and lower-slope (Flach & Heip, 1996b).

In 1994 the three stations at ~ 2200 m were clustered very closely together. Community structure and total densities at these three stations were very similar. Some small differences, however, were found in the mean individual weight (without large specimens), which was highest at station 10. At this station also two very large sipunculids were found, resulting in a total biomass of ~ 2000 mg C m^{-2}. At station 9 biomass was dominated by polychaetes (~ 75%), mainly subsurface deposit-feeding Paraonidae, Heterospionidae and Maldanidae, whereas at station F some filter- and interface-feeders were found. The median grain-size of the sediment was similar and the% org.C was also not very different. The% N at station 10, however, was very high. As N is often seen as a limiting factor for infauna (Jumars et al., 1990) the high% N at station 10 could explain high total biomass. Sediment accumulation rate was high at station 9 (~ 38 cm per 7 ka, Van Weering et al., in press). Deep-living subsurface deposit-feeders (as found at station 9) are favoured by a steady sedimentation of refractory organic material (Rice & Rhoads, 1989). The flow velocity was relatively high at station F during spring/summer, but low in autumn/winter (Thomsen & Van Weering, in press). The benthic fauna at the three ~ 2200 m stations did differ slightly in feeding-types, which could be related in differences in feeding conditions.

In 1995 the Porcupine station P2 was placed far away in the correspondence-analysis, close to station OMEX-III (Figure 2b). At these two stations the relative abundance of the two major taxa was very similar (~ 41% polychaetes and ~ 25% isopods). In absolute numbers, however, station P2 was more similar with station F and in the tree diagram they were grouped together (Figure 3b). The physical and chemical conditions at these two stations were also relatively similar. The biomass and mean individual weight was somewhat lower at station P2, as was the% org. C of the sediment.

The five stations at ~ 3500 m water depths taken in 1995, were more or less clustered together, although especially station OMEX-III was placed apart. This station, situated at the same depth as the Porcupine station P3, had the lowest number and especially low numbers of polychaetes and molluscs. The% C and N within the sediment was also much lower than at all other ~ 3500 m stations. Flow velocities were low during all seasons (Thomsen & Van Weering, in press), but the flow velocity was similar at station P3. In the tree diagram station P3 was clustered together with the along slope station A2 at 3880 m water depth and situated 330 km SSE. Whereas, the two along slope stations at similar depth, A2 and A4, were not clustered together. These two stations also differed in flow velocities, which were much higher at A2, and in the% N within the sediment, which was higher at A4. At station A2 filter-feeding taxa (Porifera and Tunicata) were found, which were not found at A4. At station A4 biomass and especially mean individual weight was much higher. In the correspondence-analysis the stations A2 and A3 were clustered together, both stations had high flow velocities and compared to A4 much lower% org. C and% N within the sediment and much lower mean individual weights of the macrofauna.

The two deepest stations P4 (4115 m) and E (4470 m) were clustered together. Both stations had low densities and besides somewhat higher numbers of ophiuroids, sponges and tunicates at P4, comparable community structure. Biomass and especially mean individual weight, however, was higher at the Porcupine Seabight station P4 than at the OMEX station E. The% of organic C and especially total N were also higher at station P4. The flow velocity at station E was very low and the sediment accumulation rate at this station very high (42 cm per 7 ka, Van Weering et al., 1997) and deep-living subsurface deposit-feeders dominated the benthic community (Flach et al., in press). Unfortunately no data on flow velocity and sediment accumulation rate for station P4 are known.

Conclusions

Stations at small geographical distances but at different depths showed larger differences in macrobenthic community structure than stations at similar depths at large distances. Thus depth is an important structuring factor. However, some regional features were observed, e.g. much higher numbers of ophiuroids at all Porcupine stations than at the OMEX stations.

An overall trend in decrease in density and biomass with increasing water depth was found, but no depth related pattern in mean individual weight could be observed. Mean individual weight, however, did show a significant negative relationship with flow velocities and a positive, but non significant relationship with the% of organic carbon within the sediment. The macrofauna could be divided into three groups: shelf, upper-slope and lower-slope communities. The shelf fauna was very different from the slope fauna, but the physical and chemical conditions were also very different. The upper-slope fauna was characterised by surface- and filter-feeders and most fauna was found in the upper 1 cm of the sediment. High % of org. C and N were found in the sediment and high flow velocities were measured, especially during autumn/winter. The lower-slope fauna was dominated by (sub)surface deposit-feeders and more animals were found in deeper sediment layers. Median grain-size was very low, as were the flow velocities and high sediment accumulation rates were observed.

Stations at similar depth, but with different physical and/or chemical conditions showed differences in density, biomass, mean individual weight and in macrobenthic community structure. More filter-feeding taxa were observed at stations with higher flow velocities, whereas more subsurface deposit-feeders were found at stations with high sedimentation rates.

Thus, besides the effect of water depth on macrobenthic community structure, other physical and chemical factors (such as flow velocities and organic matter supply) can be important structuring factors as well. Our hypothesis thus seems to be verified, macrobenthic community structure gives an indication of the quantity and quality of organic matter arriving at the sea floor and also of flow conditions over longer time scales.

Acknowledgements

The authors thank Adri Sandee for taking the macrofauna- and sediment-samples during the three cruises, and in co-operation with Joop Nieuwenhuize, for analysing the sediment composition. We would also like to thank the crew of the R. V. Pelagia, R. R. S. Charles Darwin and R. V. Meteor and the colleagues at NIOZ, Texel and GEOMAR, Kiel for the support received during the cruises. We are very grateful to Neil McAleece (Natural History Museum, London), Peter Lamont (SAMS, Oban) and Wim de Bruin (NIOZ, Texel) for providing and helping us with the BioDiversity Program. And we would also like to thank Jan Beukema (NIOZ, Texel) and two anonymous reviewers for their valuable comments on the earlier version of the manuscript. This study was financially supported through the OMEX programme (Ass. contract no. 20, Mas2-CT93-0069 and MAS3-CT96-0056), and ship-time was funded by the Netherlands Geoscience Foundation GOA, NIOZ and the German Science Foundation DFG. This is publication nr. 2346 of the Netherlands Institute of Ecology (NIOO-CEMO).

References

Billett, D. S. M., R. S. Lampitt, A. L. Rice & R. F. C. Mantoura, 1983. Seasonal sedimentation of phytoplankton to the deep sea benthos. Nature 302: 520–522.

Blake, J. A. & J. F. Grassle, 1994. Benthic community structure on the U.S. South Atlantic slope off the Carolinas: Spatial heterogeneity in a current-dominated system. Deep-Sea Res. II, 41: 835–874.

Carney, R. S., R. L. Haedrich & G. T. Rowe, 1983. Zonation of fauna in the deep sea. In G. T. Rowe (ed.), Deep-sea Biology. The Sea, Vol. 8, John Wiley & Sons, New York: 371–398.

Creutzberg, F., P. Wapenaar, G. Duineveld & N. Lopez Lopez, 1984. Distribution and density of the benthic fauna in the southern North Sea in relation to bottom characteristics and hydrographic conditions. Rapp. P.-v. Réun. Cons. int. Explor. Mer 183: 101–110.

DeMaster, D. J., R. H. Pope, L. A. Levin & N. E. Blair, 1994. Biological mixing intensity and rates of organic carbon accumulation in North Carolina slope sediments. Deep-Sea Res. II, 41: 735–753.

Diaz, R. J., G. R. Cutter & D. C. Rhoads, 1994. The importance of bioturbation to continental slope sediment structure and benthic processes off Cape Hatteras, North Carolina. Deep-Sea Res. II, 41: 719–734.

Duineveld, G. C. A., M. S. S. Lavaleye, E. M. Berghuis, P. A. W. J. de Wilde, J. van der Weele, A. Kok, S. D. Batten & J. W. de Leeuw, 1997. Patterns of benthic fauna and benthic respiration on the Celtic continental margin in relation to the distribution of phytodetritus. Int. Rev. ges. Hydrobiol. 82: 395–424.

Flach, E. & C. Heip, 1996a. Vertical distribution of macrozoobenthos along the continental slope in the Goban Spur area (NE Atlantic). Mar. Ecol. Prog. Ser. 141: 55–66.

Flach, E. & C. Heip, 1996b. Seasonal variations in faunal distribution and activity across the continental slope of the Goban Spur area (NE Atlantic). J. Sea Res. 36: 203–215.

Flach E., M. Lavaleye, H. de Stigter & L. Thomsen, in press. Feeding types of the benthic community and particle transport across the continental slope of the Goban Spur. Prog. Oceanog. (Special Issue OMEX)

Gooday, A. J. & C. M. Turley, 1990. Responses by benthic organisms to inputs of organic material to the ocean floor: a review. Phil. Trans. r. Soc., Lond. A 331: 119–138.

Grebmeier, J. M., C. P. McRoy & H. M. Feder, 1988. Pelagic-benthic coupling on the shelf of the northern Bering and Chukchi Sea I. Food supply source and benthic biomass. Mar. Ecol. Prog. Ser. 48: 57–67.

Helder, W., L. Lohse & E. Epping, 1996. Sediment-water exchange and early diagenesis. OMEX, Final Report.

Huthnance, J. M., 1995. Circulation, exchange and water masses at the ocean margin: the role of physical processes at the shelf edge. Prog. Oceanog. 35: 353–431.

Jumars, P. A., L. M. Mayer, J. W. Deming, J. A. Baross & R. A. Wheatcroft, 1990. Deep-sea deposit-feeding strategies suggested by environmental and feeding constraints. Phil. Trans. r. Soc., Lond. A, Math. Phys. Sci. 331: 85–101.

Jumars P. A. & E. D. Gallagher, 1982. Deep-sea community structure: Three playes on the benthic proscenium. In W. G. Ernst & J. G. Morin (eds), The environment of the deep sea. Prentice Hall, Inc., New Jersey: 217–255.

Lampitt, R. S., 1985. Evidence for the seasonal deposition of detritus to the deep-sea floor and its subsequent resuspension. Deep-Sea Res. 32: 885–897.

Nieuwenhuize, J., Y. E. M. Maas & J. J. Middelburg, 1994. Rapid analysis of organic carbon and nitrogen in particulate materials. Marine Cemistry 45: 217–224.

Rex, M. A., 1981. Community structure in the deep-sea benthos. Ann. Rev. Ecol. Syst. 12: 331–353.

Rice, D. L. & D. C. Rhoads, 1989. Early diagenesis of organic matter and the nutricial value of the sediment. In G. R. Lopez, G. L. Taghon, J. S. Levinton (eds), Ecology of marine deposit feeding. Springer-Verlag, New York: 59–79.

Rosenberg, R., 1995. Benthic marine fauna structured by hydrodynamic processes and food availability. Neth. J. Sea Res. 34: 303–317.

Rowe, G. T., 1983. Biomass and production of deep-sea macrobenthos. In G. T. Rowe (ed.), Deep-sea Biology. The Sea, Vol. 8, John Wiley & Sons, New York: 97–121.

Schaff, T., L. Levin, N. Blair, D. DeMaster, R. Pope & S. Boehme, 1992. Spatial heterogeneity of benthos on the Carolina continental slope: large (100 km)-scale variation. Mar. Ecol. Prog. Ser. 88: 143–160.

Sibuet, M., C. E. Lambert, R. Chesselet & L. Laubier, 1989. Density of the major size groups of benthic fauna and trophic input in deep basins of the Atlantic Ocean. J. Mar. Res. 47: 851–867.

Soetaert, K., P. M. J. Herman, J. J. Middelburg, E. Epping & W. Helder, 1995. Applying the coupled diagenetic model to OMEX sediments – preliminary results. OMEX Second Annual Report, part C: 63–64.

Soetaert, K., P. M. J. Herman, J. J. Middelburg & C. Heip, in press. Assessing organic matter mineralization, degradability and mixing rate in an ocean margin sediment (Northeast Atlantic) by diagenetic modelling. J. Mar. Res. 56.

Taghon, G. L. & R. R. Greene, 1992. Utilization of deposited and suspended particulate matter by benthic 'interface' feeders. Limnol. Oceanogr. 37: 1370–1391.

Thomsen, L., G. Graf, V. Martens & E. Steen, 1994. An instrument for sampling water from the benthic boundary layer. Cont. Shelf Res. 14: 871–882.

Thomsen, L. & Tj. C. E. van Weering, in press. Spatial and temporal variability of particulate matter in the benthic boundary layer at the North East Atlantic Continental margin (Celtic Sea). Progr. Oceanog. (Special Issue OMEX)

Van Weering, Tj. C. E., N. McCave, I. Hall & H. de Stigter, in press. Recent sedimentation in the OMEX study area: west European Margin, 47°–50° N. Progr. Oceanog. (Special Issue OMEX)

Hydrobiologia **375/376:** 287–295, 1998.
S. Baden, L. Pihl, R. Rosenberg, J.-O. Strömberg, I. Svane & P. Tiselius (eds),
Recruitment, Colonization and Physical–Chemical Forcing in Marine Biological Systems.

Interactions between two deposit feeding echinoderms: the spatangoid *Brissopsis lyrifera* (Forbes) and the ophiuroid *Amphiura chiajei* Forbes

Karin Hollertz[1], Mattias Sköld & Rutger Rosenberg
Department of Marine Ecology, Göteborg University, Kristineberg Marine Research Station, S-450 34 Fiskebäckskil, Sweden
[1] *E-mail correspondence: k.hollertz@kmf.gu.se*

Key words: competition, disturbance, food availability, behaviour, regeneration, gonad development, bioturbation, infauna

Abstract

Amphiura chiajei and *Brissopsis lyrifera* typically co-occur on some soft bottom areas of the North Sea, the Skagerrak and the Kattegat; they form the so called '*Brissopsis-chiajei* association'. Both species are deposit feeders that live partly (*A. chiajei*) or completely (*B. lyrifera*) burried in the sediment. In this association, each species is expected to affect the other one, notably through its feeding and burrowing activities. This study investigates the effects on body and gonads growth in *A. chiajei* and *B. lyrifera* as a result of their co-occurrence. The experiments were performed in aquaria with natural sediments (3 months observations) and have taken in account the population densities of both species and the availability of food. The results show that *B. lyrifera* can negatively affect the growth of body and gonads of *A. chiajei*, while *A. chiajei* seemingly has no effect on the growth of *B. lyrifera*. This situation probably results from the extensive bioturbation of the sediment by *B. lyrifera*, who also increased its surface feeding activity when food was added to the sediment surface.

Introduction

In the North Sea, the Skagerrak and the Kattegat soft-bottom areas at depths ranging from 50 to 200 m, the ophiuroid *Amphiura chiajei* and the spatangoid *Brissopsis lyrifera*, occur in high densities and biomasses. The co-occurrence of these deposit feeders has been described by Petersen (1913) as the '*Brissopsis-chiajei* association'. *A. chiajei* is buried in the sediment with its disk at 4 to 6 cm depth. One or two arms are stretched up above the sediment to collect food at the surface. Food particles are then transported along the arms to its mouth and ingested (Buchanan, 1964). *B. lyrifera* burrows down to ca 10 cm in the sediment. It ingests sediment in bulk and feeds on the associated organic matter. Spatangoid heart urchin are known to ingest both surface and deep sediment (Nichols, 1959; de Ridder & Jangoux, 1985; Lawrence, 1987). For most benthic deposit feeders, food is suggested to be a limiting factor for body and gonad growth, at least between events of sedimentation of fresh organic matter (Hargrave, 1980; Tenore, 1988). This is well illustrated by deposit feeding echinoderms which undergo body and gonad weight fluctuations related to food availability (Clements et al., 1988; Dobson et al., 1991; George, 1996; Sköld & Gunnarsson, 1996).

Through their burrowing activities, spatangoids disturb the sediment (bioturbation), and could consequently affect the co-occurring infaunal species (van Duyl, et al., 1992; Osinga et al., 1997, 1995). In the '*Brissopsis-chiajei* association', *B. lyrifera* most likely modifies the sediment grain distribution and destabilises the sediment, a situation that could affect food availability for *A. chiajei*. Other likely interactions between *B. lyrifera* and *A. chiajei* are physical disturbance and competition for food. The purpose of this work is to examine interactions between *B. lyrifera* and *A. chiajei*, manifested as effects on gonad and body growth.

Table 1. Mean (SD) density and biomass per m^2 of the ten numerically dominant species and *Brissopsis lyrifera* on the sampling location. e = echinodermata, p = polychaeta, c = crustacea, m = mollusca.

Species		Density ind. m^{-2}	Biomass g m^{-2}
Amphiura chiajei	e	481 (90)	73.5 (12.2)
Amphiura filiformis	e	307 (96)	18.4 (3.5)
Brada villosa	p	185 (61)	8.6 (4.6)
Maldane sarsi	p	133 (109)	5.8 (5.1)
Polyphysia crassa	p	78 (69)	13.4 (5.4)
Leucon nasica	c	48 (36)	0.1 (0.1)
Onoba vitrea	m	44 (68)	0.1 (0.2)
Melinna cristata	p	37 (15)	1.3 (1.0)
Abra nitida	m	30 (6)	0.3 (0.2)
Heteromastus filiformis	p	30 (13)	0.2 (0.1)
Brissopsis lyrifera	e	4 (6)	37.5 (56)

Material and methods

Two box-cosm experiments with 'undisturbed' sediment were run during three months. In the first experiment (A), we investigated how the presence or absence of *B. lyrifera* affected arm regeneration, disk area and oocyte size of *A. chiajei* with or without addition of phytoplankton. In the second experiment (B), we studied how *A. chiajei*, in two densities, affected intestine lipid content, gonad weight and oocyte size of *B. lyrifera* with or without addition of phytoplankton. During this experiment, also the behavioural response of *B. lyrifera* to the addition of organic matter was investigated.

Field sampling, sediment carbon and nitrogen content

On 19 to 21 March 1996, undisturbed sediment cores (0.25 m^2) from 52 m depth were collected with an USNEL box-corer in the Gullmarsfjord on the Swedish west coast (N 58°15.69′, E 11°27.58′). Sampling procedure followed the methods of Berge et al. (1986) and Rosenberg et al. (1997). The box-cores were connected to a deep (35 m) water flow-through system (2 l min^{-1}, 6–10 °C and 34 psu), in a thermo-constant room (6 °C).

To estimate natural macrofaunal abundance and biomass at the sampling location, five sediment cores (0.09 m^{-2}) were collected, sieved (1 mm mesh-size) and preserved (70% ethanol). Sediment profiles down to 5 cm depth were taken at the beginning and at the end of the experiment with a cut off syringe (ø: 12 mm) for analysis of total carbon (TC), total organic carbon (TOC) and nitrogen (N). The sediment profiles, sliced every cm, were freeze-dried and analysed with a NA 1500 NC Carlo Erba elemental analyzer (Fisons). TOC was achieved by treating the freeze-dried sediment samples with HCl before analysis (Hedges & Stern, 1983). Sediment in the area have been characterized as 8% sand (> 63 μm), 44% silt (> 3.9 μm) and 49% clay (< 3.9 μm) with a water content of 71% (Majoran & Agrenius (1995).

Organic matter

To verify the impact of food and feeding activity on the suggested interactions between the two species, organic matter was added to the sediment surface in half of all treatments. The organic matter was diatoms, mainly *Skeletonema costatum*, cultured or collected in the field. The first pulse was given on day 0, two weeks after density manipulations were done. Diatoms were added three more times throughout the experiment, totally 50 g DW (30 g AFDW) per m^2. When diatoms were added, the water flow was turned off for 5 h to let the algae sediment. The amount of organic matter leaving the boxes when the flow was turned on, was measured to < 5% of the amount added.

Feeding activity of A. chiajei *and* B. lyrifera

To estimate feeding activity of *A. chiajei* and *B. lyrifera*, 2.2 g fluorescently labeled polystyrene particles (Nunc Biosilon) with a diameter of 160–300 μm were added to the sediment surface of each box. The particles were added in a concentration of 1.37×10^6 particles m^{-2} at the same time as the last addition of algae was added to half of the boxes, two days before the end of the experiment. After 48 h, the particles from the stomachs of *A. chiajei* and *B. lyrifera* were counted to estimate the feeding activity. Activity of *B. lyrifera* at the sediment surface was measured daily, by counting tufts visible on the sediment surface.

Experiment A

In Experiment A, we investigated how *A. chiajei* was affected by *B. lyrifera* (present or absent), with and without addition of phytoplankton (algae added or control). These four treatments were replicated three times. The experiment ran for 55 d, from 16 April to 10 June, 1996.

Densities

To manipulate densities of *A. chiajei* and *B. lyrifera* in the boxes (0.25 m^2), the animals were forced to the sediment surface by initiating hypoxic conditions. The water flow was turned off and nitrogen gas bubbled through the water during five days. In this way, oxygen saturation decreased from 80% to about 2 to 15%. All specimens of *B. lyrifera* found in the boxes were removed and excluded from the experiment. According to calculations of *in situ* macrofaunal abundance in the boxes (Table 1), about 20% of *A. chiajei* emerged to the surface and 80% (380 ind. m^{-2}) remained in the sediment. On all specimens of *A. chiajei* that emerged to the sediment surface, the first arm clockwise (oral side) to the madreporite was cut off between the 7th and the 8th arm segments. Thereafter, a total number of 20 *A. chiajei* with an arm cut off plus 60 intact individuals, collected separately with an Agassiz trawl in the same area as the box-cores, were put in each box and an estimated total density of 700 *A. chiajei* m^{-2} was obtained. Ten specimens (mean length 47 mm, SD : 0.43) of adult *B. lyrifera*, also collected with an Agassiz trawl, were added to half of all boxes, equal to a density of 40 ind. m^{-2}. When the density manipulations were finished, the flow was turned on and the infauna that previously had emerged to the sediment surface, burrowed into the sediment.

To evaluate effects of the hypoxic treatment, 20 *A. chiajei* with one arm cut were added to each of two boxes without prior hypoxic treatment. No effect of temporary hypoxic treatment was recorded at the end of this experiment, (ANOVA; $F_{1,48} = 0.06$; $p = 0.81$). No infaunal mortality was observed.

Growth of A. chiajei

At the end of the experiment the sediment was sieved (1 mm meshes) and *A. chiajei* collected and video recorded under a dissection microscope. Disk area was measured as well as regenerated area and length of amputated arms, using image analyses, software NIH Image 1.58 (Nilsson & Sköld, 1996). Fluctuations in disk diameter may correspond to gonad size as well as amount of food in the gut (O'Connor et al., 1983; Munday & Keegan, 1992). For gonad analysis, oocyte area of *A. chiajei* was measured on 10 oocytes from 3 gonads of each individual (n = 4), using Confocal Laser Scanning Microscopy (CLSM). For laboratory procedure, see Sköld & Gunnarsson (1996). State of maturity was determined according to classification by Fenaux (1970).

Experiment B

In Experiment B, we investigated how lipid storage and gonad growth of *B. lyrifera* was affected by two densities (low and high) of *A. chiajei*, with and without addition of phytoplankton. As in Experiment A, this experiment started on 16 April 1996 and run for 55 d.

Densities

Manipulations of densities of *B. lyrifera* and *A. chiajei* followed a similar procedure as in Experiment A. Ten specimens of *B. lyrifera* were added to each box, equal to a density of 40 ind. m^{-2}, and 60 *A. chiajei* were added to 6 of 12 boxes to receive a density of 700 ind. m^{-2} (referred to as 'high density'). In the remaining 6 boxes about 20 *A. chiajei* were removed to obtain an estimated density of 380 ind. m^{-2} ('low density').

Growth of B. lyrifera

At the termination of the experiment, gut tissue and gonads of *B. lyrifera* were dissected. As echinoids may store nutrients in the gut (Lawrence et al., 1966), the intestine was weighed and stored in – 80 °C until analysing lipid content (Bligh & Dyer, 1959; modified by Silversand & Haux, 1995). Gonads were weighed immediately and stored in formalin (4%) until analysis. Gonad weight was normalised to test volume of each individual and a gonad index was obtained. Oocyte diameter was measured in a dissection microscope at 200 × magnification.

Statistics

Data in Experiment A were analysed with factorial analysis of variance (ANOVA) and regenerated arm area, disk area, large and small oocytes and fluorescent gut particle content of *A. chiajei* as dependent variables. *B. lyrifera* (present and absent) and treatment (control and algae added) were considered as fixed factors and aquarium as nested factor. For oocyte analysis the nested factors individual and gonad were also included in the model.

Data in Experiment B were analysed in a similar way but with intestine lipid content, gonad index, oocyte size and fluorescent gut particle content of *B. lyrifera* as dependent variables. For oocyte size in *B. lyrifera,* the nested factor individual was also included in the model. *A. chiajei* (low and high density) and treatment (control and algae added) were considered

Table 2. Surface activity of *B. lyrifera* (arcsin %) as dependent variable in a factorial ANOVA with treatment (low and high density of *A. chiajei*), organic enrichment (control and algae added) and before/after (B/A) algae addition as main effects and aquarium (Aq) as nested factor. The significant interaction is marked in bold.

Source of variation	df	Sum of squares	Mean square	*F*	p	Error term
Treatment	1	0.033	0.033	0.343	0.574	Aq
Organic enrichment	1	0.592	0.592	6.081	0.039	Aq
B/A	1	0.244	0.244	12.836	0.007	B/A * Aq
Aq (Treatment, Organic enrichment)	8	0.779	0.097	2.359	0.032	Residual
Treatment * Organic enrichment	1	0.013	0.013	0.129	0.729	Aq
B/A * Treatment	1	0.026	0.026	1.372	0.275	B/A * Aq
B/A * Organic enrichment	**1**	**0.340**	**0.340**	**17.915**	**0.003**	**B/A * Aq**
B/A * Aq (Treatment, Organic enrichment)	8	0.152	0.019	0.460	0.878	Residual
B/A * Treatment * Organic enrichment	1	0.072	0.072	3.768	0.088	B/A * Aq
Residual	48	1.981	0.041			

as fixed factors. Analysis of behaviour as surface activity of *B. lyrifera* was performed by evaluating the activity before and after the addition of algae (Table 2).

Pooling of nested and treatment terms was done if the nested factor was not significant ($p > 0.25$) (Underwood, 1981). When first order interaction was significant, multiple comparisons were performed using Student-Newman Keul's test (SNK-test, Sokal & Rohlf, 1995). To fulfill the criterion of homoscedasticity, gonad index and surface activity of *B. lyrifera* was corrected by arcsine transformation (Sokal & Rohlf, op. cit.). Data on fluorescent gut particle contents were transferred by log $(x+1)$. All data were tested for presence of gross heterogeneity of variances with Cochrans C-test ($\alpha = 0.05$, Snedecor & Cochran, 1989).

Trend analysis were considered significant with $\alpha = 0.05$ for the product moment correlation coefficient r (Sokal & Rohlf, 1995).

Results

Field sampling, sediment carbon and nitrogen content

Mean density and biomass of the numerically dominant macrofaunal species on the sampling location are shown in Table 1. *A. chiajei* constituted the largest density and biomass, and *B. lyrifera* the second largest biomass.

In the experiment, 125 sediment samples were analysed for TC, and N concentrations. Concentrations varied between 3.9 and 4.3% TC, and 0.3 and 0.4% N in the surface sediment down to 5 cm depth, both in the beginning and at the end of the experiment. TOC in the top 0–1 cm sediment varied between 2.9 and 3.1% in the beginning of the experiment. Thus, TC and N were rather uniform over time and between treatments.

Table 3. Mean (SD) oocyte diameter, gonad weight (wet weight) and percent of lipid in the intestine of *B. lyrifera* at the end of the experiment. Algae = Addition of phyto-plankton, C = control, High = high density of *A. chiajei*, Low = low density of *A. chiajei.* No variabel differed significantly.

Treatment		Oocyt diameter (μm)	Gonad weight (g)	Lipid content (%)
Algae	High	82.3 (12)	1.2 (0.4)	6.7 (2.3)
Algae	Low	79.4 (13)	1.2 (0.5)	6.7 (2.0)
C	High	77.8 (10)	1.3 (0.5)	6.3 (1.7)
C	Low	76.1 (9)	1.2 (0.6)	6.2 (1.8)

Experiment A

Feeding activity of A. chiajei

The feeding activity assessed by uptake of fluorescent particles was 7.6 (SD : 19) particles per individual for *A. chiajei* given phytoplankton on the sediment surface and 2.2 (SD : 6.1) for the others. The difference was not significant ($F_{1,96} = 4.42$; $p = 0.069$).

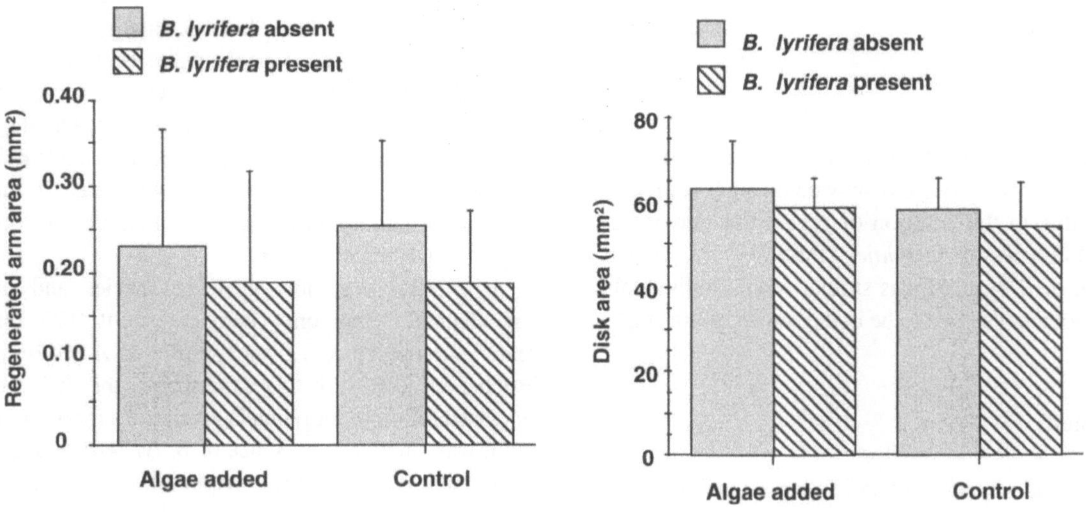

Figure 1. (A) Mean regenerated arm area (SD, $n=18$) and (B) mean disk area (SD, $n=30$) in *A. chiajei* at the end of the experiment, with and without algae added and in presence and absence of *B. lyrifera*.

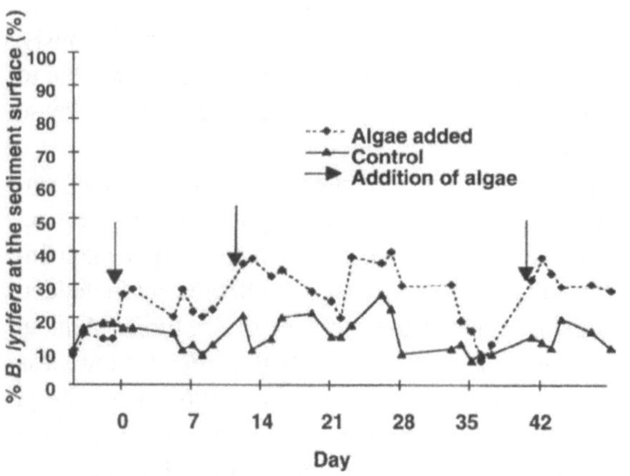

Figure 2. Percent *B. lyrifera* visible at the sediment surface in relation to supply of algae during the experiment.

Growth of A. chiajei

Regeneration of amputated arm area in *A. chiajei* decreased in the presence of *B. lyrifera* ($F_{1,68}=4.139$; $p=0.046$) (Figure 1a). No effect due to the addition of organic matter was, however, found. More than 90% of the amputated individuals were regenerating arms. Mean regenerated arm length was 1.1 mm (SD : 0.47) and 1.35 mm (SD : 0.52) in presence and absence of *B. lyrifera*, respectively. The disk area of *A. chiajei* was also significantly smaller when *B. lyrifera* was present compared to when absent ($F_{1,116}=6.700$; $p=0.011$). In addition, when organic matter was supplied, a significant increase was observed for the disk area both in the presence and the absence of *B. lyrifera* ($F_{1,116}=8.679$; $p=0.004$) (Figure 1b). At the end of the experiment, gonads of all *A. chiajei* were in the same stage of maturation, between growth and pre-maturation stage. Two different size classes of oocytes were represented in the gonads, small eggs (primary oocytes) were 52 μm (SD : 17) and large eggs (secondary oocytes) 112 μm (SD : 30) in diameter. No significant difference was found between any of the treatments for small or large oocyte-area.

Experiment B

Feeding activity of B. lyrifera

Surface activity of *B. lyrifera* increased significantly after addition of organic matter on the sediment surface (Table 2), as the apical tuft was observed more frequently shortly after addition of food (Figure 2). The organic matter added to the boxes, covered the sediment as a green carpet initially, but a few hours later a halo cleared of algae could be seen around *B. lyrifera* at the surface. Greater numbers of fluorescent particles were also found in the stomachs of *B. lyrifera* that were given organic matter than in the others ($F_{1,72} = 6{,}643$; $p = 0.033$), indicating higher feeding activity.

Growth of B. lyrifera

Intestine lipid content, mean oocyte size and total gonad weight of *B. lyrifera* showed no significant difference due to the addition of organic matter or the elevated density of *A. chiajei* (Table 3). The oocyte diameter of *B. lyrifera* was similar within individuals and all oocytes seemed to be in the same growth stage.

Discussion

The distribution pattern of soft sediment fauna is controlled by both large- and small-scale factors. Large-scale factors like water currents, salinity, temperature, food availability and sediment characteristics are relatively homogeneous over a large area (Menge & Olson, 1990) while small-scale ecological processes like competition, predation or biological disturbance are species specific (Woodin & Marinelli, 1991). Mechanisms affecting faunal distribution on a smaller scale, like interactions, need to be elaborated in controlled experiments where ambient factors can be constant and individual parameters such as body and gonad growth can be estimated. The sediment box-core system used in this study provides an excellent possibility to study animals in their natural habitat without affecting the natural sediment characteristics. Manipulative field studies with animals living at 30 m and deeper (below normal diving depths) is difficult and unpractical.

A significant part of assimilated energy in ophiuroids is allocated to regeneration of body parts as they are subject to sublethal predation, particularly arm cropping by fish (Sköld & Rosenberg, 1996). In *A. chiajei*, a high proportion of the total body weight consists of regenerated tissue (Munday, 1993). Arm regeneration in *Amphiura* has been shown to correlate positively to food supply (Sköld & Gunnarsson, 1996) and negatively to stress factors such as hypoxia (Nilsson & Sköld, 1996). Thus, the living conditions for *A. chiajei* will be reflected in arm regeneration rates, which most probably correlate to total somatic production. In this experiment, regenerated arm area of *A. chiajei* was lower when *B. lyrifera* was present, but not affected by the addition of food. This strongly suggests that *B. lyrifera* was a stress factor to *A. chiajei*. The lack of increased arm regeneration rate when organic matter was added, can be due to slow initial regeneration rates in *A. chiajei* (see below). Energy allocation for arm regeneration has been suggested to occur in two stages: an initial stage where internal reserves are used for healing the scar and a second growth stage where use of external food supply is significant (Sköld & Gunnarsson, 1996). Thus the growth phase in this experiment may not have lasted long enough or food may not have been abundant.

A. chiajei is a slow growing species and the disk diameter may only increase about 0.25 mm year^{-1}, compared to 1.6 mm year^{-1} in *A. filiformis* (Buchanan, 1967). In this experiment, the disk area increased for those individuals given extra food supply and it was smaller in presence of *B. lyrifera*. Fluctuations in disk diameter may correspond to gonad size or it may reflect the amount of food in the gut (O'Connor et al., 1983; Munday & Keegan, 1992). The increase and decrease in disk area due to algae added and presence of *B. lyrifera*, respectively, may be a result of both changes in gonad size and amount of food in the gut. As the food in the experiment was portioned over time, assimilated energy is likely to have been incorporated into body and gonad growth continuously. However, a decreased food uptake in *A. chiajei* due to disturbance of *B. lyrifera* may have caused a decrease in disk area.

In *B. lyrifera*, no changes were found in lipid content of the intestine, gonad weight or in oocyte size. Thus, the variables measured in this experiment indicates that *B. lyrifera* was not affected by the added amount of algae or the presence of *A. chiajei*. Annual test growth of juvenile or adult *B. lyrifera* is low; less than 10 mm in length (Buchanan, 1967; Rosenberg, 1972). Therefore, growth of the test is not a suitable measure in a short-term experiment like this one.

Study of the oocyte diameter showed no difference between treatments in either species. In several other echinoderm species, germinal tissue is known to increase as a response to favorable conditions which may result in an increase in total gonad weight, including protein and lipid content of the egg. These variables may be better indicators of gonad quality than oocyte size (George, 1996). All oocytes of *B. lyrifera* appeared to be of a similar size, while *A. chiajei* had a bimodal oocyte size distribution. The oocyte development of *A. chiajei* was between growth Stage 2 and prematuration Stage 3 at the end of the experiment (stages determined according to Fenaux, 1970). In the north-east Atlantic, *B. lyrifera* and *A. chiajei* build up gonads during spring and spawning occurs from late summer to late autumn (Brattström, 1945; Vasseur & Carlsén, 1948; Munday & Keegan, 1992). Accordingly, the oocytes in the present study were measured in a developmental stage towards maturity.

B. lyrifera ingested the inert fluorescent particles on the surface, which indicates that it did not select particles according to chemical composition. Feeding rate at the surface increased when food was added, which shows that *B. lyrifera* is able to localize and respond to an additional food supply. de Ridder & Jangoux (1985) found that when starved, the spatangoid *Echinocardium cordatum* selected organically rich surface sediment, while unstarved individuals alternated subsurface feeding with surface feeding. *B. lyrifera* seems to be well adapted to respond to and utilize a spring-bloom sedimentation of diatoms, as simulated in this experiment. Such events of input of high food quality may be important for growth in *B. lyrifera*. The comparatively poor food quality deeper down in the sediment, which may be the only source available during most of the year, may not have a sufficient quality for growth or gonad development. The amount of food added in this experiment (10 g AFDW m^{-2} $month^{-1}$) corresponds to approximately 5 g C m^{-2} $month^{-1}$. The sedimentation in the Skagerrak has been estimated at 4 to 12 g C m^{-2} $month^{-1}$ (Rosenberg, et al., 1997). Thus, the amount of food given in this study was in the lower range of these estimates and may have been too small to significantly increase growth.

Decreased body and gonad growth rates in *A. chiajei* in the presence of *B. lyrifera* indicate that *B. lyrifera* may be a superior competitor for food or that the activity of *B. lyrifera*, probably in combination with feeding, disturbs *A. chiajei* and affects its growth rate. Such a possible feeding disturbance was evident as *B. lyrifera* cleared the sediment surface from accumulated algae in a circular area around the test. The test area of *B. lyrifera* in this study was about 16 cm^2. This will directly affect animals close to the surface within this area, and probably also in the surrounding feeding area. In this experiment we found no effect on measured growth variables in *B. lyrifera*. The results suggest that *B. lyrifera* disturbs *A. chiajei*, most likely due to burrowing activity and while feeding. For such disturbances, densities of the disturbing species are of great importance. Densities used in this study were chosen with respect to natural densities and experimental circumstances. At various locations along the Swedish west coast, *B. lyrifera* is found in densities up to 30 ind. m^{-2} (Tunberg, 1991). However, in the North Sea densities up to 60 ind. m^{-2} have been reported (Ursin, 1960). The numbers of *B. lyrifera* found in the sampling area was low, however, densities may have been underestimated by the small (0.09 m^2) corer which is not appropriate for accurate density assessment of large individuals such as *B. lyrifera*. Although, the densities used in this study were in the higher range of the species occurrence, the recorded interaction in this study may be of significant importance even in areas with lower densities. Accordingly, the impact of different densities of both species would have been desirable to study further. Also the impact of other infaunal species that are likely to interfere with the interactions between *A. chiajei* and *B. lyrifera*, should be considered in future investigations. In this study, the effect of other infauna was assumed to be similar in each treatment and is therefore not likely to alter the outcome of the experiment.

Interactions between soft-bottom animals such as competition for food and space are important in regulating structure and function of communities. Indirect competition, e.g. by disturbance and modification of the substratum, has been suggested by Branch (1984) to be of greater importance in the organization of soft-bottom benthic communities than competition for food and space. Disruptive modifications on the substratum may result in amensalistic interactions between populations, where one species has an adverse effect on the other but is itself not affected by that species (Branch, *op. cit.*). Such indirect competition was most likely observed in the present study between *B. lyrifera* and *A. chiajei*. The large, mobile bioturbator *B. lyrifera* probably has wider effects on the properties of the sediment in the upper 10 cm, hence on the surrounding animals, than *A. chiajei* living closer to the surface. As both body and gonad growth of *A. chiajei* decreased in the presence of *B. lyrifera*, indirect competition on terms of disturbance or modification of the sediment is most likely. Generally, bioturbators modify the sediment geochemistry (Rhoads & Young, 1970; Brenchley, 1981) by their feeding, burrowing, irrigation, tunneling and biodeposition (Graf & Rosenberg, 1997). The spatangoid *E. cordatum*, with similar appearance as *B. lyrifera*, has been shown to highly affect the substratum by changing the sediment particle distribution, increasing new surfaces for bacterial growth and affecting the organic matter in the sediment (van Duyl, 1992; Osinga et al., 1997, 1995). Effects of the moving and feeding activity by *B. lyrifera* has to our knowledge not been investigated before, but impacts similar to that of *E. cordatum* can be expected.

In Peterson's pioneering work (1913), species in the '*Brissopsis-chiajei* association' were suggested to live more or less side-by-side in the sediment. The soft-bottom environment is a complex three-dimensional habitat which most likely provides space for several exquisite specialisations and complex interactions between the residents. Interaction between *B. lyrifera* and *A. chiajei* may be one among several possible examples.

Acknowledgements

The authors thank the anonymous reviewers for constructive comments and suggestions on the manuscript. We also thank Birthe Hellman, Hans Nilsson and Malin Werner for field and laboratory assistance, Paul Karila Department of Zoofysiology and Margit Fredrikson Department of Systematic Botany, Göteborg University for assistance with Confocal Laser Scanning Microscopy. Financial support for this study was provided by Swedish Natural Science Research Council (grant no. B-AA/BU 0329-312) to R. Rosenberg.

References

Berge, J. A., M. Schaanning, T. Bakke, K. Sandøy, G. M. Skeie & J. W. G. Ambrose, 1986. A soft-bottom sublittoral mesocosm by the Oslofjord: description, performance and examples of application. Ophelia 26: 37–54.

Bligh, E. G. & W. J. Dyer, 1959. A rapid method of total lipid extraction and purification. Can. J. Biochem. Physiol. 37: 911–917.

Branch, G. M., 1984. Competition between marine organisms: Ecological and evolutionary implications. Oceanogr. Mar. Biol. annu. Rev. 22: 429–593.

Brattström, H., 1945. Observations on *Brissopsis lyrifera* (Forbes) in the Gullmar Fjord. Ark. Zool. 37A: 1–27.

Brenchley, G. A., 1981. Disturbance and community structure: an experimental study of bioturbation in marine soft-bottom environments. J. mar. Res. 39: 767–790.

Buchanan, J. B., 1964. A comparative study of some features of the biology of *Amphiura filiformis* and *Amphiura chiajei* (Ophiuroidea) considered in relation to their distribution. J. mar. biol. Ass. U.K. 44: 565–576.

Buchanan, J. B., 1967. Dispersion and demography of some infaunal echinoderm populations. Symp. Zool. Soc. London 20: 1–11.

Clements, L. A. J., K. T. Fielman & S. E. Stancyk, 1988. Regeneration by an amphiuroid brittlestar exposed to different concentrations of dissolved organic material. J. exp. mar. Biol. Ecol. 122: 47–61.

de Ridder, C. & M. Jangoux, 1985. Origine des sédiments ingérés et durée du transit digestif chez l'oursin spatangide, *Echinocardium cordatum* (Pennant) (Echinodermata). Ann. Inst. océanogr., Paris. 61: 51–58.

Dobson, W. E., S. E. Stancyk, L. A. Clements & R. M. Showman, 1991. Nutrient translocation during early disc regeneration in the brittlestar *Microphiopholis gracillima* (Stimpson) (Echinodermata, Ophiuroidea). Biol. Bull. 180: 167–184.

Fenaux, L., 1970. Maturation of the gonads and seasonal cycle of the planktonic larvae of the ophiuroid *Amphiura chiajei* Forbes. Biol. Bull. 138: 262–271.

George, S. B., 1996. Echinoderm egg and larval quality as a function of adult nutritional state. Oceanol. Acta 19: 297–308.

Graf, G. & R. Rosenberg, 1997. Bioresuspension and biodeposition: a review. J. mar. Syst. 11: 269–278.

Hargrave, B. T., 1980. Factors affecting the flux of organic matter to sediments in a marine bay. In K. R. Tenore & B. C. Coull (eds), Marine Benthic Dynamics. University South Carolina Press, Columbia, SC, USA: 243–263.

Hedges, J. I. & J. H. Stern, 1983. Carbon and nitrogen determinations of carbonate-containing solids. Limnol. Oceanogr. 29: 657–663.

Lawrence, J., 1987. A functional biology of echinoderms. The John Hopkins University press, Baltimore: 340 pp.

Lawrence, J. M., A. L. Lawrence & A. C. Giese, 1966. Role of the gut as a nutrient-storage organ in the purple sea urchin (*Strongylocentrotus purpuratus*). Physiol. Zool. 39: 281–290.

Majoran, S. & S. Agrenius, 1995. Preliminary observations on living *Krithe praetexta* (Sars, 1866), *Sarsicytheridea bradii* (Norman, 1865) and other marine ostracods in aquaria. J. Micropal. 14: 96.

Menge, B. A. & A. M. Olson, 1990. Role of scale and environmental factors in regulation of community structure. Tree 5: 52–57.

Munday, B. W., 1993. Field survey of the occurrence and significance of regeneration in *Amphiura chiajei* (Echinodermata: Ophiuroidea) from Killary harbour, west coast of Ireland. Mar. Biol. 115: 661–668.

Munday, B. W. & B. F. Keegan, 1992. Population dynamics of *Amphiura chiajei* (Echinodermata: Ophiuroidea) in Killary Harbour, on the west coast of Ireland. Mar. Biol. 114: 595–605.

Nichols, D., 1959. Changes in the chalk heart-urchin *Micraster* interpreted in relation to living forms. Phil. Trans. r. Soc. Lond. B. 242: 347–437.

Nilsson, H. C. & M. Sköld, 1996. Arm regeneration and spawning in the brittle star *Amphiura filiformis* (O. F. Müller) during hypoxia. J. exp. mar. Biol. Ecol. 199: 193–206.

O'Connor, B., T. Bowmer & A. Grehan, 1983. Long-term assessment of the population dynamics of *Amphiura filiformis* (Echinodermata: Ophiuroidea) in Galway Bay (west coast of Ireland). Mar. Biol. 75: 279–28.

Osinga, R., A. J. Kop, J. F. P. Malschaert & F. Van Duyl, 1997. Effects of the sea urchin *Echinocardium cordatum* on bacterial production and carbon flow in experimental benthic systems under increasing organic loading. J. Sea Res. 37: 109–121.

Osinga, R., W. E. Lewis, J. L. M. Wopereis, C. Vreizen & F. C. van Duyl, 1995. Effects of the sea urchin *Echinocardium cordatum* on oxygen uptake and sulfate reduction in experimental benthic systems under increasing organic loading. Ophelia 41: 221–236.

Petersen, C. G. J., 1913. Valuation of the sea. 2. The animal communities of the sea bottom and their importance for marine zoogeography. Rep. Dan Biol. Ann. Rev. 16: 229–311.

Rhoads, D. C. & D. K. Young, 1970. The influence of deposit-feeding organisms on sediment stability and community trophic structure. J. mar. Res. 28: 150–178.

Rosenberg, R., 1972. Benthic faunal recovery in a Swedish fjord following the closure of a sulphite pulp mill. Oikos 23: 92–108.

Rosenberg, R., H. C. Nilsson, K. Hollertz & B. Hellman, 1997. Density-dependent migration in an *Amphiura filiformis* (Amphiuridae, Echinodermata) infaunal population. Mar. Ecol. Prog. Ser. 159: 121–131.

Silversand, C. & C. Haux, 1995. Fatty acid composition of vitellogenin from four teleost species. J. Comp. Physiol. B. 164: 593–599.

Sköld, M. & J. S. G. Gunnarsson, 1996. Somatic and germinal growth of the infaunal brittle stars *Amphiura filiformis* and *A. chiajei* in response to organic enrichment. Mar. Ecol. Prog. Ser. 142: 203–214.

Sköld, M. & R. Rosenberg, 1996. Arm regeneration frequency in eight species of Ophiuroidea (Echinodermata) from European sea areas. J. Sea Res. 35: 353–362.

Snedecor, G. W. & W. G. Cochran, 1989. Statistical methods (eight edition). Iowa state University press, Ames, Iowa, 503 pp.

Sokal, R. R. & F. J. Rohlf, 1995. Biometry (3d). W. H. Freeman & Co., New York, 887 pp.

Tenore, K. R., 1988. Nitrogen in benthic food chain. In T. H. Blackburn & J. Sörensen (eds), Nitrogen Cycling in Coastal Marine Environments. John Wiley & Sons Ltd: 191–206.

Tunberg, B., 1991. Övervakning av mjukbottenfauna längs Sveriges västkust. Swedish Environmental Protection Agency 4066.

Underwood, A. J., 1981. Techniques of analysis of variance in experimental marine biology and ecology. Oceanogr. Mar. Biol. annu. Rev. 19: 513–605.

Ursin, E., 1960. A quantitative investigation of the echinoderm fauna of the central North Sea. Meddr. Danm. Fisk. Havunders. 2: 1–204.

van Duyl, F. C., A. J. Kop, A. Kok & A. J. J. Sandee, 1992. The impact of organic matter and macrozoobenthos on bacterial and oxygen variables in marine sediment boxcosms. Neth. J. Sea Res. 29: 343–355.

Vasseur, E. & I. Carlsen, 1948. Sexual Maturity of the Sea-Urchin *Brissopsis lyrifera* (Forbes) in the Gullmar Fjord. Arkiv Zool. 41A: 1–10.

Woodin, S. A. & R. Marinelli, 1991. Biogenic habitat modification in marine sediments: the importance of species composition and activity. Symp. zool. Lond. 63: 231–250.

Hydrobiologia **375/376:** 297–305, 1998.
S. Baden, L. Pihl, R. Rosenberg, J.-O. Strömberg, I. Svane & P. Tiselius (eds), Recruitment, Colonization and Physical–Chemical Forcing in Marine Biological Systems.

Resource limitation in marine soft sediments – differential effects of food and space in the association between the brittle-star *Amphiura filiformis* and the bivalve *Mysella bidentata*?

Alf B. Josefson
National Environmental Research Institute, Dept. of Marine Ecology and Microbiology, P.O. Box 358, DK-4000 Roskilde, Denmark

Key words: macrobenthos, pelagic-benthic coupling, resource limitation, *Amphiura filiformis*, chlorophyll, Skagerrak-Kattegat

Abstract

While it may be a consensus that food is the major limiting factor for total macrofaunal standing stock in marine sediments below the euphothic zone, little is known about limiting resources on species level. This issue is addressed by analysing field population data of the facultative filter-feeding burrowing ophiuroid *Amphiura filiformis* and its coinhabitant, the bivalve *Mysella bidentata* in two gradients of pelagic productivity. In a temporal gradient, 1971–1994, with high productivity in the 80 s, abundance of *Amphiura* at 100 m depth showed logistic population growth ($R^2 = 0.90$, $n = 33$) with carrying capacity reached in 1986 (ca. 3000 ind m^{-2}). *Mysella* abundance versus time was best modelled by an exponential increase, occurring mainly during the period when *Amphiura* had reached maximum density (1986–1994) ($R^2 = 0.75$, $n = 18$). *Amphiura* showed somatic growth during the 'plateau period' of similar magnitude as predicted from independent data from the Kattegat. These observations indicate that *Amphiura* not was food limited in the plateau period. In the spatial gradient, inferred from chlorophyll concentrations in the surface sediment at 60 sites bisecting the area of the Skagerrak-Kattegat pelagic front, where primary productivity is elevated, *Amphiura* reached maximum abundance at the same level as in the temporal gradient, while *Mysella* and total benthic biomass peaked in the area of highest phytoplankton input to the sediment (Josefson & Conley, 1997). High content of the plant pigments chlorophyll-*a* (Chl-*a*) and phaeopigments in *Amphiura* suggested ingestion of labile algal matter, and a tentative chlorophyll budget indicated that *Amphiura* metabolic demand could be accounted for by ingestion of relatively fresh algal derived matter if half-lives of total chlorophyll in *Amphiura* were on the order of $^1/_2$–1 h. The results suggest that space may be a limited resource in some soft-sediment systems where filter-feeding is a dominating feeding mode, and input of fresh phytoplankton to the bottom is high i.e. where pelagic-benthic coupling is strong.

Introduction

It is a common view that food is the major limiting factor for total biomass of macrofauna in deep (below the euphotic zone) marine soft sediments (e.g. Rowe, 1971). However, on the level of individual species less is known about limiting resources. Yet, such knowledge is important for our ability to interpret changes in the benthic systems and to predict impact of environmental factors. Unlike many deposit-feeders which utilise organic matter in the surrounding sediment and living space and food therefore are difficult to separate, filter-feeders utilise a food resource that is often rapidly renewed through horizontal water movement and food capturing area is largely separated from the living space. This has the likely consequence that adjacent individuals of suspension feeders are less likely to exploit the same food resource compared to adjacent deposit feeding individuals (Levinton, 1972). The relative importance of space and food limitation in marine benthic macrofaunal invertebrates has been the subject of many experimental as well as correlative field studies especially on effects of varying resources on somatic growth (e.g. Peterson & Andre, 1980;

Olafsson, 1986; Peterson & Black, 1987; Kamermans et al., 1992), and few have considered effects on community structure (Widbom & Frithsen, 1995). In general, however, it has proven difficult to separate effects of the two different resources (Frechette & LeFaivre, 1990). In this paper I use an indirect approach using temporal and spatial field data in order to distinguish between effects of the two resources on the benthic macrofauna below the euphotic zone.

A conspicuous feature of macrobenthic soft sediment systems in the North-eastern Atlantic coastal areas is the faunal association dominated by the burrowing facultative filter-feeding ophiuroid *Amphiura filiformis* (L.), a potentially important energetic link between pelagic and benthic environments. *Amphiura* often acts as a passive suspension feeder (Loo et al., 1996) and thereby rapidly transporting organic matter from the water column down into the sediment typically to 3–6 cm depth where the mouth is situated. Parts of the organic matter not being absorbed by the ophiuroid is likely to provide food for other animals living in the *Amphiura*-burrows, such as the commensalistic bivalve *Mysella bidentata* (Montagu). Despite the likely high energetic importance of this association, little is known about what factors limit the two populations. Previous field estimates of *Amphiura filiformis* densities suggest that maximum density is attained between 2000 and 4000 ind m^{-2}, and it has been suggested that at these high densities space *per se* may be a limiting resource (Josefson, 1995; Rosenberg, 1995; Josefson & Conley, 1997). One way to test this proposition is to examine the changes of *A. filiformis* population size in gradients of increasing food availability and evaluate if density reaches an asymtotic value at the highest densities reported in the literature. As a means to indicate presence of food, the somatic growth of the ophiuroid and the density of the coinhabitant *Mysella bidentata*, which lives in the burrow of the ophiuroid, is used, as well as the content of chlorophyll in *Amphiura*-individuals and in the surface sediment.

The distributions of these two species are examined in two different organic matter (OM)-gradients, one temporal and one spatial gradient from the same geographical area, the frontal area between Skagerrak and Kattegat. Strong gradients in pelagic productivity are known to exist in this area (e.g. Heilmann et al., 1994), which also are reflected by labile organic matter distributions in the surface sediments underneath (Josefson & Conley, op. cit.). Data are reviewed and re-analysed from previously published works together with new data. I will show that the two species, behave differently in response to the gradients of productivity, and discuss this in the context of different limiting resources.

Material and methods

Temporal gradient:

Benthic faunal samples were taken at least annually over the 24 y period 1971–94, (except for in 1972 and 1983 where no samples are available), at a 100 m deep station (Vi2) positioned in the area of the Skagerrak-Kattegat pelagic front. Parts of these data are presented in Josefson (1990), Josefson & Jensen (1992) and HELCOM (1996). Densities of the species *Amphiura filiformis* and *Mysella bidentata* were estimated from quantitative Smith-McIntyre grab samples, at least 3 replicates of 0.1 m^2 each date, treated with conventional methods for benthic fauna monitoring (1 mm sieve, formalin preservation etc; details in references above).

Data on somatic growth of *Amphiura filiformis*, measured by oral width, were extracted from size distributions given in Josefson & Jensen (1992). Growth was estimated from the period 1979–1990 by pairing estimates of mean cohort oral widths measured with approximately one year intervals to give the smallest positive difference.

Considerable changes in parts of the ecosystem have been reported from this area over the past 20–25 y. Fresh-water runoff, inorganic nutrient concentrations during pre-spring bloom period, and annual mean chlorophyll concentrations in the surface waters, all showed a marked temporal increase from the 1970s to the mid 1980s (Austen et al., 1991; Josefson et al., 1993). These changes have been used to explain contemporary increases in benthic biomasses, most likely through increased sedimentation of labile organic matter to the bottom. The beginning of the 1980s also marked the start of eutrophication induced events, like hypoxia and increased frequencies of annual benthic algae, in many parts of the Kattegat (e.g. Rosenberg, 1985) also suggesting increased organic production in this period.

Spatial gradient:

Faunal and sediment data were used from 2 data sets sampled in the area of the Skagerrak-Kattegat pelagic front. Food gradient was described by the sediment

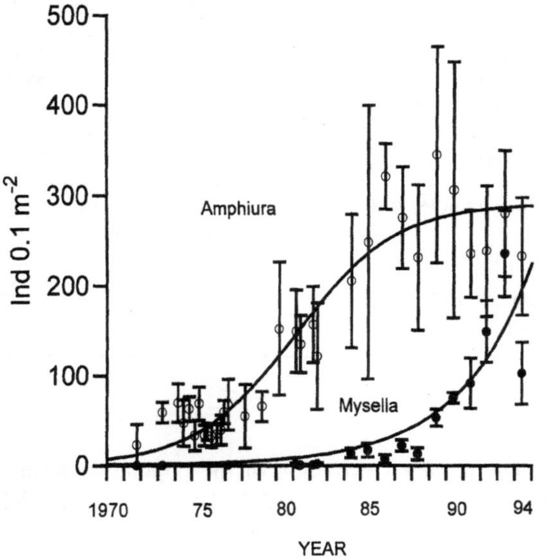

Figure 1. Abundance time series from a 24 y period of *Amphiura filiformis* and *Mysella bidentata*, means and 2 SD error bars, from each sampling date, at station Vi2 at 100 m water depth. Solid curves indicate fitted models, a logistic model $y = 292.21/(1 + e^{3.739-0.028*t})$ for *Amphiura*, and an exponential model $y = 0.6*e^{0.020*t}$ for *Mysella*, where y = the density of individuals per 0.1 m^{-2} and t = time in months, where t = 1 is January 1970.

Table 1. Results of fitting the logistic model (1) to mean densities of *Amphiura filiformis* versus time in months in the period 1971–1994 at station Vi2. A.S.E. = asymptotic standard error of the estimate, cf = confidence limit, Mean corr. R^2 = r squared corrected for means, n = number of observations.

Parameter	Estimate	A.S.E.	Lower 95% cf	Upper 95% cf
K (ind 0.1 m^{-2})	292.21	16.69	258.12	326.29
b	3.739	0.497	2.724	4.754
c	0.028	0.005	0.018	0.037
Raw R^2	0.966			
Mean corr. R^2	0.903			
$n = 33$				

distribution of the plant pigments Chl-*a* and phaeopigments which, out of several other OM-measures, has been shown to be the best predictors of total benthic biomass in the area (Josefson & Conley, 1997).

1. A spatial grid with 25 stations sampled with one 0.1 m^2 VanVeen grab per station in June and in October 1991 (Josefson, 1995). Abundance and size distribution of *Amphiura filiformis* and total faunal biomass as formalin wet weight were determined using the same extraction methods as in the temporal study above. At each station in June 1991, a separate Haps-core sample was taken for determination of organic matter in the top cm of the sediment. Chl-*a* and phaeopigments were determined spectrophotometrically chiefly in accordance with methods by Lorenzen (1967); further details in Josefson (1995).

2. A spatial grid of 35 stations bisecting the area of the Skagerrak-Kattegat pelagic front sampled for macrofauna and sediment variables with one 0.1 $m^2 \times 0.2$ m USNEL box core sample per station in June 1994 (Josefson & Conley, 1997). Data used in this paper are abundance of *Amphiura filiformis* and *Mysella bidentata* and total benthic biomass measured as AFDW of fresh deep frozen material. For comparison with biomass data from grid 1 AFDW was transformed to wet weight using conversion factors for taxonomic groups from the literature (Rumohr et al., 1987). Factors used for AFDW/WW were 0.14 for *A. filiformis*, 0.03 for remaining echinoderms (echinoids mainly), 0.085 for gastropods, 0.069 for bivalves, 0.132 for polychaetes, 0.145 for crustaceans and 0.1 for remaining fauna.

The top cm of the sediment from the June 1991 Haps samples and the June 1994 USNEL samples were analysed for Chl-*a* and phaeopigments as described in Josefson (1995) and in Josefson & Conley (1997). In addition similar data on Chl-*a* and phaeopigments were used from June 1993 from the 2nd grid of stations. Mean values of sediment chlorophyll from June 1993 and June 1994 were used to describe a spatial gradient of labile organic matter input. A data set on chlorophyll content, Chl-*a* and phaeopigments, in adult (> 1.7 mm oral width) individuals of *Amphiura filiformis* was used from the 2nd grid of stations measured in June 1994. From each station ca. 10 individuals of the ophiuroid were extracted for chlorophyll (Josefson & Conley, 1997), and analysed with same methods as for sediment samples.

Results

Temporal gradient:

Population growth

The error bars for the mean densities in the time series of *Amphiura filiformis* and *Mysella bidentata* at the 100 m station Vi2 show that the temporal variability is far greater than within-sample variability (Figure 1).

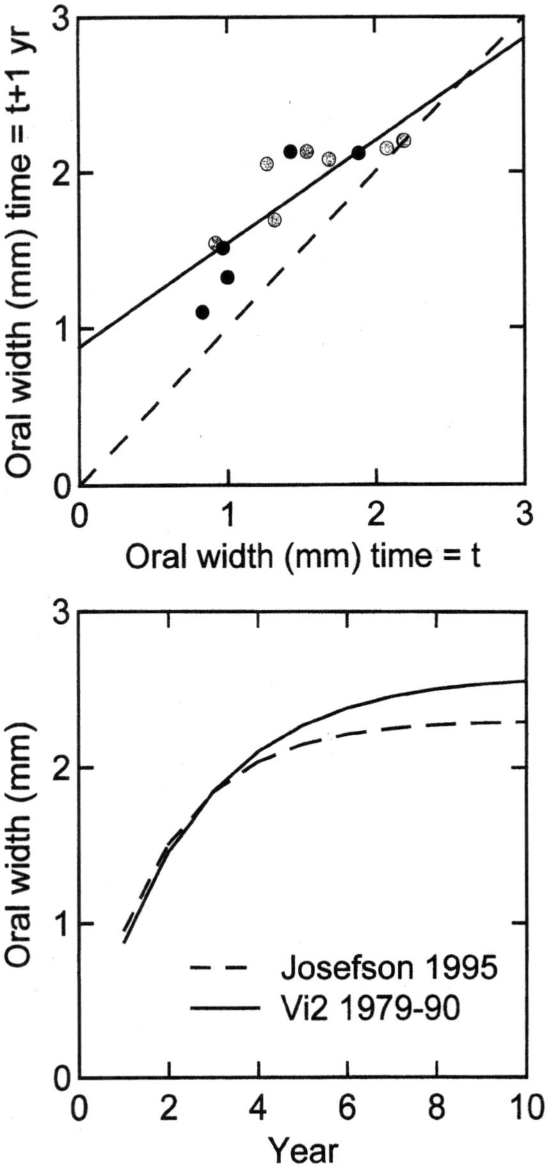

Figure 2. a) Ford-Walford plot of cohort mean oral widths of *Amphiura filiformis* (OW) with approximately 1 y intervals from the period 1979–90 at station Vi2. The fitted (solid) line follows the equation $OW_{t+1} = 0.881 + 0.661 * OW_t$ (n = 13, $R^2 = 0.73$), where OW_{t+1} = the oral width (mm) one year later than time = t. Dashed line = bisector. Grey shaded dots denote data from 1984–1990. b) Von Bertalanffi curves for growth in oral width of *Amphiura filiformis* from the time series data in 1979–90 at station Vi2, $OW_t = 2.599^* (1 - e^{-0.41^* t})$ and previously published independent data from a major part of the Kattegat based on a 4 month period, $OW_t = 2.307^* (1 - e^{-0.54^* t})$, (Josefson, 1995), where OW_t = the oral width (mm) at time t (yr).

Table 2. Results of fitting the exponential model (2) to mean densities of *Mysella bidentata* versus time in months in the period 1971–1994 at station Vi2. A.S.E. = asymptotic standard error of the estimate, cf = confidence limit, Mean corr. $R^2 = r$ squared corrected for means, n = number of observations.

Parameter	Estimate	A.S.E.	Lower 95% cf	Upper 95% cf
N_0 (ind 0.1 m^{-2})	0.600	0.746	− 0.981	2.180
b	0.020	0.005	0.010	0.029
Raw R^2	0.830			
Mean corr. R^2	0.750			
$n = 18$				

The temporal evolution of the population means of *A. filiformis* was best modelled by a logistic model (Figure 1):

$$N_t = K/(1 + e^{(b - c^* t)}), \qquad (1)$$

where N_t is the number of individuals at time t (here month number, where $t = 1$ is January 1970), K the maximum number of individuals or carrying capacity, and b and c are constants. K, b and c were estimated through iteration using the Modified Gauss-Newton algorithm in the non-linear model module of SYSTAT v. 6.0 (SPSS Inc.). Different starting values of parameters were used as well as other algorithms like Simplex and Quasi-Newton, still the same solution was obtained (Table 1). Thus K was estimated to 2922 individuals m^{-2} and the plateau was reached in 1986.

The temporal evolution of the commensal *Mysella bidentata* was best modelled by an exponential model (Figure 1, Table 2).

$$N_t = N_0 * e^{b^* t} \qquad (2)$$

where N_t and N_0 are the population densities at time t and at $t = 0$. The same methods of estimation was used as for *A. filiformis*. *M. bidentata* showed an increase, mainly during the period when the host had reached maximum population density (1986–1994).

Somatic individual growth

Mean oral width estimates for cohorts given in Josefson & Jensen (1992) were paired over approximately 1 y periods to give estimates of growth in the time period 1979–1990. The Walford plot in Figure 2a. indicates that growth occurred in this period, and that growth was not lower during the plateau period 1986–90. Estimating the growth constant by which size approaches the asymptotic maximum size, k from the

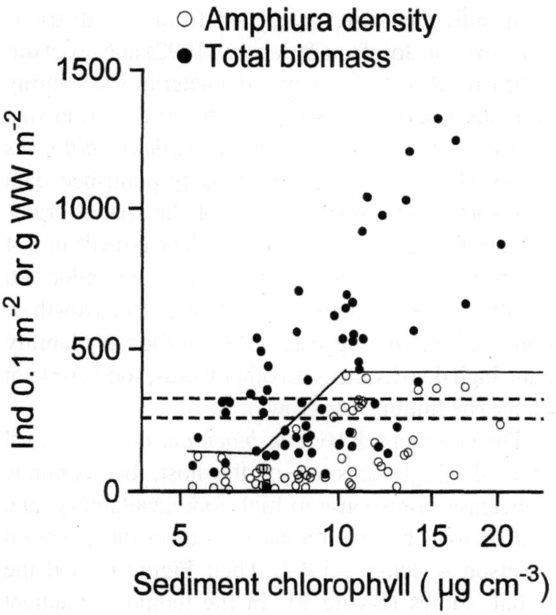

Figure 3. Distributions of *A. filiformis* abundance and total benthic biomass in wet weight along a spatial food gradient reflected by June values of sediment total chlorophyll (Chl-*a* + phaeopigments). Data from Josefson (1995) (25 stations) and Josefson & Conley (1997) (35 stations). Solid line indicating maximum densities of *A. filiformis*. Note that while the *A. filiformis* density levels off at ca. 10 μg cm^{-3}, total biomass continues to increase along the gradient. Both faunal variables are significantly positively correlated ($P < 0.01$, Pearson product moment correlation) with total sediment chlorophyll. Horizontal broken lines indicate upper and lower 95% confidence limits for carrying capacity for *Amphiura filiformis* in terms of density, estimated from the time series at station Vi2.

slope of the plot, which equals e^{-k}, and the maximum size/oral width where size at $t+1$ equals size at t, the Von Bertalanffy growth curve was constructed. In Figure 2b this curve is compared with a corresponding curve based on material from 1991 in Josefson (1995). There is a fairly good agreement between the two curves suggesting that oral width growth of *A. filiformis* at station Vi2 in the period 1979–90 not differed markedly from growth elsewhere in the area in 1991.

Spatial gradient:

Previous work has indicated that strong gradients of labile OM input to the bottom exists in the area of the Skagerrak-Kattegat pelagic front. Sediment chlorophyll distributions conform with the expected mean position of the front in the primary productive period and could explain a great part of the variation in benthic biomass in the area (Josefson & Conley, 1997). Combining total biomass and *A. filiformis* abundance data from this material with similar data from partly

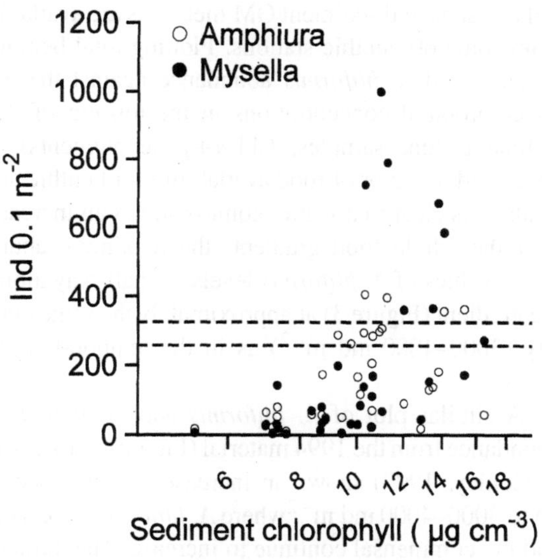

Figure 4. Abundance distributions of *Amphiura filiformis* with oral width > 1.7 mm and *Mysella bidentata* along a spatial food gradient, 32 stations, indicated by total chlorophyll concentrations in the top cm of the sediment (means for 2 cruises in June and one in October). For further information see previous figure. (data from Josefson & Conley 1997).

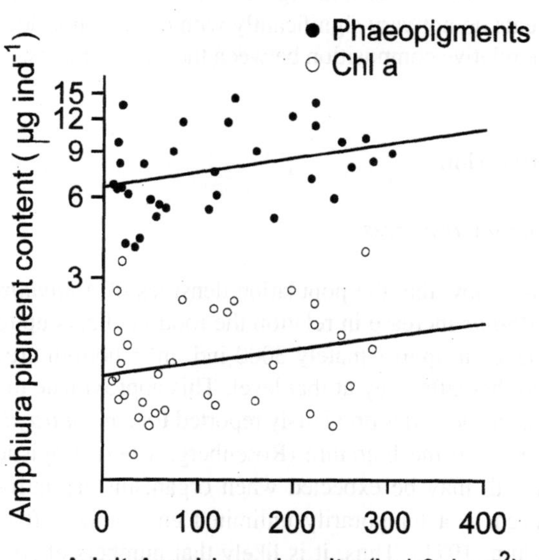

Figure 5. Chl-*a* and phaeopigments content in individuals plotted against density of adults (oral width > 1.7 mm) of *Amphiura filiformis* at 34 stations in the Skagerrak-Kattegat frontal area in June 1994. A $\log_{10}$-scale on the y-axis is used to facilitate comparison of pigment composition. The correlations between the 2 pigments and adult density based on untransformed data are not significant ($P > 0.05$, Pearson product moment correlation, $n = 34$–35).

the same area taken 3 y earlier (Josefson, 1995) material on fauna and sediment OM measures are available from some 60 benthic stations. Plotting total benthic biomass and *A. filiformis* abundance against the total chlorophyll concentrations in the top cm of the sediment (June samples, Chl-*a* + phaeopigments), a suggested reflector of food availability for benthic animals, it is clear that while biomass shows an increase over the whole food gradient, the maximum abundance values of *A. filiformis* levels off half-way along the gradient (Figure 3) at approximately the same density (3000–4000 ind m^{-2}) as in the temporal study above.

A similar plot of *A. filiformis* and *M. bidentata* abundance from the 1994 material (Figure 4) (Josefson & Conley, 1997) shows an increase of both species up to 3000–4000 ind $m^{-2}$where *A. filiformis* levels off and the commensal continue to increase. The number of commensals per burrow of *A. filiformis* peaked in the zone where the highest input of labile OM was indicated (Josefson & Conley, 1997). Again, these patterns suggest that *A. filiformis* and commensal are limited by different factors in the high input area. Chlorophyll content in *Amphiura filiformis* in June 1994 plotted against the density of adults (oral width > 1.7 mm) is shown in Figure 5. Chl-*a* and phaeopigments do not vary significantly with density, nor does the relative composition between the two pigments.

Discussion

Limiting resources

Data show that the population densities of *Amphiura filiformis* increase in relation the food gradients up to a level at approximately 3000 ind. m^{-2} bottom area and thereafter stay at that level. This approximate the highest densities previously reported for *Amphiura filiformis* in the literature (Rosenberg, 1995). Logistic growth may be expected when organisms are introduced in a temporarily unlimited environment (e.g. Odum, 1971). Thus, it is likely that numbers of *Amphiura filiformis* was limited by some factor in the 'plateau' period.

Somatic growth, measured by oral width, was not affected by the high densities in the temporal study. *Amphiura filiformis* showed somatic Bertalanffy growth during the 'plateau period' of similar magnitude as predicted from independent growth data from the whole Kattegat area (Josefson, 1995). The mean individual size, calculated from size distributions given in Josefson & Jensen (1992) and Josefson (1995), reached ca 6 mm disc diameter in 1988 during the plateau period, which is similar to the mean size in other parts the area with high *Amphiura filiformis* density (Josefson, 1995). Previously published data (Josefson, 1995) from one part of the spatial gradient area suggests density independent growth in a 4 month period. Unless resources have been allocated differentially to disc growth from, e.g. arm growth or reproduction, this suggests sufficient food availability in the high density areas. In other words, food does not seem to be limiting for growth.

The commensal *Mysella bidentata* does not level off at the high densities of the host, but continue to increase in response to high food availability, and so does total benthic biomass, both in the temporal (Josefson & Jensen, 1992, Their Figure 6) and the spatial studies (Figure 3). In the temporal gradient *Mysella* shows exponential population increase during the period when density of the host had reached carrying capacity. Assuming that the two species, at least partly, utilise similar food resources, both are known to ingest chlorophyllous material (Ockelman & Muus, 1975; Josefson & Conley, 1997). This suggests that food is available for *Mysella* in the *Amphiura* burrows. As indicated in Figure 5, chlorophyll content from the spatial gradient, and consequently ingestion of algal matter in *Amphiura* in June 1994 seem unaffected, or at least not negatively affected, by population density. This make sense if filter feeding was the dominant feeding mode and food was abundant in this period.

All these circumstances point at a limiting factor for *Amphiura* density other than food in the high density areas. One prerequisite for the alternative factor space to be important is that the individuals have the possibility to interact physically with each other. This can be evaluated by comparing the space potentially available for an individual and the space needed calculated from density and published data on burrow geometry. Adults of *A. filiformis* (disc diameter > 0.5 cm) maintain burrows penetrating 3 cm ore more into the sediment (Ockelmann & Muus, 1978). The disc is positioned in the deepest part of the burrow from which arm-burrows are radiating up to the sediment-water interface. Information on the number of arm burrows maintained and also the number of arms being active in feeding and irrigation is somewhat conflicting. Ockelmann & Muus (1978) reported 3–4 active arms per ind., while Loo et al. (1996) only observed 1–2 arms being active at a time. From

Table 3. Ranges of tissue weight specific chlorophyll contents (Chl-*a* + phaeopigments, except for Christensen & Kanneworff (1985) which gives only phaeopigments) in some benthic invertebrates in the Kattegat area. Content is given in μg chlorophyll g $AFDW^{-1}$ except for * which is μg chlorophyll g DW^{-1}.

Species	Water depth (m)	Chlorophyll content ($\mu g\ g^{-1}$)	Author
Nuculoma tenuis	30	124–1626*	Christensen & Kanneworff 1985
Abra nitida	30	33–608*	Christensen & Kanneworff 1985
Chlamys opercularis	30	20–243*	Christensen & Kanneworff 1985
Anobothrus gracilis	30	8–58*	Christensen & Kanneworff 1985
Arctica islandica	15–30	ca. 250	Josefson et al. 1995
Mya arenaria	ca. 0.5	100–500*	Josefson et al. unpublished
Amphiura filiformis	25–150	200–1200	This study, Josefson & Conley 1997

both studies, however, it seems clear that the distance between 2 arms at the sediment-water interface of adults is 2–3 cm. Assuming that the area between the arms of an individual at the sediment water interface is roughly circular, then at a density of 3000 adult ind. m^{-2} the diameter of this area being available would be ca. 2 cm. This is equal to or less than the area needed based on the published information and physical contact between individuals therefore seems possible. Rosenberg (1995) adopting a slightly different way of calculating the needed space for adult *Amphiura* arrived at similar conclusion for high density areas close to station Vi2. Thus, considering the geometry of the burrow system and the fact that other fauna utilising the sediment water interface is present, space limitation *per se* may be a realistic alternative to food limitation.

Benthic-pelagic coupling

From other studies in the area there are reasons to believe that a substantial part of plankton production sediments to the bottoms of the area at 100 m water depth. The findings that *Amphiura filiformis* contain Chl-*a* and phaeopigments suggests that the species ingests labile algal material. The stomach content of these pigments in *Amphiura* in June 1994 is of the same order of magnitude as found in both deposit and suspension feeding invertebrates at water depths ranging from 0.5 to 30 m in the Kattegat area (Table 3). Expressing the chlorophyll content in *Amphiura* as concentration on disc volume basis, and assuming a density of 3000 ind. m^{-2} with a disc diameter of 6.25 mm (equalling an oral width of 2.2 mm) and disc height of 1–1.5 mm, the concentration factor of total chlorophyll between *Amphiura* disc volume and the volume of the top cm of the ambient sediment is 20–25. Although there are no reports on chlorophyll decay rates in *Amphiura* guts, it is likely that these rates are considerably higher than decay rates in the ambient sediment. Half life of chlorophyll in bivalve guts may be an hour or less (Hawkins et al., 1986) while half-lives in sediments may be on the order of 14–25 d (Sun et al., 1991). Assuming steady state between input and decay, this suggests ingestion of relatively fresh algal material by *Amphiura* may be considerable in these high density areas i.e. the benthic-pelagic coupling is strong.

Tight benthic-pelagic coupling in parts of these areas was argued for by Josefson & Conley (1997) who calculated metabolic demand for the *Amphiura* population based on biomass and published P:B ratio and growth efficiency (Sköld et al., 1994; Loo et al., 1996). They arrived at a demand of the same order of magnitude as annual primary production in the area. The data on chlorophyll content in *Amphiura filiformis* allow for an independent way of calculating/estimating flux of algal material into the *Amphiura* population. Knowing the concentration and the decay rate of chlorophyll and assuming steady state between ingestion and loss, the ingestion of algal material can be estimated as for Chl-*a* in pelagic grazers (e.g. Mackas & Bohrer, 1976) and for algal flux into sediments (Sun et al., 1991). Conversely, if the metabolic demand is known and it is assumed that phytoplankton material is the primary food source, the decay rate of chlorophyll in *Amphiura* can be estimated.

Assuming steady state, the ingestion of chlorophyll during a time period t (I) can be expressed as

$$I = F^* C, \tag{3}$$

Table 4. Estimates of total chlorophyll (Chl-*a* + phaeopigments) decay rates (k, h^{-1}) in *Amphiura filiformis* through balancing metabolic demand and ingestion calculated from chlorophyll content of adults, assuming 100% absorption efficiency of chlorophyll, at 34 ($n=34$) stations.

Calculation mode	k (h^{-1})		Half-life (h)	
	Mean (SD)	Range	Mean	Range
P:B ratios	0.78 (0.64)	0.24–3.49	0.89	0.20–2.95
Respiration	0.68 (0.47)	0.22–2.41	1.02	0.29–3.17

where C is the amount of total chlorophyll (Chl-a+phaeopigments) in the animal and F is the fraction of chlorophyll lost (decayed) in that interval. If chlorophyll decay is exponential

$$F = (1 - e^{-k^*t}) \quad (4)$$

then for metabolic demand (MD) to balance ingestion we get

$$MD = (C * (1 - e^{-k^*t})) * AE/100, \quad (5)$$

where AE is the absorption efficiency of chlorophyll in %. Solving for k, we get the decay rate of chlorophyll needed for the absorbed part of ingestion to balance metabolic demand in a certain time interval (t).

The metabolic demand and total chlorophyll content in *Amphiura* were converted to μmol chlorophyll, μmol m^{-2} h^{-1} and μmol m^{-2} respectively.

The metabolic demand was calculated in two different ways from biomass estimates in Josefson & Conley (1997): 1. Using a respiration rate of 2.4 ml O_2 g^{-1} wet weight (WW) d^{-1} as a mean for published metabolic rates (Loo et al., 1996 and references therein). 2. Using an annual Production:Biomass ratio in *Amphiura* of 0.46 (Sköld et al., 1994) and a growth efficiency of 7% (Loo et al., 1996). It was assumed that the chemical composition of animal AFDW biomass was the same as phytoplankton AFDW biomass. Conversion factors used were: 1 ml O_2 = 20 J, AFDW/WW = 0.14, 20 kJ g^{-1} AFDW, algal g C = 0.366 g AFDW, molar ratio C/Chl-a = 60 (Sun et al., 1991).

The Chl-*a* and phaeopigment contents in *Amphiura filiformis* in June 1994 were expressed as μmol total chlorophyll m^{-2} bottom area using measured biomass and a molar weight of 893 for Chl-*a* and 593 for phaeopigments assuming most of these latter pigments was phaeophorbide.

The annual metabolic demand of *Amphiura filiformis* in the area with high standing stock (30 g AFDW m^{-2} or about 3000 ind m^{-2}) was ca. 70 g C m^{-2} yr^{-1} with both modes of calculating metabolic demand.

For the case of 100% absorption efficiency (AE), the estimated decay rates for chlorophyll corresponded to half-lives on the order of ca. 1 h (Table 4). It is likely, though, that AE is less than 100% but higher than the 70% AE for total organic matter reported by Loo & Rosenberg (1996). Using a lower AE than 100% will increase the decay rates (shorten half life) needed for the equation (5) to balance. However, if AE lies within the range 70–100%, the estimated half-lives is on the order of $^1/_2$ to 1 h. This is a very reasonable order of magnitude for chlorophyll half life in invertebrate guts such as pelagic grazers (e.g. Kiørboe & Tiselius, 1987) and filtering bivalves (Hawkins et al., 1986; Josefson et al., unpublished). Therefore it is a possibility that the *Amphiura* population to a great extent is fuelled by relatively fresh algal material.

The present comparison applies to a June situation as regards chlorophyll content in the animals, and to what degree this is representative for a whole year is unknown. Previous data from bivalves (Christensen & Kanneworff, 1985; Josefson et al., 1995; Josefson et al. unpublished), however, suggest that the June level of chlorophyll content may be intermediate between the high levels during spring bloom in March/April and the lower levels later in the year, and therefore does not represent an extreme situation.

Conclusions

The temporal and spatial distributions of *Amphiura filiformis* density and total standing stock of macrofauna in relation to the food gradients suggest that *A. filiformis* density in the area of investigation may be limited by some factor other than food when food input to the bottom is high. This is also supported by the observations that somatic growth occurs in *Amphiura* and that the population size of the coinhabitant *Mysella bidentata* continue to increase, when *Amphiura* had reached maximum densities around 3000 ind m^{-2}, and finally that chlorophyll content of *Amphiura* was not negatively correlated with density. This indicates that space *per se* may be a limiting resource also in some soft-sediment systems, where the input to the bottom of labile algal carbon is high and filter feeding is a dominant feeding mode.

Acknowledgements

Time series data used have been measured within the Swedish Environmental Monitoring Programme.

References

Austen, M. C., J. B. Buchanan, H. G. Hunt, A. B. Josefson & M. A. Kendall, 1991. Comparison of Long-term Trends in Benthic and Pelagic Communities of the North Sea. J. mar. biol. Ass. U.K. 71: 179–190.

Christensen, H. & E. Kanneworff, 1985. Sedimenting phytoplankton as major food source for suspension and deposit feeders in the Øresund. Ophelia 24: 223–244.

Frechette, M. & D. LeFaivre, 1990. Discriminating between food and space limitation in benthic suspension feeders using self-thinning relationships. Mar. Ecol. Prog. Ser. 65: 15–23.

Hawkins, A. J. S., B. L. Bayne, R. F. C. Mantoura, C. A. Llewellyn & E. Navarro, 1986. Chlorophyll degradation and absorption throughout the digestive system of the blue mussel *Mytilus edulis* L. J. exp. mar. Biol. Ecol. 96: 213–223.

Heilmann, J., K. Richardson & G. Ærtebjerg, 1994. Annual distribution and activity of phytoplankton in the Skagerrak/Kattegat frontal region. Mar. Ecol. Prog. Ser. 112: 213–223.

HELCOM, 1996. Third Periodic Assessment of the State of the Marine Environment of the Baltic Sea, 1989–1993; Background document, Balt. Sea Envir. Proc. No. 64B, Oy Edita Ab, Helsinki, 252 pp.

Josefson, A. B., 1995. Large-scale estimate of somatic growth in *Amphiura filiformis* (Echinodermata: Ophiuroidea). Mar. Biol. 124: 435–442.

Josefson, A. B. & J. N. Jensen 1992. Growth patterns of *Amphiura filiformis* support the hypothesis of organic enrichment in the Skagerrak-Kattegat area. Mar. Biol. 112: 615–624.

Josefson, A. B., J. N. Jensen & G. Ærtebjerg, 1993. The benthos community structure anomaly in the late 1970s and early 1980s – a result of a major food pulse? J. exp. mar. Biol. Ecol. 172: 31–45.

Josefson, A. B., J. N. Jensen, T. G. Nielsen & B. Rasmussen, 1995. Growth parameters of a benthic suspension feeder along a depth gradient across the pycnocline in the southern Kattegat, Denmark. Mar. Ecol. Prog. Ser. 125: 107–115.

Josefson, A. B. & D. J. Conley, 1997. Benthic response to a pelagic front. Mar. Ecol. Prog. Ser. 147: 49–62.

Kamermans, P., H. W. van der Veer, L. Karczmarski & G. W. Doeglas, 1992. Competition in deposit- and suspension-feeding bivalves: Experiments in controlled outdoor environments. J. exp. mar. Biol. Ecol. 162: 113–135.

Kiørboe, T. & P. T. Tiselius, 1987. Gut clearance and pigment destruction in a herbivorous copepod, *Acartia tonsa*, and the determination of *in situ* grazing rates. J. Plankton Res. 9: 525–534.

Levinton, J., 1972. Stability and trophic structure in deposit-feeding and suspension-feeding communities. Am. Nat. 106: 472–486.

Loo, L.-O., P. R. Jonsson, M. Sköld & O. Karlsson, 1996. Passive suspension feeding in *Amphiura filiformis* (Echinodermata: Ophiuroidea): Feeding behaviour in flume flow and potential feeding rate of field populations. Mar. Ecol. Prog. Ser. 139: 143–155.

Loo, L.-O. & R. Rosenberg, 1996. Production and energy budget in marine suspension feeding populations: *Mytilus edulis, Cerastoderma edule, Mya arenaria* and *Amphiura filiformis*. J. Sea Res. 35: 199–207.

Lorenzen, C. J., 1967. Determination of chlorophyll and phaeopigments: spectrophotometric equations. Limnol. Oceanogr. 12: 343–346.

Mackas, D. & R. Bohrer, 1976. Fluorescence analysis of zooplankton gut contents and an investigation of diel feeding patterns. J. exp. mar. Biol. Ecol. 25: 77–85.

Ockelmann, K. W. & K. Muus, 1978. The biology, ecology and behaviour of the bivalve *Mysella bidentata* (Montagu). Ophelia 17: 1–93.

Odum, E. P., 1971. Fundamentals of Ecology. W.B. Saunders Company, London, 574 pp.

Olafsson, E. B., 1986. Density dependence in suspension-feeding and deposit-feeding populations of the bivalve *Macoma balthica*: A field experiment. J. anim. Ecol. 55: 517–526.

Peterson, C. H. & S. V. Andre, 1980. An experimental analysis of interspecific competition among marine filter feeders in a soft-sediment environment. Ecology 61: 129–139.

Peterson, C. H. & R. Black, 1987. Resource depletion by active suspension feeders on tidal flats: Influence of local density and tidal elevation. Limnol. Oceanogr. 32: 143–166.

Rosenberg, R. 1985. Eutrophication – the future marine coastal nuisance? Mar. Poll. Bull. 16: 227–231.

Rosenberg, R., 1995. Benthic marine fauna structured by hydrodynamic processes and food availability. Neth. J. Sea. Res. 34: 303–317.

Rowe, G.T., 1971. Benthic biomass and surface productivity. In J. D. jr. Costlow (ed.), Fertility in the Sea, Vol. 2. Gordon and Breach Science Publisher, New York: 441–454.

Rumohr, H., T. Brey & S. Ankar, 1987. A compilation of biometric conversion factors for benthic invertebrates of the Baltic Sea. Baltic Marine Biologists Publ No. 9, 56 pp.

Sköld, M., L.-O. Loo & R. Rosenberg, 1994. Production, dynamics and demography of an *Amphiura filiformis* population. Mar. Ecol. Prog. Ser. 103: 81–90.

Sun, M., R. C. Aller & C. Lee, 1991. Early diagenesis of chlorophyll-*a* in Long Island Sound sediments: A measure of carbon flux and particle reworking. J. mar. Res. 49: 379–401.

Widbom, B. & J. B. Frithsen, 1995. Structuring factors in a marine soft bottom community during eutrophication – an experiment with radio-labelled phytodetritus. Oecologia 101: 156–168.

Hydrobiologia **375/376:** 307–315, 1998.
S. Baden, L. Pihl, R. Rosenberg, J.-O. Strömberg, I. Svane & P. Tiselius (eds),
Recruitment, Colonization and Physical–Chemical Forcing in Marine Biological Systems.

Macrobenthic metabolism as carbon and nitrogen fluxes in a coastal area exposed to strong tidal currents (Dover Strait, eastern English Channel)

Aline Migné[1] & Dominique Davoult[2]
Université des Sciences et Technologies de Lille, EP 1750 du CNRS, Station Marine, B.P. 80, F-62930 Wimereux, France
[1] *Present adress: Laboratoire d'hydrobiologie, Université P. & M. Curie, 12 rue Cuvier, F-75005 Paris, France*
[2] *e-mail: dom@loalit.univ-littoral.fr*

Key words: carbon budget, nitrogen budget, macrobenthic community, coastal area

Abstract

In view of recent studies of a Dover Strait macrobenthic community (eastern English Channel), an annual budget of carbon and nitrogen fluxes is proposed and discussed in order to understand the contribution of macrobenthic metabolism in this productive coastal area (phytoplanktonic production = 336 g C m^{-2} y^{-1}). In this area typical of coarse sediments which are exposed to strong tidal currents in the English Channel, the macrobenthic compartment is suspected to be responsible for the major part of exchanges at the bottom-boundary layer. The macrobenthic biomass is high (mean annual biomass = 281 g afdw m^{-2}) and dominated by three species (98.5%): the ophiurid *Ophiothrix fragilis* and the two cnidarians *Urticina felina* and *Alcyonium digitatum*. Carbon and nitrogen budgets including ingestion, production, respiration, excretion and egestion allow the estimation of annual fluxes from the water column to the benthic compartment (Ingestion = 820 g C m^{-2} y^{-1} and 130 g N m^{-2} y^{-1}) and from the benthic compartment to the water column (Respiration + Egestion = 710 g C m^{-2} y^{-1} and Excretion + Egestion = 100 g N m^{-2} y^{-1}). The net metabolism, as the difference between primary production and consumption of organic matter, shows that this coastal ecosystem is heterotrophic. Its functioning is based on allochtonous material and advection and it appears to be an important source of carbon dioxide.

Introduction

As a very productive area at the land-ocean interface, the coastal ocean is assumed to play an important role within the global carbon and nitrogen cycles (Smith & Hollibaugh, 1993). The topic of coastal metabolism is then of interest in defining the role of the ocean as a possible sink for CO_2 that is released by the burning of fossil fuels. Benthic metabolism is one of the relevant processes studied in the coastal zones.

The English Channel is an epicontinental sea highly governed by tidal currents which determine the distribution of sediment and benthic communities. In areas of strong hydrodynamism, the sediment is mainly composed of gravels and pebbles (about 40% of the total surface; Larsonneur et al., 1982) and colonized by a rich and diverse sessile epifauna community (Holme, 1966; Cabioch, 1968). The present paper is devoted to the understanding of the contribution of the macrobenthic metabolism in this community in the Dover Strait (eastern English Channel, study site: 50°55′5 N, 1°34′7 E), where a high pelagic primary production is observed.

Light limitation in this area (37 m deep with a high concentration of suspended material) doesn't allow the development of the macrophytobenthos. There are no field data available on the microphytobenthos (which is assumed to be absent). Phytoplankton are then assumed to be the only primary producers. Phytoplankton production is estimated to be 336 g C m^{-2} y^{-1} by Quisthoudt (1987) or 40 g N m^{-2} y^{-1} by Hoch (1995). Tidal currents can exceed 3.6 knots in mean spring tide and coarse sediments dominate. In the coastal part of the sessile epifauna community, pebbles account for more than 80% of total sediment

(Davoult, 1990). This part of the community is characterized by high species richness and biomass and by the abundance of suspension-feeders. Annual mean macrobenthic biomass was estimated to be 281.1 g m^{-2} (in ash free dry weight) with three species accounting for 98.5%: the ophiurid *Ophiothrix fragilis* (Abildgaard) and the two cnidarians *Urticina felina* (Linné) and *Alcyonium digitatum* (Linné) (Migné & Davoult, 1997a). We propose an annual budget of carbon and nitrogen fluxes at the square-metre scale according to recent results of metabolism experiments carried out on these three species.

Methods

Carbon and nitrogen budget

Assuming conservation of matter, the equation I = P + R + Ex + Eg (with I = Ingestion, P = Production, R = Respiration, Ex = Excretion and Eg = Egestion), applied to a population of a single species, includes all possible exchanges of matter in the population (Crisp, 1984). This equation was applied to the three main populations of the Dover Strait community to quantify annual exchanges of carbon and nitrogen between the community and the water column. Annual exchanges were estimated at the square-metre scale according to the population structure (density and biomass) observed during a recent quantitative survey (Migné & Davoult, 1997a). Several recent studies of the three populations (Davoult, 1989; Migné & Davoult, 1997a) and experimental measurements on the three species (Davoult et al., 1991; Davoult et al., 1992; Davoult & Gounin, 1995a; Migné, 1996; Migné et al., 1996; Migné & Davoult, 1997b,c,d) allowed us to estimate all of the terms of the equation except the ingestion (particularly difficult to measure), which was found by the difference, since the budget must balance.

Production

The annual production of the *Ophiothrix fragilis* population was estimated from the mean annual biomass, using the P:B ratio of 1.28 determined for this same population by Davoult (1989). Biomass was expressed in terms of ash free dry weight (afdw), organic carbon (C) and nitrogen (N) according to the following conversion relationships (Davoult et al., 1992) and considering the demographic structure of the population (density and size distribution: data from Migné & Davoult, 1997a):

$$B = 0.420D^{2.74} \quad (n = 198; r^2 = 0.982) \qquad (1)$$

$$C = 0.066D^{3.07} \quad (n = 74; r^2 = 0.991) \qquad (2)$$

$$N = 0.032D^{2.83} \quad (n = 71; r^2 = 0.991) \qquad (3)$$

where B = biomass as ash free dry weight in mg, C = organic carbon content in mg, N = organic nitrogen content in mg and D = disc diameter in mm. As no estimation of P:B ratio was available for the two cindarians, annual production of the *Urticina felina* and *Alcyonium digitatum* populations were calculated from mean annual biomass and mean individual (or colonial) weight according to the empirical relationship proposed by Brey (1990):

$$\log(P) = -0.473 + 1.007 \log(B) - 0.274 \log(W) \quad (4)$$

where P = annual production of the population in g afdw m^{-2}, B = mean annual biomass in g afdw m^{-2} and W = mean individual (or colonial) weight in g afdw.

Annual production of the two cnidarians populations was then expressed in terms of carbon and nitrogen using the following conversion relationships determined for the same populations (Migné et al., 1996):

$$U.\ felina \quad C = 0.508B \quad (n = 65;\ r^2 = 0.962) \qquad (5)$$

$$N = 0.113B \quad (n = 65;\ r^2 = 0.992) \qquad (6)$$

$$A.\ digitatum \quad C = 0.412B \quad (n = 109;\ r^2 = 0.970) \qquad (7)$$

$$N = 0.115B \quad (n = 111;\ r^2 = 0.980) \qquad (8)$$

where C = organic carbon content in mg, N = organic nitrogen content in mg and B = biomass in mg afdw.

Respiration

An experimental survey of respiration demonstrated a seasonal trend by sinusoidal curves fitted to the O_2-uptake data as a function of time for *O. fragilis* (Migné & Davoult, 1997b) and *A. digitatum* (Migné & Davoult, 1997c). The respiration rate of these two species was estimated in terms of oxygen consumption by integrating these curves (Euler's method):

Table 1. Known parameters of the carbon budget equation (I = P + R + Eg) of the *Ophiothrix fragilis*, *Urticina felina* and *Alcyonium digitatum* populations. P = Production, R = Respiration and Eg = Egestion (in g C m^{-2} y^{-1}), according to data from the survey of the three populations structure (Migné & Davoult, 1997a) and the studies of the three species metabolism (Respiration: Migné & Davoult, 1997b&c; Egestion: Davoult & Gounin, 1995a).

	Ophiothrix fragilis	*Urticina felina*	*Alcyonium digitatum*
P	101.62	7.45	4.80
R	175.85	22.37	1.68
Eg	482.77		

Table 2. Known parameters of the nitrogen budget equation (I = P + Ex + Eg) of the *Ophiothrix fragilis*, *Urticina felina* and *Alcyonium digitatum* populations.
P = Production, Ex = Excretion and Eg = Egestion (in g N m^{-2} y^{-1}), according to data from the survey of the three populations structure (Migné & Davoult, 1997a) and the studies of the three species metabolism (Excretion: Davoult et al., 1991 and Migné & Davoult, 1997d; Egestion: Davoult & Gounin, 1995a).

	Ophiothrix fragilis	*Urticina felina*	*Alcyonium digitatum*
P	28.17	1.66	0.47
Ex	15.59	2.70	0.69
Eg	79.85		

O. fragilis (Migné & Davoult, 1997b)

$$y = 0.310 - 0.128 sin[(2\pi/365)x + 1.094] \quad (n = 22; r^2 = 0.977) \tag{9}$$

A. digitatum (Migné & Davoult, 1997c)

$$y = 0.207 - 0.097 sin[(2\pi/365)x + 1.432] \quad (n = 16; r^2 = 0.961) \tag{10}$$

where x = Julian days and y = O_2-consumption rate in mg g^{-1} h^{-1}.

Respiration rate was then expressed in terms of carbon using the respiratory quotient of 0.69 for *O. fragilis* (according to Migné & Davoult, 1997b) and of 0.67 for *A. digitatum* (according to Migné, 1996).

As no seasonal trend was revealed in the respiration rate of *U. felina* (Migné & Davoult, 1997c), the mean O_2-consumption rate observed during the experimental survey (0.130 mg O_2 g^{-1} h^{-1}) was used for this species. Respiration rate was then expressed in terms of carbon assuming a respiratory quotient of 0.70 (according to Migné, 1996).

Excretion

An experimental survey of excretion demonstrated a seasonal trend for *O. fragilis* (Davoult et al., 1991) and *A. digitatum* (Migné & Davoult, 1997d). The excretion of these two species was estimated by integrating the sinusoidal curves fitted to the NH_4^+-release data as a function of time (Euler's method):

O. fragilis (according to data from Davoult et al., 1991)

$$y = 1.410 + 0.466 sin[(2\pi/365)x - 1.088] \quad (n = 6; r^2 = 0.983) \tag{11}$$

A. digitatum (Migné & Davoult, 1997d)

$$y = 6.480 - 2.700 sin[(2\pi/365)x + 1.100] \quad (n = 11; r^2 = 0.952) \tag{12}$$

where x = Julian days and y = NH_4^+-production rate in μg N g^{-1} h^{-1}.

As no seasonal trend was revealed in the excretion of *U. felina* (Migné & Davoult, 1997d), the mean NH_4^+-production rate observed during the experimental survey (4.120 μg N g^{-1} h^{-1}) was used for this species.

Egestion

Egestion rate (Eg) is directly related to ingestion rate (I) with absorption efficiency (Ab), which depends both on food availability and nutritional habit of the species: Eg = (1 – Ab) I.

Faeces production of *O. fragilis* has been measured in terms of carbon (unpublished data) and nitrogen (Davoult & Gounin, 1995a) at different periods of the year (egestion rate varying from 0.066 mg C ind^{-1} d^{-1} and 0.011 mg N ind^{-1} d^{-1} in November to 3.931 mg C ind^{-1} d^{-1} and 0.729 mg N ind^{-1} d^{-1} in May).

As no data on faeces production was available for the two cnidarians, egestion rate was expressed as a function of ingestion. The egestion:ingestion ratio being proposed takes into account egestion measured on *O. fragilis* and the comparative nutritional habits of the three species.

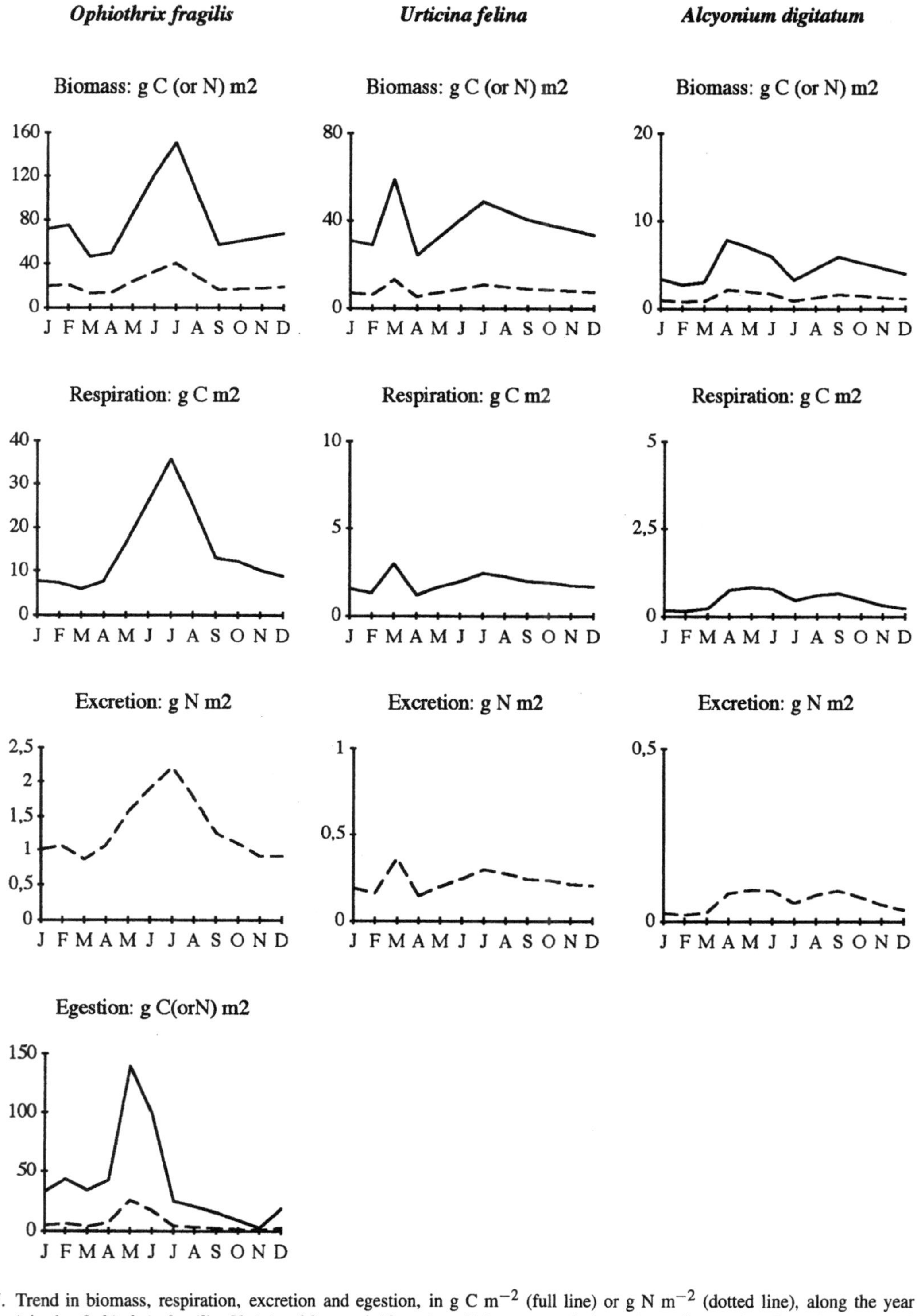

Figure 1. Trend in biomass, respiration, excretion and egestion, in g C m^{-2} (full line) or g N m^{-2} (dotted line), along the year (months in abscissa) in the *Ophiothrix fragilis*, *Urticina felina* and *Alcyonium digitatum* populations. According to data from the survey of the three populations structure (Migné & Davoult, 1997a) and the studies of the three species metabolism (Respiration: Migné & Davoult, 1997b&c; Excretion: Davoult et al., 1991 and Migné & Davoult, 1997d; Egestion: Davoult & Gounin, 1995a).

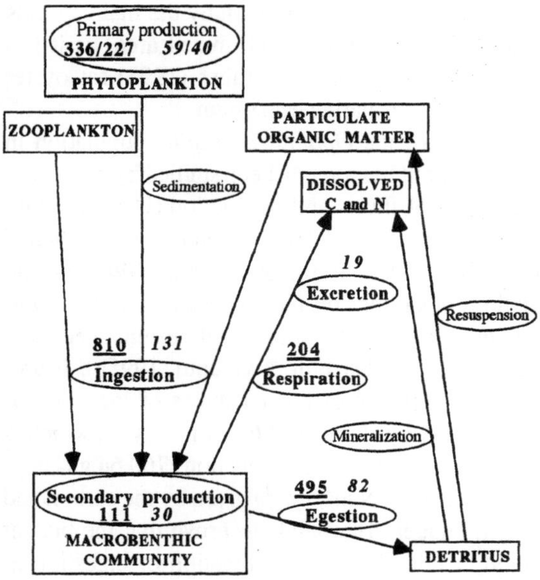

Figure 2. Annual fluxes, phytoplankton and macrobenthic-community productions of carbon (underlined) and nitrogen (italic) in g C (or N) m^{-2} y^{-1}. Phytoplankton production according to Quisthoudt (1987) and Hoch (1995); fluxes and macrobenthic production according to the present budget.

Results

Parameters of the carbon and nitrogen budget equation that could be estimated from survey of the three populations structure and studies of the three species metabolism (see Figure 1) are given in Tables 1 and 2. Annual production of the three populations was calculated from the estimation of their mean annual biomass as carbon and nitrogen stock (Figure 1):

Ophiothrix fragilis	79.39 g C m^{-2}; 22.01 g N m^{-2},
Urticina felina	38.00 g C m^{-2}; 8.45 g N m^{-2},
Alcyonium digitatum	4.80 g C m^{-2}; 1.34 g N m^{-2}.

Estimation of production according to the equation of Brey (1990) gave a P:B ratio of 0.20 for the *U. felina* population and 0.35 for the *A. digitatum* population. A P:B ratio of 0.90 was then obtained for the macrobenthic community when reduced to its three main populations.

The ingestion rate of the *O. fragilis* population was estimated to be 760.24 g C m^{-2} y^{-1} (as the sum $I = P + R + Eg$) and 123.61 g N m^{-2} y^{-1} (as the sum $I = P + Ex + Eg$). A mean Eg:I ratio of 0.65 (corresponding to an absorption efficiency of 35%) was then observed for *O. fragilis*, which is a microphagous suspension-feeder, feeding mainly and almost continuously on phytoplankton. This ratio varied from 0.15 in November (when food availability is low) to 0.85 in May (when the phytoplanktonic bloom occurs). Considering the Eg:I ratio of 0.65 observed for *O. fragilis*, we proposed an Eg:I ratio between 0.40 and 0.60 for *A. digitatum* (which is also microphagous and which feeds mainly on zooplankton) and an Eg:I ratio between 0.20 and 0.40 for *U. felina* (which is macrophagous, captures prey and feeds discontinuously). The ingestion rate of the *U. felina* population was then estimated to:

37.28 g C $m^{-2} y^{-1}$ and 5.45 g N m^{-2} y^{-1}
(with $Eg = 0.20\,I$)
49.70 g C m^{-2} y^{-1} and 7.27 g N m^{-2} y^{-1}
(with $Eg = 0.40\,I$).

And the ingestion rate of the *A. digitatum* population was estimated to:

12.12 g C m^{-2} y^{-1} and 1.93 g N m^{-2} y^{-1}
(with $Eg = 0.40\,I$)
18.18 g C m^{-2} y^{-1} and 2.90 g N m^{-2} y^{-1}
(with $Eg = 0.60\,I$).

The ingestion rate of the macrobenthic community (as the sum of the three population ingestion rates) was estimated between 809.64 and 828.12 g C m^{-2} y^{-1} and between 130.99 and 133.78 g N m^{-2} y^{-1}, with the ingestion of the *O. fragilis* population accounting for more than 90%.

Figure 2 synthetizes the estimation of annual fluxes of carbon and nitrogen related to the metabolic activity of the macrobenthic community, as well as the annual phytoplankton production (data from Quisthoudt, 1987 and Hoch, 1995). Dissolved carbon fluxes (as carbon dioxide) from the community respiration accounted for 60 to 90% of the needs for phytoplankton production (depending on the estimation of primary production). The dissolved nitrogen fluxes (as ammonium) from the community excretion accounted for 30 to 50% of the needs for phytoplankton production.

Discussion

The carbon and nitrogen budget proposed here demonstrates a very high input of organic matter to the benthic compartment (about 820 g C m^{-2} y^{-1} and 130 g N m^{-2} y^{-1}, mainly due to the dominant population) when compared with other coastal systems, Mann (1982) giving values between 100 and 300 g

C m^{-2} y^{-1}. Nevertheless, the production of the macrobenthic community is also very high (110 g C m^{-2} y^{-1} and 30 g N m^{-2} y^{-1}) and the conversion rate (production as a fraction of food ingested: about 13% in terms of carbon and about 23% in terms of nitrogen) is consistent with values found in the literature for macrobenthic communities. The carbon budgets proposed for benthic fine sand communities give macrobenthos conversion-rates of 10% (Chardy et al., 1993), 17% (Chardy & Dauvin, 1992) and 19% (Améziane et al., 1996) in the Western English Channel and about 24% in the Northern Wadden Sea (Asmus, 1982); the nitrogen budget proposed for a South-African surf zone gives a macrobenthos conversion-rate of 20% (Cockcroft & McLachlan, 1993). The high input demonstrated in the Dover Strait suggests that the functioning of the macrobenthic system is widely dependant on allochtonous sources of organic matter and on advection processes. The same trend was shown by a model developed for the study of a coarse sediment benthic system of the English Channel (Chardy, 1987), sensitivity analysis emphazising the major role of allochtonous matter in the maintenance of the filter-feeding compartment. The ecological importance of horizontal advective processes for energy transfer from the pelagic to the benthic system was also emphasized in the energy budget performed on some marine suspension-feeding populations by Loo & Rosenberg (1996) in the Skagerrak and in the Kattegat. In the Dover Strait, this relationship can be explained by the physical characteristics of the area. Terrestrial inputs and advection are important but inshore-offshore diffusion remains low (Pingree et al., 1975). Turbulence due to tidal currents and wind mixes the water column, as a result phytoplankton are then accessible to benthic filter-feeders. The important role of dense populations of suspension-feeders in controlling phytoplankton biomass was demonstrated in similar well-mixed coastal areas: Cloern (1982) and Officer et al. (1982) in the San Francisco Bay, Hily (1991) in the Bay of Brest. These organisms may not only act as filters using water-column organic matter but could also contribute, by their metabolism, to nutrient supplies and an enhanced primary production. We estimate, by the excretion and egestion rates of the three populations, an annual flux of nitrogen (element often limiting phytoplankton primary production) of about 100 g m^{-2} with nearly 20 g as ammonium, that constitutes most of the nitrogen uptake by phytoplankton (Matsakis, 1992; L'Helguen et al., 1993). This value supports the hypothesis of a positive feedback to the high phytoplankton consumption by the main species of the community in this well-mixed area, as it has been suggested by Dame & Libes (1993) for oyster reefs within an estuarine ecosystem.

The role of the *Ophiothrix fragilis* population in the Dover Strait has already been studied by constructing a model of fluxes and production (Davoult et al., 1998). The model took into account an experimental function of ingestion (Davoult et al., 1994) and the variations of suspension-feeding behaviour of *O. fragilis* with variations of bottom tidal current speed and food availability (Davoult & Gounin, 1995b). It provided an estimation of annual fluxes of ingestion of 634.5 g C m^{-2} y^{-1} and 120.6 g N m^{-2} y^{-1}, which is comparable to the present estimation (760.64 g C m^{-2} y^{-1} and 123.61 g N m^{-2} y^{-1}) given by the carbon and nitrogen budget. That tends to prove that the budget method is suitable for the estimation of ingestion of the *O. fragilis* population (with previous studies permitting to estimate all the other items of the equation). The C:N ratio of ingested matter here obtained for this phytoplankton-feeding species (C:N = 6.150) is very close to the Redfield ratio assumed for phytoplankton (C:N = 6.625); that also contributes to validate the estimation of annual fluxes of *O. fragilis* ingestion.

The energy or matter budget approach is widely used at a population scale, in particular for exploited marine organisms (Rodhouse, 1978; Barkai & Griffiths, 1988). Some budgets have also been compiled for communities (Cockcroft & McLachlan, 1993) or to compare benthic populations dominated by suspension-feeders in different habitats (Loo & Rosenberg, 1996). The aim of this paper was to extend the carbon and nitrogen budget to the macrobenthic community, considering that the sum of the three main populations (which biomass accounts for 98.5% of total macrobenthic biomass) is representative of the whole community. The major difficulty was to obtain values of production and egestion for the two cnidarians populations. The empirical relationship proposed by Brey (1990) allowed us to estimate a production from available parameters (mean annual population biomass and mean individual biomass), but resulted in only the somatic production. Values of production used in our budget for the cnidarian populations are then underestimated. Nevertheless, the P:B ratio obtained for the community (as the sum of the three populations), equal to 0.9, is consistent with values found in the literature for macrobenthic communities. George & Warwick (1985) estimated a P:B ratio of 1.4 for a hard-bottom reef macrobenthic community of the Bristol Channel. They explained this 'high value' by the dominance of *O. fragilis* (70% of total biomass), for which they calculated a P:B ratio of 1.8, the remaining species giving a P:B of only 0.37. In the English Channel, the model proposed by Chardy (1987) gave a P:B ratio of 0.9 for a macrobenthic community of coarse sediment, the one of Chardy & Dauvin (1992) a P:B of 1.84 and those of Améziane et al. (1996) P:B ratios between 0.4 and 2.3 for soft-bottom macrobenthic communities. The egestion of the two cnidarians was expressed as a function of ingestion. This procedure was used by various authors (notably Chardy et al., 1993 and Améziane et al., 1996) with different egestion rates for the different macrobenthic trophic-compartments. We proposed different rates for *A. digitatum* and *U. felina* in view of the rate obtained for *O. fragilis* and our knowledge on nutritional habits of the three species. *O. fragilis* is a microphage that feeds primarily and almost continuously on phytoplankton (Warner & Woodley, 1975). During nutritional experiments, ingestion rate first increased with the phytoplankton flux and then at a maximum rate rapidly stagnated (Davoult et al., 1994). *A. digitatum* is also microphagous but feeds discontinuously and mainly on zooplankton (Matthews, 1917; Roushdy, 1962). During nutritional experiments, the maximum ingestion rate was never reached in spite of the high concentration of food (Migné, 1996). *U. felina* is mainly macrophagous, sporadically captures prey and may undergo long periods of fasting (Stephenson, 1928). We supposed *O. fragilis* (being often in saturated conditions) to have a low absorption efficiency, *U. felina* (having to undergo fasting) to have a high absorption efficiency and *A. digitatum* to have some absorption efficiency between that of *O. fragilis* and *U. felina*. The annual Eg:I ratio of 0.65 (that is an absorption efficiency of 35%) obtained in the present study for *O. fragilis* is consistent with these assumptions; it is higher than the ratio of 0.40 used by Chardy et al. (1993) for suspension-feeders. Furthermore, the seasonal variation of this ratio (from 0.15 to 0.85) confirms the relationship between absorption efficiency and food availability. The proposed proportion of ingested matter which is egested for *U. felina* (between 0.20 and 0.40) is close to the 0.30–0.40 measured on the sea anemone *Metridium senile* (Linné) (Lesser et al., 1994).

An assessment of benthic respiration and its partitioning in the size range of organisms (from micro-organisms to megafauna) was attempted in two benthic habitat types (differing in depth, bottom current velocities and sediment texture) of the northwestern Barents Sea (Piepenburg et al., 1995). This study underlines the influence of both environmental setting and community structure on the partitioning of the metabolism in the benthic compartment (suggesting that benthic respiration could be dominated by macro-organisms when the seabed is composed of coarse sediments and stones). Because of the sediment characteristics (pebbles accounting for more than 80% of total sediment) and the high macrobenthic biomass (281 g m^{-2}), the study of benthic-system of the Dover Strait was based on the assumption that the macrobenthic compartment was responsible for the major part of the fluxes at the water-boundary layer. On the contrary, investigations of energy flow in sand-bottom marine benthic communities often show a dominating contribution of bacteria or meiofauna to the system (Gerlach, 1971). Nevertheless, the macrobenthic contribution varies considerably from one site to another and can even dominate: we have noticed in relevant literature the contribution of macrofauna to benthic respiration to be from 6% to 84%. A higher amount of respiration has been estimated for macrobenthos than for micro- and meiofauna in some sand communities characterized by their high macrobenthic biomass: in the cold polar waters of the Bering Shelf (Grebmeier et al., 1988), where bacterial activity may be repressed, but also in the temperate waters of the Wadden sea (Asmus, 1982). In the Dover Strait, we assume that both sediment characteristics and macrobenthos biomass support the hypothesis of the leading role of the macrobenthos in benthic metabolism.

Even if the present approach is not entirely satisfactory, the gross estimation obtained permits us to propose an hypothesis on the functioning of this complex coastal system. A system is said to be 'net autotrophic' (sink of CO_2) if it produces more organic matter than it consumes and 'net heterotrophic' (source of CO_2) if organic consumption exceeds production. Assuming that phytoplankton is the only primary producer, the Dover Strait system appears to be highly heterotrophic as one heterotrophic compartment (and even a single population of this compartment) consumes more organic matter than is produced. This conclusion is based both on our estimation of macrobenthic consumption (that has been discussed) and on the primary production estimation. Recently, the photoadaptation of phytoplankton has been studied in the eastern English Channel (Lizon et al., 1995). The estimates of daily primary production (computed without and with the observed daily and depth variability of photosynthetic parameters of the phytoplankton populations) showed large differences between them (from 1.2 to 2.1 g C m^{-2} d^{-1}, i.e. prediction of primary production may be underestimated by about 40%). By increasing the value of primary production used in this study by 40%, the consumption of the system remains higher than its production. It can then be said that this system constitutes a carbon-dioxide source for the surrounding water and adjacent systems, as has been emphasized for some coastal systems (Smith & Hollibaugh, 1993).

Acknowledgements

This work was supported by the PNOC (Programme National d'Océanographie Côtière, Chantier Manche).

References

Améziane, N., P. Chardy & J. C. Dauvin, 1996. Modelling carbon flows in soft-bottom communities from the bay of Morlaix, western English Channel. In Eleftheriou et al. (eds), Biology and Ecology of Shallow Coastal Waters. 28 EMBS Symposium, Danemark: 215-224.

Asmus, H., 1982. Field measurements on respiration and secondary production of a benthic community in the northern Wadden sea. Neth. J. Sea Res. 16: 403–413.

Barkai, R. & C. L. Griffiths, 1988. An energy budget for the south african abalone *Haliotis midae* Linnaeus. J. moll. Stud. 54: 43–51.

Brey, T., 1990. Estimating productivity of macrobenthic invertebrates from biomass and mean individual weight. Meeresforsch. 32: 329–343.

Cabioch, L., 1968. Contribution à la connaissance des peuplements benthiques en Manche Occidentale. Cah. Biol. mar. IX 5 (suppl.): 493–720.

Chardy, P., 1987. Modèle de simulation du système benthique des sédiments grossiers du golfe normand-breton (Manche). Oceanol. Acta 10: 421–434.

Chardy, P. & J. C. Dauvin, 1992. Carbons flows in a subtidal fine sand community from the western English Channel: a simulation analysis. Mar. Ecol. Prog. Ser. 81: 147–161.

Chardy, P., P. Gros, H. Mercier & Y. Monbet, 1993. Benthic carbon budget for the Bay of Saint-Brieuc (Western Channel). Application of inverse method. Oceanol. Acta 16: 687–694.

Cloern, J. E., 1982. Does the benthos control phytoplankton biomass in south San Francisco bay? Mar. Ecol. Prog. Ser. 9: 191–202.

Cockcroft, A. C. & A. McLachlan, 1993. Nitrogen budget for a high-energy ecosystem. Mar. Ecol. Prog. Ser. 100: 287–299.

Crisp, D. J., 1984. Energy flow measurements. Chapter 9. In Holme, N. A. & A. D. McIntyre (eds), Methods for the Study of Marine Benthos. International Biological Program Handbook N° 16, 2nd ed. Blackwell Scientific Publications, Oxford, 387 pp.

Dame, R. & S. Libes, 1993. Oyster reefs and nutrient retention in tidal creeks. J. exp. mar. Biol. Ecol. 171: 251–258.

Davoult, D., 1989. Structure démographique et production de la population d'*Ophiothrix fragilis* (Abildgaard) du détroit du Pas de Calais, France. Vie mar. hors série 10: 116–127.

Davoult, D., 1990. Biofaciès et structure trophique du peuplement des cailloutis du détroit du Pas de Calais (France). Oceanol. Acta 13: 335–348.

Davoult, D., N. Degros, M. A. Janquin & B. Soyez, 1992. Biometrics, carbon and nitrogen content in the ophiuroid *Ophiothrix fragilis*. J. mar. biol. Ass. U.K. 72: 915–918.

Davoult, D., F. Dewailly & A. Migné, 1998. Carbon and nitrogen budget of a dense population of the suspension feeding ophiuroid *Ophiothrix fragilis* in a coastal ecosystem. In Mooi & Telford (eds), Echinoderms: San Francisco. Balkema, Rotterdam: 337–342.

Davoult, D. & F. Gounin, 1995a. Nitrogen excretion by a dense population of *Ophiothrix fragilis* (Abildgaard): Role in the exchanges at the water-sediment interface. In Emson, Smith & Campbell (eds), Echinoderm Research, Balkema, Rotterdam, 65–69.

Davoult, D. & F. Gounin, 1995b. Suspension-feeding activity of a dense *Ophiothrix fragilis* (Abildgaard) population at the water-sediment interface: time coupling of food availability and feeding behaviour of the species. Estuar. Coast. shelf Sci. 41: 567–577.

Davoult, D., F. Gounin & M. A. Janquin, 1991. Ammonium excretion by the ophiurid *Ophiothrix fragilis* as a function of season and tide. Mar. Biol. 111: 425–429.

Davoult, D., F. Gounin & M. A. Janquin, 1994. Experimental nutrition in the suspension-feeding ophiurid *Ophiothrix fragilis* (Abildgaard) as a function of chlorophyll *a* flux. In David, Guille, Féral & Roux (eds), Echinoderms through Time, Balkema, Rotterdam: 399–403.

George, C. L. & R. M. Warwick, 1985. Annual macrofauna production in a hard-bottom reef community. J. mar. biol. Ass. U.K. 65: 713–735.

Gerlach, S. A., 1971. On the importance of marine meiofauna for benthos communities. Oecologia 6: 176–190.

Grebmeier, J. M., C. P. McRoy & H. M. Feder, 1988. Pelagic-benthic coupling on the shelf of the northern Bering and Chukchi Seas. I. Food supply source and benthic biomass. Mar. Ecol. Prog. Ser. 48: 57–67.

Hily, C., 1991. Is the activity of benthic suspension feeders a factor controlling water quality in the bay of Brest? Mar. Ecol. Prog. Ser. 69: 179–188.

Hoch, T., 1995. Modélisation du cycle biogéochimique des éléments limitant la production biologique en Manche (N, P, Si). Thèse de Doctorat, Université de Paris VII, 213 pp.

Holme, N. A., 1966. The bottom fauna of the English Channel, part II. J. mar. biol. Ass. U.K. 46: 401–493.

Larsonneur, C., P. Bouysse & J. P. Auffret, 1982. The superficial sediments of the English Channel and its Western approaches. Sedimentology 29: 851–864.

Lesser, M. P., J. D. Witma & K. P. Sebens, 1994. Effects of flow and seston availability on scope for growth of suspension-feeding invertebrates from the Gulf of Maine. Biol. Bull. 187: 319–335.

L'Helguen, S., C. Madec & P. Le Corre, 1993. Nutrition azotée du phytoplancton dans les eaux brassées de la Manche occidentale. Oceanol. Acta. 16: 653–660.

Lizon, F., Y. Lagadeuc, C. Brunet, D. Aelbrecht & D. Bentley, 1995. Primary production and photoadaptation of phytoplankton in relation with tidal mixing in coastal waters. J. Plankton Res. 17: 1039–1055.

Loo, L. O. & R. Rosenberg, 1996. Production and energy budget in marine suspension feeding populations: *Mytilus edulis*, *Cerastoderma edule*, *Mya arenaria* and *Amphiura filiformis*. J. Sea Res. 35: 199–207.

Mann, K. H., 1982. Ecology of coastal waters. A system approach. Studies in Ecology. University of California Press, Los Angeles, 322 pp.

Matsakis, S., 1992. Ammonia excretion rate of *Clytia* spp. hydromedusae (Cnidaria, Thecata): effects of individual dry weight, temperature and food availability. Mar. Ecol. Prog. Ser. 87: 55–63.

Matthews, A., 1917. The development of *Alcyonium digitatum* with some notes on the early colony formation. Quart. J. Micro. Sci. 62: 43–94.

Migné, A., 1996. Rôle des organismes suspensivores dans les transferts pelago-benthiques d'une zone de fort hydrodynamisme (Détroit du Pas de Calais, Manche orientale): flux et bilans d'azote et de carbone. Thèse de Doctorat, Université de Lille I – Station Marine de Wimereux, 181 pp.

Migné, A. & D. Davoult, 1997a. Distribution quantitative de la macrofaune benthique du peuplement des cailloutis dans le détroit du Pas de Calais (Manche Orientale, France). Oceanol. Acta 20: 453–460.

Migné, A. & D. Davoult, 1997b. Carbon-dioxide production and metabolic parameters in the ophiurid *Ophiothrix fragilis*. Mar. Biol. 127: 699–704.

Migné, A. & D. Davoult, 1997c. Oxygen consumption in two benthic cnidarians: *Alcyonium digitatum* (Linnaeus, 1758) and *Urticina felina* (Linnaeus, 1767). Proc. 6th int. Conf. on Coelenterate Biology, 1995: 321–328.

Migné, A. & D. Davoult, 1997d. Ammonium excretion in two benthic cnidarians: *Alcyonium digitatum* (Linnaeus, 1758) and *Urticina felina* (Linnaeus, 1767). J. Sea Res. 37: 101–107.

Migné, A., D. Davoult, M. A. Janquin & A. Kupka, 1996. Biometrics, carbon and nitrogen content in two cnidarians: *Urticina felina* and *Alcyonium digitatum*. J. mar. biol. Ass. U.K. 76: 595–602.

Officer, C. B., T. J. Smayda & R. Mann, 1982. Benthic filter feeding: a natural eutrophication control. Mar. Ecol. Prog. Ser. 9: 203–210.

Piepenburg, D., T. H. Blackburn, C. F. von Dorrien, J. Gutt, P. O. J. Hall, S. Hulth, M. A. Kendall, K. W. Opalinski, E. Rachor & M. K. Schmid, 1995. Partitioning of benthic community respiration in the Arctic (northwestern Barents Sea). Mar. Ecol. Prog. Ser. 118: 199–213.

Pingree, R. D., L. Pennycuick & B. Gaw, 1975. A time-varying temperature model of mixing in the English Channel. J. mar. biol. Ass. U.K. 55: 975–992.

Quisthoudt, C., 1987. Production primaire phytoplanctonique dans le détroit du Pas de Calais (France): variations spatiales et annuelles au large du Cap Gris-Nez. C. r. Acad. Sci., Paris 304 série III: 245–250.

Rodhouse, P. G., 1978. Energy transformations by the oyster *Ostrea edulis* L. in a temperate estuary. J. exp. mar. Biol. Ecol. 34: 1–22.

Roushdy, H. M., 1962. Expansion of *Alcyonium digitatum* L. (Octocorallia) and its significance for the uptake of food. Vidensk. Medd. Dan. Naturhist. Foren. Khobenhavn 124: 409–419.

Smith, S. V. & J. T. Hollibaugh, 1993. Coastal metabolism and the oceanic organic carbon balance. Rev. Geophys. 31: 75–89.

Stephenson, T. A., 1928. The british sea anemones, volume 1. London Ray Society, 148 pp.

Warner, G. F. & J. D. Woodley, 1975. Suspension-feeding in the brittle-star *Ophiothrix fragilis*. J. mar. biol. Ass. U.K. 55: 199–210.

Hydrobiologia **375/376:** 317–329, 1998.
S. Baden, L. Pihl, R. Rosenberg, J.-O. Strömberg, I. Svane & P. Tiselius (eds),
Recruitment, Colonization and Physical–Chemical Forcing in Marine Biological Systems.

Long-term changes in the diversity and faunal structure of benthic communities in the northern North Sea: natural variability or induced instability?

T. H. Pearson & H-P. Mannvik
Akvaplan-niva AS, 9005, Tromsø, Norway

Key words: North Sea, benthic communities, variability, natural disturbance, dominance, diversity

Abstract

Changes in benthic sedimentary communities in many areas of the northern North Sea in the vicinity of oil and gas fields have been monitored intermittently over the past three decades, in most cases triennially but in some areas annually for short periods. Accumulating evidence from these surveys suggests that large scale temporal variability occurs in both the diversity and structure of these communities on time scales varying from years to decades. A recent change to regional, rather than field based, surveys in the Norwegian sector of the North Sea has provided evidence that spatially such changes occur on broad scales and are not local responses to field based environmental disturbance. The possible factors inducing such changes are reviewed and the relative temporal and spatial stability of benthic infaunal communities in response to fluctuating pelagic conditions and levels of sedimentary disturbance are discussed. It is suggested that broad scale temporal and spatial variability in these benthic communities is driven by climatic forces influencing the overlying water masses but that there may have been some increase in pelagic productivity and/or in benthic pelagic coupling in the area in recent years.

Introduction

Monitoring of physical, chemical and biological variables in the sediments around oil and gas installations in the Norwegian sector of the northern North Sea has been undertaken for the past three decades. The impact of hydrocarbon extraction activities is focused on the sedimentary environment, primarily through the discharges of drilling cuttings and fluids from the drilling and extraction platforms (Davies et al., 1984; Kingston, 1992; Olsgard & Gray, 1995). The effects of these inputs on the sedimentary benthos may be major up to a radius of 1000 m from production platforms and discernible up to 5/6000 m (GESAMP, 1993; Olsgard & Gray, 1995). However, such field related effects may be modified by natural changes in benthic communities in response to environmental variables fluctuating on much wider spatial and longer temporal scales than are field related variables. Lack of sufficiently wide ranging temporal and spatial information has generally prevented any examination of the relative roles of natural and anthropogenic factors in forcing benthic community dynamics. However a recent change to regional rather than field based surveys in the Norwegian sector of the North Sea has provided information on benthic population levels over extensive areas including many individual oil and gas fields and intermediate regional sampling stations. This, together with the history of changes recorded at individual fields and references areas in the past, now allows some assessment to be made of the relative importance of the natural and anthropogenic factors influencing benthic populations in the area on both temporal and spatial scales.

Study area

The information discussed is drawn from a monitoring survey undertaken in 1996 in the area designated as Region 1, which comprises the southern part of the Norwegian sector in the North Sea. It extends from

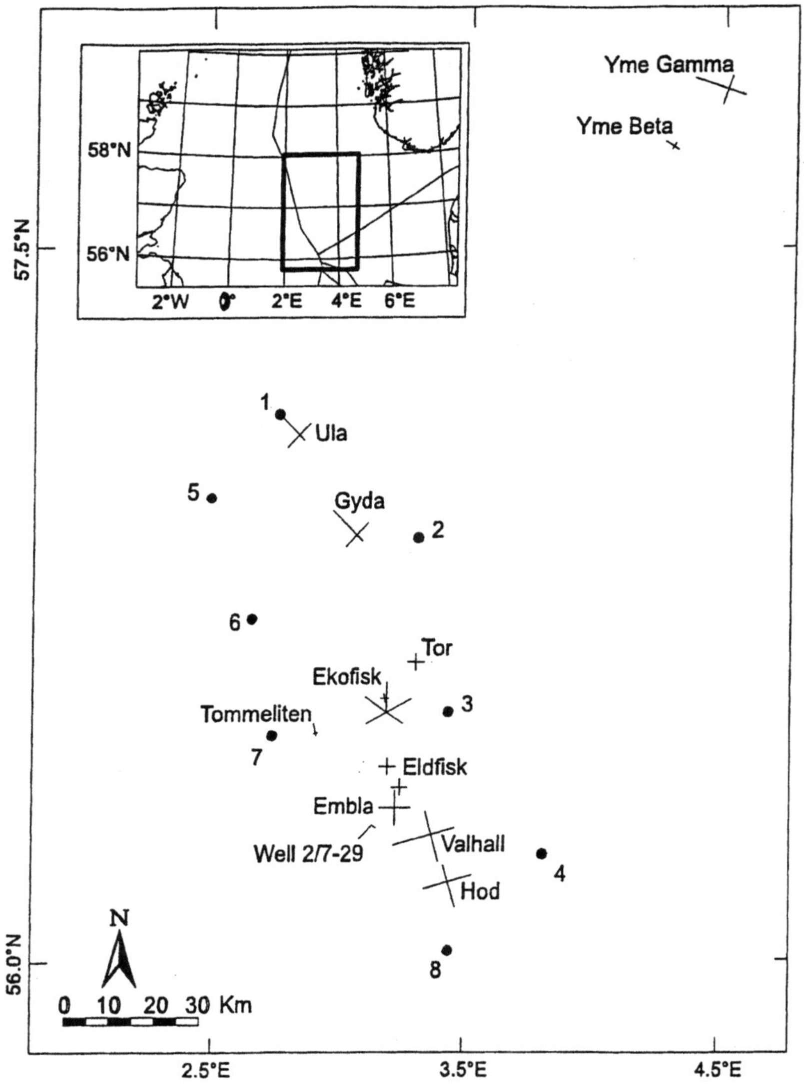

Figure 1. Map showing the position of the region, the regional and reference station positions within the region and the centre positions of the individual fields.

Table 1. Region 1: summary of industrial activity, 1993–96.

	No. of fields	No. of wells drilled	Tones of hydrocarbons discharged	Production water (tonnes) discharged
Regional total	14	80	2554	500
Range of field totals		0–15	0–254	0–297

approximately 56.0° to 57.5° N and 2.5° to 4.5° E (see Figure 1), covering an area of some 14 000 km^2 where sandy-silt sediments predominate at depths varying from about 65 m in the south to 90 m in the north. Tidal currents in the area run in a NE-SW direction while the residual current is directed to the SE. The region includes fourteen active oil and gas fields from which discharges of drilling fluids and production waters contaminated with a variety of hydrocarbons and minerals take place. A summary of the industrial discharges made in the area for the period 1993–96 is given in Table 1.

Sampling locations

For the purposes of the regional survey sampling stations were located in groups at varying distances from individual field platforms and installations (field stations). To permit comparability with earlier monitoring surveys field station positions were chosen from the suite of stations in each field which had been sampled in previous surveys in each area. These always included the reference station designated for each field and positioned in an area outside the influence of field related activities. In addition eight stations were positioned at various points throughout the region beyond the apparent influence of field activities and representative of varying geographic areas and sedimentary topographies (regional stations). Three of these regional stations coincided in position with previously designated field reference stations. The positions of the individual fields within the region and of the regional stations are shown in Figure 1. The number of stations sampled in each field in the 1996 survey and over the preceding 8 y are detailed in Table 2.

Methods

The sampling and analytical methodology used followed the protocols laid down in the guidelines issued for the conduct of monitoring surveys in the vicinity of petroleum installations in the North Sea (SFT, 1990). Five individual 0.1 m^2 van Veen grab samples for benthic fauna were taken at each sampling station and sieved on a 1 mm screen. Additional grab samples were subsampled for material for the analysis of physical and chemical sedimentary characteristics. The population density and the species richness data reported here are based on the cumulative totals of the five faunal samples (0.5 m^2). In one or two instances data from earlier surveys were based on slightly different sample sizes, in which case such data have been normalised to 0.5 m^2 in the comparisons given here. General spatial and temporal comparisons of population changes are based on simple univariate statistics, but where appropriate, multivariate analyses have been used, including hierarchical clustering based on the Bray-Curtis dissimilarity index and group average sorting (Rohlf, 1991) and ordinations using Correspondence Analysis (CA) and Canonical Correspondence Analysis (CCA) (Greenacre, 1993).

Table 3. Region 1: Number of individuals and taxa per 0.5 m^2 for each of the Regional and field reference stations (Ref. st.), 1996. SD., standard deviation: C, coefficient of variation.

St. no. (degree/distance)	No. of individuals	No. of taxa
1 (Ref. st. Ula)	516	65
2 (56°55′/03°20′)	622	72
3 (Ref. st. Ekofisk)	510	66
4 (56°15′/03°50′)	508	87
5 (57°00′/02°30′)	864	85
6 (56°45′/02°40′)	605	68
7 (56°30′/02°45′)	593	68
8 (Ref. st. Hod)	931	79
Ref. Yme Gamma	476	84
Ref. Gyda	462	72
Ref. Valhall	523	72
Ref. Tommeliten	503	78
Mean	592	75
S.D.	151	9
C (%)	25	11

Results

Comparison of spatial variability at regional and reference stations

In the 1996 survey the eight regional stations and four additional field reference stations were positioned in areas considered to be beyond the influence of field related industrial activities. Faunal variations between these twelve stations can be considered, therefore, to be the result of natural factors. Table 3 gives the total number of individuals and number of taxa recorded at each of these stations, together with their mean and standard deviation. The number of individuals found

Table 2. Region 1: summary of sampling activity, 1988–96.

Sampling area	Number of stations sampled								
	1988	1989	1990	1991	1992	1993	1994	1995	1996
Region	1	1	1	2	2	1		1	8
Yme Gamma							14		11
Ula		16			18				9
Gyda			17			19			10
Tor					20				9
Ekofisk 2/4 B&K			15	15	15	12			8
Ekofisk Centre			22	22	22	22			10
Eldfisk 2/7A/FTP			20	20	20	16		9	9
Eldfisk 2/7B									9
Embla			20	20	20	20			9
Tommeliten		10							8
Valhall	20			25					12
Hod	30			23					12
Total	51	27	95	127	117	90	14	10	124

Table 4. Region 1: the two most abundant species at each of the Regional stations (RS) and field reference stations (FR). G, Gyda; V, Valhall; T, Tommeliten; Y, Yme.

Species	RS1	RS2	RS3	RS4	RS5	RS6	RS7	RS8	FRG	FRV	FRT	FRY
Chaetozone setosa	1				1	2			2			
Amphiura filiformis	2	1		1	2		2	2	1		1	
Phoronis sp.		2		2								
Mysella bidentata			1									
Spiophanes kroyeri			2									
Eudorellopsis deformis						1				2		
Spiophanes bombyx							1				2	1
Myriochele oculata								1				
Ophiura affinis										1		
Hippomedon denticulatus												2

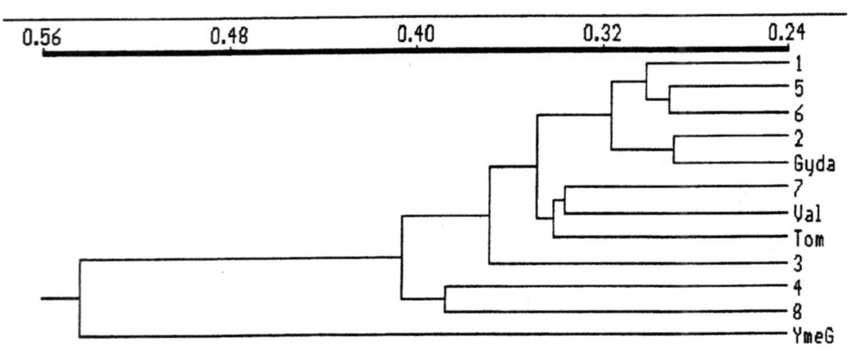

Figure 2. Dendrogram for cluster analysis carried out on whole station data from the regional and reference stations in Region I, 1996.

Table 5. Region 1: comparison of the relative abundance (no. per 0.5 m^2) of the dominant species at each of the regional and reference stations. The stations are ordered from left to right in relation to increasing sediment grain size. Station designations as in Figure 4; SD, standard deviation; C, coefficient of variation.

Station	FRT	RS6	RS5	RS8	RS7	RS3	RS2	FRG	FRV	RS1	RS4	FRY	Mean	SD	C
Chaetozone setosa	24	103	353	17	37	3	55	50	18	165	10	2	70	101	144
Amphiura filiformis	48	65	62	87	66	51	77	56	51	58	101	18	62	21	34
Phoronis sp.	32	8	27	24	43	20	62	38	23	29	31	0	28	16	57
Mysella bidentata	11	41	32	1	22	73	24	31	58	22	4	1	27	22	81
Spiophanes kroyeri	60	25	31	9	30	61	25	5	14	11	9	31	26	19	73
Eudorellopsis deformis	26	110	55	37	59	34	39	46	52	22	26	0	42	27	64
Spiophanes bombyx	37	29	24	8	76	57	24	5	25	0	18	96	33	29	88
Myriochele oculata	11	12	15	439	25	10	4	7	18	15	16	2	48	123	256
Ophiura affinis	16	23	10	27	18	24	23	9	65	29	19	9	23	15	65
Hippomedon denticulatus	2	2	1	0	0	6	0	3	1	0	2	34	4	9	237
Sediment% < 0.25 mm	97.6	97.5	96.6	96.2	95.2	95.1	93.3	92.5	92.4	91.3	90.1	5.9			
Depth (m)	70	67	68	67	67	67	64	65	67	71	64	87			

varied from 462 at the Gyda reference station to 931 at the Hod reference station, whereas the number of taxa recorded ranged from 65 at the Ula reference station (Regional station 1) to 87 at Regional station 4. The standard deviation of faunal density was 25% of the mean and that of species richness was 11% of the mean. These coefficients of variation (C) can be considered, then, as the statistical limits to natural variability in the macrobenthic populations of the region as a whole.

The two most abundant species at each of the regional and reference stations are listed in Table 4, this demonstrates a considerable degree of variation between stations throughout the region. Only one species, the predominantly suspensivorous brittle star *Amphiura filiformis*, occurs as a top dominant in the majority of these stations. Otherwise only the deposit feeding polychaete worm *Chaetozone setosa,* was a dominant in more than three of the twelve stations. These two species were the most abundant at Regional station 1 and the Gyda reference station, which are geographically close to each other in the north of the region, but also at Regional station 4 in the south east of the area sampled. One or other of these two species recurred as top dominants in nine of the twelve stations. Otherwise eight different species occurred amongst the two most abundant across the region, including *Phoronis* sp., a suspensivore, *Myriochele oculata, Spiophanes kroeyeri, S. bombyx*, and *Hippomedon denticulatus*, surface detritivores and the epifaunal carnivore *Ophiura affinis*. This degree of difference suggests some considerable patchiness in the distribution of dominant species across the region as a whole. The relative abundance of these predominant species, given as the mean and standard deviation of the populations in five replicate samples taken from each of the twelve stations is compared in Table 5. This reinforces the impression of patchiness in the faunal dominance across the region. Thus *C. setosa* populations have a coefficient of variation (SD as a percentage of the mean) of 136 and the value of C for *M. oculata* is 256. Only one, *A filiformis*, of the nine dominant species listed had a value for C of less than 40.

The table also includes the proportion of sediment with particle diameters < 0.25 mm (fine sand/silt/clay) and the depth at each station. These factors were closely similar in all areas but the Yme field, which is isolated in the north-eastern corner of the region where the water is slightly deeper (≅ 15 m) and medium sands form the bulk of the sediments. The Yme reference station was dominated by *Spiophanes bombyx,* which predominated at only two of the other stations and the amphipod *H. denticulatus* which was absent or present in only low numbers elsewhere, suggesting that the coarser sediments and deeper water had influenced the dominance distribution in that area. These differences suggest that the populations in the Yme area differ markedly from those in the rest of the region.

Multivariate statistical analyses carried out on the regional and reference station data can quantify inter-

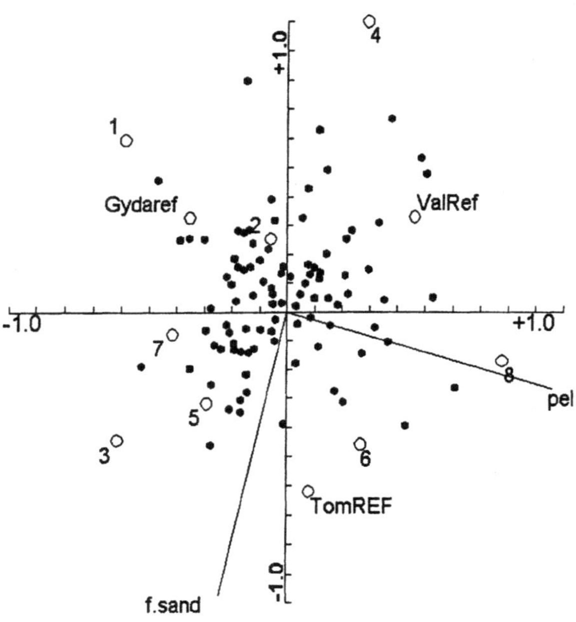

Figure 3. A triplot of the CCA ordination of the data from the regional stations (numbered 1–8 on the plot and individual field reference stations (Gydaref, reference station for the Gyda field; Valref, reference station for the Valhall field; Tomref, reference station for the Tommeliten field). Open circles represent the station positions within the ordination space: black dots represent the positions of individual taxa contributing significantly to the biological variance. The vector lines represent the relationship of significant environmental variables to the ordination axes and their length is proportional to their relative significance.

station similarities and relationships with measured environmental variables. Figure 2 shows the dendrogram from a hierarchical cluster analysis of the data. The cluster analysis separated out the reference station at Yme Gamma at a dissimilarity level of just below 55%, whilst dissimilarities amongst the other stations were below 42% further emphasising the faunal differences in the Yme area. Product Moment correlation coefficient used to test the goodness of fit of the analytical outcome to the Bray-Curtis data matrices showed a very good fit to the data ($r = 0.95$).

A CCA ordination of the full data sets from the regional stations and the individual field reference stations was carried out to examine the association between the biological data and the measured environmental variables in these areas. Because of the less than 50% similarity between the Yme reference station and the other eleven stations it was omitted from this analysis to allow meaningful separations among the other stations in the ordination. Through the forward selection procedure in CANOCO one of the ten fitted variables, Pelite, gave the best fit and was significant. 30% of the biological variance was explained by the first two axes of the ordination space which is constrained by the environmental variables Pelite and fine sand.

Figure 3 shows a triplot from the analysis using Pelite and Fine sand as the constraining environmental variables. Note that fine sand, although the next best fitting variable to Pelite, was non-significant in the analysis. The plot shows vectors for the most significant environmental variables against the first two axes representing the variance in the biological data, i.e. the groups of faunal samples are spread along the axes in relation to their relative degrees of similarity; the further apart they are the less similar they are.

The first canonical axis shows a gradient from Regional station 8 at the positive end of the axis to Regional station 3 at the negative end and is positively correlated with the pelite content of the sediments. It separates Regional stations 4, 6 and 8 and the Valhalla and Tommeliten reference stations which have positive values from Regional stations 1, 2, 3, 5, and 7 with the Gyda reference station which all have negative values. This is effectively a north-south gradient across the region associated with slightly higher silt/clay content in the sediments at the more southerly stations. The second axis shows a gradient from Regional station 4 at the positive end to the Tommeliten reference station at the negative end. This axis has a weak negative association with fine sand. Since the more southerly and easterly stations have the more positive values on this axis it probably reflects the east west depth gradient across the region.

Comparison of spatial variability at all stations

In addition to the twelve regional and reference stations 112 field stations were sampled in the 1996 Region 1 monitoring survey (Table 2). The comparative influence of the measured environmental factors on the entire set of stations sampled, with the exception of those in the Yme area, was assessed through a CCA ordination, shown in Figure 4. The forward selection procedure in CANOCO was used to find those environmental variables which gave the best fit to the biological data. These are drawn as vectors in the ordination plots. The relative length of the vectors drawn denotes the comparative importance of each variable. 25% of the species variance can be explained by the first three axes of the ordination space which is constrained by the environmental variables Zinc, Lead,

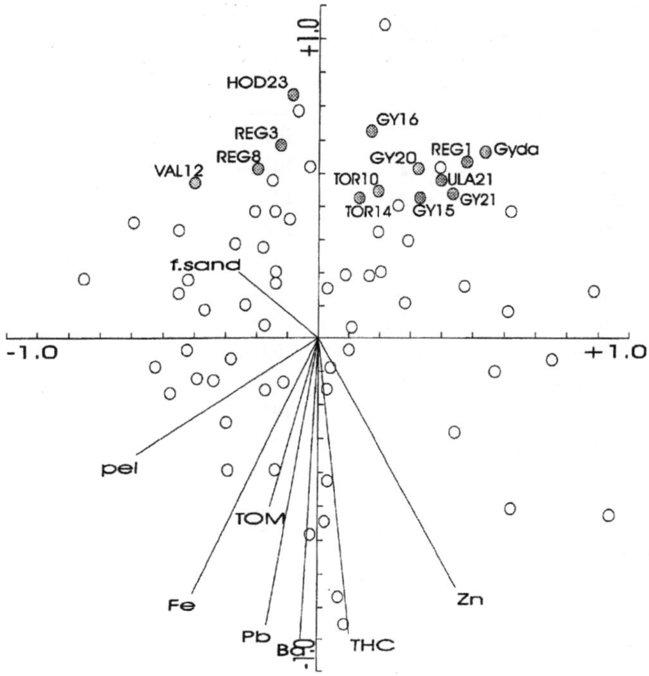

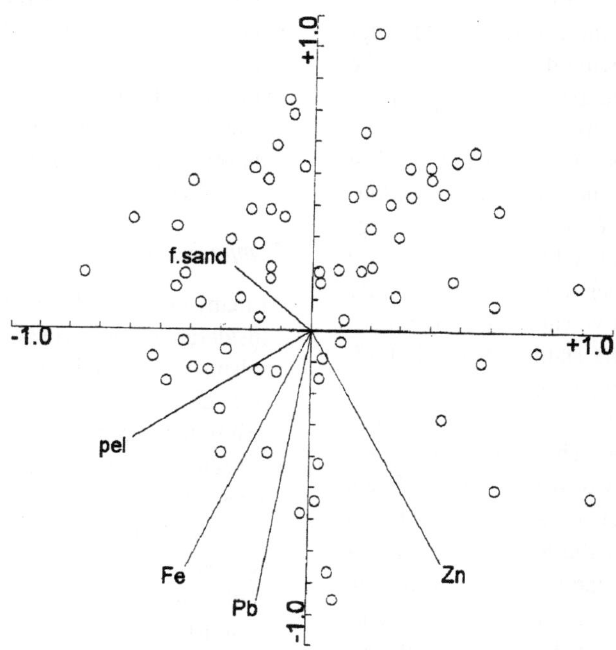

Figure 4. Triplots for the CCA ordination of the data from all the stations sampled during the 1996 Region 1 survey. See caption to Figure 3 for description of the plot symbols.

a. Triplot showing the stations with the highest scores (±) on the two axes and the significant environmental vectors. f. sand, Fine sand; pel, Pelite; Fe, Iron; TOM, Total Organic Matter; Pb, Lead; Ba, Barium; THC, Total Hydrocarbons; Zn, Zinc.

b. Triplot showing the positions of the regional and reference stations within the ordination and the vectors of the most significant environmental variables. Station designations as in Figure 3.

Table 6. Percentage change since the previous survey in total faunal density and in species numbers at stations showing no relationship with field related disturbance in the CCA analysis.

Field	Station No.	CCA score on axis 2	Data of previous survey	% change in total abundance	% change in species numbers
Gyda	15	+1.2	1993	+133	+29
Tor	14	+1.2	1992	+116	+46
Gyda	21	+1.2	1993	+8	+5
Tor	10	+1.2	1992	+117	+13
Valhall	12	+1.3	1991	+81	+37
Ula	21	+1.4	1992	+73	0
Gyda	20	+1.5	1993	+113	+8
Region	8 (Hod Ref.)	+1.5	1991	−42	+1
Region	1 (Ula Ref.)	+1.5	1992	−9	−9
Gyda	Reference	+1.6	1993	+5	+16
Region	3 (Ekofisk Ref.)	+1.7	1993	+60	+35
Gyda	16	+1.8	1993	+315	+16
Hod	23	+2.1	1991	+78	+97
Number of stations exceeding normal limits of C				10	8

Iron, Fine Sand, Barium, Total hydrocarbons (THC), Pelite (silt/clay) and Total organic matter (TOM). Figure 4a presents a triplot showing the stations with the highest scores (±) on the two axes and the significant environmental vectors. This first canonical axis separates stations Gyda 9, Ula 8, 16 and 20 and Tor 5 at the positive end of the axis and a number of stations from Hod, Valhall and the Ekofisk area at the negative end. This axis is correlated with the grain size variables, with positive values indicating the slightly coarser sediments and negative values denoting muddier conditions. Many of the stations (e.g. Hod 15, Valhall 11, Ekofisk 11) show a poor fit to the environmental space (< 20%) however, suggesting that this is a relatively weak gradient. The second axis has a strong negative correlation with Iron, Zinc, and Lead. These are all variables associated with industrial field activities. It separates all of the Regional and reference stations at the positive end of the axis from, e.g. Ekofisk 44 and Valhall 16 at the negative end. This is thus a gradient strongly related to field disturbance since the latter stations are sited close to the field installations and have been described as having moderately disturbed faunas (Mannvik et al., 1997). Many of the stations with high negative scores on this axis show a good fit (> 50%) to the environmental space, e.g. Ula 8, Ekofisk 19, indicating a close relationship with the environmental variables. Stations high on the positive axis however show a poorer fit (< 20%) in general, e.g. Hod 23, Gyda 16, suggesting the influence of other non field related environmental factors in these areas. Faunal data from these stations with high positive axis 2 scores will be used, therefore, to assess the level of temporal variability in the populations attributable to natural factors.

Comparison of temporal variability over the region

At many of the fields sampled in 1996 there had been considerable changes in both the total abundance of organisms found and in the number of species recorded since the previous survey in the area. Table 7 summarises the percentage change in faunal density and in species richness for the stations with high positive scores on axis 2 of the CCA ordination and which are deemed, therefore, to be undisturbed by field related factors. The five regional stations which were not also field reference stations had not been sampled previously, thus no estimates of temporal change are available from these areas. Changes at these undisturbed stations show that increases in faunal densities and species richness considerably greater than the established normal variance limits had occurred in recent years at most fields in the region, with the exception of Yme, Ula, Hod and Tommeliten. This pattern suggests that over much of the central and southern part of the region a considerable increase in the populations had

Table 7. Region 1: temporal changes in some of the commonest species found at different reference stations.

	1988	1989	1990	1991	1992	1993	1994	1995	1996	X	s.d.
Area: Ekofisk											
Spiophanes kroeyeri			0	1	7	1			61	14	26
Spiophanes bombyx			1	0	14	4			57	15	24
Chaetozone setosa			1	1	3	2			3	2	1
Myriochele oculata			0	0	12	2			10	5	6
Eudorellopsis deformis			6	3	15	6			34	13	13
Mysella bidentata			1	65	43	24			73	41	29
Amphiura filiformis			82	73	65	54			51	65	13
Ophiura affinis			0	0	0	1			24	5	11
Area: Valhall											
Spiophanes kroeyeri	1			0					14	5	8
Spiophanes bombyx	1			0					3	1	1
Chaetozone setosa	5			13					18	12	7
Myriochele oculata	60			49					18	42	22
Eudorellopsis deformis	11			12					52	25	23
Mysella bidentata	2			18					58	26	29
Amphiura filiformis	18			61					51	43	22
Ophiura affinis	2			0					65	22	37
Area: Hod											
Spiophanes kroeyeri	1			0					9	3	5
Spiophanes bombyx	1			0					8	3	4
Chaetozone setosa	5			7					17	10	6
Myriochele oculata	60			213					439	237	191
Eudorellopsis deformis	11			19					37	22	13
Mysella bidentata	2			74					1	25	42
Amphiura filiformis	18			87					87	64	40
Ophiura affinis	2			1					27	10	15
Area: Gyda											
Spiophanes kroeyeri			11			15			5	10	5
Spiophanes bombyx			17			7			5	10	6
Chaetozone setosa			1			4			50	18	27
Myriochele oculata			0			1			7	3	4
Eudorellopsis deformis			54			37			46	46	8
Mysella bidentata			3			79			31	38	38
Amphiura filiformis			68			70			56	65	8
Ophiura affinis			5			1			9	5	4
Area: Ula											
Spiophanes kroeyeri		14			10				11	12	2
Spiophanes bombyx		366			10				0	125	208
Chaetozone setosa		3			10				165	59	92
Myriochele oculata		3			15				15	11	7
Eudorellopsis deformis		109			51				22	61	44
Mysella bidentata		0			128				22	50	68
Amphiura filiformis		21			51				58	43	20
Ophiura affinis		2			1				29	11	16

taken place in areas considered to be uninfluenced by industrially related disturbances. On the other hand in the north and east of the region populations in such areas had declined or only increased marginally.

At some fields in the region monitoring surveys have been taking place for over twenty years, e.g. Ekofisk Centre, but because of changing sampling techniques and non-standardisation in the early years not all data is comparable. The data from the period 1990–96 at the Ekofisk reference area was obtained using standard methodology, however, and can be used to obtain a measure of temporal variance in the population statistics over a seven year period. Comparable data on some of the most abundant species is also available from four other fields in the region (Ula, Gyda, Valhall and Hod) at varying times between 1988 and 1996. Population changes in a suite of species common to the reference stations in all these areas which have dominated the populations in at least one of the sampling periods are illustrated in Table 9. In 1996 two species of surface deposit feeding polychaetes, *S. kroeyeri* and *S. bombyx* had high populations at Ekofisk although throughout the period 1990–93 their populations had been low. Very high numbers of *S. bombyx* were recorded in 1989 at Ula but few were present there in 1992 and it was absent in 1996. *S. kroeyeri* numbers were low throughout the period in other areas. The deposit feeding polychaete *C. setosa* had high populations at Ula and Gyda in 1996 although having only minor populations in the more southerly fields at that time and previously. Another small deposit feeding polychaete, *M. oculata,* had very numbers in 1996 at Hod, following moderately high populations in previous years. This species was also numerous in 1988 and 1991 at Valhall but present only in low numbers there in 1996 and throughout the period at the other fields. The cumacean *E. deformis* was very numerous at Ula in 1988 but declined to moderate levels in subsequent surveys. At Hod its populations were relatively high throughout the period but elsewhere it was numerous only in 1996. In 1996 numbers of this species were low at Ekofisk but remained high in all other areas. The bivalve *M. bidentata* had high numbers at all fields sampled in 1991–93 but was numerous only at Ekofisk, Valhall and Gyda in 1996. At the Ula field numbers had declined markedly and at Hod only one specimen was found in 1996. *A. filiformis*, a suspensivorous brittle star, had high populations at all five fields in the period 1990–96 although at the fields sampled prior to that date numbers were considerably lower. Another brittle star, the carnivore *O. affinis*, was numerous at four of the five fields in 1996 but scarce in all areas prior to that date.

Discussion

Relative levels of natural variability

The regional and reference stations were positioned in areas considered to be outside the influence of any disturbances emanating from the industrial activities in the oil field areas. If it is assumed, therefore, that changes at those stations are the result of natural factors rather than associated with field related disturbances, then the biological variance across this group of stations at any one time may be taken to represent the normal limits to naturally induced change in the region as a whole. It has been shown that the coefficient of variation (C) for population abundance at these stations was 25% and that for species richness was 11%. The values of C for samples from the Ekofisk reference area over a period of seven years were comparable thus any variation beyond these limits will be taken as being abnormal for these populations.

The data in Table 7 showed that amongst the thirteen stations suggested by the CCA analysis to be least influenced by industrial disturbance and which could be compared with previous data 10 (77%) had increases in population density and 8 (61%) increases in species richness greater than these limits. The highest increases in abundance were at Gyda and Tor, two fields to the north of the region whilst fewer stations from fields in the south were above the established limits. However individual stations from the most southerly stations, Valhall and Hod showed large increases as did the reference stations from fields in the north, south and centre.

It is apparent, therefore, that in these undisturbed areas of the region temporal changes in population levels are far from uniform. However the extent of increases in species richness and population density exceeded that of decreases in 1996 in those areas where they occurred. In general then in all the undisturbed or relatively undisturbed areas sampled the trend in 1996 was for strong, but patchily distributed increases in density and diversity of the benthic fauna of the region.

The comparison of changing dominance of some species common to a group of fields sampled between 1988 and 1996 allowed a more detailed examination of temporal fluctuations within the benthic communities.

Table 8. Region 1: comparison of spatial and temporal variation in populations (measured as C, the coefficient of variation) of a suite of dominant species found throughout the area.

Species	Spatial variability across the regional stations	Temporal variability in the Ula area	Temporal variability in the Gyda area	Temporal variability in the Ekofish area	Temporal variability in the Valhall area	Temporal variability in the Hod area	No. of fields with higher temporal variability
S. kroeyeri	68	18	48	189	156	148	3
S. bombyx	85	166	66	158	115	143	4
C. setosa	128	154	150	50	55	64	2
M. oculata	225	63	141	121	51	80	0
E. deformis	60	73	18	99	94	60	3
M. bidentata	82	137	102	72	111	63	3
A. filiformis	23	45	12	20	52	62	3
O. affinis	28	149	80	212	116	147	5
No. of species with higher temporal variability		6	3	5	6	4	

This suggested that there was considerable temporal and spatial variance in dominance across the region as a whole which is reasonably represented by the five fields for which data is available. Thus these range from Ula and Gyda in the north through Ekofisk in the centre to Valhall and Hod in the south of the region. Only one species *A. filiformis*, showed consistently high populations in all areas for most of the period. The other seven species examined all showed high populations at some fields in some periods but few general patterns are apparent. The bivalve *M. bidentata* had high populations at one time or another in all areas but never for successive surveys in the same area. *E. deformis* and *S. bombyx* predominated more frequently in the north and centre of the region than in the southern fields, where *M. oculata* and *C. setosa* were more frequent. These types of distributions may be linked to the north-south gradient in grain size, since the sediments of the northern fields are slightly coarser. *M. oculata* and *C. setosa* are both known to be favoured by organic enrichment (Pearson & Rosenberg, 1978). High populations of *M. oculata* have been associated with moderately enriched conditions in the vicinity of oil platforms in the area (Mannvik et al., 1993) and it is possible that there may be some additional detrital inputs affecting this supposedly undisturbed area at those times when such populations increase. Similarly the higher populations of the deposit feeding polychaete *C. setosa* than would be expected in an undisturbed area recorded at Ula in 1996 were also recorded from Regional stations 5 and 6 situated between Ula and Ekofisk. Numbers of the filter feeding brittle star *A. filiformis* at that time were relatively high throughout the area and particularly so at Regional stations 4 and 8 in the south. Similarly high numbers of these species in parts of the region in 1993 were attributed to natural fluctuations resulting from long-term changes in the benthic faunal dominants of the area driven by climatic fluctuations acting on recruitment and food availability (Mannvik et al., 1994).

Table 8 compares changes in the coefficient of variance (C) for these species at the five fields with the C values recorded for the same species at the eight regional stations. Thus a comparison is given of the spatial variance in each species abundance with temporal variance in the same populations at the five fields. In the Ula and Valhall fields six and at the Ekofisk field five of the eight species show higher temporal C values than the spatial values. Of the eight species examined only one, *O. affinis,* had consistently higher temporal than spatial variance. *S. bombyx* had higher temporal variance at four of the five fields. Conversely *M. oculata* had a higher spatial than temporal coefficient of variance at all five fields. In general the analysis suggests that temporal changes in the populations of the majority of dominant species in the region are considerably higher than the spatial variation in those species with the notable exception of the small detritivorous polychaete *M. oculata*.

Causes of natural variability

The suggestion referred to above, that detrital fluctuations caused by changes in the primary production in the overlying water column as a result of climatic influences on the movement of water masses, seems the most likely general explanation for these elevated populations in the industrially undisturbed areas of the region. A similar explanation has been advanced to explain large diversity and dominance differences in other areas in the North Sea. Creutzenberg et al., (1984) described large population differences locally on either side of a frontal area in the southern N. Sea and on a regional scale changing benthic diversity and dominance have been related to variation in detrital deposition near topographically controlled gyres in the Dogger Bank area (Künitzer, 1990 a & b, Kröncke, 1990). Low benthic diversities in stratified areas of the north central North Sea have been contrasted with the higher benthic populations beneath adjacent mixed water bodies (Künitzer et al., 1992; Heip et al., 1992). Strong temporal fluctuations in the population dominance and density of macrobenthos in an area off the north-east coast of England have been attributed by Buchanan & Moore (1986) to changing detrital inputs driven by inter-annual differences in local plankton blooms perhaps influenced by the changing balance of Atlantic water inflows to the northern North Sea. These cause the movement of waters of differing temperatures and nutrient content through the area. However such general hydrographic factors under climatic or tidal control seem inadequate to explain the spatial and temporal variability in the high dominance and diversity levels recorded within the region. Other factors which might have influenced some of the changes observed include disturbance by bottom trawling, localised eutrophication and the storm driven advection of sedimentary material. Fishing intensity in the Region 1 is quite high and, although no fishing is allowed within the 500 m exclusion zones drawn around each oil installation bottom trawling occurs throughout the rest of the area. Trawling is known to have a considerable impact on benthic communities, differentially removing larger epifauna and smothering or damaging many infaunal species (Bergman & Hup, 1992). This tends to promote the dominance of small opportunist mobile or semi-mobile species, particularly polychaetes and can produce to some extent the type of changes that have been discussed. However such fishing activity has been carried out in the area since well before the development of hydrocarbon extraction and it is thought unlikely that the type of large scale changes observed in both time and space over the area could have been generated only by fishing related disturbance, although it cannot be dismissed as a potential contributory factor.

Eutrophication has been suggested as the cause of increasing benthic populations in many coastal areas of the southern and central North Sea in recent years (Beukema & Cadée, 1986; Rachor, 1990; Reise & Schubert, 1987) but has been thought to be unlikely to have any major impact on the deeper and more extensive northern areas because of summer stratification and consequent reduction in benthic-pelagic coupling (Heip et al., 1992). However there is increasing evidence that coccolithophorid blooms occur frequently in stratified areas of the northern North Sea and that there is a high downward particulate flux from such blooms through the formation of macroaggregates (van der Wal, et al., 1995), thus stratification may be no bar to a rapid benthic response to increased surface production in such areas. The Norwegian coastal current impinges on the northern edge of Region 1 and has been increasingly enriched by enhanced nutrient loading from the Baltic outflow through the Skagerrak coastal current (Andersson, 1996). It is possible that this could have had some role in increasing primary production rates in recent years, particularly in the south of the area. The advection of sedimentary material from other areas through storm driven bottom water surges has been advanced as a possible explanation for large scale population changes elsewhere on the Norwegian shelf in the period 1990–1992 (Mannvik et al., 1993). In that case decreases in fine material on the upper slopes of the Norwegian trench coincided with increase in fines over areas of the adjacent central plateau following an exceptionally stormy winter. It was postulated that subsequent dominance and diversity changes in the benthos of the plateau area could have been a response to the redistribution of sediment. This or similar events might have had an influence on the populations in Region 1, but no direct evidence is available. Rosenberg (1995) suggested that high benthic populations, particularly of *A. filiformis*, in deep areas of the Skagerrak were sustained by the advection of detrital material into the area. Any or all of these factors may have had an influence on the recorded population changes. In the absence of any direct information on primary production and detrital sedimentation in the region it is not possible at present to differentiate between these factors and assess their reality or relative impact.

Conclusions

It seems probable that over a large part of Region 1 there was an increase in the detrital Carbon supply to the benthos between 1993 and 1996 sufficient to increase the diversity of the macrofaunal populations and to change the dominance structure of the communities. This effect was patchily distributed but most evident in the south east of the region and did not occur in the north east at the Yme field. Evidence from earlier field surveys in different parts of the region suggests that such fluctuations, attributable to natural forces, occur at varying spatial and temporal scales through out the region. On the other hand population changes in response to industrial disturbance are localised to areas in the vicinity of installations. In these areas density changes may be up to ten times greater and diversity fluctuations six times higher than those attributable to natural causes in undisturbed areas.

As the Regional survey strategy develops in the Norwegian sector of the North Sea the accumulation of subsequent data sets from Region 1 and from the other designated survey regions in the northern North Sea should provide sufficient data to explore the causes and full extent of these broad scale changes and allow a better assessment of the factors driving them.

References

Andersson, L., 1996. Trends in nutrient and oxygen concentrations in the Skagerrak-Kattegat. J. Sea Res. 35: 63–71.

Bergman, M. J. N. & M. Hup, 1992. Direct effects of beam trawling on macrofauna in a sandy sediment in the southern North Sea. ICES J. mar. Sci. 49: 5–11.

Beukema, J. J. & G. C. Cadée, 1986. Zoobenthos responses to eutrophication of the Dutch Wadden Sea. Ophelia 26: 55–64.

Buchanan, J. B. & J. J. Moore, 1986. long-term studies at a benthic station off the coast of Northumberland. Hydrobiologia 142: 121–127.

Creutzenberg, F., P. Wapenaar, G. Duineveld & N. Lopez Lopez, 1984. Distribution and density of the benthic fauna in the southern North Sea in relation to bottom characteristics and hydrographic conditions. Rapp. P.-v. Réun. Cons. int. Explor. Mer 183: 101–110.

Davies, J. M., J. M. Addy, R. Blackman, J. R. Blanchard, D. C. Moore, H. J. Sommerville, A. Whitehead & T. Wilkinson, 1984. Environmental effects of oil based mud cuttings. Mar. Pollut. Bull. 15: 363–370.

GESAMP (IMO/FAO/UNESCO/WMO/IAEA/UN/UNEP Joint Group of Experts on the Scientific Aspects of Marine Pollution), 1993. Impact of oil and related chemicals and wastes on the marine environment. GESAMP Rep. Stud. 50: 1–180.

Greenacre, M. J., 1993. *Correspondence analysis in practice.* Academic Press, London, 193 pp.

Heip, C., 1995. Eutrophication and zoobenthos dynamics. Ophelia 41: 113–136.

Heip, C., D. Basford, J. A. Craeymeersch, J. M. Dewarumez, J. Dörges, P. de Wilde, G. C. A. Duineveld, A. Eleftheriou, P. M. J. Herman, U, Niermann, P. Kingston, A. Künitzer, E. Rachor, H. Rumohr , K. Soetaert & T. Soltwedel, 1992. Trends in biomass, density and diversity of North Sea: macrofauna. ICES J. mar. Sci. 49: 13–22.

Kingston, P. F. 1992. Impact of offshore oil installations on the benthos of the North Sea. ICES J. mar. Sci. 49: 45–53.

Kröncke, I., 1990. Macrofauna standing stock of the Dogger Bank. A comparison: II. 1951–52 versus 1985–87. Are changes in the community of the north-eastern part of the Dogger Bank due to environmental changes. Neth. J. Sea Res. 25: 189–198.

Künitzer, A., 1990a. The infauna and epifauna of the central North Sea. Meeresforsch. 33: 23–37.

Künitzer, A., 1990b. A comparison of the *Amphiura filiformis*-associations north-east of the Dogger Bank and of the German Bight. Neth. J. Sea Res. 25: 199–208.

Künitzer, A., D. Basford, J. A. Craeymeersch, J. M. Dewarumez, J. Dörges, G. C. A. Duineveld, A. Eleftheriou, C. Heip, P. Herman, P. Kingston, U, Niermann, E. Rachor, H. Rumohr & P. A. J. de Wilde, 1992. The benthic infauna of the North Sea: species distribution and assemblages. ICES J. mar. Sci. 49: 127–143.

Mannvik, H-P., R. Fieler, T. H. Pearson, K. Hansen, L. K. Sydnes, R. Evans & S. Falk-Petersen, 1993. Environmental Monitoring Survey of the Ekofisk, Eldfisk, Embla and Tor Fields, May 1992, 274 pp + appendices.

Mannvik, H-P., S. Cochrane, T. H. Pearson, E. Jørgensen, K. Hansen & R. Fieler, 1994. Environmental Monitoring Survey of the Ekofisk, Eldfisk and Embla Fields, May 1993, 329 pp + appendices.

Olsgard, F. & J. S. Gray, 1995. A comprehensive analysis of the effects of offshore oil and gas exploration and production on the benthic communities of the Norwegian continental shelf. Mar. Ecol. Prog. Ser. 122: 277–306.

Pearson T. H. & R. Rosenberg. 1978. Macrobenthic succession in relation to organic enrichment and pollution of the marine environment. Oceanogr. mar. Biol. Ann. Rev. 16: 229–311.

Rachor, E., 1990. Changes in sublittoral zoobenthos in the German Bight with regard to eutrophication. Neth. J. Sea Res. 25: 209–214.

Reise, K. & A. Schubert, 1987. Macrobenthic turnover in the subtidal Wadden Sea: the Norderaue revisited after 60 y. Helgoländer Meeresunters. 41: 69–82.

Rohlf, F. J., 1991. NTSYS-pc. Numerical Taxonomy and Numerical Analysis System. Exeter Publishing, Ltd, New York.

Rosenberg, R., 1995. Benthic marine fauna structured by hydrodynamic processes and food availability. Neth. J. Sea Res. 34: 303–317.

SFT, 1990. Manual for overvåkningsundersøkelser rundt petroleumsinstallasjoner I norske havområder. Statens Forurensningstilsyn (SFT). 90.01. 29 pp.

Van der Wal, P., R. S. Kempers & M. J. W. Veldhuis, 1995. Production and downward flux of organic matter and calcite in a North Sea bloom of the coccolithophore *Emiliania huxleyi*. Mar. Ecol. Prog. Ser. 126: 247–265.

Hydrobiologia **375/376:** 331–341, 1998.
S. Baden, L. Pihl, R. Rosenberg, J.-O. Strömberg, I. Svane & P. Tiselius (eds),
Recruitment, Colonization and Physical–Chemical Forcing in Marine Biological Systems.

Vertical migratory behaviour of the euphausiid, *Meganyctiphanes norvegica*, and its dispersion in the Kattegat Channel

G. A. Tarling[1], J. B. L. Matthews[1], R. Saborowski[2] & F. Buchholz[2]
[1] *Scottish Association for Marine Science, PO Box 3, Oban, Argyll, PA34 4AD, U.K*
[2] *Biologische Anstalt Helgoland, Meeresstation, Postfach 180, D-27483 Helgoland, Germany*

Abstract

The euphausiid *Meganyctiphanes norvegica* (Northern Krill) is predominantly an oceanic species common to the North Atlantic and adjacent seas. In the Kattegat the species concentrates in a series of depressions in the Kattegat Channel east of the island of Læsø which represent havens of marine conditions beneath the low salinity Baltic outflow.

The vertical migratory behaviour of *Meganyctiphanes* results in it encountering considerable contrasts in physical conditions during its diurnal cycle. This behaviour and the resulting physical experience of the animals were investigated by means of a combination of net sampling and ADCP transects across the Alkor Deep (131 m) in summer (18–19/7/96) and winter (9–10/3/97).

In both summer and winter the krill tended to concentrate within the basin during daytime and to disperse in the upper layers at night. The period of dispersion was longer in winter (17.00–05.00 h) than in summer (23.00–02.00 h). The complex layering of different wind-induced and tidal current systems acted to advect the krill away from the basin during their upward phase at night. ADCP measurements showed that in summer the krill would be advected SSW at an average rate of 3 cm s^{-1}, but that in winter they would be advected WSW at 3.2 cm s^{-1}. Calculations show that the krill would be capable of swimming against such currents without increasing their standard metabolism and that their distribution in the vicinity of the basin seems to be determined more by biotic than by abiotic factors.

Introduction

Meganyctiphanes norvegica is an important part of the Atlantic pelagic community as it is the largest euphausiid species in the northern hemisphere, eats predominantly zooplankton but also phytoplankton as energy sources and is itself a staple food of vertebrates, particularly fish, birds and whales (Pearcy et al., 1979). The species is found within a temperature range of 2–15 °C (Einarsson, 1945; Lindley, 1982) and occurs within a wide geographic area from the east coast of Canada to western Europe, including the Mediterranean and regions of the Arctic. Accordingly, it is expected that the species exhibits specific behavioural and physiological adaptations that allow it to inhabit a wide range of different oceanographic regimes. The Kattegat represents an extremity in the species' distributional range because it is influenced by two contrasting water bodies, a brackish input from the Baltic and a marine input from the Skagerrak. The physiological tolerances and behavioural adaptations exhibited by krill in this region would be expected to be close to their limits of tolerance, and study of this population will produce insights into which processes are able to adapt and in what ways.

Within the northern Kattegat, the species is found mainly in the basins which represent havens of marine conditions beneath freshwater outflow in the surface layers. As with many euphausiid species (Mauchline & Fisher, 1969; Mauchline, 1980), *M. norvegica* typically exhibits a pronounced diurnal vertical migratory pattern in the Kattegat (Buchholz et al., 1995) where the species ascends to upper layers during the night and descends during the day. Given the contrasting directions of the marine inflow and the freshwater outflow within the region as well as the variable effect

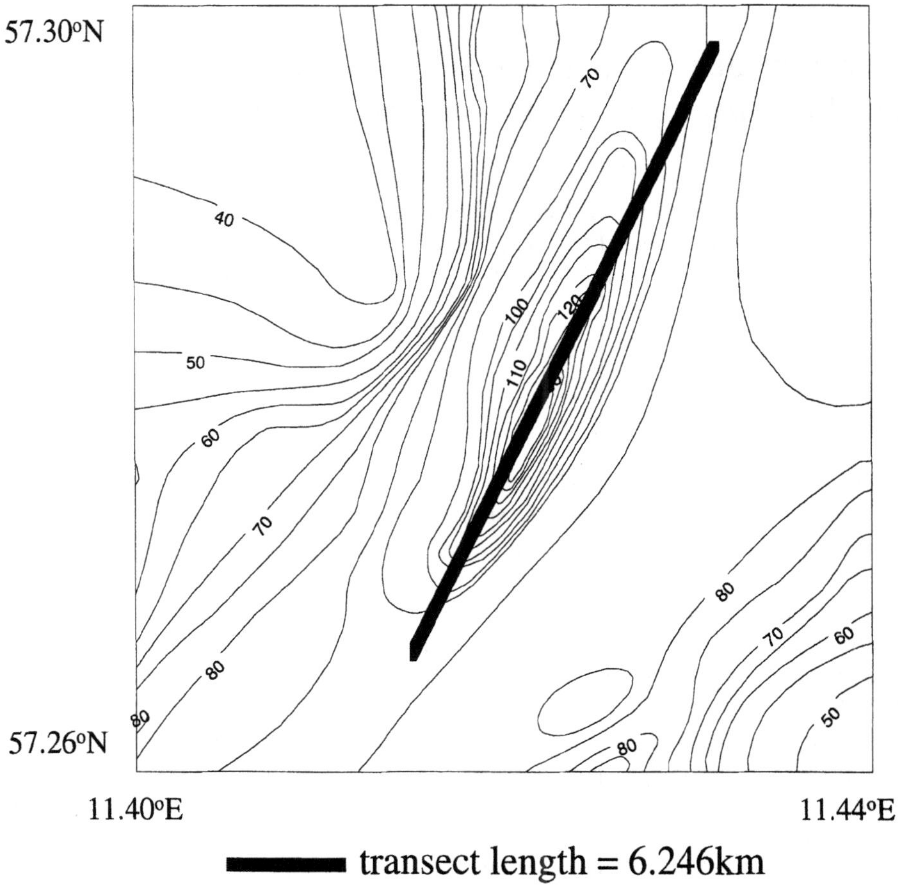

Figure 1. The Alkor Deep showing the net sampling transect.

of tide and wind on current direction, this behavioural pattern results in the species being subjected to changeable and often considerable advective forces which may act to carry the organisms away from the basins. Nevertheless, Boysen & Buchholz (1984) showed that the population within a particular basin was autonomous and exhibited an uninterrupted life-history pattern, suggesting that the krill were able to counter the disruptive advective influences and maintain their preferred geographic position. However, a certain degree of immigration and emigration from the basins is still to be expected as a result of advective forces.

This study examines the vertical migration of the krill population within the Alkor Deep during a summer and a winter situation, primarily to characterise the relationship between these patterns and the physical environment and, secondarily, to consider the role of advective forces on distribution patterns. This approach allows advective displacement from a typical deep to be estimated and hence addresses the question of the metabolic cost of maintaining a preferred geographic location.

Hydrography and bathymetry

Circulation and currents – The Kattegat can be considered a large estuary with outflow of hyposaline Baltic water along the surface and inflow of seawater of nearly full salinity confined to the lower part of the water column in the deeper portion of the Kattegat. The strength and direction of currents are regulated by large-scale weather conditions and the distribution of barometric pressure over the North Sea and Baltic Sea. Under most wind conditions, there is a general northward flow at the surface and southward flow is confined to the deepest passages. Only when there are strong westerly and northerly winds are surface currents deflected southwards.

Salinity – The circulation pattern normally results in Kattegat water being stratified with a halocline at ca. 10 m in the vicinity of the basin system. Above the halocline, values seldom exceed 20 psu and are generally lower during summer months. Below the halocline, salinities range between 30 and 34 psu for most of the year but are higher during the windy months of October and November (Buchholz & Boysen, 1988).

Temperature – Temperature is extremely variable with depth and season. Surface waters range from more than 16 °C during summer to less than 1 °C during January and February (Buchholz & Boysen, 1988). Steep temperature gradients and a difference of ca. 10 °C exist between the surface and 80m between June and November. Below 80 m, seasonal temperatures are more constant and range between 5 and 10 °C.

Bathymetry – The Kattegat is mostly quite shallow, the average water depth being 27 m. The Læsø Deep consists of a series of four depressions < 130 m deep running north to south immediately to the east of the island of Læsø (Ulrich, 1983). This study concentrates on the Alkor Deep, which is 3.5 km long, 1 km wide and runs in a NNE-SSW direction.

Methods

Sampling was carried out on board the R/V *Heincke*, a 55 m vessel equipped with a starboard winch and conducting cable for net deployments. This was used to operate a 1 m^2 MOCNESS system with 2 mm mesh, flowmeter and temperature and salinity sensors. The net was deployed every 3 h over a 30 h period in a double oblique fashion. The ship passed over the Alkor Deep in a north-south plane at 2–3 knots, the direction depending on prevailing wind conditions.

Net 1 was sent down open and arrived at 80 m when the ship was over the rim of the basin. Net 2 was open between 80 and a maximum of 115 m, depending on the exact course of the ship over the basin. Nets 3 to 8 were opened and closed every 20 m from 100 m to the surface, corresponding to ca. 5 min opening time. The last 20 m were sampled in 10 m intervals. Total deployment time was ca. 45 min which corresponded with the time taken for the ship to travel from one rim of the basin to the other.

Animals were immediately frozen at – 20 °C, then thawed on return to the laboratory where they were sexed, staged (Makarov & Denys, 1980) and measured in terms of carapace length. Carapace lengths were converted to total length and dry weight using calibrations taken from the same samples.

A 153 kHz narrow-band ADCP (RD Instruments, San Diego, California) was deployed through the oceanographic well of the ship during each 30-h sampling period. The ADCP was set to collect data from 30 depth 'bins', each with a depth of 8 m. Raw data were collected every ca. 1.5 sec and averaged over six minutes to obtain current measurements, giving a standard deviation for each measurement of 0.83 cm s^{-1}. The data were re-averaged over 30-sec intervals for relative backscatter patterns in order to gain higher temporal resolution. Relative backscatter anomaly for each depth bin was calculated by subtracting an average value for the 30 h period from the mean value observed during each 3-hourly sampling period.

All times are given in local time (UTC + 1 in winter; UTC + 2 in summer).

Results

Description of prevailing weather

Summer situation – During the three days before sampling, wind direction was west-northwesterly, between 6 and 11 m s^{-1}. During the 30-h sampling period, wind strength decreased from 12 to 2.7 m s^{-1} and shifted from northwesterly to northeasterly. As a result, surface currents would have been influenced to flow south against the prevailing northward current, slowing the outward-flowing Baltic current.

Winter situation – For two days prior to sampling, wind direction was northwesterly at ca. 12 m s^{-1}. This shifted to south-southwesterly on the day before sampling at a speed of 11 m s^{-1}. During sampling, wind direction drifted from west-northwesterly to south-southwesterly and varied between 4 and 8.5 m s^{-1}. The westerly wind would have acted against the northward flowing surface currents, although the effect would have been small given the low strength of the winds.

Physical conditions over the Alkor Deep

In the summer situation, there was a 3-layered temperature structure (Figure 2) which can be divided into surface, mid-layer and bottom. Between 0 and 16m, temperatures were approximately 15 °C. In the mid-layer (17–50 m) there was a steep temperature gradient

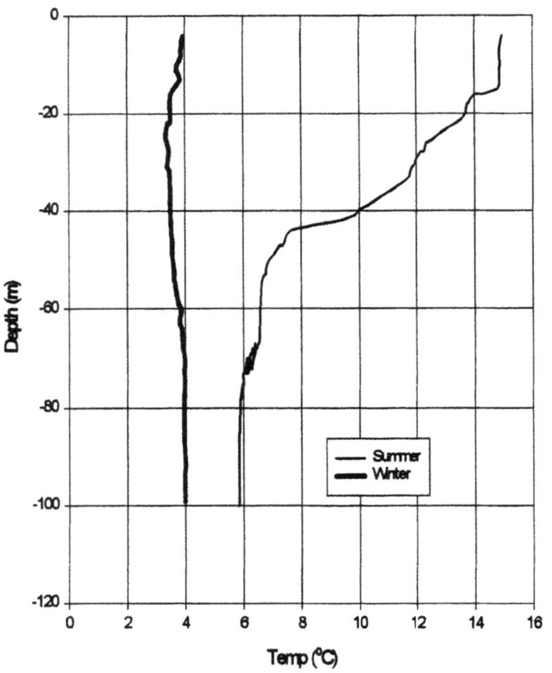

Figure 2. Temperature profile for the Alkor Deep in summer (18/7/96) and winter (8/3/97).

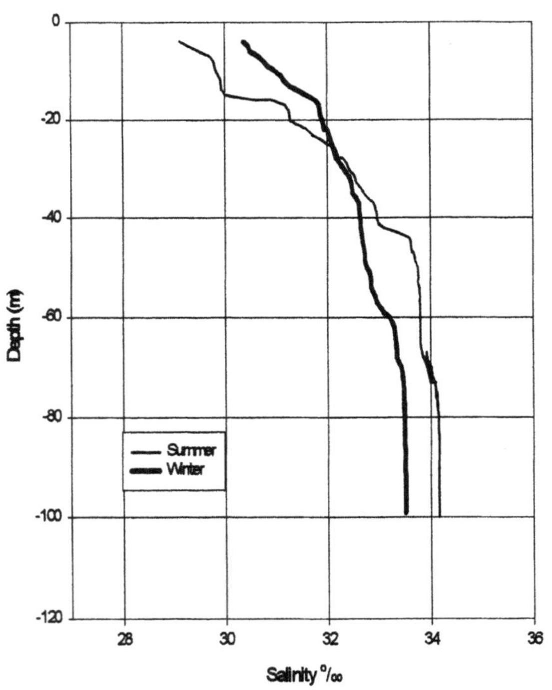

Figure 3. Salinity profile for the Alkor Deep summer (18/7/96) and winter (8/3/97).

where temperature dropped from 15 °C to 7 °C. Below 50 m, temperature was quite stable, falling gradually from 7 °C to 6 °C at 100m. By contrast, temperatures during the winter situation were virtually uniform throughout the water column at around 4 °C.

The 3-layered structure in summer temperatures was mirrored in the salinity profiles (Figure 3), with quite stable values at the surface and bottom and a steep gradient in the mid-layer. Typical surface salinities were around 29 psu, rising to 33.5 psu down through the mid-layer and remaining around 33.5–34 psu between 50 and 100 m. During the winter situation, salinities were generally higher than in summer between 0 and 30 m (30.4 psu as opposed to 29.1 psu at 4 m in summer) but lower than in summer below 30 m (33.5 psu as opposed to 34.2 psu at 90 m in summer); this illustrates a greater degree of mixing between fresh and marine inputs during the winter months. Nevertheless, there was still a 3-layered pattern as observed during summer with stable layers at the surface and bottom and a gradient in the mid-layer.

Figures 4 and 5 illustrate the strength and direction of currents with depth. Overall, there were three general patterns corresponding to the three layers identified in the T/S plots. In the summer situation, surface currents (at 16 m) varied between N and NE and were between 5 and 15 cm s^{-1}. In the mid-layer (40 m), currents were weaker (5 to 10 cm s^{-1}) and more variable in direction compared with the upper layer, with a certain degree of tidal influence. For instance, between 15.00 h and 21.00 h on 18/7 currents were to the east, corresponding with the low to high tide interval, whereas between 09.00 h and 15.00 h on 19/7 currents varied between NW and SW, corresponding with the high to low tide interval. Currents at 80 m were S to SSW in the range of 15–20 cm s^{-1}.

In the winter situation, surface currents (at 16 m) were N to NW between 15 and 30 cm s^{-1}. The mid-layer (at 40 m) exhibited currents that were generally W to SW between 10 to 15 cm s^{-1} but there was also a weaker S to SE component with speeds around 5 cm s^{-1}, which corresponded with the high to low tide interval. Currents at 80 m were variable, generally S to SW between 5 and 10 cm s^{-1} at most times but N to NE at 5 to 10 cm s^{-1} between 19.00 h and 02.00 h on 18/7.

Overall, surface currents were strong and predominantly to the north throughout both sampling periods, corresponding to the brackish Baltic outflow. The flow of this surface current was probably retarded in summer and in winter by a northerly and a westerly current

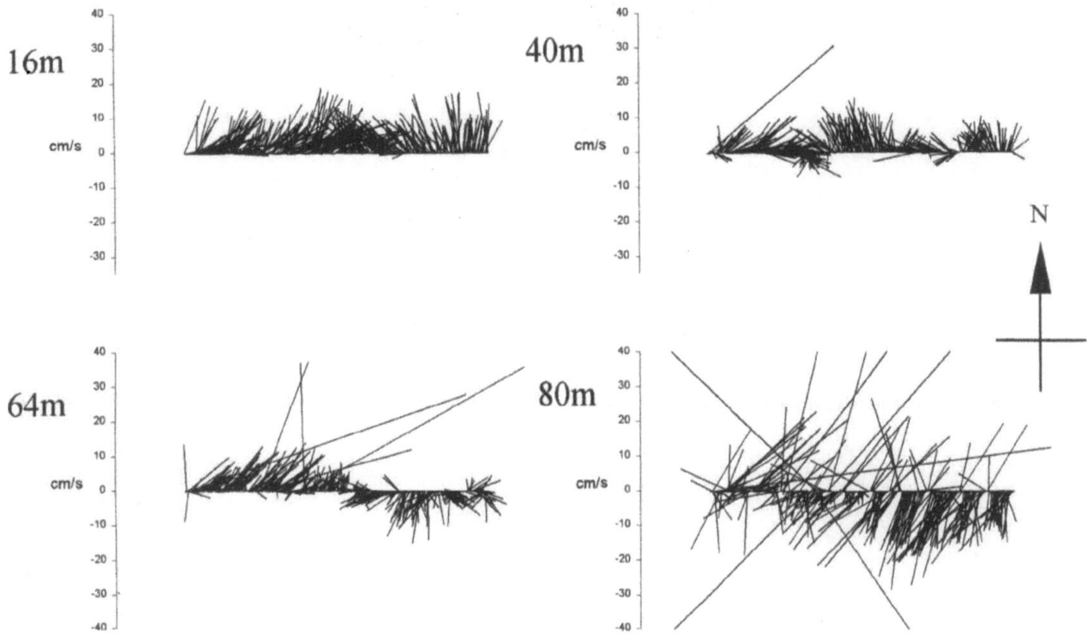

Figure 4. Current strength and direction during a 30 h period over the Alkor Deep in summer (18–19/7/96).

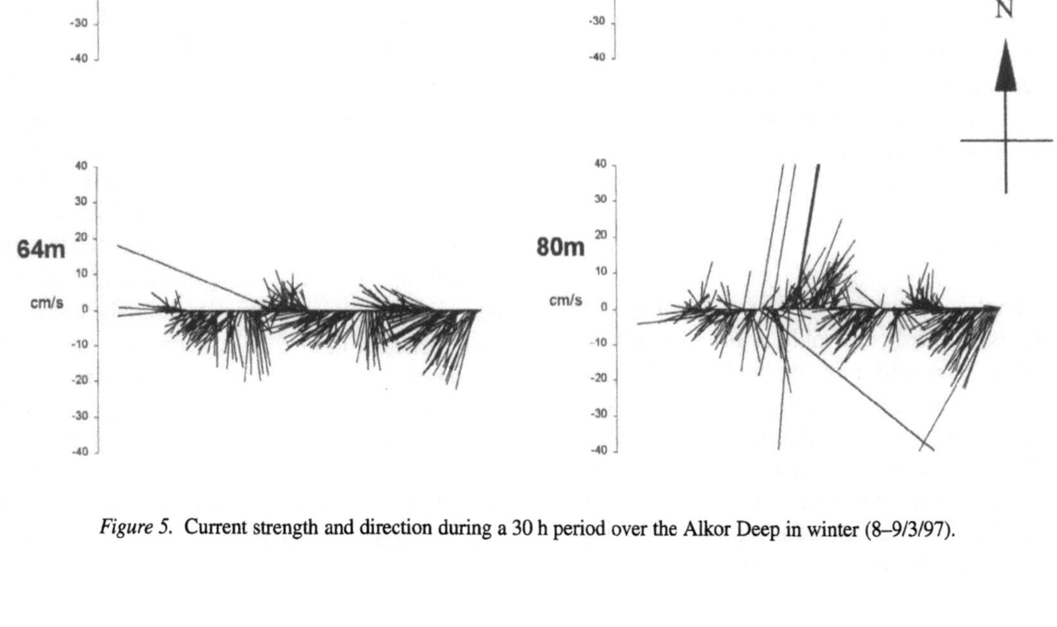

Figure 5. Current strength and direction during a 30 h period over the Alkor Deep in winter (8–9/3/97).

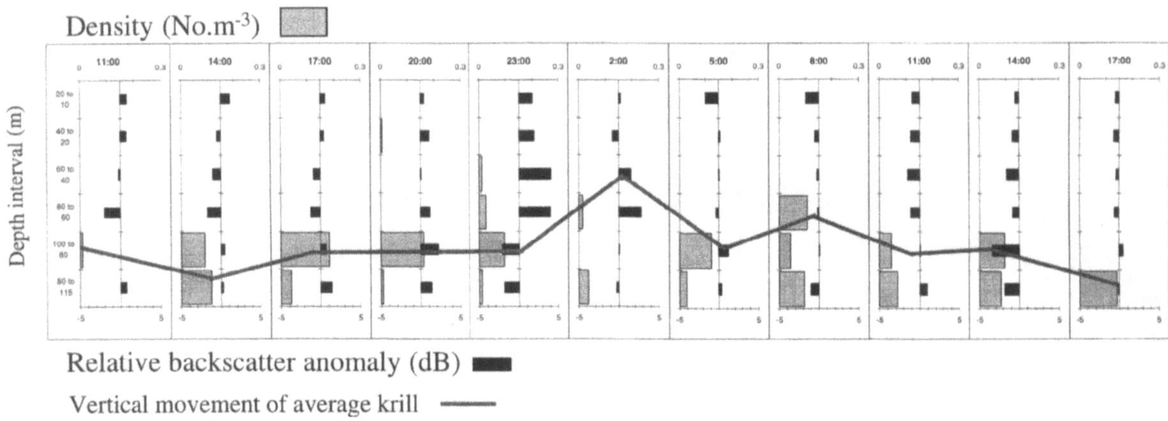

Figure 6. Density (grey) and relative backscatter anomaly (black) of krill through the water column during a 30 h period in summer.

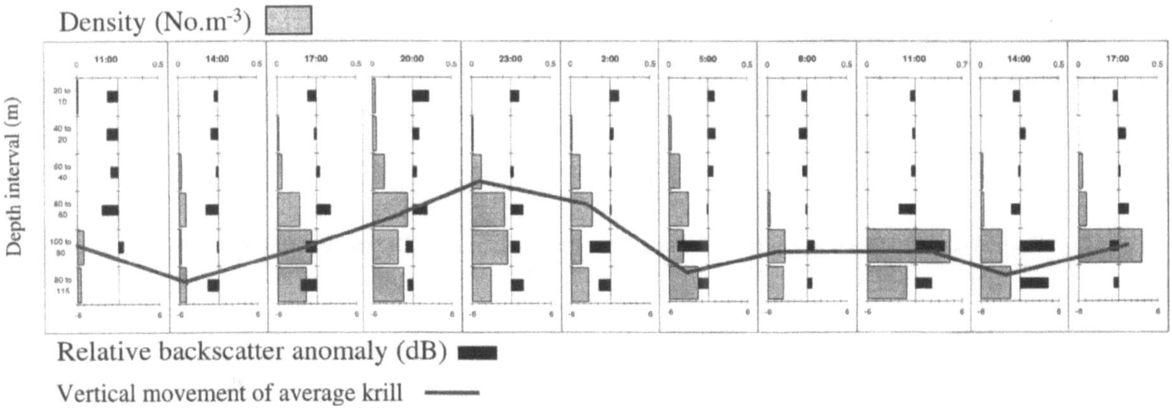

Figure 7. Density (grey) and relative backscatter anomaly (black) of krill through the water column during a 30 h period in winter.

respectively. Bottom currents were predominantly to the south and conformed to the marine inflow from the Skagerrak. Greater variability of the bottom current in the winter may have been a result of increased mixing between layers during the winter months. Currents in the intermediate layer were variable and tidally influenced, representing the transition between the contrasting flow regimes.

Biological patterns

A full description of the net catches is given by Tarling et al. (submitted) and only a brief resumé of the general patterns will be given here. The densities of krill were lower during the summer period compared with winter, (0.0267–3.46 ind m^{-2} compared with 0.82–10.05 ind m^{-2}), but in both seasons the krill characteristically concentrated within the basin during the day and ascended during the night (Figures 6 and 7). In summer, low densities were detected between 40 and 80 m at 23.00 and 02.00 h whilst densities within the basin decreased. In winter, density decreases within the basin were observed over a longer period (02.00 h to 11.00 h) and animals were regularly found between 20 and 40 m from 17.00 h to 05.00 h. The size distribution of the populations found during summer and winter corresponded with a previous study of the annual cycle of the Kattegat population (Boysen & Buchholz, 1984), with a unimodal distribution (peak at 30 mm total length) in summer and a bimodal distribution (peaks at 23 and 30 mm total length) in winter. Smaller animals tended to migrate vertically more than larger animals in winter but no such tendency was evident in summer.

Backscatter patterns (see Tarling et al., submitted) and net catches were used jointly to interpret the typical vertical migration of the krill population over a diurnal cycle in summer and in winter. Figures 6 and 7 show the percentage distribution in the net samples (grey histograms) and the backscatter anomalies for each ADCP depth bin (black histograms) throughout the sampling period; the line superimposed on this double set of histograms is an attempt to estimate the 'typical' vertical movement of a hypothetical krill through the whole sampling period. In most cases, the position of this line has largely been determined by the distribution in the net samples, but at night, when there were high positive ADCP anomalies at 40–60 and 60–80 m and net catches were small (presumably the result of dispersion away from the vicinity of the basin) more weight has been given to the backscatter results. The vertical movement so derived has then been related to ADCP data on the strength and direction of currents through the water column, in order to estimate the influence of horizontal advective forces on the krill. Figure 8 depicts the horizontal displacement, at 6-min intervals, of a passive particle following the same vertical movements of a hypothetical krill. During each 6-min period, the particle is advected by the horizontal current values for the appropriate depth and time, and moved to a new co-ordinate. This co-ordinate then becomes the starting point for the next 6-min period such that each movement is cumulative. The starting point for the particle was the centre of the Alkor Deep and the calculation was carried out for the 30-h period covered by the net sampling series.

The calculation showed that a passive particle in summer would be advected towards the S or SW, with the deep southward flowing Skaggerak water being a major influence. After 30 h, the particle would have travelled 3230 m from the middle of the basin. This would mean that an individual would have to maintain an average speed of 3.0 cm s^{-1} in order to remain within the Alkor Deep. During winter, nett advection would be to the WSW, because of the greater influence of the mid-layer current on the population as a result of the longer period of time it spends in that layer. The nett distance transported from the deep would have been 3440 m in 30 h, implying that an individual would have to maintain a speed of 3.2 cm s^{-1} in order to retain its original position.

Cost of a pelagic lifestyle in the Kattegat

Work carried out by Kils (1981) showed that, in the case of *Euphausia superba*, animals may swim between speeds of between 0 and 15 cm s^{-1} without affecting their 'standard metabolism'. This is because, within this range, krill regulates its speed mainly by

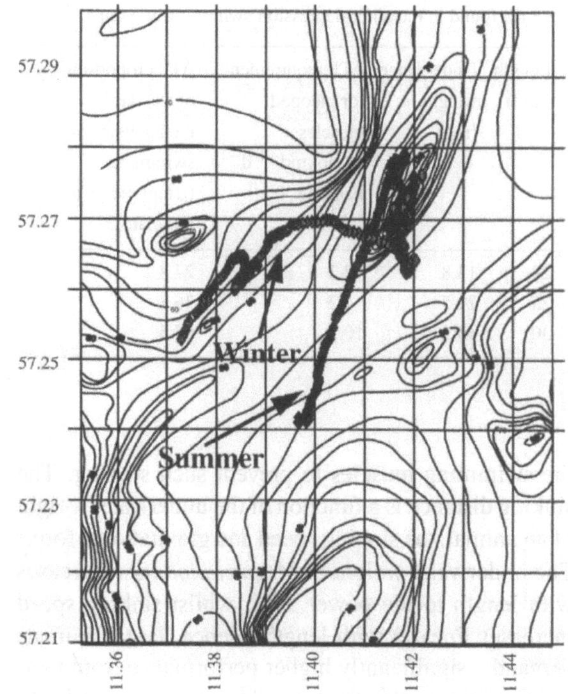

Figure 8. Horizontal displacement of a passive particle following the same vertical movements as a hypothetical krill over 30 h in summer and winter.

changes in the execution of the pleopod swimming stroke. At 15 cm s^{-1}, the pleopod stroke is optimally executed. Between 15 and 40 cm s^{-1}, the frequency of strokes controls the speed which would in turn increase the metabolic rate to an 'active metabolism'. Although similar experiments have not been carried out as comprehensively on *Meganyctiphanes*, it is believed that similar values would apply because the reduction in propulsive force from the pleopods would be counteracted by a corresponding decrease in the resistance of the body moving through water, since both parameters are a function of surface area. For the purposes of this investigation, therefore, one can conclude that the speeds *Meganyctiphanes* needs to achieve in order to remain within the vicinity of the Alkor Deep over a diurnal cycle are well within its capabilities and do not require the animal to increase its metabolic rate above the 'standard' level.

Further data provided by Kils (1981) allow the approximate daily costs of *Meganyctiphanes* maintaining a pelagic lifestyle to be calculated. One way this can be estimated is by considering what distance an organism would have sunk in a given time and hence, what 'performance' (P, J d^{-1}) is demanded of

Table 1. Comparative metabolic requirements for different sizes of krill found in the alkor depression swimming at 3 cm s^{-1}.

Length (mm)	Underwater weight (mg)	ATP requirement for pleopod muscles (μmol ind^{-1} d^{-1}) Estimate 1	ATP proportion of standard metabolism for swimming (μmol ind^{-1} d^{-1}) Estimate 2
22	13.8	2.3	27.8
25	22.5	4.2	36.8
30	45.2	10.2	67.5

the swimming muscles to prevent such sinking. The sinking distance is a function of the underwater weight of an animal, the sinking speed and gravitational force. The underwater weight of *Meganyctiphanes* increases with length to the power 3.76, whilst sinking speed increases linearly with length. Hence, larger animals demand a significantly higher performance from their swimming muscles than smaller animals in order to remain buoyant. To give some idea of the comparative costs within a typical krill population found in the Alkor Deep, the metabolic demand for maintaining buoyancy has been calculated for three common size classes at the time of the winter survey, 22 mm, 25 mm and 30 mm:

$$P = m \cdot g \cdot h \cdot t^{-1}, \qquad (1)$$

where:
P = performance (J d^{-1})
m = underwater weight (kg)
g = earth acceleration (m s^{-2})
h = sinking distance per second (m)
t = time unit (s)

From equations given in Kils (1981), this can be reduced to

$$P = 6.54 \cdot 10^{-9} \cdot L^{-4.83}, \qquad (2)$$

where:
L = length (mm)

Performance (J d^{-1}) can be converted into the amount of ATP that must be provided to the pleopod swimming muscles by assuming that the conversion efficiency of muscles is 25% and that 1 mol ATP⁼35 kJ (see Table 1, Estimate 1).

Another approach is to compare the metabolic rates of an animal during swimming and resting. Since krill do not exhibit typical resting behaviour comparisons have to be made with a species that does. The shrimp, *Crangon crangon*, shows resting, burrowing and swimming (hovering) behaviour, and it has been shown that its metabolic rate increases by 60% whilst hovering (Kils, 1981). To apply to *Meganyctiphanes* however, further parameters need to be applied to this value, as will be outlined. Within the Alkor Deep, krill must swim at 3 cm s^{-1} in order to maintain their geographic position (see above). Swimming at this speed creates a significant amount of hydrodynamic uplift which would counteract the underwater weight and hence the sinking force. In *Euphausia superba*, animals swimming at 3 cm s^{-1} had a body angle of 30^0 which created an uplift that was 29% of the sinking force. Observations of *Meganyctiphanes* made by Kils (1981) suggest that the relationship between body angle and speed are approximately the same but with higher standard deviation. The proportion of krill metabolism used in swimming would therefore be approximately 40% of the total metabolism (i.e. (100–29)% of 60%). Respiration rates for 30 mm krill from the Alkor deep were measured at 1.1 mg $O_2 \cdot {g_d}^{-1} \cdot h^{-1}$ (Saborowski, unpubl.data; g_d – dry weight (g), h – hour) which can, in turn, be converted to ATP production during swimming (see Table 1, Estimate 2).

The first estimate of swimming represents an underestimate of true swimming costs, since during the passing on of energy through the pleopod muscles to the water, a certain amount will be lost. The second estimate is certainly higher than reality since krill are far more efficient swimmers than *Crangon crangon*. Hence the true value for carrying out a pelagic lifestyle in the Kattegat should be between these two values.

Discussion

Comparing MOCNESS net catches with ADCP backscatter patterns may be subject to two main errors (Pieper, 1979): variability in backscattering strength for a given euphausiid biomass, and differences inherent in the two sampling methods. As Cochrane et al. (1994) discovered when using a bottom-mounted ADCP over a 49-d period, day and night backscattering strength may vary, probably as a result of different angles of orientation of the animals. When the decrease in backscatter during the period of upward swimming was related to target-strength models by Cochrane et al. (1994), results indicated that the

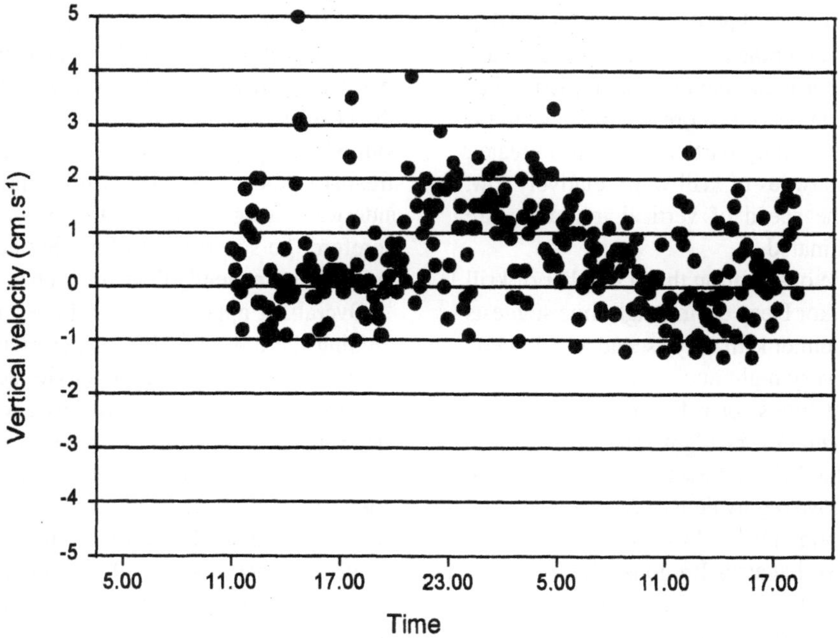

Figure 9. Vertical velocities (cm s^{-1}) measured at 80 m over 30 h during summer (18–19/7/96) [-ve downwards; +ve upwards].

organisms probably changed their orientation from horizontal to 30 °. This fits well with the prediction by Kils (1981) that the propulsive force necessary for upward lift is minimised at body angles of around 20–30 °. In relation to the present study, such variability in backscattering is small compared with the larger distributional changes seen diurnally and probably does not mask any general pattern. Correspondence between the ADCP and the MOCNESS is more of a problem since krill populations are known to be patchy and currents may deflect the path of the net over the basin in relation to the swath of the shipboard ADCP. Tarling et al. (submitted) consider the correlation between net samples and relative backscatter in both the summer and winter situations and conclude that backscatter data may be used to complement information gained from net catches. For instance, very little krill was caught in the upper layers, or indeed in the lower layers, during night-time in the summer, but the ADCP did show higher backscatter from the upper layers at this time, indicating that krill did occur there and may have dispersed horizontally.

Data collected by ADCPs allow the vertical velocity of backscattering particles to be measured directly. Assuming there is no significant upwelling or downwelling, this can provide estimates of vertical migration speeds that can be compared with data from net catches and the movement of backscattering bands over time. The nature of krill distributions during this study made estimates of vertical migration speed from net catches and backscattering data difficult, because the krill did not ascend in discrete bands but dispersed during the night. Therefore, the vertical velocity component derived from the ADCP is the most valid method of estimating vertical migration speeds in this instance. Figure 9 shows vertical velocities observed in the layer of highest krill densities (80 m) in summer. Although there is variability in the measurements, a higher proportion of upward velocities were observed between 23.00 and 02.00 h whilst a greater proportion of downward velocities occurred during daytime. Since *Meganyctiphanes* is one of the main backscattterers (Figures 6 and 7), it is plausible that this species was a major contributor to this diurnal pattern. Upward velocities averaged around 2 cm s^{-1} whilst downward velocities averaged around 1 cm s^{-1}. These velocities are slightly below estimates from other studies of the vertical velocity of *Meganyctiphanes*, both from ADCP data (Buchholz et al., 1995 – 3 cm s^{-1}, Cochrane et al., 1994 – 2.8 cm s^{-1}, Cochrane et al., 1991 – 4 cm s^{-1}, Heywood, 1996 – 2 to 6 cm s^{-1}) and from observations made in the laboratory (Hardy & Bainbridge, 1954 – 2.5 cm s^{-1} over a 100 m interval upwards, 3.6 cm s^{-1} downwards over 130 m). The ADCP averages any detected movement over 6-min periods in order to reduce the otherwise large standard deviation; thus the speed of migrating organisms will be combined with those of stationary particles within a particular depth/time bin. Estimates of vertical migration become more accurate as the ratio of moving to stationary particles becomes greater. In this instance, densities of krill were relatively low, so estimates of the speed of vertical migration are probably underestimated.

The winter-time observation that a number of krill remained in the Alkor Deep during nighttime suggests one of two things: either krill may ascend and descend for brief periods every night and different individuals ascend at different times, or a fraction of the population does not ascend every night. The first scenario is feasible given that krill can ascend at a rate of 2 cm s^{-1} at least (see above) equivalent to 72 m in one hour. Furthermore, Gibbons (1992) has shown by analysis of stomach contents that asynchronous migration does occur in *Euphausia lucens*. However, this behaviour has yet to be demonstrated in *M. norvegica*. In the Skagerrak, Bergström & Strömberg (1996) found high numbers of *Meganyctiphanes* remaining at depth at night, suggesting that the second scenario applied in that case. Assuming this can also occur in the Kattegat, one could predict the frequency of ascent to the upper levels over a period of several days, by comparing the relative densities of the day and night populations within the basin. For example, during winter, when catches were larger, the ratio of day to night densities was 3 to 1, suggesting that one third of the population remained at depth every night and that perhaps any one individual made an ascent on two out of every three days. To gain a more accurate estimate, one would need to consider the variability in day and night densities over an extended period.

Any differences in the frequency of ascent and the depth distribution of the various size classes may be related to the disproportionate increase in underwater weight with size, as described by Kils (1981) and illustrated in Table 1: a 25 mm krill only has to combat 50% of the underwater weight of a 30 mm krill and a 22 mm krill only 30.5%. Converting this into ATP requirements, a 25 mm krill would need to provide between 41 and 54% of that necessary for a 30 mm krill if it is to maintain a pelagic lifestyle and a 22 mm krill between 22 and 40%. These values represent the costs of hovering, not of vertical migration; in order to ascend, an individual must change its orientation so that a greater propulsive force goes into the vertical vector rather than the horizontal. This, in turn, would result in a decrease in the horizontal vector and hence a loss of speed. In order to maintain location within the Alkor Deep, therefore, both a change in angle and an increase in propulsive force would be necessary. The cost of these actions would also vary with size, both with respect to the resistance of a body facing a stream of water, which scales as L^2 and with respect to underwater weight, which scales as $\sim L^4$. It would not, therefore, be surprising to find that large individuals ascend less frequently than small ones.

Overall, it appears that krill are capable of withstanding the advective forces experienced within the Læsø Deep system and their activity does not necessitate raising their metabolic activity above a 'standard' rate. Behavioural changes with size would be expected from calculations made of the relative costs of swimming behaviour, such that larger individuals would migrate over a smaller vertical range and/or less frequently. Further analyses of the population dynamics and size-specific depth distributions are necessary to address this hypothesis.

Acknowledgements

We would like to thank the following people for their assistance in collecting net samples during the summer and winter missions: S. Bröhl, P. Schumann, K. Grau, M. Salomon, Dr D. Müller-Navarra, E. Alberssard, Dr J.-P. Labat, Dr S. Dallot, Dr A. Bedo, Dr J. Cuzin-Roudy, Dr P. Mayzaud, Dr P David and Dr O. Guerin. In particular, we are grateful for the efforts of S. Bröhl in collecting length-weight measurements and helping in respiration experiments. Finally, we would like to acknowledge the support and efficiency of the crew of the *R/V Heincke* during both missions. This project was funded by EU-MAST III (MAS3-CT-0013), The PEP-Programme.

References

Bergström, B. & J.-O. Strömberg, 1997. Behavioural differences in relation to pycnoclines during vertical migration of the euphausiids *Meganyctiphanes norvegica* (M. Sars) and *Thysanoessa raschii* (M. Sars). J. Plankton Res. 19: 255–261.

Boysen, E. & F. Buchholz, 1984. *Meganyctiphanes norvegica* in the Kattegat. Mar. Biol. 79: 195–207.

Buchholz, F. & E. Boysen-Ennen, 1988. *Meganyctiphanes norvegica* (Crustacea: Euphausiacea) in the Kattegat: studies on the horizontal distribution in relation to the hydrography and zooplankton. Ophelia 29: 71–82.

Buchholz, F., C. Buchholz, J. Reppin & J. Fisher, 1995. Diel vertical migrations of *Meganyctiphanes norvegica* in the Kattegat: Comparison of net catches and measurements with Acoustic Doppler Current Profilers. Helgolander wiss. Meeresunters. 49: 849–866.

Cochrane, N. A., D. Sameoto, A. W. Herman & J. Neilson, 1991. Multiple-frequency acoustic backscattering and zooplankton aggregations in the inner Scotian Shelf basins. Can. J. Fish. aquat. Sci. 48: 340–355.

Cochrane, N. A. & D. D. Sameoto, 1994. Temporal variability of euphausiid concentrations in a Nova Scotia shelf basin using a bottom-mounted acoustic Doppler current profiler. Mar. Ecol. Prog. Ser. 107: 55–66.

Einarsson, H., 1945. Euphausiacea I. Northern Atlantic Species. Dana Report 27: 1–184.

Gibbons, M. J., Vertical migration and feeding of *Euphausia lucens* at two 72 h stations in the southern Benguela upwelling region. Mar. Biol. 116: 257–268.

Hardy, A. C. & R. Bainbridge, 1954. Experimental observations on the vertical migrations of planktonic animals. J. mar. biol. Ass. U.K. 33: 409–448.

Heywood, K. J., 1996. Diel vertical migration of zooplankton in the Northeast Atlantic. J. Plankton Res. 18: 163–184.

Kils, U., 1981. Swimming behaviour, swimming performance and energy balance of Antarctic krill *Euphausia superba*. BIOMASS Scientific Series 3: 121 pp.

Lindley, J. A., 1982. Population dynamics and production of euphausiids III. *Meganyctiphanes norvegica* and *Nyctiphanes couchii* in the north Atlantic ocean and the North Sea. Mar. Biol. 66: 37–46.

Makarov, R. R. & C. J. Denys, 1980. Stages of sexual maturity of *Euphausia superba*. BIOMASS Scientific Series 11: 13 pp.

Mauchline, J. & L. R. Fisher, 1969. The biology of euphausiids. Adv. Mar. Biol. 7: 1–454.

Mauchline, J., 1980. The biology of mysids and euphausiids. Adv. Mar. Biol. 18: 1–681.

Nicol, S., 1986. Shape, size and density of daytime swarms of the euphausiid *Meganyctiphanes norvegica* in the Bay of Fundy. Mar. Ecol. Prog. Ser. 18: 241–251.

Pearcy, W. G., C. C. E. Hopkins, S. Grønvik & R. A. Evans, 1979. Feeding habits of cod, capelin and herring in Balsfjorden, northern Norway, July-August 1978: the importance of euphausiids. Sarsia 67: 269–277.

Pieper, R. E., 1979. Euphausiid distribution and biomass determined acoustically at 102 kHz. Deep Sea Res. 26: 687–702.

Pleuddemann, A. J. & R. Pinkel, 1989. Characterization of the patterns of diel migration using a Doppler sonar. Deep Sea Res. 36: 509–530.

Ulrich, V. J., 1983. Zur Bathymetrie und Topographie der nördlichen Kattegat-Rinne. Meeresforsch. 30: 61–68.

Hydrobiologia **375/376:** 343–351, 1998.
S. Baden, L. Pihl, R. Rosenberg, J.-O. Strömberg, I. Svane & P. Tiselius (eds),
Recruitment, Colonization and Physical–Chemical Forcing in Marine Biological Systems.

Vertical distribution, grazing and egg production of calanoid copepods during winter–spring in Gullmarsfjorden

Josefin Titelman* & Peter Tiselius
Kristineberg Marine Research Station, S-450 34 Fiskebäckskil, Sweden
* *Present address: Danish Institute for Fisheries Research, Department of Marine and Coastal Ecology, Kavalergården 6, DK-2920 Charlottenlund, Denmark*

Abstract

The vertical distribution of copepods in relation to their potential food was examined in Gullmarsfjorden, Sweden (58°15.6′ N, 11°27.2′ E). Plankton distributions were determined from bottle samples at 5 m intervals on four occasions; 30 January, 28 February, 7 March and 10 April 1996. Potential food was crudely divided into chlorophyll containing protists (generally phytoplankton) and ciliates. Weak correlations between copepods and phytoplankton were found in January and April, while there were no correlations during the diatom spring bloom (February-March) when phytoplankton were superfluous throughout the water column. Copepods and ciliates were never correlated. Ingestion as determined from gut fluorescence and egg production analyses suggested a higher degree of herbivory during the spring bloom than before and after. There was potential for copepod predation control of ciliate biomass throughout the study. Due to low copepod abundance in January through March estimated grazing pressure was not sufficient to control phytoplankton biomass until after the diatom spring bloom. Egg production rates were constant for *Centropages hamatus* ($\sim$ 20 eggs female^{-1} day^{-1}) on all dates but more variable (1–26 eggs female^{-1} day^{-1}) for the other species. *Temora longicornis* and *Acartia* sp. both revealed their lowest fecundity during the bloom.

Introduction

Distributions of copepods and their potential food are horizontally and vertically patchy at varying scales. Vertical patchiness occurs on all scales, from the smallest caused by minute eddies and individual behaviours (10 cm–1 m, Owen, 1989; Davis et al., 1992; Tiselius et al., 1994) to slightly larger due to wind mixing and Langmuir cells (1–10 m, e.g. Davis et al., 1992) to the largest associated with internal waves and density structure (10–100 m, e.g. Haury, 1976). Patch formation and dissipation of both copepods and their prey are governed by a range of physical factors, such as currents, turbulence (Rothschild & Osborn, 1988; Davis et al., 1991) and light (Stearns & Forward, 1984), and by biological or behavioral factors such as predator evasion (Bollens & Stearns, 1992) and food abundance (Uchima & Hirano, 1988).

In such a variable environment the average food concentrations are frequently too low to support maximum copepod growth (Kiørboe, 1991). Thus, adaptations to respond quickly to evanescent or thin patches of food and to remain in such patches are crucial for copepod survival in food limited environments (Dagg, 1977; Tiselius, 1992). Laboratory studies of zooplankton have revealed that animals use different swimming behaviors during food search and feeding. (e.g. Williamson, 1981; Tiselius, 1992). For example, the copepod *Acartia tonsa* keeps swimming in search for food when food is scarce, but switches behavior to more closed feeding bouts when it encounters a food patch of sufficiently high concentration. The trigger for this type of behavioral change probably depends on the degree of food limitation and may vary between species and developmental stages (Frost, 1985). The reduced motility which generally accompanies the change in behavior results in longer residence time in favorable food patches. If there was no other reason than food limitation for this type of behavior, we would expect zooplankton to accumulate in food patches. Deviations from a concurrent distribution of zooplankton and food could be interpreted as a predator avoidance behavior or a result of physical disturbances.

On a vertical scale of about 10 m or more it is well known that copepods aggregate at or above the chlorophyll maximum (Longhurst, 1976; Ortner et al., 1980). Small species that do not store food long and may be affected by even short periods of starvation can be expected to have the ability to locate and remain in food layers at even smaller vertical scales.

The aim of this study was to investigate if small calanoid copepods aggregate at the locations of their potential food. We examined this by high resolution vertical sampling in three different food environments with varying degrees of food limitation. Adaptations for survival in food limited environments often include an ability to switch between different feeding modes (Saiz & Kiørboe, 1995; Kiørboe et al., 1996) and food types which also ensures a nutritionally complete diet. Since copepods are omnivores (e.g. Cowles et al., 1988; Støttrup & Jensen, 1990; Stoecker & Egloff, 1987) the choice of proxy for food is not obvious. In this study we crudely separated the potential food into chlorophyll containing protists (generally phytoplankton) and ciliates and estimated their relative contribution to the diet by combining gut fluorescence and egg production data.

Methods

Sampling took place at the entrance of Gullmarsfjorden, Sweden (58°15.6′ N, 11°27.2′ E, water depth = 50 m) on 30 January, 28 February, 7 March and 10 April 1996. Salinity and temperature profiles were obtained with a ADM TERMI 120 portable CTD, and profiles of relative fluorescence with a Q-Instruments *in situ* fluorometer.

Plankton distribution

Water was collected with a 5 liter GO Flo bottle at 5, 10, 15 and 25 m depths. Sampling on 30 January consisted of 2 replicates from each depth and included samples from 20 m. On consecutive dates 3 replicates were collected. Only 5 and 10 m depths were sampled on 7 March.

Phytoplankton

Samples of ~ 100 ml of unfiltered seawater were preserved in 1% acid Lugol's and refrigerated until analysis. Subsamples of 50 ml were allowed to settle for 24 h in sedimentation chambers before phytoplankton were identified, counted and measured under an inverted microscope (Leica LEITZ DMIL). We considered the entire chamber area at 100 X for cells > 50 μm and a fraction corresponding to 1.23 ml at 400 X for cells < 50 μm.

Plasma volumes were calculated (Thomsen, 1992) and averaged from a minimum of 60 cells species^{-1}. For species found in lower numbers or with complicated geometrical shapes, plasma volumes from Edler (1977) and Thomsen (1992) were used. We applied carbon to plasma volume ratios of 0.13 pg C μm^{-3} for thecate dinoflagellates (Thomsen, 1992) and 0.11 for other phytoplankton (Strathmann, 1967). Only phytoplankton > 10 μm in diameter or length were analyzed (unless in chains), since this fraction has revealed a good representative of the phytoplankton food available to copepods (Kiørboe & Nielsen, 1994).

Ciliates

Samples of ~ 900 ml unfiltered seawater were preserved and stored as described for phytoplankton. Subsamples of 50 ml sedimented for 24 h before ciliates (50–250 sample^{-1}) were counted and measured under an inverted microscope (Leica LEITZ DMIL) considering the entire chamber area at 100 X. To get a sufficient number of ciliates for analysis from January 30, the entire sample was counted. Biovolumes were calculated assuming conical shapes with length to diameter ratios of 1.25 for ciliates < 50 μm and 2 for ciliates > 50 μm (Tiselius, 1989). A spheroid shape was used for *Mesodinium rubrum*. We applied a conversion factor of 0.071 pg C μm^{-3} (Fenchel & Finlay, 1983) to obtain biomass.

Copepods

About 4 l of seawater were screened on a 90 μm sieve and retained zooplankton were preserved in 4% formaldehyde. During the spring bloom (28 February and 7 March) samples were screened (250 μm) again to remove phytoplankton. Copepods were identified, counted and measured under a dissecting microscope (Leica Wild M8). Biomass was calculated using length-weight regressions in Bottrell & Robins (1984) (*Calanus finmarchicus*) and in Klein Breteler et al. (1982) (other copepods) and converted to carbon assuming 0.45 μgC μgdw^{-1} (Kiørboe et al., 1985). Vertical distribution of copepods was also examined by vertical tows with a WP-2 (90 μm) net between the pycnocline and surface and between the bottom and pycnocline. On 7 march a 250 μm mesh was used to avoid phytoplankton clogging. Samples were treated and preserved as described above. Subsamples of 3/1000–1/10 were analyzed considering > 200 individuals.

Egg production

Live animals for egg production and gut fluorescence analyses were caught in oblique tows of a 250 μm net in the upper 10 m. Measurements of egg production were generally conducted according to Kiørboe et al. (1985). One to 4 bottles, each with 1–3 females, were incubated per species. Females were incubated at 5 °C for 22–24 h in standing 620 ml glass bottles containing screened (63 μm) surface water from the site of zooplankton collection. Egg production was converted to amount of ingested carbon by assuming a gross growth efficiency of 0.33 and species specific egg carbon of 37, 50, 222, and 36 ngC egg^{-1} for *Centropages hamatus*, *Temora longicornis*, *C. finmarchicus*, and *Acartia* sp. respectively, estimated from data in Kiørboe et al. (1985). Species analyzed depended on their availability in the upper 10 m.

Gut fluorescence

Aliquots of 100–250 animals were concentrated on MG/C filters (47 mm Ø, 1.2 μm nominal pore size) and instantaneously frozen. On 7 March, circular pieces of 200 μm nylon mesh replaced the filters to reduce phytoplankton coating. Three replicates of 2–20 stage C5–C6 copepodids were analyzed per species. After over night extraction in 90% acetone in darkness at 8 °C, chlorophyll *a* and phaeopigment *a* fluorescence were determined with a Turner Designs 10-AU fluorometer. We applied a carbon to pigment ratio of 40 (Kiørboe et al., 1985) and specific gut evacuation rates determined from regressions of gut evacuation rates and temperature (Dam & Peterson, 1988) in converting gut fluorescence to ingested carbon. In determining the evacuation rates we used the respective temperatures at 5 m (middle of layer where copepods were caught).

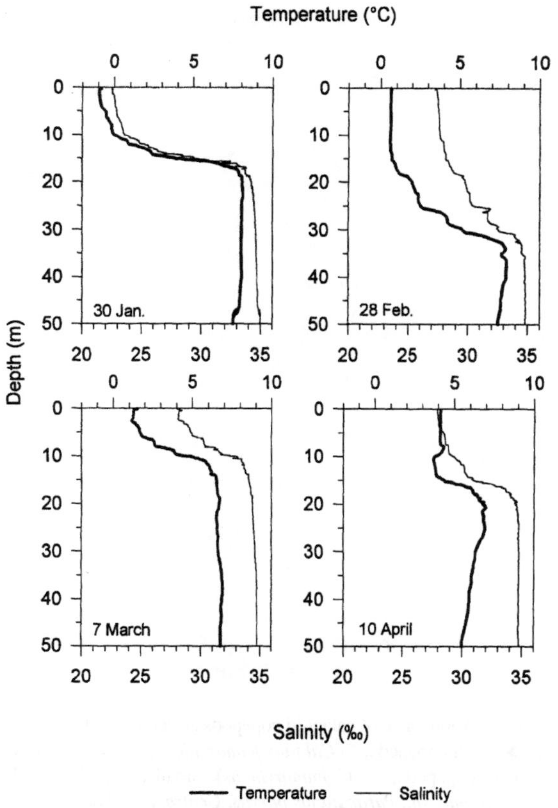

Figure 1. Vertical profiles of temperature and salinity.

Statistical analyses

Product-moment correlation coefficients for copepods and ciliates and for copepods and phytoplankton were calculated and tested for significance using a Student's t-test. Only bottles where copepods were found were included in correlations. *Mesodinium rubrum* was not included in estimates of ciliate biomass for correlations because its motility pattern protects it from copepod predation (Jonsson & Tiselius, 1990).

Results

Plankton distribution and biomass

Generally, copepods were most abundant above the pycnocline and not strongly associated to either pycnoclines or fluorescence peaks (Figures 1 & 2). Copepods dominated the total plankton biomass except during the diatom spring bloom (Figure 3). The diatoms *Thalassiosira nordenskioeldii* and several *Chaetoceros* species, including *C. laciniosus*, *C.*

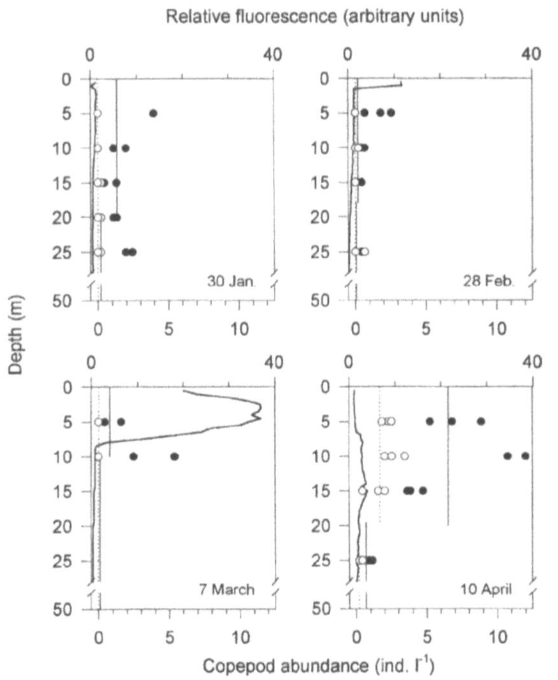

Figure 2. Abundance of calanoid copepods as determined from bottles (● small copepods; ○ *Calanus finmarchicus*) and vertical tows (— small copepods; ··· *C. finmarchicus*). Small copepods include *Pseudocalanus sp., Paracalanus parvus, Centropages hamatus, Microcalanus pusillus, Temora longicornis and Acartia* spp. Relative fluorescence (thick line) is included for comparison. Some data points are overlaid by others.

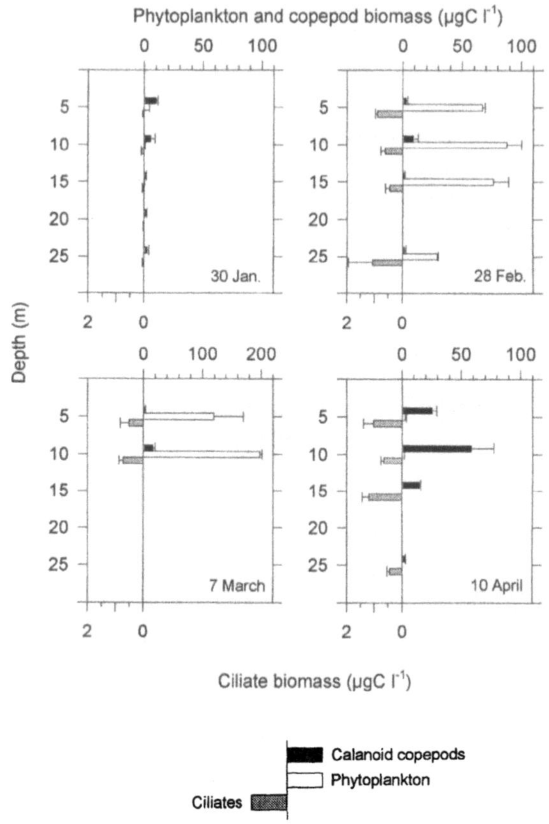

Figure 3. Biomass of plankton as determined from bottles. Note different scale for 7 March and reversed scale for ciliates. Error bars = SE.

debilis and *C. diadema*, dominated phytoplankton biomass during the bloom while *Dinophysis norvegica* dominated before and after the bloom.

Abundance and biomass determined from vertical tows generally agreed with those derived from the bottles regarding both *C. finmarchicus* and small copepods (Figure 2). However, on 28 February during the spring bloom, vertical tows above the pycnocline underestimated abundance of small calanoid copepods as compared with data from the bottles.

Pseudocalanus sp. dominated copepod abundance and biomass in the upper mixed layer on all sampling occasions (Table 1). *C. finmarchicus* was not present in January but gained importance for total biomass with time. *Paracalanus parvus*, *C. hamatus*, *T. longicornis, Acartia* spp. and *Microcalanus pusillus* appeared throughout the study and all but *M. pusillus* increased in abundance with time.

Correlations

Copepod abundance exhibited weak or no correlations with phytoplankton biomass (Figure 4). The strongest relationships were found in January when the correlation was significant ($r=0.86$, $p < 0.01$, $n=10$, Student's *t*-test) and in April when there was a tendency ($r=0.60$, $0.05 < p < 0.10$, $n=11$) for a correlation.

Egg production

Egg production of *C. hamatus* was around 20 eggs female^{-1} d^{-1} throughout the study (Table 2). Egg production of the other small copepods was more variable and ranged between 1–11 eggs female^{-1} d^{-1}. Both *T. longicornis* and *Acartia* sp. revealed their lowest fecundity during the bloom. Females of *C. finmarchicus* were present in the upper mixed layer only during the bloom with a production of ~ 25 eggs female^{-1} d^{-1}. Egg sacs were present on *Pseudocalanus* sp. through-

Table 1. Abundance and biomass of calanoid copepods integrated from vertical net tows above and below the pycnocline. Actual depth intervals of tows are indicated in the tables. Biomasses are calculated based on length-weight regressions (see methods)

A. Copepod abundance.

Species	Abundance (ind. m^{-3})							
	30 January		28 February		7 March		10 April	
	0–20 m	20–50 m	0–18 m	18–50 m	0–10 m	10–50 m	0–20 m	19–50 m
Centropages hamatus	70	17	44	8	108	4	2664	119
Temora longicornis	30	0	2	0	4	0	133	42
Pseudocalanus sp.	800	37	89	25	548	75	2398	133
Paracalanus parvus	120	20	20	24	32	9	1332	400
Microcalanus pusillus	290	148	11	9	56	24	200	0
Acartia spp.	20	1	7	1	40	2	1332	14
Calanus finmarchicus	0	0	18	1	2	2	1665	204
Total	1330	223	191	68	790	116	8192	912

B. Copepod biomass.

Species	Biomass (μgC m^{-3})							
	30 January		28 February		7 March		10 April	
	0–20 m	20–50 m	0–18 m	18–50 m	0–10 m	10–50 m	0–20 m	19–50 m
Centropages hamatus	373	25	90	27	142	6	6502	186
Temora longicornis	21	0	17	0	47	0	1016	311
Pseudocalanus sp.	1977	59	348	130	2413	296	9552	589
Paracalanus parvus	156	33	30	50	749	21	2971	817
Microcalanus pusillus	363	83	15	11	81	38	920	0
Acartia spp.	54	3	17	1	0	2	1854	11
Calanus finmarchicus	0	0	470	51	44	6	6273	487
Total	2945	203	987	271	2128	368	29087	2402

out the study, but incubations in January and February of females with egg sacs showed no production.

Gut fluorescence

Total gut pigment content in the small copepods increased from 0.2–0.3 ng ind.$^{-1}$ on 30 January to 0.2–0.7 ng ind.$^{-1}$ after the spring bloom (Table 2). Gut pigment content of *C. finmarchicus* showed highs during the spring bloom followed by a decline after the bloom. Gut fluorescence of *Acartia* sp. was measured only after the bloom and revealed very low pigment concentrations. During the bloom, phytoplankton smeared the filters during the collection of copepods. Despite the use of 200 μm screens, the phytoplankton interfered with the analysis as it was difficult to pick the small copepods without including surrounding phytoplankton. This problem was less severe for *C. finmarchicus*. Consequently, gut pigment content was unrealistic for small copepods during the bloom (Table 2), and therefore no ingestion rates based on gut fluorescence were calculated.

Discussion

Abundance and biomass determined from vertical tows generally agreed with those derived from the bottles regarding both *C. finmarchicus* and small copepods (Figure 2). Although a volume of only 5 l is small to estimate copepod abundance, the average of 2–3 bottles apparently gives reliable results. We feel confident that our sampling method was adequate to

Table 2. Total gut pigment (chlorophyll *a* and phaeopigment *a*) contents and egg production rates of the dominant copepods. Gut pigment contents are averages of 3 samples each of 1–20 individuals. Egg production rates are averages of 1–4 incubation bottles (n) with 1–3 adult females. – = no animals were found

Species	Date	Total gut pigment (ng ind.$^{-1}$ ± SE)	Egg production (eggs female^{-1} d^{-1} ± SE)	*n*
Centropages hamatus	30 January	0.26 ± 0.03	22.6 ± 10.1	2
	28 February	–	18.2 ± 4.9	4
	7 March	6.4 ± 1.3	21.0	1
	10 April	0.37 ± 0.13	19.6 ± 4.1	4
Temora longicornis	30 January	0.22 ± 0.08	2.8 ± 1.5	2
	28 February	11.5 ± 5.0	1.3 ± 0.3	3
	7 March	–	–	–
	10 April	0.72 ± 0.28	10.7 ± 4.2	4
Calanus finmarchicus	30 January	–	–	–
	28 February	24.3 ± 9.7	26.4 ± 5.9	4
	7 March	12.3 ± 2.3	24.0 ± 1.9	4
	10 April	0.70 ± 0.28	–	–
Acartia sp.	30 January	–	–	–
	28 February	–	–	–
	7 March	–	2.3	1
	10 April	0.048 ± 0.013	6.0 ± 1.7	3
Pseudocalanus sp.	30 January	0.20 ± 0.03	–	
	28 February	17.6 ± 2.6	–	
	7 March	4.4 ± 0.6	–	
	10 April	0.21 ± 0.11	–	

Table 3. Estimated carbon ingestion (μgC ind.$^{-1}$ d^{-1} ± SE) calculated from egg production (I_E) and from gut fluorescence (I_H) (For calculations see methods). The ratio of I_H/I_E is a crude measure of the relative degree of herbivory. Values < 1 suggest omnivory with falling numbers approaching total carnivory, while values ≥ 1 suggest herbivory (Peterson et al., 1990)

Species	Date	I_H	I_E	I_H/I_E
Centropages hamatus	30 January	0.161 ± 0.019	2.51 ± 1.11	0.06
	10 April	0.400 ± 0.141	2.18 ± 0.457	0.18
Temora longicornis	30 January	0.134 ± 0.047	0.421 ± 0.223	0.32
	10 April	0.784 ± 0.303	1.58 ± 0.620	0.50
Calanus finmarchicus	7 March	10.2 ± 1.88	16.0 ± 1.283	0.64
Acartia sp.	10 April	0.0518 ± 0.0079	0.648 ± 0.187	0.08

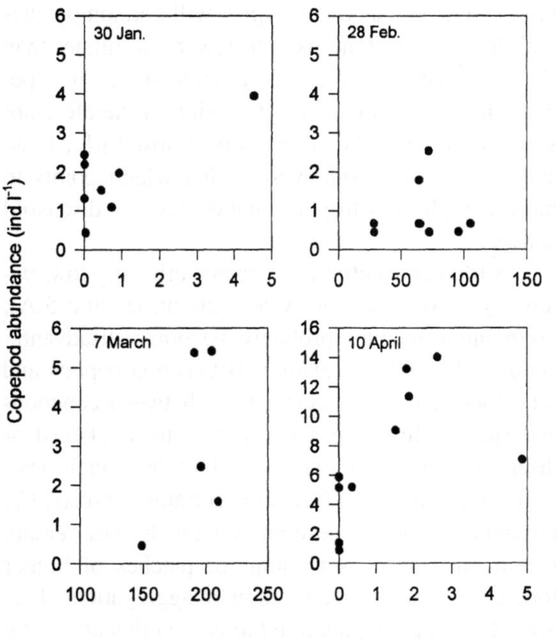

Figure 4. Copepod abundance versus phytoplankton biomass. Each dot represents one 5 l sample. Bottles with no copepods are not included in plots. Both scales vary between dates.

assess the distribution of the copepods in relation to their food.

There was only one significant correlation between copepods and food. However, tendencies for a relationship between copepods and phytoplankton were evident before and especially after the bloom (Figure 4). At these occasions, total food concentrations were generally low (Figure 3). The weak correlations between copepods and food during periods of food limitation can be interpreted in several ways. First, it is possible that the copepods may not have been feeding, or that they may have only been feeding on a portion of what we have defined as available food. In an attempt to tackle this issue, we estimated the degree of herbivory from data of egg production and gut fluorescence for January 30 and April 10 (Table 3). Following Peterson et al. (1990) we estimated carbon ingestion from egg production (I_E) and gut fluorescence (I_H). A comparison of I_H and I_E provides a rough estimate of the degree of herbivory, where values $I_H/I_E > 1$ suggest herbivory and values < 1 imply omnivory approaching carnivory in the lower end. Although a crude tool, the I_H/I_E ratio is a useful complement to the copepod v/s food correlations as the ratio hints towards what major food type is actually ingested. The I_H/I_E revealed that all small copepods were feeding omnivorously both before and after the bloom suggesting that food abundance was indeed limited. Food sources without chlorophyll were especially important for *Centropages hamatus* before the spring bloom and for *Acartia* sp. after the spring bloom.

Secondly, grazing or predation pressure may obscure our ability to detect patterns of aggregation by food. If ingestion is high in a food limited environment we would expect a negative correlation between grazers and their food. Using ingestion rates determined from egg production (Table 3) and biomasses of adult females from length-weight regressions (see methods), mean weight specific ingestion rates were calculated. Multiplying weight specific ingestion with total biomass (Table 1:B), assuming that all copepod stages have similar weight specific ingestion, yields a rough estimate of potential grazing and/or predation (from here on referred to as grazing) ability of the calanoid copepod community in the upper mixed layer. Potential copepod grazing was then compared with the standing stock of phytoplankton and ciliates. For the comparisons, we used the mean biomass of phytoplankton and ciliates (excluding *Mesodinium rubrum*) combined from bottles at 5 and 10 m (Figure 3).

On 30 January the potential copepod grazing of 0.4 μgC l^{-1} d^{-1} was not sufficient to consume the available food biomass of 1.9 μgC l^{-1} in a day because of low copepod abundance. Potential grazing ability during the spring bloom was similar to that on 30 January while the available food was 80–190 μgC l^{-1} suggesting no possibility of copepods to control phytoplankton biomass. On 10 April, copepod grazing potential had increased to 4.9 μgC l^{-1} d^{-1} while the available food concentration was 2.6 μgC l^{-1} connoting that copepods were abundant enough to consume the ciliate and phytoplankton biomass in a day. The low I_H/I_E (Table 3) suggests that predation on ciliates was high and that all or most of the ciliates present could have been consumed that day. Since specific growth rates of both phytoplankton (Banse, 1995) and ciliates (Nielsen & Kiørboe, 1994) are of order 1 d^{-1}, there was potential for control of both ciliates and phytoplankton in April. On the same assumptions, total copepod biomass was sufficient to control ciliate biomass but not phytoplankton biomass on all sampling dates. The indicated correlation between phytoplankton and copepods in April together with the lack of correlation with ciliates suggests a situation where copepods exert strong predation pressure on ciliates found in strata of higher phytoplankton abundance.

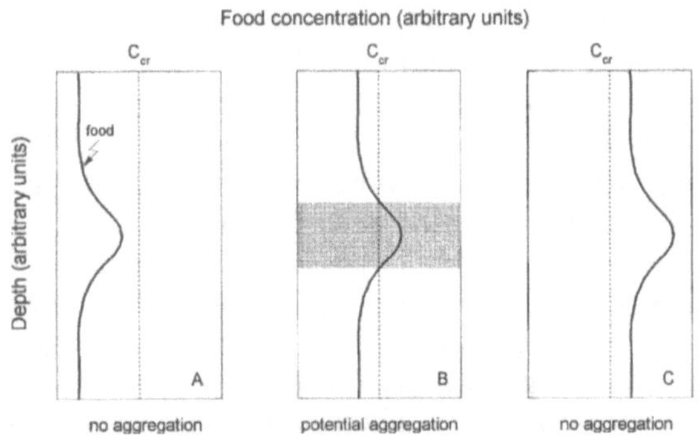

Figure 5. Model of threshold response hypothesis (modified after Tiselius et al., 1993) at (A) food limited, (B) moderately food limited, (C) non food limited conditions. Food concentration (solid line) varies with depth with a sub surface maximum. Food can be any type of food or a combination of food items. C_{cr} = Critical concentration (indicated by dotted line) when copepods switch behavior from food search to feeding. Threshold responses will lead to aggregations of copepods in food patches only during moderately food limited conditions where patches are within the copepods' migration range and patch concentration higher than C_{cr}. Aggregations in non food limited environments will be governed by factors other than food.

The above points towards a complicated interaction between biological interactions and physics for the distributions of plankton. Can we then make any predictions of the distributions of copepods and their prey in the field? Laboratory studies have revealed that copepods have one swimming behavior when searching for food and change behavior when they encounter a food patch of sufficiently high concentration (e.g. Williamson, 1981; Tiselius, 1992). The adaptive value of this behavior was examined in a model by Tiselius et al. (1993). In the model the zooplankter alters behavior at a critical concentration of food (C_{cr}) and reduces its motility when in food patches of sufficient concentrations. This model may help explain some of the patterns found in this study. Since the critical factor for the change in behavior is the threshold food concentration, we term this the threshold hypothesis.

Assuming that copepods behave similarly in the field as in the laboratory studies, copepod aggregation to food is expected only during periods of moderate food limitation (Figure 5). Theoretically, copepods will search for food patches of concentrations exceeding C_{cr} and after encounter with such a patch, switch behavior to feeding. Allowing sufficient time, such behavior would result in aggregations of copepods in patches of higher food abundance and thus, optimally, generate correlations between copepods and potential food items. A moderately food limited environment, such as on 10 April, will contain patches with food concentrations both lower and higher than C_{cr}. The tendency for a correlation between copepods and phytoplankton on 10 April may therefore not seem surprising. The comparatively low I_H/I_E, however, indicates an omnivorous diet which points to more complicated trophic relationships as discussed above.

As food concentrations become critically low, below C_{cr} throughout the water column (Figure 5:A), preferential feeding probably becomes disadvantageous. Although correlations between copepods and total food may exist, correlations between copepods and specific food types likely become weakened or disappear during completely food limited conditions.

Conversely, when food concentrations exceed C_{cr} throughout a water column, copepods will remain feeding instead of searching for patches of higher food concentrations resulting in no aggregations (Figure 5:C). As phytoplankton biomass builds up during bloom events ciliates become proportionally less significant, potentially causing a shift towards herbivory for all copepods (Gifford & Dagg, 1988; Fessenden & Cowles, 1994). The complete lack of correlations between copepods and food in combination with the larger degree of herbivory (i.e. I_H/I_E approaching 1) observed here during the spring bloom suggest unlimited food supply.

Although our field data are not sufficient, data from laboratory experiments (Tiselius, 1992) and computer simulations (Tiselius et al., 1993) support the threshold hypothesis. Because copepods vary their degree of omnivory and food preference depending on a multitude of factors, what is considered food and thereby C_{cr} may have to be redefined for each field situation. Food types will need to be crudely grouped together, as in this study. Our field data are not adequate, partly due to low copepod abundance in January through March, to fully test the threshold hypothesis. However, as suggested by the April data, the hypothesis remains appealing for explaining field data and could serve as a framework for designing future field investigations. Combining laboratory determined behavioral responses with *in situ* prey distributions may be a means of interpreting predator-prey relationships in the field.

Acknowledgements

We are grateful to the crew of R/V 'Oscar von Sydow' and Carina Johanson for help with field work. P.T. was supported by the Swedish Natural Sciences Research Council, grant B-AA/BU 09889-316.

References

Banse, K., 1995. Zooplankton: Pivotal role in the control of ocean production. ICES J. Mar. Sci. 52: 265–277.

Bollens, S. M. & D. E. Stearns, 1992. Predator induced changes in the diel feeding cycle of a planktonic copepod. J. exp. mar. Biol. Ecol. 156: 179–186.

Bottrell, H. H. & D. B. Robins, 1984. Seasonal variations in length, dry weight, carbon and nitrogen of *Calanus helgolandicus* from the Celtic Sea. Mar. Ecol. Prog. Ser. 14: 259–268.

Cowles, T. J., R. J. Olson & S. W. Chisholm, 1988. Food selection by copepods: discrimination on the basis of food quality. Mar. Biol. 100: 41–49.

Dagg, M. J., 1977. Some effects of patchy food environments on copepods. Limnol. Oceanogr. 22: 99–107.

Dam, H. G. & W. T. Peterson, 1988. The effect of temperature on the gut clearance rate constant of planktonic copepods. J. exp. mar. Biol. Ecol. 123: 1–14.

Davis, C. S., G. R. Flierl, P. H. Wiebe & P. J. S. Franks, 1991. Micro patchiness, turbulence and recruitment in plankton. J. Mar. Res. 49: 109–151.

Davis, C. S., S. M. Gallager & A. R. Solow, 1992. Microaggregations of oceanic plankton observed by towed video microscopy. Science 257: 230–232.

Edler, L., 1977. Phytoplankton and primary production in the Sound. Ph.D. dissertation, Göteborg University, Göteborg, Sweden.

Fenchel, T. & B. J. Finlay, 1983 Respiration rates in heterotrophic free-living protozoa. Microb. Ecol. 9: 99–122.

Fessenden, L. & T. J. Cowles, 1994. Copepod predation on phagotrophic ciliates in Oregon coastal waters. Mar. Ecol. Prog. Ser. 107: 103–111.

Frost, B. W., 1985. Food limitation of the planktonic marine copepods *Calanus pacificus* and *Pseudocalanus* sp. in a temperate fjord. Arch. Hydrobiol. (Beih.) Ergeb. Limnol. 21: 1–13.

Gifford, D. J. & M. J. Dagg, 1988. Feeding of the estuarine copepod *Acartia tonsa* Dana: Carnivory vs. herbivory in natural microplankton assemblages. Bull. Mar. Sci. 43: 458–468.

Haury, L. R., 1976. Small-scale patterns of a California zooplankton assemblage. Mar. Biol. 37: 135–157.

Jonsson, P. R. & P. Tiselius, 1990. Feeding behaviour, prey detection and capture efficiency of the copepod *Acartia* tonsa *feeding* on planktonic ciliates. Mar. Ecol. Prog. Ser. 60: 35–44.

Kiørboe, T., 1991. Pelagic fisheries and spatio-temporal variability in zooplankton productivity. Bull. Plankton Soc. Japan, Spec. Vol. 229–249.

Kiørboe, T., F. Møhlenberg & H. U. Riisgård, 1985. *In situ* feeding rates of planktonic copepods: A comparison of four methods. J. exp. mar. Biol. Ecol. 88: 67–81.

Kiørboe, T. & T. G. Nielsen, 1994. Regulation of zooplankton biomass and production in a temperate coastal ecosystem. 1. Copepods. Limnol. Oceanogr. 39: 493–507.

Kiørboe, T., E. Saiz & M. Viitasalo, 1996. Prey switching behaviour in the planktonic copepod *Acartia tonsa*. Mar. Ecol. Prog. Ser. 143: 66–75.

Klein Breteler, W. C. M., H. G. Fransz & S. R. Gonzalez, 1982. Growth and development of four calanoid copepod species under experimental and natural conditions. Neth. J. Sea Res. 6: 195–207.

Longhurst, A. R., 1976. Interactions between zooplankton and phytoplankton profiles in the eastern tropical Pacific Ocean. Deep-Sea Res. 23: 729–754.

Nielsen, T. G. & T. Kiørboe, 1994. Regulation of zooplankton biomass and production in a temperate, coastal ecosystem. 2. Ciliates. Limnol. Oceanogr. 39: 508–519.

Ortner, P. B., P. H. Wiebe & J. L. Cox, 1980. Relationships between oceanic epizooplankton distributions and the seasonal deep chlorophyll maximum in the Northwestern Atlantic Ocean. J. Mar. Res. 38: 507–531.

Owen, R. W., 1989. Microscale and finescale variations of small plankton in coastal and pelagic environments. J. Mar. Res. 47: 197–240.

Peterson, W. T., S. J. Painting & L. Hutchings, 1990. Diel variations in gut pigment content, diel vertical migration and estimates of grazing impact for copepods in the southern Benguela upwelling region in October 1987. J. Plankton Res. 12: 259–281.

Rothschild, B. J. & T. R. Osborn, 1988. Small-scale turbulence and plankton contact rates. J. Plankton Res. 10: 465–474.

Saiz, E. & T. Kiørboe. 1995. Predatory and suspension feeding of the copepod *Acartia tonsa* in turbulent environments. Mar. Ecol. Prog. Ser. 122: 147–158.

Stearns, D. E. & R. B. J. Forward, 1984. Copepod photobehavior in a simulated natural light environment and its relation to nocturnal vertical migration. Mar. Biol. 82: 91–100.

Stoecker, D. K. & D. A. Egloff, 1987. Predation by *Acartia tonsa* Dana on planktonic ciliates and rotifers. J. exp. mar. Biol. Ecol. 110: 53–68.

Støttrup, J. G. & J. Jensen, 1990. Influence of algal diet on feeding and egg production of *Acartia tonsa* Dana. J. exp. mar. Biol. Ecol. 141: 87–105.

Strathmann, R. R., 1967. Estimating the organic content of phytoplankton from cell volume or plasma volume. Limnol. Oceanogr. 12: 411–418.

Thomsen, H. A., 1992. Plankton i de indre danske farvande. Havforskning fra Miljøstyrelsen (11). Miljøstyrelsen, Denmark.

Tiselius, P., 1989. Contribution of aloricate ciliates to the diet of *Acartia clausi* and *Centropages hamatus* in coastal waters. Mar. Ecol. Prog. Ser. 56: 49–56.

Tiselius, P., 1992. Behavior of *Acartia tonsa* in patchy food environments. Limnol. Oceanogr. 37: 1640–1651.

Tiselius, P., P. J. Jonsson & P. G. Verity, 1993. A model evaluation of the impact of food patchiness on foraging strategy and predation risk in zooplankton. Bull. Mar. Sci. 53: 247–264.

Tiselius, P., G. Nielsen & T. G. Nielsen, 1994. Microscale patchiness of plankton within a sharp pycnocline. J. Plankton Res. 16: 543–554.

Uchima, M. & R. Hirano, 1988. Swimming behavior of the marine copepod *Oithona davisae*: internal control and search for environment. Mar. Biol. 99: 47–56.

Williamson, C. E., 1981. Foraging behavior of a freshwater copepod: Frequency changes in looping behavior at high and low prey densities. Oecologia 50: 332–336.

Hydrobiologia **375/376:** 353–361, 1998.
S. Baden, L. Pihl, R. Rosenberg, J.-O. Strömberg, I. Svane & P. Tiselius (eds),
Recruitment, Colonization and Physical–Chemical Forcing in Marine Biological Systems.

An in situ experiment to investigate the modification of particulate matter and urea above a benthic sandy silt community in the Baltic Sea

Laurenz Thomsen[1] & Sabine Jähmlich[2]
[1] *GEOMAR, Wischhofstr. 1–3, 24148 Kiel, Germany, Tel.: [+49] 431 600 2123; fax: [+49] 431 600 2928*
[2] *Universität Rostock, Freiligrathstr. 7–8, 18055 Rostock, Germany*

Abstract

During two cruises to the Baltic (Mecklenburg Bight) in September 1993 and November 1994 bottom water and sediment samples were taken from 5 stations on a 2.0 km long transect above a benthic sandy silt community. Profiles of total particulate matter, particulate organic carbon, chlorophyll equivalents and urea were taken in the benthic boundary layer (5–40 cm height above sea floor) on the downstream stations across an area occupied by macrofauna feeding at the sediment-water interface at 26 m water depth. Particulate matter concentration profiles varied under the two different flow conditions in September ($u_* = 0.7$ cm s^{-1}) and November ($u_* = 0.2$ cm s^{-1}). In September 1993 resuspension of total particulate matter (TPM) of 22 to 130 mg m^{-2} h^{-1} occurred while particulate organic carbon (POC) and chlorophyll (CPE) were deposited with a rate of 9 mg m^{-2} h^{-1} and 0.11 mg m^{-2} h^{-1} respectively at the station of highest macrofauna abundance. In November 1994 physical sedimentation and biological deposition of up to 388 mg m^{-2} h^{-1} TPM, 7.4 mg m^{-2} h^{-1} POC and 0.07 mg m^{-2} h^{-1} CPE occurred. Urea was released into the water column. Data suggest that in shallow water environments local sediment and benthic boundary layer characteristics prevent large scale calculations of fluxes of particulate matter.

Introduction

Recent attention to the ecological significance of the benthic boundary layer (BBL) shows that the near seabed region is characterised by strong vertical gradients of flow velocity and particle concentration (Nowell & Jumars, 1984; Sternberg et al., 1986). Benthic organisms often rely on advective sources for their food supply. They are dependent on the fluid medium for transport of this material and can in many cases influence these transport processes (Grant, 1983; LaBarbera, 1984; Eckmann & Nowell, 1984; Jumars & Nowell, 1984; Miller et al., 1984). The importance of biodeposition and bioerosion has been summarised by Graf and Rosenberg (1997). The data base obtained from studies on feeding behaviour of many macrobenthic species (Miller et al., 1992) show that the modification of particulate matter in the benthic boundary layer (BBL) can be more important as a food source of the benthos than material deposited by sedimentation.

Experiments on biodeposition at continental margin environments showed that under constant deep sea conditions fluxes of particulate matter to the sediment could be determined (Thomsen et al., 1995; Thomsen & Flach, 1997).

The aim of this study was to investigate the modification of particulate matter and urea above a benthic sandy silt community in a shallow water environment under changing environmental conditions. It was expected that these changes in macrofaunal composition, sediment -and flow conditions would locally alter the fluxes into and out of the sediments.

Methods

The study was carried out during two RV 'ALKOR' cruises in September 1993 and November 1994 to the Mecklenburg Bight (Western Baltic, Figure 1). The western Baltic is the southernmost part of the transition area between North Sea and Baltic Proper. The relatively strong stratified region is subject to

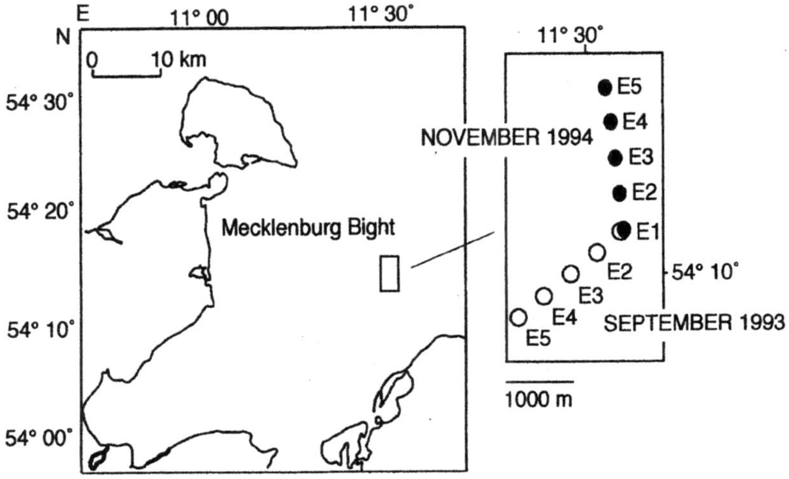

Figure 1. Location of stations of the *in-situ* experiments in Mecklenburg Bight, Baltic Sea.

three kinds of external forcing: local winds, barotropic pressure gradients (sea level differences) and baroclinic pressure gradients (freshwater outflow). The time scale of weather patterns is 2 to 4 days and therefore the circulation pattern is mostly governed by transient rather than by permanent currents. Westerly winds generate eddy patterns rather than coastal jets. The circulation of the western Baltic Sea is characterised in most cases by adjustment processes rather than by permanent flows (Fennel & Sturm, 1992).

For each experiment samples of bottom water and sediments were taken along 5 stations on a 2.0 km long transect starting at 54°13′ N, 11°33′ E at a water depth of 26 m and then extended down current.

Sampling

The BIOPROBE system, an in situ bottom water sampler (Thomsen et al., 1994) was used to collect 12 l of water at 5, 10, 20 and 40 cm height above the bottom [h.a.b]. Current velocity was measured by a thermistor flow meter at 40 cm height above the sea floor or by means of a particle camera (Thomsen et al., 1996). The flow direction was obtained by videopictures of the position of a current-vane relative to a compass-data logger. After deployment of the first station, the water masses were followed and the remaining stations were approached in downstream direction every 500 m.

Replicate samples of the sediment surface (0–10 cm depth) were taken with a multicorer (Barnett et al., 1984) modified to attach eight tubes of 10 cm in diameter and with a large Boxcorer (0.25 m^2). Macrofauna samples of the top 30 cm of sediment were sieved through 500 μm mesh and were preserved in borax-buffered 3% Formalin. Feeding types of macrofauna were determined according to Fauchald & Jumars (1979), Taghon & Greene (1992), Aller & Yingst (1985) and Levinton (1991).

The bottom water samples [5–40 cm h.a.b.] and sediment samples at each station were analysed for the following parameters: particulate organic carbon [POC] of water was measured with a Carlo-Erba -NA-1500 Analyser after a 3-hour HCl vapour treatment to dissolve carbonate. Chlorophyll equivalents (CPE) were analysed with a Turner Fluorometer by using the equations of Edler (1979). TPM (total particulate matter) was determined using the method of Bodungen et al. (1991). Urea and oxygen was measured according to Grasshoff et al. (1976). Statistics were evaluated using the statistical software 'Stat View 4.05' for Macintosh computers.

Flow velocity spot measurements during 15 minutes each were log-transformed and used to calculate the shear velocity (u_*) from the Karman-Prandtl log-profile relationship via $u_* = \Delta u / 5.75 \times \Delta \log z$ using a z_0 of 0.1 cm (Middleton & Southard, 1984). Turbulence intensity values (standard deviation of the instantaneous velocities, Middleton & Southard, 1984) from the flow measurements and weather data on wind speed and wave height were gathered from cruise log data to consider possible steady/unsteady flow regimes under wave generated fluid movements (Miller et al., 1992).

Results

Surface sediments

Although the flow direction and thus the experimental transect varied considerably between the two experiments, the sediment characteristics on both transects were similar. Sediments at the experimental stations at 26 m water depth consisted of silt (90% < 63 μm). During both cruises the CPE content in the sediments integrated over the top 10 cm was similar and varied between 510 and 570 mg m^{-2} 10 cm^{-1}.

Macrofaunal composition differed between the cruises and the stations. Figure 2 shows the different feeding types at the stations. In September 1993 the mean macrofauna abundance varied between 944 and 1440 ind. m^{-2} with a maximum value at station E3, 1000 m away from the leading edge of the experiment. Juvenile *Macoma baltica*, *Arctica islandica* and *Scoloplos armiger* were the dominant species but the distribution was extremely patchy and differed between the box corer samples at the stations by one order of magnitude. The percentage of interface feeders and filter feeders increased from 60% to 82% between station E1 and E3 with lowest values of 51% found at station E5 where deposit feeders accounted for 38% of the four feeding types (Figure 2).

A similar patchiness was found in November 1994, when macrofauna was less abundant and strongly decreased in flow direction from 824 to 144 to 72 ind m^{-2}. At all stations in November *Polydora ciliata* was the dominant species and accounted for 61 to 87% of total abundance, followed by *Arctica islandica* and *Mytilus edulis*, which accounted for 6 to 31% of the total abundance. Deposit feeders were almost absent and carnivores were rarely found (< 10%) at the stations (Figure 2).

Hydrographic conditions

During both experiments (8–10 hours each and 24 hours before) the weather conditions were calm with moderate wind and wave conditions. In September 1993 a persistent pycnocline occurred at 16 m water depth. Bottom water temperature was 11.5 °C, salinity was 23 ppt. 36 hours before and during the experiment wind speed increased from 2 to 5 m s^{-1} and the wave height was <1 m. In November 1994 no pycnocline could be detected. Bottom water temperature was 9 °C and salinity was 17.6 ppt. 36 hours before and during the experiment wind speed increased from 0 to maximal 6 m s^{-1} at the end of the bottom water sampling and wave height was 0 to <1 m. Bottom currents during both experiments were always directed in one direction.

Bottom water

In September 1993 average flow velocity at 40 cm height above the sea floor was 11 cm s^{-1} with an estimated bottom shear velocity of 0.7 cm s^{-1}. Turbulence intensities ranged between 5 and 10% during the 15 minutes of measurement at each station. TPM varied between 5.5 and 7.1 mg dm^{-3} and showed maximum values at 10 to 20 cm h.a.b. between stations E3 and E5 (Figure 3A–D). POC ranged between 0.07 and 0.12 mg dm^{-3} and showed minimum values where maximum TPM concentrations occurred (sta. E3). Highest values of POC were found between 20 and 40 cm h.a.b. and the average POC/PON ratio was 11. CPE was uniformly distributed in the bottom water, varied between 0.38 and 0.78 μg dm^{-3} and showed minimum values at station E3 (1000 m). Urea decreased into the water column and in flow direction from ≈ 0.38 μm between stations E2 and E3 to less than 0.1 μm between 30 and 40 cm h.a.b. Oxygen concentrations in the bottom water was < 20 μm, ie. less than 10% oxygen saturation.

In November 1994 average flow velocity at 40 cm height above the sea floor was 3.5 cm s^{-1} and the calculated u_* was 0.23 cm s^{-1}. Turbulence intensities ranged between 8 and 12% during the 15 minutes of measurement at each station. Under these flow conditions TPM decreased both into the lowermost 10 cm of the water column and in flow direction from 40 to < 20 mg dm^{-3} (Figure 3E–H). POC increased into the water from less than 0.1 to 0.3 mg dm^{-3} with lowest values of 0.05 mg dm^{-3} at station E5. The average POC/PON ratio was 7. Like TPM, CPE decreased in both the flow direction and into the water column from > 4 to less than 3 μg dm^{-3}. Urea was uniformly distributed up to station E3. At that station high amounts of urea dispersed into the water column and in flow direction from 0.6 to 0.4 μm. Oxygen concentration in the bottom water was around 220 μm at that time, i.e. 65% oxygen saturation.

Discussion

The flow conditions during the two experiments differed clearly, resulting in two different ways of particle transport. Generally the main factor that determines

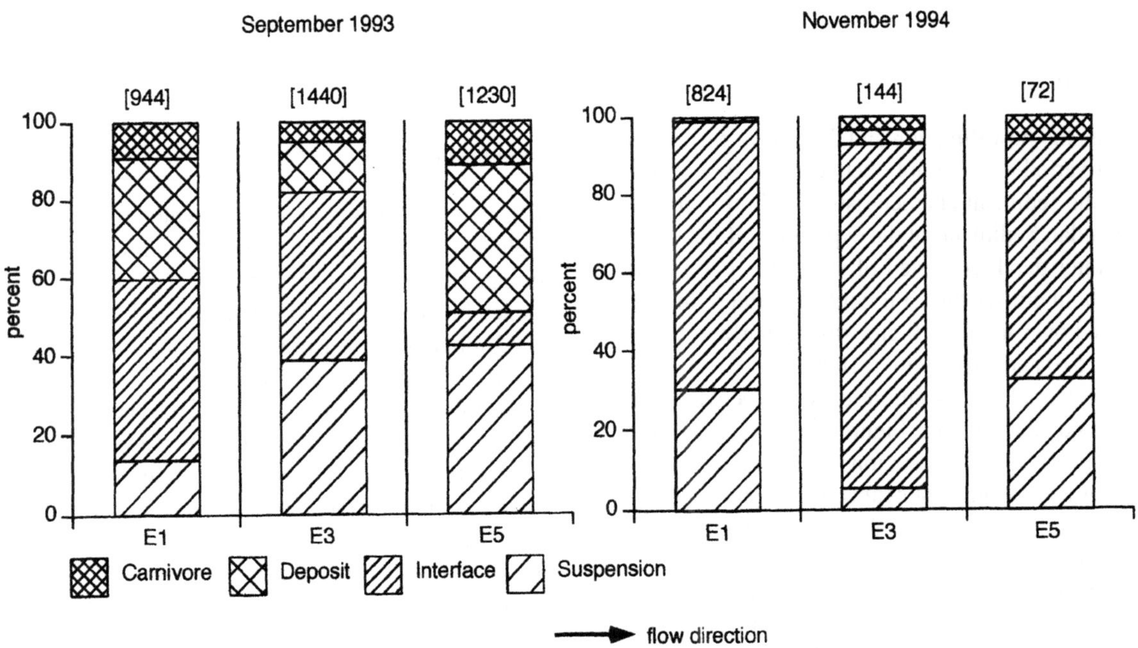

Figure 2. Macrofauna feeding type distributions [%] and abundance [in brackets] in sediments of the experimental stations.

the distribution of suspended sediment with height above the bed is the ratio of the settling velocity to the shear velocity (ws/κ u_*, where κ is the v. Karman constant 0.41 and ws is the settling velocity) called Rouse number [ps]. If ps is of order of 0.1 to 1, increased settling is expected with increasing particle concentrations towards the sea floor. If ps is $\ll 0.1$, material will stay in suspension (Middleton & Southard, 1984). Once resuspended, particles of ws < 0.64 u_* are transported further downcurrent (Allen, 1971). According to Stoke's Law (Middleton & Southard, 1984) the calculated settling velocities for single silty sediment particles at the experimental stations were in the order of 0.4 to 2×10^{-2} cm s^{-1}, those of aggregates measured in the BBL with a particle camera were ≥ 0.1 cm s^{-1} (Jähmlich, 1996). Thus fine sediments would show a uniformly distributed particle concentration profile in September 1993; whereas under the low flow conditions in November 1994 particles were concentrated towards the lowermost 10 cm, indicating particles of settling velocities of more than 2×10^{-2} cm s^{-1}. During both cruises, the low POC /TPM values indicate the 'resuspended' origin of the suspended carbon fraction. The shear velocity needed to erode the silty sediments found at the stations would be in the order of 1 cm s^{-1} (Thomsen & v. Weering, 1998). This could only be the case in the September 1993 study, if a different z_0 of 0.5 to 1 cm (instead of $z_0 = 0.1$ cm) was used for the log-transformed single velocity measurements. This would result in a u_* of approximately 1 cm s^{-1} (Middleton & Southard, 1984; Gust & Weatherly, 1985). However these high z_0 values should only be applied if increased roughness elements were found on the sediment surface. Video pictures of the sediment surface on the transects indeed revealed the existence of sediment disturbances of up to 20 cm height from bottom-trawling (Springer, pers. com.). Possible effects of wave generated fluid movements resulting in large oscillatory components and unsteady flow conditions can be neglected. Miller et al. (1992) showed that in a water depth of 30 m water, surface waves of 1 m height can generate a peak of u_* of 1.3 cm s^{-1} and Bock and Miller (1995) pointed out the influence of waves on the particle transport, even from the previous day. However both the data on weather conditions (36 hour data) and turbulence intensities do not reflect any influence of waves and steady/unsteady flow conditions during the time of experiments. Thus, beside the TPM profiles in September 1993 and November 1994 the determined organic particle distributions in the BBL at the experiments do not reflect the theoretical particle flow. Decreasing particle concentrations of POC and CPE during both cruises and of CPE at station E3 during the September cruise towards the sea floor, respectively, indicate

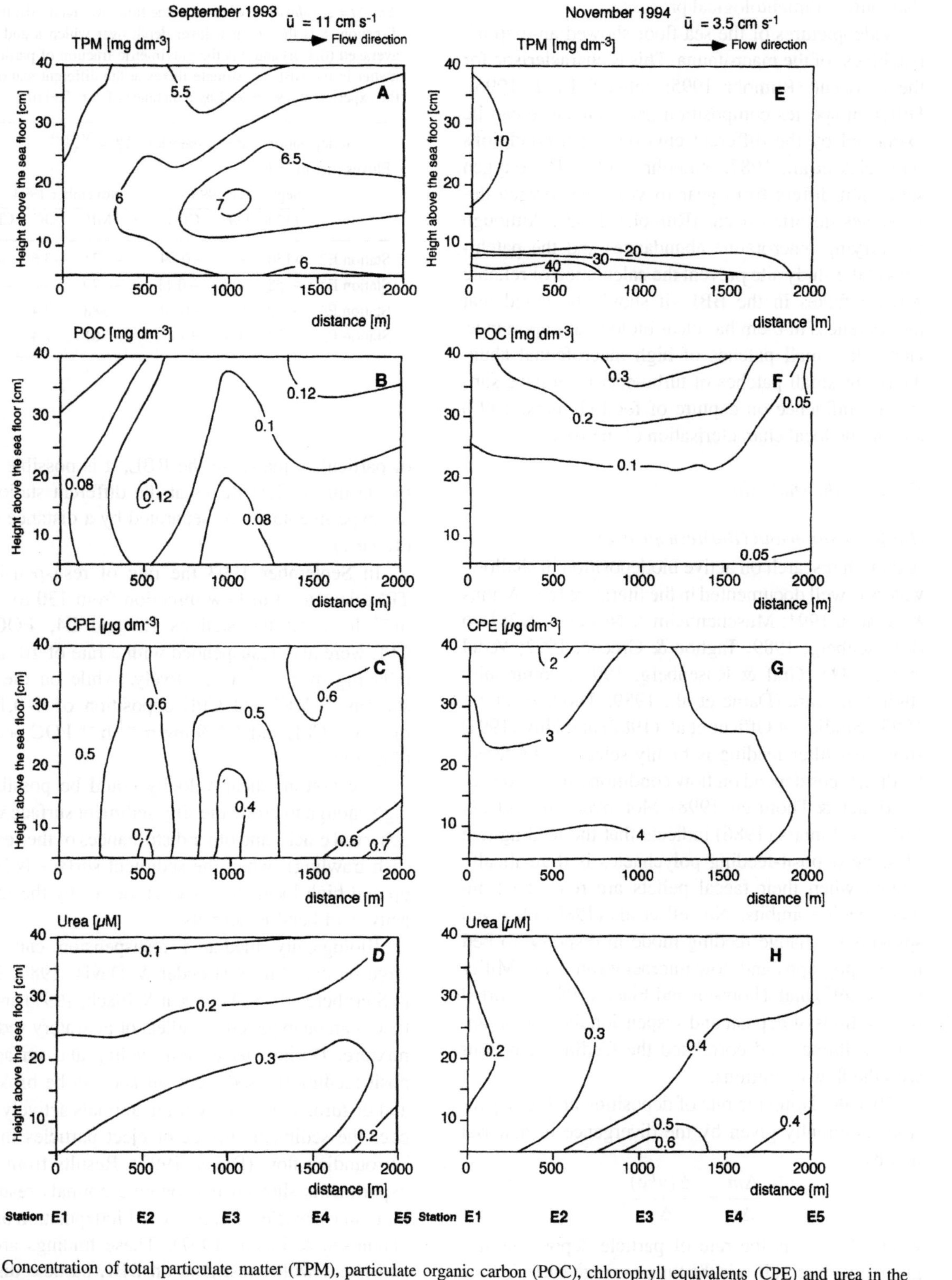

Figure 3. Concentration of total particulate matter (TPM), particulate organic carbon (POC), chlorophyll equivalents (CPE) and urea in the near-bottom water of the stations of the experiments.

that additional factors control the particle distributions than pure sedimentological processes.

Videopictures of the sea floor showed an extreme patchiness of the macrofauna. This is characteristic for the study site (Rumohr, 1995; Zeller & Blick, 1996). Different species composition and abundance can be explained by the different environmental conditions (Gosselck et al., 1987; Rumohr, 1993). The oxygen saturation differs from year to year and oxygen deficiencies do often occur (Rumohr, 1993). Although the varying macrofauna abundances and the patchiness at the study site prevent the calculation of reliable particle fluxes in the BBL, it should be noted that the spikeness in them has clear biological importance. Generally small patches of high macrofaunal abundance (or small patches of turbulence) can have substantial influence on capture of food (Jumars, 1993) and on the local characterisation of the BBL.

Table 1. Calculation of the rate of deposition and resuspension, nominally given by the divergence in material flux by $\Delta m/\Delta t = \Delta\,(uh\o)/\Delta x$, where u is the time-averaged drift in flow direction, h is the nominal layer depth over which u and $\o$ are averaged (0.4 m) and $\o$ is the gravimetric fraction of particulate matter in the BBL to estimate fluxes at the different stations of the experiments, separated by a distance of Δx (500 m)

Rate of deposition and resuspension $\frac{\Delta m}{\Delta t} = \frac{\Delta(uh\o)}{\Delta x}$
Fluxes [mg m^{-2} h^{-1}]

	September 1993			November 1994		
	TPM	POC	CPE	TMP	POC	CPE
Station E2	+130	+11	+0.09	− 21	−3.6	−0.05
Station E3	+ 58	− 9	−0.11	− 79	−3.6	−0.05
Station E4	+ 22	+10	+0.09	−388	−2.4	−0.05
Station E5	+ 22	−1.4	±0	−381	−7.4	−0.07

Fluxes at the study site

(Bio)deposition and (Bio)resuspension

Although research on active biodeposition in shallow-waters is well documented in the literature (e.g. Asmus & Asmus, 1991; Muschenheim & Newell, 1992; Loo & Rosenberg, 1989; Taghon & Greene, 1992; Vedel et al., 1994; Graf & Rosenberg, 1997), community studies are rare (Dame et al., 1989; Thomsen et al., 1995). Studies of Officer et al. (1982) and Hily (1991) show that filter feeding is highly selective. Different feeding types depend on flow conditions and food supply (Flach & Thomsen, 1998). Moreover, studies from Miller & Jumars (1986) indicate that the feeding rate of some deposit-feeding polychaetes is dramatically altered when their faecal pellets are removed from their feeding ambits. Nowell et al. (1989) observed spionids to change feeding mode in response to bed microtopography and flow microenvironments. Miller et al. (1992) and Thomsen and Flach (1997) reported observations of deposit and suspension feeding organisms in flumes and correlated the feeding behaviour with the flow conditions.

By calculating the rate of deposition and resuspension, nominally given by the divergence in material flux by

$$\frac{\Delta m}{\Delta t} = \frac{\Delta(uh\o)}{\Delta x},$$

where $\Delta m/\Delta t$ is the rate of particle deposition at a given time intervall, u is the time-averaged drift in flow direction, h is the nominal layer depth over which u and $\o$ are averaged and $\o$ is the gravimetric fraction of particulate matter in the BBL, it is possible to estimate the particle fluxes at the different stations on the experimental sites, separated by a distance of Δx (Table 1).

In September 1993 the rate of resuspension of TPM decreased in flow direction from 130 to 22 mg m^{-2} h^{-1}. At the stations E2 and E4, POC and CPE were also resuspended with a rate of 10–11 and 0.09 mg m^{-2} h^{-1} respectively, while on the other stations a POC and CPE deposition of 0–0.11 mg m^{-2} h^{-1} CPE and 1.4–9 mg m^{-2} h^{-1} POC occurred (Table 1).

The bottom shear velocity could be possibly be high enough to erode the silty sediment surface via the observed cracks and other disturbances of the sea floor (fish trawling), when the sediment surface is broken up and high local shear can erode; or by the feeding activity of benthic animals.

Biologically mediated resuspension can create large particle fluxes (Bender & Davis, 1984; Miller & Sternberg, 1988; Thomsen & Flach, 1997) and particles can be expelled as pellets or as watery sediment mixture. During suspension feeding and surface deposit feeding the sediment surface can be broken up and disturbed, especially if the animals actively move over the sediment surface or eject particles into the surrounding flow (Davis, 1993). Results from flume experiments showed that benthic animals resuspend inorganic particles during active biodeposition of POC (Thomsen & Flach, 1997). These findings are also consistent with latest data on BBL particle distributions at continental margins, where higher amounts of lithogenic matter were found in the BBL during times of enhanced benthic activity in summer (Thomsen & v. Weering, 1998).

In September 1993, minimum organic particle concentrations with maximum TPM concentrations were found around station E3. At that station, highest macrofauna abundance of 1440 ind m^{-2} was determined of which 39% were suspension feeders and 43% were surface deposit feeders. The dominant deposit feeders *Macoma baltica* and *Polidora ciliata,* however, are able to switch over to suspension feeding under distinct flow conditions and act as interface feeders (Michaelis, 1978; Levinton, 1991). Despite the erosion of TPM at that station POC and CPE were (bio)deposited and urea was released. Biodeposition of POC also occurred at station E5, where 1230 ind m^{-2} were found. The urea flux was lower but at that station interface feeders and suspension feeders penetrating into the BBL only accounted for 51% of total abundance. However, at all stations on the transect a correlation between faunal abundance, feeding types and particle concentrations seems evident but can not be properly statistically evaluated.

Erosion of all particle fractions at stations E2 and E4 suggest that pure physical erosion occurred, perhaps due to sediment disturbances form bottom trawling. However only limited data on bottom topography with sediments exposed to the flow can support these ideas.

Despite lower flow conditions in November 1994 the TPM and CPE concentrations in the BBL were 5 to 10 times higher (Figure 3). The CPE content in the bottom water was one order of magnitude higher suggesting that fresh phytodetritus arrived in the BBL (POC/PON ratio $\approx$ 7) and had not been bioturbated into the sediments. The CPE concentrations in the first 10 cm of sediments were similar to the September 1993 values and deposit feeders were almost absent. Macrofauna abundance was up to one order of magnitude lower in November 1994 but adult animals dominated. Pure physical resuspension of the silty sediments ($u_{*c} > 1$ cm s^{-1}) of organic/inorganic aggregates ($u_{*c} > 0.4$ cm s^{-1}, Thomsen & v. Weering, 1997) or via cracks of the sediment surface can be neglected under the low flow conditions, even when a z_0 of 0.5–1 cm was used to calculate u_*. Thus, bioresuspension of mainly inorganic particles of low settling velocities or a progressive physical sedimentation under low flow conditions of lateral advected particles are possible explanations. Within the high particle loads, particle deposition at all stations occurred, ranging between 21 to 388 mg m^{-2} h^{-1} for TPM, 2.4 to 7.4 mg m^{-2} h^{-1} for POC and 0.05 to 0.07 mg m^{-2} h^{-1} for CPE (Table 1). Within this progress mainly driven by physical sedimentation, the low POC values between 5 and 20 cm h.a.b. can not be explained by sedimentological processes and the reduction of CPE in flow direction indicate additional active or passive biodeposition, mainly by the interface feeder *Polydora ciliata*. Thus, a combination of both, physical sedimentation and selective biodeposition of settling particles can explain the modification of particulate matter during that experiment.

The calculated local particulate organic carbon deposition of 0.03 to 0.22 g m^{-2} d^{-1} are in the same order of magnitude as reported for mixed benthic populations in shallow waters with 0.2–0.23 g m^{-2} d^{-1} being deposited (Hily, 1991, Loo & Rosenberg, 1995) and at continental margin environments with a biodeposition of 0.01 to 0.04 g m^{-2} d^{-1} (Thomsen et al., 1995; Thomsen & Flach, 1997).

The calculated particle entrainment of 0.5 to 1.4 g m^{-2} d^{-1} (0.2 to 0.5 kg m^{-2} d^{-1}) dry sediment at station E3 and E5 in September 1993 are more than one order of magnitude lower that reported for other shallow water environments (Bender & Davis, 1984; Rowden & Jones, 1994) and could be the result of both physical and biological resuspension.

Effluxes of urea

Urea was dispersed into the water column during both cruises. In September 1993 highest effluxes originate at stations of highest abundances of interface- and suspension feeders. Due to the enhanced flow conditions in September 1993, urea diffused upward and showed lower concentrations in the water column than in November 1994, when macrofauna was less abundant and no direct link between macrofauna abundances, feeding types and urea efflux was evident. The extreme patchiness at that time might be responsible for this. An urea efflux into the water column can be estimated via the calculation of the vertical eddy diffusion coefficient K_v. K_v for the constant stress layer can be calculated via $K_v = \kappa\ u_*\ z$, where κ is the von Karman constant (0.41), and z is the height above the sea floor. The calculated K_v for the BBL [5–40 cm h.a.b.] during the September and November cruise are in the order of $0.03–5 \times 10^{-3}$ m^2 s^{-1} and $0.001–0.9 \times 10^{-3}$ m^2 s^{-1} respectively and would be reduced to 10^{-4} m^2 s^{-1} into the euphotic zone. For the case of turbulent eddy diffusion through the 26 m water layer at that station the approximate length of time the diffusion takes over the distance L [26 m] is calculated from L^2/K_v (Mann & Lazier, 1991). For both cases it would take too long (more than 10 days) for the urea to diffuse from the sediments into the upper water layers, even in September 1993, because a pycnocline with an expected K_v of 10^{-5} m^2 s^{-1} would further reduce the efflux. However, after storm events with K_v values of up to 10^{-2} m^2 s^{-1} (Mann & Lazier, 1991), when the mixed layer reaches down to the sediments, it would only take an hour or less for the urea to diffuse into the euphotic zone and support primary production. Thus beside a fast pelagic benthic coupling under turbulent flow conditions in shallow waters, when surface production becomes available to the benthos within a short period of time, this process can be reversed.

Conclusions

In the shallow waters environment of the Mecklenburg Bight with its changing macrofaunal compositions and flow conditions, local modifications in the transport of particulate matter occur. Contrary to deep water environments, these modifications only allow calculations of local fluxes within the BBL and reflect the high variability and importance of local processes. Small scale temporal variations due to changing weather conditions, the generation of waves and resulting unsteady flow conditions could also drastically influence any kind of mass balance calculations. In any case the study points to the importance of benthic activity for the processes in the BBL.

Acknowledgements

The authors wish to thank the crew of RV ALKOR for their help. We thank Michael Bock, Peter Tiselius and one anonymous reviewer for constructive criticism of the manuscript. This study was supported by the 'BMBF' and is a publication of GEOMAR and Rostock University.

References

Allen, J. R. L., 1971. A theoretical and experimental study of climbing-ripple cross lamination, with a field application to the Uppsoka esker. Geogr. Ann. 53A: 157–187.

Aller, R. & J. Yingst, 1985. Effects of the marine deposit-feeders *Herteromastus filiformis* Polychaeta, Macoma baltica Bivalvia and Tellina texana Bivalvia on averaged sedimentary solute transport, reaction times and microbial distributions. J. Mar. Res. 433: 615–645.

Asmus, R. M. & H. Asmus, 1991. Mussel beds: limiting or promoting phytoplankton J. exp. mar. Biol. Ecol. 148: 215–232.

Barnett, P. R. O., J. Watson & D. Conelly, 1984. A multiple corer for taking virtually undisturbed samples from, shelf, bathyal and abyssal sediments. Oceanol. Acta 7: 339–409.

Bender, K. & W. R. Davis, 1984. The Effect of Feeding by Yoldia Limatula on Bioturbation. Ophelia 23: 91–100.

Bock M. & D. C. Miller, 1995. Storm effects on particulate food resources on an intertidal sandflat. J. exp. mar. Biol. Ecol. 187: 81–101.

Bodungen, B. V., M. Wunsch & H. Fürderer, 1991. Sampling and analysis of suspended and sinking particles in the North Atlantic. Geophys. Monogr. 63: 47–56.

Dame, R. F., J. D. Spurrier & T. G. Wolaver, 1989. Carbon, nitrogen and phosphorus processing by an oyster reef. Mar. Ecol. Prog. Ser. 54: 249–256.

Davis, W. R., 1993. The role of bioturbation in sediment resuspension and its interaction with physical shearing. J. exp. mar. Biol. Ecol. 171: 187–200.

Eckman, J. E. & A. R. M. Nowell, 1984. Boundary skin friction and sediment transport about an animal tube mimic. Sedimentology 31: 851–862.

Edler, L., 1979. Recommendation methods for marine biological studies in the Baltic Sea – Phytoplankton and Chlorophyll. BMB Publications 5: 25–31.

Fauchald, K. & P. A. Jumars, 1979. The diet of worms: A study of polychaete feeding guilts. Oceanogr. Mar. Biol. Ann. Rev. 17: 193–284.

Fennel, W. & M. Sturm, 1992. Dynamics of the western Baltic. J. Mar. Syst. 3: 183–205.

Flach, E. & L. Thomsen, 1998. Do physical and chemical factors structure the macrobenthic community at a continental slope in the NE Atlantic. Hydrobiologia 375/376: 265–285.

Gosselck, F., F. Doerschel & T. Doerschel, 1987. Further developemnts of macrozoobenthos in Lübeck Bay, followimg recolonisation in 1980/81. Int. Rev. Ges. Hydrobiol. 725: 631–638.

Graf, G. & R. Rosenberg, 1997. Bioresuspension and Biodeposition – A Review. J. Mar. Systems 11: 269–278.

Grant, J., 1983. The relative magnitude of biological and physical sediment reworking in an intertidal community. J. Mar. Res. 41: 673–689.

Grasshoff, K., K. Ehrhard & K. Kremling, 1976. Methods of seawater analysis. Verlag Chemie, Weinh. 317 pp.

Gust, G. & G. L. Weatherly, 1985. Velocities, turbulence and skin friction in a deep-sea logarithmic layer. J. Geophys. Res. 90: 4779–4792.

Hily, C., 1991. Is the activity of benthic suspension feeders a factor controlling water quality in the Bay of Brest? Mar. Ecol. Prog. Ser. 69: 179–188.

Jähmlich, S., 1996. Untersuchungen zur Partikeldynamik in der Bodengrenzschicht der Mecklenburger Bucht, GEOMAR Reports, Universität Kiel. 85 pp.

Jumars, P. A. & A. R. M. Nowell, 1984. Effects of benthos on sediment transport: difficulties with functional grouping. Cont. Shelf Res. 3: 115–130.

Jumars, P. A., 1993. Concepts in biological oceanography, Oxford University Press. 348 pp. LaBarbera, M., 1984. Feeding currents and particle capture mechanisms in suspension feeding animals. Am. Zool. 24: 71–84.

Levinton, J. S., 1991. Variable feeding behaviour in three species of Macoma as a response to water flow and sediment transport. Mar. Biol. 110: 375–383.

Loo, L.-O. & R. Rosenberg, 1989. Bivalve suspension feeding dynamics and benthic-pelagic coupling in an eutrophicated marine bay. J. exp. mar. Biol. Ecol. 130: 253–276.

Mann, K. & J. Lazier, 1991. Dynamics of marine ecosystems. Blackwell Scientific Publications. 306 pp.

Michaelis, H., 1978. Zur Morphologie und Ökologie von Polydora cilliata und P. Ligni Polychaete, Spionodae. Helg. Wiss. Unters. 31: 102–116.

Middleton, G. V. & J. B. Southard, 1984. Mechanics of sediment movement. S.E.P.M. Short course Number 3, 2nd Edition, Tulsa, USA.

Miller, D. C., P. A. Jumars & A. R. M. Nowell, 1984. Effects of sediment transport on deposit feeding: scaling arguments. Limn. Oceanogr. 29: 1202–1217.

Miller, D. C. & P. A. Jumars, 1986. Pellet accumulation sediment supply and crowding as determinants of surface deposit-feeding rate in Pseudopolydora kempi japonica. J. exp. mar. Biol. Ecol. 99: 1–17.

Miller, D. C. & R. W. Sternberg, 1988. Field measurements of the fluid and sediment-dynamic environment of a benthic deposit feeder. J. Mar. Res. 46: 771–796.

Miller, D. C., M. J. Bock & E. J. Turner, 1992. Deposit and suspension feeding in oscillary flows and sediment fluxes. J. Mar. Res. 50: 489–520.

Muschenheim, D. K. & C. R. Newell, 1992. Utilization of seston over a mussel bed. Mar. Ecol. Prog. Ser. 85: 131–136.

Nowell, A. R. M. & P. A. Jumars, 1987. Flumes: theoretical and experimental considerations for simulation of benthic environments. Oceanogr. Mar. Bio. Ann. Rev. 25: 91–112.

Nowell, A. R. M., P. A. Jumars, R. L. Self & J. B. Southard, 1989. The effects of sediment transport and deposition on infauna: results obtained in a specifically designed flume. In S. R. Lopez, G. Taghon & J. Levinton (eds), Ecology of Marine Deposit Feeders. Springer Verlag: 215–228.

Officer, C. B, T. J. Smayda & R. Mann, 1982. Benthic filter feeding: a natural eutrophication control. Mar. Ecol. Prog. Ser. 2: 203–210.

Rowden, A. A. & M. B. Jones, 1994. A contribution to the biology of the burrowing mud shrimp, *Callianasssa subteranea*. J. Mar. Biol Assoc. U.K. 74: 623–635.

Rumohr, H., 1993. Erfahrungen und Ergebnisse aus 7 Jahren Benthosmonitoring in der südlichen Ostsee. In P. DF. J. Duinker (ed.), Das biologische Monitoring der Ostsee im Insitut für Meereskunde Kiel, 1985–1992. IFM Report, 240, Inst. F. Meereskunde, Kiel.

Rumohr, H., 1993. Monitoring the marine environment with imaging methods. Sci. Mar. 95: 129–138.

Sternberg, R. W., R. V. Johnson, D. A. Cacchione & D. E. Drake, 1986. An instrument system for monitoring and sampling suspended sediment in the benthic boundary layer. Mar. Geol. 71: 187–199.

Thagon, G. L. & R. R. Greene, 1992. Utilisation of deposited and suspended particulate matter by benthic interface feeders. Limnol. Oceanogr. 67: 1370–1391.

Thomsen, L., G. Graf, V. Martens & E. Steen, 1994. An instrument for sampling water from the benthic boundary layer. Cont. Shelf Res. 14: 871–882.

Thomsen, L., G. Graf, K. V. Juterzenka & U. Witte, 1995. An in situ experiment to investigate the depletion of seston above a suspension feeder field on the continental slope of the western Barents Sea. Mar. Ecol. Prog. Ser. 123: 295–300.

Thomsen, L., S. Jähmlich, G. Graf, M. Friedrichs, B. Springer & S. Wanner, 1996. An instrument for aggregate studies in the benthic boundary layer. Mar. Geol. 135: 153–157.

Thomsen, L. & E. Flach, 1997. Mesocosm observations of fluxes of particulate matter within the benthic boundary layer. J. Sea Res. 37: 67–79.

Thomsen, L. & T. C. v. Weering, 1998. Spatial and temporal variability of particulate matter in the benthic boundary layer at the North East Atlantic Continental Margin (Goban Spur). Prog. Oceanogr. (in press)

Turley, C., 1980. Distribution and biodegradation of urea in coastal waters. Dissertation, University of Wales.

Vedel, A., B. B. Andersen & T. Riisgård, 1994. Field investigations of pumping activity of the facultativ filter-feeding polychaete *Nereis diversicolor* using an improved infrared phototransducer system. Mar. Ecol. Progr. Ser. 103: 91–101.

Zettler, M. L. & A. Bick, 1996. The analysis of small- and mesoscale dispersion pattern *of Marenzelleria viridis* Polychaeta: Spionidae in acoastal water of the southern Baltic. Helgol. Meeresunters. 502: 265–286.

Hydrobiologia **375/376:** 363–367, 1998.
S. Baden, L. Pihl, R. Rosenberg, J.-O. Strömberg, I. Svane & P. Tiselius (eds),
Recruitment, Colonization and Physical–Chemical Forcing in Marine Biological Systems.

Small-scale plankton patchiness in the Black Sea euphotic layer

Yu. N. Tokarev[1], R. Williams[2]* & S. A. Piontkovski[1]
[1] *Institute of Biology of the Southern Seas, 335011 Sevastopol, Nakhimov av., 2, Ukraine*
[2] *Plymouth Marine Laboratory, Prospect Place, Plymouth PL1 3DH, United Kingdom*
* *Corresponding author: e-mail bw@wpo.nerc.ac.uk; fax (44) 01752 633101*

Key words: plankton, bioluminescence, acoustic field, temperature, patchiness

Abstract

Small-scale horizontal heterogeneites or patches of bioluminescence intensity and sound volume back scattering strength (SVBS) were studied in the central part of the Black Sea in October, 1989. Bathyphotometric casts and measurements of SVBS, were taken every 3 minutes in the surface layer (0–20 m) over a period of several hours from a drifting vessel. Casts were accompanied by temperature profiles and zooplankton sampling. Measurements were carried out during the day and night periods. From adaptive spectral analysis it was shown that the horizontal dimensions of the bioluminescent and SVBS heterogeneities were between 75 to 1500 m and 20 to 443 m respectively. In the case of zooplankton biomass the dimensions were 85 to 818 m and 80 to 1800 m for the temperature field. Physical and dynamical factors dominate in the formation of small-scale spatial heterogeneity of the studied fields. Biological factors, such as diel vertical migrations, trophic interactions etc. will also contribute to the formation of patches at night.

Introduction

Investigation of the factors influencing the spatial structure of pelagic communities has posed one of the major problems in modern marine biology. It must also be realised that plankton organisms can influence the physical properties of water masses (Tokarev, 1996) and that plankton is the most important biological product of the pelagic waters (Vinogradov et al., 1996).

In the various spatial-temporal scales of structure within plankton communities, from thousands of kilometers to several centimeters, it is the smaller scale and microscales, which are poorly understood. Various International meetings over the last decade (Marine Zooplankton Colloquium, 1989; Behavioural mechanisms of food selection, 1990, etc.) have emphasised the necessity to study these smaller scale ranges. This is because the heterogeneity of plankton distributions in the small- and microscale are thought to be important in determining population and ecosystem dynamics (Mackas & Boyd, 1979). Simulation models have shown that microscale patchiness of prey density can greatly enhance predator population growth. The half-maximum growth of planktonic Copepoda (Crustacea) is reached in several hours in patches of food with a dimension of approximately 10 meters (Davis et al., 1991). Field sampling and quantification of small-scale plankton variability, is still a difficult technical problem. One of the possible ways to estimate smaller scale variability is to apply biophysical methods to evaluate plankton communities (Bityukov et al., 1996; Pieper et al., 1990; Tokarev et al., 1995).

The main goals of our study were:
(1) to assess daily dynamics of the small-scale spatial structure or patches of the plankton community in the Black Sea euphotic layer, utilising a combination of two biophysical methods:

(a) measurements of the bioluminescent signals of the plankton community;

(b) measurements of the sound volume back scattering strength (SVBS);

(2) to investigate the role of temperature in the formation of the small-scale structure of the plankton community at different times of the day.

Materials and methods

A field survey was carried out in October at a drift station in the centre of the western circulation of the Black Sea (43°10′ N, 31°00′ E) during the 116th cruise of R/V 'Akademic A. Kovalevsky', 1989. Investigations were conducted in the form of two time series, each 4 h duration, in the day (10.00–14.00 h) and night (22.00–02.00 h). Determination of the speed and direction of drift of the vessel was an important element of the study. From the beginning of each experiment the drift was measured every 20 min in relation to a moored buoy. The vessel location was determined using the ship's navigation system. The average value determined over the experiments was 4.5 m min^{-1}, which was adopted as the ship's vessel drift and the speed of the horizontal transfer of the studied layer. From the literature (Sipilov & Altman, 1991) the flow speeds for this region of the Black Sea in the upper isothermal layer do not exceed the wind drift and agree with our measured wind drift.

The bioluminescence intensity was measured by means of vertical casts every 3 min with the bathyphotometer 'SALPA' (Vasilenko et al., 1997). This instrument has a dynamic band of 57 db. The sensitivity of the bioluminescence sensor was in a range from 10^{-13} to 10^{-8} W cm^{-2} l^{-1} with a band of spectral sensitivity from 160 to 600 nm. The repeat time of the bioluminescence, temperature and conductivity sensors was 180 ms. The bathyphotometer was deployed at a velocity of 1.0–1.2 m s^{-1}. The depth of the submersible gear was estimated by means of a pressure sensor DDV (precision 0.01 MPa). The absolute errors of the other sensors were 0.01 °C for temperature and 0.0025 mho m^{-1} for the conductivity channel.

'Plankton-2' (MHI-5102), an automatic hydroacoustic device was built at IBSS and has an echolocation frequency of 80 kHz. This equipment was used to obtain the spatial-temporal variability of SVBS (Sokolov & Tokarev, 1987). The range of sound pressure measurements was 0.15–10^5 Pa. A series of consecutive measurements were made at intervals of 52.5 s. The wavelength of 80 kHz frequency, although not ideal for investigating mesozooplankton, was chosen because we were initially studying patchiness of accumulations of plankton, which were represented by quite large forms such as *Mnemiopsis leidyi* and *Aurelia aurita*.

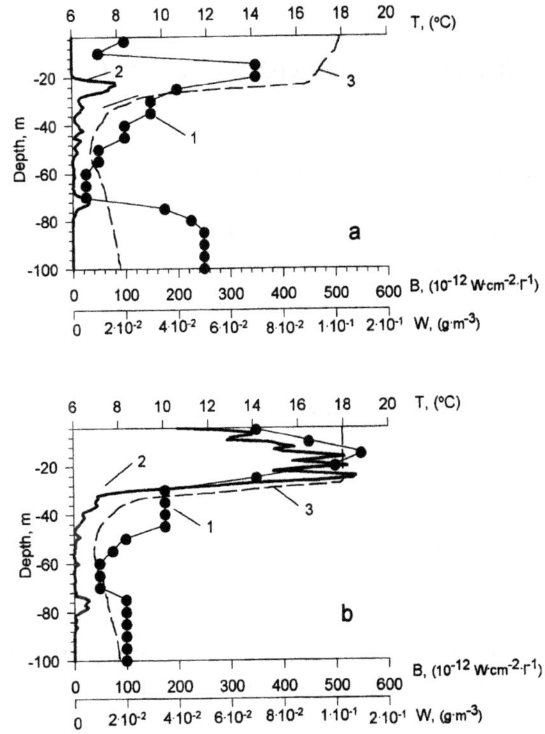

Figure 1. Mean profiles of vertical structure of plankton biomass (1), bioluminescence (2) and temperature (3) in the Black Sea photic layer during the day (a) and night (b).

Zooplankton sampling was carried out in the upper 20 m layer every 3 mins, using a small Juday Net (36 cm mouth diameter) with 112 μm mesh towed at 0.5 m s^{-1}, simultaneously with the bathyphotometer casts. At this time of the year the water in the upper 20 m is isothermal and lies above the thermocline. The total plankton biomass (wet weight) caught by net sampling was determined in the laboratory and some of the samples were processed to species level. A further series of net sampling was carried out to 100m with a Juday net fitted with an opening and closing device together with the bathyphotometer every hour over a 36 h period. This sampling was carried out either side of the two 4 h detailed 3 minute sampling in the upper 20 m. Mean vertical distributions of zooplankton, bioluminescence and temperature from this sampling programme are shown in Figure 1.

Autoregression spectral analysis (Marple, 1987) was applied to the time series data of bioluminescence intensity, temperature, SVBS and biomass of zooplankton samples. The peaks of spectral density obtained at the $n/2$ to $n/6$ order of the autoregression model and which statistically exceeded the noise variance were interpreted as plankton patches or spatial heterogeneities. It is known that the minimum correctly estimated temporal periodicity (and in our case the plankton patch) is pre-ordinated by the ratio $4t \leq P$, where 't' is the resolution of measurements, 'P' is the studied period. During the deployment of the over-side equipment the speed of the vessel drift was 4 to 5 m min^{-1}. This means that peaks of spectral density less than 10–12 mins were omitted.

Results

Analysis of the vertical profiles of temperature, bioluminescence intensity, and plankton showed a highly stratified structure in the Black Sea over a 24 h period, especially during the night (see Figure 1). The diel changes of bioluminescence were much greater than in the plankton biomass. The day/night differences of bioluminescence intensity in the upper layer (0–30 m) varied from approximately 8 to 50 times, while the plankton biomass only doubled (see Figure 1).

To quantify the horizontal plankton heterogeneity the upper 20 m layer was chosen for analysis, as this layer contained the majority of the plankton. The dimensions of the spatial heterogeneities of the bioluminescent field obtained in the Black Sea in day time varied from 76 to 345 m. In comparison, the patches of zooplankton biomass changed from 86 to 450 m, which was in good agreement with those of the temperature heterogeneities (Table 1). At night, the dimensions of the heterogeneities of the studied fields at the Black Sea were enhanced. The maximal dimensions were 1500 m, 818 m, 1800 and 443 m for bioluminescence, plankton, temperature and SVBS respectively. The number of the spectral density maxima for bioluminescence, plankton and SVBS increased 1.5 times, but decreased from 8 to 6 peaks for temperature (Table 1).

The larger SVBS patches in the day time (146 m) corresponded to the size of mesoplankton heterogeneities described in the literature (Smith et al., 1976; Omori & Hamner, 1982). Smaller aggregations in sound reverberation (20 m) corresponded to the assumed common sizes of macroplankton patches, which are not efficiently caught by Juday nets (Kovalev et al., 1993) and for this reason are not displayed in the spectral density of the plankton field.

From analysis of species in the zooplankton samples it can be seen that the general increase in zooplankton abundance and SVBS in the surface layer at night is accompanied by changes in species diversity. For instance, the copepod *Pseudocalanus elongatus* was not observed in the day time, while it was dominant (by the numbers) at night.

Discussion

The general agreement between the dimensions of spatial heterogeneities of the plankton biomass and temperature indirectly implies that the distribution of plankton was responding to the small-scale fluctuations in the temperature regime. Indeed, in some cases direct correlation between zooplankton abundance and temperature have been noted for the upper layers (Cassie, 1963). However, the dimensions of the heterogeneities of plankton biomass and bioluminescence intensity can differ, as was noted above. This might be due to the fact that only a section of the plankton community bioluminesce, and it is this part which is associated with diel rhythms in bioluminescence in the studied regions (Bityukov et al., 1993).

The increase of spatial heterogeneities at night could be based upon vertical migration and the behavioural mechanisms of the planktonic organisms forming the patches. Patches of organisms can be maintained by visual contacts, chemoreception and vibroreception between and among species (Newbury, 1972; Omori and Hamner, 1982).

Our observation that the number of spectral peaks of the bioluminescent, plankton and acoustic fields change during the day cycle, on a background of more stable statistical structure of the temperature field heterogeneity, also supports the idea that biological mechanisms can be powerful at certain times (at night for instance) in the formation of these heterogeneites. The biological mechanisms acting on the meso and microscales should interact with the other biological factors and mechanisms acting on larger scales. In the upper layers of the Black Sea the fluctuations in the diel rhythm of zooplankton abundance can be affected by the synchronised rhythms of predator/prey interactions and the reproduction rate of zooplankton populations (Kovalev, 1993; Zagorodnaya and Kovalev, 1988). In general, zooplankton reproduction rate acts as an ecologically important factor in the formation of spatial heterogeneity in the microscale (Owen, 1989).

Table 1. The typical size of patch (m) corresponding to the maximum of the spatial spectra of plankton biomass (*P*), bioluminescence (*B*), temperature (*T*) and SVBS (*A*) in the near surface layer (0–20 m) of the Black Sea

No. of max.	*P*		*B*		*T*		*A*	
	Day	Night	Day	Night	Day	Night	Day	Night
1	450	818	345	1500	450	1800	146	443
2	300	600	214	643	250	562	59	291
3	188	428	132	450	173	392	40	129
4	155	321	100	346	141	251	30	90
5	137	250	76	290	122	200	25	70
6	86	200	–	225	99	164	20	60
7	–	184	–	183	80	–	–	–
8	–	164	–	171	–	–	–	–

On the other hand, the general agreement between the bioluminescence, plankton and temperature spectral functions indirectly indicate that spatial-temporal variability of these fields can be influenced by other factors causing this similarity. Previous field studies of microscale variability of zooplankton abundance together with turbulent diffusion have indicated good agreement of spectral functions with the concentration of tracer dye and species abundance in the upper layers of the Black Sea (Nemirovski et al., 1991).

A number of maxima in the spectral analysis of the SVBS distribution in the surface layer remained constant throughout the 24 hr period (as shown in Table 1). A level of SVBS at 80 kHz frequency in the Black Sea is connected with the abundance of macroplankton populations (Medusae, Ctenophora) and fish larvae which do not extensively migrate out of the upper isothermal layer in daytime (Tokarev & Sokolov, 1993). There is a considerable increase (in 2–4 times) in sound reverberation in patch sizes at night, which can be connected (besides the increase in zooplankton abundance) with the change in spatial orientation of organisms. The results of studies of the SVBS oscillations of *Euphausia superba*, euphausiids and some fish demonstrate that the change of target strength of these organisms at night was connected with changes in their orientation during trophic interactions and in relation to light intensity (Inigo, 1982; Roe, 1983).

Conclusions

I. Physical factors dominate in the formation of small-scale spatial heterogeneity of the studied plankton fields. The general agreement obtained between the dimensions of the spatial heterogeneities of the plankton biomass and temperature indirectly implies that the plankton distributions were adapted during daylight hours to the small-scale fluctuations in the temperature regime.

II. Biological factors, such as diel vertical migrations, trophic interactions etc. will also contribute to the formation of patches of plankton biomass at night.

III. The method of high frequency profiling of bioluminescent and acoustic fields together with temperature profiles provide a fast technique in obtaining quantitative estimates of the zooplankton small-scale spatial-temporal variability.

Acknowledgements

The authors are grateful to E. P. Bityukov, B. Sokolov and V.Vasilenko for the technical assistance with the treatment of the data and useful comments. This work was supported by a grant from ONR N00014-95-1-0089 for which we are grateful.

References

Behavioral mechanisms of food selection. 1990. NATO ASI-Series, A-Life Sciences. Springer. Berlin.

Bityukov, E. P., P. V. Evstigneev & Yu. N. Tokarev, 1993. Bioluminescent Dinoflagellata in the Black Sea under anthropogenic impact. Hydrobiol. J. (Gidrobiologicheskii Jurnal) 30: 74–81 (in Russian).

Bityukov, E. P., V. I. Vasilenko, I. M. Serikova & Yu. N. Tokarev, 1996. Results and prospects of the bioluminescent investigations at the Black Sea. Ecology of the Sea (Ecologiya Morya) 46: 19–24 (in Russian).

Cassie, R. M., 1963. Microdistribution of plankton. In Barnes, H. (ed.), Oceanography and marine biology. George Allen & Unwin Ltd., London 1: 223–252.

Davis, C. S., G. R. Flierl, P. H. Wiebe & P. J. S. Franks, 1991. Micropatchiness, turbulence and recruitment in plankton. J. Mar. Res. 49: 109–151.

Inigo, E., 1982. Duirnal variations in mean volume backscattering strength of an Antarctic krill (Euphausia superba). J. Plankton Res. 4: 155–162.

Kovalev, A. V., 1993. Zooplankton. 1. Mesozooplankton. In A. V. Kovalev & Z. Z. Finenko (eds), Plankton of the Black Sea. Naukova Dumka, Kiev: 144–164 (in Russian).

Kovalev, A. V., V. V. Melnikov, N. A. Ostrovskaya & I. Yu. Prusova, 1993. Zooplankton. 3. Macroplankton. In A. V. Kovalev & Z. Z. Finenko (eds), Plankton of the Black Sea. Naukova Dumka, Kiev: 183–193 (in Russian).

Mackas, D. L. & C. M. Boyd, 1979. Spectral analysis of zooplankton spatial heterogeneity. Science 204: 62–64.

Marple, S. L., Jr., 1987. Digital spectral analysis with applications. Prentice-Hall Inc., New Jersey. 584 pp.

Nemirovskij, M. S., S. A. Piontkovskij, S. A. Seregin & A. N. Korchenko, 1991. Microdistribution of zooplankton: advective-diffusive processes and behaviour of individuals. In V. I. Kuzmicheva (ed.), Fisheries related studies of plankton. Part 1. Ocean and marginal seas. Nauka, Moscow: 100–111.

Newbury, T. K., 1972. Vibration perception by Chaetognatha. Nature 236: 459–460.

Omori, M. & W. M. Hamner, 1982. Patchy distribution of zooplankton: behavior, population assessment and sampling problems. Mar. Biol. 72: 193–200.

Owen, R. W., 1989. Microscale and finescale variations of small plankton in coastal and pelagic environments. J. Mar. Res. 47: 197–240.

Pieper, R. E., D. V. Holliday & G. S. Kleppel, 1990. Quantitative zooplankton distributions from multifrequency acoustics. J. Plankton Res. 12: 433–441.

Silopov, A. I. & E. N. Altman (eds), 1991. Hydrometeorology and hydrochemistry of the USSR Seas. V. 4. The Black Sea. N. 1. The hydrometeorological conditions. Gidrometeoizdat, Sakt-Petersburg. 357 pp.

Smith, L. R., C. B. Miller & R. L. Holton, 1976. Small-scale horizontal distribution of coastal copepods. J. exp. mar. Biol. Ecol. 23: 241–253.

Sokolov, B. G. & Yu. N. Tokarev, 1987. Investigation of the sound volume backscattering strength spatial-temporal heterogeneities at the upper productive layer of the Atlantic ocean tropical zone by means of the hydroacoustic complex MHI-5102. Marine Hydrophysical Institute. Sevastopol. Division of VINITI. No. 6825–87: 227–235 (in Russian).

Tokarev Yu. N., 1996. Plankton as a factor modifying the physical characteristics in the marine environment. Mar. Ecol. 17: 187–196.

Tokarev, Yu. N. & B. G. Sokolov, 1993. Hydroacouctic characteristics of active layer and its relationship with composition and abundance of the plankton. In A. V. Kovalev & Z. Z. Finenko (eds), Plankton of the Black Sea. Naukova Dumka, Kiev: 205–215 (in Russian).

Tokarev, Yu. N., E. P. Bityukov & B. G. Sokolov, 1995. Finescale spatial structure of the acoustic field of the upper productivity layer of the Atlantic ocean. Hydrobiological J. (Gidrobiologicheskii Jurnal) 31: 79–86 (in Russian).

Vasilenko, V. I., E. P. Bitykov, B. G. Sokolov & Yu. N. Tokarev, 1997. Hydrobiophysical device 'Salpa' of the Institute of Biology of the Southern Seas used for bioluminescent investigation of the upper layers of the ocean. In J. W. Hastings, L. J. Kricka & P. E. Stanley (eds), Bioluminescence and Chemiluminescence. Molecular Reporting with Photons. John Wiley & Sons, Chichester: 549–552.

Vinogradov, M. E., A. L. Vereshchaka & E. A. Shushkina, 1996. Vertical structure of the zooplankton communities in the oligotrophic areas of the Northern Atlantic, and influence of the hydrothermal vent. Oceanology (Okeanologiya) 36: 71–79.

Zagorodnyaya, Yu. A. & A. V. Kovalev, 1988. Daily rhythm of marine copepods reproduction and ontogenetic transmutations. Hydrobiol. J. (Gidrobiologicheskii Jurnal) 24: 20–23 (in Russian).

Hydrobiologia **375/376:** 369–380, 1998.
S. Baden, L. Pihl, R. Rosenberg, J.-O. Strömberg, I. Svane & P. Tiselius (eds),
Recruitment, Colonization and Physical–Chemical Forcing in Marine Biological Systems.

Pelagic–benthic coupling in the Bay of Fundy

D. J. Wildish[1] & G. B. J. Fader[2]
[1] *Fisheries and Oceans Canada, Marine Environmental Sciences, Biological Station, St. Andrews, NB, Canada E0G 2X0*
[2] *Geological Survey of Canada (Atlantic), Bedford Institute of Oceanography, Dartmouth, NS, Canada B2Y 4A2*

Key words: Bay of Fundy, pelagic–benthic coupling

Abstract

Bivalve reefs are important in pelagic–benthic coupling and, in order to assess these linkages quantitatively, it is necessary to obtain accurate maps of reef biomass and production. Presented here is a preliminary attempt to use modern surficial geological techniques, including sidescan sonar and a high resolution seismic reflection system to map the distribution of bivalve reefs in the Bay of Fundy. Acoustic techniques were groundtruthed with a 0.5-m^2 video grab. The grab jaws could be hydaulically closed from the deploying vessel on the basis of a video view of the sediment from a camera focused between the grab jaws. Sampling was limited to the upper Bay of Fundy where five geological provinces were recognized. Preliminary results show that horse mussels are limited to the harder, more stable sedimentary provinces (gravel/cobble and mottled gravel), but also to an intermediate province: sand with bioherms. The latter are raised features formed by the horse mussels on megarippled sand which are long and thin, and flow-parallel structures covered with epifauna. Although positive, the present data are insufficient to conclude whether acoustic methods can significantly improve on conventional benthic grab sampling methods to map benthic production.

Introduction

Bivalve reefs are considered to be universal features of continental shelf benthic environments where substrates, phytoplankton and water movement patterns converge at the same place and time (Wildish & Kristmanson, 1997). The ecological importance of bivalve reefs in pelagic-benthic coupling is well established, and published work shows that water movement transport of sestonic food is a requirement to sustain these highly productive reefs (e.g. Bernard, 1974; Dame et al., 1980, 1984; Cohen et al., 1984; Nichols, 1985; Emerson et al., 1986; Dame & Dankers, 1988; Asmus & Asmus, 1990).

We present a review of the knowledge relating specifically to pelagic–benthic coupling within the Bay of Fundy. This shows that three main deficiencies are present in work completed to date, and the main purpose here is to address one of them: insufficient benthic sampling coverage to map bivalve reef production. We determine whether modern acoustic, surficial geological mapping techniques enable benthic biologists to accurately map the distribution of the dominant horse mussel, *Modiolus modiolus*, reefs at the head of the Bay of Fundy (see Wildish et al., 1986; Wildish et al., 1998).

Physical description of the Bay of Fundy

The Bay of Fundy opens to the Gulf of Maine (Figure 1) and is situated at ~45 °N, having a cold, temperate climate. Seawater mean temperatures at the mouth of the Bay range seasonally from near 0 to 12 °C. The Gulf Stream warm currents run too far to the southeast to ameliorate local climate of the Bay. The system is a large, macrotidal estuary, the principal freshwater discharge being from the Saint John River, with an annual total discharge of 550 000 $m^3 \cdot s^{-1}$ in 1995. Many smaller rivers also discharge into the Bay. Salinities in the main part of the Bay range from 28–32 psu.

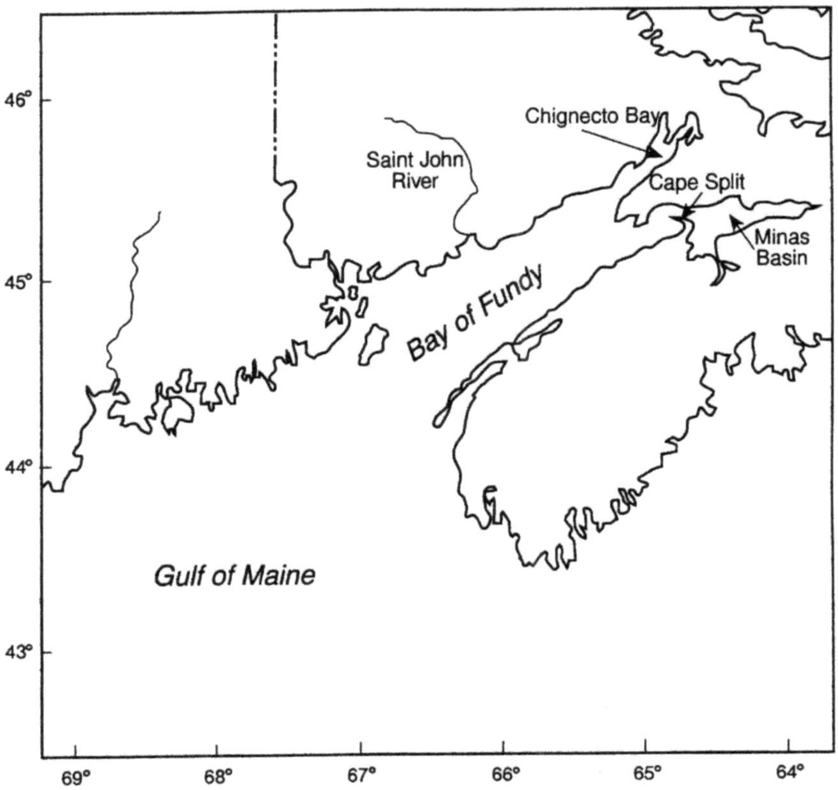

Figure 1. Map showing the Bay of Fundy/Gulf of Maine.

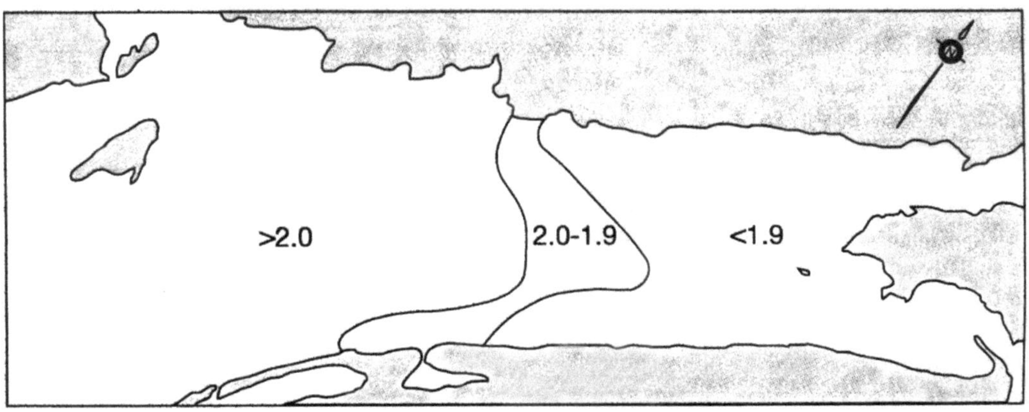

Figure 2. Calculated values of $\log_{10}$ (H/D) for the Bay of Fundy based on V_{av} from the Greenberg (1979) model.

The total area of the Bay is 11 149 km^2, inclusive of an intertidal area of 1437 km^2, and two upper arms, Chignecto Bay and Minas Basin, of 2010 km^2 combined (Wildish et al., 1986).

In the Bay of Fundy, depths exceed 200 m near the mouth on the southeast side. The tidal prism here is 8.3 m at maximum spring tides, whereas in Minas Basin, at the head of the Bay, it may exceed 16 m (Trites & Garrett, 1983). Depth-integrated tidal current speeds in 1978 for each of the grid squares in the finite element tidal model of Greenberg (1979) show that maximum tidal velocities, V_{max}, is an inverse function of depth within the Bay (Wildish & Peer, 1983). The range of V_{max} is $\sim$25–150 cm $\cdot$ s^{-1} for the grid squares of the 98 benthic stations worked in Wildish & Peer (1983), although $V_{max} > 400$ cm $\cdot$ s^{-1}

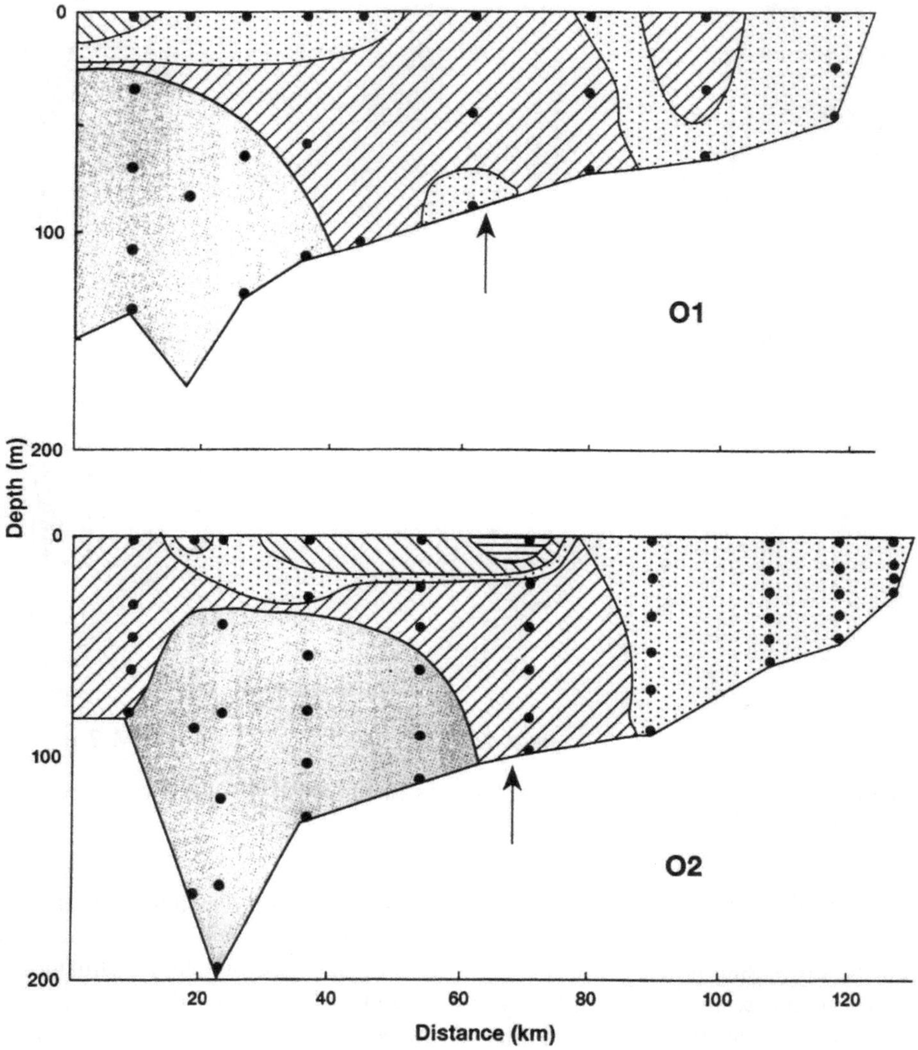

Figure 3. NE-SW profiles of Chl a during 22 August-2 September 1977 in the Bay of Fundy. Transect 01 (upper panel) is an offshore transect on the northwest side of the Bay, 02 is an offshore transect on the southeast side of the Bay. Shading categories: stippled, < 0.2; slash up, 0.2–1.0; dots, 1.0–2.5; slash down, 2.5–5.0; and horizontal lines, > 5.0 μg Chl $a \cdot L^{-1}$. Arrow indicates the position of the deepest horse mussel reef recognized by Wildish et al. (1986).

has been recorded at Cape Split in the Minas Basin (Wildish et al., 1986). Because of the strong tidal currents and rapid shallowing along the length of the Bay, it is likely that stratified seawater becomes well mixed near its head. We have calculated the parameter: $\log_{10}$ (*H/D*), for a number of stations in the Bay, where *H* is depth in m, *D* is tidal energy dissipation, derived from V_{av} from the Greenberg (1979) model. Garrett et al. (1978) consider that the transition between stratified and well mixed seawater occurs when $\log_{10}$ (*H/D*) = 1.9–2.0, with < 1.9 as well mixed and > 2.0 as stratified (see Figure 2).

As a consequence of the hydrodynamics, bed shear stresses of sediments are higher towards the head of the Bay, as the water column shallows. Fader et al. (1977) described the surficial geology of the Bay, using conventional methods, as consisting of three major types. Sambro Sand, located at the head of the Bay, is undergoing intensive tidal reworking, Scotian Shelf Drift represents the original till material available after the retreat of the last Ice Age, whilst the LaHave Clay is a net depositional sediment found in the deeper parts of the mouth of the Bay, and where tidal currents are at their lowest.

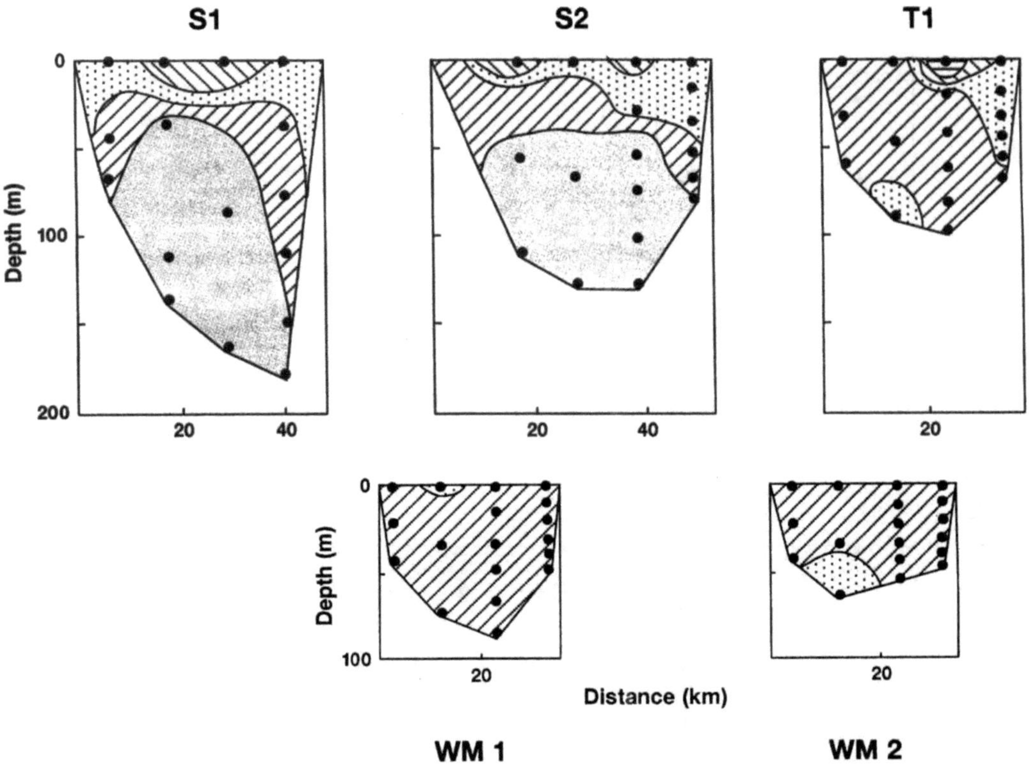

Figure 4. NW-SE profiles of Chl *a* (across the long axis) during 22 August-2 September 1977 in the Bay of Fundy. S = stratified, T = transitional and WM = well mixed water masses. Shading categorization as in Figure 3.

Plant pigments in seawater of the Bay

Published information regarding primary production or plant pigment distribution within the Bay of Fundy by Prouse et al. (1984), Daborn (1986), and Emerson et al. (1986) is too limited seasonally, geographically and with depth strata sampled to be of much use in pelagic-benthic coupling studies. However, some previously obtained data from fieldwork aboard CSS Dawson in 1977 is available (Wildish, Hargrave, Kranck & Milligan, unpublished). This work meets criteria for adequate geographic coverage and depth strata sampled, although no seasonal changes of plant pigments were studied since the fieldwork was all completed within an 11-d period beginning 22 August 1977.

The chlorophyll *a* (Chl *a*) profiles along two NE-SW offshore transects on the northwest side and southeast side of the Bay are shown in Figure 3. Each dot represents a sampling station taken at various tidal times and the arrow indicates the origin of the largest horse mussel reef area, as in Wildish & Peer (1983). Figure 3 clearly shows the stratified nature of the water column towards the mouth of the Bay, a transitional zone at depths down to 100 m where live phytoplankton, as indicated by Chl *a* distribution, reaches the bottom, and a well mixed zone in the upper Bay. This biological characterization accords quite well with the physical characterization shown in Figure 2 and suggests that downwelling is the mechanism of the downward transport of live phytoplankton. Transects across the Bay are shown in Figure 4, identified as occurring in stratified, transitional and well mixed water masses.

It is of interest that the sediments of the Bay associated with high biomasses or production of horse mussels occur only where the water column is transitional or well mixed.

Pelagic-benthic coupling

A partial carbon budget for the Bay of Fundy is presented in Wildish & Kristmanson (1997) based on the work of Emerson et al. (1986). In stratified, deeper water at the mouth of the Bay, the phytoplankton-copepod-herring food web is dominant with the eu-

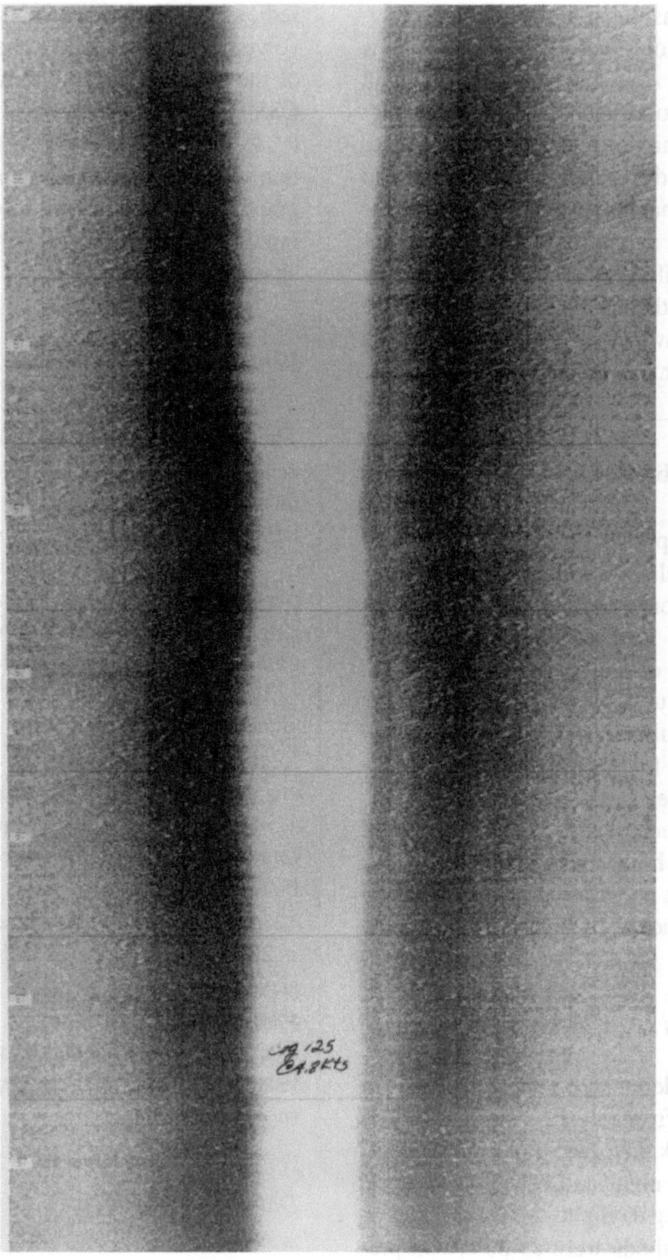

Figure 5. Simrad 300 kHz sidescan sonogram from the Bay of Fundy of the gravel/cobble province. Track distance shown is ~600 m. Horizontal distance apart of the lines is 15 m. Range ~75 m on either side of the unsurveyed portion, directly under the ship's hull. Same scale in Figure 5–8.

photic zone limited to the surface 20 m or so. Energy reaches the benthos in the LaHave Clay part of the Bay mainly as POM, although not all of it, 60 g $C \cdot m^{-2} \cdot yr^{-1}$, is accounted for by macrofaunal utilization or sediment burial rates (Emerson et al., 1986). It is proposed by Emerson et al. (1986) that the excess is tidally transported by excursion distances of 20 km to horse mussel reefs on shallower, harder substrates farther towards the head of the Bay of Fundy.

The euphotic zone becomes increasingly narrowed at the head of the Bay due to the increased load of inorganic and organic particulates carried down from the

Chignecto Bay and Minas Basin. Plant pigment analyses reviewed in the previous section show that Chl *a* is reduced in the euphotic zone, but present throughout the water column due to efficient tidal mixing, suggesting that live phytoplankton reach the bottom due to downwelling. This provides circumstantial evidence to support the hypothesis that horse mussels in the upper Bay may be trophically supported directly from primary production. Upwelling processes at some stages of the tide could return the nitrogen wastes of the bivalves to the surface seawater where its utilization by phytoplankton can occur.

Spatial distribution of bivalve reefs

We have previously reported on bioherms in sand which are horse mussel, *Modiolus modiolus*, reefs identified by sidescan sonar and high resolution seismic reflection systems, and confirmed visually by video camera mounted on an ROV (Wildish et al., 1998). In the present study, we have used the same surficial geological techniques but a new video grab to groundtruth the acoustic findings.

Our purpose was to determine whether modern surficial sedimentary characterization techniques could improve the mapping of horse mussel reefs in the Bay of Fundy. Such maps are necessary if an accurate estimate of secondary production by bivalve reefs is to be made.

Acoustic survey methods

All field work was undertaken on the first leg of a CSS Hudson mission from 16–22 June 1997. The collaborative field work involved surficial geological methods (Fader) at night and benthic biological methods (Wildish) during daylight.

Acoustic survey methods were as described previously (Wildish et al., 1998) with acoustic pulses sent and received from a tow fish. Two sidescan sonar systems were employed: a Bedford Institute of Oceanography (BIO) designed 70 kHz system, and a Simrad dual frequency system operated at 100 kHz and 300 kHz. The sonograms obtained reflected acoustic energy from the seabed on gray-scale thermal recorders and showed textural changes of the seabed as well as identified certain topographic features defined by an increase in reflected energy and associated acoustic shadows. These records were qualitatively analyzed each morning to identify surficial sedimentary provinces, although a full analysis involved digitization and further lab analysis. The range of the sidescan sonar systems during passage of the CSS Hudson could be varied from 50–1000 m.

In addition, sleeve gun and Huntec high resolution seismic reflection systems provided seabed topographic information, and subsurface sediment stratigraphy and structure, in particular, the presence or absence of bioherms.

Benthic sampling methods

The video grab used is described in Schwinghamer et al. (1996) and was developed specifically for the purpose of quantifying groundfish trawling impacts on benthic macrofaunal communities.

The 0.5-m^2 grab is supported on a large metal, support frame and is closed hydraulically from the mother ship by remote control. A video camera and a still camera were mounted on the grab frame. The video camera is focussed through the grab jaws and can be switched on and off by the operator. It can be used to choose a suitable site for sampling. The still camera was focussed on the substrate near the grab, from 1.5 m above it, and could take single exposures (to a total of 250). The shutter speed was set at 1/250 s and F stop of 11.5.

All grab samples with equal to, or more than, $\sim 25\%$ of the expected volume were quantitatively sieved on a mesh screen of 5×5 mm and all mussels, *Modiolus modiolus* or *Mytilus edulis* retained. The mussels were shucked in the field and both wet tissues and valves stored deep frozen for further study (not reported here).

Data interpretation

We offer a preliminary analysis of the acoustic data and a later publication will provide a detailed surficial geological interpretation, inclusive of cruise track details, which are omitted here. Field sampling was confined to the head of the Bay of Fundy where previous work suggested that horse mussels were abundant (Wildish and Peer, 1983; Wildish et al., 1986, Wildish et al., 1998).

A preliminary interpretation of the gray-scale sonograms obtained in the period up to 22 June 1997 suggested that five major surficial sedimentary provinces were present as follows:

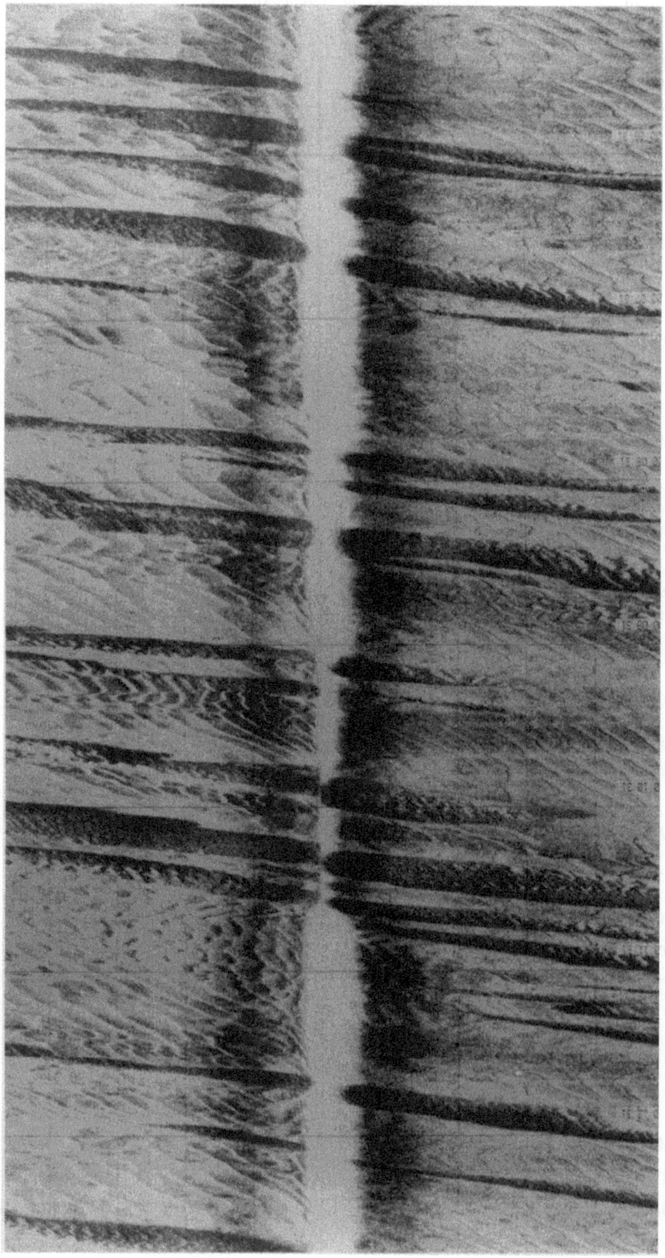

Figure 6. Simrad 300 kHz sidescan sonogram from the Bay of Fundy of the sand with bioherm province. The ship's track is at right angles to the flow-parallel bioherms or horse mussel reefs.

Gravel/cobble – A solid, dark, often featureless relief with a smooth or slightly cobbly surface, but lacking bioherms (Figure 5).

Mottled gravel – Similar to the above, but with patches where the dark background is broken up by lighter areas of irregular shape (not shown).

Sand with bioherms – A sandy substrate has a much lighter background with sand megaripples evident and darker strips running parallel to the prevailing tidal flow. The latter are raised features (bioherms) of up to 3 m in height from the seabed previously identified as horse mussel reefs (Figure 6).

Table 1. Quantitative video grab samples and the density of horse mussels in the Bay of Fundy.

Geological province	Grab number	Depth (m)	Coordinates North	West	Density (m^2)
Gravel/cobble	13	86	45°00.66′	65°30.66′	4
	14	86	45°00.81′	65°30.63′	48
	15	81	45°00.75′	65°30.68′	32
	16	79	45°00.72′	65°30.75′	50
	22	68	44°54.29′	65°38.49′	90
	23	70	44°54.37′	65°38.44′	76
	24	72	44°57.30′	65°30.59′	0
Mottled gravel	18	79	44°54.67′	65°36.88′	106*
	20	74	44°56.13′	65°33.76′	158
	21	74	44°56.06′	65°33.90′	98
Sand with bioherms	01	84	45°06.07′	65°15.58′	4
	03	88	45°05.62′	65°17.06′	78
	06	83	45°05.62′	65°17.07′	64
	08	84	45°06.13′	65°15.50′	18
	09	87	45°06.20′	65°15.44′	18
Rippled sand	17	77	45°00.68′	65°31.05′	0
	31	81	44°59.53′	65°30.30′	0
	32	76	45°59.81′	65°29.89′	0
	33	77	44°59.82′	65°29.87′	0
	34	59	45°08.51′	65°14.48′	0
Sand with comet marks	26	75	45°09.70′	65°31′48′	0

* Also 8 *Mytilus edulis* per m^2.

Rippled sand – Shows up as a light relief with ripple marks and darker strips running in the same direction as the tidal flow, but normal to the megaripple marks (Figure 7).

Sand with comet marks – A light relief background with dense spots, indicating cobbles around which dark shadows stretch either in one or both tidal directions (Figure 8). The shadows, or comet marks, are indications of downstream erosion with the relative length of each indicating the amount of residual flow in either direction.

Of 35 video grab deployments undertaken, 21 were successful, semi- or fully quantitative samples. Of the latter, we have classified where the samples were taken according to the surficial geological province in which they occur. The results (Table 1) suggest that horse mussel reefs are not just associated with bioherms, but also with harder substrates (mottled gravel and gravel/cobble provinces). They seem to be absent from mobile sandy substrates (rippled sand province), although only one satisfactory sample was obtained in the sand with comet mark province.

We have not yet analyzed the video footage taken during deployment of the video grab, although still photographs taken just before grab closure are used to further characterize the acoustically identified geological provinces as follows:

Gravel cobble – The surface of this substrate may have the best developed epifauna, with the presence of *Flustera* sp., various sponges, hydroids and worm tubes, although this is in some cases absent (e.g. grab #13, but still with a few mussels present). In other cases, e.g. grab #24, a little epifauna was present, but no mussels were found.

Mottled gravel – All three of the Table 1 stations show horse mussels (live and dead) at the surface, with a sparse epifauna associated with the valves of the mussels. Fine scale rippling of the adjacent sediments is present, indicating tidal reworking.

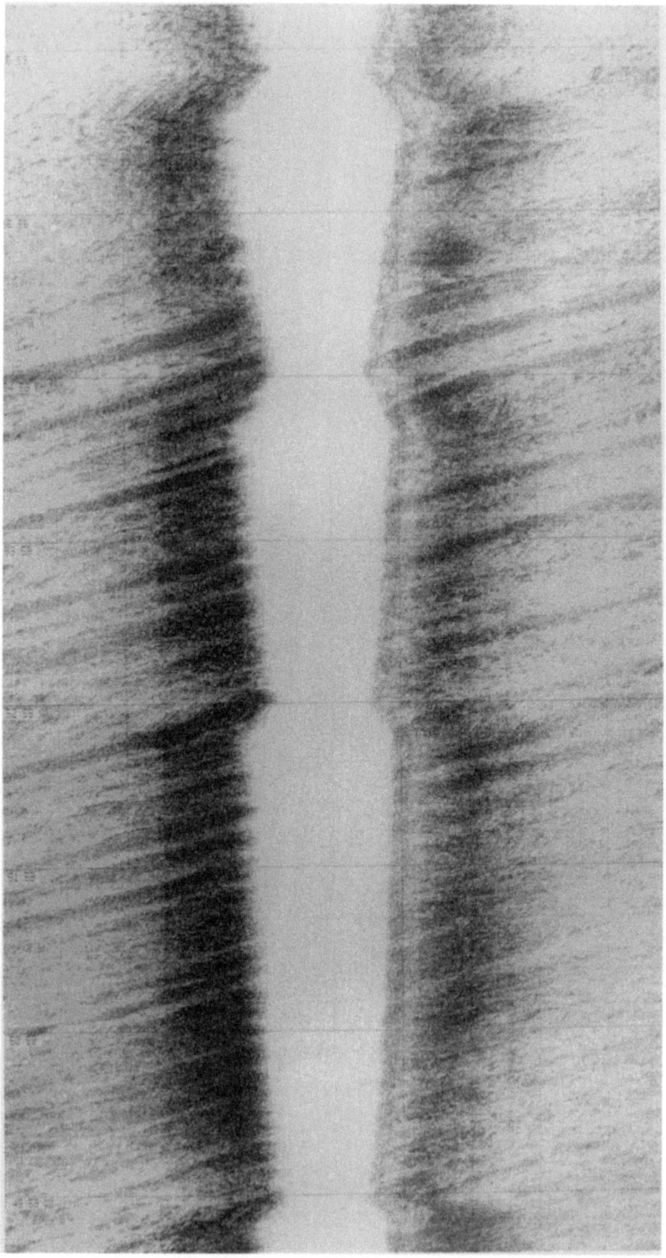

Figure 7. Simrad 300 kHz sidescan sonogram from the Bay of Fundy of the rippled sand province. The ship's track is nearly at right angles to the flow-parallel sand strips and darker gravel/cobble.

Sand with bioherms – Sand rippled surfaces were present with fine, white shell gravel on the slopes of the ripples. We found that it was necessary to wait until the grab drifted over an area with dense epifauna (*Sertularia* sp., sponges and hydroids), or occasionally with live and dead mussels visible, to ensure collecting mussels from this province.

Rippled sand – Sand rippled surfaces with white shell gravel on the slopes of the ripples was present uniformly in all photographs of this substrate. However, despite considerable searching, no epifauna was seen and no horse mussels collected in five satisfactory grab samples (Table 1).

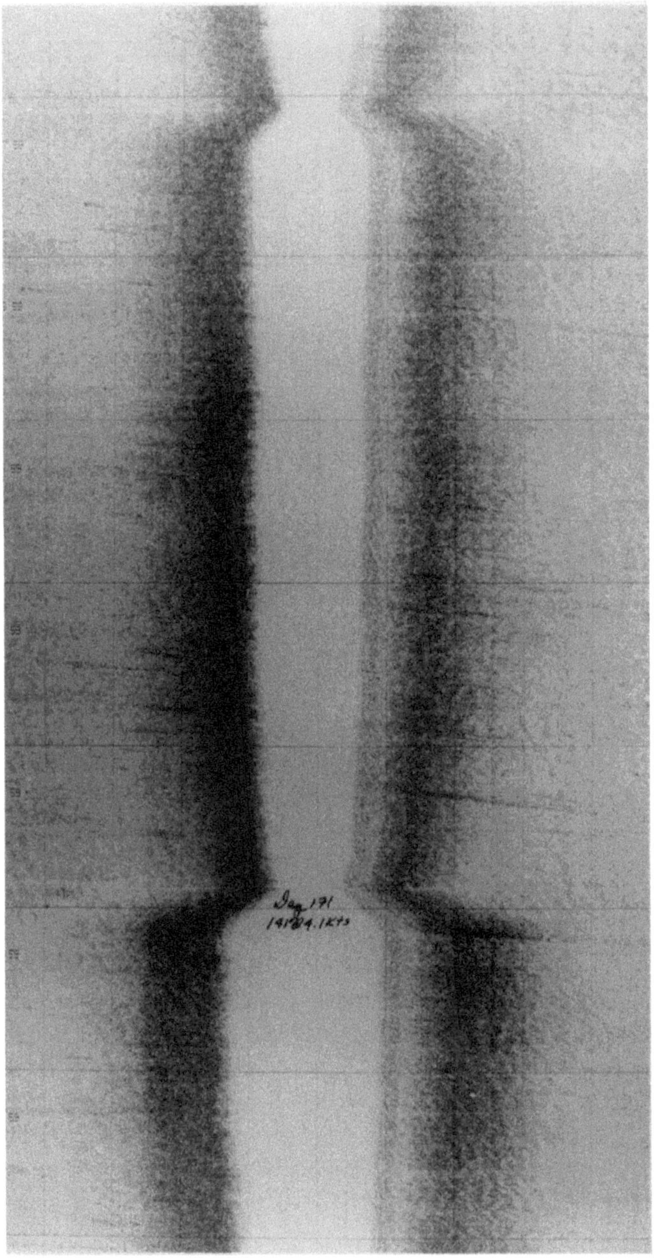

Figure 8. Simrad 300 kHz sidescan sonogram from the Bay of Fundy of the sand with comet marks province. The ship's track is at right angles to the flow-parallel comet marks.

Sand with comet marks – Sand rippled surfaces were present as above, but the shell gravel and inorganic particulates were larger and not water-movement sorted as in other sand substrates. There were a few small epifauna present but no direct photographic evidence of a comet mark was found. Only one satisfactory grab sample was collected, and so further sampling is required to determine if horse mussels are present or absent.

Discussion

Our brief review of the scientific knowledge of the marine disciplines relating to pelagic-benthic coupling in the Bay of Fundy reveals at least three major deficiencies. The first concerns physical oceanography and the lack of detailed knowledge regarding the tidal and seasonal behavior of fronts and up- and downwelling processes that are involved. Nevertheless, the general area of the fronts is associated spatially with high benthic production areas where horse mussel reefs are concentrated, suggestive of a linkage. The front is also associated with the highest Chl a levels (Figure 3) and the up/downwelling can provide a mechanism to transport phytoplankton to the mussels and relay excreted ammonia to the euphotic zone as a plant nutrient.

Secondly, our knowledge of primary production, or Chl *a* distribution, for the Bay is inadequate. This remains particularly true for temporal coverage within the Bay. Results of a later summer sampling (Wildish, unpublished) suggest that the head of the Bay has a well mixed water column, and that adjacent to it is a transitional water mass. In both these areas live phytoplankton, as indicated by Chl *a*, reach the seabed to ~ 100 m depth.

Previous estimates of secondary benthic production of horse mussel reefs using conventional methods (Wildish et al., 1986) relied on 266, 0.1- or 0.5-m^2 grab samples, which represents a very small percentage of the total Bay area sampled. Because of the contagious distribution of bivalve reefs, it is likely that estimates based on such a small sample will be misleading. We have attempted to see whether surficial sedimentary acoustic techniques might improve on the spatial coverage and thus improve on the mussel reef charts currently available. This would lead to a more accurate estimate of secondary benthic production for the Bay.

We offer here only a preliminary analysis of the acoustic sonograms and a later publication will provide a more detailed surficial geological interpretation. Our field sampling was confined to that area of the Bay where conventional grab sampling suggested that horse mussels were concentrated (Wildish et al., 1986). Five surficial sedimentary provinces were recognized in this region (see Table 1). Horse mussels were generally limited to the harder, more stable, gravel/cobble sediments. Here, a solid substratum is available for bysuss threads to attach. This contrasts with sand sediments where solid substrates were generally absent and horse mussels also were absent. Upper Bay of Fundy sands are mobile and may be physically abrasive to epifauna as well as causing poor quality food for suspension feeders by diluting seston in the lower part of the benthic boundary layer (Muschenheim (1987).

One interesting intermediate province, sand with bioherms, also had horse mussels present (Wildish et al., 1998). The mussels were associated only with the bioherms, which were raised up to 3 m higher than the adjacent sand and, consequently, protected from sand blasting and sestonic food dilution by sand grains. The bioherms were recognizable from the video camera images as discreet flow parallel structures with epifauna, in contrast to the bare adjacent sand. The origin and genesis of these structures is of considerable interest to both geologists and benthic biologists.

We believe that we have made a promising start in deciding whether surficial sedimentary acoustic techniques can be used to chart horse mussel reefs, but a final conclusion on whether the acoustic methods employed here are as useful to benthic biologists as they are to surficial geologists must await:

- further detailed analysis of the sidescan sonograms and seismic reflection data;
- additional geological field surveys of the same kind so that a greater proportion of the Bay is covered, and to determine whether additional categories of sediments can be identified;
- further benthic biological groundtruthing with the video grab to determine the probability of finding horse mussels on a particular geological province; and
- further benthic biological grab sampling to determine the typical densities, biomasses and secondary production estimates associated with particular geological provinces.

Acknowledgements

We wish to thank the Captain, officers and crew of the CSS Hudson who made the success of this fieldwork possible. Peter Vass and Kelly Bentham kept the video grab operating in difficult sampling conditions.

References

Asmus, H. & R. M. Asmus, 1990. Trophic relationships in tidal flat areas: to what extent are tidal flats dependent on imported food? Neth. J. Sea Res. 27: 93–99.

Bernard, F. R., 1974. Annual biodeposition and gross energy budget of mature Pacific oysters, *Crassostrea gigas*. J. Fish. Res. B. Can. 31: 185–190.

Cohen, R. R. H., P. V. Dresler, E. J. P. Phillips & R. L. Cory, 1984. The effect of the Asiatic clam, *Corbicula fluminea*, on phytoplankton of the Potomac River, Maryland. Limnol. Oceanogr. 29: 170–180.

Daborn, G. R., 1986. Effects of tidal mixing on the plankton and benthos of estuarine regions of the Bay of Fundy. Lecture notes on Coastal and Estuarine Studies No. 17.

Dame, R. F. & N. Dankers, 1988, Uptake and release of materials by a Wadden Sea mussel bed. J. exp. mar. Biol. Ecol. 118: 207–216.

Dame, R. F., R. G. Zingmark & E. Haskin, 1984. Oyster reefs as processors of estuarine materials. J. exp. mar. Biol. Ecol. 83: 239–247.

Dame, R., R. Zingmark, H. Stevenson & D. Nelson, 1980. Filter feeder coupling between the water column and benthic subsystem. In V. S. Kennedy (ed.), Estuarine Perspectives. Academic Press: 521–526.

Emerson, C. W., J. C. Roff & D. J. Wildish, 1986. Pelagic-benthic energy coupling at the mouth of the Bay of Fundy. Ophelia 26: 165–180.

Fader, G. B. J., L. H. King & B. MacLean, 1977. Surficial geology of the eastern Gulf of Maine and Bay of Fundy. Can. Hydrog. Sci. Pap. 19, and Geol. Survey Can. Pap. 76–17: 23 pp.

Garrett, C. J. R., J. R. Keelay & D. A. Greenberg, 1978. Tidal mixing versus thermal stratification in the Bay of Fundy and Gulf of Maine. Atmosphere-Ocean 16: 403–423.

Greenberg, D. A., 1979. A numerical model investigation of tidal phenomena in the Bay of Fundy and Gulf of Maine. Geodosey 2: 161–187.

Muschenheim, D. K., 1987. The role of hydrodynamic sorting of seston in the nutrition of a benthic suspension feeder, *Spio setosa* (Polychaeta, Spionidae). Biol. Oceanogr. 4: 265–288.

Nichols, F. H., 1985. Increased benthic grazing: an alternative explanation for low phytoplankton biomass in northern San Francisco Bay during the 1976–1977 drought. Estuar. Coast. Shelf Sci. 21: 379–388.

Prouse, N. J., D. C. Gordon, B. T. Hargrave, C. J. Bird, J. McLachlan, J. S. S. Lakshiminarayana, J. Sita Devi & M. L. H. Thomas, 1984. Primary production: organic matter supply to ecosystem in the Bay of Fundy. In D. C. Gordon & M. J. Dadswell (ed.), Update on the Marine Environmental Consequences of Tidal Power in the Upper Reaches of the Bay of Fundy. Can. Tech. Rep. Fish. aquat. Sci. 1236: 65–96.

Schwinghamer, P., J. Y. P. Guigné & W. G. Siu, 1996. Quantifying the impact of trawling on benthic habitat structure using high resolution acoustics and chaos theory. Can. J. Fish. aquat. Sci. 53: 288–296.

Trites, R. W. & C. J. R. Garrett, 1983. Physical oceanography of the Quoddy Region. In M. L. H. Thomas (ed.), Marine and Coastal Systems of the Quoddy Region, New Brunswick. Can. Spec. Publ. Fish. aquat. Sci. 64: 9–34.

Wildish, D. J. & D. Peer, 1983, Tidal current speed and production of benthic macrofauna in the lower Bay of Fundy. Can. J. Fish. aquat. Sci. 40: 309–321.

Wildish, D. J., G. B. J. Fader, P. Lawton & A. J. MacDonald, 1998. The acoustic detection and characteristics of sublittoral bivalve reefs in the Bay of Fundy. Con. Shelf Res. 18: 105–113.

Wildish, D. J. & D. D. Kristmanson, 1997. Benthic Suspension Feeders and Flow. Cambridge University Press, 409 pp.

Wildish, D. J., D. L. Peer & D. A. Greenberg, 1986. Benthic macrofaunal production in the Bay of Fundy and possible effects of a tidal power barrage at Economy Point-Cape Tenny. Can. J. Fish. aquat. Sci. 43: 2410–2417.

MIX
Papier aus verantwortungsvollen Quellen
Paper from responsible sources
FSC® C105338

If you have any concerns about our products,
you can contact us on
ProductSafety@springernature.com

In case Publisher is established outside the EU,
the EU authorized representative is:
Springer Nature Customer Service Center GmbH
Europaplatz 3, 69115 Heidelberg, Germany

Printed by Libri Plureos GmbH
in Hamburg, Germany